CHEMISTRY OF DISCOTIC LIQUID CRYSTALS

FROM MONOMERS TO POLYMERS

T0141259

THE LIQUID CRYSTALS BOOK SERIES

Edited by

Virgil Percec

Department of Chemistry
University of Pennsylvania
Philadelphia, PA

The Liquid Crystals book series publishes authoritative accounts of all aspects of the field, ranging from the basic fundamentals to the forefront of research; from the physics of liquid crystals to their chemical and biological properties; and from their self-assembling structures to their applications in devices. The series will provide readers new to liquid crystals with a firm grounding in the subject, while experienced scientists and liquid crystallographers will find that the series

PUBLISHED TITLES

Introduction to Liquid Crystals: Chemistry and Physics
By Peter J. Collings and Michael Hird

The Static and Dynamic Continuum Theory of Liquid Crystals:
A Mathematical Introduction
By Iain W. Stewart

Crystals That Flow: Classic Papers from the History of Liquid Crystals
Compiled with translation and commentary by Timothy J. Sluckin, David A. Dunmur, and Horst Stegemeyer

Nematic and Cholesteric Liquid Crystals: Concepts and Physical Properties
Illustrated by Experiments
By Patrick Oswald and Pawel Pieranski

Alignment Technologies and Applications of Liquid Crystal Devices
By Kohki Takatoh, Masaki Hasegawa, Mitsuhiro Koden, Nobuyuki Itoh, Ray Hasegawa, and Masanori Sakamoto

Adsorption Phenomena and Anchoring Energy in Nematic Liquid Crystals
By Giovanni Barbero and Luiz Roberto Evangelista

Chemistry of Discotic Liquid Crystals: From Monomers to Polymers
By Sandeep Kumar

THE LIQUID CRYSTALS BOOK SERIES

CHEMISTRY OF DISCOTIC LIQUID CRYSTALS

FROM MONOMERS TO POLYMERS

Sandeep Kumar

CRC Press
Taylor & Francis Group
Boca Raton   London   New York

CRC Press is an imprint of the
Taylor & Francis Group, an **Informa** business

CRC Press
Taylor & Francis Group
6000 Broken Sound Parkway NW, Suite 300
Boca Raton, FL 33487-2742

First issued in paperback 2022

© 2011 by Taylor and Francis Group, LLC
CRC Press is an imprint of Taylor & Francis Group, an Informa business

No claim to original U.S. Government works

ISBN 13: 978-1-4398-1143-6 (hbk)
ISBN 13: 978-1-03-240259-8 (pbk)

DOI: 10.1201/b10457

Visit the Taylor & Francis Web site at
http://www.taylorandfrancis.com

and the CRC Press Web site at
http://www.crcpress.com

To

Navita

Navdeep and Sudeep

My mother, family members, and the memory of my father

Contents

Preface

Although 32 years have elapsed since the discovery of discotic liquid crystals (DLCs) by Chandrasekhar and coworkers, our knowledge of DLCs is still limited as compared to calamitic liquid crystals. The hierarchical self-assembly of disk-shaped molecules leads to the formation of DLCs. The *self-assembled* columns of DLCs *self-organize* in different 2D lattices, which possess two very important and attractive properties: *self-processing* on/between substrates and *self-healing* of structural defects in the columnar phase. All these *self-contained* properties render them as potential functional materials for many semiconducting device applications and models for energy and charge migration in self-organized dynamic functional soft materials. The negative birefringence films formed by polymerized nematic DLCs have been commercialized as compensation foils to enlarge the viewing angle of commonly used twisted nematic liquid crystal displays. Moreover, a liquid-crystal-display device with wide and symmetrical viewing angle has been demonstrated by using nematic DLCs. The past three decades have seen tremendous interest in this area, fueled primarily by the possibility of creating a new generation of organic semiconductors and wide viewing displays using DLCs.

Researchers working on DLCs need to have an up-to-date source of reference material to establish a solid foundation of understanding. It is extremely important that students and researchers in the field of liquid crystals have ready access to what is known and what has already been accomplished in the field. Prior to this book, the only way this could be achieved was by conducting extensive searches through the literature. While a number of books on classical calamitic liquid crystals are available, there are no books that are dedicated exclusively to the basic design principles, synthesis, and physical properties of DLCs. Therefore, this is as good a time as any to examine work in the area of DLCs and make a compilation of the scattered literature.

The book deals mainly with the chemistry and thermal behavior of DLCs. It is divided into six chapters and is targeted at a wide readership. Chapter 1 provides a basic introduction to liquid crystals. It describes the molecular self-assembly and the types of liquid crystals, provides their classification, and covers their history and general applications. It then focuses on DLCs and describes their discovery, structure, characterization, and alignment. Chapter 2 deals with the chemistry and physical properties of various monomeric DLCs. It consists of 25 sections describing the synthesis and mesomorphic properties of monomeric DLCs formed by different cores. Chapters 3 through 5 cover the chemistry and mesomorphism of discotic dimers, oligomers, and polymers. Chapter 6 presents some applicable properties of DLCs. Each chapter begins with a general description, which provides the necessary background and context for the uninitiated reader to understand the concepts involved. The remainder of the section is a comprehensive review of work. Researchers working in the field of discotics would find this a comprehensive, up-to-date source of work. The extensive reference list will also help the reader to pursue further investigations. I have primarily covered literature that appeared in scientific journals up to the end of 2008. Though a large number of patents, including authors' patents, are available on DLCs, I have avoided using any of this material. It is likely that some interesting materials may lie buried within the patent and remote journals literature. There is a great deal of physics associated with DLCs. I have only summarized the physical properties of these materials, as a detailed account is beyond the scope of this book. However, efforts have been made to provide important references dealing with the physical properties of DLCs.

This book is the first reference book that covers the various aspects of DLCs. Hopefully, it would become a valuable addition not only to the bookshelves of all those who are linked to the field of liquid crystals, but also for those in the fields of supramolecular chemistry, polymer chemistry, supramolecular materials, organic electronics, and complex soft condensed matter. I hope that this book will be helpful not only to students and researchers but also to the directors and principal investigators working in this field. Moreover, this first book on DLCs will lead to further advances in this fast-growing technological field.

It is my pleasure to thank Professor V. Percec for inviting me to write this book. I enjoyed working with Hilary Rowe, chemistry editor, and David Fausel, project coordinator, Dr. Vinithan Sedumadhavan, project manager, and Richard Tressider, project editor, and I would like to express my sincere gratitude for their constant support.

I would also like to thank various publishers and authors for their permission to reproduce figures from previous publications. These figures, taken in part or adapted from other sources, are acknowledged in their respective legends.

It is my privilege to express my gratitude to my colleague, Professor Lakshminarayanan, for many helpful discussions. I would like to express my sincere gratitude to my students Hari K. Bisoyi, Satyam Gupta, Avinash B. S., and Swaminathan K. for their interest and help throughout the preparation of this book. Thanks are also due to my colleagues Srinivasa, Jayshankar, Indu, Thanigaivelan, and Shadakshari for their assistance in organizing the references. I would also like to thank my former students Dr. S. K. Pal and Dr. Jaishri Naidu for providing some literature, and the staff at Raman Research Institute (RRI) library for providing me a lot of literature.

This book could not have been written without the support of my family members. I would like to thank my wife Navita Rani and sons Navdeep Kumar and Sudeep Kumar for their patience, encouragement, cooperation, and moral support during the course of this work.

Author

Sandeep Kumar received his PhD in chemistry/medicinal chemistry from Banaras Hindu University, Varanasi, India, in 1986, under Professor A. B. Ray. He then worked with Dr. Sukh Dev at Malti Chem Research Centre, Vadodara, India, for about two years on the synthesis of food-flavoring agents. He has worked as a postdoctoral research fellow at the Hebrew University of Jerusalem, Israel; at Technion, Israel Institute of Technology, Haifa, Israel; at the Scripps Research Institute, La Jolla, San Diego, California; and at the University of Mainz, Mainz, Germany, with Professors E. Glotter, E. Keinen, and H. Ringsdorf. During his postdocs, he worked on natural cairomones, organometallic chemistry, catalytic antibodies, and liquid crystals. He was a visiting research professor at the Naval Research Laboratory, Washington, District of Columbia, during 1999–2000 and at the National Dong Hwa University, Hualien, Taiwan, during 2008. He has also visited many other countries such as the United Kingdom, France, Ireland, Japan, China, Korea, Malaysia, Slovenia, and Italy to deliver lectures.

Dr. Kumar joined the Centre for Liquid Crystal Research, Bangalore, India, to start a new chemistry lab in 1995. In 2002, he moved to the Raman Research Institute, Bangalore, India, where he currently holds the position of Professor and Coordinator, Soft Condensed Matter Group. He has published over 140 research papers in peer-reviewed international journals including over 110 papers on discotic liquid crystals. He also has a few patents to his credit. Several of his papers were among the most cited and accessed on the Web. The Royal Society of Chemistry (RSC) awarded him a journals grant for international authors in 2001 for his significant publications in RSC journals. He was awarded the inaugural LG Philips Display Mid-Career Award by the International Liquid Crystal Society in 2008. His current research interests include design, synthesis and applications of liquid crystals, conducting polymers, green chemistry, and nanotechnology.

1 Introduction

1.1 SELF-ORGANIZATION OF MOLECULES AND LIQUID CRYSTALS

Life on earth begins with the self-organization of molecules. No life would be possible without the self-assembly of lipids into bilayers within the cell membrane. Molecular self-assembly and self-organization are nature's elegant and effective tools/strategies for the dynamic functional materials of life. The *supramolecular engines of creation*, that is, DNA, proteins, enzymes, etc., are created by the hierarchical organization of small prototype discrete molecular building blocks by using molecular recognition and supramolecular interactions [1,2]. Nature utilizes supramolecular interactions, that is, non-covalent intermolecular interactions otherwise known as molecular information, such as hydrogen bonding, π-stacking, polar–nonpolar interactions, metal coordination, charge transfer complex, ionic interactions, etc., to build dynamic functional soft materials (α-helix and β-pleated structures of polypeptides, formation of double helix of nucleic acids, etc.) by the process of self-assembly and self-organization at different molecular levels and accomplish the desired biological functions (DNA replication, reversible binding of oxygen to hemoglobin, molecular motors, ion pumps, etc.) that are vital for life [3]. The often-cited beauty of self-assembly and self-organization is their spontaneity [4]. Spontaneous self-assembly and self-organization to supramolecular functional nanostructures is nature's solution to vital biological processes. The supramolecular organization, while dynamic in nature, is stable enough to small environmental perturbations. The stability of supramolecular aggregates is determined by the number density of a particular interaction and the number of different supramolecular interactions involved in the self-organization process [2–4]. The double helix of DNA reveals and represents one of the most essential and stable supramolecular structures in which single building blocks can organize. Supramolecular systems are scientifically intriguing and challenging because they involve the rational design and development of large-scale structures, leading to molecular materials of dimension similar to those of complex systems found in nature. One of the hallmarks of many self-assembled systems is the presence of liquid crystalline phases [5–16]. In a broad sense, liquid crystals (LCs) can be considered as prototypical self-organizing molecular materials of today [17–21]. Liquid crystals stand between the isotropic liquid and the strongly organized solid state. Similarly, life stands between complete disorder which is death, and complete rigidity, which is death again [22]. In materials science, non-covalent interactions have been used to obtain well-defined, self-assembled architectures in neat systems as well as in solvents. LCs belong to one of such systems. Supramolecular interactions such as van der Waals forces, dipolar and quadrupolar interactions, charge transfer interactions and hydrogen bonding, etc., play a crucial role in the formation of LCs and in the determination of their mesomorphic properties.

LCs are unique functional soft materials that combine both *order* and *mobility* on a molecular, supramolecular, and macroscopic level [23–53]. Hierarchical self-assembly in LCs offers a powerful strategy for producing nanostructured mesophases. Molecular shape, microsegregation of incompatible parts, specific molecular interaction, self-assembly, and self-organization are important factors that drive the formation of various LC phases. LCs are accepted as the fourth state of matter after solid, liquid, and gas. They form a state of matter intermediate between the solid and the liquid states. For this reason, they are referred to as intermediate phases or *mesophase*. However, these are true thermodynamic stable states of matter. The constituents of the mesophase are called *mesogens*. Mesogens can be organic (forming thermotropic and lyotropic phases), inorganic (metal oxides forming lyotropic phases) [54], or organometallic (metallomesogens) [55]. LCs are equally important in materials science as well as in life science. Important applications of thermotropic LCs are electro-optic displays, temperature sensors, and selective reflecting pigments. Lyotropic systems are incorporated in cleaning processes (soap, detergent, etc.), and are important in cosmetic and food industries [56–60]. They are used as templates for the preparation of mesoporous materials and also serve as model systems for biomembranes [61]. LCs are important in living matter. Most important are biological membranes, DNA, etc. [62,63]. Anisotropic fluid states of rigid polymers are used for the processing of high-strength fibers like Kevlar [64]. LCs can potentially be used as new functional materials for electron and ion transportation and as sensory, catalytic, optical, or bioactive material [65]. They are extremely diverse since they range from DNA to high-strength synthetic polymers like Kevlar (used for bulletproof vests, protective clothing, high-performance composites for aircraft and automotive industries) and from small organic molecules like alkyl and alkoxycyanobiphenyls used in liquid crystal displays (LCDs) to self-assembling amphiphilic soap molecules. Recently, their biomedical applications such as in controlled drug delivery, protein binding, phospholipid labeling, and microbe detection have been demonstrated [66–71]. Apart from materials science and bioscience, LCs are now playing a significant role in nanoscience and nano-technology, such as the synthesis of nanomaterials using LCs as templates [72], the design of LC nanomaterials [73], alignment and self-assembly of nanomaterials using LC phases [74–76], and so on. Owing to their dynamic nature and photochemically, thermally, or mechanically induced structural changes, LCs can be used for the construction of stimuli-responsive materials [65]. Although LCs have diverse applications such as temperature sensing and solvents in chemical reactions, chromatography, spectroscopy, holography, etc., they are primarily known for their extensive exploitation in electro-optical display devices such as watches, calculators, telephones, personal organizers, laptops, flat panel televisions, etc.

1.1.1 Liquid Crystals as an Intermediate Phase (Mesophase) of Matter

The distinction between solid, liquid, and gas, the common states of matter, is evident even to the nonscientist. In a solid, the constituents (molecules, atoms, or ions) are rigidly fixed and, therefore, it has a definite shape and a definite volume. On the other hand, in a gas, the molecules have random motion and therefore it has neither a definite shape nor a definite volume; it takes the shape of the container and occupies the volume of the container. In a liquid, the molecules are not as rigidly fixed as in solid; they have some freedom of motion, which is, however, much more restricted than that in a gas (Figure 1.1). Therefore, a liquid has a definite volume although not a definite shape; it takes the shape of the container. It is much less compressible and far denser than a gas. Solids are characterized by incompressibility, rigidity, and mechanical strength. This indicates that the constituents are closely packed. They are held together by strong cohesive forces and cannot move at random. Some solids, like sodium chloride, besides being incompressible and rigid also possess characteristic

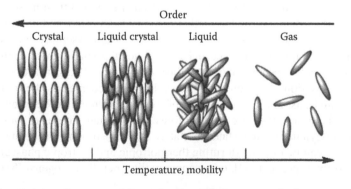

FIGURE 1.1   Different states of matter and the molecular ordering present in them.

geometrical forms. Such substances are said to be crystalline solids. X-ray studies reveal that their ultimate particles are arranged in a definite pattern throughout the entire three-dimensional (3D) network of a crystal. This is termed as *long-range order*. There is another category of solids, such as glass, which possesses properties of incompressibility and rigidity to a certain extent but they do not have definite geometrical forms. Such substances are called amorphous solids and possess *short-range order* of the constituents. Similarly, liquids exhibit only short-range order while gases show no order at all. The order in a crystal is usually both *positional* and *orientational*, that is, the molecules are constrained both to occupy specific sites in a lattice and to point their molecular axes in specific directions. The molecules in liquids, on the other hand, diffuse randomly though the sample container with the molecular axes tumbling wildly, which imparts fluidity. Crystalline solids are *anisotropic*, that is, their physical properties are different in different directions. Amorphous solids, liquids, and gases are *isotropic*, that is, all directions are identical and all properties are alike in all directions.

Consider a molecular crystal that is being heated. With increase in temperature, the molecular vibrations increase and ultimately become so high that molecules break away from their fixed positions and lose their specific orientations. They then begin to move more freely and have rotational motion as well. The solid now changes into the liquid state where the molecules are neither fixed in specific sites nor oriented to specific directions in the sample. In other words, an anisotropic solid becomes an isotropic liquid. If, however, the molecules of the crystal have pronounced geometric shape anisotropy like a rod or a disk, then, in certain compounds, the isotropic liquid state is preceded by another intermediate state in terms of molecular ordering. This state possesses some degree of orientational ordering and sometimes some positional ordering of the anisotropic molecules, that is, the molecules diffuse throughout the sample but while doing so they maintain some *orientational* and *positional* ordering albeit short-ranged. This is the mysterious and fascinating liquid crystalline state. Here, the anisotropic solid changes into a stable anisotropic fluid before turning into an isotropic liquid. This has been recognized as a true thermodynamically stable state of matter. Owing to the diffusion of the molecules, this state of matter is fluid in nature, and owing to the orientational ordering of the molecules, this state of matter possesses anisotropic physical properties. Hence, LCs possess both the fluidity of liquids and anisotropic properties of crystals. Since this anisotropic ordered fluid lies between the crystalline solid state and the isotropic liquid state and possesses properties of both, it has been referred to as an intermediate phase or *mesophase*. This has also been called as the fourth state of matter. The order (positional and orientational) present in the mesophase is much less than the crystalline phase. There is only a slight tendency for the molecules to point more in one direction than others or to spend more time in some positions than others. There is another condensed phase that exhibits intermediate order, in which the molecules are generally fixed at lattice points but in addition to vibration, molecules can freely rotate. This phase is referred to as *plastic crystal*. In this phase, unlike the LC phase, the molecules do not diffuse at all. This is also the case with the isotropic amorphous solids. There exists another state of matter, also known as the fourth (distinct) state of matter, called the plasma phase. It has nothing to do with LCs, but it is a true state of matter just as the solid, liquid, and gaseous states are. If a substance is heated to a very high temperature, the random motion becomes so violent that the electrons that are normally bound to the atoms get knocked off. This phase of matter is composed of positively charged ions and negatively charged electrons, which normally attract each other so strongly that the ions and electrons bind together. However, the temperature is so high that the rate at which the ions and electrons bind together is equal to the rate at which the electrons are being knocked off the atoms. Thus, the substance exists in this state with unbound electrons and ions. It is a new phase of matter that normally exists in and around stars. Moreover, scientists presently create plasma in their experiments on nuclear fusion.

In the most simple LC phase, one molecular axis tends to point along a preferred direction as the molecules undergo diffusion. The preferred direction is called the *director* and is denoted by **n**. To specify the amount of orientational order in such a liquid crystalline phase, an *order parameter* is defined. This can be expressed in many ways, but the most useful formulation for calculating the order parameter is as follows:

$$S = \left\langle \frac{(3\cos^2\theta - 1)}{2} \right\rangle$$

where the bracket denotes an average over many molecules at the same time or the average over time for a single molecule. "θ" is the angle between the molecular axis and the director (Figure 1.2). S is the order parameter. The order parameter is defined such that $S = 1$ for perfectly crystalline solid and $S = 0$ for an isotropic liquid; obviously, S lies in between these two states, that is, $0 < S < 1$ for LCs.

FIGURE 1.2 Molecular order in a nematic LC phase.

1.2   BRIEF HISTORY OF LIQUID CRYSTALS

The serendipitous discovery of LCs in 1888 marks an important milestone in the history of scientific discoveries. When determining the melting point of cholesteryl benzoate, **1** (Figure 1.3), Friedrich Reinitzer, an Austrian botanist, noticed the unusual melting behavior of this compound. It melts at 145.5°C to form a cloudy liquid. This opaque liquid then appears to melt again at 178.5°C to a clear transparent liquid [77]. He also observed some unusual color behavior upon cooling. He could not explain the phenomenon he observed; however, he was aware of the work of a German physicist, Otto Lehmann, who used to study, under microscope, how substances crystallize on a heating stage. So he sent some of the samples to Otto Lehmann. Lehmann performed many experiments on these samples with his heating stage microscope and explained the phenomenon of existence of "double melting" [78]. He first referred to them as "soft crystals"; later he used the term "crystalline fluids." As he became more convinced that the opaque phase was a homogeneous phase of

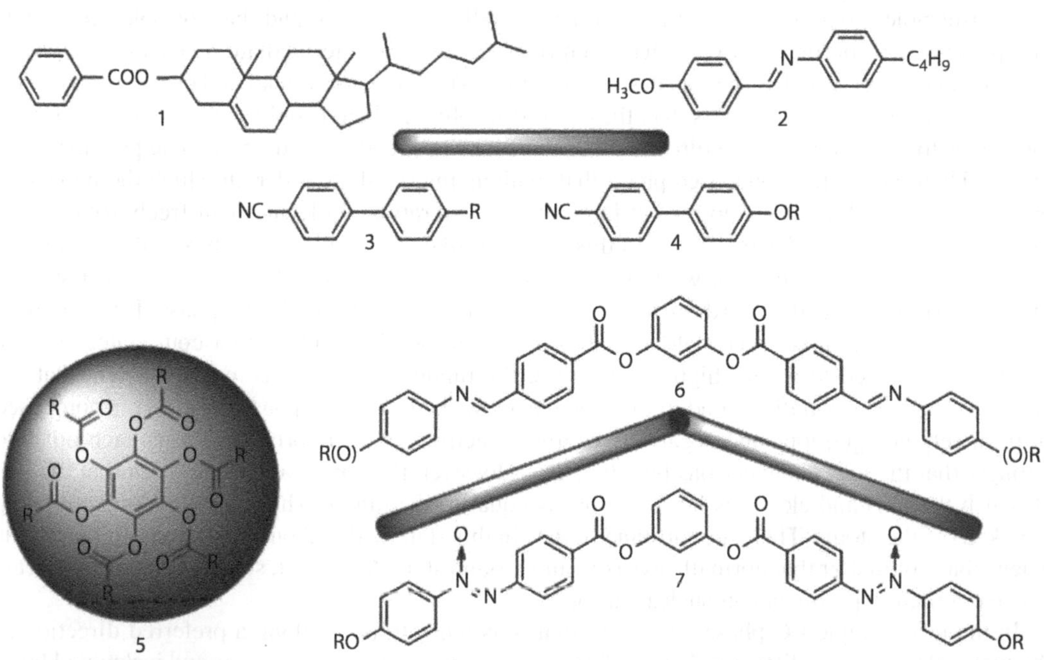

FIGURE 1.3   Historically important LC molecules.

matter sharing properties of both liquids and solids, he began to call them LCs. In spite of many subsequent arguments over nomenclature, this was the name that eventually survived. The term is in widespread use today. As it carries two contradictory terms, it possesses an element of mystery and attraction. The discovery of LCs itself was a multinational and multidisciplinary task, and so also the present-day science and technology of LCs embraces many branches of science. Reinitzer is usually called the discoverer of LCs [79]. It should be noted that researchers as early as the 1850s actually dealt with LCs but did not realize the uniqueness of the phenomena [80]. It was observed that the outer coverings of a nerve fiber forms soft and flowing forms when left in water, and these forms produce unusual effects when polarized light was used.

Lehmann was the dominating figure in LC research around that time. He not only dealt with LCs derived from natural products but also with synthetic organic molecules. He also observed that a solid surface in contact with a liquid crystalline substance causes the LC to orient in a certain direction. This is of great practical importance today with LCDs. Though all of Lehmann's ideas were not accepted by other investigators, simultaneous developments in LC research occurred both in Germany and France. An important contributor at that time was the German chemist Daniel Vorlander, who worked in Halle. He and his coworkers synthesized many new liquid crystalline substances and were the first to observe a single substance that possessed more than one LC phase [81]. Out of his work, Vorlander was able to identify what kinds of substances were likely to be liquid crystalline. He laid down the foundation of the relationship between molecular structure and LC properties. In 1907, he remarked that "...the crystalline-liquid state results from a molecular structure which is as linear as possible" [82]. It means all compounds exhibiting LC behavior had elongated (rod-like) molecules, now called *calamitic* molecules. Following this, the progress was both swift and substantial [79]. In 1922, Georges Freidel published the first classification of LCs into nematic, smectic, and cholesteric [83]. In the early years, he objected to Lehmann's term LC on the basis that LCs were neither true liquids nor true crystals. He preferred the term meso-morphic to describe the LC state and the associated term *mesophase* reflecting the intermediate nature of these phases between the crystal and isotropic liquid states. These terms are widely used today and coexist happily with the Lehmann terminology. A useful term from Freidel's nomen-clature is the word *mesogen* (*nematogen* and *smectogen*) used to describe a material that is able to produce mesophase. If the compound does really form a mesophase, the description of it as *mesomorphic* is perfectly adequate. Friedel gave us today's terms *smectic* and *nematic* with their well-known Greek derivations. He also understood that an LC could be oriented by an electric field. The effect of electric and magnetic fields later became the subject of great attention. X-ray experiments in France and Germany revealed in a most unambiguous way that LCs possess more order than liquids but less than solids. Gradually, the field developed, and in 1957 Glenn Brown, an American chemist, published an extensive and informative review article "The Mesomorphic State" on LC phases [84]. In 1958, the Faraday Society of London organized a conference on LCs. In 1962, George Gray, a British chemist, published a full-length book describing the molecular structure and properties of LCs [85]. In 1965, Glenn Brown instituted the first *International Liquid Crystal Conference* (ILCC), where the application of cholesteric LCs in thermography was pre-sented. In 1968, at the second ILCC, a group of researchers from Radio Corporation of America (RCA) gave the first indication for an application of LCs in electro-optical display technology. This report increased the interest in LC research exponentially, which continues even today. In 1969, an important advancement was the synthesis of the first moderately stable room temperature LC *p*-methoxybenzylidene-*p*-*n*-butylaniline (MBBA, **2**) for display applications [86]. However, the introduction of stable room temperature LCs, 4-alkyl, and 4-alkoxy-4'-cyanobiphenyls (**3,4**) by Gray and coworkers in 1973 provided a secure basis for LC research [87]. Following this, progress was made in the direction of their technological applications in subsequent years. Fundamental research, therefore, moved forward rapidly with very good readily available materials and funding, which was released due to the potential for technological applications. Research on LCs exploded during the 1970s and 1980s.

In 1977, when the rod-like LCs started to revolutionize commercial display technologies and the general belief that only rod-like (calamitic) molecules can form LCs was prevailing, Chandrasekhar and coworkers in India reported that not only rod-like molecules but also compounds with disk-like molecular shape are able to form LC phases [88]. They prepared a number of benzene hexa-*n*-alkanoates **5** and from optical, thermodynamic, and x-ray studies established that these materials form a new class of LCs. This opened up a whole new field of fascinating LC research. However, it is interesting to note that Vorlander in 1924 supposed the possibility of the existence of mesophases in leaf-shaped molecules, but his attempts to realize any example with this behavior had been unsuccessful probably because the molecules, he looked at were devoid of flexible alkyl chains [89]. He mentioned in his article that leaf-shaped molecules do not form any LCs at all. Of course, the same molecules surrounded by long aliphatic chains are now well known for forming columnar mesophases.

About two decades later, in 1996, when the concept that chiral molecules can form both chiral and achiral mesophases but achiral molecules cannot form chiral mesophases by themselves was becoming generalized, Niori and coworkers discovered ferroelectricity in non-chiral banana-shaped molecules **6** [90]. This led to a very intense research activity involving bent-shaped molecules, which provide access to mesophases with polar order and superstructural and supramolecular chirality despite the molecules being achiral. It is interesting to note that the first banana-shaped LCs **7** were prepared in the research group of Vorlander in 1929, but the type of mesophase was not reported [91]. However, with the present state of knowledge, the liquid crystalline behavior of these compounds have been reinvestigated and it has been found that some of them form banana phases.

The last two decades have seen so many developments in different directions that a simple pattern of evolution simply does not exist. Developments have occurred in an explosive way emanating outward from the core of fundamental knowledge. Now, the LC research field has its own dedicated international scientific journals and scientific meetings. Research funding is flowing from both public and private agencies for carrying out cutting edge research on LCs, unlike the early days of LC research. In addition, the International Liquid Crystal Society (ILCS) encourages and values the contributions of researchers toward the development and understanding of LC science and technology by conferring awards and honors to scientists of all age and from various disciplines. Glenn Brown prizes are conferred on entry-level researchers, while scientists toward the end of their career are conferred with honorary memberships for their significant contributions. Recently, a mid-career award was introduced to encourage scientists in the middle of their scientific career, the author being the first recipient of such a mid-career award. One can be certain that many more encouraging steps will be taken in the future for driving the field at a high pace. Moreover, irrespective of funding, awards, and honors, the driving thrust for LC research is always there since most of the time it is driven by curiosity, and curiosity never dies. It is very much evident that LC research has progressed from curiosity to commodities. LCs have become a part of our daily lives, ranging from wristwatches and pocket calculators to portable computers and televisions. Progress in our understanding of LC phases has also aided our understanding of the cell membrane and of certain diseases. There seems to be no end in our progress to understand LCs. So, we can be certain that just as many new developments will be made by scientists in the future.

1.3 CLASSIFICATION OF LIQUID CRYSTALS

There are various ways of classifying LCs based on the molar mass of the constituent molecules, that is, low molar mass (monomeric and oligomeric) and high molar mass (polymeric) LCs; based on how the liquid crystalline phase has been obtained, that is, by adding solvent (*lyotropic*) or by varying the temperature (*thermotropic*); based on the nature of the constituent molecules (organic, inorganic, and organometallic); based on the geometrical shape of the molecules (rod-like, disk-like, banana-like); and based on the organization of the molecules in the

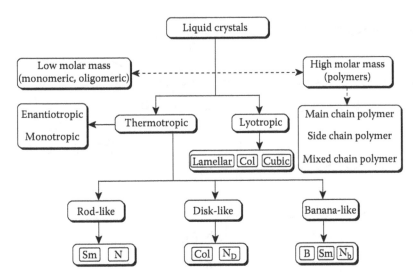

FIGURE 1.4 Classification of LCs.

liquid crystalline phase (nematic, smectic, columnar, helical, B phases, etc.). The classification of LCs is shown in Figure 1.4.

However, the most widely recognized and used classification of LCs is into two major categories: (a) thermotropic LCs (mesophase formation is temperature dependent) and (b) lyotropic LCs (mesophase formation is solvent and concentration dependent). If a compound displays both thermotropic and lyotropic liquid crystalline phases, then it is called *amphotropic* LC [92].

1.4 LYOTROPIC LIQUID CRYSTALS

While thermotropic LCs are obtained by the effect of temperature on pure compounds or a mixture of compounds, lyotropic LCs, otherwise known as anisotropic solutions, are formed by dissolving amphiphilic compounds in suitable solvents under appropriate conditions of concentration and temperature [56–60]. The amphiphilic compounds are characterized by two distinct parts of contrasting character: a hydrophilic polar head and a hydrophobic nonpolar tail. Apart from temperature, both the concentration of the solute and the solvent (most often water) also play a very significant role in lyotropic LC systems. Typical examples of lyotropic LCs are soaps in water and various phospholipids. Just as there are different types of structural modifications for thermotropic LCs, there are several different types of lyotropic LC phases. Each of these different types has a different extent of molecular ordering within the solvent matrix and also various kinds of molecular aggregates varying in shape. Lyotropic LC phases can have positional order in one (lamellar), two (columnar hexagonal), or three (cubic) dimensions. Lyotropic LCs possess both industrial and biological significance. Lyotropic LCs are very important in everyday cleaning processes (soaps and detergents) and foods. Of far greater importance is the occurrence of lyotropic LC phases in biological membranes and DNA, and their potential applications in drug delivery and gene therapy [58]. The cell membranes in the body are a result of the lyotropic LC phase that is generated from the dissolution of phospholipids in water. Therefore, life itself critically depends on lyotropic LC phases.

Most surfactants (surface active agents) in water form lyotropic LC phases. Surfactants are amphiphilic materials whose constituent molecules have a molecular structure that includes a polar head group and a nonpolar chain. There are various kinds of surfactant molecules such as cationic **8, 9**, anionic **10, 11**, nonionic **12–14**, and zwitterionic **15** (Figure 1.5).

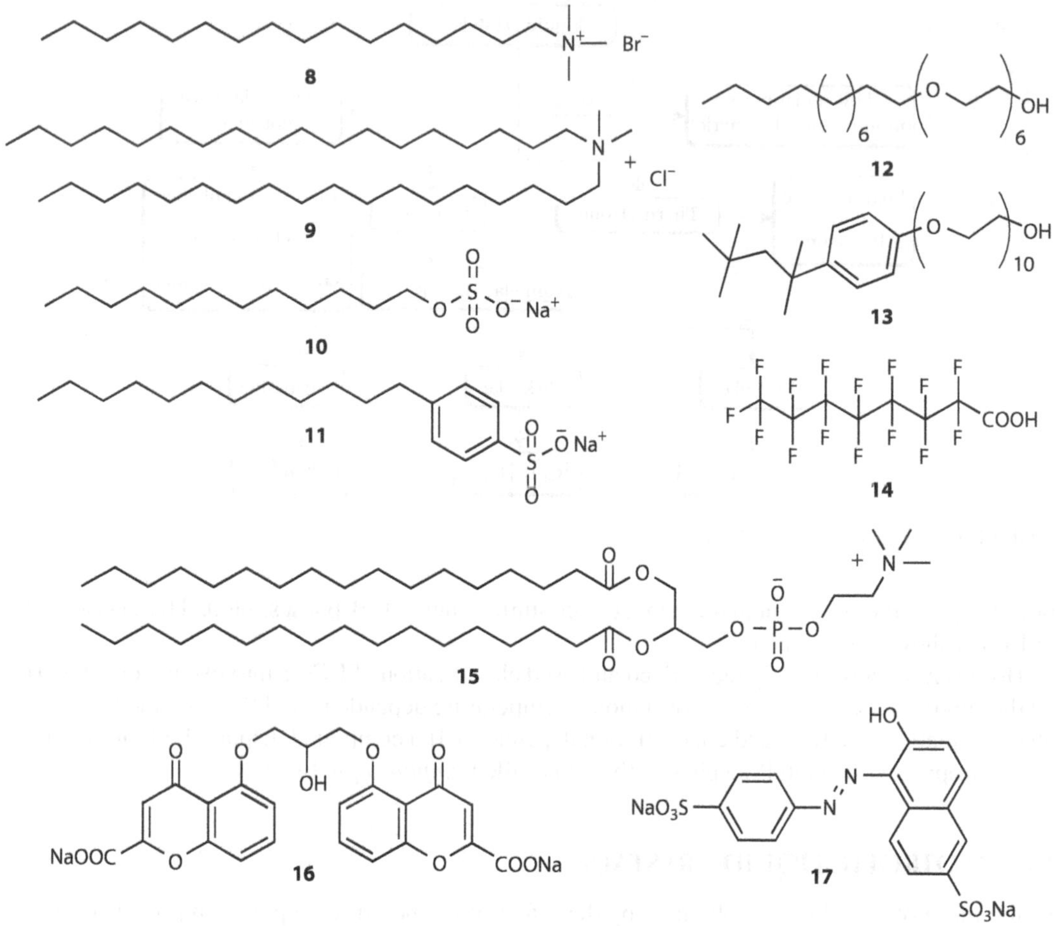

FIGURE 1.5 Different surfactant molecules forming lyotropic liquid crystalline phases and chromonic LCs.

When a small amount of amphiphilic material is dissolved in a polar solvent such as water, it goes into the solution. As the concentration of the material is increased, the hydrophobic tails assemble together and present the hydrophilic polar heads to the solvent, thereby arranging themselves into spheres called *micelles* (Figure 1.6). So the polar head groups are on the surface of the *micelle* and the nonpolar hydrocarbon chains are toward the center. In phospholipids, the structures are called *vesicles*. The micelles are stable as long as the amount of amphiphilic material is above a certain concentration called *critical micelle concentration* (CMC). If the concentration of the material is further increased, more micelles are formed. In some cases, the size and shape of the micelles remain fairly constant as the number of micelles increase. In other cases, the shape of the micelles changes from spherical to cylindrical. It should be noted that similar structures begin to form if amphiphilic material is added to nonpolar liquid such as oil. In this case, the micelles form with the polar head groups toward the inside and the nonpolar end chains toward the outside. These are referred to as inverted (reverse) structures (Figure 1.6). These structures also change in shape and size as the amount of amphiphilic material is increased.

If the concentration of the material is increased, a point is reached where the micelles combine to form larger structures and these structures are liquid crystalline. Three different classes of lyotropic LC phase structures are widely recognized. These are the lamellar, the hexagonal columnar, and the cubic phases, and their structures have each been classified by x-ray diffraction techniques (Figure 1.6). The structure of a classical lamellar phase consists of a stacked array of amphiphilic

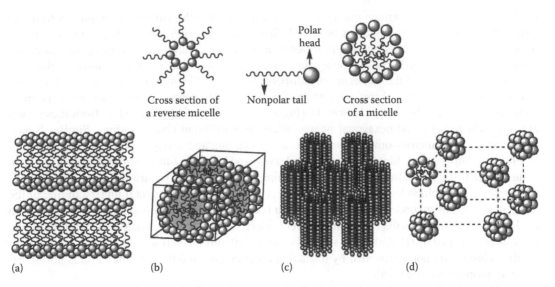

FIGURE 1.6 Cross sections of micelles and reverse micelles formed by surfactant molecules and lyotropic liquid crystalline phases: (a) lamellar, (b) bicontinuous cubic, (c) hexagonal columnar, and (d) micellar cubic.

bilayer sheets separated by a solvent. The hexagonal columnar phase consists of long cylindrical rods of amphiphilic molecules arranged on a hexagonal lattice. There are two different kinds of cubic phases exhibited by surfactants: one lies between the lamellar and hexagonal phase, the other appears between the hexagonal phase and the micellar solutions. In the latter phase, the surfactant molecules arrange themselves in spherical or nonspherical micelles, which in turn form a cubic lattice called the micellar cubic phase. The former phase, which appears between the lamellar and the hexagonal phase, is called the bicontinuous cubic phase and consists of an interwoven network of branched columns. The cubic phases are optically isotropic and possess high viscosity.

Temperature has a significant effect on the stability of the lyotropic phases. In fact none of the phases can form if the temperature is not high enough. All LC structures require that the amphiphilic molecules be able to move relative to one another. If the temperature is too low, the molecules tend to form rigid crystalline structures. So the temperature above which mesophases form is called the *Kraft temperature*. The Kraft temperature increases slightly as the concentration of the amphiphile increases.

The above-described LC phases occur in binary systems of solvent and surfactant. Under suitable conditions, however, in a mixture of a highly polar liquid, a slightly polar liquid, and an amphiphilic molecule (ternary system), there exist three nematic phases with distinct physical signature: two uniaxial nematic phases and a biaxial nematic phase. The uniaxial nematic phases are formed from disk-like and rod-like micellar aggregates; however, each of these phases can undergo a transition to a biaxial nematic phase as a function of concentration or temperature [93,94]. It should be noted that the biaxial nematic phase was first detected in lyotropic LC systems, which possess two orthogonal optic axes and hence two orthogonal directors. Lyotropic liquid crystallinity is not only limited to flexible surfactants but also to inorganic metal oxides, and some polymers like Kevlar also exhibit lyotropic mesomorphism.

Another interesting class of lyotropic LCs is called chromonic LCs (chromonics). Chromonics is a novel class of lyotropic LCs exhibited by rigid aromatic compounds surrounded by polar groups at their periphery [95–98]. These are formed by the self-organization of aromatic compounds with ionic or hydrophilic groups in aqueous solutions. Chromonic LCs represent an intersection of two very active fields of research, supramolecular assembly and ordered complex fluids. In the chromonic LC phase, an aqueous solution of a dye, drug, or nucleic acid assembles to form aggregates that are anisotropic in shape. If the concentration of these aggregates is high enough and the shape of the aggregates is anisotropic enough, a nematic LC phase (the chromonic N phase) forms in which the aggregate axes possess a preferred direction as the aggregates diffuse, break up, and reform. At higher concentrations and/or lower temperatures, the aggregates form hexagonal arrays known as the chromonic M phase. The aggregates can be of monomolecular or multimolecular wide stacks and the lengths of the stacks are polydisperse in nature. The most widely studied chromonic system to date is disodium chromoglycate **16** (DSCG), an anti-asthmatic drug sold under the trade name Intal, followed by Sunset Yellow **17** (Figure 1.5), a food-coloring azodye. Both these compounds form nematic and hexagonal phases although at different concentrations. Similar behavior is shown by numerous other systems, including cyanine and azodyes. Though their potential as optical compensators, biosensors, patterned dyes, and as other functional materials have been demonstrated, the structure–property relationships of chromonic LCs are not understood to the same extent as amphiphile-based lyotropic LCs. A much better fundamental understanding of the molecular mechanism underlying the aggregation is required for the full potential of chromonics to be developed. Sometimes the chromonics are considered as the lyotropic analogues of thermotropic discotic liquid crystals (DLCs), but it should be noted that, unlike thermotropic DLCs, in chromonics the columns are not surrounded by insulating alkyl chains, and the column width is not necessarily of monomolecular width.

1.5 THERMOTROPIC LIQUID CRYSTALS

When the liquid crystalline phases are obtained by varying the temperature of the compounds, they are called thermotropic LCs. The mesophase can be obtained either by heating a solid or by cooling an isotropic liquid. The transition temperature from the crystal to the mesophase is called the *melting point*, while the transition temperature from the mesophase to the isotropic liquid is called the *clearing point*. When thermodynamically stable mesophases are obtained both on heating and cooling, the phases are called *enantiotropic*. If the mesophase is obtained only while cooling the isotropic liquid, it is called the *monotropic* phase and is a metastable mesophase since the mesophase transition occurs below the melting point. The essential requirement for a molecule to be a thermotropic LC is a structure consisting of a central core (often aromatic) and a flexible peripheral moiety (generally aliphatic chains). Moreover, the geometric anisotropy, interaction anisotropy, self-assembly, self-organization, and microsegregation are the driving parameters for mesophase formation. Based on the shape of the mesogenic molecules, thermotropic LCs are classified into three main groups: (a) calamitic (rod-like), (b) bent-core (bad rods, boomerang, banana-like), and (c) discotic (disk-like) LCs.

1.6 CALAMITIC LIQUID CRYSTALS

The most common type of molecules that form thermotropic mesophase are rod-like molecules. These molecules possess an elongated shape, that is, the molecular length (l) is significantly greater than the molecular breadth (b), as depicted in Figure 1.7.

The geometric shape anisotropy in combination with interaction anisotropy and microsegregation of incompatible parts in calamitic LCs leads to a number of mesophase morphologies. Most of

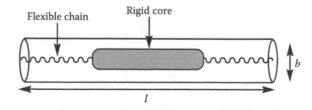

FIGURE 1.7 Representation of a calamitic LC molecule where $l \gg b$.

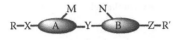

FIGURE 1.8 General template for calamitic LC molecules.

the calamitic liquid crystalline compounds consist of two or more ring structures, bonded together directly or via linking groups. They usually have terminal hydrocarbon chains and sometimes lateral substituents as well. The typical chemical structure of these molecules can be represented by the general template, as shown in Figure 1.8, where A and B are core units (benzene, naphthalene, biphenyl, etc.), R and R′ are flexible moieties such as normal and/or branched alkyl groups, M and N are generally small lateral substituents (–Cl, –Br, –NO_2, –CH_3, –OCH_3, –CN, etc.). Y is a linking group to the core units and X and Z are linking groups of terminal chains and core units.

The nature of the central core, linking groups, and lateral substituents impart significant effect on the mesophase morphology and physical properties of calamitic LCs. Calamitic LCs generally exhibit two types of mesophases: (a) nematic (from the Greek word *nematos* meaning "thread") and (b) smectic (from the Greek word *smectos* meaning "soap").

1.6.1 NEMATIC PHASE

The nematic phase has a high degree of long-range orientational order of the molecules, but no long-range positional order. This is the least-ordered mesophase (closest to the isotropic liquid state). It is denoted by "N." It differs from the isotropic liquid in that the molecules are spontaneously oriented with their long axes approximately parallel to each other (Figure 1.9). The term nematic is derived from the Greek word for thread and relates to the thread-like texture formed by defects between ordered regions visible under the polarizing optical microscope. The preferred direction of orientation of the molecules, depicted as a long arrow, is called the *director*.

Nematic LCs are primarily known for their extensive application in LCD devices. Due to their high fluidity (less viscosity), the effects of electric and magnetic fields are substantial on nematic LCs. They can be easily aligned and reoriented by the application of relatively small electric and magnetic fields, which is the basis for their practical use in LCD devices.

1.6.2 CHIRAL NEMATIC PHASE

The first thermotropic liquid crystalline material, discovered in 1888 by the Austrian botanist Reinitzer, exhibits what is now known as the chiral nematic phase (N*). Historically, the chiral nematic phase was called the *cholesteric phase* because the first materials exhibiting this phase were cholesterol derivatives. Today, this is not the case; there are many different types of chiral materials that exhibit the chiral nematic (cholesteric) phase and these have no resemblance to cho-

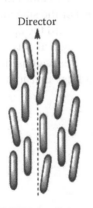

Director

FIGURE 1.9 Schematic representation of calamitic nematic LC phase.

lesterol. In fact, a chiral nematic phase can be generated by adding a small quantity of chiral material (not necessarily liquid crystalline) to a nematic material. It is thought that this is achieved by the chiral dopant creating a chiral environment for all of the other achiral molecules, and hence a helical macrostructure is generated. The asymmetry of the constituent molecules of the chiral nematic phase causes a slight and gradual rotation of the director. This gradual change describes a helix that has a specific, temperature-dependent pitch (Figure 1.10). The pitch of a chiral nematic phase is the distance along the helix over which the director rotates by 360°. The helical structure has the ability to selectively reflect light of a wavelength equal to that of the helical pitch length. If the pitch length is of the order of the wavelength of visible light, then colors are selectively reflected. The pitch length is temperature dependent and so is the color of the reflected light. This is the basis behind the commercially successful use of chiral nematic materials in thermochromic thermometer devices and other devices that change color with temperature. A racemic mixture (equal parts of each enantiomer) possesses an infinite pitch and, therefore, behaves as a normal nematic phase. However, if one enantiomer is in excess, then the mixture will exhibit the chiral nematic phase. Mixing the two optical isomers in various proportions allows the pitch of the chiral nematic phase to be changed from the pitch of either of the pure optical isomers. Two enantiomers of the same material will describe opposite twist senses for the helix.

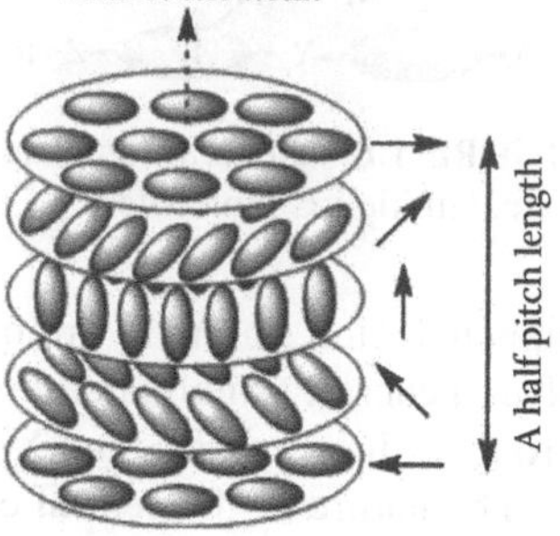

1.6.3 SMECTIC PHASES

Rod-like (calamitic) molecules are able to form liquid crystalline phases, where, in addition to the orientational order, the molecules possess short-range positional order as well. The molecules are arranged in layers with well-defined layer spacing or periodicity

FIGURE 1.10 Schematic representation of the cholesteric phase.

[50,99]. The interlayer attractions are weak as compared to the lateral forces between the molecules, and the layers are able to slide over one another relatively easily. This gives rise to the fluid property in the system but with higher viscosity than nematic phase. The smectic phase is denoted by the symbol "Sm." There are several types of smectic mesophases, characterized by a variety of molecular arrangements within and between the layers. Smectic LCs can be classified depending on whether the director is parallel or tilted with respect to the layer normal and depending on the in-plane and in-between-plane correlations. Accordingly, there are orthogonal and tilted smectic phases with structured and unstructured layers. Although the total number of smectic mesophases cannot be specified, the following types have been defined: SmA, SmB_{hex}, SmC, SmF, and SmI. However, SmA and SmC mesophases are more commonly encountered (Figure 1.11).

In the SmA phase, the molecules form layers with their long molecular axes orthogonal to the layer planes and hence the director is parallel to the layer normal. In the layers, the molecules exhibit very short-range positional ordering and there is no positional correlation between the layers. The SmC phase has the same layer structure of the SmA, but the molecules are tilted with respect to the layer normal. In this phase too, there is no positional correlation either within the layers or between the layers. Both smectic A and smectic C phases possess unstructured layers, that is, the layers correspond to that of a two-dimensional (2D) liquid. However, they possess one-dimensional (1D) layer periodicity (periodical stacking of the 2D fluid layers). SmB, SmF, and SmI phases are more ordered smectic phases in which the molecules possess hexagonal order within the layers (Table 1.1). However, the layers do not possess positional order between them. The molecules in the smectic B (hexatic) phase are orthogonal to the layer planes and located at the corners and centers of a network of hexagons. The smectic F and smectic I phases are tilted analogues of the smectic B phase, in which the molecules are tilted toward the side of the hexagonal net and toward the apex of the hexagonal net, respectively. Although there is little correlation between the hexagonal networks from one smectic plane to another, the orientation of the hexagons in different planes is consistent. This type of order is called *bond-orientational order*, with "bonds" describing lines that join the

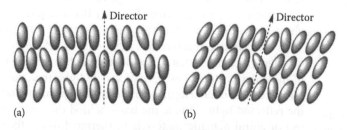

FIGURE 1.11   Side views of (a) smectic A and (b) smectic C LC phases.

TABLE 1.1
General Characteristic Structural Features of Achiral Smectic and Soft Crystal Phases

Phase Type	Orientation of Long Molecular Axis	Plane and Side Views	Positional Ordering	Distribution of Center of Mass
SmA	Orthogonal		Short range	Isotropic
SmC	Tilted		Short range	Isotropic
SmB (hexatic)	Orthogonal		Short range	Hexagonal
SmI	Tilted to apex		Short range	Pseudo-hexagonal
SmF	Tilted to side		Short range	Pseudo-hexagonal
Crystal B	Orthogonal		Long range	Hexagonal
Crystal J	Tilted to apex		Long range	Pseudo-hexagonal
Crystal G	Tilted to side		Long range	Pseudo-hexagonal
Crystal E	Orthogonal		Long range	Orthorhombic
Crystal K	Tilted to side *a*		Long range	Monoclinic
Crystal H	Tilted to side *b*		Long range	Monoclinic

centers of mass of nearest neighbor molecules and are not to be confused with chemical bonds. There are also smectic phases, both tilted and not tilted, that display strong correlation between the preferred positions of the molecules in the different smectic planes. These include the crystal B phase and its tilted analogues crystal G and crystal J phases, and the crystal E phase and its tilted analogues crystal H and crystal K phases. In crystal B and its tilted analogues, the molecules lie on hexagonal packing networks. In the crystal E phase, the molecules are orthogonal to the layer plane and produce an orthorhombic lattice; however, its tilted analogues crystal H and crystal K phases produce a monoclinic lattice with the tilt being toward the shorter edge of the packing net for the H phase and toward the longer edge of the packing array for the K phase (Table 1.1).

Thus, smectic A, C, B_{hex}, I, and F are essentially smectic LCs, whereas smectic B, E, J, G, H, and K are crystal phases. There is also the smectic D phase, which seems to have some type of cubic arrangement of the molecules. There is little evidence for a layered structure, and the phase seems to be optically isotropic. Apart from the non-chiral smectic phases, chiral analogues of the tilted phases also exist [50]. It is interesting to note that the original classification of the smectic modifications was not made through structural investigations of the phases, but rather by miscibility studies using phase diagrams. Phases that were found to be miscible over the entire concentration range of the phase diagram for binary systems were said to belong to the same miscibility group.

More recently, a frustrated smectic LC phase has been discovered that occurs above the temperature of the normal smectic phases in some chiral materials; this phase is called the *twist-grain boundary* (TGB) phase [100,101]. Similarly, another frustrated phase exhibited by materials that are highly chiral and occurs at a temperature above a chiral nematic phase is called the *blue phase* [102–104]. This phase exists for only a few degrees Celsius before the material clears to the isotropic liquid.

1.6.4   Smectic C* Phase

When the constituent molecules of the smectic C phase are chiral, the phase is called smectic C*. The phase structure is basically the same except that the molecular chirality causes a slight and gradual change in the direction of the molecular tilt (there is no change in the tilt angle with respect to the layer normal). This change in tilt direction from layer to layer gradually describes a helix (Figure 1.12). Since the tilted molecules of the smectic C* phase are chiral, the symmetry of the layers is reduced to a twofold the axis of rotation.

Even though the molecules are undergoing rapid reorientational motion, the overall result of the reduced symmetry is to create an inequivalence in the dipole moment along the C_2 axis of the layers. Such dipole inequivalence generates a spontaneous polarization (P_S) along the C_2 axis, the direction of which changes with the changing tilt direction of the helix. Accordingly, P_S is reduced to zero throughout a bulk sample, which is not influenced by external forces; hence the SmC* phase is truly defined as *helielectric*. However, a single layer of the chiral smectic C phase is ferroelectric, and so by unwinding the helix (by external forces such as surface interactions), a true ferroelectric phase is generated [52,53]. In the chiral smectic C phase, the molecular chirality in conjunction with molecular tilt makes the smectic layers polar and the polarity can be switched by the application of an appropriate electric field.

1.6.5   Ferro-, Antiferro-, and Ferrielectric Chiral Smectic C Phases

Ferroelectricity arises when the naturally helielectric smectic C* phase is unwound. If this unwinding of the helix allows the layer polarizations to point in the same direction, then the phase

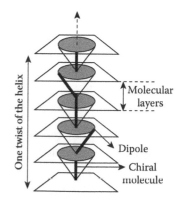

FIGURE 1.12 Helielectric structure of smectic C* LC.

One twist of the helix

Molecular layers

Dipole

Chiral molecule

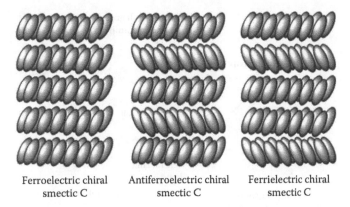

Ferroelectric chiral   Antiferroelectric chiral   Ferrielectric chiral
smectic C   smectic C   smectic C

FIGURE 1.13 Schematic representation of ferro-, antiferro-, and ferrielectric chiral smectic phases.

is called ferroelectric. This phase possesses a net bulk polarization. However, the unwinding of the helix can lead to two other phases known as antiferroelectric and ferrielectric phases (Figure 1.13).

The suggested structure of the antiferroelectric phase shows that the molecular layers are arranged in such a way that the polarization directions in subsequent layers point in opposite directions, which results in an averaging of the spontaneous polarization to zero. Evidence to support this hypothesis comes from the fact that when a strong electric field is applied to this phase, the layer ordering is perturbed and the phase returns to a normal ferroelectric phase. Removal of the field will generate the antiferroelectric phase. In the ferrielectric phase, the layers are stacked in such a way that there is a net overall spontaneous polarization. In this phase, the number of layers of opposite polarization is not equal, that is, more layers are tilted in one direction than the other.

1.7 BENT-CORE LIQUID CRYSTALS

Vorlander, a pioneer in the field of LC synthesis, has synthesized several bent-core LCs and reported the mesogenic properties in 1929, mentioning that the thermal stability of the mesophase is low compared to straight core analogues [91,105]. Matsunaga and coworkers also synthesized bent-core mesogens in the early 1990s [106,107] but did not realize the physical importance of the molecules before the discovery of polar switching in these compounds in 1996 [90]. Bent-shaped molecules have been thought of as "bad" molecules for forming LC phases, because if such molecules freely rotate about their long molecular axes, the excluded volume becomes large and violates the ability to form mesophase; therefore, only a few bent-core molecules were synthesized. However, in 1996 Niori et al. reinvented these molecules as banana LCs and showed ferroelectric switching in these achiral molecules [90]. The discovery of polar order in non-chiral banana-shaped molecules has led to very intense research activity in the field. Several hundred bent molecular shape compounds have been synthesized so far. Bent-shaped molecules provide access to mesophase with polar order and supramolecular chirality despite the mesogens being achiral [108–115]. Typically, their molecular structure can be regarded as being composed of three units: an angular central core, two linear rigid cores, and terminal chains (Figure 1.14).

There are numerous new mesophases that have been detected with the bent-core molecules. Most of them have no analogues in LC systems formed by conventional calamitic molecules. There are eight phases designated as B1, B2 to the most recent B8 according to the sequence of their discovery, where B stands for "bent-core," "banana," "bow," etc., and not to be confused with the smectic B phases of calamitic LCs. The distinctions between the phases are made on the basis of optical textures, the different characteristic x-ray diffraction diagrams, and electro-optical responses. Now, the number of new phases has risen and it turned out that some of the B phases have several subphases, that is, these phases actually represent families of related phases. There are new phases whose

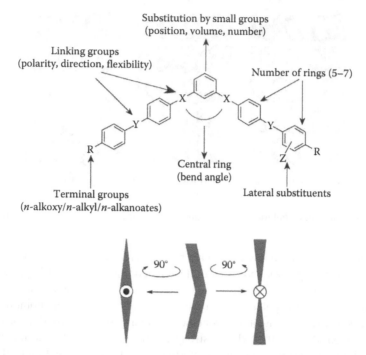

FIGURE 1.14   General template for banana LC molecules.

phase structures are not sufficiently well characterized at present. Moreover, some of the B phases appear to be soft crystals. The structures proposed for some of the mesophases are characterized by columnar order and others by lamellar order. In addition to the fascinating B phases, bent-core LCs also form conventional nematic and smectic phases. Bent-core mesogens are the first thermotropic LCs for which a biaxial nematic (N_b) phase has been unambiguously determined [116–119]. These compounds also exhibit biaxial and polar smectic A phases. None of the B phases are miscible with any smectic phase of calamitic compounds. Bent-core LCs open the door to novel complex types of molecular self-organization and to the new field of supramolecular stereochemistry. Various new application potentials of these materials include nonlinear optics (NLO), flexoelectricity, photoconductivity and the design of biaxial nematic phase, etc.

Simply bending molecules from rod-like shapes brings about new properties that cannot be imagined in calamitic LCs, particularly with regard to polarity and chirality. The bend in the rigid cores of the banana LC compounds leads to a reduction of the rotational disorder of the molecules about their long axes. The reduced symmetry of the rigid segments of such molecules leads to a directed packing of the molecules within layers. The important consequence of the directed packing of such molecules is the occurrence of polar order parallel to the smectic layers, provided that the molecules possess a lateral dipole moment. In the polar SmA (designated as SmAP) phase, the molecules are arranged, on an average, perpendicular to the layer planes, whereas in the most frequently observed polar SmC (designated as SmCP) phase, the molecules are tilted with respect to the layer normal. If the polar axes are the same in the adjacent layers, the SmAP or the SmCP phase are ferroelectric. However, if the polar axes alternate from layer to layer, the ground state is antiferroelectric. In case of the SmCP phase, four distinct structural variants can be distinguished, depending on the stacking of molecules in adjacent layers [110]. A synclinic or anticlinic arrangement of the molecules can occur, which is indicated in the code letter by suffixes "S" or "A" after C. On the other hand, ferroelectric structures (polar axes parallel) or antiferroelectric structures (polar axes antiparallel) are designated by the suffixes "F" and "A," respectively, after P. Thus, the structural variants are represented by SmC_SP_A, SmC_AP_A, SmC_SP_F, and SmC_AP_F (Figure 1.15). It should be emphasized

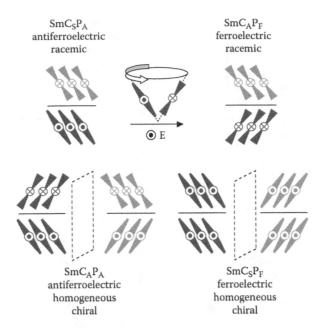

FIGURE 1.15 Supramolecular arrangements resulting from the combination of the different tilt directions and polar directions in bent-core LCs. (Reproduced from Reddy, R.A. and Tschierske, C., *J. Mater. Chem.*, 16, 907, 2006. With permission from the Royal Society of Chemistry.)

that the ground state is antiferroelectric in order to escape from a bulk polarization. In addition to the polarity, another aspect of fundamental interest is chirality of the layers. The combination of director tilt and polar order in the SmCP phase leads to a chirality of the smectic layers, even though the constituent molecules are achiral; this can be regarded as a "superstructural chirality." The three vectors, layer normal, tilt direction, and polar axis of a layer, can make up an orthogonal coordinate system, which may be left- or right-handed, and since left- and right-handed systems are mirror images of one another, the layers are chiral (Figure 1.16). Macroscopic samples are achiral. In the so-called racemic states, the chirality alternates from layer to layer whereas in the so-called homochiral states, the chirality is the same in adjacent layers. Since the chirality is *geometrical* in nature and does not result from a molecular chirality, both enantiomeric structures are equally probable. Moreover, in some cases macroscopic regions of opposite handedness are spontaneously formed. These dark conglomerates or dark racemates are optically isotropic phases with a local SmCP order. The occurrence of superstructural chirality in the mesophase of bent-core compounds without having any chiral auxiliary in the molecules is not only of fundamental scientific interest but also of industrial application, as this chirality can be switched in external electric fields. The formation of helical super structures in order to escape from a macroscopic polarization is another way to chirality in such supramolecular systems composed of achiral bent-core molecules. This has been referred to as "supramolecular chirality."

In order to escape from a macroscopic polarization, the layer structures are modified, and this leads to new mesophase morphologies in bent-core LCs. The new mesophases are interesting not only from structural point of view but also with respect to some unusual physical properties. Phase chirality is the consequence of the polar order in combination with the tilt of the bent-core molecules. Please remember that the polarity of chiral smectic C phase results from the combination of molecular chirality and layer tilt in calamitic LC systems. Alignment techniques developed for classical LCs are not applicable to these systems. Sample alignment is not only important for applications but also for basic research, as monodomain samples will allow better characterization and improved comparative studies. This is a big challenge. Even though some progress has been made

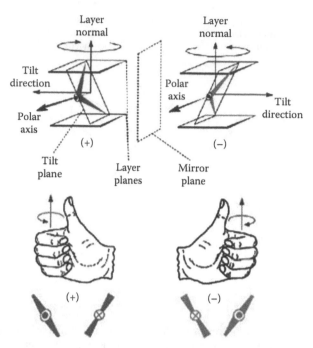

FIGURE 1.16   Origin of the superstructural (layer) chirality within the smectic phases of bent-core molecules. (Reproduced from Reddy, R.A. and Tschierske, C., *J. Mater. Chem.*, 16, 907, 2006. With permission from the Royal Society of Chemistry.)

to decrease the transition temperatures of these compounds, efforts to prepare bent-core LCs that satisfy the targeted commercial range are rare. Although significant progress has been made in this field, the literature shows that there are still many open questions that remain to be answered.

1.8   DISCOTIC LIQUID CRYSTALS

In September 1977, Chandrasekhar and colleagues at the Raman Research Institute, India reported "...what is probably the first observation of thermotropic mesomorphism in pure, single component systems of relatively simple plate like, or more appropriately disk-like molecules" [88]. They designed and synthesized a number of benzene hexa-*n*-alkanoates and investigated possible mesomorphism in this novel molecular architecture. By thermodynamic, optical, and x-ray studies, they could establish that these materials form a new class of LCs in which molecules are stacked one on top of the other in columns, and the columns in turn constituted a 2D hexagonal arrangement.

Self-organization of disk-like molecules forms an entirely new class of LC, quite different from the classical LCs formed by rod-like molecules that had been known since they were observed by Friedrich Reinitzer in 1888. The disk-like molecules spontaneously self-assemble into 1D stacks, which in turn self-organize on various 2D lattices; the third dimension has no translational order (Figure 1.17).

Mesophases formed by disk-shaped molecules are primarily of four types: (1) nematic, (2) smectic, (3) columnar, and (4) cubic. The columnar phase is ubiquitous in discotics followed by the nematic phase, whereas the other phases are rarely observed. Most of the discotics exhibit only one type of mesophase but a few examples are known to exhibit polymorphism [120–133]. A single discotic compound exhibiting all phase structures is not known in the literature. It is important to note that though columnar phases are most characteristic in discotic (disk-shaped) mesogens, they are by no means unique to discotics. Surfactants that aggregate into cylindrical micelles to form lyotropic

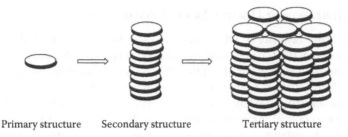

Primary structure   Secondary structure   Tertiary structure

FIGURE 1.17 Self-assembly and self-organization of discotic liquid crystalline molecules into columnar phases.

columnar phases are known for a long time, and even some rod-like molecules (polycatenar molecules), dendrimers, and bent-core mesogens are found to exhibit thermotropic columnar phases. Soon after the discovery of DLCs, terms such as "discotic phases" or "discotics" were used to denote the disk-like molecules as well as the (but not always) columnar phases formed by them. As this incorrect nomenclature for the structure of a phase and the shape of its constituting molecules created a source of ambiguity, it was strictly pointed out that "it is the molecule that is discotic and not the mesophase, which may be columnar, nematic, or lamellar."

1.8.1 STRUCTURE OF THE DISCOTIC MESOGENS

Since the discovery of the first discotic mesogens, most efforts have been aimed at understanding the nature of the molecular parameters that favor the formation of mesophases and control their transition temperatures. It is now well established that molecules forming DLCs are typically made of a central discotic core substituted by 3–12 saturated chains of three or more carbon atoms. These materials often have two-, three-, four-, or six-fold rotational symmetry. However, there are many exceptions, and materials with low symmetry, a nonplanar, nonaromatic core having shorter number of chains are also documented. The liquid crystallinity results from the microsegregation of the two constituents: the crystalline character is promoted by the interaction between the conjugated cores while the liquid character originates from the melting of the saturated alkyl chains in the mesophase. Such discotic molecules organize spontaneously in the form of 1D columns, which can be oriented easily and possess self-healing properties, that is, the capacity of repairing structural defects in contrast to crystalline materials. The search for such mesophases is mostly ruled by subtle changes in the number, size, and nature of the lateral chains in addition to the central core. A general template for discotic mesogens is shown in Figure 1.18. By tailoring the shape, size, and nature of the central core as well as the type of the attached side chains, compounds with different abilities to self-organize into different mesophase morphologies can be obtained.

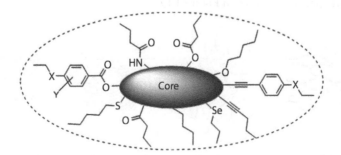

FIGURE 1.18 General template for DLC molecular architecture.

1.8.2 Characterization of Discotic Liquid Crystal Phases

The thermotropic phase behavior of DLCs, like any other LC, is usually studied by using differential scanning calorimetry (DSC), polarized optical microscopy (POM), small- and wide-angle x-ray scattering [134], and solid-state NMR [135,136]. DSC is used to determine the temperatures of phase transitions and enthalpy changes related to each transition. The fluid character of mesophases and in many cases the characteristics textures are easily observed by POM. DSC is always employed as a complementary tool to optical microscopy and reveals the presence of LC phases by detecting the enthalpy change that is associated with a phase transition. However, this technique cannot identify the type of LC phase, but the magnitude of enthalpy change does give some information about the degree of molecular ordering within a mesophase. The supramolecular organization and the corresponding packing parameters in each phase can be studied in detail by x-ray diffraction, in particular by 2D x-ray scattering of macroscopically orientated samples. This technique allows one to obtain a detailed insight into the intra- and intercolumnar order. It is not only possible to determine the intercolumnar spacings but also to obtain information about the arrangement of disks within the columns, such as tilting and helical packing, and also to provide deeper insight into the various microstructures adopted in the self-assembly of the mesogens in the mesophase. Solid state NMR is one of the most powerful tools for the study of the molecular dynamics in DLCs [135,136]. This technique allows one to derive independent conclusions about the rotation of the core or about the peripheral mobility of side chains. Moreover, because of the different electronic environments of the aromatic protons in the intracolumnar packing, the tilted arrangement of the disks in the solid phase can be determined as well. In general, it is necessary to apply all these complementary experimental methods in order to obtain a clear, comprehensive, and unambiguous picture of the bulk behavior of discotic mesogens.

The phase transition mechanism in DLCs is as follows: at the melting point of the molecular crystals, it is believed that the alkyl chains melt and provide fluidity to the mesophase, and at the clearing point, the unstacking of the central cores occurs leading to an isotropic liquid state. The phase transition from crystalline to the LC phase is accompanied by a significant increase in molecular dynamics such as the rotation of the disks around the columnar axis, lateral and axial displacement of the disks, etc. The centers of gravity of the molecules in the columnar LC mesophases are positioned along the column axis, and columns can slide relative to each other giving rise to the fluid character of the phase. It is important to stress that the molecular fluctuations in the liquid crystalline mesophases support the self-healing of structural defects and hence improve various physical properties such as the charge carrier mobilities and photoconductivity along the columnar stacks. The following section is devoted to the detailed structures of the various liquid-crystal phases of discotic molecules and their experimental identification and characterization by x-ray diffraction and polarizing optical microscopy.

1.9 STRUCTURE OF THE NEMATIC PHASES OF DISCOTIC MESOGENS

The nematic phases of disk-like molecules can be subdivided into four types: (1) discotic nematic (N_D), (2) chiral nematic (N_D^*), (3) columnar nematic (N_{Col}), and (4) nematic lateral (N_L). The structure of these nematic phases is shown in Figure 1.19.

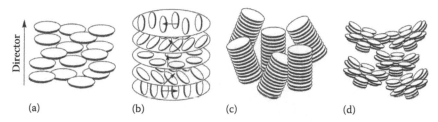

FIGURE 1.19 Structure of various nematic phases exhibited by discotic mesogens: (a) discotic nematic, (b) chiral nematic, (c) columnar nematic, and (d) nematic lateral.

Discotic nematic (N_D) phase is the least ordered, least viscous, and more symmetric mesophase among the other nematic phases [137]. In a discotic nematic mesophase, the discotic molecules possess full translational and rotational freedom around their short molecular axis (disk normal) but on an average, the short molecular axes are oriented in a preferred direction called the director **n**. In other words, the short molecular axes of the molecules orient more or less parallel to each other, while their centers of mass are isotropically distributed in the nematic phase. N_D has the same symmetry as that of calamitic nematic but both are not miscible with one another and hence phase separation occurs owing to the fundamental structural differences. Like chiral calamitic or cholesteric phase chiral discotic nematic phase N_D^* also exists [138]. The mesophase occurs in mixtures of discotic nematic and mesomorphic or non-mesomorphic chiral dopants as well as in pure chiral discotic molecules [139]. The helical structure of the chiral discotic nematic phase is shown in Figure 1.19b. The columnar nematic phase (N_{Col}) is characterized by a columnar stacking of the molecules. However these columns do not form 2D lattice structures [140–144]. They display a positional short-range order and an orientational long-range order. The columns behave like supramolecular rods and can be regarded as building blocks of the N_{Col} phase instead of single molecules. Recently, another nematic phase has been reported, where the disk-shaped molecules aggregate into large superstructure, and these supramolecular aggregates show a nematic arrangement. The phase is referred to as the nematic lateral phase (N_L) due to the strong lateral interactions [145–147].

All nematic discotic phases are fluid and exhibit typical *schlieren* textures (Figure 1.20) with two and four brush defects and marbled textures similar to that of a nematic phase of calamitic molecules. Similar to the textures of cholesteric phases of calamitic molecules, oily streaks, fingerprint, and polygonal textures are observed for the N_D^* phase. Both nematic columnar and nematic lateral phases, generally exhibited by charge transfer (CT) interactions, show deep colors under the microscope and their viscosities are higher than normally observed for discotic nematic phase. Transition enthalpies of different phase transitions depend on the amount of order present in the system. Since the discotic nematic phase is least ordered, the N_D–I transition enthalpy is usually less than $1\,kJ\,mol^{-1}$ whereas the isotropic transition enthalpies of nematic columnar and nematic lateral phases are $1–3\,kJ\,mol^{-1}$ and about $5\,kJ\,mol^{-1}$, respectively.

The x-ray diffraction profile of an N_D phase resembles that of an isotropic phase. The wide-angle diffraction peak is related to the lateral distance between the cores, while the small angle diffraction peak is attributed to the diameter of the core. In an aligned sample of the nematic phase, the small-angle reflections are normal to the reflections in the wide angle region (Figure 1.20). Owing to the negative diamagnetic anisotropy of discotic nematic LCs, the director of the discotic nematic phase will always be orthogonal to the applied magnetic field. An x-ray pattern of a columnar nematic phase shows relatively sharp reflection in the wide-angle regime that corresponds to the regular stacking of the discotic molecules on top of each other. The reflections in the small-angle regime are rather diffuse and broadly related to the liquid-like arrangement of the columns. Therefore, the

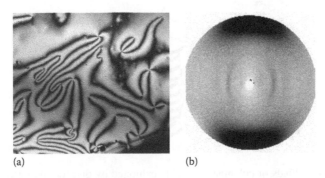

(a)   (b)

FIGURE 1.20 (See color insert following page 240.)   Typical (a) polarizing optical microscopic Schlieren texture and (b) x-ray diffraction pattern of an aligned sample of discotic nematic LCs.

columns are oriented more or less parallelly and have only short-range positional order, which is characteristic for a nematic mesophase. In the N_L phase, relatively sharp reflections are found in both the small-angle and the wide-angle regions of the diffraction pattern. The correlation length is a measure of spatial order in terms of the molecular dimensions. The correlation length shows low value for the N_D phase; in the N_{Col} phase, the correlation length increases in the columnar direction, and in the N_L phase, it increases in both columnar and lateral directions. Hence, XRD is a powerful technique to distinguish between these nematic phases.

1.10 SMECTIC PHASES OF DISCOTIC MESOGENS

When there is an uneven distribution of the peripheral chains or there is a reduced number of peripheral chains, the discotic mesogens exhibit smectic mesophase, as shown in the Figure 1.21. Like the calamitic smectic mesophases, in discotic smectic mesophases the disks are arranged in a layered fashion separated by sublayers of peripheral chains [148–155]. The smectic mesophases of discotic mesogens are otherwise known as discotic lamellar (D_L) phases. Since the molecular rotations about their long molecular axes will be restricted in the layers, they are expected to exhibit biaxial smectic phases. Characteristic optical textures have not been observed for the smectic phases; however, the x-ray diffraction pattern of the mesophase gives sharp diffraction rings in the small-angle region whose spacings are in the ratio 1:2:3, which suggests a lamellar organization of the molecules. Inside the layers, the molecules do not form any columnar aggregates. Unlike columnar phases, smectic phases are rare in DLCs.

1.11 COLUMNAR PHASES OF DISCOTIC MESOGENS

In columnar mesophases, molecules self-assemble one on top of the other in columns and these so-formed columns (Figure 1.22) self-organize in various 2D lattices. The molecules may be arranged in a regular ordered manner or aperiodically (disordered). Depending on the degree of order in the molecular stacking, the orientation of molecules along the columnar axis (orthogonal, tilted, helical, etc.), the dynamics of the molecules within the columns, and the 2D lattice symmetry of the columnar packing, the columnar mesophases may be classified into seven classes: (1) columnar hexagonal mesophase (Col_h), (2) columnar rectangular mesophase (Col_r), (3) columnar oblique phase (Col_{ob}), (4) columnar plastic

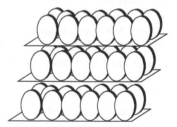

FIGURE 1.21 Smectic (discotic lamellar) phase of discotic mesogens.

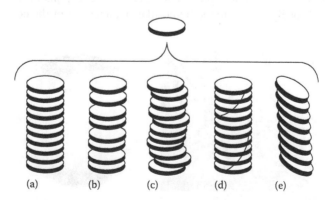

(a)   (b)   (c)   (d)   (e)

FIGURE 1.22 Different kinds of columnar assembly exhibited by discotic mesogens: (a) ordered column, (b) and (c) disordered columns, (d) helical column, and (e) tilted column. These self-assembled columns self-organize into different types of 2D columnar lattices.

phase (Col$_p$), (5) columnar helical phase (H), (6) columnar square (tetragonal) phase (Col$_{tet}$ or Col$_t$), and (7) columnar lamellar phase (Col$_L$).

1.11.1 HEXAGONAL COLUMNAR MESOPHASE

Columnar hexagonal mesophase is characterized by a hexagonal packing of the molecular columns. Hexagonal mesophases are often denoted as Col$_{ho}$ (Figure 1.23) or Col$_{hd}$ where "h" stands for hexagonal and "o" and "d" for ordered or disordered stacking of the molecules. In both cases, fluidity exists; only the correlation lengths are different and, therefore, it is recommended to discontinue o and d subscripts. The recommended abbreviation for columnar hexagonal phase is "Col$_h$." The planar space group of a hexagonal columnar mesophase is *P6/mmm*.

The 2D diffraction pattern and the x-ray scattering profiles of an unoriented (powder) sample of a columnar hexagonal phase are shown in Figure 1.24 [156]. In the small-angle region, the columnar hexagonal phase generally exhibits four diffraction rings whose spacings are in the ratio $1:1/\sqrt{3}:1/\sqrt{4}:1/\sqrt{7}$ along with two broad peaks in the wide-angle region. However, geometric considerations suggest that the Col$_h$ phase can, in principle, display more reflections in the small-angle region. Out of the two wide-angle reflections, one corresponds to the liquid-like packing of flexible alkyl chains and the other one, which is relatively narrow, corresponds to the intracolumnar stacking of discotic cores. Further evidence of the hexagonal arrangement of the columns is obtained by a 2D diffraction pattern of a monodomain sample. For a powder sample of a columnar hexagonal mesophase, scattering rings are observed; however, a monodomain sample results in six point-shaped reflections arranged in a perfect hexagon in the small-angle regime [157]. This can be obtained by sending the x-ray beam along the columnar axis (Figure 1.25). However, obtaining a monodomain aligned sample is not trivial.

It is possible to draw aligned fibers from a Col$_h$ mesophase, either by shearing or extrusion. All small-angle reflections related to the 2D hexagonal lattice are located on the equator of the x-ray pattern and the diffuse halo in the wide-angle regime is found on the meridian (Figure 1.25). While the reflections in the small-angle regime belong to the large periods of the hexagonal lattice, short distances correlate to the reflections in the wide-angle regime. For Col$_h$ mesophase focal conic, fan-shaped, mosaic, and dendritic textures are commonly observed (Figure 1.26).

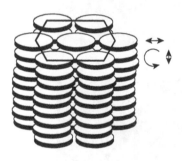

FIGURE 1.23 Structure of the hexagonal columnar phase of DLCs. The arrows indicate the lateral and axial displacements and rotational disorder of the molecules in the columns.

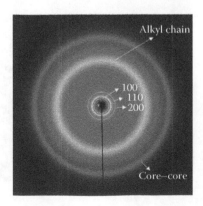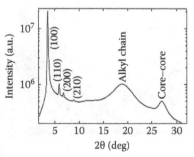

FIGURE 1.24 Typical x-ray powder diffraction pattern of columnar hexagonal phase and its intensity vs. diffraction angle profile. (Reproduced from Prasad, S.K. et al., *Mol. Cryst. Liq. Cryst.*, 396, 121, 2003. With permission.)

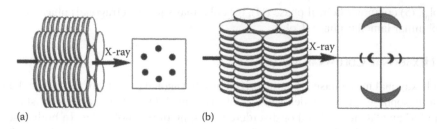

FIGURE 1.25 Schematic representation of x-ray diffraction patterns of (a) a homeotropic sample and (b) aligned fibers and illustration of columnar alignments toward the incident x-ray beam.

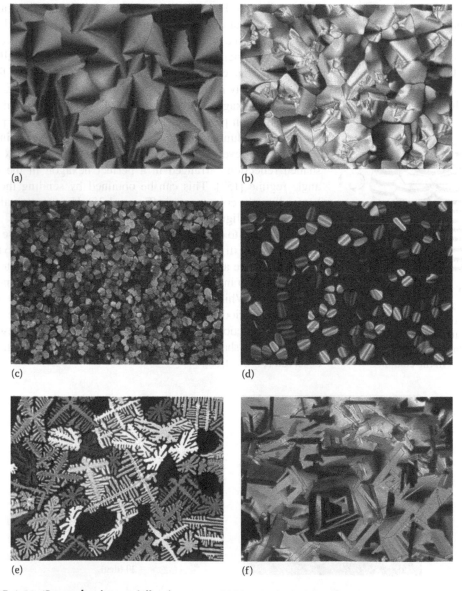

FIGURE 1.26 (See color insert following page 240.) Optical photomicrographs of different columnar phases under crossed polarizers: (a) focal conic, (b) pseudo focal conic, (c) mosaic, (d) texture with rectilinear defects of Col_h phases, (e) dehydritic texture, and (f) texture of helical (H) phase with crystalline order.

1.11.2 RECTANGULAR COLUMNAR MESOPHASE

The columnar rectangular mesophase consist of the stacking of the aromatic cores of the molecules in columns surrounded by disordered aliphatic chains and packed in a rectangular fashion. Three different types of rectangular mesophase Col_r have been identified (Figure 1.27). In general, the molecules are tilted with respect to the column axis [158–162], whereby the cross section, orthogonal to the axis of a column, is elliptic.

Depending on the mutual orientation of the molecules (ellipses) and the number of columns per unit cell in the lattice, Col_r phases have been divided into three different types. The symmetry of the 2D lattices are specified by three different planar space group, that is, $P2_1/a$, $P2/a$, and $C2/m$ (Figure 1.27), belonging to the subset of space groups without any translational order in the direction of the principal symmetry axis, that is, the direction of the columns. In the lattice with space group $P2_1/a$, the ellipses are oriented alternatively along two different directions (herringbone packing of elliptical columns); the long axes of the ellipses are orthogonal to the column axis and the lattice contains two columns per unit cell. The $P2/a$ lattice has elliptical columns having three different orientations and has four columns per unit cell, whereas in the $C2/m$ lattice the ellipses are oriented along a unique direction and the unit cell possesses two columns. Stronger core–core interactions are needed for the formation of the rectangular phases because the cores of one column have to "know" how they must be tilted with respect to the cores of the neighboring columns. Therefore, crossover from columnar rectangular to hexagonal mesophases with increasing side-chain lengths has often been observed. Two strong reflections (results by the splitting of the (10) reflection of the hexagonal lattice) in the small angle region of x-ray diffraction pattern are characteristic of a rectangular columnar phase (Figure 1.28). The indexation of a columnar rectangular mesophase and the determination of the lattice structure are complex

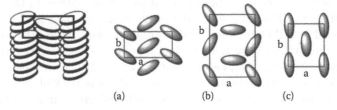

(a)   (b)   (c)

FIGURE 1.27 Structure of columnar rectangular phase and different types of rectangular columnar phases: (a) Col_r ($P2_1/a$), (b) Col_r ($P2/a$), and (c) Col_r ($C2/m$).

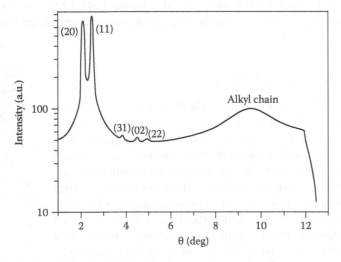

FIGURE 1.28 Typical x-ray diffraction profile of columnar rectangular phase.

and often not completely unambiguous. The clear discrimination between the different lattice structures requires the observation of large number of peaks in the x-ray pattern. Since only a limited number of reflections are actually observed, in practice an unequivocal determination of the symmetry is not reliable. Two-dimensional x-ray patterns of aligned samples can be most helpful to further clarify the symmetry of rectangular phases. As a result of the minor differences in the structures, textures known for Col_h phases can also be observed for Col_r phases. However, broken fan-shaped textures are more common for columnar rectangular mesophases than for columnar hexagonal ones.

FIGURE 1.29 Structure of columnar oblique phase.

1.11.3 COLUMNAR OBLIQUE MESOPHASE

Figure 1.29 shows the arrangement of the columns in a columnar oblique (Col_{ob}) mesophase, in which the tilted columns are represented by elliptic cross sections. The symmetry of this 2D lattice corresponds to the space group P_1. Examples of columnar oblique mesophases are rare because of strong core–core interactions [134,162,163]. Since $P1$ is a primitive planar space group, there are no reflection conditions like columnar hexagonal and rectangular phases and therefore all peaks are allowed. Hence, the assignment of the oblique mesophase by x-ray diffraction study is not so straightforward. Fan-shaped textures and spiral textures are characteristic for Col_{ob} phase.

1.11.4 COLUMNAR PLASTIC MESOPHASE

Columnar plastic phase, denoted as Col_p, has been identified recently in DLCs [156,164]. This phase is characterized by 3D crystal-like order of the center of mass of the molecules, but the columns are arranged in a 2D hexagonal lattice, while the disks within the columns are able to rotate about the column axis (Figure 1.30). In the case of Col_h phase, structural disorders such as nonparallel arrangement of the disks, longitudinal and lateral displacements, and rotation around the columnar axis occur, while the motional freedom of disks in the Col_p phase is restricted. The x-ray diffraction pattern of Col_p phase exhibits low-angle reflections that can be indexed to a 2D hexagonal lattice. In addition, the profile for the plastic phase has reflections having mixed indices. The presence of the diffuse peak due to the alkyl chains differentiates this phase from a truly crystalline phase. The wide angle core–core reflection, which is usually diffuse in the columnar phase, becomes very sharp and splits into two peaks (Figure 1.31). The texture of the Col_p phase is very similar to the textures observed for the hexagonal columnar phase, but the dendritic texture (Figure 1.26e) is often observed.

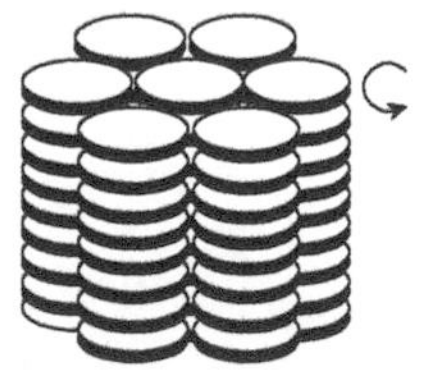

1.11.5 COLUMNAR HELICAL (H) PHASE

An exceptional mesophase structure with helical order has been demonstrated for triphenylene derivatives, namely, hexahexylthiotriphenylene (HHTT) [165] and a triphenylene ester derivative [166]. In this so-called H phase, there is a helicodal stacking of the triphenylene cores within each column, the helical period being incommensurate with the intermolecular (intracolumnar) spacing. In addition, a three-column superlattice develops as a result of the frustration caused by molecular interdigitation in triangular symmetry. In the superlattice, the third column is displaced by half an intracolumnar distance vertically with respect to the other two columns. The H phase found in HHTT is illustrated in Figure 1.32a. The x-ray diffraction profile obtained in the helical (H) phase is shown

FIGURE 1.30 Structure of columnar plastic phase— the curved arrow indicates the rotational disorder of the molecules in the columns. Note that the axial and lateral displacements of the molecules are not present in this phase.

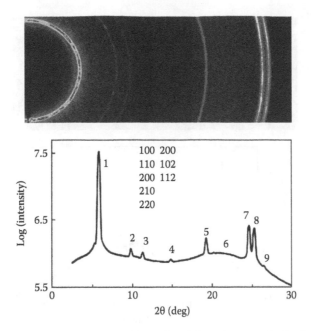

FIGURE 1.31 X-ray diffraction pattern and I–2θ profile of the columnar plastic phase. (Reproduced from Prasad, S.K. et al., *Mol. Cryst. Liq. Cryst.*, 396, 121, 2003. With permission.)

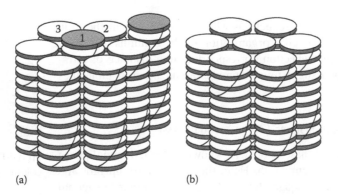

FIGURE 1.32 Structures of columnar helical phases: (a) helical phase with crystalline order and three-column superlattice and (b) helical phase with liquid crystalline order (this phase is a columnar hexagonal phase with intracolumnar helical order of the discotic molecules).

in Figure 1.33 and clearly indicates the 3D positional order of the H phase [156]. Figure 1.26f shows the texture of the H phase. There is another helical columnar liquid crystalline phase wherein the discotic molecules possess intracolumnar helical ordering and unlike the H phase of HHTT, this phase does not involve superlattice formation, as shown in Figure 1.32b [167].

1.11.6 COLUMNAR LAMELLAR MESOPHASE

A layered structure with columnar organization is known to exist for mesophases of certain discotic compounds [168–172]. Such a columnar lamellar mesophase, which is denoted by the symbol Col_L, is shown in Figure 1.34. In this phase, discotic molecules stack to form columns and these columns are arranged in layers, where the columns in layers can slide but the columns in different layers do not possess any positional (translational) correlation. The x-ray profile of this phase in the

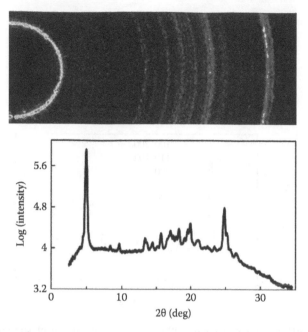

FIGURE 1.33 X-ray diffraction pattern and *I*–2θ profile of the columnar helical phase of HHTT. (Reproduced from Prasad, S.K. et al., *Mol. Cryst. Liq. Cryst.*, 396, 121, 2003. With permission.)

low angle region displays reflections whose spacings are in the ratio 1:2:3 suggesting the lamellar organization of the columns and there appears a wide angle reflection corresponding to a distance of intra-columnar separation suggesting columnar organization of molecules in the layers.

1.11.7 COLUMNAR SQUARE (TETRAGONAL) PHASE

The columnar square phase, otherwise known as the tetragonal phase (Col$_{tet}$), is shown in Figure 1.35. Here, the columns are upright and they are arranged in a square lattice. Like the columnar hexagonal phase, this phase also exhibits spontaneous homeotropic alignment of the columns. It is exhibited by some sugar molecules and phthalocyanine derivatives, etc. [173–178]. The dendritic texture [173] of the columnar tetragonal phase is very similar to the textures observed for other columnar phases. Therefore, on the basis of optical texture alone it is not possible to assign the phase structure and hence, x-ray diffraction studies have always been used to ascertain the phase structure where the diffraction pattern of the phase displays (10) and (11) reflections in the small angle region with a reciprocal spacing ratio of 1:√2 [173,174].

FIGURE 1.34 Structure of columnar lamellar phase.

FIGURE 1.35 Structure of columnar tetragonal (square) phase.

1.12 CUBIC PHASE

Cubic phases are ubiquitous in lyotropic LC systems; however, some discotic phthalocyanine derivatives [176,177] exhibit cubic phase, which seems to be a bicontinuous cubic phase. This phase consists of branched interwoven columns of discotic molecules (Figure 1.36), and the phase is optically isotropic in nature. The XRD pattern of the phase is consistent with a simple cubic lattice.

1.13 ALIGNMENT OF DISCOTIC LIQUID CRYSTALS

At the beginning of the DLC research, more attention was devoted to reveal and evaluate structure–property relationships in this new class of LCs. It was just a scientific curiosity to investigate new aromatic cores, which would provide DLCs and the effect of various substituents on the mesomorphism of DLCs. Accordingly, a variety of DLCs were prepared, and the relationship between the molecular structure and mesophase properties was revealed to some extent. However, the situation changed

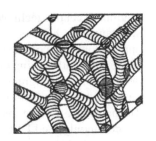

FIGURE 1.36 Structure of the cubic phase exhibited by DLCs.

when Haarer et al., in 1994, demonstrated that the columnar phases possess very high charge carrier mobility along the column axes [179]. It was realized that the supramolecular order of the columnar phases of the DLCs has the potential to act as active functional materials in organic electronic and optoelectronic devices. Then the attention was slightly diverted from structure–property relationships to structure–performance relationships of DLCs. Just after a couple of years in 1997, it was again demonstrated that the negative birefringence films formed by polymerized discotic nematic LCs can act as optical compensation films to enlarge the viewing angle and to increase the contrast ratio of thin film transistor liquid crystal display (TFT-LCD) devices [180–182]. Following these developments, an unprecedented growth of interest during the last decade has been motivated by the scientific interest in self-organizing systems and, as demonstrated [183], for technological applications of DLCs.

In order to take advantage of tailor-made discotic molecules for devices, it is of paramount importance to control the supramolecular structure of the molecules in functional structures and materials since functions such as electrical conductivity and charge migration depend crucially on the supramolecular arrangement of the material. The challenge is hence to control not only nearest neighbor interactions but also the nano- and microscale morphology of the sample by proper molecular design and choice of sample preparation method. Moreover, the molecules should be coded in such a way as to allow them to find their way from a disordered state in solution or isotropic melt into a highly ordered columnar state through proper recognition of their nearest neighbors.

DLCs are flexible enough to be spread on a plate and between two plates either from solution or from isotropic melt. However, uniform spreading is not achieved over a macroscopic area, which is desirable for the performance of electronic devices. Charge carrier mobilities in the discotic liquid crystalline system are still slower than those in organic single crystals. However, it is very difficult to prepare single organic crystals large enough to be used for electronic devices. DLCs are relatively easily processable and possess the property of self-healing of defects owing to their dynamic nature. Therefore, in this respect, DLCs are superior for low-cost solution and melt processing on plastic substrates for electronic and optoelectronic device fabrication.

The intrinsic high viscosity of columnar phases of DLCs owing to the strong π–π interaction is a major problem for their alignment and subsequent reorientation. Unlike calamitic LCs, which can be easily aligned and reoriented by small external stimuli such as electric and magnetic fields, DLCs cannot be easily aligned and reoriented because of their high viscosity. Moreover, the alignment methods developed for calamitic LCs hardly affect the alignment of DLCs. Aligned samples are required for various purposes such as to understand the detailed supramolecular order in the columnar phase (ordered, disordered, orthogonal, and tilted arrangement of disks, etc.), to evaluate the electronic properties and charge migration efficiency of the columns and conductivity anisotropy of the columnar phase, which are very important for their potential use in electronic devices as organic semiconductors. Very little work has been devoted in designing alignment layers for discotic columnar phase and very few systematic investigations on their surface alignment have been published. Presumably, the perceived lack of application for discotics is one of the reasons why the alignment control technology of discotics was lagging behind the calamitic LCs. DLCs can align spontaneously but the alignment is neither uniform nor monodomain in nature. From a practical applications point of view, it is highly desirable to have uniform uniaxial alignment of the columnar phases. Recently, owing to their application potential in organic semiconducting devices and display devices, diverse alignment control techniques for both columnar and nematic phases have been developed and now almost any kind of alignment can be achieved for discotic materials [184–188]. Triphenylene, truxene, phthalocyanine, porphyrin, and hexabenzocoronene (HBC) derivatives are the commonly used DLCs for alignment and property evaluation. They can be aligned on a single substrate, in between two substrates, or in microscopic and nanoscopic pores by processing either from solution owing to their solubility in common organic solvents or from isotropic liquid due to their low isotropization temperatures promoted by the radial flexible alkyl chains surrounding the central discotic cores.

1.13.1 ALIGNMENT CONTROL TECHNIQUES FOR DISCOTIC NEMATIC LIQUID CRYSTALS

The orientational control of LCs is highly required not only to comprehend the correlation of their orientational order with emerged anisotropic properties but also to optimize the performances of molecular devices using LC systems. There have been extensive reports on the orientational control of calamitic LCs by mechanical rubbing and external field effects as well as photo-alignment techniques, leading to their versatile applications such as in display devices. In contrast, the alignment technology for DLCs was not well developed in general, in particular the alignment of discotic nematic LCs. The alignment of discotic nematic (N_D) is hindered by the rare and high-temperature occurrence of the nematic phase and very few compounds are known to exhibit stable room temperature discotic nematic phase [189]. Moreover, alignment technology developed for calamitic nematic LCs are not very effective for the alignment of discotic nematic phase. However, recently, significant work has been done for the alignment control of discotic nematic LCs owing to their practical commercial application as negative optical compensation films in TFT-LCDs for widening the viewing angle and increasing the contrast ratio [182]. Discotic nematic LCs were discovered during the time when its calamitic counterpart was revolutionalizing the LCD technology. Obviously, it immediately caught the attention for the evaluation of various display parameters and the development of novel alignment techniques, then, being a new potential contender for LCD devices with negative optical anisotropy. Attempts have been made to align discotic nematic phase both horizontally and vertically between two coated glass plates to evaluate various physical parameters significant for display applications.

The first report dealing with the alignment of DLCs is by Vouchier et al. Glass surfaces and glass surfaces coated with disk-shaped aromatic molecules such as hexaphenol, mellitic acid, hexahydroxytriphenylene, and rufigallol promoted a homeotropic alignment of the nematic phase [190]. Similarly, freshly cleaved surfaces of crystals such as apophyllite and muscovite aligned the mesophase homeotropically. Uniform homogeneous alignment was achieved for the nematic mesophase on slides coated with obliquely deposited silicon oxide.

Nematic DLCs in sandwiched LC cells have been aligned by a 5T magnetic field and the quality of homogeneously aligned films was found to be dependent on the type of alignment layer [191]. Uniform alignment was achieved when a magnetic field was applied parallel to the cell surface during the cooling process. The director was perpendicular (homeotropic) to the surface when the cell was coated with polyimide alignment layer and parallel (homogeneous) when the surface was uncoated. The application of magnetic field was not effective to obtain uniformly aligned cells when the magnetic field was perpendicular to the cell surface. DLCs usually show negative magnetic susceptibility so that the disk-shaped cores align parallel to the magnetic field while the orientation of the director is confined to the plane perpendicular to the magnetic field. Thus, a monodomain can only be obtained if the sample is spun in a magnetic field to control the in-plane orientation of the director.

The alignment of discotic nematic LCs is also achieved by applying alignment layers [192]. Most of the rubbed alignment layers (polyvinyl alcohol, polyimides) provide flat adhesion of the molecular disks yielding splay or homeotropic alignment of the molecules. Two chemically modified polyimides (fluorinated polyimide) were found to provide planar orientation of the optic axis and the molecular disks adhere with their edges to the substrate. Using the chiral discotic compound for the fabrication of planar structure, films with twisted optic axis were obtained.

Polarized infrared irradiation technique has been used for the alignment control of columnar phases; however, this technique is not good enough for the alignment control of nematic phase as the nematic phase shows only dynamic turbulence of the texture and goes back to the original alignment [193]. The discotic nematic mesophase shows a strong tendency for homeotropic alignment between BaF_2 substrates. The alignment behavior of discotic nematic LC on self-assembled monolayers (SAM) of alkane thiol and asymmetrical alkyl disulfide has been studied by Monobe et al. [194]. The N_D phase shows typical *Schlieren* texture without any preferred alignment on gold and alkane thiol SAM, while the larger area of planar alignment was observed on asymmetrical alkyl disulfide SAM when thin films were used. However, nematic textures were observed for thicker films probably as a result of the hybrid alignment of discotic molecules. The alignment behavior of N_D phase has also been studied on polyimide and cetyltrimethylammoniumbromide (CTAB)-coated substrates [195]. The orientational behavior of the nematic phase on substrates coated with a polyimide and CTAB was investigated by POM, and average order parameter was evaluated by the infrared dichroic method. The discotic nematic mesophase exhibits a homeotropic alignment on a polyimide film and a tilted or planar homogeneous alignment on a CTAB-coated substrate. The order parameter is higher on a polyimide film than on a CTAB-coated substrate. In the case of the polyimide film, the molecular core is uniformly parallel to the substrate, while on CTAB, the discotic core is perpendicular to the substrate, surface. Recently, Ichimura et al. have reported a "command surface" effect in which nematic LCs homeotropically aligned on an azobenzene monolayer film change their alignment homogeneously by *trans–cis* photoisomerization [196,197] (Figure 1.37). Polymethacrylate films containing photoactive *p*-cyanoazobenzene groups have been used for the orientational control of DLCs. Conoscopic observations revealed that a film of polymer irradiated with linearly polarized light from surface normal induces solely homeotropic alignment with an orientational director of DLCs perpendicular to the surface plane. Oblique irradiation of the polymer film by nonpolarized light induces tilted alignment of the discotic molecules in the nematic phase or hybrid alignment containing a distribution of the DLC director. An average pretilt angle of 70° from the substrate plane was estimated by birefringence measurement. This is a novel procedure to control the alignment of DLC materials based on a surface-assisted photo-alignment technique.

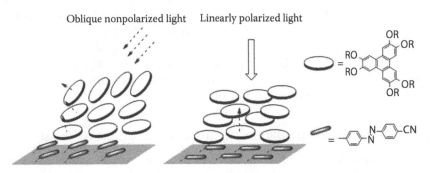

FIGURE 1.37 Tilted and homeotropic alignment of DLC molecules in the N_D phase on a thin film of a polymer containing *p*-cyanoazobenzene groups, exposed to irradiation obliquely with nonpolarized light and perpendicularly to linearly polarized light, respectively. (Adapted from Ichimura, K. et al., *Adv. Mater.*, 12, 950, 2000.)

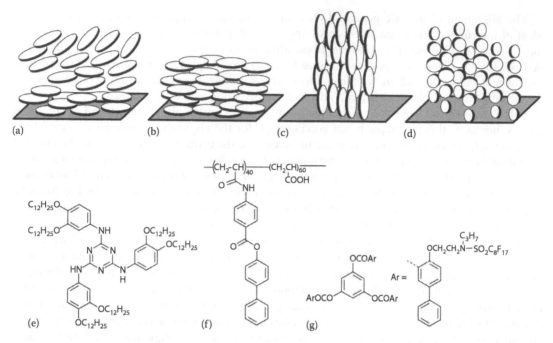

FIGURE 1.38 Various alignment structures of discotic nematic LCs: (a) hybrid, (b) homeotropic, (c) planar, and (d) vertical twisted. The alignment promoters used were (e) the horizontal alignment promoter from the air interface side, (f) the vertical alignment promoter from the substrate side, and (g) the vertical orientation promoter from the free air interface.

After theoretical optimization [181] and successful practical demonstration [182] of the utilization of hybrid aligned and photo-polymerized discotic nematic films in commercial TFT-LCDs, deliberate attempts have been made in Fuji photo film laboratory to achieve different kinds of alignment of discotic nematic phase, which are retained upon photo-polymerization and can be used as optical compensation films for any kind of LCD modes to widen the viewing angle and increase the contrast ratio of the device. In the hybrid alignment thin film, which is very effective in expanding the viewing angle, the molecules in the neighborhood of the orientation film adopt horizontal orientation whereas the molecules are tilted near the free interface, that is, the hybrid alignment structure is a mixture of splay and bend deformations [182]. The obtained thin films with appropriate alignment are durable and do not interfere with the transmittance and image quality of the display. They have established hybrid, horizontal, vertical, and vertical-twisted alignments of discotic nematic LCs on web-coated thin films with the help of their newly developed alignment promoters (Figure 1.38). Polymerizable hexabenzoates of triphenylene **18–20** (Figure 1.39) are the revolutionary materials, which have made the less abundant discotic nematic liquid crystalline materials more ubiquitous in a world with more LCDs than people by acting as the compensation films for widening the viewing angle. Among the above polymerizable compounds, the acryloyl derivative **18** is compatible for photopolymerization speed and thermal stability.

1.13.2 Alignment Control Techniques for the Discotic Columnar Phase

The columnar phases of DLCs can be considered as active functional components for electronic and optoelectronic devices, such as light-emitting diodes (LED), field-effect transistors (FET), and photovoltaic solar cells. The key physical processes for the optoelectronic applications are the formation, transport, and recombination/collection of electrical charges. In particular, the mobility of

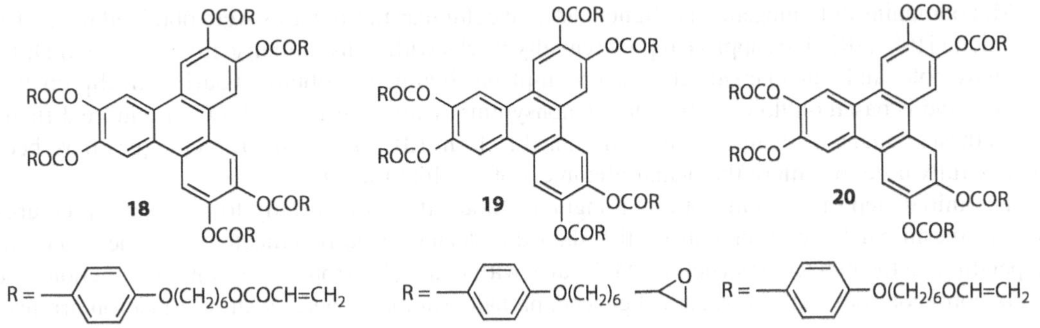

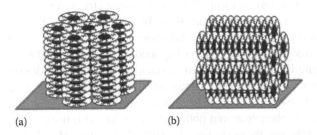

FIGURE 1.39 Chemical structures of the most important and revolutionary DLC compounds, which have found real practical application in LCD devices.

FIGURE 1.40 Schematic representation of (a) homeotropic and (b) planar alignment configurations of discotic columnar phases.

charge carriers in solid state materials is one of the most important parameters determining device performance. These electronic mobilities not only depend on the intrinsic electronic properties of the materials but also on the macroscopic order of the molecules in thin and thick films. Therefore, the control of their supramolecular order over macroscopic dimensions is a key issue to obtain optimized performance. The control of the molecular alignment on/between the surfaces is also an important issue during the discotic material processing. Columnar phases can exhibit two different alignments: the homeotropic alignment, where the molecules adopt face-on orientation on the substrate (Figure 1.40a), and the planar alignment, where the molecules adopt edge-on orientation on the substrate surface (Figure 1.40b). For FET, an edge-on organization (Figure 1.40b) of the discotics in uniaxially oriented columns is required. In this arrangement, charge carriers drift through the columns from the source electrode to the drain electrode, under controlled gate voltage. In contrast, the large monodomain face-on arrangement (Figure 1.40a) of the disks leads to a homeotropic alignment, which allows faster charge transport between the top and bottom electrodes, and favors the photovoltaic and light-emitting performance of DLCs. So detailed studies on the control of the supramolecular order and alignment on/between the surfaces leading to device applications are considered.

1.13.2.1 Planar Alignment of Discotic Columnar Phase
DLCs can form uniaxial alignment of the columns lying horizontal on different types of substrates by using appropriate processing techniques. These techniques include zone casting, friction transferred, and preoriented poly(tetrafluoroethylene) PTFE, Langmuir–Blodgett (LB) technique, crystalline surfaces, zone melting, magnetic field, field-force alignment, circularly polarized infrared radiation, etc.

Monodomains of homogeneous aligned discotic columnar mesophases were obtained by the LB technique [198–201]. This approach preferentially works with nonsymmetrically substituted DLCs that have polar and nonpolar side chains. The in-plane alignment is obtained during the dip-coating process and is based on flow fields. Though nonsymmetrically substituted DLCs are more difficult to synthesize, many of them have been designed to form LB films, and LB techniques have been successfully used to achieve the planar alignment of amphiphilic DLCs.

The initial step in the study of any Langmuir–Blodgett film is usually to measure a pressure–area isotherm. Such measurements of disk-shaped molecules usually indicate one of the two forms depending on the molecular structure. Molecules with relatively strong core–water interactions and weak core–core interactions, such as benzene, phthalocyanine, generally show a face-on arrangement with the cores cofacial to the water surface and the aliphatic chains extending away from the surface [198]. In contrast, molecules with stronger core–core attractions such as truxene, triphenylene show an edge-on arrangement. After transfer to solid substrates via LB techniques, such films have been studied by various structural probes. Those films that are edge-on at the water interface are most often found to form edge-on columnar structures, similar to those in the 3D columnar LC after transfer to the solid substrate. In contrast, the face-on molecules are believed to form columns extending away from the substrate but not necessarily always. Langmuir films of disk-shaped molecules provide an interesting model system for a 2D system with anisotropic interactions. In Langmuir monolayers, where π-stacking is allowed to dominate over alkyl chains packing, the structures formed at the air–water interface resemble bulk structures of similar molecule [199,200]. The Langmuir–Blodgett technique, which is one of the most frequently applied alignment methods for triphenylene and phthalocyanine, has also been established for the orientation of HBCs [202–204]. For these studies, the HBC derivatives **21, 22** (Figure 1.41) were asymmetrically substituted and terminated by a carboxylic acid group to provide the desired amphiphilic character [202]. The molecules formed well-defined monolayer when spread from a solution at the air–water interface. Efficient transfer of the monolayer to substrate by vertical dipping gave well-defined multilayer films (Figure 1.42). Columns were oriented along the dipping direction with disk planes perpendicular to the columnar axes and stacked in a cofacial manner, as has been observed for triphenylene derivatives **40** (Figure 1.43). Further improvement of the LB alignment was found when poly(ethylene imine) functionalized substrates were used as anchor points for the first layer [203]. The amphiphilic HBC disks formed a macroscopic in-plane orientation of the columns with their main axes parallel to the dipping direction.

Self-assembled monolayers (SAM) of DLCs on gold surfaces displayed properties similar to LB films. Planar alignment has been achieved by forming SAM of thiol-terminated discotics **36, 37** (Figure 1.43) and disulfide-substituted triphenylenes **39** on gold surface [205]. The cores of triphenylenes **36, 37, 41** with 1–2 thiols attached align homogeneously and formed domains of columnar stacks with different in-plane orientation [206]. A face-on alignment was found for the derivative **32** with thioethers attached to the core [206]. Terminal thiols attached to more than two side chains per molecule **42** resulted in the formation of nonuniform layers. Substrate surfaces modified by SAM of alkanethiol and disulfides have been found to promote the planar alignment of discotic columnar phases when spin coated from solutions [207]. Planar alignment has also been achieved in a metal–SAM–metal junction with an HBC derivative **23** bearing a dithiolane functionality [208]. The tunneling of electrons across the columns indicate that HBC units are transparent to electrons as compared to the aliphatic chains.

Discotics can be successfully aligned by the zone-crystallization technique. Amorphous or polycrystalline films of DLCs have been recrystallized into single-crystal-like thin films of micrometer thickness by a zone-melting technique [209], in which an electrically heated wire generates a narrow molten zone on the organic layer sandwiched between two pieces of glass or indium tin oxide-coated glass. When the molten zone was moved slowly across the layer from one end of the cell to the other, a single-crystal-like film was produced after a single pass. Accordingly, an HBC derivative **24** was aligned by the zone melting method [210]. The sample was moved at a defined velocity from a hot plate with a temperature above the isotropic phase to a cold plate with a temperature lower than

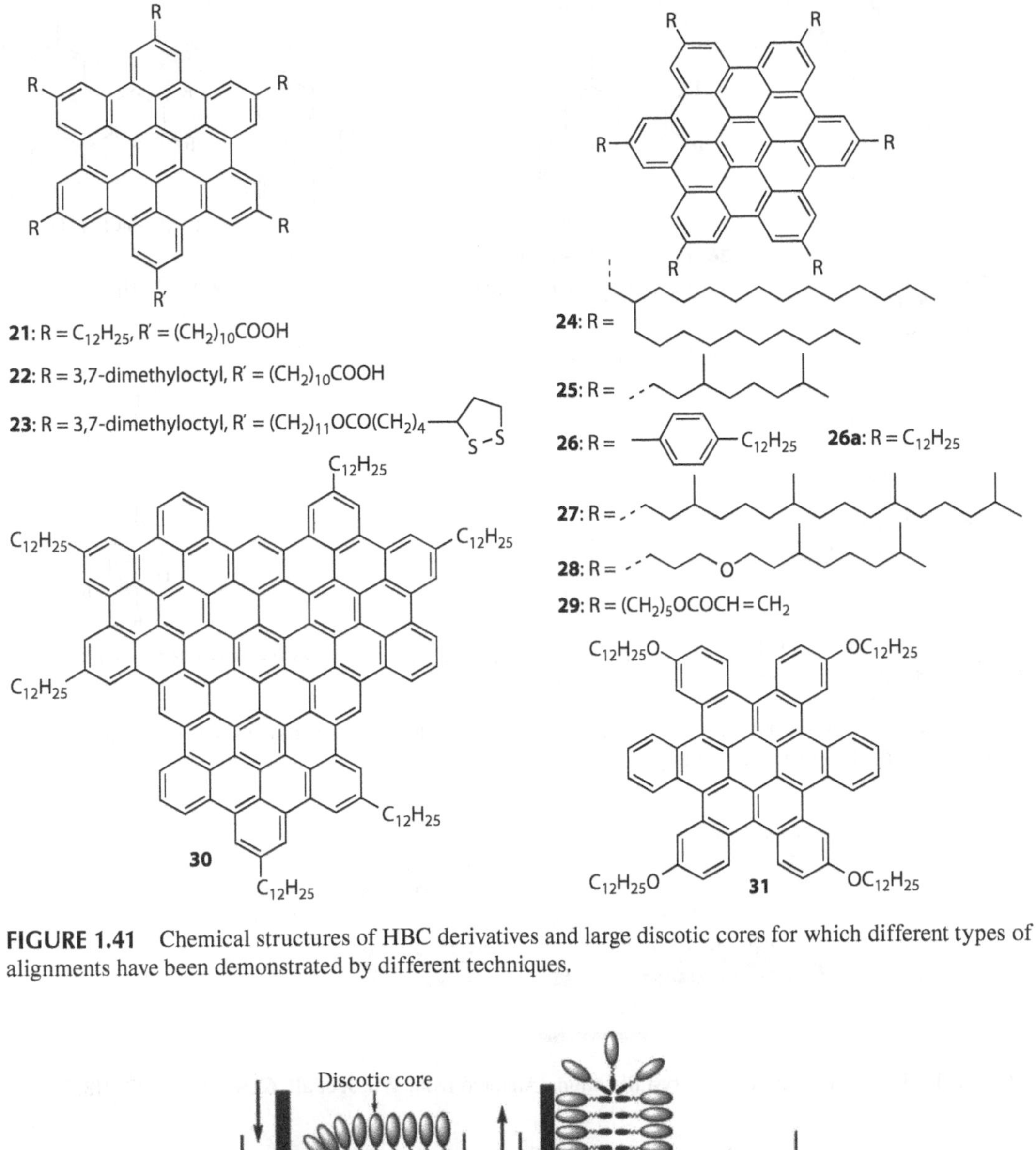

21: R = $C_{12}H_{25}$, R′ = $(CH_2)_{10}COOH$
22: R = 3,7-dimethyloctyl, R′ = $(CH_2)_{10}COOH$
23: R = 3,7-dimethyloctyl, R′ = $(CH_2)_{11}OCO(CH_2)_4$
24: R =
25: R =
26: R = $C_{12}H_{25}$
26a: R = $C_{12}H_{25}$
27: R =
28: R =
29: R = $(CH_2)_5OCOCH{=}CH_2$
30
31
$C_{12}H_{25}$
$C_{12}H_{25}O$
$OC_{12}H_{25}$

FIGURE 1.41 Chemical structures of HBC derivatives and large discotic cores for which different types of alignments have been demonstrated by different techniques.

Discotic core
Polar group

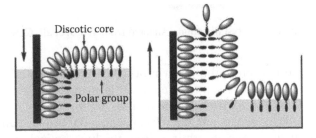

FIGURE 1.42 Schematic illustration of the deposition of discotic molecules onto a substrate with the edge-on arrangement using the Langmuir–Blodgett technique. (Adapted from Wu, J. et al., *Chem. Rev.*, 107, 718, 2007.)

the isotropic temperature. Between these two plates the material crystallizes along the temperature gradient as an aligned film (Figure 1.44). After zone crystallization, the sample revealed a high homogeneous structural order with columns oriented uniaxially, edge-on, and aligned in the moving direction. Two-dimensional x-ray scattering revealed the structure and POM revealed the high optical anisotropy. A closer inspection by AFM showed the lamella morphology of the film.

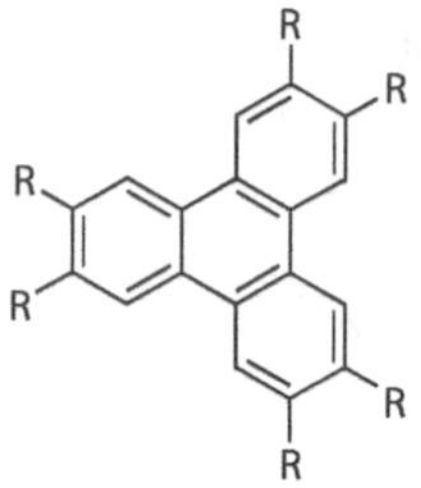

32: R = SC$_6$H$_{13}$
33: R = OC$_5$H$_{11}$
34: R = OC$_6$H$_{13}$
35: R = O(CH$_2$)$_3$C$_4$F$_9$

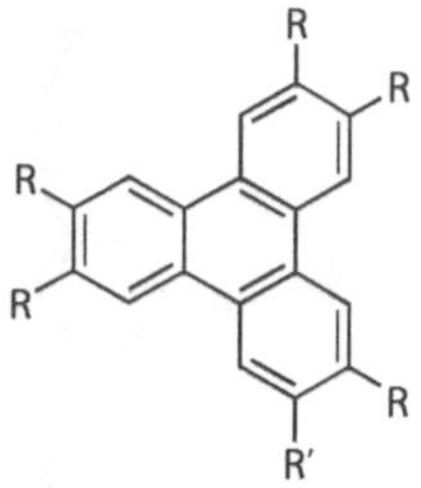

36: R = OC$_5$H$_{11}$, R′ = O(CH$_2$)$_{10}$SH
37: R = OC$_6$H$_{13}$, R′ = O(CH$_2$)$_5$SH

38: R = OC$_4$H$_9$, R′ =

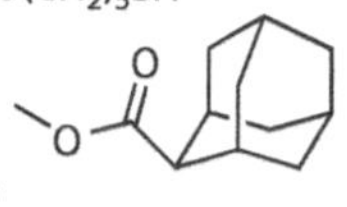

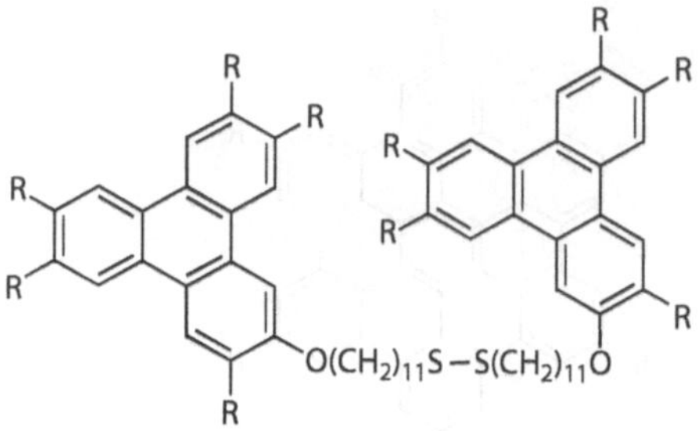

39: R = OC$_5$H$_{11}$

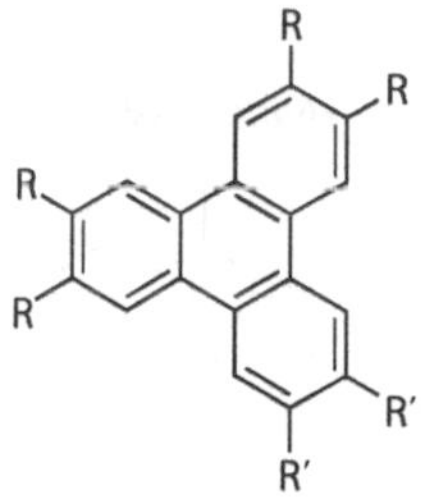

O(CH$_2$)$_{11}$S–S(CH$_2$)$_{11}$O

40: R = OC$_5$H$_{11}$, R′ = O(CH$_2$)$_6$OH
41: R = OC$_6$H$_{13}$, R′ = O(CH$_2$)$_5$SH

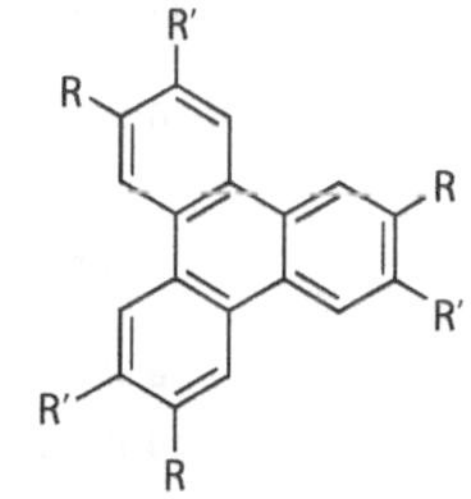

42: R = OC$_6$H$_{13}$, R′ = O(CH$_2$)$_5$SH

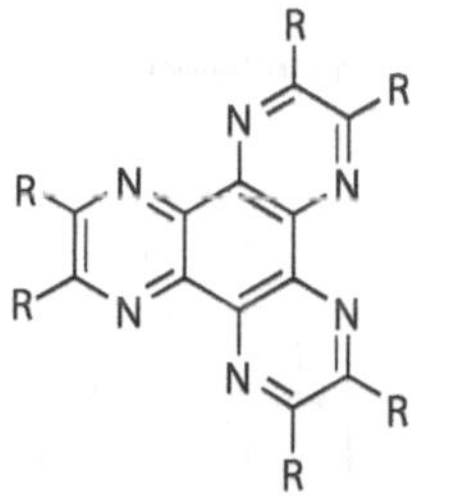

43: R = CONHCH$_2$CH(C$_2$H$_5$)C$_4$H$_9$

FIGURE 1.43 Chemical structures of triphenylene derivatives for which different kinds of alignments have been demonstrated by different techniques.

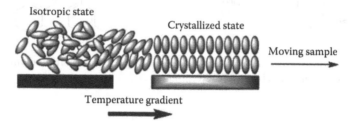

Isotropic state

Crystallized state

Moving sample

Temperature gradient

FIGURE 1.44 Alignment by zone crystallization. (Adapted from Wu, J. et al., *Chem. Rev.*, 107, 718, 2007.)

Another way to achieve a uniaxial alignment of columns lying parallel to the substrate is the use of an alignment layer to benefit from epitaxial growth. There are various examples of PTFE-aligned DLCs [211–215]. A triphenylene derivative **38** was spin coated onto the PTFE orientation layer, and after evaporation of the solvent, a uniaxial in-plane alignment of columns along the orientation direction was determined from the high optical anisotropy observed in POM [214]. The optical absorption and photoluminescence spectra were significantly polarized. Annealing the film at higher temperature further increased the degree of orientation. Similarly, an identical alignment was observed for HBC derivatives **25, 26** that were deposited from solution onto the PTFE film [211–213]. The structure evaluated by grazing-incidence x-ray diffraction and high optical anisotropy in polarized absorption indicated well-aligned HBC films with a parallel orientation of the columnar stacks on the underlying PTFE chains. The implementation of the HBC layers in an FET showed promising device performance with high mobility and a high anisotropy in conductivity with maximum values obtained for the directions along the columnar alignment.

Another innovative way to macroscopically align discotic HBC as a thin film by solution processing is the application of a strong magnetic gradient [216]. Using this technique with a horizontal 20T gradient, a solution of HBC **26** was cast onto FET wafers. This resulted in large-area monodomain films as proven by POM and x-ray experiments. The HBC molecules were found to align edge-on with their planes along the applied magnetic field, whereas the close-packed columns were oriented ~40° with respect to the gradient. FETs with high charge carrier mobilities have been demonstrated, which are significantly enhanced with respect to the unaligned materials. Recently, field-force alignment [217,218] of HBC **26** has been successful for obtaining planar alignment. A solution of HBC was applied to a glass surface by drop casting and the molecules were oriented into highly ordered structures by an electric field during solvent evaporation. AFM, SEM, and TEM showed a long-range alignment where the disk-like molecules were organized in columns perpendicular to the direction of the imposed electric field. The high anisotropy of the uniaxially aligned films was characterized by cross-polarized light microscopy.

Zone-casting technique has been shown to be very efficient in forming thin films with a higher macroscopic order from solution [219–226]. As shown in Figure 1.45, a solution of the organic material is deposited by a nozzle onto a moving support. Here, a concentration gradient is formed between the nozzle and the support. At a critical concentration, the material is nucleated from the solution onto the moving support as an aligned thin layer. In case of DLCs capable of π-stacking, the optimal processing conditions closely related to the self-aggregation of the molecules in solution. Using the zone-casting technique, HBC derivatives **24**, **26**, **27** were successfully aligned into highly ordered surface layers. The success of the technique comes from the strong self-aggregation and pronounced size of the preaggregates already existing in solution before the deposition. AFM-tapping mode revealed pair-packed columnar structures uniaxially oriented in the deposition direction with exceptionally high column length. HR-TEM displayed a perfect orientation suggesting a high supramolecular order with edge-on arranged molecules [224]. Additionally, the filtered inverse fast Fourier transform image disclosed an obvious intracolumnar periodicity corresponding to the characteristic stacking of the disks. X-ray diffraction of the thin layer showed that the disks were arranged in a herringbone order with respect to the stacking axis. However, heating the zone-cast film to the mesophase induces a significant change in the optical behavior from near net zero optical anisotropy in POM to highly anisotropic medium above the phase transition temperature. This change was related to the rotation of the disks from a herringbone arrangement to the cofacial packing. FET devices based on zone cast films displayed superior device performance owing to the highly ordered and uniformly oriented HBC **26a** disks (HBC-C_{12}). Planar alignment was achieved for spin-coated Hexa-*cata*-hexabenzocoronene **31**, which showed promising field effect transistor properties [225]. HBCPhC$_{12}$ **26** and a larger discotic compound **30** (C_{96}–C_{12}) possess an apparently

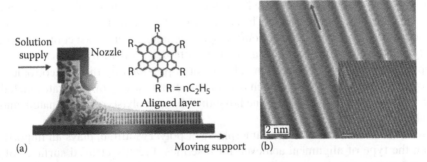

FIGURE 1.45 (a) Schematic illustration of zone-cast HBC-C12 and (b) filtered inverse FFT image showing the intermolecular periodicity within the columns; the arrows in (b) indicate the deposition direction. (Reproduced from Pisula, W. et al., *Adv. Mater.*, 17, 684, 2005. With permission. Copyright Wiley-VCH Verlag GmbH & Co.)

lower self-organization in drop cast films indicating a poorer self-aggregation in solution [221]. However, zone-cast films of both compounds displayed high macroscopic uniaxial orientation of the columns with a molecular edge-on arrangement on the support. Electron diffraction studies implied a high intracolumnar periodicity but a low intercolumnar correlation of the disks due to increased molecular dynamics in the liquid crystalline phase.

1.13.2.2 Homeotropic Alignment of Discotic Columnar Phases

The face-on homeotropic alignment of DLCs has been achieved either in between two substrates or on a single substrate, and the alignment can be obtained from isotropic melt or from solution by different solution processing techniques [188]. This is a thermodynamically preferable alignment and spontaneously occurs for different DLCs during cooling of the isotropic melt. Spontaneous homeotropic alignment of DLCs on substrates or between two electrodes is termed as "self-processing." It is noteworthy that the *self-assembled* columns of DLCs *self-organize* in different 2D lattices, which possess two very important and attractive properties that is *self-processing* on/between substrates and *self-healing* of structural defects in the columnar phase. All these *self-contained* properties render them as potential functional materials for many semiconducting device applications and models for energy and charge migration in self-organized dynamic functional soft materials. Annealing is used to improve molecular ordering and promote non-tilted homeotropic orientation in thin films of discotics. It has been shown that an improved homeotropic order improves the photoconductivity of a sample **27** as compared to poorly oriented sample [227].

Homeotropic aligned samples do not show birefringence in the POM between crossed-polarizers, since the optical axis in this case coincides with the columnar axes. In order to distinguish between the homeotropically aligned columnar and the isotropic phase, additional experiments are necessary; x-ray diffraction is one of the best methods, which also precisely confirms the alignment.

By changing the chemical nature of the aromatic core and/or of the side chains, it is possible to modify the surface affinity of the molecules. Moreover, the surface affinity can also be modified by treating (modifying) the substrate surfaces as well. Homeotropic alignment between two solid substrates has been reported for many different discotic molecules and is independent of the core size and film thickness. However, there are essential differences between the alignments on one or between two solid substrates. Only very limited examples of successful homeotropic alignment on one solid substrate have been described [228], where the self-alignment of the disks was strongly dependent on the film thickness. For homeotropic alignment, the first layer of discotics in contact with the surface presumably acts as the nucleation center and the sample grows vertically. It has been speculated that some specific substrate–molecule interactions are involved in homeotropic alignment but there are no systematic studies to uncover this situation. It has been observed that all-hydrocarbon discotics **27** exhibit spontaneous homeotropic alignment like heteroaromatic systems **28** [227]. While homeotropic alignment is in fact the thermodynamically preferred orientation and does not require heteroatoms for specific molecule–surface interactions, it is greatly enhanced by their presence. The heteroatoms can interact with substrates such as glass and aluminum surfaces, and this allows homeotropic alignment even at fast cooling rates and more significantly on a single surface, which has not been attained with all-hydrocarbon discotics. The more complex mechanism of homeotropic alignment is provided by the homeotropic alignment of phthalocyanine **44** (Figure 1.46), which occurs between a variety of substrates including gold, glass, ITO, PEDOT-PSS (poly(3,4-ethylenedioxy-thiophene)-poly(styrenesulfonate)), and polyisobutylene rubber [229].

The structure of the substrate, either amorphous or polycrystalline, plays an important role in determining the type of alignment achieved for discotics. Freshly cleaved surfaces of apophyllite (a lamellar tetragonal silicate) and muscovite (a monoclinic mica) promote the homeotropic alignment of discotics [190]. Like the discotic nematic phase, the columnar phase of discotics also exhibits face-on orientation between two substrates coated with discotic surfactants like hexaphenol, hexahydroxytriphenylene, and rufigallol. Experiments for the alignment of several

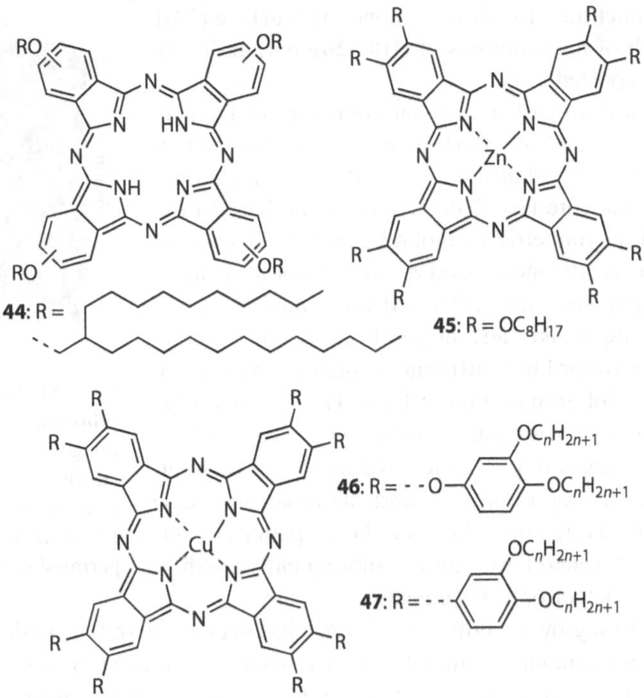

FIGURE 1.46 Chemical structures of discotic liquid crystalline phthalocyanine derivatives for which homeotropic (face-on) alignment has been demonstrated by different techniques.

discotic triphenylenes **33**, **34** have been carried out on a number of different substrates. These are Si, CaF_2, ZnSe, ZnS, and BaF_2. The results experiments show that in most cases (depending on the cleanness, flatness and crystallinity, amorphous/isotropic), the alignment achieved is homeotropic for discotics in the columnar phase if the sample is cooled slowly from the isotropic phase between the two substrates [188,230]. However, if the substrate surface is uneven like ZnSe and has a crystalline structure, the alignment achieved is planar or edge-on. Nevertheless, if the sample is cooled to the glassy phase and left at a lower temperature for several hours even on a Si substrate, then the alignment changes from homeotropic to planar heterogeneous. In other words, the column axes get parallel to the surface of the substrate and these are more or less randomly distributed along the surface. This shows that planar heterogeneous alignment has the lowest energy especially in the crystalline state. When the sample is slowly heated, the alignment continues to be planar, but if the sample is kept for several hours (annealed) in the columnar phase, then the alignment changes slowly from planar to homeotropic. This is termed as "anchoring transition," where the column axes go from being parallel to the substrate to being normal to it [188]. This implies that the homeotropic alignment is thermodynamically more stable. The anchoring transition from planar to homeotropic alignment is usually observed on heating the sample and bringing it close to the isotropic temperature. The anchoring transition is dependent on the temperature and the substrate.

Face-on self-assembled monolayers (SAM) of discotics were studied on highly oriented pyrolytic graphite (HOPG). Monolayers of discotic molecules lying flat on the substrate can be obtained by the deposition of dilute solution onto HOPG, as reported for triphenylene **33** [231–233], hexaazatriphenylene **43** [234], and HBC **26a** [235–238]. In this case, the main driving force for the face-on organization is the interaction between the disk and the surface. Graphite surfaces generally promote the homeotropic alignment of columnar mesophases (Figure 1.47). Hexaalkoxy triphenylenes are also known to form face-on alignment layers on HOPG, and surface-induced chirality has been observed for nonchiral discotics on a nonchiral surface [231]. Recently, the growth of nanocolumns of HBC **26a** perpendicular to HOPG has been reported [238].

As has been mentioned earlier, face-on alignment of triphenylenes can also be obtained by SAM technique on gold surface by controlling the number and position of the thiol anchor groups attached to the discotic core [205,206]. Triphenylene-based thioethers **32** [205] and a symmetric trithiol **42** [206] lie face-on on gold substrate; however, the mono- and *ortho* di-thiol derivatives display edge-on alignment with highly ordered planar columns. This alignment technique takes advantage of the specific interaction of gold substrate toward the sulfur atoms present in the thioethers and terminal thiol groups (Figure 1.48). The self-assembly of disks at interfaces is also of great importance because it relates to the charge carrier transport at organic molecule/substrate interfaces and also opens the opportunity to fabricate molecular scale electronics. Alternatively, the monolayer on these "patterned surfaces" can act as the "ground floor" for the subsequent growth of vertical columns for homeotropic alignment.

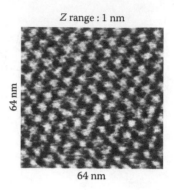

Z range : 1 nm

64 nm

64 nm

FIGURE 1.47 (See color insert following page 240.) STM image of a triphenylene discotic on an HOPG surface. (Reproduced from Gupta, S.K. et al., *J. Phys. Chem. B*, 113, 12887, 2009. With permission from ACS.)

Similarly, the face-on growth of discotics **25**, **26a** has been observed on MoS$_2$ surface also [239]. Highly ordered, several nanometer thick films have been grown on MoS$_2$ substrate. The films are characterized by a 2D lateral arrangement of columns standing at the surface on a macroscopic scale. The self-assembly of such insulated columns of face-to-face disks with surface-induced vertical alignment has been achieved directly from solution processing.

Some Cu-phthalocyanine derivatives **46** exhibit spontaneous uniform homeotropic alignment between two glass plates in the columnar tetragonal phase [176]. These DLCs also exhibit the homeotropic alignment behavior between two ITO-coated plates, which allowed to measure the charge carrier mobility in these compounds by TOF analysis [240]. It is interesting to note that the analogous compound **47** without oxygen atoms between the phthalocyanine core and peripheral phenyl groups do not exhibit homeotropic alignment, which again prompts one to think of the involvement of specific molecule–surface interactions for homeotropic alignment.

Like the epitaxial planar alignment of DLCs with an alignment layer, a Zn-phthalocyanine **45** has been homeotropically aligned on an ITO surface modified by pyridine-functionalized siloxane. By the coordination of the pyridine to the zinc centers, the epitaxial growth of highly ordered perpendicular stacks of Zn-Pc has been obtained from their chloroform solution [241]. Interestingly, the height of the columns can be easily controlled by varying the immersion time.

It has recently been observed that homeotropic alignment between two solid surfaces of different natures can be induced by the introduction of appropriate partially fluorinated chains at the periphery of triphenylene molecules **35** [242]. These observations seem to indicate that homeotropic alignment is achieved between two solid substrates regardless of the chemical nature of the surfaces.

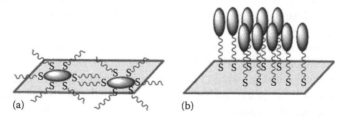

(a)   (b)

FIGURE 1.48 (a) The cores of discotic thioethers adsorbed to a Au surface in *face-on orientation* and (b) monofunctionalized thiol discotics adsorbed to a Au surface in *edge-on orientation*.

Linearly polarized IR-FEL irradiation

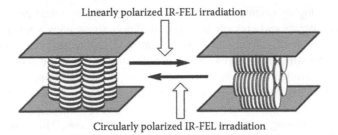

Circularly polarized IR-FEL irradiation

FIGURE 1.49 Schematic representations of alignment change of discotic triphenylene on infrared irradiation with linearly polarized light and on irradiation with circularly polarized light. (Adapted from Monobe, H. et al., *Adv. Mater.*, 18, 607, 2006.)

Homeotropically aligned hexaalkoxytriphenylene **34**, between two BaF_2 substrates, can be reoriented to homogeneous alignment by vibrational excitation with the help of infrared radiation obtained from free-electron laser [243]. Recently, the alignment control and reorientation of the columnar phase have been obtained by using polarized infrared irradiation. Circularly polarized infrared irradiation provides an alignment change, forming a new domain with a uniform homeotropic alignment. Interestingly, switching between edge-on and face-on arrangements of a columnar phase was observed by vibrational excitation using a combination of linearly and circularly polarized infrared irradiations (Figure 1.49) [244].

Oligomers of DLCs have certain advantages over monomeric DLCs as far as alignment is concerned [185]. Contrary to linear oligomers, star-branched oligomers have been described, which form macroscopic monodomains of columnar stacks that self-align homeotropically on a glass substrate [245].

1.13.2.3 Alignment of Discotic Liquid Crystals in Pores

To understand the effect of geometric confinement on DLCs and to produce highly ordered carbonaceous nano objects by subsequent carbonization of the ordered structures, DLCs have been aligned in nanoscopic and microscopic porous templates. Wendorff and coworkers have used porous alumina templates to produce aligned liquid crystalline triphenylene nanowires (Figure 1.50) [246]. When the pores with a diameter of a few hundred nanometers were filled with the molten triphenylene derivative **38**, which possesses a highly ordered plastic columnar phase, only the pore walls were wetted by the material. It has been observed that various parameters such as pore geometry, interfacial phenomena, and the thermal history influence the order of the disks within the pores.

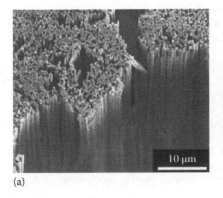
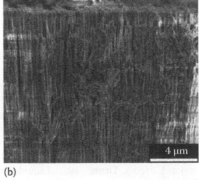

(a)   (b)

FIGURE 1.50 Scanning electron microscopy (SEM) images of released, aligned discotic triphenylene wires with diameters of (a) 400 and (b) 60 nm. (Reproduced from Steinhart, M. et al., *Nano Lett.*, 5, 429, 2005. With permission from ACS.)

The LC columns are highly oriented along the long axis of the template pores, but the LC–pore wall interactions counteract and promote the growth of columns having a homeotropic orientation with respect to the pore walls. A further development of the templating method was the preparation of nanotubes consisting of a polymer layer outside the tube and a discotic triphenylene layer inside [247]. Under controlled annealing in the mesophase, the disks arrange to produce a columnar alignment along the axes of the tubes. Recently, an HBC derivative **24** was successfully templated in nanoscopic pores and in macroscopic glass capillary by melt processing [248]. In both cases, the columnar structures were long-range aligned along the template axis. This behavior was explained by the pronounced directional self-assembly of the molecules, while the influence of the template curvature was negligible. An alternative to produce stable graphitic nanotubes was presented for discotic HBC derivative **26**. The filling of membrane pores with a solution of material resulted in a wetting of the walls. After carbonization at high temperature and removal of the inorganic template, carbon nanotubes with ordered graphene sheets were obtained [249]. Interestingly, a polymerizable HBC derivative **29** has been aligned in the pores of an inorganic template, and subsequent annealing at high temperature in the mesophase resulted in cross-linked HBCs forming nanotubes [250]. These nanostructures were collected by the careful dissolution of the template, and characterized by microscopic techniques, which suggest a high degree of order with columnar structure. This method provides promising donor materials for application in the field of organic electronics, where an established morphology can be fixed to ensure the mechanical and thermal stability of the device.

REFERENCES

1. Brown, G. H. and Wolken, J. J. *Liquid Crystals and Biological Structures*, Academic Press, New York, 1979.
2. Ringsdorf, H., Schlarb, B., and Venzmer, J. *Angew. Chem. Int. Ed.* 27, 113–158, 1988.
3. Lehn, J. M. *Supramolecular Chemistry*, VCH, Weinheim, Germany, 1995.
4. Whitesides, G. M. and Grzybowski, B. *Science* 295, 2418–2421, 2002.
5. Goodby, J. W., Saez, I. M., Cowling, S. J., Gasowska, J. S., MacDonald, R. A., Sia, S., Watson, P., Toyne, K. J., Hird, M., Lewis, R. A., Lee, S. E., and Vaschenko, V. *Liq. Cryst.* 36, 567–605, 2009.
6. Goodby, J. W., Saez, I. M., Cowling, S. J., Gortz, V., Draper, M., Hall, A. W., Sia, S., Cosquer, G., Lee, S. E., and Raynes, E. P. *Angew. Chem. Int. Ed.* 47, 2754–2787, 2008.
7. Saez, I. M. and Goodby, J. W. *J. Mater. Chem.* 15, 26–40, 2005.
8. Goodby, J. W., Mehl, G. H., Saez, I. M., Tuffin, R. P., Mackenzie, G., Velty, R. A., Benvegnu, T., and Plusquellec, D. *Chem. Commun.* 2057–2070, 1998.
9. Tschierske, C. *J. Mater. Chem.* 8, 1485–1508, 1998.
10. Tschierske, C. *J. Mater. Chem.* 11, 2647–2671, 2001.
11. Tschierske, C. *Chem. Soc. Rev.* 36, 1930–1970, 2007.
12. Kato, T. *Science* 295, 2414–2418, 2002.
13. Kato, T., Mizoshita, N., and Kishimoto, K. *Angew. Chem. Int. Ed.* 45, 38–68, 2006.
14. Percec, V., Dulcey, A. E., Balagurusamy, V. S. K., Miura, Y., Smidrkal, J., Peterca, M., Nummelin, S., Edlund, U., Hudson, S. D., Heiney, P. A., Duan, H., Magonov, S. N., and Vinogradov, S. A. *Nature* 430, 764–768, 2004.
15. Percec, V., Glodde, M., Bera, T. K., Miura, Y., Shiyanovskaya, I., Singer, K. D., Balagurusamy, V. S. K., Heiney, P. A., Schnell, I., Rapp, A., Spiess, H. W., Hudson, S. D., and Duan, H. *Nature* 419, 384–387, 2002.
16. Chandrasekhar, S. and Kumar, S. *Sci. Spectra.* 8, 66–72, 1997.
17. Bruce, D. W., Goodby, J. W., Sambles, J. R., and Coles, H. J. *Phil. Trans. R. Soc. A* 364, 2567–2571, 2006.
18. Seddon, J. M. *Curr. Opin. Colloid. Interface Sci.* 7, 296–297, 2002.
19. Goodby, J. (ed.). Thematic issue: Liquid crystals, *Chem. Soc. Rev.* 36, 1845–2128, 2007.
20. Tschierske, C. (ed.). Theme issue: Liquid crystals beyond display applications, *J. Mater. Chem.* 18, 2857–3060, 2008.
21. Goodby, J. W., Bruce, D. W., Hird, M., Imrie, C., and Neal, M. (eds.). Molecular topology in liquid crystals, *J. Mater. Chem.* 11, 2631–2886, 2001.
22. Dervichian, D. G. *Mol. Cryst. Liq. Cryst.* 40, 19–31, 1977.

23. Demus, D., Goodby, J., Gray, G. W., Spiess, H. W., and Vill, V. (eds.). *Handbook of Liquid Crystals*, Vol. 1–3, Wiley VHC, Weinheim, Germany, 1998.
24. Collings, P. J. and Patel, J. S. (eds.). *Handbook of Liquid Crystals Research*, Oxford University Press, Oxford, NY, 1997.
25. Kelker, H. and Hatz, R. *Handbook of Liquid Crystals*, Verlag Chemie, Weinheim, Germany, 1980.
26. Bahadur, B. (ed.). *Liquid Crystals: Applications and Uses*, Vols. 1–3, World Scientific, Singapore, 1990.
27. de Gennes, P. G. and Prost, J. *The Physics of Liquid Crystals*, Clarendon Press, Oxford, NY, 1993.
28. Collings, P. J. *Liquid Crystals: Nature's Delicate Phase of Matter*, Princeton University Press, Princeton, NJ, 2002.
29. Collings, P. J. and Hird, M. *Introduction to Liquid Crystals: Chemistry and Physics*, Taylor & Francis, London, U.K., 1997.
30. Chandrasekhar, S. *Liquid Crystals*, Cambridge University Press, Cambridge, U.K., 1992.
31. Demus, D., Goodby, J., Gray, G. W., Spiess, H. W., and Vill, V. (eds.). *Physical Properties of Liquid Crystals*, Wiley VCH, Weinheim, Germany, 1999.
32. Dunmur, D. A., Fukuda, A., and Luckhurst, G. R. (eds.). *Physical Properties of Liquid Crystals: Nematics*, INSPEC-IEE, London, U.K., 2001.
33. Ramamoorthy, A. (ed.). *Thermotropic Liquid Crystals: Recent Advances*, Springer, Dordrecht, the Netherlands, 2007.
34. Singh, S. *Liquid Crystals: Fundamentals*, World Scientific, Singapore, 2002.
35. Vertogen, G. and de Jeu, W. H. *Thermotropic Liquid Crystals, Fundamentals*, Springer-Verlag, Berlin, Germany, 1988.
36. Oswald, P. and Pieranski, P. *Nematic and Cholesteric Liquid Crystals: Concepts and Physical Properties Illustrated by Experiments*, Taylor & Francis, CRC Press, Boca Raton, FL, 2005.
37. Oswald, P. and Pieranski, P. *Smectic and Columnar Liquid Crystals: Concepts and Physical Properties Illustrated by Experiments*, Taylor & Francis, CRC Press, Boca Raton, FL, 2005.
38. Gray, G. W. and Luckhurst, G. R. (eds.). *The Molecular Physics of Liquid Crystals*, Academic Press, London, U.K., 1979.
39. Kumar, S. (ed.). *Liquid Crystals: Experimental Study of Physical Properties and Phase Transitions*, Cambridge University Press, Cambridge, U.K., 2001.
40. Khoo, I. C. and Wu, S. T. *Optics and Nonlinear Optics of Liquid Crystals*, World Scientific, Singapore, 1993.
41. Jakli, A. and Saupe, A. *One- and Two-Dimensional Fluids: Properties of Smectic, Lamellar and Columnar Liquid Crystals*, CRC Press, Boca Raton, FL, 2006.
42. Gray, G. W. and Winsor, P. A. *Liquid Crystals and Plastic Crystals*, Vols. 1–2, Ellis Horwood Ltd., Chichester, U.K., 1974.
43. Gray, G. W. *Thermotropic Liquid Crystals*, John Wiley, New York, 1987.
44. Fisch, M. R. *Liquid Crystals, Laptops and Life*, World Scientific, Singapore, 2004.
45. Rasing, T. and Musevic, I. (eds.). *Surfaces and Interfaces of Liquid Crystals*, Springer-Verlag, Berlin Heidelberg, Germany, 2004.
46. Crawford, G. P. and Zumer, S. (eds.). *Liquid Crystals in Complex Geometries: Formed by Polymers and Porous Networks*, Taylor & Francis, London, U.K., 1996.
47. Carroll, P. L. *Cholesteric Liquid Crystals: Their Technology and Applications*, Ovum Ltd., London, U.K., 1973.
48. Demus, D. and Richter, L. *Textures of Liquid Crystals*, Verlag Chemie, Weinheim, Germany, 1978.
49. Dierking, I. *Textures of Liquid Crystals*, Wiley VCH Verlag, Weinheim, Germany, 2003.
50. Gray, G. W. and Goodby, J. W. *Smectic Liquid Crystals: Textures and Structures*, Leonard Hill, Glasgow, U.K., 1984.
51. Goodby, J. W. *Ferroelectric Liquid Crystals: Principles, Properties and Applications*, Gordon and Breach, Philadelphia, PA, 1991.
52. Lagerwall, S. T. *Ferroelectric and Antiferroelectric Liquid Crystals*, Wiley-VCH, Weinheim, Germany, 1999.
53. Musevic, I., Blinc, R., and Zeks, B. *The Physics of Ferroelectric and Antiferroelectric Liquid Crystals*, World Scientific, Singapore, 2000.
54. Sonin, A. S. *J. Mater. Chem.* 8, 2557–2574, 1998.
55. Serrano, J. L. (ed.). *Metallomesogens: Synthesis, Properties and Application*, VCH, Weinheim, Germany, 1996.
56. Friberg, S. In *Lyotropic Liquid Crystals*, Gould, R. F. (ed.), Advances in Chemistry Series, American Chemical Society, Washington, DC, 1976.

57. Mingos, D. M. P. and Kato, T. (eds.). *Structure and Bonding*, Vol. 128, Springer-Verlag, Berlin, Germany, 2008.
58. Toledano, P. and Neto, A. M. F. (eds.). *Phase Transitions in Complex Fluids*, World Scientific, Singapore, 1998.
59. Petrov, A. G. *The Lyotropic State of Matter: Molecular Physics and Living Matter Physics*, Gordon & Breach Science Pub., Amsterdam, the Netherlands, 1999.
60. Lynch, M. L. and Spicer, P. T. (eds.). *Bicontinuous Liquid Crystals*, Taylor & Francis/CRC Press, Boca Raton, FL, 2005.
61. Koltover, I., Salditt, T., Radler, J. O., and Safinya, C. R. *Science* 281, 78–81, 1998.
62. Nakata, M., Zanchetta, G., Chapman, B. D., Jones, C. D., Cross, J. O., Pindak, R., Bellini, T., and Clark, N. A. *Science* 318, 1276–1279, 2007.
63. Zanchetta, G., Nakata, M., Buscaglia, M., Bellini, T., and Clark, N. A. *Proc. Natl. Acad. Sci. USA* 105, 1111–1117, 2008.
64. O'Connor, I., Hayden, H., O'Connor, S., Coleman, J. N., and Gun'ko, Y. K. *J. Mater. Chem.* 18, 5585–5588, 2008.
65. Kato, T., Yasuda, T., Kamikawa, Y., and Yoshio, M. *Chem. Commun.* 729–739, 2009.
66. Woltman, S. J., Jay, G. D., and Crawford, G. P. (eds.). *Liquid Crystals: Frontiers in Biomedical Applications*, World Scientific, Singapore, 2007.
67. Woltman, S. J., Jay, G. D., and Crawford, G. P. *Nat. Mater.* 6, 929–938, 2007.
68. Watson, S. J., Gleeson, H. F., D'emanuele, A., Serak, S., and Grozhik, V. *Mol. Cryst. Liq. Cryst.* 331, 375–382, 1999.
69. Gupta, V. K., Skaife, J. J., Dubrovsky, T. B., and Abbott, N. L. *Science* 279, 2077–2080, 1998.
70. Brake, J. M., Daschner, M. K., Luk, Y. Y., and Abbott, N. L. *Science* 302, 2094–2097, 2003.
71. Shiyanovskii, S. V., Schneider, T., Smalyukh, I. I., Ishikawa, T., Niehaus, G. D., Doane, K., Woolverton, C. J., and Lavrentovich, O. D. *Phys. Rev. E* 71, 020702, 2005.
72. Hegmann, T., Qi, H., and Marx, V. M. *J. Inorg. Organomet. Polym. Mater.* 17, 483–508, 2007.
73. Cseh, L. and Mehl, G. H. *J. Mater. Chem.* 17, 311–315, 2007.
74. Kumar, S. and Bisoyi, H. K. *Angew. Chem. Int. Ed.* 46, 1501–1503, 2007.
75. Kumar, S., Pal, S. K., Kumar, P. S., and Lakshminarayanan, V. *Soft Matter.* 3, 896–900, 2007.
76. Kumar, S. and Lakshminarayanan, V. *Chem. Commun.* 1600–1601, 2004.
77. Reinitzer, F. *Monatsh. Chem.* 9, 421–441, 1888, for English translation see *Liq. Cryst.* 5, 7–18, 1989.
78. Lehmann, O. *Z. Phys. Chem.* 4, 462–467, 1889.
79. Sluckin, T. J., Dunmur, D. A., and Stegemeyer, H. *Crystals That Flow: Classic Papers from the History of Liquid Crystals*, Taylor & Francis, London, U.K., 2004.
80. Heintz, W. *J. Prakt. Chem.* 66, 1–51, 1855.
81. Vorlander, D. *Kristallinisch-Flussige Substanzen,* Enke, Stuttgart, Germany, 1908.
82. Vorlander, D. *Ber. Dtsch. Chem. Ges.* 40, 1970–1972, 1907.
83. Friedel, G. *Ann. Phys.* 18, 273–274, 1922.
84. Brown, G. H. and Shaw, W. G. *Chem. Rev.* 57, 1049–1157, 1957.
85. Gray, G. W. *Molecular Structure and the Properties of Liquid Crystals*, Academic Press, London, U.K., 1962.
86. Kelker, H. and Scheurle, B. *Angew. Chem. Int. Ed.* 8, 884–885, 1969.
87. Gray, G. W., Harrison, K. J., and Nash, J. A. *Electron. Lett.* 9, 130–131, 1973.
88. Chandrasekhar, S., Sadashiva, B. K., and Suresh, K. A. *Pramana* 9, 471–480, 1977.
89. Vorlander, D. *Chemische Kristallographic der Flussigkeiten*, Akademische Verlagsgesellschaff, Leipzig, Germany, p. 34, 1924.
90. Niori, T., Sekine, T., Watanabe, J., Furukawa, T., and Takezoe, H. *J. Mater. Chem.* 6, 1231–1233, 1996.
91. Fuchs, K. and Katscher, E. E. *Ber. Dtsch. Chem. Ges.* 62, 2381–2386, 1929.
92. Tschierske, C. *Curr. Opin. Colloid Interface Sci.* 7, 355–370, 2002.
93. Yu, L. J. and Saupe, A. *Phys. Rev. Lett.* 45, 1000–1003, 1980.
94. de Melo Filho, A. A., Laverde Jr., A., and Fujiwara, F. Y. *Langmuir* 19, 1127–1132, 2003.
95. Tam-Chang, S. W. and Huang, L. *Chem. Commun.* 1957–1967, 2008.
96. Lydon, J. *Curr. Opin. Colloid Interface Sci.* 8, 480–490, 2004.
97. Attwood, T. K., Lydon, J. E., and Jones, F. *Liq. Cryst.* 1, 499–507, 1986.
98. Lydon, J. E. *Curr. Opin. Colloid. Interface Sci.* 3, 458–466, 1998.
99. Lagerwall, J. P. F. and Giesselmann, F. *ChemPhysChem* 7, 20–45, 2006.
100. Goodby, J. W., Waugh, M. A., Stein, S. M., Chin, E., Pindak, R., and Patel, J. S. *Nature* 337, 449–452, 1989.
101. Goodby, J. W. *Curr. Opin. Colloid. Interface Sci.* 7, 326–332, 2002.

102. Crooker, P. P. *Liq. Cryst.* 5, 751–775, 1989.
103. Wright, D. C. and Mermin, N. D. *Rev. Mod. Phys.* 61, 385–432, 1989.
104. Coles, H. J. and Pivnenko, M. N. *Nature* 436, 997–1000, 2005.
105. Vorlander, D. and Apel, A. *Ber. Dtsch. Chem. Ges.* 65, 1101–1109, 1932.
106. Matsunaga, Y. and Miyamoto, S. *Mol. Cryst. Liq. Cryst.* 237, 311–317, 1993.
107. Matsuzaki, H. and Matsunaga, Y. *Liq. Cryst.* 14, 105–120, 1993.
108. Tschierske, C. and Dantlgraber, G. *Pramana* 61, 455–481, 2003.
109. Takezoe, H. and Takanishi, Y. *Jpn. J. Appl. Phys.* 45, 597–625, 2006.
110. Reddy, R. A. and Tschierske, C. *J. Mater. Chem.* 16, 907–961, 2006.
111. Pelzl, G., Diele, S., and Weissflog, W. *Adv. Mater.* 11, 707–724, 1999.
112. Ros, M. B., Serrano, J. L., de la Fuente, M. R., and Folcia, C. L. *J. Mater. Chem.* 15, 5093–5098, 2005.
113. Etxebarria, J. and Ros, M. B. *J. Mater. Chem.* 18, 2919–2926, 2008.
114. Weissflog, W., Murthy, H. N. S., Diele, S., and Pelzl, G. *Phil. Trans. R. Soc. A* 364, 2657–2679, 2006.
115. Link, D. R., Natale, G., Shao, R., Maclennan, J. E., Clark, N. A., Korblova, E., and Walba, D. M. *Science* 278, 1924–1927, 1997.
116. Madsen, L. A., Dingemans, T. J., Nakata, M., and Samulski, E. T. *Phys. Rev. Lett.* 92, 145505, 2004.
117. Acharya, B. R., Primak, A., and Kumar, S. *Phys. Rev. Lett.* 92, 145506, 2004.
118. Luckhurst, G. R. *Angew. Chem. Int. Ed.* 44, 2834–2836, 2005.
119. Hessel, F. and Finkelmann, H. *Polym. Bull.* 15, 349–352, 1986.
120. Kumar, S. *Chem. Soc. Rev.* 35, 83–109, 2006.
121. Kumar, S. *Liq. Cryst.* 36, 607–638, 2009.
122. Kumar, S. *Liq. Cryst. Today* 18, 2–27, 2009.
123. Laschat, S., Baro, A., Steinke, N., Giesselmann, F., Hagele, C., Scalia, G., Judele, R., Kapatsina, E., Sauer, S., Schreivogel, A., and Tosoni, M. *Angew. Chem. Int. Ed.* 46, 4832–4887, 2007.
124. Sergeyev, S., Pisula, W., and Geerts, Y. H. *Chem. Soc. Rev.* 36, 1902–1929, 2007.
125. Wu, J., Pisula, W., and Mullen, K. *Chem. Rev.* 107, 718–747, 2007.
126. Boden, N., Bushby, R. J., Clements, J., and Movaghar, B. *J. Mater. Chem.* 9, 2081–2086, 1999.
127. Bushby, R. J. and Lozman, O. R. *Curr. Opin. Solid State Mater. Sci.* 6, 569–578, 2002.
128. Bushby, R. J. and Lozman, O. R. *Curr. Opin. Colloid Interface Sci.* 7, 343–354, 2002.
129. Ohta, K., Hatsusaka, K., Sugibayashi, M., Ariyoshi, M., Ban, K., Maeda, F., Naito, R., Nishizawa, K., van de Craats, A. M., and Warman, J. M. *Mol. Cryst. Liq. Cryst.* 397, 25–45, 2003.
130. Cammidge, A. N. and Bushby, R. J. In *Handbook of Liquid Crystals*, Vol. 2B, Demus, D., Goodby, J., Gray, G. W., Spiess, H. W., and Vill, V. (eds.), Wiley-VCH, Weinheim, Germany, Chap. VII, pp. 693–748, 1998.
131. Chandrasekhar, S. *Liq. Cryst.* 14, 3–14, 1993.
132. Chandrasekhar, S. and Ranganath, G. S. *Rep. Prog. Phys.* 53, 57–84, 1990.
133. Donnio, B., Guillon. D., Deschenaux, R., and Bruce, D. W. In *Comprehensive Coordination Chemistry II*, Vol. 7, McCleverty, J. A. and Meyer, T. J. (eds.), Elsevier, Oxford, U.K., pp. 357–627, 2003.
134. Destrade, C., Foucher, P., Gasparoux, H., Nguyen, H. T., Levelut, A. M., and Malthete, J. *Mol. Cryst. Liq. Cryst.* 106, 121–146, 1984.
135. Fischbach, I., Ebert, F., Spiess, H. W., and Schnell, I. *ChemPhysChem* 5, 895–908, 2004.
136. Lehmann, M., Fischbach, I., Spiess, H. W., and Meier, H. *J. Am. Chem. Soc.* 126, 772–784, 2004.
137. Bisoyi, H. K. and Kumar, S. *Chem. Soc. Rev.* 39, 264–285, 2010.
138. Praefcke, K. In *Physical Properties of Liquid Crystals: Nematics*, Dunmur, D. A., Fukuda, A., and Luckhurst, G. R. (eds.), INSPEC, London, U.K., pp. 17–35, 2001.
139. Langner, M., Praefcke, K., Kruerke, D., and Heppke, G. *J. Mater. Chem.* 5, 693–699, 1995.
140. Ringsdorf, H., Wustefeld, R., Zerta, E., Ebert, M., and Wendorff, J. H. *Angew. Chem. Int. Ed.* 28, 914–918, 1989.
141. Grafe, A., Janietz, D., Frese, T., and Wendorff, J. H. *Chem. Mater.* 17, 4979–4984, 2005.
142. Kouwer, P. H. J., Jager, W. F., Mijs, W. J., and Picken, S. J. *Macromolecules* 33, 4336–4342, 2000.
143. Stracke, A., Wendorff, J. H., Janietz, D., and Mahlstedt, S. *Adv. Mater.* 11, 667–670, 1999.
144. Mahlstedt, S., Janietz, D., Stracke, A., and Wendorff, J. H. *Chem. Commun.* 15–16, 2000.
145. Kouwer, P. H. J., Jager, W. F., Mijs, W. J., and Picken, S. J. *Macromolecules* 34, 7582–7584, 2001.
146. Kouwer, P. H. J., Jager, W. F., Mijs, W. J., and Picken, S. J. *Macromolecules* 35, 4322–4329, 2002.
147. Kouwer, P. H. J., van den Berg, O., Jager, W. F., Mijs, W. J., and Picken, S. J. *Macromolecules* 35, 2576–2582, 2002.
148. Sakashita, H., Nishitani, A., Sumiya, Y., Terauchi, H., Ohta, K., and Yamamoto, I. *Mol. Cryst. Liq. Cryst.* 163, 211–219, 1988.

149. Shimizu, Y., Miya, M., Nagata, A., Ohta, K., Yamamoto, I., and Kusabayashi, S. *Liq. Cryst.* 14, 795–805, 1993.
150. Ohta, K., Yamaguchi, N., and Yamamoto, I. *J. Mater. Chem.* 8, 2637–2650, 1998.
151. Ohta, K., Ando, N., and Yamamoto, I. *Liq. Cryst.* 26, 663–668, 1999.
152. Akai, T. and Shimizu, Y. *Liq. Cryst.* 27, 437–441, 2000.
153. Bruce, D. W., Dunmur, D. A., Santa, L. S., and Wali, M. A. *J. Mater. Chem.* 2, 363–364, 1992.
154. Alameddine, B., Aebischer, O. F., Amrein, W., Donnio, B., Deschenaux, R., Guillon, D., Savary, C., Scanu, D., Scheidegger, O., and Jenny, T. A. *Chem. Mater.* 17, 4798–4807, 2005.
155. Kadam, J., Faul, C. F. J., and Scherf, U. *Chem. Mater.* 16, 3867–3871, 2004.
156. Prasad, S. K., Rao, D. S. S., Chandrasekhar, S., and Kumar, S. *Mol. Cryst. Liq. Cryst.* 396, 121–139, 2003.
157. Bisoyi, H. K. and Kumar, S. *J. Phys. Org. Chem.* 21, 47–52, 2008.
158. Levelut, A. M. *J. Chim. Phys.* 80, 149–161, 1983.
159. Frank, F. C. and Chandrasekhar, S. *J. Phys.* 41, 1285–1288, 1980.
160. Morale, F., Date, R. W., Guillon, D., Bruce, D. W., Finn, R. L., Wilson, C., Blake, A. J., Schroder, M., and Donnio, B. *Chem. Eur. J.* 9, 2484–2501, 2003.
161. Billard, J., Dubois, J. C., Vaucher, C., and Levelut, A. M. *Mol. Cryst. Liq. Cryst.* 66, 115–122, 1981.
162. Donnio, B., Heinrich, B., Allouchi, H., Kain, J., Diele, S., Guillon, D., and Bruce, D. W. *J. Am. Chem. Soc.* 126, 15258–15268, 2004.
163. Destwde, C., Tinh, N. H., Mamlok, L., and Malthete, J. *Mol. Cryst. Liq. Cryst.* 114, 139–150, 1984.
164. Simmerer, J., Glusen, B., Paulus, W., Kettner, A., Schuhmacher, P., Adam, D., Etzbach, K. H., Siemensmeyer, K., Wendorff, J. H., Ringsdorf, H., and Haarer, D. *Adv. Mater.* 8, 815–819, 1996.
165. Fontes, E., Heiney, P. A., and de Jeu, W. H. *Phys. Rev. Lett.* 61, 1202–1205, 1988.
166. Voigt-Martin, I. G., Garbella, R. W., and Schumacher, M. *Liq. Cryst.* 17, 775–801, 1994.
167. Wu, J., Watson, M. D., Zang, L., Wang, Z., and Mullen, K. *J. Am. Chem. Soc.* 126, 177–186, 2004.
168. Mery, S., Haristoy, D., Nicoud, J. F., Guillon, D., Diele, S., Monobe, H., and Shimizu, Y. *J. Mater. Chem.* 12, 37–41, 2002.
169. Davidson, P., Levelut, A. M., Strzelecka, H., and Gionis, V. *J. Phys. Lett.* 44, 823–828, 1983.
170. Ziessel, R., Douce, L., El-Ghayoury, A., Harriman, A., and Skoulios, A. *Angew. Chem. Int. Ed.* 39, 1489–1493, 2000.
171. Haristoy, D., Mery, S., Heinrich, B., Mager, L., Nicoud, J. F., and Guillon, D. *Liq. Cryst.* 27, 321–328, 2000.
172. Steinke, N., Frey, W., Baro, A., Laschat, S., Drees, C., Nimtz, M., Hagele, C., and Giesselmann, F. *Chem. Eur. J.* 12, 1026–1035, 2006.
173. Yasuda, T., Ooi, H., Morita, J., Akama, Y., Minoura, K., Funahashi, M., Shimomura, T., and Kato, T. *Adv. Funct. Mater.* 19, 411–419, 2009.
174. Bubulak, T. V., Buchs, J., Kohlmeier, A., Bruma, M., and Janietz, D. *Chem. Mater.* 19, 4460–4466, 2007.
175. Ohta, K., Watanabe, T., Hasebe, H., Morizumi, Y., Fujimoto, T., Yamamoto, I., Lelievre, D., and Simon, J. *Mol. Cryst. Liq. Cryst.* 196, 13–26, 1991.
176. Hatsusaka, K., Ohta, K., Yamamoto, I., and Shirai, H. *J. Mater. Chem.* 11, 423–433, 2001.
177. Ichihara, M., Suzuki, A., Hatsusaka, K., and Ohta, K. *Liq. Cryst.* 34, 555–567, 2007.
178. Praefcke, K., Marquardt, P., Kohne, B., Stephan, W., Levelut, A. M., and Watchtel, E. *Mol. Cryst. Liq. Cryst.* 203, 149–158, 1991.
179. Adam, D., Schuhmacher, P., Simmerer, J., Haussling, L., Siemensmeyer, K., Etzbachi, K. H., Ringsdorf, H., and Haarer, D. *Nature* 371, 141–143, 1994.
180. Mori, H., Itoh, Y., Nishiura, Y., Nakamura, T., and Shinagawa, Y. *Jpn. J. Appl. Phys.* 36, 143–147, 1997.
181. Mori, H. *Jpn. J. Appl. Phys.* 36, 1068–1072, 1997.
182. Kawata, K. *Chem. Rec.* 2, 59–80, 2002.
183. Mende, L. S., Fechtenkotter, A., Mullen, K., Moons, E., Friend, R. H., and MacKenzie, J. D. *Science* 293, 1119–1122, 2001.
184. O'Neill, M. and Kelly, S. M. *Adv. Mater.* 15, 1135–1146, 2003.
185. Eichhorn, S. H., Adavelli, A., Li, H. S., and Fox, N. *Mol. Cryst. Liq. Cryst.* 397, 47–58, 2003.
186. Eichhorn, H. *J. Porphyr. Phthalocyanines* 4, 88–102, 2000.
187. Yoshio, M., Kagata, T., Hoshino, K., Mukai, T., Ohno, H., and Kato, T. *J. Am. Chem. Soc.* 128, 5570–5577, 2006.
188. Vij, J. K., Kocot, A., and Perova, T. S. *Mol. Cryst. Liq. Cryst.* 397, 231–244, 2003.
189. Kumar, S. and Varshney, S. K. *Angew. Chem.* 112, 3270–3272, 2000.
190. Vauchier, C., Zann, A., Le Barny, P., Dubois, J. C., and Billard, J. *Mol. Cryst. Liq. Cryst.* 66, 103–114, 1981.

191. Ikeda, S., Takanishi, Y., Ishikawa, K., and Takezoe, H. *Mol. Cryst. Liq. Cryst.* 329, 589–595, 1999.
192. Sergan, T., Sonpatki, M., Kelly, J., and Chien, L. C. *Mol. Cryst. Liq. Cryst.* 359, 245–257, 2001.
193. Shimizu, Y., Monobe, H., Heya, M., and Awazu, K. *Mol. Cryst. Liq. Cryst.* 441, 287–295, 2005.
194. Monobe, H., Terasawa, N., Kiyohara, K., Shimizu, Y., Azehara, H., Nakasa, A., and Fujihira, M. *Mol. Cryst. Liq. Cryst.* 412, 229–236, 2004.
195. Monobe, H., Mima, S., Sugino, T., Shimizu, Y., and Ukon, M. *Liq. Cryst.* 28, 1253–1258, 2001.
196. Ichimura, K., Furumi, S., Morino, S., Kidowaki, M., Nakagawa, M., Ogawa, M., and Nishiura, Y. *Adv. Mater.* 12, 950–953, 2000.
197. Furumi, S., Kidowaki, M., Ogawa, M., Nishiura, Y., and Ichimura, K. *J. Phys. Chem. B* 109, 9245–9254, 2005.
198. Bjornholm, T., Hassenkam, T., and Reitzel, N., *J. Mater. Chem.* 9, 1975–1990, 1999.
199. Karthaus, O., Ringsdorf, H., Tsukruk, V. V., and Wendorff, J. H. *Langmuir* 8, 2279–2283, 1992.
200. Maliszewskyj, N. C., Heiney, P. A., Josefowicz, J. Y., Plesnivy, T., Ringsdorf, H., and Schuhmacher, P. *Langmuir* 11, 1666–1674, 1995.
201. Mindyuk, O.Y. and Heiney, P. A. *Adv. Mater.* 11, 341–344, 1999.
202. Reitzel, N., Hassenkam, T., Balashev, K., Jensen, T. R., Howes, P. B., Kjaer, K., Fechtenkotter, A., Tchebotareva, N., Ito, S., Mullen, K., and Bjornholm, T. *Chem. Eur. J.* 7, 4894–4901, 2001.
203. Kubowicz, S., Pietsch, U., Watson, M. D., Tchebotareva, N., Mullen, K., and Thunemann, A. F. *Langmuir* 19, 5036–5041, 2003.
204. Laursen, B. W., Norgaard, K., Reitzel, N., Simonsen, J. B., Nielsen, C. B., Nielsen, J. A., Bjornholm, T., Solling, T. I., Nielsen, M. M., Bunk, O., Kjaer, K., Tchebotareva, N., Watson, M. D., Mullen, K., and Piris, J. *Langmuir* 20, 4139–4146, 2004.
205. Schonherr, H., Kremer, F. J. B., Kumar, S., Rego, J. A., Wolf, H., Ringsdorf, H., Jaschke, M., Butt, H. J., and Bamberg, E. *J. Am. Chem. Soc.* 118, 13051–13057, 1996.
206. Boden, N., Bushby, R. J., Martin, P. S., Evans, S. D., Owens, R. W., and Smith, D. A. *Langmuir* 15, 3790–3797, 1999.
207. Monobe, H., Azehara, H., Shimizu, Y., and Fujihira, M. *Chem. Lett.* 30, 1268–1269, 2001.
208. Duati, M., Grave, C., Tcbeborateva, N., Wu, J., Mullen, K., Shaporenko, A., Zharnikov, M., Kriebel, J. K., Whitesides, G. M., and Rampi, M. A. *Adv. Mater.* 18, 329–333, 2006.
209. Liu, C. Y. and Bard, A. J. *Chem. Mater.* 12, 2353–2362, 2000.
210. Pisula, W., Kastler, M., Wasserfallen, D., Pakula, T., and Mullen, K. *J. Am. Chem. Soc.* 126, 8074–8075, 2004.
211. van de Craats, A. M., Stutzmann, N., Bunk, O., Nielsen, M. M., Watson, M., Mullen, K., Chanzy, H. D., Sirringhaus, H., and Friend, R. H. *Adv. Mater.* 15, 495–499, 2003.
212. Piris, J., Debije, M. G., Stutzmann, N., van de Craats, A. M., Watson, M. D., Mullen, K., and Warman, J. M. *Adv. Mater.* 15, 1736–1740, 2003.
213. Bunk, O., Nielsen, M. M., Solling, T. I., van de Craats, A. M., and Stutzmann, N. *J. Am. Chem. Soc.* 125, 2252–2258, 2003.
214. Zimmermann, S., Wendorff, J. H., and Weder, C. *Chem. Mater.* 14, 2218–2223, 2002.
215. Brinkmann, M., Wittmann, J. C., Barthel, M., Hanack, M., and Chaumont, C. *Chem. Mater.* 14, 904–914, 2002.
216. Shklyarevskiy, I. O., Jonkheijm, P., Stutzmann, N., Wasserberg, D., Wondergem, H. J., Christianen, P. C. M., Schenning, A. P. H. J., de Leeuw, D. M., Tomovic, Z., Wu, J., Mullen, K., and Maan, J. C. *J. Am. Chem. Soc.* 127, 16233–16237, 2005.
217. Cristadoro, A., Lieser, G., Rader, H. J., and Mullen, K. *ChemPhysChem* 8, 586–591, 2007.
218. Cristadoro, A., Ai, M., Rader, H. J., Rabe, J. P., and Mullen, K. *J. Phys. Chem. C* 112, 5563–5566, 2008.
219. Tang, C., Tracz, A., Kruk, M., Zhang, R., Smilgies, D. M., Matyjaszewski, K., and Kowalewski, T. *J. Am. Chem. Soc.* 127, 6918–6919, 2005.
220. Breiby, D. W., Bunk, O., Pisula, W., Solling, T. I., Tracz, A., Pakula, T., Mullen, K., and Nielsen, M. M. *J. Am. Chem. Soc.* 127, 11288–11293, 2005.
221. Pisula, W., Tomovic, Z., Stepputat, M., Kolb, U., Pakula, T., and Mullen, K. *Chem. Mater.* 17, 2641–2647, 2005.
222. Tracz, A., Jeszka, J. K., Watson, M. D., Pisula, W., Mullen, K., and Pakula, T. *J. Am. Chem. Soc.* 125, 1682–1683, 2003.
223. Piris, J., Pisula, W., Tracz, A., Pakula, T., Mullen, K., and Warman, J. M. *Liq. Cryst.* 31, 993–996, 2004.
224. Pisula, W., Menon, A., Stepputat, M., Lieberwirth, I., Kolb, U., Tracz, A., Sirringhaus, H., Pakula, T., and Mullen, K. *Adv. Mater.* 17, 684–689, 2005.

225. Xiao, S., Myers, M., Miao, Q., Sanaur, S., Pang, K., Steigerwald, M. L., and Nuckolls, C. *Angew. Chem. Int. Ed.* 44, 7390–7394, 2005.
226. Kastler, M., Pisula, W., Wasserfallen, D., Pakula, T., and Mullen, K. *J. Am. Chem. Soc.* 127, 4286–4296, 2005.
227. Liu, C. Y., Fechtenkotter, A., Watson, M. D., Mullen, K., and Bard, A. J. *Chem. Mater.* 15, 124–130, 2003.
228. Pisula, W., Tomovic, Z., El Hamaoui, B., Watson, M. D., Pakula, T., and Mullen, K. *Adv. Funct. Mater.* 15, 893–904, 2005.
229. de Cupere, V., Tant, J., Viville, P., Lazzaroni, R., Osikowicz, W., Salaneck, W. R., and Geerts, Y. H. *Langmuir* 22, 7798–7806, 2006.
230. Perova, T. S. and Vij, J. K. *Adv. Mater.* 7, 919–922, 1995.
231. Charra, F. and Cousty, J. *Phys. Rev. Lett.* 80, 1682–1685, 1998.
232. Xu, S., Zeng, Q., Lu, J., Wang, C., Wan, L., and Bai, C. L. *Surf. Sci.* 538, L451–L459, 2003.
233. Gupta, S. K., Raghunathan, V. A., Lakshminarayanan, V., and Kumar, S. *J. Phys. Chem. B* 113, 12887–12895, 2009.
234. Palma, M., Levin, J., Lemaur, V., Liscio, A., Palermo, V., Cornil, J., Geerts, Y., Lehmann, M., and Samori, P. *Adv. Mater.* 18, 3313–3317, 2006.
235. Wasserfallen, D., Fischbach, I., Chebotareva, N., Kastler, M., Pisula, W., Jackel, F., Watson, M. D., Schnell, I., Rabe, J. P., Spiess, H. W., and Mullen, K. *Adv. Fun. Mater.* 15, 1585–1594, 2005.
236. Samori, P., Yin, X., Tchebotareva, N., Wang, Z., Pakula, T., Jackel, F., Watson, M. D., Venturini, A., Mullen, K., and Rabe, J. P. *J. Am. Chem. Soc.* 126, 3567–3575, 2004.
237. Samori, P., Severin, N., Simpson, C. D., Mullen, K., and Rabe, J. P. *J. Am. Chem. Soc.* 124, 9454–9457, 2002.
238. Piot, L., Marie, C., Feng, X., Mullen, K., and Fichou, D. *Adv. Mater.* 20, 3854–3858, 2008.
239. Friedlein, R., Crispin, X., Simpson, C. D., Watson, M. D., Jackel, F., Osikowicz, W., Marciniak, S., de Jong, M. P., Samori, P., Jonsson, S. K. M., Fahlman, M., Mullen, K., Rabe, J. P., and Salaneck, W. R. *Phys. Rev. B* 68, 195414, 2003.
240. Fujikake, H., Murashige, T., Sugibayashi, M., and Ohta, K. *Appl. Phys. Lett.* 85, 3474–3476, 2004.
241. Hoogboom, J., Garcia, P. M. L., Otten, M. B. J., Elemans, J. A. A. W., Sly, J., Lazarenko, S. V., Rasing, T., Rowan, A. E., and Nolte, R. J. M. *J. Am. Chem. Soc.* 127, 11047–11052, 2005.
242. Teresawa, N., Monobe, H., Kiyohara, K., and Shimizu, Y. *Chem. Commun.* 1678–1679, 2003.
243. Monobe, H., Awazu, K., and Shimizu, Y. *Adv. Mater.* 12, 1495–1499, 2000.
244. Monobe, H., Awazu, K., and Shimizu, Y. *Adv. Mater.* 18, 607–610, 2006.
245. Bisoyi, H. K. and Kumar, S. *Tetrahedron Lett.* 49, 3628–3631, 2008.
246. Steinhart, M., Zimmermann, S., Goring, P., Schaper, A. K., Gosele, U., Weder, C., and Wendorff, J. H. *Nano Lett.* 5, 429–434, 2005.
247. Steinhart, M., Murano, S., Schaper, A. K., Ogawa, T., Tsuji, M., Gosele, U., Weder, C., and Wendorff, J. H. *Adv. Funct. Mater.* 15, 1656–1664, 2005.
248. Pisula, W., Kastler, M., Wasserfallen, D., Davies, R. J., Gutierrez, M. C. G., and Mullen, K. *J. Am. Chem. Soc.* 128, 14424–14425, 2006.
249. Zhi, L., Wu, J., Li, J., Kolb, U., and Mullen, K. *Angew. Chem. Int. Ed.* 44, 2120–2123, 2005.
250. Kastler, M., Pisula, W., Davies, R. J., Gorelik, T., Kolb, U., and Mullen, K. *Small* 3, 1438–1444, 2007.

2 Monomeric Discotic Liquid Crystals

The majority of discotic liquid crystals belong to the monomeric class. Since the discovery of the first discotic liquid crystals derived from benzene core, most efforts have been geared toward understanding the nature of the molecular parameters that favor the formation of discotic liquid crystals and control their transition temperatures. Accordingly, a large number of monomeric discotic liquid crystals have been designed, synthesized, and studied for their mesomorphism and to comprehend the chemical structure–physical property relationships. To date, the number of discotic liquid crystals derived from more than 60 different cores comes to about 3000. This chapter presents the synthesis and mesomorphic behavior of discotic monomers derived from these cores. The chapter is divided into 25 sections. Sections 2.1 through 2.14 deal with the chemistry and thermal behavior of discotic liquid crystals derived from aromatic cores. Discotic liquid crystals generated from heterocyclic cores are covered in Sections 2.15 through 2.24. Section 2.25 presents the chemistry and mesomorphism of discotic liquid crystals derived from saturated cores.

Each section begins with a general introduction of the core followed by the synthesis and thermal behavior of liquid crystalline compounds derived from it. In addition to a brief description of the mesomorphic properties of important compounds given in the text, the thermal behavior of all the liquid crystalline materials is summarized in different tables. In all the tables, mainly liquid crystalline derivatives have been tabulated. The thermal behavior of a few relevant non-liquid crystalline final molecules has also been given, but most of the non-liquid crystalline intermediates have not been covered. The transition temperatures are given in degree centigrade. Multiple crystal–crystal transitions have not been given. Monotropic transitions are placed in brackets. In the tables, Cr=crystal; Ss=semi solid; Col=columnar phase (undefined symmetry); Col_h=hexagonal columnar phase (please note that the suffixes "o" and "d" have been removed); Col_r=rectangular columnar phase; Col_{ob}=columnar oblique phase; Col_t=columnar tetragonal phase; Col_p=columnar plastic phase; D_L=discotic lamellar phase; N=nematic phase; N_D=nematic discotic phase; N_c=nematic columnar phase; N*=chiral nematic; BP=blue phase; g=glass transition; Sm=smectic phase; M, LC, and X=unknown mesophase; poly=polymer; dec=decompose; and I=isotropic phase. A question mark (?) in the table means that the exact transition temperature is not observed.

2.1 BENZENE CORE

Benzene and its derivatives are not only dominating in organic chemistry textbooks as model aromatic compounds and solvents in chemical reactions, but they also serve as very important first-aid pharmaceuticals and are suspect carcinogens as well. Benzene made a giant leap into materials science in 1977 when its hexaesters were reported to form columnar liquid crystalline phases in disk-like molecules by Chandrasekhar and coworkers [1–3]. Hence, benzene is the first discotic core and its hexaesters are the prototype discotic liquid crystalline materials discovered. Ever since the discovery, a large variety of benzene derivatives have been designed, synthesized, and studied for their mesomorphism and to comprehend the chemical structure–physical property relationships [4–12]. Recently, the studies of discotic benzene derivatives have moved from structure–property relationship to device performance evaluation. Both thermotropic and lyotropic liquid crystalline phase behaviors of benzene derivatives have been studied. Moreover, the effect of symmetry on the phase behavior of benzene-based liquid crystals has also been reported. Furthermore, alignment and reorientation of columnar phases have been demonstrated, and photoconductivity, ionic conductivity, and nonlinear optical properties have been evaluated. More interestingly, a display device has been demonstrated with benzene multiynes [13,14]. To gain greater insight into the phase behavior and useful physical properties, many theoretical and simulation studies have been accomplished on benzene-based discotic compounds [15–24]. The theoretical studies include molecular interactions and dynamics in the mesophase, thermodynamic properties, pattern and texture formation, phase morphology and orientation dependence of viscosity, etc.

Chandrasekhar and coworkers prepared a series of hexaesters of hexahydroxybenzene, and from optical, calorimetric, and x-ray studies, they established the columnar liquid crystalline nature of these novel class of materials [1]. The synthetic route used to prepare these compounds is shown in Scheme 2.1. Hexahydroxybenzene **3** is prepared from glyoxal **1** and treated with an acid chloride to give the esters **4**. Yields of the hexaesters are improved if an inert atmosphere is used. Typically, no solvent is used for these reactions. The mesophase behavior of benzene hexaesters is summarized in Table 2.1 [25–28]. Almost all the mesomorphic derivatives exhibit a columnar hexagonal

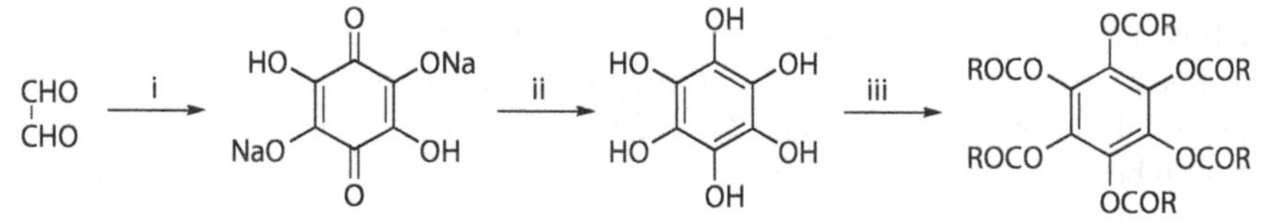

SCHEME 2.1   Synthesis of benzene hexaesters: (i) Na_2SO_3/O_2; (ii) $SnCl_2/HCl$; (iii) RCOCl.

TABLE 2.1
Transition Temperatures of Hexaesters of Benzene

Structure	R	Phase Transition	Ref.
4a	C_6H_{13}	Cr 68 Col$_h$ 86 I	[1]
4b	C_7H_{15}	Cr 80 Col$_h$ 84 I	[1]
4c	OC_6H_{13}	Liquid	[25]
4d	$CH_2OC_5H_{11}$	Cr 68 Col$_h$ 94 I	[25]
4e	$C_2H_4OC_4H_9$	Cr 30 Col$_h$ 32 I	[25]
4f	$C_3H_6OC_3H_7$	Cr 43 Col$_h$ 44 I	[25]
4g	$C_5H_{10}OCH_3$	Cr 16 Col$_h$ 25 I	[25]
4h	$CH_2OC_3H_7$	Cr 110 Col$_h$ 112 I	[25]
4i	$CH_2OC_7H_{15}$	Cr 74 Col$_h$ 94 I	[25]
4j	$CH_2SC_5H_{11}$	Cr 66 Col$_h$ 67 I	[25]
4k	$C_2H_4SC_4H_9$	Cr 101 I	[27]
4l	$C_3H_6SC_3H_7$	? 53 Col 57 I	[27]
4m	$C_5H_{10}SCH_3$	Cr 32 Col 37 I	[27]
4n	$CH(Me)C_6H_{13}$	Liquid	[28]
4o	$CH_2CH(Me)C_5H_{11}$	Cr 96 I	[28]
4p	$C_2H_4CH(Me)C_4H_9$	M 96 I	[28]
4q	$C_3H_6CH(Me)C_3H_7$	M 102 I	[28]
4r	$C_4H_8CH(Me)C_2H_5$	Cr 61 M 87 I	[28]
4s	$C_5H_{10}CH(Me)CH_3$	Cr 85 M 87 I	[28]
4t	$C_4H_8CH(Et)C_2H_5$	Cr 60 M$_1$ 86 M$_2$ 87 I	[28]
4u	$C_5H_{10}CH(Me)C_2H_5$	Cr 52 M 75 I	[28]

mesophase. The stability of the mesophase is sensitive to the presence of heteroatoms in the peripheral tails. Significant stabilization is observed for derivatives bearing a β-oxygen atom in the side chain. However, when the position of this oxygen atom is moved toward or away from the core, the mesophase stability is either reduced or the mesophase is suppressed. Branched side chains stabilize the mesophase if the branch point is close to the middle, but destabilization is noticed as it approaches the core. However, terminal branching points can have either an effect on the mesophase stability depending on whether they exert steric effect or space filling effect, which contrast to each other.

Being the first examples of liquid crystalline materials of its kind, the benzene hexaesters have been studied for their physical properties to gain more insight into the mesophase structure and molecular interactions involved in stabilizing discotic columnar phases [29–41]. Accordingly, solute–solvent interaction in the mesophase, heat capacity studies in the mesophase, infrared characterization of the relative orientation of molecular parts, phase transitions and molecular dynamics by nuclear magnetic resonance, electric birefringence, Langmuir–Blodgett films, etc., have been studied. The possible formation of lyotropic liquid crystalline phases has also been tested on some of these hexaester derivatives.

Tetraesters of 2,3,5,6-tetrahydroxy-1,4-benzoquinone 6 and their corresponding hydroquinones are also found to exhibit columnar mesomorphism, though over a very narrow range of temperature [42–44]. The synthesis of these compounds is shown in Scheme 2.2 and their phase behavior is shown in Table 2.2.

Phloroglucinol-derived C_3 symmetric liquid crystals 9 have been reported by Lehmann et al. These materials can be easily prepared via esterification of phlorogucinol with functionalized acid chlorides (Scheme 2.3). Their thermal data are listed in Table 2.3 [45–47]. These compounds adopt an E-shaped conformation in the columnar phase rather than a star-shaped conformation in order to avoid the free space in the columnar phase of some of the derivatives. Interestingly, the columnar

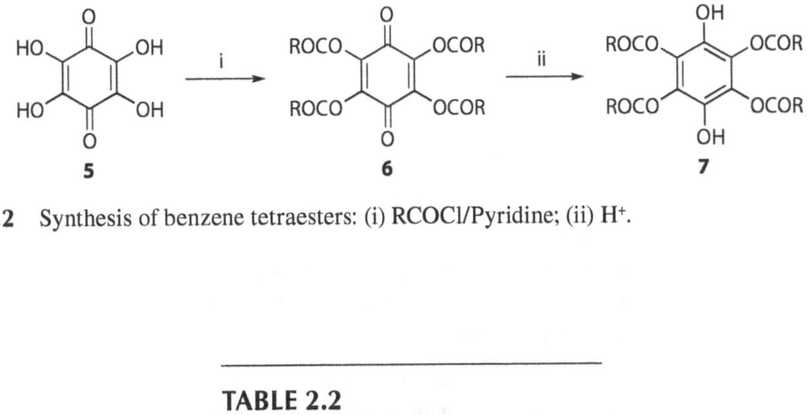

SCHEME 2.2 Synthesis of benzene tetraesters: (i) RCOCl/Pyridine; (ii) H⁺.

TABLE 2.2
Transition Temperatures of Benzene Tetraesters

Structure	R	Phase Transition	Ref.
6a	C_6H_{13}	Cr 55 Col 57 I	[42]
6b	C_7H_{15}	Cr 68 Col 70 I	[42]
7a	C_6H_{13}	Cr 80 Col 85 I	[44]
7b	C_7H_{15}	Cr 79 Col 82 I	[44]

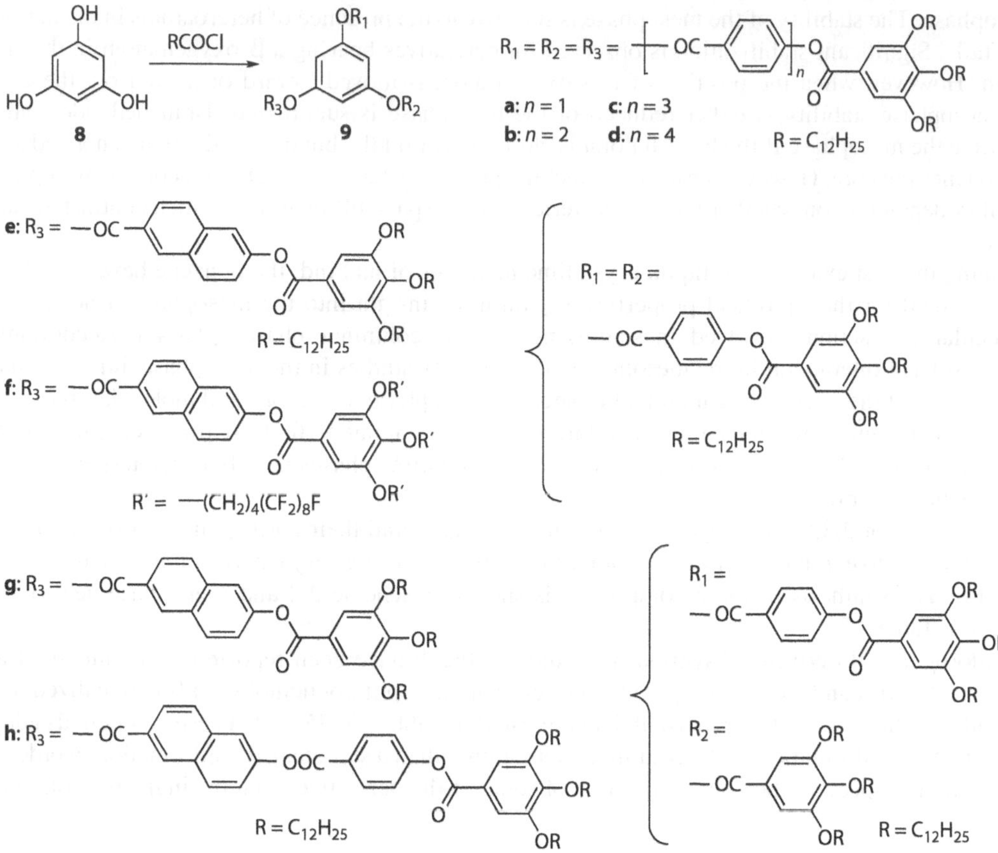

SCHEME 2.3 Phloroglucinol-derived discotic liquid crystals.

TABLE 2.3
Phase Transition Temperatures of C₃ Symmetrical and Unsymmetrical Triesters of Benzene

Structure	Phase Transition Temperatures	Ref.
9a	g 20 Col$_h$ 53 I	[45]
9b	Col$_1$ 79 Col$_2$ 98 I	[45]
9c	Col$_1$170 Col$_2$172 I	[45]
9d	Col$_1$ 240 Col$_2$ 249 dec	[45]
9e	Col$_h$ 65.1 I	[47]
9f	Col$_r$ 151.9 SmA 190.9 I	[47]
9g	Col$_h$ 43.2 I	[47]
9h	Cub 60.6 I	[47]

phase of some of the derivatives serve as templates for crystal growth, and it has been revealed that the crystal growth is one dimensional in nature and preserves the orientation of the parent mesophase. Non-symmetric mesogens containing naphthalene chromophore have also been studied. Structural investigations reveal a rich mesomorphism from columnar, lamellar, to cubic phases depending on the type of oligobenzoate arms.

Diacetylene-armed C_3-symmetric esters 12 and amides 13 derived from benzene tricarbonyl chloride have been studied by Chang et al. These derivatives exhibit a rather ill-defined mesomorphism and undergo thermal polymerization [48,49]. The photopolymerization of these compounds has also been carried out in the solid as well as liquid crystalline state. The chemical structures of these compounds are shown in Scheme 2.4 and their phase transition temperatures are listed in Table 2.4. It is interesting to note that the phloroglucinol analogue 14 is not mesomorphic, which demonstrates the effect of changing the linkage to central benzene core that might act against the π-stacking of the molecules.

Matsunaga and coworkers have replaced some of the ester linkages of the benzene hexaesters with amide groups. The heavily substituted members of this class of compounds are prepared by the total acylation of 3,6-diamino-1,2,4,5-tetrahydroxybenzene [50]. The synthetic route to prepare these compounds is shown in Scheme 2.5, and their mesophase behavior is summarized in Table 2.5. It is very clear that the introduction of two amide groups stabilizes the columnar mesophase substantially with respect to the parent hexaesters indicating the role of hydrogen bonding in the mesophase stability. The hydrogen bonding–induced phase stability of the above amides has prompted them to synthesize a variety of mixed ester amides carrying even lesser number of

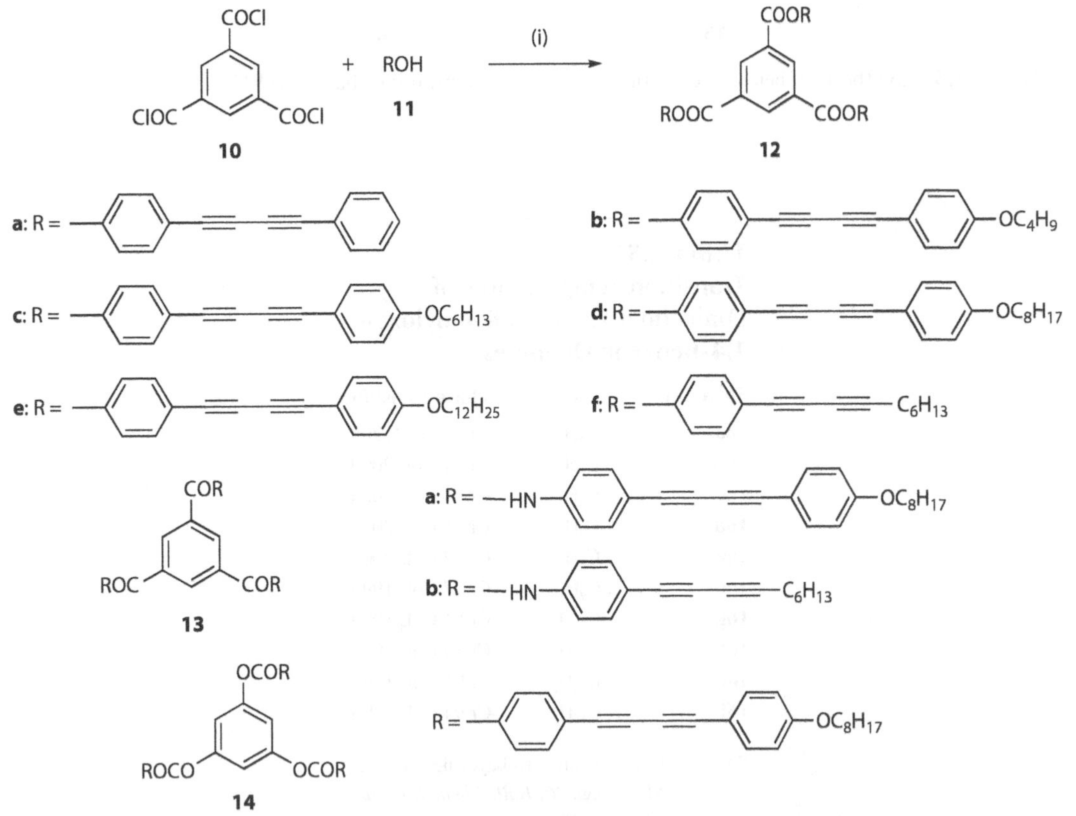

SCHEME 2.4 Diacetylene-armed symmetric benzene esters: (i) NaH, THF, reflux.

TABLE 2.4
Phase Transition Temperatures of Diacetylene-Armed Benzene Derivatives

Structure	Phase Transition	Ref.
12a	Cr 200 I > 250 poly	[49]
12b	g120 M 135 poly	[49]
12c	g 140 M	[49]
12d	Cr (137 Col$_h$) 146 I	[48]
12e	Non LC	[49]
12f	Non LC	[49]
13a	Cr 190 N$_D$ 204 I	[48]
13b	Cr 120 M >200 poly	[48]
14	Non LC	[49]

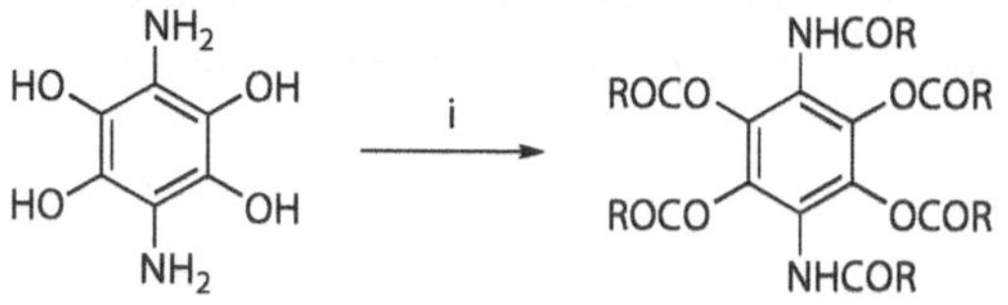

SCHEME 2.5   Synthesis of benzene discotics with mixed ester–amide chains: (i) RCOCl/pyridine.

TABLE 2.5
Transition Temperatures of Dialkanoyl Tetrakis(alkanoyloxy)-1,4-Benzene Diamines

Structure	R	Phase Transition
16a	C$_4$H$_9$	Cr 134 Col$_h$ 208 I
16b	C$_5$H$_{11}$	Cr 71 Col$_h$ 209 I
16c	C$_6$H$_{13}$	Cr 64 Col$_h$ 208 I
16d	C$_7$H$_{15}$	Cr 75 Col$_h$ 205 I
16e	C$_8$H$_{17}$	Cr 77 Col$_h$ 200 I
16f	C$_9$H$_{19}$	Cr 77 Col$_h$ 199 I
16g	C$_{10}$H$_{21}$	Cr 81 Col$_h$ 198 I
16h	C$_{11}$H$_{23}$	Cr 89 Col$_h$ 197 I
16i	C$_{13}$H$_{27}$	Cr 92 Col$_h$ 189 I
16j	C$_{15}$H$_{31}$	Cr 94 Col$_h$ 190 I

Source: Data from Kobayashi, Y. and Matsunaga, Y., *Bull. Chem. Soc. Jpn.*, 60, 3515, 1987.

peripheral alkyl chains [51–53]. The synthesis of such materials is shown in Schemes 2.6 through 2.8, and their thermal phase behavior properties are given in Tables 2.6 through 2.8. The C_3 symmetric benzene tricarboxamides, which can be prepared from 1,3,5-benzene tricarbonyl chloride and alkyl amines as shown in Scheme 2.9, also form mesophases [54]. However, their columnar structure has not been confirmed unequivocally. The phase transition temperatures of these compounds are collected in Table 2.9.

The existence of both thermotropic and lyotropic mesomorphisms in disk-shaped compounds is rather unusual. However, the latter can be stabilized by hydrogen bonding, as has been shown by Meijer and coworkers [55–63]. They have prepared and studied mesomorphism of diamino-bipyridine-based

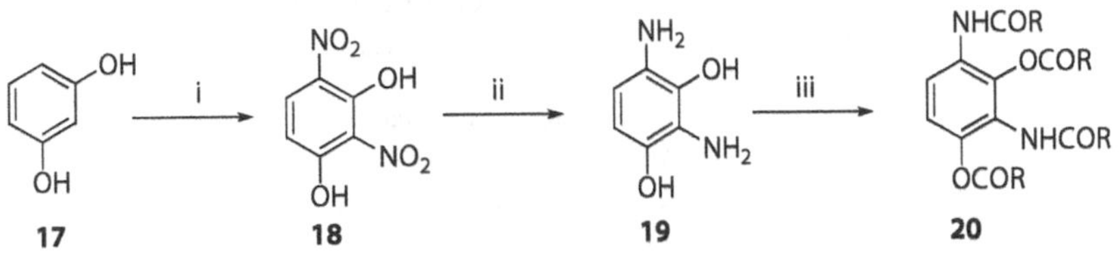

SCHEME 2.6 Synthesis of tetrasubstituted benzene discotics with mixed ester-amide chains: (i) NaNO$_2$, (ii) SnCl$_2$/HCl, (iii) RCOCl/pyridine.

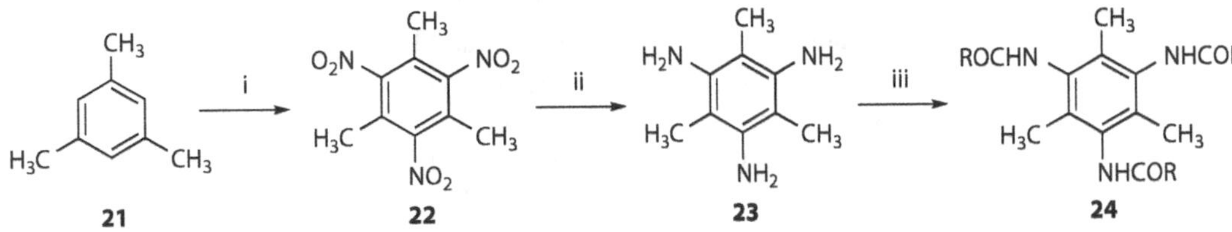

SCHEME 2.7 Synthesis of mesitylene-derived discotics: (i) HNO$_3$/H$_2$SO$_4$; (ii) Sn/HCl; (iii) RCOCl/pyridine.

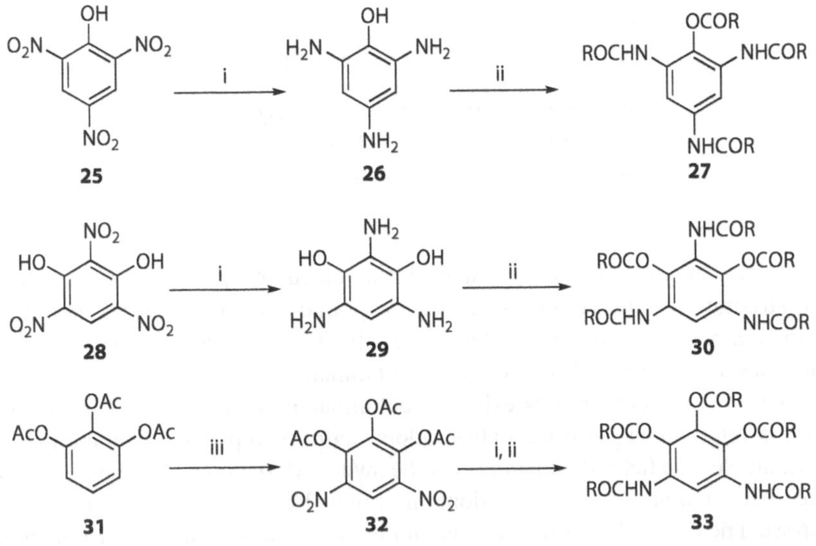

SCHEME 2.8 Synthesis of mixed ester–amide derivatives of benzene: (i) Sn/HCl; (ii) RCOCl/pyridine; (iii) HNO$_3$.

TABLE 2.6
Transition Temperatures of Benzene
1,3-Diamide-2,4-Diesters

Structure	R	Phase Transition
20a	C_6H_{13}	Cr 89 Col$_h$ 124 I
20b	C_7H_{15}	Cr 87 Col$_h$ 122 I
20c	C_8H_{17}	Cr 90 Col$_h$ 121 I
20d	C_9H_{19}	Cr 93 Col$_h$ 122 I
20e	$C_{10}H_{21}$	Cr 95 Col$_h$ 120 I
20f	$C_{11}H_{23}$	Cr 97 Col$_h$ 117 I
20g	$C_{13}H_{27}$	Cr 101 Col$_h$ 114 I
20h	$C_{15}H_{31}$	Cr 103 Col$_h$ 114 I

Source: Data from Kawada, H. and Matsunaga, Y.,
Bull. Chem. Soc. Jpn., 61, 3083, 1988.

TABLE 2.7
Transition Temperatures of
1,3,5-Trimethyl Benzene Triamides

Structure	R	Phase Transition
24a	C_3H_7	Cr 380 Col$_h$ 410 I
24b	C_4H_9	Cr 315 Col$_h$ 380 I
24c	C_5H_{11}	Cr 300 Col$_h$ 357 I
24d	C_6H_{13}	Cr 257 Col$_h$ 357 I
24e	C_7H_{15}	Cr 239 Col$_h$ 346 I
24f	C_8H_{17}	Cr 200 Col 222 Col$_h$ 340 I
24g	C_9H_{19}	Cr 189 Col 206 Col$_h$ 342 I
24h	$C_{10}H_{21}$	Cr 185 Col 191 Col$_h$ 338 I
24i	$C_{11}H_{23}$	Cr 183 Col 189 Col$_h$ 342 I
24j	$C_{13}H_{27}$	Cr 118 Col 182 Col$_h$ 338 I
24k	$C_{15}H_{31}$	Cr 120 Col 175 Col$_h$ 328 I

Source: Data from Harada, Y. and Matsunaga, Y., *Bull.
Chem. Soc. Jpn.*, 61, 2739, 1988.

C_3 symmetric extended discotics stabilized by intramolecular hydrogen bonding. The synthesis and chemical structures of the liquid crystalline compounds are shown in Scheme 2.10, and their phase transition temperatures are collected in Table 2.10. The large core induces strong interactions between molecules and hence leads to mesophases of enhanced thermal stability.

In dilute solutions, the compounds exhibit a columnar nematic phase owing to the hydrogen bonding and π–π interactions. This mesophase adopts a uniaxial planar orientation in between two glass plates. In an electric field, the columns can be switched to a homeotropic alignment and such voltage-induced formation of large monodomains might be useful for one-dimensional ion and charge transport. The chiral derivatives self-assemble into dynamic chiral helix in apolar solvents. The chirality at the supramolecular level in the stacks can be tuned by temperature and solvent, as has been revealed by circular dichroism spectroscopy.

TABLE 2.8
Transition Temperatures for Mixed Ester and Amide Derivatives of Benzene

Structure	R	Phase Transition		
		Structure 27	Structure 30	Structure 33
A	C_4H_9	Cr 144 Col$_h$ 222 I	Cr 100 Col$_h$ 241 I	Cr 113 Col$_h$ 196 I
B	C_5H_{11}	Cr 133 Col$_h$ 225 I	Cr 55 Col$_h$ 238 I	Cr 67 Col$_h$ 208 I
C	C_6H_{13}	Cr 129 Col$_h$ 229 I	Cr 63 Col$_h$ 241 I	Cr 88 Col$_h$ 208 I
D	C_7H_{15}	Cr 123 Col$_h$ 227 I	Cr 49 Col$_h$ 242 I	Cr 77 Col$_h$ 208 I
E	C_8H_{17}	Cr 109 Col$_h$ 226 I	Cr 73 Col$_h$ 227 I	Cr 93 Col$_h$ 206 I
F	C_9H_{19}	Cr 111 Col$_h$ 227 I	Cr 65 Col$_h$ 225 I	Cr 92 Col$_h$ 203 I
G	$C_{10}H_{21}$	—	—	Cr 96 Col$_h$ 202 I
H	$C_{11}H_{23}$	Cr 107 Col$_h$ 225 I	Cr 78 Col$_h$ 224 I	Cr 101 Col$_h$ 203 I
I	$C_{13}H_{27}$	Cr 112 Col$_h$ 225 I	Cr 84 Col$_h$ 224 I	Cr 103 Col$_h$ 198 I
J	$C_{15}H_{31}$	Cr 113 Col$_h$ 218 I	Cr 66 Col$_h$ 225 I	Cr 103 Col$_h$ 193 I

Source: Data from Kawamata, J. and Matsunaga, Y., *Mol. Cryst. Liq. Cryst.*, 231, 79, 1993.

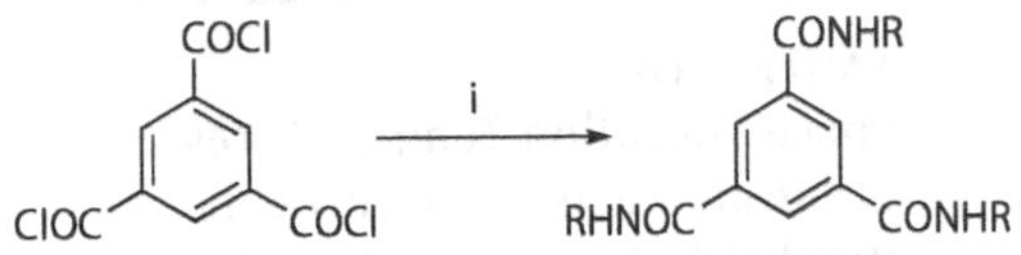

SCHEME 2.9 Synthesis of benzene triamides: (i) RNH$_2$/pyridine.

TABLE 2.9
Transition Temperatures for *N,N′,N″*-1,3,5-Benzene Tricarboxamides

Structure	R	Phase Transition
34a	C_5H_{11}	Cr 119 M 206 I
34b	C_6H_{13}	Cr 99 M 205 I
34c	C_7H_{15}	Cr 116 M 208 I
34d	C_8H_{17}	Cr 102 M 204 I
34e	C_9H_{19}	Cr 65 M 215 I
34f	$C_{10}H_{21}$	Cr 49 M 208 I
34g	$C_{11}H_{23}$	Cr 72 M 216 I
34h	$C_{12}H_{25}$	Cr 88 M 212 I
34i	$C_{13}H_{27}$	Cr 81 M 216 I
34j	$C_{14}H_{29}$	Cr 61 M 209 I
34k	$C_{15}H_{31}$	Cr 88 M 214 I
34l	$C_{16}H_{33}$	Cr 73 M 205 I
34m	$C_{17}H_{35}$	Cr 87 M 211 I
34n	$C_{18}H_{37}$	Cr 78 M 206 I

Source: Data from Matsunaga, Y. et al., *Bull. Chem. Soc. Jpn.*, 61, 207, 1988.

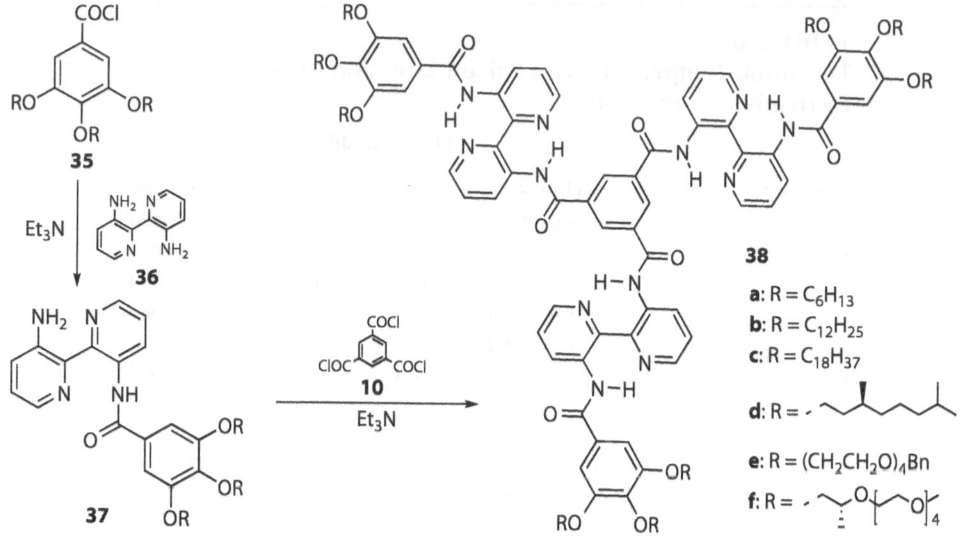

SCHEME 2.10 Synthesis of intramolecular hydrogen bonded benzene-based discotic liquid crystals.

TABLE 2.10
Phase Transition Temperatures
of Intramolecularly Hydrogen
Bonded C$_3$ Symmetric Benzene
Based Discotic Liquid Crystals

Structure	Phase Transition	Ref.
38a	Col$_h$ 383 I	[55]
38b	Cr 9 Col$_h$ 355 I	[55]
38c	Cr 62 Col$_h$ 308 I	[55]
38d	Col$_h$ 373 I	[60]
38e	g −50 Col$_h$ 244 I	[58]
38f	g −74 Col$_h$ 270 I	[59]

Recently, Nuckolls and coworkers designed and synthesized new benzene derivatives **43** forming columnar phases that are held together by hydrogen bonding and π–π interactions [64–68]. These hexa-substituted benzene derivatives contain three amides in the 1,3,5-positions that are flanked by substituents other than hydrogen at each of the remaining positions. The first series of compounds are substituted by amides and alkoxy groups alternatively (Scheme 2.11A), whereas the second series of compounds are substituted by amides and ethynyl substituents (Scheme 2.11B). The key to the synthesis of the first series of amides is the discovery that 1,3,5-tribromo-2,4,6-trialkoxy benzene undergoes a triple lithium/halogen exchange reaction, as shown in Scheme 2.11A. These compounds in the bulk state self-assemble to columnar superstructures. The phase transition temperatures of both the series of compounds are collected in Table 2.11. The key step in the synthesis of the second series of compounds is the palladium-mediated coupling between an alkynyl subunit and the trisbromotriester (Scheme 2.11B). Sonogashira coupling in this reaction was very sluggish and produced a mixture of products even after an extended reaction time from which the desired product could not be isolated. However, a modified Stille coupling furnished the desired product.

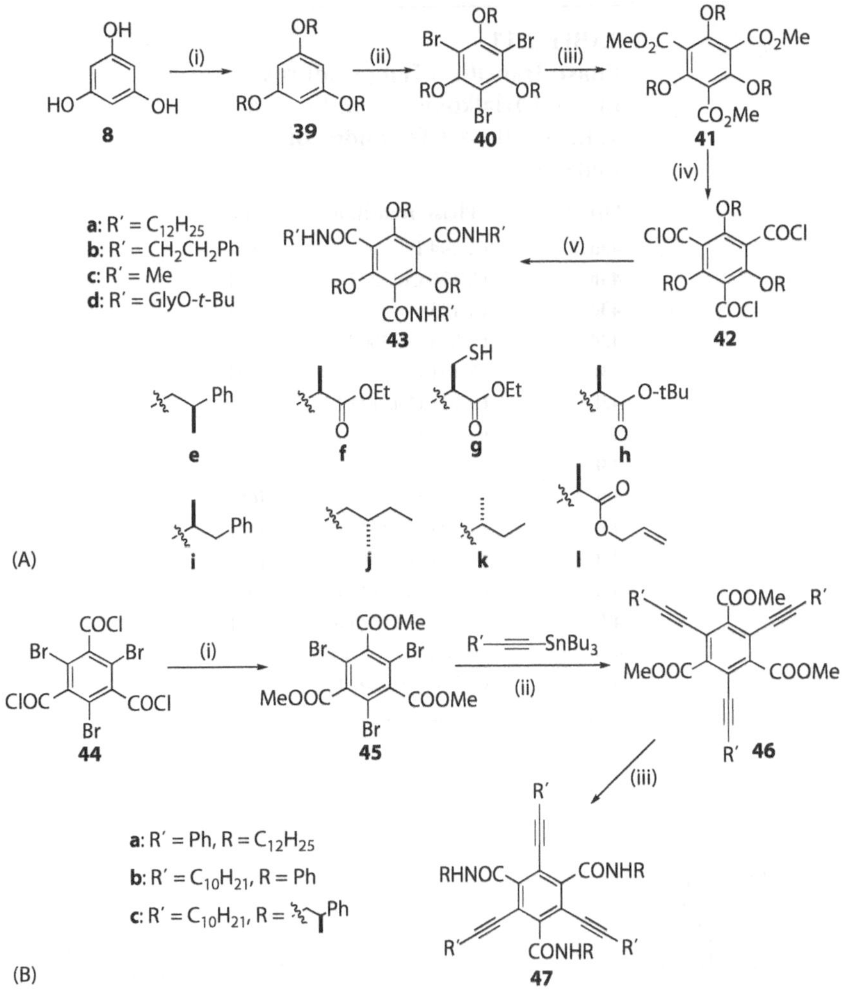

SCHEME 2.11 (A) Synthesis of trialkoxy-triamides of benzene: (i) K_2CO_3, $C_{12}H_{25}I$; (ii) Br_2, $FeCl_3$; (iii) t-BuLi; $ClCO_2Me$; (iv) NaOH; $SOCl_2$; (v) RNH_2, Et_3N. (B) Synthesis of amide-alkyne-substituted benzene discotics: (i) MeOH, pyridine; (ii) $Pd(PPh_3)$, $AsPh_3$ (catalyst); (iii) KOH, i-PrOH; $SOCl_2$; RNH_2.

These compounds also exhibit columnar mesomorphism in the bulk state. The compounds of both the series also exhibit lyotropic mesophases and hence they are amphotropic in nature. In solution, depending on the solvent, concentration, temperature, and side chains, they form helical columnar aggregates of various sizes that can be transferred to substrates. At lower concentrations, they exhibit columnar nematic phase, but at higher concentrations, they exhibit chiral nematic phases of helical columns, that is, the helical columns form a super-helix in solution, which reflects circularly polarized light. Since the molecules possess dipole moment, on aggregation the columns possess polar order and interestingly these polar aggregates can be directed by electric fields after transferring from solutions to substrates. Electric field–induced reorientation of the columns in these amides is very promising from the device point of view.

Another class of discotic benzene derivatives consists of hexakis[(alkoxyphenoxy)methyl] derivatives **49** [69]. These compounds are simply prepared from hexakis(bromomethyl)-benzene **48** and an excess of p-alkoxyphenoxide, as shown in Scheme 2.12. The thermal behavior of these materials is summarized in Table 2.12. The mesophase of the compounds are ill characterized

TABLE 2.11
Phase Transition Temperatures of 1,3,5-Trialkoxy Triamides and 1,3,5-Triethynyl Triamides of Benzene

Structure	Phase Transition	Ref.
43a	Cr 294 I	[64]
43b	Cr 176 Col$_h$ 232 I	[64]
43c	Cr 123 I	[64]
43d	Cr 85 Col$_h$ 200 I	[64]
43e	Cr 191 Col$_h$ 233 I	[65]
43f	Cr 141 Col$_h$ 172 I	[65]
43g	Cr 164 I	[65]
43h	Cr 103 I	[65]
43i	Cr 249 I	[65]
43j	Cr 90 Col$_h$ 254 I	[65]
43k	Cr 98 Col$_h$ 248 1	[65]
43l	Cr 113 Col$_h$ 174 I	[65]
47a	Cr 141 Col$_h$ 225 dec	[67]
47b	Cr 96 Col$_h$ 225 dec	[67]
47c	Cr 110 Col$_h$ 207 I	[67]

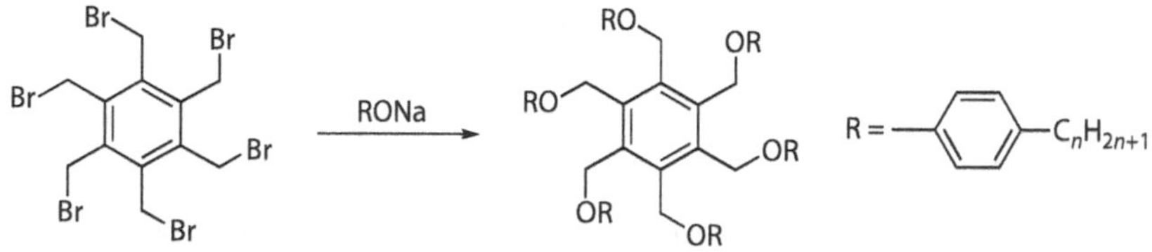

SCHEME 2.12   Synthesis of hexakis[(alkoxyphenoxy) methyl] derivatives of benzene.

TABLE 2.12
Transition Temperatures for Hexakis [(Alkoxyphenoxy) Methyl] Benzenes

Structure	R	Phase Transition
49a	p-C$_5$H$_{11}$ OC$_6$H$_4$	Cr 68 Col$_r$ 97 I
49b	p-C$_6$H$_{13}$ OC$_6$H$_4$	Cr 76 Col$_r$ 83 I
49c	p-C$_7$H$_{15}$ OC$_6$H$_4$	Cr 67 Col$_r$ 71 I

Source: Data from Kok, D.M. et al., *Mol. Cryst. Liq. Cryst.*, 129, 53, 1985.

since the authors were unable to confirm unambiguously the presence of a discotic mesophase and pointed out that the mesophase might be a special type of crystal.

Praefcke and coworkers have discovered that hexakis(alkylsulfone) derivatives **52** also form columnar mesophases [70–73]. The synthesis of the compounds is straightforward and is depicted in Scheme 2.13. The thermal behavior of the compounds is summarized in Table 2.13. It appears that the compounds exhibit disordered hexagonal columnar phase with significant supercooling.

Benzene-based multiynes (hexaynes [74–81] and pentaynes [82–107]) are the most widely studied discotic liquid crystals among the benzene derivatives. Praefcke introduced these fascinating discotic materials, and groups around Janietz, Kumar, and Mehl have developed different strategies to synthesize and study the important physical properties of these materials. Most of these materials provide and stabilize the rarely occurring discotic nematic phase. Of particular significance is the demonstration of a liquid crystal display device having wide and symmetrical viewing angle characteristics by using some of these materials as the active switching components [13], which was further advanced [14] by the introduction of a room temperature discotic nematic material by Kumar and coworkers [78]. Both symmetrical **56** and unsymmetrical **57** derivatives have been designed and studied. The synthesis and structures of the hexaynes are shown in Scheme 2.14 and their phase transition temperatures are shown in Table 2.14. Pentaynes were introduced by Praefcke et al., whereas functionalized and amphiphilic pentaynes were introduced by Janietz and coworkers [83] from which a variety of dimers, oligomers, and polymers have been synthesized and studied.

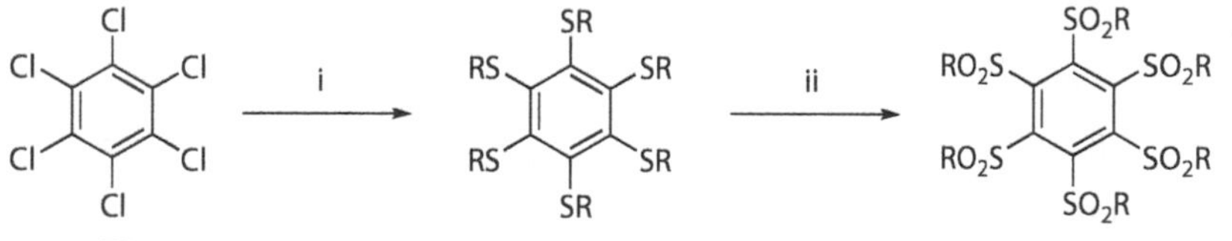

SCHEME 2.13 Synthesis of hexkisalkylsulfone derivatives of benzene: (i) RSNa/HMPT, (ii) *m*-CPBA/CHCl$_3$.

TABLE 2.13
Transition Temperatures for Hexakis(Alkylsulfone) Benzene Derivatives

Structure	R	Phase Transition
52a	C_6H_{13}	Cr 164 I
52b	C_7H_{15}	Cr 120 Col$_h$ 138 I
52c	C_8H_{17}	Cr 108 Col$_h$ 134 I
52d	C_9H_{19}	Cr 87 Col$_h$ 130 I
52e	$C_{10}H_{21}$	Cr 77 Col$_h$ 125 I
52f	$C_{11}H_{23}$	Cr 65 Col$_h$ 115 I
52g	$C_{12}H_{25}$	Cr 45 Col$_h$ 103 I
52h	$C_{13}H_{27}$	Cr 44 Col$_h$ 90 I
52i	$C_{14}H_{29}$	Cr 42 Col$_h$ 77 I
52j	$C_{15}H_{31}$	Cr 49 Col$_h$ 63 I
52k	$C_{16}H_{33}$	Cr 69 I

Source: Data from Praefcke, K. et al., *Z. Naturforsch.*, 39b, 950, 1984.

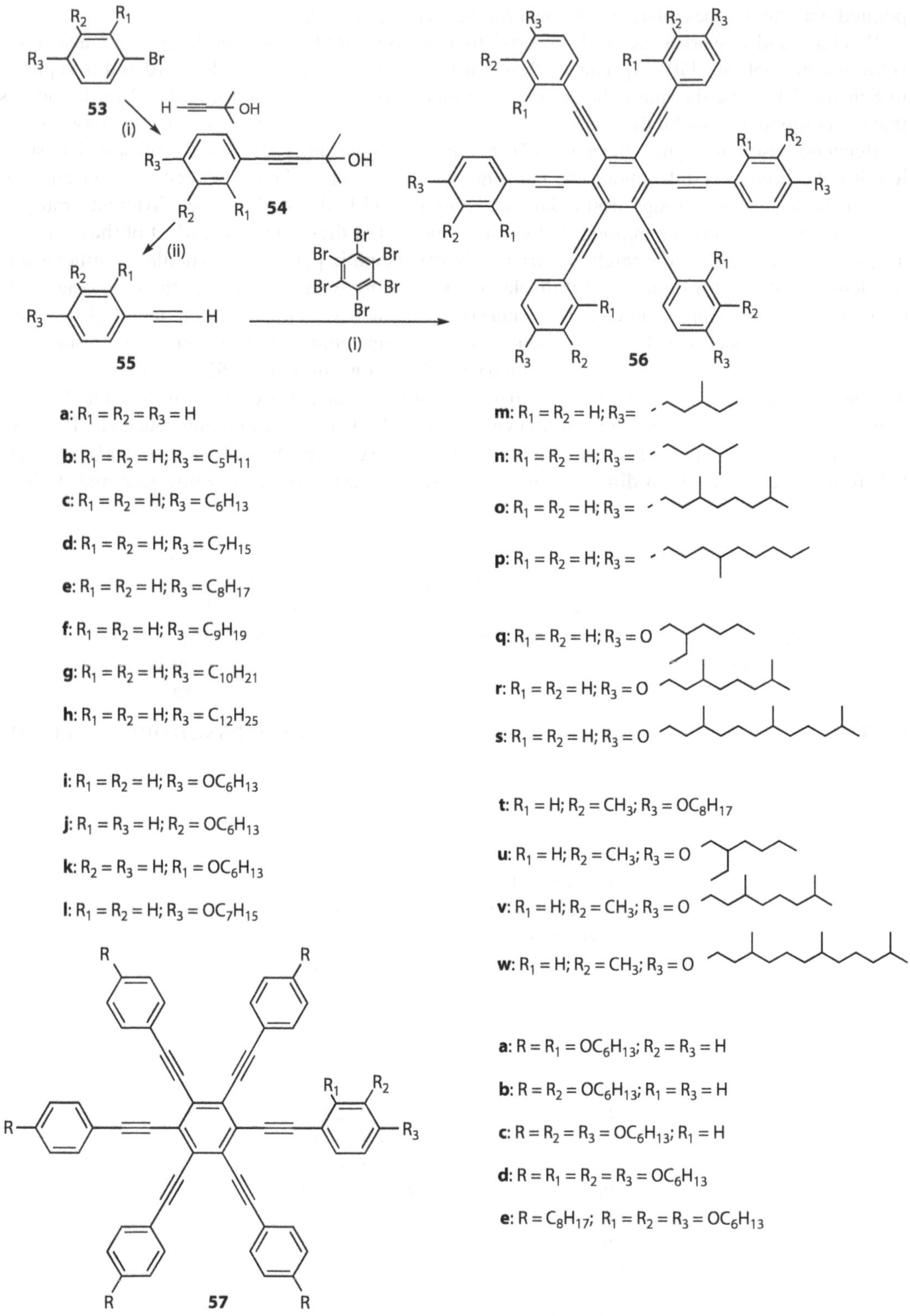

a: $R_1 = R_2 = R_3 = H$

b: $R_1 = R_2 = H; R_3 = C_5H_{11}$

c: $R_1 = R_2 = H; R_3 = C_6H_{13}$

d: $R_1 = R_2 = H; R_3 = C_7H_{15}$

e: $R_1 = R_2 = H; R_3 = C_8H_{17}$

f: $R_1 = R_2 = H; R_3 = C_9H_{19}$

g: $R_1 = R_2 = H; R_3 = C_{10}H_{21}$

h: $R_1 = R_2 = H; R_3 = C_{12}H_{25}$

i: $R_1 = R_2 = H; R_3 = OC_6H_{13}$

j: $R_1 = R_3 = H; R_2 = OC_6H_{13}$

k: $R_2 = R_3 = H; R_1 = OC_6H_{13}$

l: $R_1 = R_2 = H; R_3 = OC_7H_{15}$

m: $R_1 = R_2 = H; R_3 =$

n: $R_1 = R_2 = H; R_3 =$

o: $R_1 = R_2 = H; R_3 =$

p: $R_1 = R_2 = H; R_3 =$

q: $R_1 = R_2 = H; R_3 = O$

r: $R_1 = R_2 = H; R_3 = O$

s: $R_1 = R_2 = H; R_3 = O$

t: $R_1 = H; R_2 = CH_3; R_3 = OC_8H_{17}$

u: $R_1 = H; R_2 = CH_3; R_3 = O$

v: $R_1 = H; R_2 = CH_3; R_3 = O$

w: $R_1 = H; R_2 = CH_3; R_3 = O$

a: $R = R_1 = OC_6H_{13}; R_2 = R_3 = H$

b: $R = R_2 = OC_6H_{13}; R_1 = R_3 = H$

c: $R = R_2 = R_3 = OC_6H_{13}; R_1 = H$

d: $R = R_1 = R_2 = R_3 = OC_6H_{13}$

e: $R = C_8H_{17}; R_1 = R_2 = R_3 = OC_6H_{13}$

SCHEME 2.14   Synthesis of benzene hexaalkynes: (i) Pd(0), CuI, Et$_3$N; (ii) KOH, toluene, reflux.

TABLE 2.14
Phase Transition Temperatures of Benzene Hexaalkynes

Structure	Phase Transition	Ref.
56a	Cr 310 I	[74]
56b	Cr 169.5 N_D 187 I	[75]
56c	Cr 123.7 N_D 142 I	[75]
56d	Cr 98.2 N_D 131.2 I	[75]
56e	Cr 80.1 N_D 95.7 I	[74]
56f	Cr 70.4 N_D 87.3 I	[77]
56g	Cr (54 N_D) 71 I	[79]
56h	Cr 57 I	[79]
56i	Cr 143.5 N_D 215.9 I	[81]
56j	Cr 87 I	[81]
56k	Cr 63 I	[81]
56l	Cr 108.8 N_D 192.5 I	[76]
56m	Cr > 250 dec	[79]
56n	Cr > 195 dec	[79]
56o	Non-LC	[79]
56p	Cr 6.1 N_D 69.3 I	[78]
56q	Cr 78.3 Col 231 I	[80]
65r	Cr 80 N_D 124 I	[79]
56s	Liquid	[79]
56t	Cr 95 N_D 176 I	[79]
56u	Cr 64.3 Col 199.9 I	[80]
56v	Cr 71 N_D > 147 dec	[79]
56w	Liquid	[79]
57a	Cr 87.3 N_D 100 I	[81]
57b	Cr 108.8 N_D 178.5 I	[81]
57c	Cr 93.7 N_D 195 I	[81]
57d	Cr 83 N_D 120 I	[81]
57e	Cr 30.2 N_D 60.8 I	[81]

Pentaynes containing chiral branched chains exhibit chiral discotic nematic phase, whereas those containing racemic side chains furnish room temperature discotic nematic phase [86,79]. Mehl and coworkers have introduced microsegregating (siloxane and perfluorinated) side chains to the pentaynes to study the effect of these polyphilic chains on the mesomorphism [93,94]. The chemical structures of the compounds are shown in Scheme 2.15, and the phase behavior of these compounds is summarized in Table 2.15.

Compounds combining discotic and calamitic nematogens have been designed and studied as model compounds for the much sought after biaxial nematic materials [97,98,103]. The chemical structures of the compounds are shown in Scheme 2.16 and their thermal data are collected in Table 2.16. Although these compounds possess a significant molecular biaxiality and avoid the phase separation problem encountered in rod disk mixtures, they fail to stabilize the biaxial nematic phase. Some of the compounds exhibit metastable nematic phase while others furnish stable nematic phase. Interestingly, some of the derivatives exhibit multiple mesophases with multiple levels of phase-segregated molecular organization in the liquid crystalline phase [105,106].

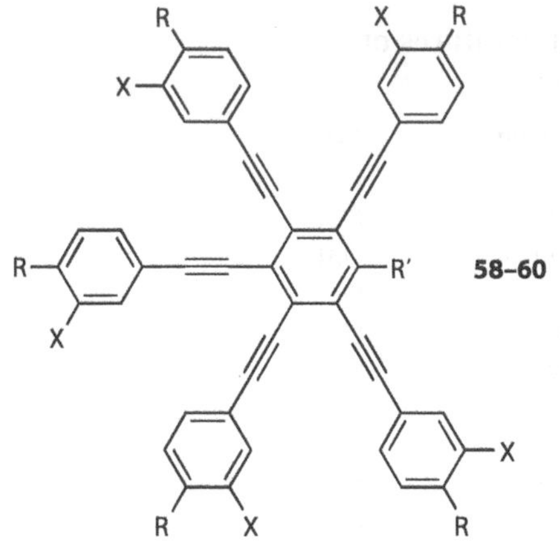

58a–c: X = R = H

59a–f: X = H, R = C_nH_{2n+1}

60a–r: X = H, R = OC_nH_{2n+1}/other group

60s: X = CH_3, R = OC_nH_{2n+1}

58–60

58a: R′ = $O(CH_2)_{11}OH$
58b: R′ = $O(CH_2)_{10}COOEt$
58c: R′ = $O(CH_2)_{10}COOH$

59a: R = CH_3, R′ = $O(CH_2)_{11}OH$
59b: R = CH_3, R′ = $O(CH_2)_{11}OCOCH_2CH_3$
59c: R = CH_3, R′ = $O(CH_2)_{10}COOH$
59d: R = CH_3, R′ = $O(CH_2)_{10}COOCH_3$
59e: R = CH_3, R′ = $O(CH_2)_{10}CONHC_3H_7$
59f: R = CH_3, R′ = $O(CH_2)_{10}COOCH_2C_6F_{13}$
59g: R = CH_3, R′ = $O(CH_2)_{10}COOCH_2C_{11}F_{23}$
59h: R = CH_3,
    R′ = $O(CH_2)_{11}OCO(CH_2)_4Si(CH_3)_2OSi(CH_2)_3$

59i: R = CH_3, R′ = $O(H_2C)_{10}OCO$

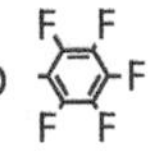

59j: R = CH_3, R′ = $O(H_2C)_{10}OCN$

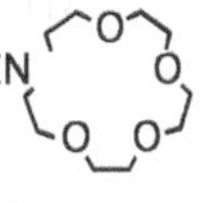

59k: R = C_2H_5, R′ = $O(CH_2)_{11}OH$
59l: R = C_3H_7, R′ = $O(CH_2)_{11}OH$
59m: R = C_5H_{11}, R′ = $O(CH_2)_{11}OH$
59n: R = C_5H_{11}, R′ = $O(CH_2)_{10}COOEt$
59o: R = C_5H_{11}, R′ = $O(CH_2)_{10}COOH$
59p: R = C_5H_{11}, R′ = OC_9H_{19}
59q: R = C_5H_{11}, R′ = $OC_{10}H_{21}$
59r: R = C_5H_{11}, R′ = $OC_9H_{18}CH=CH_2$
59s: R = C_5H_{11}, R′ = $OC_{11}H_{23}$
59t: R = C_7H_{15}, R′ = $O(CH_2)_{11}OH$

59u: R = C_4H_9, R′ = O

59v: R = C_5H_{11}, R′ = O

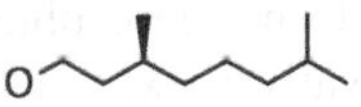

60a: R = OCH_3, R′ = $O(CH_2)_{11}OH$
60b: R = OCH_3, R′ = $O(CH_2)_{11}OCOPhO(CH_2)_{10}CH_3$
60c: R = OCH_3, R′ = $O(CH_2)_{11}OCO(CH_2)_9CH=CH_2$
60d: R = OCH_3, R′ = $O(CH_2)_{11}OCO(CH_2)_6CH_3$
60e: R = OC_6H_{13}, R′ = $O(CH_2)_{11}OH$
60f: R = OC_6H_{13},
    R′ = $O(CH_2)_{11}OCO(CH_2)_4Si(CH_3)_2OSi(CH_2)_3$

60g: R′ = $OC_{16}H_{33}$, R = O

60h: R′ = $OC_{16}H_{33}$, R = O

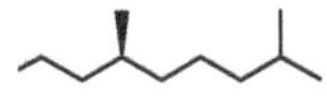

60i: R′ = $OC_{16}H_{33}$, R = O

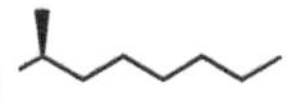

60j: R′ = $OC_{10}H_{21}$, R = O

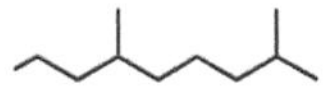

60k: R′ = R = O

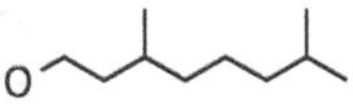

60l: R′ = CN, R = - -

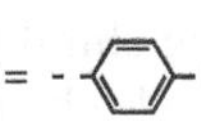

60m: R′ = CN, R 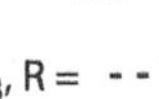 = - C_5H_{11}

60n: R′ = $OC_{16}H_{33}$, R = - -

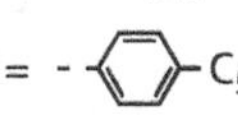

60o: R′ = $OC_{16}H_{33}$, R 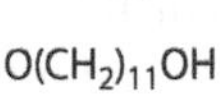 = - C_5H_{11}

60p: R = SC_6H_{13}, R′ = $O(CH_2)_{11}OH$

60q: R = $SO_2C_6H_{13}$, R′ = $O(CH_2)_{11}OH$

60r: R = $NHCOC_6H_{13}$, R′ = $O(CH_2)_{11}OH$

60s: X = CH_3, R′ = R = O

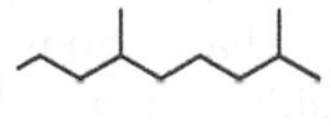

SCHEME 2.15   Pentaalkynylbenzene-based discotic liquid crystals.

TABLE 2.15
Phase Transition Temperatures of Benzene Pentaynes

Structure	Phase Transition	Ref.
58a	Cr 118 I	[84]
58b	Cr 90.2 I	[83]
58c	Cr 140.4 I	[83]
59a	Cr 171.6 N_D 226.6 I	[91]
59b	Cr 155 N_D195 I	[101]
59c	Cr 187 N_D 244 I	[94]
59d	Cr 167 N_D 201 I	[94]
59e	Cr 168 N_D 189 I	[94]
59f	Cr165 N_D 176 I	[94]
59g	Cr 181 N_D 192 I	[95]
59h	Cr 135 I	[93]
59i	Cr 150 N_D 196 I	[94]
59j	Cr 151 I	[94]
59k	Cr 122.2 N_D 147.3 I	[82]
59l	Cr 90 N_D 135.5 I	[82]
59m	Cr 67 N_D 94.2 I	[82]
59n	Cr 56.6 N_D 69 I	[83]
59o	Cr 83.2 N_D 91.2 I	[83]
59p	Cr 86.4 N_D 109.9 I	[90]
59q	Cr 82 N_D 105.5 I	[97]
59r	Cr 76.7 N_D 101.4 I	[88]
59s	Cr 75 N_D 101 I	[89]
59t	Cr (39 N_D) 57 I	[102]
59u	Cr (117 N_D^*) 118 I	[86]
59v	Cr 77.8 N_D^* 98.4 I	[86]
60a	Cr 132 N_D 246 I	[100]
60b	Cr (81 N_D) 92 I	[96]
60c	Cr 80 N_D 140 I	[96]
60d	Cr 95 N_D 157 I	[96]
60e	Cr 71 N_D 118 I	[102]
60f	Cr 11 N_D 24 I	[93]
60g	Cr 87.7 N_D^* 100.1 I	[86]
60h	g −36 N_D^* 23.4 I	[86]
60i	Non-LC	[87]
60j	N_D 40 I	[79]
60k	Non-LC	[79]
60l	Cr 246 Col > 260 dec	[85]
60m	Cr 62 Col > 270 dec	[85]
60n	Cr 171 Col 230 N_D 250 I	[85]
60o	Cr 106 Col 285 I	[85]
60p	Cr 56 Col_h 167 I	[102]
60q	g 115 Col_h > 260 dec	[102]
60r	Cr 215 I	[102]
60s	N_D 73 I	[79]

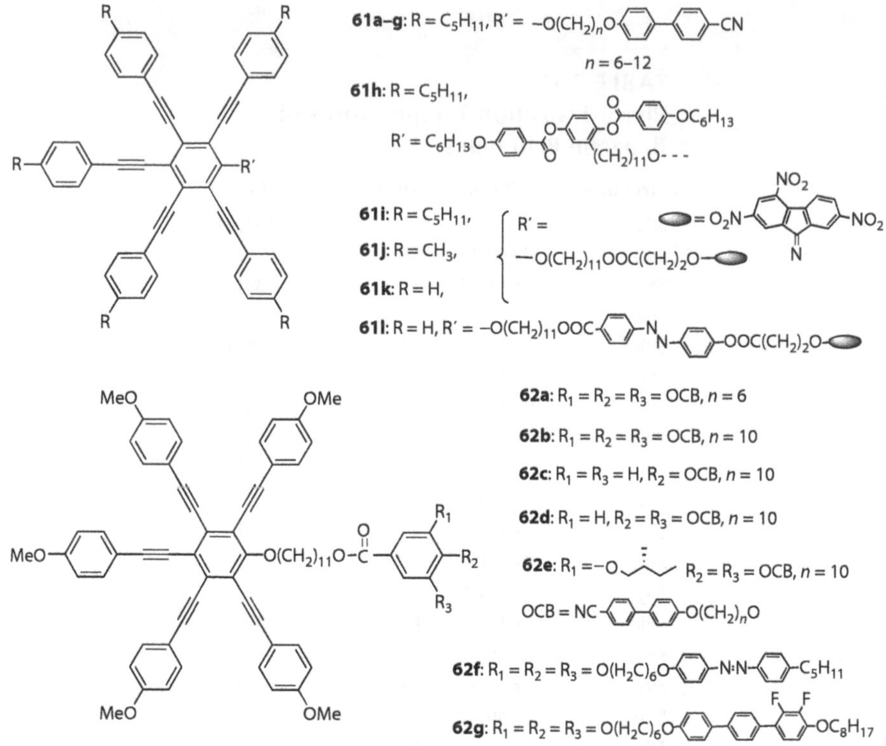

61a–g: R = C_5H_{11}, R′ = $-O(CH_2)_nO$—⬡—⬡—CN

$n = 6–12$

61h: R = C_5H_{11},

R′ = $C_6H_{13}O$—structure—OC_6H_{13} ... $(CH_2)_{11}O$- - -

61i: R = C_5H_{11},

61j: R = CH_3,   R′ = $-O(CH_2)_{11}OOC(CH_2)_2O$—⬬   ⬬ = O_2N—fluorenone— NO_2

61k: R = H,

61l: R = H, R′ = $-O(CH_2)_{11}OOC$—⬡—N≡N—⬡—$OOC(CH_2)_2O$—⬬

62a: $R_1 = R_2 = R_3 = OCB, n = 6$

62b: $R_1 = R_2 = R_3 = OCB, n = 10$

62c: $R_1 = R_3 = H, R_2 = OCB, n = 10$

62d: $R_1 = H, R_2 = R_3 = OCB, n = 10$

62e: $R_1 = -O$—structure $R_2 = R_3 = OCB, n = 10$

OCB = NC—⬡—⬡—$O(CH_2)_nO$

62f: $R_1 = R_2 = R_3 = O(H_2C)_6O$—⬡—N·N—⬡—$C_5H_{11}$

62g: $R_1 = R_2 = R_3 = O(H_2C)_6O$—⬡—⬡—⬡—$OC_8H_{17}$

SCHEME 2.16   Benzene-based disk–rod oligomers.

TABLE 2.16

Phase Transition Temperatures of Disk–Rod Oligomers

Structure	Phase Transition	Ref.
61a	Cr (N 52) 150 I	[103]
61b	Cr 114 I	[103]
61c	Cr (52 N) 122 I	[103]
61d	Cr 88 I	[103]
61e	Cr 125 I	[103]
61f	Cr (33 N) 102 I	[103]
61g	Cr (54 N) 100 I	[103]
61h	Cr (28.5 N) 74 I	[97]
61i	Col 67.4 I	[91]
61j	Cr 224.6 I	[91]
61k	Cr 118 Col 124.5 N_{Col} 134 I	[91]
61l	g 68.5 N_{Col} 105.4 I	[92]
62a	g 10 N 41 I	[104]
62b	g 8 N 52 I	[104]
62c	Cr 22.5 N 69.1 I	[99]
62d	g 31 N 77 I	[98]
62e	g 23 N* 50 I	[107]
62f	Cr 72 N 81 I	[105]
62g	Cr 115 SmC 134 SmA 135 N 159 I	[106]

The benzene-based multiynes have also been studied for their various physical properties [108–125]. Apart from the LCD device, the hexaynes have been studied for their bend, splay elastic constants, selective reflection in the chiral nematic phase, electrooptic effects, and nematic–nematic transitions by optical, diamagnetic anisotropy, and XRD studies. The amphiphilic pentaynes have been studied for their organization at the air–water interface and Langmuir–Blodgett films. Pentaynes form a two-dimensional nematic phase at the air–water interface. Their alignment and reorientation have been studied on patterned surfaces. Photophysical properties and the effect of molecular symmetry on spectroscopic properties of these derivatives have also been studied. The non-mesomorphic pentaynes exhibit mesomorphism by charge-transfer interaction with non-mesomorphic electron acceptor molecules.

Recently, two series of electron-deficient tris(N-salicylideneanilines) that form multiple columnar mesophases have been reported [126–129]. The chemical structures of the compounds are shown in Scheme 2.17 and their thermal phase behavior is collected in Table 2.17. These compounds possess the unique property of proton-transfer accompanied by a configurational change of π-electrons. In the columnar phase of these compounds, the proton and electron interact with each other through the H-bonding environment, which is expected to exhibit high charge-carrier mobility. Although the two geometric forms of these systems, that is, enol-imine and keto-enamine are expected to be in equilibrium with each other, interestingly, it is the keto-enamine form that exists exclusively. Moreover, the structural characterization of the keto-enamine forms indicates that it exists with C_{3h} and C_s symmetric tautomeric forms. The compounds derived from single chain aliphatic alkyl or aromatic alkyl and alkoxy amines do not furnish any mesomorphism; however, when two or three chain aromatic alkoxy amines are used, they provide stable mesophases, suggesting the importance of the space-filling effect in discotic liquid crystals. Compounds containing chiral tails self-organize into a room temperature helical columnar mesophase resulting from the chiral stacking of the constituent molecules.

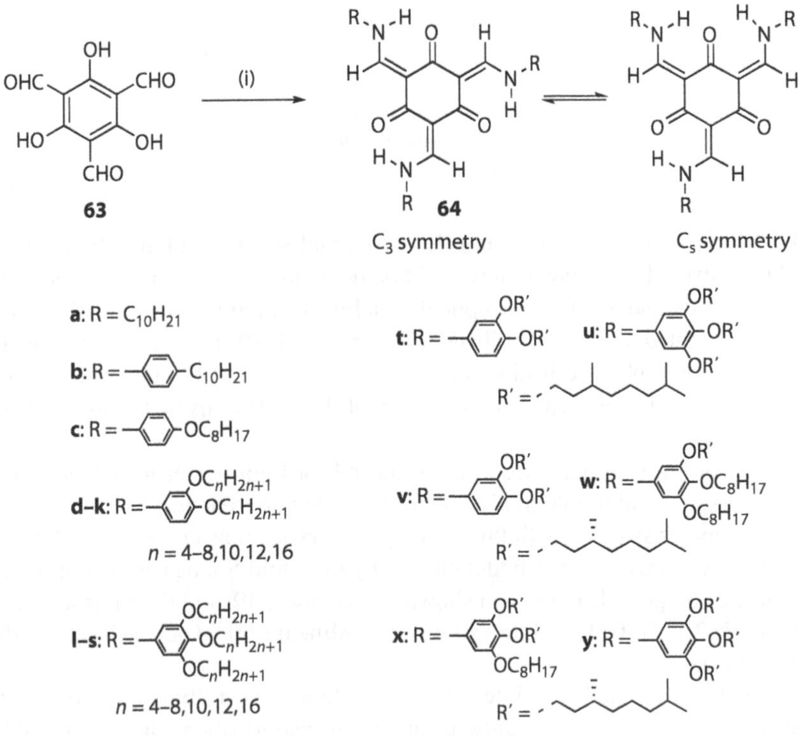

SCHEME 2.17   Synthesis of tris-N-salicylideneanilines: (i) RNH_2, EtOH, reflux.

TABLE 2.17
Phase Transition Temperatures of Tris-*N*-Salicylideneanilines

Structure	Phase Transition	Ref.
64a	Cr 74 I	[128]
64b	Liquid	[128]
64c	Non-LC	[128]
64d	Cr 128.5 I	[126]
64e	Cr 104 Col_h 119.3 I	[126]
64f	Cr 84.9 Col_r 131.7 I	[126]
64g	Cr 75.3 Col_r 138 I	[126]
64h	Cr 75.6 Col_r 136.4 I	[126]
64i	Cr 82.6 Col_r 141.9 I	[126]
64j	Cr 82.6 Col_r 133.9 I	[126]
64k	Cr 73.4 Col_r 127.3 I	[126]
64l	Cr 176.1 Col_r 219.1 Col_h 246 I	[126]
64m	Cr 168.5 Col_r 186.5 Col_h 228.6 I	[126]
64n	Cr 55.4 Col_r 173.4 Col_h 214.6 I	[126]
64o	Cr 54.2 Col_r 156.2 Col_h 197 I	[126]
64p	Cr 57.2 Col_r 164.4 Col_h 186.2 I	[126]
64q	Cr 64.2 Col_r 158.5 Col_h 175.2 I	[126]
64r	Cr 52.9 Col_r 148.5 Col_h 163.5 I	[126]
64s	Cr 59.9 Col_r 96.9 Col_h 128.1 I	[126]
64t	Col_h 109.9 I	[126]
64u	Col_r 68 Col_h 109.9 I	[126]
64v	Col_h 117.2 I	[127]
64w	Cr 64.3 Col_r 136 Col_h 160.8 I	[127]
64x	Col_r 133.3 Col_h 144.4 I	[127]
64y	Col_r 53.9 Col_r 108.2 I	[127]

Note: All the compounds are mixtures of C_{3h} and C_s symmetric keto-enamine tautomeric forms.

Stilbenoid compounds play a prominent role in material science and are already used as optical brightners and laser dyes. These compounds might be potentially interesting for light-emitting diodes, photoconductive devices, and imaging and optical switching techniques. Accordingly Meier, Lehmann, and coworkers investigated a series of stilbenoid dendrimers [130–135]. The synthesis and chemical structures of the mesomorphic derivatives are shown in Scheme 2.18, and their phase behavior is collected in Table 2.18. The key step in the synthesis of these all-trans materials involves a threefold Wittig–Horner olefination.

Materials with nonlinear optical effects are required for high-performance electro-optic modulators, frequency doublers, and holographic memories. One approach toward the nonlinear optical properties uses octopolar systems with threefold symmetry such as compound **72**, which has been obtained from 1,3,5-trialkoxy-2,4,6-triiodobenzene by threefold Sonogashira coupling. The chemical structures of the compounds have been shown in Scheme 2.19, and their phase transition behavior is shown in Table 2.19 [136–138]. Interestingly, a nonlinear optical device has been demonstrated by using these materials.

Kato and coworkers reported very interesting ionic liquid crystalline derivatives **74–76** of benzene (Scheme 2.20), which exhibit one-dimensional ion transport when aligned uniaxially [139,140]. They could measure the ion transport anisotropy by photopolymerizing aligned columnar phase

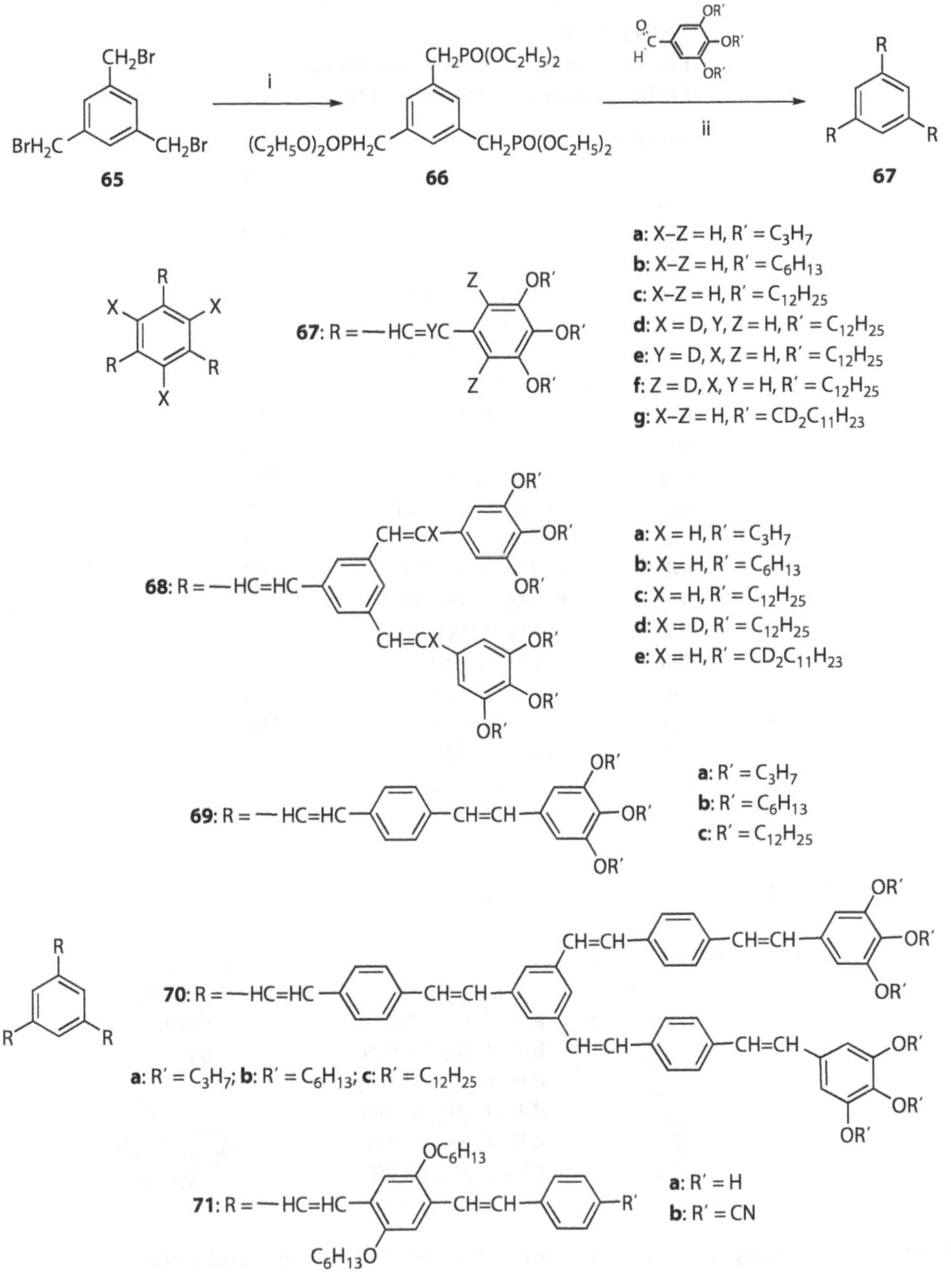

SCHEME 2.18 Synthesis of stilbenoid discotics: (i) P(OC$_2$H$_5$)$_3$, 160°C; (ii) KOC(CH$_3$)$_3$, THF.

of polymerizable ionic derivatives. The ionic conductivities parallel to the column axis are higher than those that are perpendicular to it. Moreover, the films with columns oriented vertically to the surface show anisotropy of ionic conductivities higher than that of the films with columns aligned parallel to the surface.

Discotic liquid crystals for polyelectrochromic materials and multielectron redox behavior may exhibit multiple electron transfer and significant absorption in the visible range. In this context, the use of azulene as redox active chromophores is interesting because their electrochemical reduction

TABLE 2.18
Phase Transition Temperatures of Stilbenoid-Armed Benzene Derivatives

Structure	Phase Transition	Ref.
67a	Cr 189 I	[131]
67b	g −15 Col_h 74 I	[131]
67c	Cr 38 Col_h 75 I	[131]
67d	Cr 36.7 Col_h 75.4 I	[133]
67e	Cr 37.7 Col_h 74.6 I	[133]
67f	Cr 37.3 Col_h 73.1 I	[133]
67g	Cr 37.3 Col_h 73.7 I	[133]
68a	Cr 161 I	[131]
68b	Col_h? Col_{ob} 144 I	[131]
68c	Cr 11 Col_h 32 Col_{ob} 99 I	[131]
68d	Cr 4.1 Col_h 29.7 Col_{ob} 98.3 I	[133]
68e	Cr 2.2 Col_h 27.5 Col_{ob} 94.6 I	[133]
69a	Cr 216 I	[135]
69b	g −42 g 24 D_L 129 I	[135]
69c	Cr −35 g 21 Col_h 108 I	[135]
70a	g 102 Col 254 I	[135]
70b	g −61 Col_h 235 I	[135]
70c	Cr −27 Col_{ob} 195 I	[135]
71a	g 8 N_D 114 N_D 126 I	[134]
71b	Cr 209 N_D 232 I	[134]

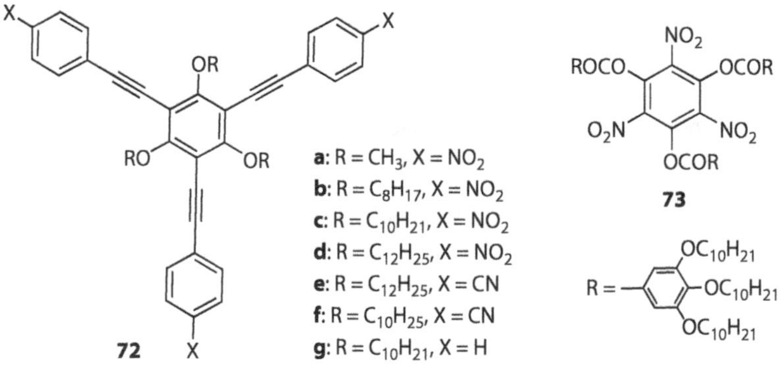
X
X
OR
RO
OR
X
72
a: R = CH3, X = NO2
b: R = C8H17, X = NO2
c: R = C10H21, X = NO2
d: R = C12H25, X = NO2
e: R = C12H25, X = CN
f: R = C10H25, X = CN
g: R = C10H21, X = H
NO2
ROCO
OCOR
O2N
NO2
OCOR
73
R =
OC10H21
OC10H21
OC10H21

SCHEME 2.19 Trialkoxy-triethynyl and trinitrotrialkoxybenzoyloxy-benzene discotics.

is strongly facilitated by the formation of the cyclopentadienide subunit. Hence, combining multiple redox behavior with liquid crystalline properties is very fascinating [141,142]. Accordingly, an azulene-substituted benzene derivative **77** has been prepared by the reaction of substituted 6-bromoazulene with hexaethynylbenzene prepared *in situ* from hexakis(trimethylsilyethynyl)benzene [141]. Compound **77** with *n*-hexyloxycarbonyl chains forms columnar phases in contrast to the hexaalkynylbenzenes, which generally furnish nematic phase. This could be because of the strong interaction of carbonyl substituents on the azulenes. The tetraazulene-substituted benzene derivative **78** also behaves similarly. The other azulene-substituted benzene derivative **80**, which can be conveniently prepared from diazulene-substituted tolane **79** by cobalt-mediated cyclotrimerization,

TABLE 2.19
Phase Transition Temperatures of Triethynyl Trialkoxy Benzene Derivatives

Structure	Phase Transition	Ref.
72a	Non-LC	[136]
72b	Cr 120.3 N_D 143.7 I	[136]
72c	Cr 59.9 Col_h 68.5 N_D 111.9 I	[136]
72d	Cr 78 Col_h-N_D-I	[136]
72e	Cr 77 Col 83 N_D 98 I	[137]
72f	Cr 97 Col_r 128 N_D 134 I	[137]
72g	Cr 42 I	[137]
73	Cr 46.9 Col_h 93.5 I	[138]

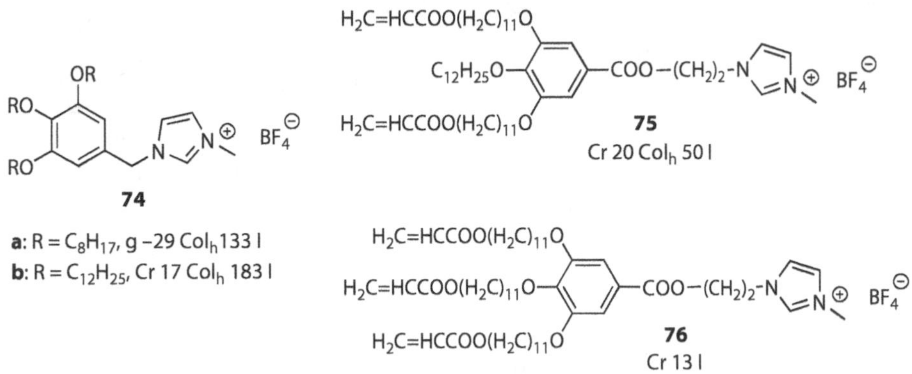

74

a: R = C_8H_{17}, g −29 Col_h 133 I
b: R = $C_{12}H_{25}$, Cr 17 Col_h 183 I

75
Cr 20 Col_h 50 I

76
Cr 13 I

SCHEME 2.20 Ionic liquid crystalline derivatives of benzene.

exhibits multiple columnar hexagonal phases [142]. The chemical structures and their phase behavior are depicted in Scheme 2.21. This represents, like the alkyl-substituted benzenehexaynes, an all-hydrocarbon discotic liquid crystal.

Other benzene-based discotic liquid crystals that form either columnar or nematic mesophase are shown in Scheme 2.22, and their thermal phase behavior is collected in Table 2.20 [143–150]. The dendronized C_3 symmetric esters derived from phloroglucinol exhibit monotropic and enantiotropic columnar hexagonal phase [143,144]. The oxadiazole-armed compound exhibits columnar phase; however, when the oxadiazole arms are connected to the central unit via phenylethynyl spacer, a nematic phase is stabilized [145,146]. The hydrazone derivatives exhibit both columnar and nematic phases, depending on the number of peripheral chains [147]. The hexaarylbenzenes substituted with alkyl or different aryl (phenyl or thienyl) arms exhibit columnar phases [148]. The alkoxyphenyl-substituted C_3 symmetric triesters seem to form a metastable nematic phase at or below room temperature [149], whereas the sugar-substituted triester stabilizes a columnar phase [150].

In addition to the above-mentioned discotics, a large number of di-, tri-, tetra-, penta-, and hexa-substituted benzene derivatives are known to exhibit classical calamitic phases. The chemistry and physical properties of these derivatives have not been covered in this section.

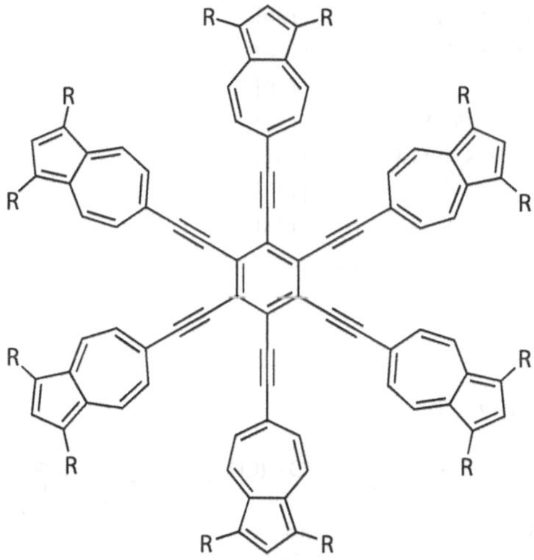

77: R = COOC$_6$H$_{13}$

Cr 77.3 Col$_h$ 173.3 Col$_r$ > 270 I (dec.)

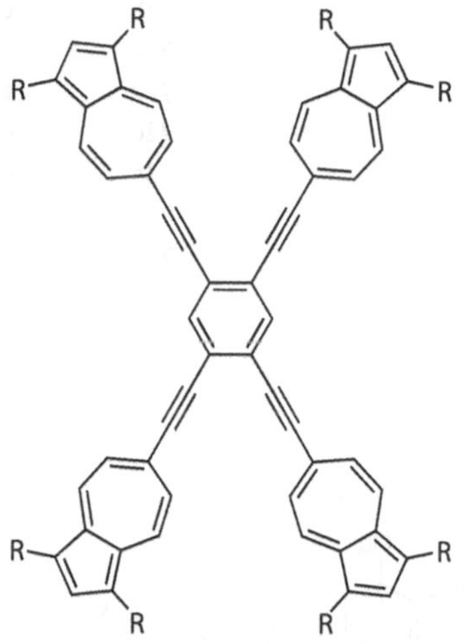

78: R = COOC$_6$H$_{13}$

Cr 215 Col$_r$ 244.4 I (dec.)

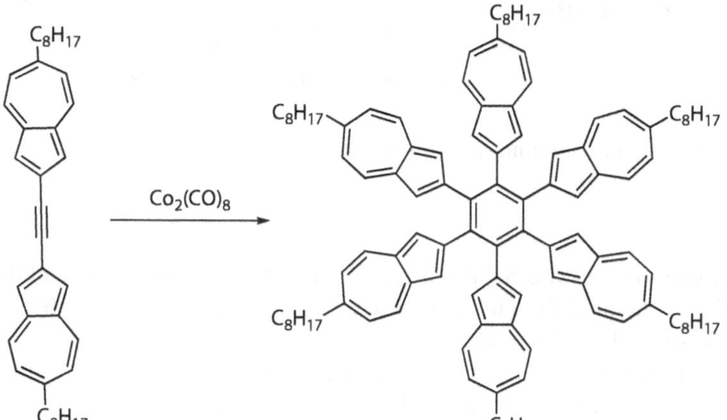

Co$_2$(CO)$_8$

79

80

Cr 115.5 Col$_{h1}$ 125.3 Col$_{h2}$ 137.5 Col$_{h3}$ 143.5 Col$_{h4}$ 183.4 M 199.9 I

SCHEME 2.21 Azulene-substituted benzene discotics.

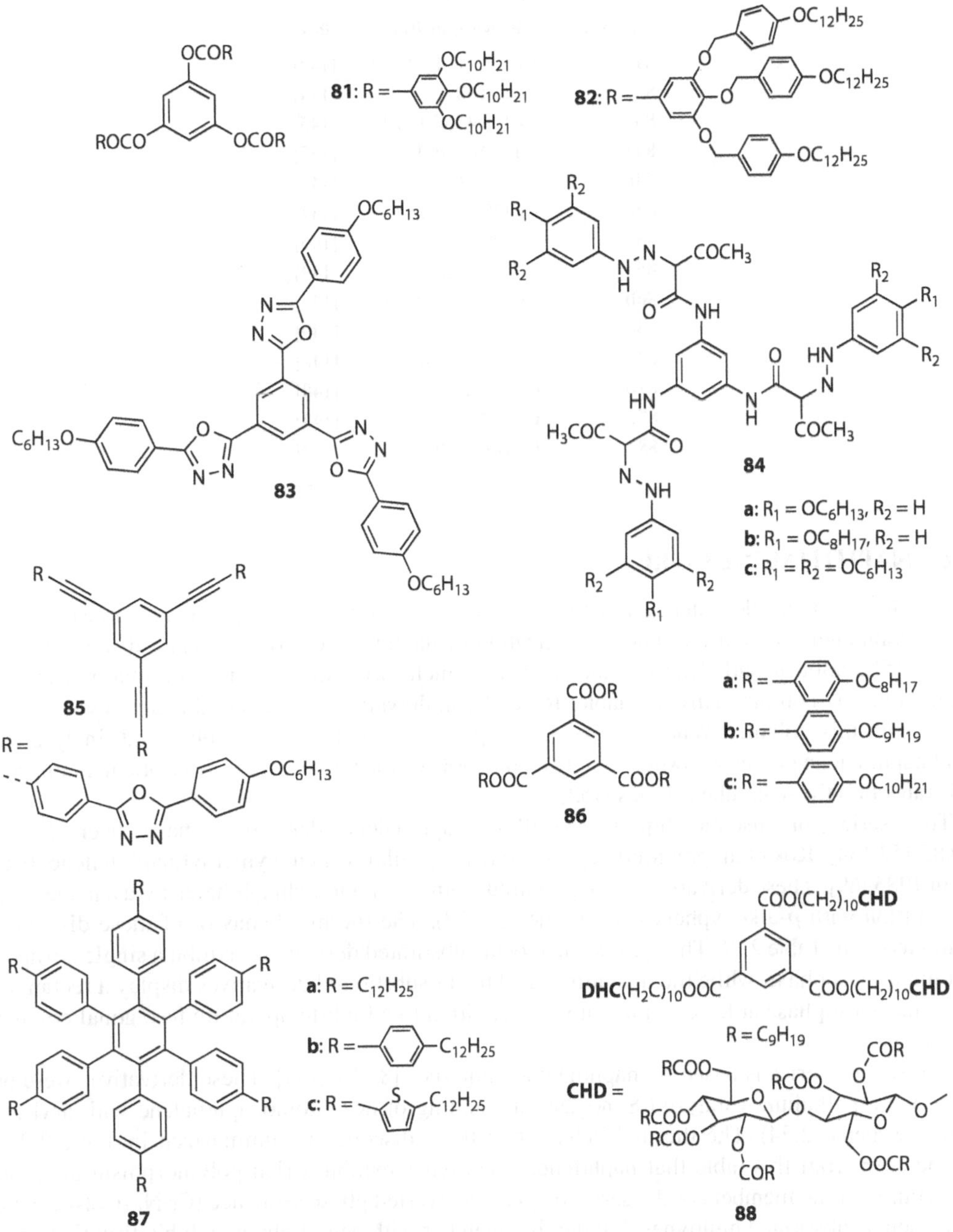

SCHEME 2.22 Molecular structure of miscellaneous benzene-based discotic liquid crystals.

TABLE 2.20
Phase Transition Temperatures
of Miscellaneous Benzene-Based
Discotic Liquid Crystals

Structure	Phase Transition	Ref.
81	Cr (27.3 Col$_h$) 29.5 I	[143]
82	Col$_h$ 159 I	[144]
83	Cr 123.4 Col 128.2 I	[145]
84a	g 105 N$_D$ 119 I	[147]
84b	g 69 N$_D$ 82 I	[147]
84c	Cr 120 Col$_h$ 175 I	[147]
85	Cr 189 N$_D$ 200 I	[146]
86a	Cr (23.5 N$_D$) 72.2 I	[149]
86b	Cr (22.5 N$_D$) 78.3 I	[149]
86c	Cr (25.5 N$_D$) 57.5 I	[149]
87a	Cr 47.2 Col 66.6 I	[148]
87b	Col$_h$ 83.4 I	[148]
87c	Cr 29.2 Col$_h$ 155.5 I	[148]
88	Cr 41 Col 104 I	[150]

2.2 NAPHTHALENE CORE

A naphthalene molecule is derived by the fusion of a pair of benzene rings. It is the most abundant single component of coal tar. The carbonization of naphthalene proceeds via a discotic nematic phase [151]. Solid naphthalene polymer powder, which forms discotic nematic phase at elevated temperature, is commercially available. Recently, multiwall carbon nanotube cavities were infiltrated with this polymer. When a mixture of tip-opened multiwall nanotubes and finely ground solid naphthalene polymer powder were heated to 300°C, the polymer in the discotic nematic phase infiltrates into the nanotube cavities [152].

Two series of discotic liquid crystalline naphthalene derivatives have been realized [69,85,153,154]. Kok et al. prepared hepta- and octa-(p-alkoxyphenoxymethyl)naphthalene discotics in 1985 [69]. These derivatives were prepared from bromomethylnaphthalene via a nucleophilic substitution with p-alkoxyphenoxides (Scheme 2.23). The thermal behavior of these discotics is summarized in Table 2.21. The symmetrical octa-substituted derivatives exhibit a single hexagonal columnar mesophase, while the unsymmetrical hepta-substituted derivatives display a rectangular columnar mesophase at lower temperature in addition to a high-temperature hexagonal columnar mesophase.

The second series is based on naphthalene multiynes [85,153,154]. These derivatives were prepared via a palladium-catalyzed Sonogashira coupling of hexabromonaphthalene with aryl acetylenes (Scheme 2.24). The thermal behavior of these discotics is summarized in Table 2.22. It can be seen from this table that naphthalene multiynes exhibit a rich polymorphism and, more importantly, some members of the series display an inverted phase sequence (Cr-N$_D$-Col-I). All the alkylphenyl-substituted multiynes, but the first member with pentyl-chain, exhibit discotic nematic and columnar phases. Linking the alkyl chain via oxygen atom increases the melting and clearing transition temperatures significantly. Similarly, the addition of another phenyl ring in the periphery (3g–3j) also enhances the clearing temperatures. The multiynes that do not have any phenyl rings do not show any mesomorphism. Thus it may be interpreted that mesophases are induced due to the formation of a "super-disk" structure composed of central naphthalene, acetylene bridge, and peripheral phenyl rings.

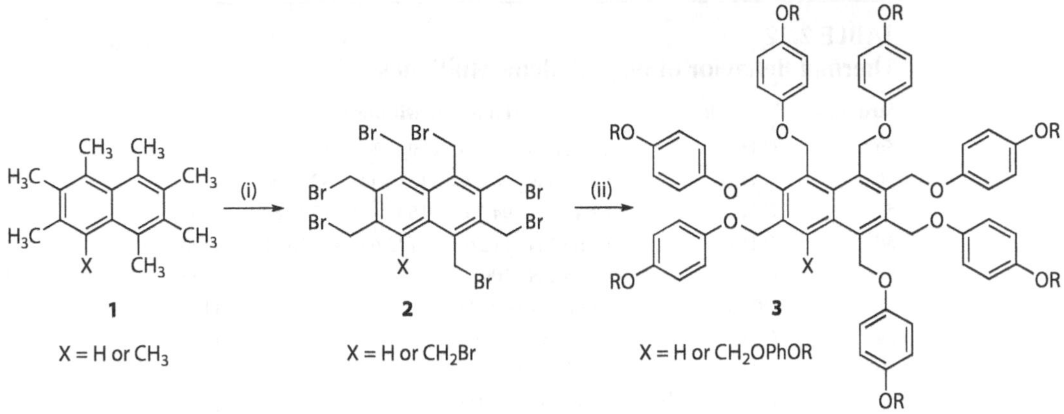

SCHEME 2.23 Synthesis of (*p*-alkoxyphenoxymethyl)naphthalenes: (i) Br$_2$, hv; (ii) NaOC$_6$H$_4$OR, DMF, 80°C.

TABLE 2.21
Thermal Behavior of (*p*-Alkoxyphenoxymethyl)-naphthalene Discotics

Structure	R	X	Phase Transitions
3a	C$_5$H$_{11}$	CH$_2$OPhOC$_5$H$_{11}$	Cr 162 Col$_h$ 164 I
3b	C$_6$H$_{13}$	CH$_2$OPhOC$_6$H$_{13}$	Cr 135 Col$_h$ 167 I
3c	C$_7$H$_{15}$	CH$_2$OPhOC$_7$H$_{15}$	Cr 105 Col$_h$ 167 I
3d	C$_8$H$_{17}$	CH$_2$OPhOC$_8$H$_{17}$	Cr 78 Col$_h$ 167 I
3e	C$_9$H$_{19}$	CH$_2$OPhOC$_9$H$_{19}$	Cr 62 Col$_h$ 165 I
3f	C$_6$H$_{13}$	H	Cr 65 Col$_r$ 147 Col$_h$ 184 I
3g	C$_7$H$_{15}$	H	Cr 79 Col$_r$ 92 Col$_r$ 138 Col$_h$ 182 I

Source: Data from Kok, D.M. et al., *Mol. Cryst. Liq. Cryst.*, 129, 53, 1985.

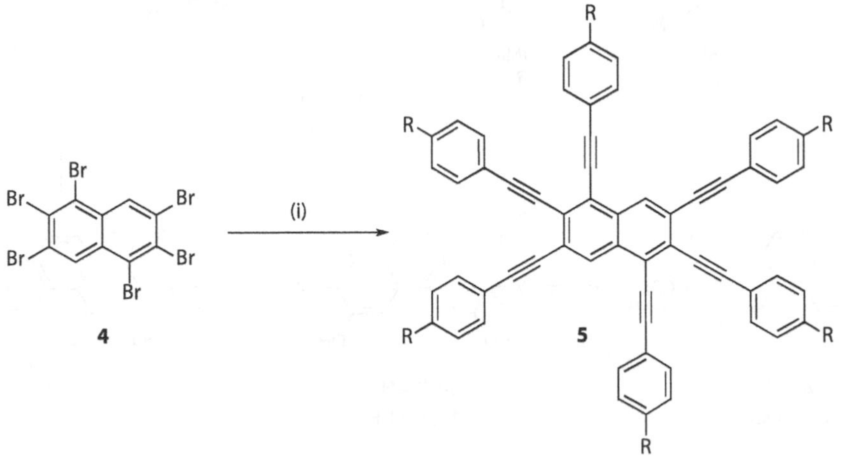

SCHEME 2.24 Synthesis of naphthalene multiynes: (i) HC≡CC$_6$H$_4$R, Pd(PPh$_3$)$_4$, CuI, Et$_3$N.

TABLE 2.22
Thermal Behavior of Naphthalene Multiynes

Structure	R	Phase Transitions	Ref.
5a	C_5H_{11}	Cr 121.5 Col_r 156.3 Col_h ~260 dec	[153]
5b	C_6H_{13}	Cr 69.3 N_D 98.4 Col_r 133.9 Col_h ~245 dec	[153]
5c	C_7H_{15}	Cr 48.5 N_D 94.7 Col_r 159.0 Col_h ~230 dec	[153]
5d	C_8H_{17}	Cr 60.3 N_D 112.6 Col_r 136.6 Col_h ~200 I	[153]
5e	C_9H_{19}	Cr 68.2 N_D 109.9 I	[153]
5f	OC_5H_{11}	Cr 68.2 N_D ~250 dec	[153]
5g	C_6H_5	Cr 240 Col > 260 dec	[85]
5h	C_6H_4-(p)-C_5H_{11}	Cr 122 Col > 240 dec	[85]
5i	C_6H_4-(p)-C_8H_{17}	Cr 75 Col > 240 dec	[85]
5j	C_6H_4-(p)-C_9H_{19}	Cr 58 Col ~180 N_D 222 I	[85]

2.3   PHENANTHRENE CORE

Phenanthrene is a tricyclic aromatic hydrocarbon composed of three fused benzene rings. It is omnipresent in the environment as a product of incomplete combustion of fossil fuels and wood and has been identified in air, surface, drinking water, and in foods. Phenanthrene can be obtained by fractional distillation of high-boiling coal tar oil and is employed in the manufacturing of dyes, explosives, drugs, and phenanthrenequinone. Like naphthalene, phenanthrene also displays a discotic nematic carbonaceous mesophase during the carbonization process [151,155,156].

A number of synthetic methods are available for the assembly of phenanthrene nucleus [157]. Discotic liquid crystalline derivatives of phenanthrene were first reported by Scherowsky and Chen in 1994 [158]. They synthesized two 1,2,3,6,8-penta-, **12**, and six 1,2,3,6,7,8-hexa-substituted phenanthrene, **13**, esters as shown in Scheme 2.25. The bromination of 2,3,4-trimethoxybenzyl alcohol **1** with phosphorous tribromide, yielded **2**, which on reaction with triphenylphosphine furnished the phosphonium salt, **3**. Wittig reaction of **3** with 2,4-dimethoxy benzaldehyde **4** or

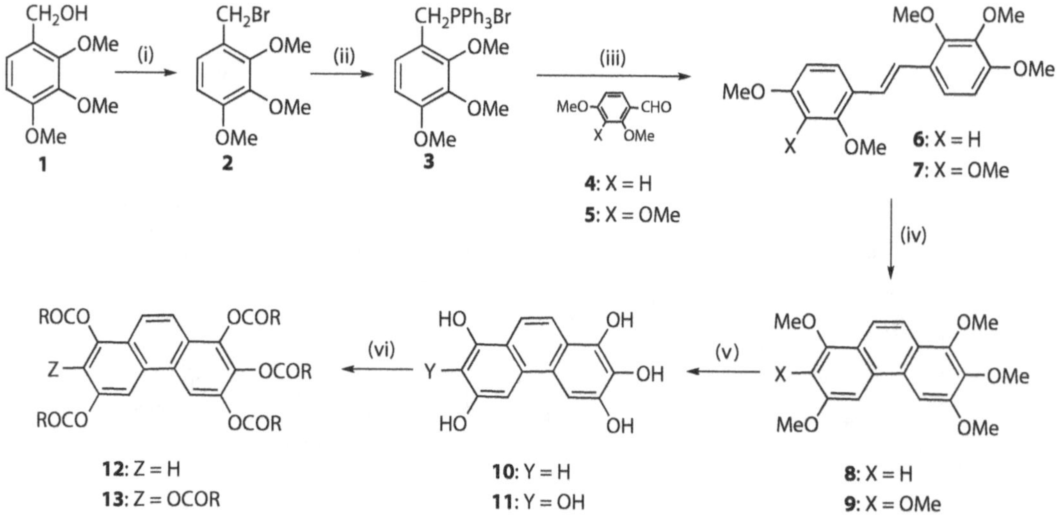

SCHEME 2.25   Synthesis of penta- and hexa-substituted phenanthrene esters: (i) PBr_3; (ii) PPh_3; (iii) LiOMe, MeOH; (iv) hv, I_2; (v) BBr_3; (vi) DCC, DMAP, RCOOH.

TABLE 2.23
Thermal Behavior of Penta and
Hexaalkanoyloxyphenanthrenes

Structure	R	Phase Transition	Ref.
12a	(S)C*H(CH$_3$)OC$_7$H$_{15}$	Col 51 I	[158]
12b	(R)CH$_2$OC*H(CH$_3$)C$_6$H$_{13}$	Col 75 I	[158]
13a	C$_6$H$_{13}$	Cr (96 Col) 117.5 I	[158]
13b	C$_7$H$_{15}$	Cr (79 Col) 107.3 I	[158]
13c	C$_8$H$_{17}$	Cr (89 Col) 110 I	[158]
13d	C$_{10}$H$_{21}$	Cr 112 I	[158]
13e	(S)C*H(CH$_3$)OC$_7$H$_{15}$	Cr 77.1 I	[158]
13f	(R)CH$_2$OC*H(CH$_3$)C$_6$H$_{13}$	Cr −12.1 Col 145 I	[158]
20a	(S)C*H(CH$_3$)OC$_7$H$_{15}$	Col$_h$ 131 I	[159]
20b	(S)CH$_2$OC*H(CH$_3$)C$_6$H$_{13}$	Col$_h$ 138 I	[159]
20c	C$_8$H$_{17}$	Col$_h$ 125 I	[159]
20d	(S)CH$_2$C*H(CH$_3$) C$_7$H$_{15}$	Col$_h$ 107 I	[159]

2,3,4-trimethoxybenzaldehyde **5**, gives the penta- or hexamethoxy-stilbene **6** and **7**, respectively. The photocyclization of polymethoxylated stilbene **6** and **7** furnished polymethoxylated phenanthrene **8** and **9**. Boron tribromide–mediated demethylation of **8** and **9** provided penta- and hexahydroxy-phenanthrene **10** and **11**. Finally, the penta- or hexa-esters of phenanthrene **12** and **13** were achieved by DCC, DMAP-mediated esterification of **10** and **11** with appropriate fatty acids. Chiral acids have also been utilized to visualize ferroelectric switching in these chiral discotic esters.

The thermal behavior of penta- and hexa-alkanoyloxy phenanthrene discotics is summarized in Table 2.23. Hexaesters of phenanthrene having normal alkyl chains in the periphery exhibit only monotropic columnar phases. On heating these, phenathrene derivatives **13a–d** clear in a temperature range of 107°C–118°C. The columnar mesophase appears in between 79°C and 96°C on cooling. On the other hand, chiral penta- and hexaesters display enantiotropic columnar phases. While the hexaesters show strong tendency to crystallize, the pentaesters do not show any sign of crystallization even at low temperature. A helical phase structure was proposed for these chiral phenanthrene discotics. These derivatives exhibit strong electro-optical switching properties.

Another series of 2,3,6,7,9,10-hexa-substituted phenanthrene derivatives **20**, which are structural isomers of the above-mentioned 1,2,3,6,7,8-hexa-substituted phenanthrene derivatives, is reported by the same authors [159]. The synthesis of these derivatives is presented in Scheme 2.26. The benzoin condensation of 3,4-dimethoxybenzaldehyde **14** in the presence of sodium cyanide furnished tetramethoxy benzoin derivative **15**. The oxidation of **15** with copper (II) sulfate in pyridine yielded veratril **16**. The oxidative coupling of **16** with aluminum trichloride in nitrobenzene yields 2,3,6,7-tetramethoxyphenanthrene-9,10-dione **17**, which on reductive acetylation with zinc in acetic anhydride afforded **18**. The hexahydroxyphenanthrene **19** was obtained by the boron tribromide–mediated demethylation of **18** in dichloromethane. The esterification of **19** with chiral acids furnished the target hexaester **20**.

All these hexaesters exhibit very broad enantiotropic columnar phases (Table 2.23). The incorporation of oxygen atom into the side chain stabilizes the mesophase (compare **20b** vs. **20d**). On the other hand, branching of the side chain destabilizes the mesophase (compare **20d** vs. **20c**). The extent of destabilization increases as the branching point approaches the phenanthrene core (compare **20a** vs. **20b**). Unlike chiral 1,2,3,6,7,8-hexa-substituted phenanthrenes **13**, the 2,3,6,7,9,10-hexa-substituted phenanthrene derivatives **20**, bearing the same chiral chains do not possess tilted columnar phase and, therefore, do not display any electro-optical effect. However, a mixture of **20a**

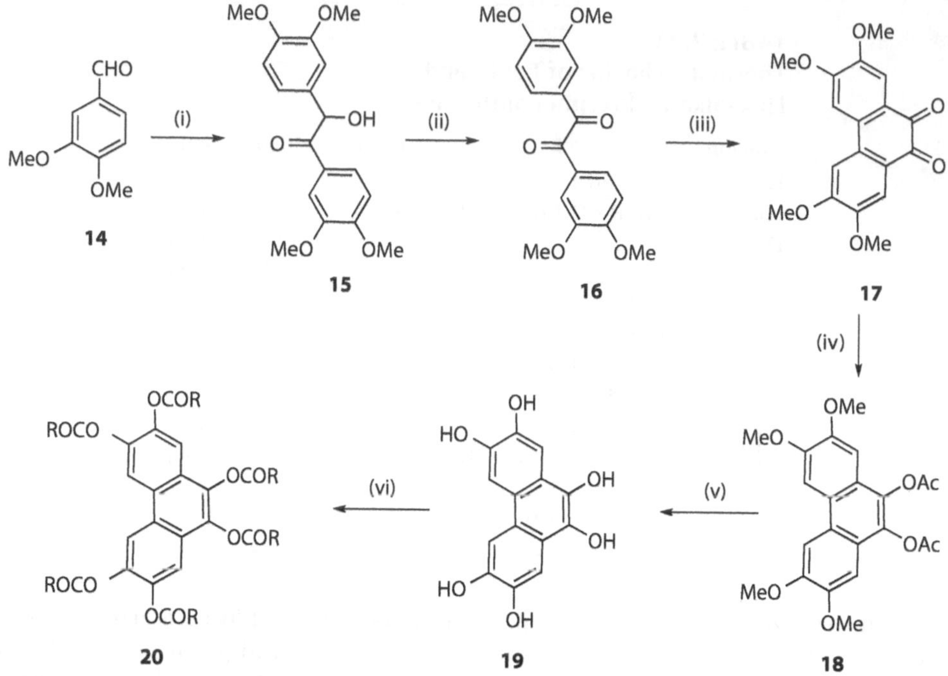

SCHEME 2.26 Synthesis of hexa-substituted phenanthrene esters **20**: (i) NaCN; (ii) CuSO$_4$, pyridine; (iii) AlCl$_3$, nitrobenzene; (iv) Zn, Ac$_2$O; (v) BBr$_3$, CH$_2$Cl$_2$; (vi) DCC, DMAP, Zn, RCOOH.

and **13e** [15:85 ratio] showed ferroelectric switching. The switching time decreases on increasing temperature (at constant electric field) as well as on increasing electric field (at constant temperature), but temperature has relatively small effect on switching time as compared to electric field.

2.4 ANTHRAQUINONE CORE

Anthraquinone belongs to one of the most widely occurring compounds in nature. A number of anthraquinones have found applications as eco-friendly natural colorants in food, cosmetics, and textiles [160,161]. They have also been exploited as dichroic dyes in host–guest displays [162], chemical sensors [163], organogelators [164], and anticancer drugs [165].

The 1,2,3,5,6,7-hexahydroxy-9,10-anthraquinone, commonly known as rufigallol (**1**), is a molecule of both biological and materials science interest. It is one of the first compounds to be synthesized in the history of organic compounds by Robiquet in 1836 by the dimerization of gallic acid, as described by Grimshaw and Haworth in 1956 [166]. Rufigallol has been reported as a novel oxidant drug [167–169]. A remarkable synergistic antimalarial interaction between rufigallol and the structurally similar compound exifone has been described. It is believed that rufigallol acts in a pro-oxidant fashion to produce oxygen radicals inside parasitized erythrocytes [167–169]. Rufigallol has also been recognized for its vitamin K activity [170].

Rufigallol has been found to function as the core fragment for a remarkable family of discotic liquid crystals. 1,2,3,5,6,7-hexahydroxyanthraquinone derivatives are one of the earliest systems reported to form columnar mesophases. They are interesting materials as these molecules have an elongated core with a twofold symmetry axis; they are colored, they exhibit an important polymorphism, the core is electron-deficient in nature, they are thermally stable, and their chemistry is fairly easy. Billard and coworkers reported the first discotic liquid crystalline hexaesters of rufigallol in 1980 [171], and since then more than 100 different discotic liquid crystalline derivatives of this molecule have been prepared and studied for various physical properties [172–213].

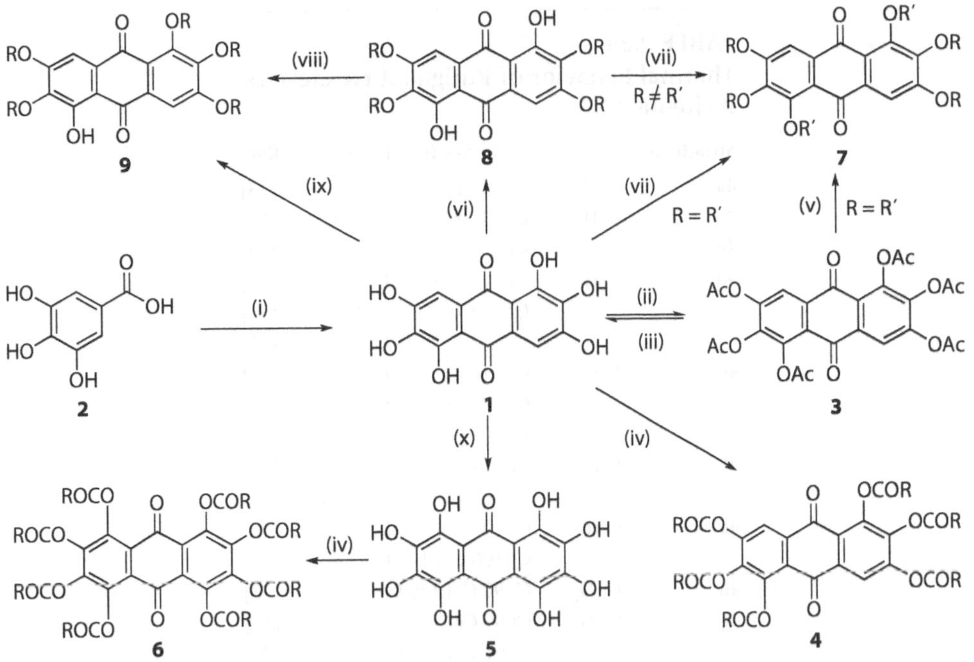

SCHEME 2.27 Synthesis of hexa- and octa-substituted rufigallol discotics: (i) H$_2$SO$_4$, heat; (ii) Ac$_2$O, H$_2$SO$_4$; (iii) Aq NaOH; (iv) Py-acetone, RCOCl; (v) DMF, Na$_2$CO$_3$, RBr, 160°C; (vi) DMSO, NaOH, RBr, TBAB, 80°C; (vii) DMF, K$_2$CO$_3$, RBr, TBAB, ~100°C; (viii) DMF, Cs$_2$CO$_3$, RBr; (ix) ROH, TPP, DEAD; (x) H$_3$BO$_3$, HgO, H$_2$SO$_4$.

Rufigallol (**1**) is easily obtained by the action of sulfuric acid on gallic acid (**2**) (Scheme 2.27) [166]. Thus, heating gallic acid in 98% H$_2$SO$_4$ at 100°C gives rufigallol in about 20% yield. The crude product is difficult to filter as it clogs badly. The purification of rufigallol can be achieved by centrifuging the crude product followed by the acetylation of the dried product in acidic conditions [166]. The hexaacetate **3** separates as yellow plates and can be recrystallized from acetic anhydride. Very recently, the use of microwave heating was realized to prepare rufigallol efficiently and quickly. The self-condensation of gallic acid in the presence of sulfuric acid can be achieved in high yield in about 1 min by using microwave heating [205]. However, the procedure is applicable only for a small-scale synthesis. The product was isolated simply by adding water to the reaction mixture followed by the filtration of the solid product that was converted to a hexaacetate by treatment with acetic anhydride in 60% yield. The hydrolysis of the pure hexaacetate furnished pure rufigallol. The overall yield of pure rufigallol from gallic acid was about 50%. Rufigallol-based monomeric DLCs may be broadly classified into the following classes: (1) hexaesters of rufigallol, (2) octaesters of rufigallol, (3) hexaethers of rufigallol having all the six identical chains, (4) mixed chain hexaethers, (5) pentaethers of rufigallol, (6) rufigallol-cyanobiphenyl-based discotic-calamitic hybrids, and (7) rufigallol-based discotic metallomesogens. The synthesis and mesomorphic properties of all these derivatives are described in the following sections.

2.4.1 Rufigallol-Hexa-N-Alkanoates

Esterification of rufigallol with an excess of appropriate acid chloride in acetone-pyridine or toluene-pyridine at elevated temperature yields hexaesters **4** in low (23%) yield (Scheme 2.27) [171]. No further effort has been made to improve the synthesis of these materials. Eleven members of the hexa-*n*-alkanoyloxy-9,10-anthraquinone series have been prepared. Their thermal behavior is presented in Table 2.24.

TABLE 2.24
Thermal Behavior of Rufigallol Hexaesters
(Scheme 2.27)

Structure	R	Phase Transition	Ref.
4a	C_3H_7	Cr 216 I	[173]
4b	C_4H_9	Cr 170 I	[173]
4c	C_5H_{11}	Cr 152 I	[173]
4d	C_6H_{13}	Cr 112 Col 134 I	[173]
		Cr 98 Col 108 Col 127 I	[176]
		Cr 104 Col 133 I	[176]
4e	C_7H_{15}	Cr 110 Col 133 I	[173]
		Cr 108 Col 128 I	[176]
		Cr 106 Col 126 I	[176]
4f	C_8H_{17}	Cr 91 Col 128 I	[173]
		Cr 80 Col 106 Col 123 I	[176]
4g	C_9H_{19}	Cr 109 Col 128 I	[173]
		Cr 106 Col 128 I	[176]
4h	$C_{10}H_{21}$	Cr 92 Col 126 I	[173]
4i	$C_{11}H_{23}$	Cr 99 Col 124 I	[173]
		Cr 91 Col 106 I	[176]
4j	$C_{12}H_{25}$	Cr 96 Col 120 I	[173]
4k	$C_{13}H_{27}$	Cr 103 Col 117 I	[173]

While lower homologues (up to hexanoate) of the series do not show liquid crystallinity, all the higher members of the series display enantiotropic mesophases. Moreover, three members (heptanoloxy- to decanoyloxy-) also exhibit monotropic columnar phase. X-ray diffraction studies indicate that both the phases are columnar rectangular. They differ only in the size of lattice and molecular orientations in the columns.

Corvazier and Zhao investigated the confinement effect on the phase behavior of rufigallol-hexa-*n*-octanoate [183]. Millipore membranes of various pore sizes were taken as the confining media. The suppression of the monotropic phase inside the membrane of smaller pore size was observed. However, the confinement does not impart any effect on the overall crystallization mechanism. This material was also dispersed in different polymers like polystyrene, poly(methyl methacrylate), and poly(ethyl methacrylate). It is reported that the orientation of discotic molecules can be achieved by the stretching of polymer-dispersed discotic liquid crystal films [182].

2.4.2 OCTA-ALKANOYLOXY-9,10-ANTHRAQUINONES

The oxidation of 1,2,3,5,6,7-hexahydroxyanthraquinone yields octahydroxy-anthraquinone **5** (Scheme 2.27). Thus, heating rufigallol with boric acid and mercuric acid in concentrated H_2SO_4 at 250°C furnished 1,2,3,4,5,6,7,8-octahydroxy-9,10-anthraquinone in about 35% yield. The esterification of this octahydroxy-9,10-anthraquinone with alkanoyl chlorides provides octa-alkanoyloxyanthraquinone discotics **6** in about 60% yield [176]. Nine members of the series ranging from octa-*n*-octanoyloxy- to octa-*n*-hexadecanoyloxy-9,10-anthraquinone have been prepared, and all are reported to be liquid crystalline. Their thermal behavior is presented in Table 2.25.

The first four members of the series display only a single optically biaxial columnar phase, while higher members exhibit two or three columnar mesophases. Compared with the hexaalkanoyloxy-9,10-anthraquinone, the octa-alkanoyloxyanthraquinone possesses larger mesomorphic range and the higher homologues exhibit a uniaxial columnar phase that was absent in the hexa-substituted

TABLE 2.25
Thermal Behavior of Octa-Alkanoyloxy-9,10-Anthraquinones (Scheme 2.27)

Structure	R	Phase Transition
6a	C_8H_{17}	Cr 85.5 Col 150.1 I
6b	C_9H_{19}	Cr 86.5 Col 148.0 I
6c	$C_{10}H_{21}$	Cr 98.7 Col 147.2 I
6d	$C_{11}H_{23}$	Cr 102.0 Col 143.4 I
6e	$C_{12}H_{25}$	Cr 101.0 Col 104.6 Col 140.1 I
6f	$C_{13}H_{27}$	Cr 99.2 Col 104.5 Col 138 I
6g	$C_{14}H_{29}$	Cr 104.0 Col 110.1 Col 134.6 Col$_h$ 135.9 I
6h	$C_{15}H_{31}$	Cr 102.5 Col 109.5 Col 121.9 Col$_h$ 132.3 I
6i	$C_{16}H_{33}$	Cr 107.3 Col 113.0 Col 121.2 Col$_h$ 131.1 I

Source: Data from Billard, J. et al., *Liq. Cryst.*, 16, 333, 1994.

derivatives. Though a variety of novel mesogens can be derived from octahydroxy-9,10-anthraquinone, surprisingly, no further work has been carried out on this molecule.

2.4.3 Hexa-*n*-Alkoxyrufigallols

Ten aliphatic ethers of rufigallol having identical peripheral chains (R = R′) have been prepared by Carfagna et al. [174]. These rufigallol-ethers **7** were prepared directly from rufigallol-hexaacetate. Thus, alkylation of hexaacetoxy-rufigallol with an excess of appropriate alkyl halide in DMF at reflux temperature afforded 1,2,3,5,6,7-hexaalkoxy-9,10-anthraquinones (Scheme 2.27). Recently, it has been observed that the alkylation of rufigallol under microwave heating produced these ethers in a few minutes [205,210].

The thermal data for rufigallol hexaethers with identical peripheral chains are listed in Table 2.26. As generally observed in a homologous series, the temperature for the mesophase to the isotropic

TABLE 2.26
Thermal Behavior of Symmetrical Rufigallol-Hexaethers (Scheme 2.27)

Structure	R	Phase Transition	Ref.
7a	C_4H_9	Cr 105 Col$_h$ 131 I	[174]
		Cr (122.5 Col$_h$) 123.7 I	[179]
7b	C_5H_{11}	Cr 83 Col$_h$ 117 I	[174]
7c	C_6H_{13}	Cr 53 Col$_h$ 105 I	[174]
		Cr 20 Col$_r$ 49 Col$_h$ 106 I	[203]
7d	C_7H_{15}	Cr 49 Col$_h$ 102 I	[174]
7e	C_8H_{17}	Cr 37 Col$_h$ 96 I	[174]
		Cr 7 M 22 M 35 Col$_h$ 97 I	[186]
		Cr 2 Cr 19 M 31 Col$_h$ 96 I	[196]
7f	C_9H_{19}	Cr 32 Col$_h$ 92 I	[174]
		Cr 41 Col$_h$ 92 I	[203]
7g	$C_{10}H_{21}$	Cr 48 Col$_h$ 87 I	[174]
7h	$C_{11}H_{23}$	Cr 26 Col$_h$ 84 I	[174]
7i	$C_{12}H_{25}$	Cr 37 Col$_h$ 79 I	[174]
7j	$C_{13}H_{27}$	Cr 62 Col$_h$ 72 I	[174]

phase decreases smoothly with increasing length of the alkyl chains. On the other hand, the melting transitions do not follow a regular pattern. It was also observed that solution-crystallized and melt-crystallized samples display different phase behaviors [203]. X-ray diffraction studies indicate the formation of hexagonal columnar phase in all the hexaalkoxy-rufigallols. A few members also exhibit rectangular columnar phase at lower temperature [174,176,203].

Maeda et al. studied the phase behavior of three hexaalkoxy-rufigallols under hydrostatic pressure using a high pressure differential thermal analyzer [203]. Under pressure, on one hand, induction of Col_r phase in hexaoctyloxy-rufigallol was observed. On the other hand, the stable Col_r phase of hexahexyloxy-rufigallol has a decreased temperature range with increasing pressure and then the Col_r phase disappears under higher pressure. The temperature-dependent dielectric spectroscopy of four homologous hexaalkoxy-rufigallols in the frequency range of 10 Hz to 10 MHz has recently been carried out by Gupta et al. [204]. The dielectric anisotropy $\left(\Delta\varepsilon' = \varepsilon'_{\parallel} - \varepsilon'_{\perp}\right)$ has been found to be positive throughout the entire range of the Col_h phase for all of the four compounds of this series. No relaxation phenomenon is found in the frequency range of measurement, that is, 10 Hz to 10 MHz. They have also measured the DC conductivity of these compounds, which was found in the range of 10^{-10}–10^{-11} S m^{-1}. Chandrasekhar et al. measured AC conductivity of hexapentyloxy-rufigallol doped with a small amount of an electron donor anthracene [191]. The conductivity was found to be nearly seven orders of magnitude higher in the columnar phase relative to that in the isotropic phase, as well as that in the columnar phase in the undoped state. Thermoelectric power studies confirmed the nature of the charge carrier as electrons in anthracene-doped hexapentyloxy-rufigallol.

2.4.4 Mixed Tail Hexaalkoxyrufigallols

The nature of aliphatic chains around the periphery of the core plays an important role in deciding the thermal behavior of DLCs. Generally, the introduction of dissymmetric side chains does not affect the nature of the mesophase but results in the reduction of mesophase stability (lower clearing temperature). As melting temperature decreases significantly, often the mesophase range increases in dissymmetric DLCs. In the hexahydroxy-rufigallol, the hydroxyl groups at 1- and 5-positions are hydrogen bonded and, therefore, are less reactive. Under milder etherification conditions, the hydrogen bonded 1- and 5-positions do not get alkylated and thus 1,5-dihydroxy-2,3,6,7-tetraalkoxy-9,10-anthraquinone **8** forms (Scheme 2.27). All the 1,5-dihydroxy-2,3,6,7-tetraalkoxy-9,10-anthraquinones having normal alkyl chains were reported to be non-liquid crystalline; however, very recently it has been observed that a branched-chain substituted tetraalkoxy derivative **8a** displays a broad room-temperature mesophase [205]. This could be because the branched chains fill the necessary space around the core to induce mesomorphism. Further substitution on this compound at 1- and 5-positions may lead to many new DLCs.

8a: Col$_h$ 115.7 I

The unequal reactivity of hydroxyl groups leads to the preparation of several mixed-tail rufigallol derivatives **7** (R≠R'). These unsymmetrical rufigallols can be subdivided into six groups (Figure 2.1): (a) compounds having two different *n*-alkyl chains at 1- and 5-positions (Structure **7a**), (b) compounds having branched alkyl chains at 1- and 5-positions (Structure **7b**), (c) compounds in which the two H-bonded positions are having *n*-alkyl chains but the other four positions are bearing branched alkyl chains (Structure **7c**), (d) compounds in which the two H-bonded positions were substituted by benzyl groups (Structure **7d**), (e) compounds in which the two H-bonded positions are having *n*-alkyl chains but the other four positions are bearing benzyl groups (Structure **7e**), and (f) compounds in which the two H-bonded positions were substituted by hydroxyethyl or THP-protected hydroxyethyl groups (Structure **7f**). Thermal data of all these materials are listed in Tables 2.27 through 2.32. Most of these materials exhibit hexagonal columnar phase. Their thermal behavior largely depends on the number and nature of substitutions. An important observation is that the introduction of two branched alkyl chains reduces the melting point significantly and these mixed tail derivatives display very wide temperature range columnar phase that is stable at room temperature [205]. While rufigallols with two or four branched alkyl chains are room-temperature liquid crystals, the substitution of all the six alkyl chains by branched alkyl chains destroys the mesomorphism completely [205].

2.4.5 MONO-HYDROXY-PENTAALKOXYRUFIGALLOLS

A few monohydroxy-pentaalkoxy-rufigallols have been prepared to realize their dimers and oligomers (Chapters 3 and 4). In principle, monoalkylation of 1,5-dihydroxy-2,3,6,7-tetraalkoxy-9,

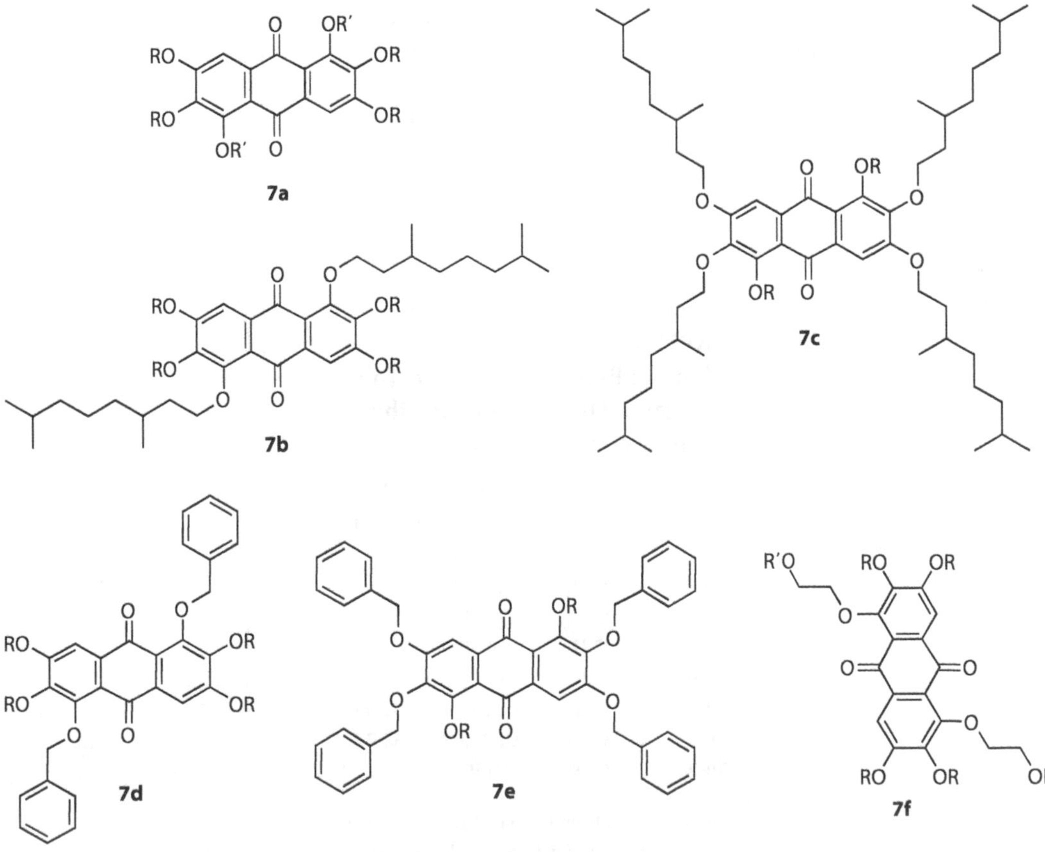

FIGURE 2.1 Structural types of mixed tail rufigallol discotics.

TABLE 2.27
Thermal Behavior of Mixed-Tail Anthraquinone Hexaethers with the Structure 7a (Figure 2.1)

Structure	R	R'	Phase Behavior	Ref.
7a-1	C_4H_9	C_3H_7	Cr 133.4 I	[179]
7a-2	C_4H_9	C_5H_{11}	Cr 105.5 Col$_h$ 123.2 I	[179]
7a-3	C_4H_9	C_6H_{13}	Cr 84 Col$_h$ 118.7 I	[179]
7a-4	C_4H_9	C_7H_{15}	Cr 82.8 Col$_h$ 108 I	[179]
7a-5	C_4H_9	C_8H_{17}	Cr 77.2 Col$_h$ 92.1 I	[179]
7a-6	C_4H_9	C_9H_{19}	Cr 77.3 Col$_h$ 79.5 I	[179]
7a-7	C_4H_9	$C_{10}H_{21}$	Cr (38.6) Col$_h$ 59.8 I	[179]
7a-8	C_4H_9	$C_{11}H_{23}$	Cr 34.1 I	[179]
7a-9	C_4H_9	$C_{12}H_{25}$	Cr 39 I	[179]
7a-10	C_8H_{17}	C_3H_7	Cr 80.2 I	[179]
7a-11	C_8H_{17}	C_4H_9	Cr 80.7 I	[179]
7a-12	C_8H_{17}	C_5H_{11}	Cr 55.4 Col$_h$ 73.9 I	[179]
7a-13	C_8H_{17}	C_6H_{13}	Cr 53.9 Col$_h$ 86.5 I	[179]
7a-14	C_8H_{17}	C_7H_{15}	Cr 39.9 Col$_h$ 91.4 I	[179]
			Cr −8.6 M 44.8 Col$_h$ 93.4 I	[186]
7a-15	C_8H_{17}	C_9H_{19}	Cr 51.1 Col$_h$ 83.4 I	[179]
			Cr 53.9 Col$_h$ 96.6 I	[186]
7a-16	C_8H_{17}	$C_{10}H_{21}$	Cr 32.2 Col$_h$ 57.9 I	[179]
			Cr 48.5 Col$_h$ 95.3 I	[186]
7a-17	C_8H_{17}	$C_{11}H_{23}$	Cr 38.2 Col$_h$ 56.5 I	[179]
			Cr 22.2 Col$_h$ 91.9 I	[186]
7a-18	C_8H_{17}	$C_{12}H_{25}$	Cr 50 Col$_h$ 70.1 I	[179]

TABLE 2.28
Thermal Behavior of Mixed Tail Rufigallol Hexaethers with the Structure 7b (Figure 2.1)

Structure	R	Phase Transition
7b-1	C_4H_9	Cr 54.7 Col$_h$ 65.8 I
7b-2	C_5H_{11}	Col$_h$ 75.2 I
7b-3	C_6H_{13}	Col$_h$ 77.7 I
7b-4	C_7H_{15}	Col$_h$ 78.7 I
7b-5	C_8H_{17}	Col$_h$ 72.3 I
7b-6	C_9H_{19}	Col$_h$ 67.4 I
7b-7	$C_{10}H_{21}$	Cr 37.6 Col$_h$ 61.0 I
7b-8	$C_{11}H_{23}$	Cr 42.0 Col$_h$ 54.7 I
7b-9	$C_{12}H_{25}$	Cr (30.5 Col$_h$) 39.8 I

Source: Data from Bisoyi, H.K. and Kumar, S., *New J. Chem.*, 32, 1974, 2008.

TABLE 2.29
Thermal Behavior of Mixed Tail Rufigallol Hexaethers with the Structure 7c (Figure 2.1)

Structure	R	Phase Transition
7c-1	C_5H_{11}	Cr 52.6 I
7c-2	C_6H_{13}	Cr 57.2 I
7c-3	C_7H_{15}	Col_h 52.6 I
7c-4	C_8H_{17}	Col_h 56.8 I
7c-5	C_9H_{19}	Col_h 59.7 I
7c-6	$C_{10}H_{21}$	Col_h 59.6 I
7c-7	$C_{11}H_{23}$	Col_h 56.1 I
7c-8	$C_{12}H_{25}$	Col_h 52.7 I
7c-9	$C_{14}H_{29}$	Col_h 39.2 I

Source: Data from Bisoyi, H.K. and Kumar, S., *New J. Chem.*, 32, 1974, 2008.

TABLE 2.30
Thermal Behavior of Mixed Tail Rufigallol Hexaethers with the Structures 7d (Figure 2.1)

Structure	R	Phase Transition
7d-1	C_5H_{11}	Cr 171.9 I
7d-2	C_6H_{13}	Col 135.8 Col_h 151 I
7d-3	C_7H_{15}	Cr 77 Col 128.9 Col_h 143.5 I
7d-4	C_8H_{17}	Col 121.4 Col_h 137 I
7d-5	C_9H_{19}	Cr 84.2 Col 104.8 Col_h 132.1 I
7d-6	$C_{10}H_{21}$	Col 92.8 Col_h 129.6 I

Source: Data from Prasad, V. et al., *Liq. Cryst.*, 27, 1075, 2000.

10-anthraquinone, which can be easily prepared from rufigallol, should give the monohydroxy-pentaalkoxy-rufigallol, but in practice the alkylation of rufigallol-tetraethers under different reaction conditions results in the formation of a mixture of unreacted (tetraalkylated), pentaalkylated, and hexaalkylated products. All efforts to isolate pure monohydroxy-pentaalkoxyanthraquinone were futile. However, the problem was solved by acetylating the crude product as R_f values of the tetraalkoxydiacetoxyrufigallol and pentaalkoxymonoacetoxyrufigallol, and hexaalkoxyrufigallol were significantly different on a chromatographic column and thus, all three products could be separated readily by column chromatography. Pure monoacetoxy-pentaalkoxyrufigallol on hydrolysis afforded the desired monofunctionalized anthraquinone derivatives [200]. It has recently been observed that alkylation of rufigallol with aliphatic alcohol under Mitsunobu reaction conditions predominantly produces monohydroxy-pentaalkoxyrufigallol in good yield [206]. All the monohydroxy-pentaalkoxyrufigallols **9** (Scheme 2.27) display hexagonal columnar phase over a wide temperature range. Their thermal data are summarized in Table 2.33.

TABLE 2.31
Thermal Behavior of Mixed Tail Rufigallol Hexaethers with the Structures 7e (Figure 2.1)

Structure	R	Phase Transition
7e-1	C_5H_{11}	Cr 138 Cr 167.5 I
7e-2	C_6H_{13}	Cr (140 Col$_h$) 158.5 I
7e-3	C_7H_{15}	Cr (141 Col$_h$) 156.5 I
7e-4	C_8H_{17}	Cr (134.5 Col$_h$) 143.5 I
7e-5	C_9H_{19}	Cr (128 Col$_h$) 132.5 I
7e-6	$C_{10}H_{21}$	Cr (115.5 Col$_h$) 118.5 I
7e-7	$C_{12}H_{25}$	Cr 120 I

Source: Data from Prasad, V., *Liq. Cryst.*, 28, 647, 2001.

TABLE 2.32
Thermal Behavior of Mixed Tail Rufigallol Hexaethers with the Structure 7f

Structure	R	R'	Phase Transition
7f-1	C_4H_9	H	Cr 162 I
7f-2	C_6H_{13}	H	Cr 140 I
7f-3	C_8H_{17}	H	Cr 72.5 Col 130.5 I
7f-4	$C_{10}H_{21}$	H	Cr 82 Col 107.7 Col 122.5 I
7f-5	C_3H_7	THP	Cr 126 I
7f-6	C_4H_9	THP	Cr (93 Col$_h$) 103 I
7f-7	C_5H_{11}	THP	Cr 79.5 Col$_h$ 101.5 I
7f-8	C_6H_{13}	THP	Cr 83.5 Col$_h$ 91 I
7f-9	C_7H_{15}	THP	Cr 80.5 Col$_h$ 85.5 I
7f-10	C_8H_{17}	THP	Cr 64 Col$_h$ 76 I
7f-11	C_9H_{19}	THP	Cr 58.5 Col$_h$ 70 I
7f-12	$C_{10}H_{21}$	THP	Cr 52 Col$_h$ 62.5 I

Source: Data from Prasad, V. and Rao, D.S.S., *Mol. Cryst. Liq. Cryst.*, 350, 51, 2000.

2.4.6   RUFIGALLOL-BASED DISCOTIC-CALAMITIC HYBRIDS

Recently, Pal and Kumar prepared a number of alkoxycyanobiphenyl-substituted rufigallols **10–14** by systematically replacing one, two, four, five, and six alkyl chains by cyanobiphenyl-tethered alkoxy chains (Figure 2.2). Commercially available 4′-hydroxy-4-biphenylcarbonitrile was alkylated under classical conditions with an excess of the appropriate α,ω-dibromoalkane to obtain the ω-brominated product. The bromo-terminated alkoxycyanobiphenyls were reacted with monohydroxyanthraquinone under microwave heating to give the desired rod–disk dimers **10**. When the bromo-terminated alkoxycyanobiphenyls were reacted with dihydroxy-tetraalkoxy-9,10-anthraquinone under microwave heating in the presence of cesium carbonate as base and NMP as solvent, the rod–disk–rod trimers **11** formed within 2 min. The tetraalkoxycyanobiphenyl-substituted

TABLE 2.33
Thermal Behavior of Monohydroxy-Pentaalkoxy-Rufigallols (Scheme 2.27)

Structure	R	R′	Phase Transition	Ref.
9a	C_3H_7	C_3H_7	Cr 68.4 Col 106.6 I	[200]
9b	C_4H_9	C_4H_9	Cr 57.7 Col 127.6 I	[200]
9c	C_5H_{11}	C_5H_{11}	Cr 32.4 Col 125.3 I	[200]
9d	C_6H_{13}	C_6H_{13}	Col 113 I	[200]
9e	C_7H_{15}	C_7H_{15}	Col 104.8 I	[200]
9f	C_8H_{17}	C_8H_{17}	Col 90.4 I	[200]
9g	C_9H_{19}	C_9H_{19}	Col 86.4 I	[200]
9h	$C_{10}H_{21}$	$C_{10}H_{21}$	Col 80.6 I	[200]
9i	$C_{12}H_{25}$	$C_{12}H_{25}$	Col 67.5 I	[200]
9j	C_5H_{11}	CH_2PhNO_2	M 71.9 Col_h 171.6 I	[184]

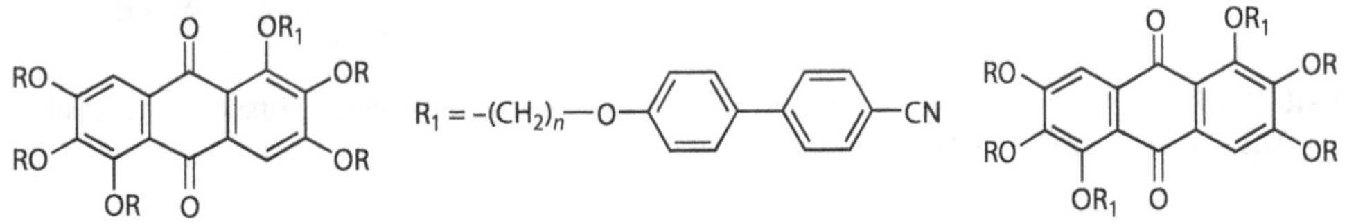

$R_1 = -(CH_2)_n-O-$

10a: R = $C_{10}H_{21}$, n = 4;
10b: R = $C_{10}H_{21}$, n = 5
10c: R = $C_{10}H_{21}$, n = 10
10d: R = $C_{10}H_{21}$, n = 14

11a: R = C_6H_{13}, n = 4
11b: R = C_6H_{131}, n = 9
11c: R = C_6H_{13}, n = 12
11d: R = $C_{12}H_{25}$, n = 4

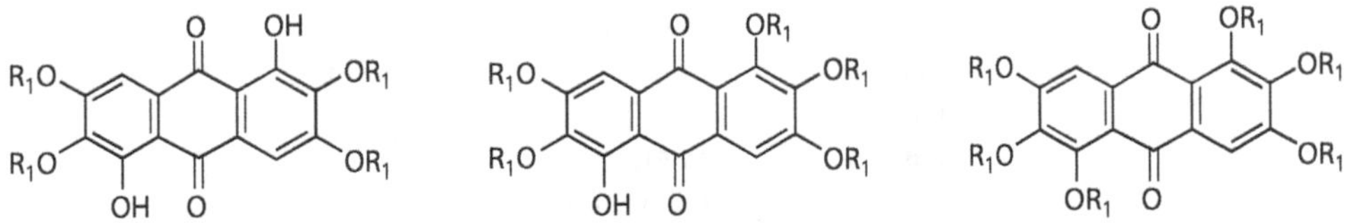

12: n = 9

13a: n = 6
13b: n = 8

14a: n = 6
14b: n = 9
14c: n = 12

FIGURE 2.2 Chemical structures of rufigallol-cyanobiphenyl hybrids.

rufigallol **12** was prepared by reacting bromo-terminated alkoxycyanobiphenyl with rufigallol under microwave heating within 4–5 min. It should be noted that attempted tetramerization under classical reaction conditions did not produce any desired product. A tiny amount of pentaalkoxycyanobiphenyl-substituted rufigallol **13** also formed in the reaction, but it could not be isolated in pure form. The desired pentaalkoxycyanobiphenyl-substituted rufigallols were obtained via Mitsunobu reaction. When rufigallol was reacted with bromo alcohols in the presence of DEAD, TPP to get the corresponding ethers, exclusively pentasubstituted product **15** is formed as shown in Scheme 2.28. The corresponding pentasubstituted product on reaction with commercially available 4′-hydroxy-4-biphenylcarbonitrile gives the pentaalkoxycyanobiphenyl-substituted rufigallols **13** in 15%–20% yield. While the classical alkylation of rufigallol with bromo-terminated

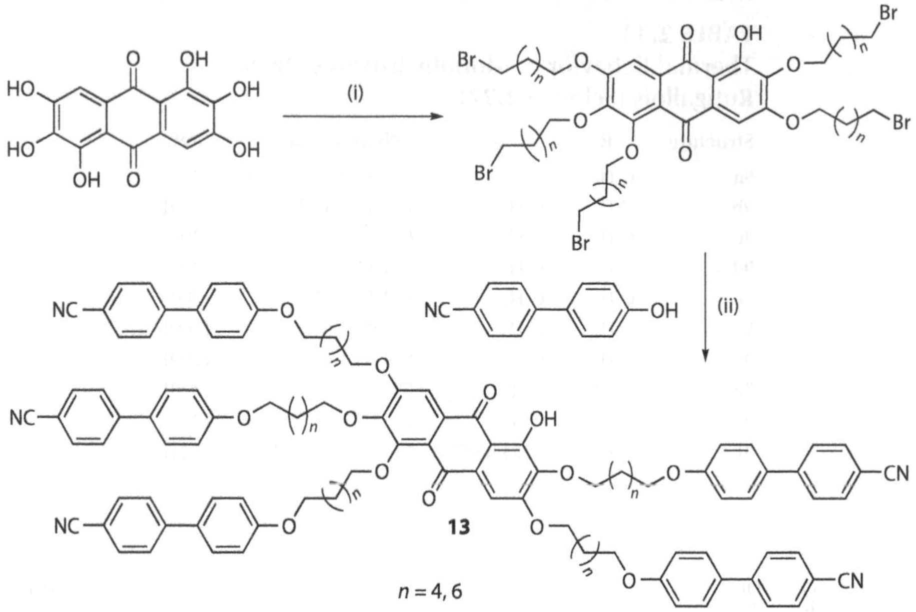

SCHEME 2.28 Synthesis of pentacyanobiphenyl-substituted rufigallols under Mitsunobu reaction conditions: (i) DEAD, TPP, Br(CH$_2$)$_n$OH; (ii) Cs$_2$CO$_3$, acetone, reflux.

TABLE 2.34
Thermal Behavior of Alkoxycyanobiphenyl-Substituted Rufigallols (Figure 2.2)

Structure	Phase Transition
10a	Col$_h$ 40 I
10b	Cr 65 I
10c	Cr 55 I
10d	Cr 61.5 I
12	Cr 85 N 138 I
13a	Cr 143 SmA 166 N 187 I
13b	Cr 183 SmA 188 N 203 I
14a	Cr (126 N) 183 I
14b[a]	Cr 94 SmA 116 N 125 I
14c	Cr 143 (130 SmA) I

Sources: Data from Pal, S.K. and Kumar, S. (unpublished result); Pal, S.K. et al., *Liq. Cryst.*, 35, 521, 2008.

[a] On cooling another Sm phase at 110°C and a re-entrant nematic phase at 79°C appear.

alkoxycyanobiphenyls failed to produce the hexaadduct **14**, it could be obtained under microwave heating in 8–10 min.

The thermal behavior of cyanobiphenyl-substituted rufigallols is presented in Table 2.34. The dimer **10a** having four methylene units, showed columnar liquid crystalline property at room temperature. On heating, it transforms to the isotropic phase at about 43°C. Interestingly, increasing the length of spacer connecting the cyanobiphenyl unit to rufigallol destroys the liquid crystalline property of the material. Thus, compounds **10b** and **10c** with 5 and 10 methylene units do not show any liquid crystalline phase. They melted at 65°C and 55°C, respectively, to the isotropic phase. Similarly, increasing the number of atoms in the periphery also destabilized the columnar phase. Compound **10d** with 12 alkyl chains in the rufigallol moiety melted at 61.5°C to the isotropic phase. None of the trimer **11** exhibited liquid crystalline properties. The tetraalkoxycyanobiphenyl-substituted rufigallol **12** exhibited an enantiotropic nematic phase. On first heating, compound **12** melted at about 85°C to a nematic phase, which clears at about 138°C. On cooling, the nematic phase appears at 137°C and it stays down to room temperature. The pentaalkoxycyanobiphenyl-substituted rufigallol **13b** exhibited an SmA phase at 143°C on heating, which transforms to a nematic at about 166°C and finally clears at 187°C. Compound **13b** displayed similar thermal behavior. Compound **14a** showed only a monotropic nematic phase. The crystalline compound **14a** on heating melts at 183°C to the isotropic phase. However, on cooling, the nematic phase appears at 176°C, crystallizing at 148°C. Compound **14b** displayed interesting thermal behavior. On heating, it melts at 94°C to an SmA phase that transforms to a nematic phase at 116°C and finally clears at 125°C. Although its DSC trace shows only a single isotropic-nematic transition at 124°C, POM reveals two other second-order transitions. On cooling, the appearance of an SmA phase at 110°C and a re-entrant nematic phase at 79°C was observed under POM. Compound **14c**, having longer spacer, shows only a monotropic SmA phase. On heating, it melts at 143°C to the isotropic phase. On cooling, the SmA phase appears at 130°C, which crystallizes at about 82°C.

2.4.7 RUFIGALLOL-BASED DISCOTIC METALLOMESOGENS

The molecular architecture of monohydroxy-pentaalkoxy-rufigallol (1-hydroxy-2,3,5,6,7-pentakis(alkoxy)-9,10-anthraquinone) is similar to the β-diketonate system (Figure 2.3). It may be noted that β-diketonates were the first disk-like metal complexes reported to exhibit mesomorphism in the pure state. They are among the most widely synthesized and studied metallomesogens. They exhibit either calamitic or discotic mesomorphism, depending on subtle differences in the molecular structure. A number of β-diketone derivatives are known to show nematic, smectic, columnar, and lamellar mesophases. β-Diketone complexes having both calamitic and discotic features are also known.

Due to the similarity between monohydroxy-pentaalkoxy-rufigallol and the β-diketonate system, it is expected that the chemistry that has been well developed to the latter can be applied to the former, and thus, a family of metallomesogens can be realized. Kumar and coworkers prepared a number of metal-bridged rufigallol dimers (Scheme 2.29) by refluxing monohydroxy-pentaalkoxy-rufigallols with metal acetate in acetonitrile-pyridine.

Two series of complexes, one with copper and other with Pd, were prepared in this way [192–194]. While the Pd complexes were found to be thermally unstable at higher temperature, the Cu complexes of the same ligand were stable. The thermal data of these discotic metallomesogens are listed in Table 2.35. The higher homologues of both the series form columnar phase, but the exact nature of the columnar phase has not been revealed. Lower members of both the series are not liquid crystalline,

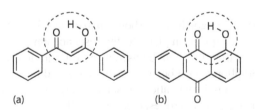

FIGURE 2.3 Molecular structures showing similarity between (a) β-diketone (keto-enol form) and (b) 1-hydroxyanthraquinone systems.

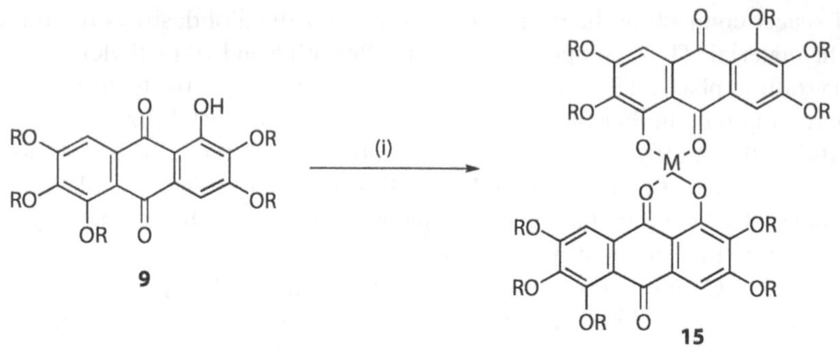

SCHEME 2.29   Synthesis of rufigallol-based metallomesogens: (i) metal acetate, acetonitrile-pyridine.

TABLE 2.35
Thermal Behavior of Rufigallol-Based
Metallomesogens (Scheme 2.29)

Structure	R	R′	Phase Transition
15a	C_4H_9	Cu	Cr 184 I
15b	C_5H_{11}	Cu	Cr 175 I
15c	C_6H_{13}	Cu	Cr 170 I
15d	C_8H_{17}	Cu	Cr 152 Col 156 Col 166 I
15e	$C_{12}H_{25}$	Cu	Cr 72 Col 121 Col 146 I

Source: Data from Naidu, J.J. and Kumar, S., *Mol. Cryst. Liq. Cryst.*, 397, 17, 2003.

but columnar mesophase can be induced by doping them with an electron acceptor, trinitrofluorenone [192–194]. All the Pd complexes start decomposing on heating and, therefore, their thermal data are not reproducible. It is imperative to prepare metal complexes of branched-chain rufigallol discotics having room-temperature mesophase stability and low clearing temperature.

As described above, a variety of discotic liquid crystals having very wide mesophase range and room-temperature stability have already been derived from rufigallol, and it can be expected that many more new materials like side-chain polymers, oligomers, donor–acceptor–donor or acceptor–donor–acceptor triads, discotic-calamitic hybrids, etc., would be realized in future. Unfortunately, not too many physical studies have been carried out on these interesting materials to uncover the full potential of these interesting mesogenic materials. Efforts have to be made to utilize rufigallol-based DLCs in fabricating devices like photovoltaic cells, TFTs, LEDs, etc.

2.5   PYRENE CORE

Pyrene is a planar polycyclic aromatic hydrocarbon consisting of four fused benzene rings. It is well known for its high fluorescence quantum yield. Its fluorescence emission spectrum is very sensitive to the solvent's polarity and, therefore, can be used as a probe to determine solvent environments. Pyrene derivatives as fluorophores have been extensively studied in various areas of chemical biology. Its derivatives have also found applications as sensors, organic light emitting diodes, photoconductors, fluorescent polymers, genetic probes, etc. [214–224].

Pyrene hydrocarbon can be easily substituted at 1-, 3-, 6-, and 8-positions to generate discotic liquid crystals. Halogenation (bromination or chlorination) of pyrene **1** provides 1,3,6,8-tetrahalopyrene

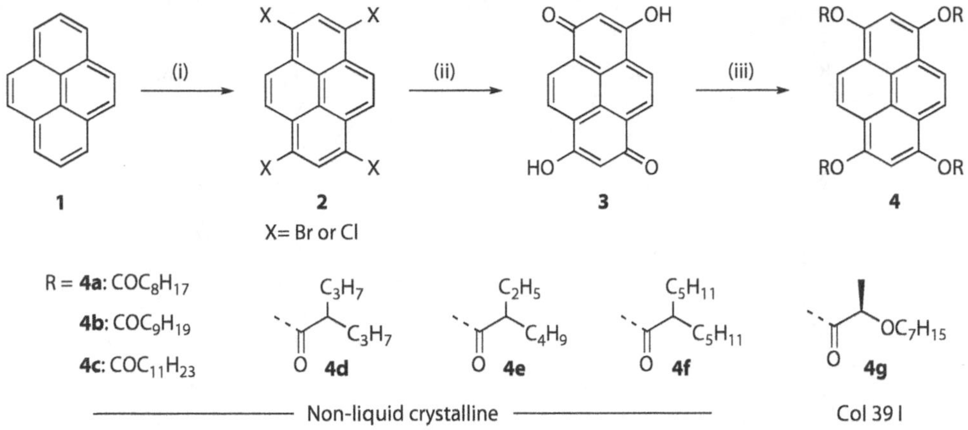

SCHEME 2.30 Synthesis of pyrene tetraesters: (i) Br$_2$, nitrobenzene or Cl$_2$, C$_2$H$_2$Cl$_4$; (ii) 25% H$_2$SO$_4$-SO$_3$; 40% H$_2$SO$_4$, H$_2$O; (iii) RCOCl, Zn, DMAP, THF, pyridine.

2 in excellent yield (Scheme 2.30) [225,226]. Treatment of **2** with fuming sulfuric acid followed by hydrolysis yields 3,6-dihydroxypyrene-1,6-quinone **3**. The reductive esterification of this quinone in the presence of different acid chlorides offers various 1,3,6,8-tetraalkanoyloxypyrene derivatives **4**.

Pyrene tetraesters with long linear alkyl chains **4a–c** and branched chains **4d–f** are reported to be non-mesomorphic, but columnar phases can be induced via charge-transfer complexation with TNF. Conversely, the chiral tetraester **4g** exhibits a monotropic columnar phase on slow cooling below 34°C [225]. The mesophase clears at 39°C on heating. This material displays ferroelectric switching, albeit very slow. Interestingly, short-chain normal alkyl esters **5a**, **5b** and racemic 2-ethylhexyl ester, **5h**, of pyrene 1,3,6,8-tetracarboxylic acid are reported to be liquid crystalline [227–229]. These tetraesters have been used to fabricate an organic light emitting diode device [228].

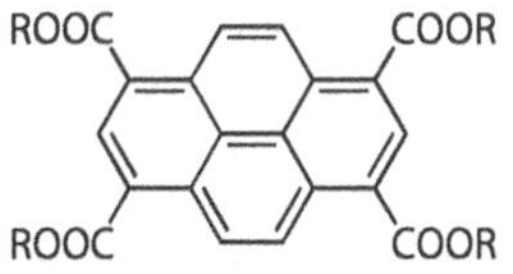

5a: CH$_3$, Cr 266 [Col 262] I
5b: C$_2$H$_5$, Cr 190 Col 204 I
5c: C$_3$H$_7$, Cr 181 I
5d: C$_4$H$_9$, Cr 101 I

5e: C$_5$H$_{11}$, Cr 101 I
5f: C$_6$H$_{13}$, Cr 104 I
5g: C$_8$H$_{17}$, Cr 88 I
5h: , Col 94 I

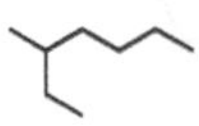

The tetrabromopyrene **2** is a valuable precursor for a variety of discotics (Scheme 2.31). Thus, the Suzuki coupling in between **2** and commercially available 1-methylphenylboronic acid and 1-fluorophenylboronic acid yields tetraphenyl derivatives **6** and **9**, respectively. The demethylation of **6** followed by alkylation of the resultant 1,3,6,8-tetrakis(4-hydroxyphenyl)-pyrene **7** with different alkyl halides yields tetraethers **8a–e**. On the other hand, the esterification of the tetraphenol **7** with acid chloride or benzoic acid affords tetraesters **8f**, **8g**, or benzoate **8h**, respectively [230]. Similarly, the Suzuki coupling reaction of tetrabromopyrene **2** and appropriate boronic acid ester yields tetraaryl-pyrenes **11** [231,232]. On the other hand, Sonogashira–Hagihara reaction in between **2** and 3,4,5-tridodecyloxyphenylacetylene produces 1,3,6,8-tetrakis(3,4,5-trisdodecyloxy-phenylethynyl)pyrene **12** [233].

All the ethers, thioethers, esters, and benzoates **8a–h**, **10a**, and **10b** were found to be non-mesomorphic. The monoalkoxyphenyl-substituted derivative **11a** was also not liquid crystalline; however, dialkoxyphenyl-substituted derivatives **11b**, **11c** and trialkoxyphenylethynyl-substituted derivative **12** exhibit columnar phases (Table 2.36). This clearly demonstrates the importance of space filling around the core in inducing the mesomorphism.

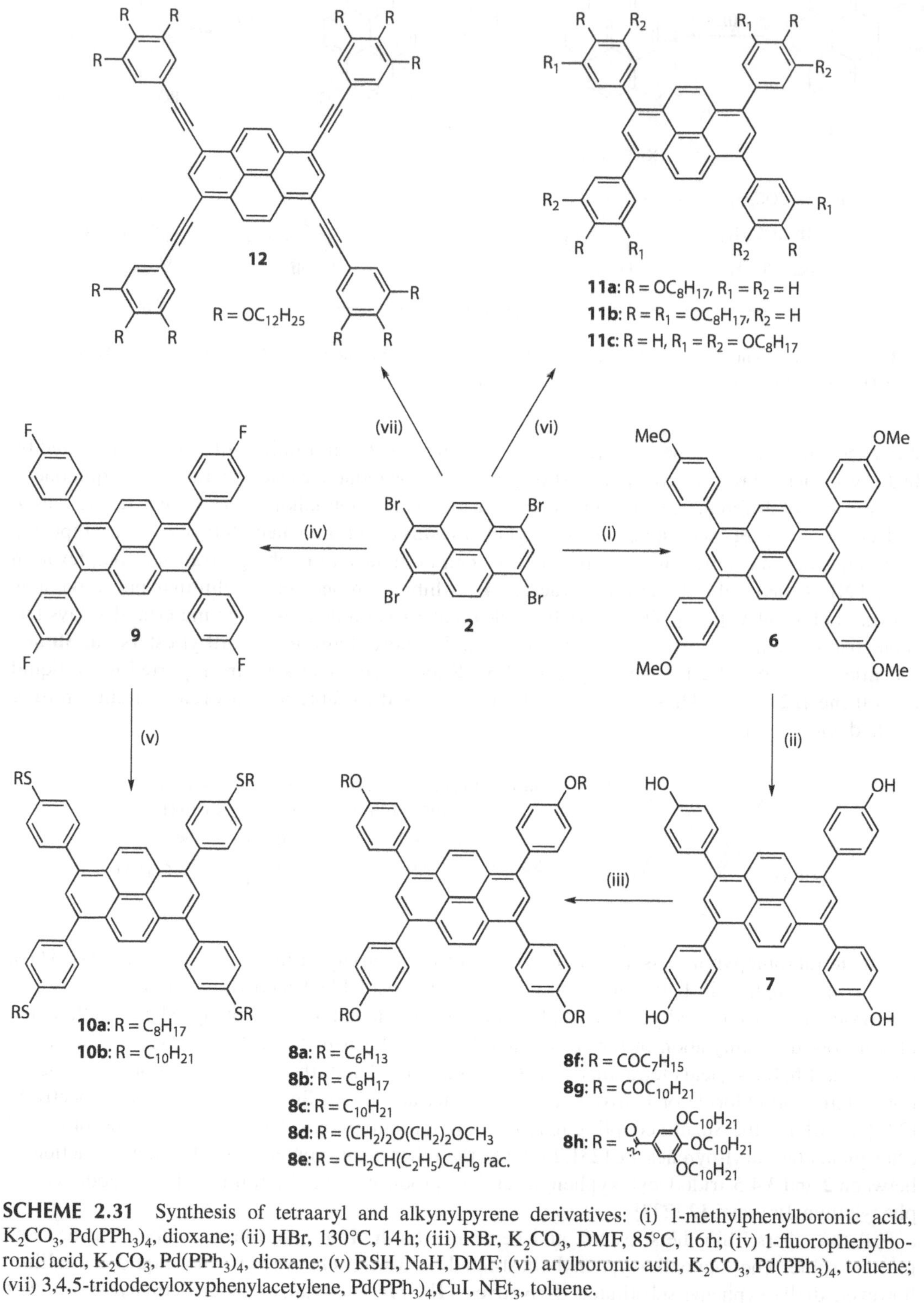

SCHEME 2.31 Synthesis of tetraaryl and alkynylpyrene derivatives: (i) 1-methylphenylboronic acid, K$_2$CO$_3$, Pd(PPh$_3$)$_4$, dioxane; (ii) HBr, 130°C, 14 h; (iii) RBr, K$_2$CO$_3$, DMF, 85°C, 16 h; (iv) 1-fluorophenylboronic acid, K$_2$CO$_3$, Pd(PPh$_3$)$_4$, dioxane; (v) RSH, NaH, DMF; (vi) arylboronic acid, K$_2$CO$_3$, Pd(PPh$_3$)$_4$, toluene; (vii) 3,4,5-tridodecyloxyphenylacetylene, Pd(PPh$_3$)$_4$, CuI, NEt$_3$, toluene.

TABLE 2.36
Thermal Phase Behavior of
Mesomorphic Tetrasubstituted
and Disubstituted Pyrenes

Structure	Phase Transition	Ref.
4g	Col 39 I	[225]
5a	Cr (262 Col) 266 I	[227]
5b	Cr 190 Col 204 I	[227]
5h	Col 94 I	[227]
11b	Cr 73 Col$_h$ 87 I	[232]
11c	Cr (25 Col$_h$) 81 I	[232]
12	Cr 51.5 Col$_h$ 95.5 I	[233]
15c	I 208 D$_L$ 184 Cr	[234]
15d	I 184 D$_L$ 74 Cr	[234]
16b	I 170 D$_L$ 134 Cr	[234]
16c	I 148 D$_L$ 106 Cr	[234]
16d	I 139 D$_L$ 105 Cr	[234]
17	Cr −35 Cubic 175 I	[235]

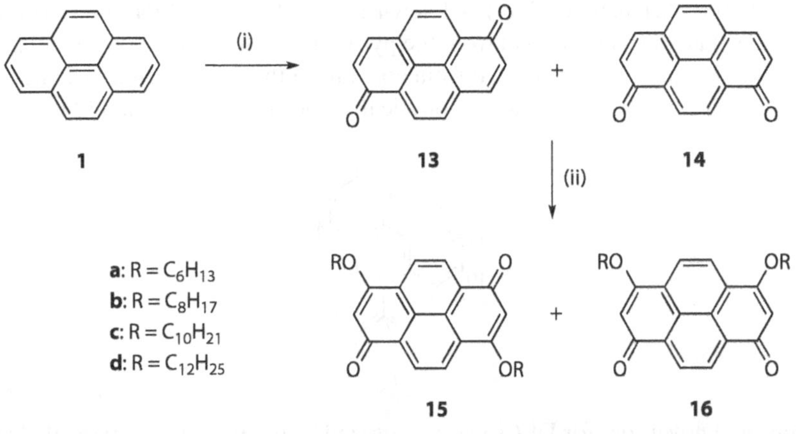

SCHEME 2.32 Synthesis of dialkoxy *syn*- and *anti*-pyrenediones: (i) $Na_2Cr_2O_7$, 3M H_2SO_4; (ii) ROH, $FeCl_3$.

Oxidation of pyrene with sodium dichromate in sulfuric acid yields a mixture of pyrene-1,6-dione **13** and pyrene-1,10-dione **14** in equal amount (Scheme 2.32) [234]. Treatment of this mixture with an alcohol in the presence of $FeCl_3$ generates corresponding alkoxypyrene derivatives **15** and **16** in low yield. While short chain derivatives **15a**, **15b**, and **16a** are non-mesomorphic, longer alkoxy chain derivatives **15c**, **15d**, **16b**, **16c**, and **16d** exhibit discotic lamellar phases (Table 2.36).

Sagara and Kato reported the liquid crystalline nature of disubstituted pyrene **17** prepared from 1,6-diethynylpyrene and the fan-shaped dendron by Sonogashira coupling [235]. Compound **17** exhibits a liquid crystalline optically isotropic cubic phase from −35°C to 175°C. Similarly, a number of dendrimers derived from monosubstituted pyrenes have been reported to form columnar phases [236–239]. These dendrons form a helical superstructure around the columnar axis. A detailed discussion of such supramolecular liquid crystals is out of the scope of this chapter.

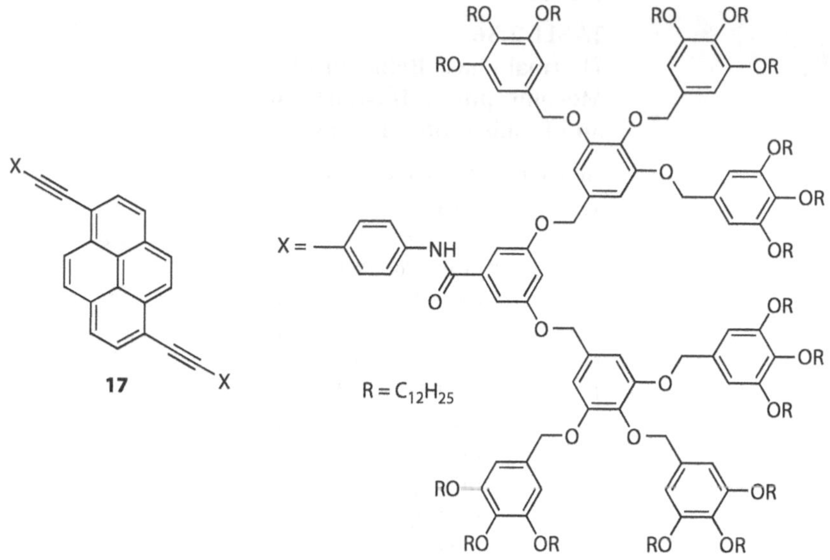

17   X = —⟨⟩—NH

R = C₁₂H₂₅

2.6   TRIPHENYLENE CORE

The best studied core system in the field of DLCs is triphenylene (TP). The symmetrical fused aromatic hydrocarbon "triphenylene" (**1**) is known in the chemistry literature since more than a century. Schultz isolated this compound from the pyrolytic products of benzene and gave the name triphenylene [240]. It can be easily isolated from coal tar. In the early twentieth century, it was synthesized from cyclohexanone and its various physical properties were studied [240].

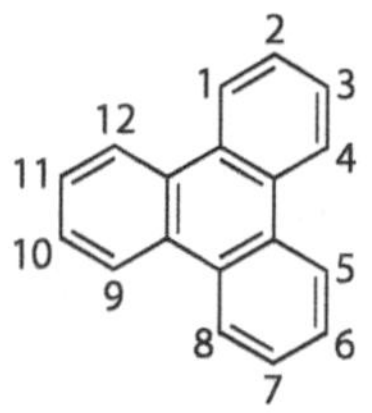

1

Triphenylene as a novel core for DLCs was recognized by the French group in 1978 [241,242] just a few months after of the discovery of mesomorphism in disk-shaped molecules. Since then, triphenylene has remained the focus of considerable attention of liquid crystal scientists around the world for a number of reasons: thus, its derivatives are thermally and chemically stable, their chemistry is fairly accessible, they show a variety of mesophases, and their one-dimensional charge and energy migration properties offer tremendous potential applications [243,244]. This stimulated numerous synthetic efforts to prepare a variety of triphenylene-based DLCs. In addition to their importance as DLCs, triphenylene derivatives have also been recognized as interesting materials for molecular scale devices [245–249], molecular receptors [250–254], etc. More than 500 monomeric triphenylene derivatives have been prepared to investigate their mesomorphic properties. Additionally, a large number of triphenylene-based discotic dimers, oligomers, and polymers have also been realized (Chapters 3 through 5). This section presents the chemistry and thermal behavior of monomeric triphenylene discotics. These discotics may be subdivided into the following categories: (1) symmetrical triphenylene hexaethers, (2) symmetrical triphenylene hexaesters, (3) unsymmetrical triphenylene discotics, (4) alkoxy-hydroxy triphenylenes, (5) discotics derived from alkoxy-hydroxy triphenylenes, (6) discotics derived from symmetrical hexaalkoxy-TPs, (7) discotics derived from symmetrical hexabromo-TP, and (8) trisubstituted TP discotics. The synthesis and mesomorphic properties of these derivatives are presented in the following sections.

2.6.1 SYMMETRICAL TRIPHENYLENE HEXAETHERS

Hexaalkoxy-TPs are the most widely synthesized and studied discotic mesogens, and a number of methods have been developed for their synthesis. Traditionally, these molecules were prepared by the alkylation of 2,3,6,7,10,11-hexahydroxy-TP, which in turn was prepared by the demethylation of hexamethoxy-TP (Scheme 2.33). Oxidative coupling is a classical method for the formation of aryl–aryl bonds and has been extensively used for the synthesis of natural products, polymers, and many other low molar mass compounds of materials science interest [255–257]. The oxidation of 1,2-dialkoxybenzene to hexaalkoxy-TP is one of the unusual cases of Scholl reaction [258,259], where more than one aryl–aryl bonds are formed. In this reaction, linear coupling products have not been realized. The cyclization is enthalpically favored by the aromatic character of the central ring and the greater conjugation of the planar triphenylene relative to nonplanar acyclic oligomers and polymers [260]. 1,2-Dimethoxybenzene (veratrole) 2 is oxidatively trimerized using chloranil or $FeCl_3$ in concentrated H_2SO_4 at room temperature to obtain hexamethoxy-TP, 3 [261–267]. The trimerization involves three consecutive Scholl reactions, but surprisingly under normal Scholl conditions using aluminum chloride and nitrobenzene, this does not work well [266]. Triphenylene derivatives show mesomorphism only when the six peripheral alkoxy chains have a minimum of three carbon atoms, that is, propyloxy chains. The trimerization using chloranil is limited to the preparation of hexamethoxy-TP; with higher homologues such as 1,2-dihexyloxybenzene it gives only a poor yield of hexahexyloxy-TP with many side products. Therefore, to prepare these long-chain derivatives, the methyl groups of hexamethoxy-TP are dealkylated with boron tribromide or HBr, and the resulting hexaphenol is then alkylated with an appropriate alkyl halide to obtain different hexaalkoxy-TP discotics (Scheme 2.33). As polyphenols are generally sensitive to air oxidation and are difficult to store for a long time, the hexahydroxy-TP may be preserved as hexaacetate, which can be directly converted to hexaalkoxy-TP discotics by heating with alkyl halide in DMSO in the presence of powdered KOH [268]. Similarly, treatment of the hexamethoxy-TP with trimethylsilyl iodide gives the hexatrimethylsilyloxy-TP 5, which can be directly alkylated with alkyl halide to hexaalkoxy-TP discotics via in situ desilylation with tetrabutylammonium fluoride (Scheme 2.33) [269,270]. A very clean synthesis of hexaalkoxytriphenylenes in moderate yield is the cyclic anodic trimerization of dialkoxybenzene followed by a chemical or electrochemical reductive work-up [271–275]. Though the reaction is not limited to the synthesis of hexamethoxy-TP, this slow reaction is only applicable to a very small scale, and thus limited in its use.

The use of $FeCl_3$ as oxidant was further explored by the Ringsdorf group, which reported the synthesis of hexahexyloxy-TP by the trimerization of 1,2-dihexyloxybenzene 7 ($R = C_6H_{13}$) in 70% H_2SO_4 at 80°C using iron chloride in 20% yield [276]. Naarmann et al. performed the reaction at room temperature and realized higher yields [277]. A dramatic advancement in the synthesis of triphenylene hexaethers was achieved by the Leeds group, who used only a catalytic amount of H_2SO_4 (0.3%) in dichloromethane followed by a reductive work-up using methanol (Scheme 2.33) [278–281]. Care must be taken during the reductive workup as it is highly exothermic, and a large amount of formaldehyde is liberated. Pouring the reaction mixture in a large amount of cold methanol is recommended. The preparation of different alkoxytriphenylenes in 55%–86% yields was reported using this method. Later, two other reagents, molybdenum pentachloride [282] and vanadium oxytrichloride [283,284], were reported to be highly efficient for the preparation of triphenylene hexaethers. It may be noted that the oxidative coupling can be accomplished by using a number of other reagents such as $SbCl_5$, VOF_3, $K_3Fe(CN)_6$, $CuCl_2$ or $Cu(OT_f)_2$ and $AlCl_3$ in CS_2, $Pb(OAc)_4$/ BF_3-ET_2O in MeCN, $Tl(OCOCF_3)_2$ in CF_3COOH, etc.; however, the potential of these oxidants in the synthesis of triphenylene-based DLCs has not yet been explored.

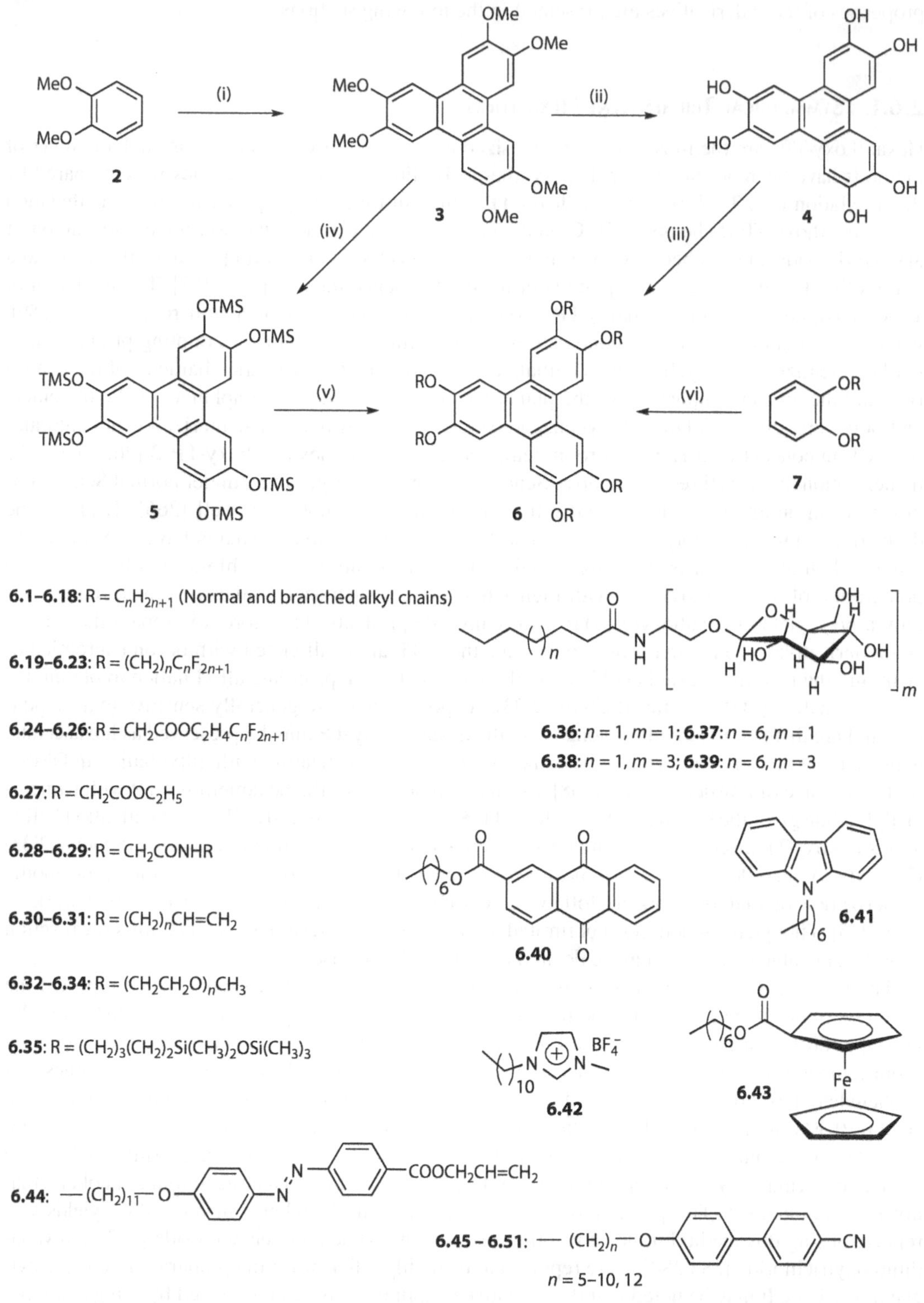
MeO
MeO
2
(i)
OMe
OMe
MeO
MeO
OMe
3 OMe
(ii)
OH
OH
HO
HO
OH
4 OH
(iv)
(iii)
OTMS
OTMS
TMSO
TMSO
OTMS
5 OTMS
(v)
OR
OR
RO
RO
OR
6 OR
(vi)
OR
OR
7
6.1–6.18: R = C$_n$H$_{2n+1}$ (Normal and branched alkyl chains)
6.19–6.23: R = (CH$_2$)$_n$C$_n$F$_{2n+1}$
6.24–6.26: R = CH$_2$COOC$_2$H$_4$C$_n$F$_{2n+1}$
6.27: R = CH$_2$COOC$_2$H$_5$
6.28–6.29: R = CH$_2$CONHR
6.30–6.31: R = (CH$_2$)$_n$CH=CH$_2$
6.32–6.34: R = (CH$_2$CH$_2$O)$_n$CH$_3$
6.35: R = (CH$_2$)$_3$(CH$_2$)$_2$Si(CH$_3$)$_2$OSi(CH$_3$)$_3$
6.36: n = 1, m = 1; 6.37: n = 6, m = 1
6.38: n = 1, m = 3; 6.39: n = 6, m = 3
6.40
6.41
BF$_4^-$
6.42
Fe
6.43
6.44: —(CH$_2$)$_{11}$—O— —N=N— —COOCH$_2$CH=CH$_2$
6.45 - 6.51: —(CH$_2$)$_n$—O— —CN
n = 5–10, 12

SCHEME 2.33 Synthetic routes to hexaalkoxytriphenylenes: (i) chloranil, H_2SO_4; (ii) BBr_3 or HBr; (iii) RBr, base; (iv) TMSI; (v) RBr, TBAF; (vi) $FeCl_3$ or $MoCl_5$ or $VOCl_3$, CH_2Cl_2.

In the beginning, it was considered necessary to add a little acid to the reaction mixture; however, later it was observed that even the catalytic amount of H_2SO_4 is not required for this trimerization [282]. The reagent $VOCl_3$ is of particular importance as unlike other reagents, which are solid and insoluble in organic solvents, it can be readily handled using standard syringe techniques or by using an addition funnel. $VOCl_3$ has a high solubility in various organic solvents, which could be a reason for the almost spontaneous trimerization of dialkoxy benzenes into hexaalkoxy-TPs. The easy handling of the reagent, convenient work-up of the reaction, and high product yields make $VOCl_3$ a very attractive reagent for the preparation of triphenylene discotics, albeit it is expensive and more sensitive to moisture than $FeCl_3$. Dichloromethane was found to be the only solvent that gives a good yield of the trimerization product [284]. The reason for the poor performance of the reaction in various other solvents is still not clear. Although only 2 equivalents of $VOCl_3$ are required to generate diradical cation species, the best yield of hexaalkoxy-TPs was realized using about 2.5 equivalents of $VOCl_3$. This could be due to the impurities present in the commercial reagent or non-perfect anhydrous reaction conditions. As expected, reducing the amount of the reagent decreases the product yield, but increasing the reagent concentration also decreases the yield of hexaalkoxy-TPs. This could be due to the formation of side products. Therefore, it is generally safer to use little excess of the reagent but not too much. Two main side products are formed in this reaction. These are mono-hydroxy-pentaalkoxy-TP (major) and α–chloro-hexaalkoxy-TP (minor). If the oxidizing reagent is not used in large excess and the reaction is not left for a longer period, only partially dealkylated product is observed. Under normal reaction conditions, about 2%–3% of these products are formed. Both can be separated by efficient column chromatography. To obtain the pure hexaalkoxy-TP, it is beneficial to realkylate the crude reaction mixture before purification. The trimerization reaction works well at ambient temperature. Performing the reaction at low temperature (0°C) reduces the yield of the product as a lot of unreacted starting material remains, but efforts have not been made to optimize the reaction conditions at low temperature. The yield of the product also decreases when the reaction is carried out at elevated temperature.

The mesophase behavior of symmetrical hexaalkoxy-TPs (Scheme 2.33) is summarized in Table 2.37. Please note that many other non-liquid crystalline derivatives are also known, but in this table as well as in all other tables, only discotic liquid crystalline compounds and some non-mesomorphic but closely related to liquid crystalline materials are presented. All derivatives with peripheral chains longer than propyloxy exhibit hexagonal columnar mesophases [285–288]. Both the transition temperatures and the mesophase range decrease on increasing the length of the side chains. The hexapropyloxy-TP was originally described to be non-liquid crystalline but recently its mesomorphic nature was reported by Bushby et al. [288]. Further, the nature of the mesophase of the hexabutyloxy-TP was originally classified as hexagonal columnar, but recently it was revealed that the material possesses a more ordered columnar plastic mesophase [289].

The concept of using branched chains instead of normal alkyl chains to modify thermal properties of calamitic as well as discotic materials is well documented. The technique is particularly very useful in bringing down the melting and clearing points of large-core discotics [78,210,290–293]. Compound 6.13 having peripheral butyloxy chains attached with a methyl branch at the terminal does not display mesomorphic behavior (Table 2.37) [287]. The high melting point of 6.13 indicates that the space-filling effect of the branched chain has increased the melting point in comparison to normal butyloxy chain derivative 6.4 with the same chain length. The melting point of 6.13 is also higher than the melting point of compound 6.5 having the same mass unit that can again be attributed to the space-filling effect of the branched chains. The position of branching imparts significant effect on mesomorphism. Branching near the core generally reduces the possibility of forming liquid crystalline phases. Bulkier branched unit, such as ethyl instead of methyl, generally decreases the melting point drastically due to the steric effect of the branch. Overall, the mesomorphism in such materials depends on the π–π interaction of the core, number, and nature of the branch exerting space-filling and steric effects.

TABLE 2.37
Thermal Behavior of Symmetrical Hexasubstituted Triphenylene Ethers Having Normal or Branched Peripheral Alkyl Chains (Scheme 2.33)

Structure	R	Phase Transition	Ref.
6.1	CH_3	Cr 317 I	[242]
6.2	C_2H_5	Cr 247 I	[242]
6.3	C_3H_7	Cr 177 I	[242]
		Cr 100 Col$_p$ 174.5 I	[289]
6.4	C_4H_9	Cr 88.6 Col$_h$ 145.6 I	[242]
6.5	C_5H_{11}	Cr 69 Col$_h$ 122 I	[242]
6.6	C_6H_{13}	Cr 68 Col$_h$ 97 I	[242]
		Cr 67.6 Col$_h$ 100.1 I	
6.7	C_7H_{15}	Cr 68.6 Col$_h$ 93 I	[242]
6.8	C_8H_{17}	Cr 66.8 Col$_h$ 85.6 I	[242]
6.9	C_9H_{19}	Cr 57 Col$_h$ 77.6 I	[242]
6.10	$C_{10}H_{21}$	Cr 58 Col$_h$ 69 I	[242]
6.11	$C_{11}H_{23}$	Cr 54 Col$_h$ 66 I	[286]
6.12	$C_{13}H_{27}$	(49 Col$_h$) I	[242]
6.13	$(CH_2)_2CH(CH_3)_2$	Cr 130.3 I	[287]
6.14	$(CH_2)_3CH(CH_3)_2$	Cr 91.8 Col$_h$ 122.1 I	[287]
6.15	$CH_2CH(CH_2CH_3)_2$	Cr 29.5 Col$_h$ 120.0 I	[287]
6.16	$(CH_2)_2CH(CH_3)(CH_2)_3CH(CH_3)_2$	Cr 19 Col$_h$ 36 I	[294]
6.17	$[CH_2CH(CH_3)]_3CH_2CH_3$	Cr −36 I	[295]
6.18	$[CH_2CH(CH_3)]_4CH_2CH_3$	Cr −37 I	[295]
6.19	$(CH_2)_3CF_3$	Cr 132 Col$_h$ 171 I	[299]
6.20	$(CH_2)_3C_2F_5$	Cr 126 Col$_h$ 150 I	[299]
6.21	$(CH_2)_3C_3F_7$	Cr 130 Col$_h$ 149 I	[299]
6.22	$(CH_2)_3C_4F_9$	Cr 116 Col$_h$ 157 I	[299]
6.23	$(CH_2)_3C_6F_{13}$	Cr 89 Col$_h$ 183 I	[299]
6.24	$CH_2COOC_2H_4C_6F_{13}$	Cr 30 Col$_h$ 210 dec	[296]
6.25	$CH_2COOC_2H_4C_8F_{17}$	Cr 47 Col$_h$ 217 dec	[296]
6.26	$CH_2COOC_2H_4C_{10}F_{21}$	Cr 115 Col 215 dec	[296]
6.27	$CH_2COOC_2H_5$	Cr 122 Col$_h$ 172 I	[296]
6.28	$CH_2CONHC_4H_9$	Col	[300]
6.29	$CH_2CONHC_{12}H_{25}$	Col	[300]
6.30	$(CH_2)_3CH=CH_2$	Cr 78.3 Col$_h$ 97.1 I	[305]
6.31	$(CH_2)_8CH=CH_2$	Cr 55.9 Col 100.4 Col 198.7 I	[268]
6.32	$CH_2CH_2OCH_3$	Lyotropic nematic phases	[307]
6.33	$(CH_2CH_2O)_2CH_3$	Lyotropic nematic phases	[307]
6.34	$(CH_2CH_2O)_3CH_3$	Lyotropic nematic phases	[307]
6.35	$(CH_2)_5Si(CH_3)_2OSi(CH_3)_3$	Cr −19 Col$_h$ 52.6 I	[305]
6.36		g 74 Col$_h$ 127 I	[306]
6.37		g 51 Col$_r$ 165 I	[306]
6.38		g 130	[306]
6.39		g 104 Col$_h$ 225 I	[306]
6.40		Cr 70 Cr 98	[303]

TABLE 2.37 (continued)
Thermal Behavior of Symmetrical Hexasubstituted Triphenylene Ethers Having Normal or Branched Peripheral Alkyl Chains (Scheme 2.33)

Structure	R	Phase Transition	Ref.
6.41		Cr 89 I	[302]
6.42		g 8 M 47 Col 111 I	[310]
6.43		—	[304]
6.44		Cr 81.1 Col 104.6 I	[331]
6.45	$n = 5$	g 50 I	[335]
6.46	$n = 6$	g 49 N 109 I	[335]
6.47	$n = 7$	g 33 N 95 I	[335]
6.48	$n = 8$	g 30 N 116 I	[335]
6.49	$n = 9$	g 31 N 115 I	[335]
6.50	$n = 10$	g 32 N 123 I	[335]
6.51	$n = 12$	g 38 N 130 I	[333]

While hexaalkoxy-TP **6.16** having chiral 2,7-dimethyloctyl chains exhibits a hexagonal columnar phase to very low temperature [294], triphenylene derivatives with peripheral alkyl chains having three methyl (trimethyloctane), **6.17**, and four methyl (tetramethyldecane), **6.18**, branches are as such not liquid crystalline. However, their mixtures with hexakis(octyloxy)triphenylene or hexakis(decyloxy)triphenylene exhibit columnar mesophases [295]. An equimolar mixture of compound **6.17** with 2,3,6,7,10,11-hexafluoro-TP shows highly stabilized hexagonal columnar phase due to arene–perfluoroarene interactions [294].

Recently, a number of symmetrical hexaalkoxy-TPs having peripheral chains other than normal or branched alkoxy chains (Scheme 2.33) have been realized [296–305]. Most of these materials were prepared by alkylating hexahydroxy-TP with functionalized alkyl bromides. The concept of mesophase stabilization by the introduction of semifluorinated alkyl chains, well studied in calamitic systems, has been extended to triphenylene-based DLCs [296–299]. The nature of the mesophase structure does not usually change with the introduction of fluorinated chains, but the thermal behavior is strongly affected. In the case of butyloxy chains, **6.19**, the hexagonal plastic phase was converted to a disordered columnar phase upon terminal fluorination [298]. It has been reported recently that triphenylene discotics with such terminal fluoroalkylated chains show a strong tendency toward homeotropic alignment of the columnar phase [299]. The symmetrical triphenylenes **6.24–6.26** having terminal fluorinated chains and ester groups in the peripheral chains also exhibit stabilized hexagonal columnar phase but all the derivatives decompose at the clearing temperature [296]. The triphenylene derivative **6.30** having long spacer with terminal double bond exhibits highly stabilized columnar mesophases [268], but similar architecture with a short spacer displays a narrow columnar phase [305]. The terminal double bonds can be converted to siloxane derivative **6.35**, which exhibits a broad room temperature columnar mesophase [305]. Such materials are also interesting to create discotic networks.

Compounds **6.28** and **6.29** prepared from hexahydroxy-TP and 2-bromo-N-alkylacetamides were found to be efficient gelators of some hydrocarbon solvents. The formation of the hexagonal columnar phase and rectangular columnar phase in the cyclohexane gels of **6.28** and **6.29**, respectively, was observed by x-ray diffraction studies. Such organogelators have recently been reported to be useful as photo- and electrochemical materials [300,301].

Carbazole derivatives are well known for their photoconducting properties. The incorporation of the carbazole moiety in the supramolecular structure of triphenylene discotics may enhance their physical properties. To investigate this possibility, a number of triphenylene derivatives with pendent carbazole groups **6.41** were prepared but none of them show mesomorphic properties in the pure state. However, upon doping with the electron-acceptor trinitrofluorenone (TNF), the induction of a columnar phase due to charge-transfer (CT) complex formation was observed [302]. Similarly, triphenylene derivatives having other bulky units such as anthraquinone, ferrocene, etc., also do not self-assemble to form mesophases [303,304].

Sugar-coated triphenylene-based discotic liquid crystals have been reported by the Stoddort group [306]. In compounds **6.36–6.39** bulky carbohydrate units are attached to the ends of the alkyl chains. These amphiphilic materials, having a central hydrophobic part and peripheral hydrophilic units, exhibit both thermotropic and lyotropic mesophase behavior. On the other hand, the attachment of ethyleneoxy chains around the triphenylene periphery generates amphiphilic discotics **6.32–6.34** showing lyotropic phases [307–309].

The incorporation of ionic imidazolium units into the alkyl side-chain termini of a triphenylene derivative resulted in a waxy solid **6.42** that exhibits a hexagonal columnar phase at 111°C on cooling from the isotropic phase [310]. It goes to a mesophase transition at 47°C and vitrifies at about 8°C. On heating, the glassy phase changes to an unidentified phase at 10°C, which transforms into a columnar phase at 70°C and finally clears at 114°C. The ionic compound **6.42** can be easily prepared by N-alkylation of 1-methylimidazole with hexakis(10-bromodecyloxy)triphenylene followed by anion exchange from Br^- to BF_4^- using $AgBF_4$. It may be noted that the corresponding non-ionic imidazole derivative does not show liquid crystalline properties. Thus, the mesomorphism is induced by the ionic self-assembly.

Attaching calamitic liquid crystals to a discotic core is an interesting approach to obtain biaxial phases. The search for the elusive biaxial nematic phase [311,312], where the unique axes of the molecules are arranged not only in a common direction (known as the director) but also there is a correlation of the molecules in a direction perpendicular to the director, is one of the most active areas of research in liquid crystal science in recent years. The hunt for this new liquid crystal phase began more than 30 years ago, when it was recognized that the molecules forming liquid crystals deviate from their presumed cylindrical shape [313]. In fact, the molecules are more lozenge like, and it is because of this lowering of the molecular symmetry that two nematic phases should be possible. Indeed, the first claimed discovery of a biaxial nematic [314] was for a compound formed of spoon-like molecules, and similar claims soon followed for cross-shaped [315] and bone-shaped [90] molecules. Till now, there have been many claims [316,317] for its discovery in low molar mass thermotropic liquid crystals, although experimental difficulties in unambiguously identifying the symmetry of these phases raise questions concerning these assignments. Recently, a biaxial nematic phase was reported in bent-core thermotropic mesogens [318].

Theoretical studies and mean field calculations [319,320] have shown that the biaxial nematic phase (N_b) is obtained by changing a shape biaxiality parameter (η) between a rod at one extreme ($\eta = 0$) and a disk ($\eta = 1$) at the other. The N_b phase exists over ranges such as $0.2 \leq \eta \leq 0.8$, but is most stable at $\eta = 0.4$. Such a structure is then properly intermediate between a rod and a disk, and this led to proposals that the N_b phase might be realized in rod–disk mixtures. In such a mixture, the optimum packing arrangement has the long axes of the rods arranged perpendicularly to the short axes of the disks, and hence the system has two directors. This idea was investigated in theoretical approaches to the biaxial nematic phase formed from binary mixtures of rod- and disk-like molecules [321]. The situation with respect to physical mixtures is not so straightforward because a mixture of rods and disks should [322], and indeed does [22,323,324], separate into two uniaxial nematic phases, one rich in rods and the other rich in disks. The inter-mixing of triphenylene hexabenzoates and calamitic molecule 4-(*trans*-4-*n*-propylcyclohexyl)phenyl-4-*n*-propylbenzoate [323] or phenyl biphenyl carboxylate [324] have been studied. The systems were found to be immiscible. However, theoretical work by Sharma et al. [325] and by Vanakaras et al. [326] has shown that rod–disk mixtures can lead to N_b phases if the rod and disk are attracted more to one another than to each other.

One way to overcome this problem is to attach rod-like and disk-like units covalently via flexible alkyl spacers, so that they cannot phase separate. Recently, several efforts have been made in this direction [96–99,103,106,207,327–335]. Some of the triphenylene-based discotic-calamitic systems are discussed here. The triphenylene derivative **6.44** (Scheme 2.33) possesses six polymerizable azobenzene-based calamitic units linked to the triphenylene core via ethereal linkage [331]. Surprisingly, a normal hexagonal columnar phase was reported in this material (Table 2.37). As the mesophase characterization is based only on textural studies, detailed x-ray investigations may be required to confirm this observation. Jeong et al. attached six alkoxycyanobiphenyls to the triphenylene core via ether linkages [333]. Compound **6.51** exhibits a nematic to the isotropic phase transition at 130°C on heating, while cooling the nematic phase goes to a glassy state at about 38°C. However, the glass transition temperature depends on the rate of cooling and other exothermic and endothermic peaks appear on slow cooling and heating rates. A biaxial molecular arrangement of rod–disk molecule under an electric field was reported [334]. Imrie and coworkers extended this series by preparing another six lower homologues **6.45–6.50** [335]. All the members of the series exhibit nematic phases. While the lowest homologue **6.45** with pentamethylene spacer exhibits a monotropic nematic phase, all other higher homologues exhibit enantiotropic nematic phase.

2.6.2 SYMMETRICAL TRIPHENYLENE HEXAESTERS

Hexaalkanoates of triphenylene are prepared from hexahydroxy-TP (Scheme 2.34). As mentioned earlier, the hexahydroxy-TP can be prepared by demethylating hexamethoxy-TP (Scheme 2.33). The esterification of the hexahydroxy-TP with normal acid chlorides, substituted benzoyl chlorides, or alkylcyclohexylcarbonyl chlorides under classical reaction conditions furnished the triphenylene

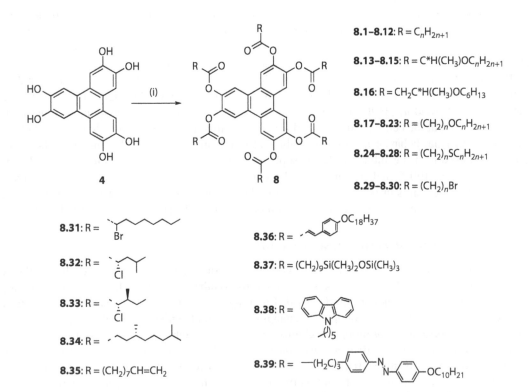

SCHEME 2.34 Synthesis of triphenylene hexaalkanoates: (i) RCOCl, pyridine.

esters, benzoates, and cyclohexanoates, respectively. These derivatives can also be prepared by condensing hexahydroxy-TP with various acids in the presence of DCC and DMAP.

The mesophase behavior of the discotic triphenylene hexaesters **8** [27,242,285,286,302,332, 336–344] is summarized in Table 2.38. The first four members of simple straight-chain series are not liquid crystalline; other members form a rectangular mesophase and the higher homologues also form a hexagonal columnar phase. The introduction of a methyl branch in the alkyl chain, as expected, reduces the melting and clearing temperatures. The effect of heteroatom addition in the peripheral chains depends on its nature and position. The incorporation of oxygen or sulfur atom into the side chains generally decreases the melting points and improves the mesophase range. The introduction of bulky bromine atom in the side chains or at the terminal position of side chains imparts divesting effect on mesomorphism and none of these compounds show liquid crystalline properties. However, α-chloro-substituted compounds exhibit columnar phases, albeit monotropic in nature.

Similar to hexaalkoxy-TP, a few triphenylene hexaester with terminally functionalized groups have also been prepared. The presence of large terminal group, such as carbazole group, destroys mesomorphism. The discotic–calamitic hybrid molecule **8.39** is particularly interesting as it shows kinetically controlled monotropic bimesomorphism, with a metastable smectic and a stable hexagonal columnar phase [332].

Among the hexaesters of triphenylene, hexabenzoates (Figure 2.4) have received much attention as they display a nematic phase [285,345–361]. However, most of the triphenylene benzoates, **9**, show the nematic phase at very high temperatures (Table 2.39), thus limiting their use. In order to reduce their transition temperatures, lateral substitution in the inner and outer sides of the peripheral phenyl groups was systematically investigated. Lateral substitution, in general, suppresses columnar phases, but the stability of the nematic phase remains relatively unaffected (Table 2.39). While lateral alkyl group substitution does not change the nature of the mesophase, lateral polar group substitution significantly suppresses the nematic phase. One of the chiral alkoxy chain-substituted hexabenzoates is reported to exhibit a discotic blue phase.

A number of hexanaphthoates, **10** (Figure 2.5), were also found to display the discotic nematic phase at elevated temperatures in addition to a columnar rectangular phase [362]. The mesophase behavior of these materials is summarized in Table 2.40. In comparison to triphenylene hexabenzoates, these materials exhibit much broader nematic mesophase range.

Beattie et al. prepared a series of hexa-*n*-alkylcyclohexanoates of triphenylene, **11** (Figure 2.5), by the esterification of hexahydroxy-TP with a variety of 4-*n*-alkylcyclohexylcarbonyl chlorides in the presence of pyridine [349]. These compounds are related to the above-mentioned triphenylene hexabenzoates, but the benzene ring was reduced to a saturated *trans*-1,4-disubstituted cyclohexane ring. Their thermal behavior is shown in Table 2.40. In comparison with the analogous benzoate esters, cyclohexanoates exhibit a wider mesophase range. The clearing temperatures of cyclohexanoates are about 100°C higher than those of the analogous *n*-alkyl esters and about 50°C higher than the benzoate esters of triphenylene. In an earlier report, compound **11.4** was reported to be liquid crystalline at room temperature and decomposes before clearing at about 345°C [285].

2.6.3 Unsymmetrical Triphenylene Derivatives

Whereas the synthesis of symmetrically substituted hexakis(alkoxy)triphenylenes by oxidative trimerization of 1,2-dialkoxybenzenes is quite easy, the synthesis of well-defined unsymmetrically substituted derivatives is not straightforward. Breaking of symmetry in a hexaalkoxy-TP can be achieved in different ways, for example, by changing the chain length of some of the side chains, by changing the nature of one or more side chains, or by low and high degrees of substitution.

To examine the effect of dissymmetry on the mesophase, Tinh et al. prepared several dissymmetrical hexa-substituted triphenylenes by putting different alkyl chains in the periphery and found that an introduction of non-symmetric side chains does not affect the nature of Col$_h$ phase but results in the reduction of the mesophase stability [363]. Traditionally, these materials

TABLE 2.38
Thermal Behavior of Symmetrical Hexasubstituted
Triphenylene Esters (Scheme 2.34)

Structure	R	Phase Transition	Ref.
8.1	C_2H_5	Cr 296 I	[242]
8.2	C_3H	Cr 230 I	[242]
8.3	C_4H_9	Cr 193 I	[242]
8.4	C_5H_{11}	Cr 146 I	[242]
8.5	C_6H_{13}	Cr 108 Col$_r$ 120 I	[242]
8.6	C_7H_{15}	Cr 66 Col 126 I	[242]
		Cr 64 Col$_r$ 130 I	[286]
8.7	C_8H_{17}	Cr 62 Col$_r$ 125 I	[242]
8.8	C_9H_{19}	Cr 75 Col$_t$ 125 I	[342]
		Cr 75 Col$_r$ 125.5 I	[242]
8.9	$C_{10}H_{21}$	Cr 67 Col$_r$ 108 Col$_h$ 121.5 I	[242]
8.10	$C_{11}H_{23}$	Cr 79 M$_1$ 109 M$_2$ 120 I	[242]
		Cr 80 Col 93 Col$_r$ 111 Col$_h$ 122 I	[286]
8.11	$C_{12}H_{25}$	Cr 80 M$_1$ 92 M$_2$ 103 I	[242]
		Cr 83 Col$_r$ 99 Col$_h$ 118 I	[286]
8.12	$C_{13}H_{27}$	M$_1$ 55.5 M$_2$ 106 I	[242]
		Cr 87 Col$_r$ 96 Col$_h$ 111 I	[286]
8.13	$C^*H(CH_3)OC_4H_9$	Cr? Col 93 I	[340]
8.14	$C^*H(CH_3)OC_6H_{13}$	Cr 84 Col$_h$ 124 I	[340]
8.15	$CH(CH_3)OC_7H_{15}$	Cr 49 Col 90 I	[342]
8.16	$CH_2C^*H(CH_3)C_6H_{13}$	Cr 59 Col$_r$ 64.5 Col$_h$ 97.5 I	[337]
8.17	CH_2OCH_3	Col	[338]
8.18	$CH_2OC_3H_7$	Col	[338]
8.19	$CH_2OC_5H_{11}$	Col	[338]
8.20	$CH_2OC_7H_{15}$	Cr −14 Col 227 I	[338]
8.21	$CH_2OC_9H_{19}$	Col	[338]
8.22	$CH_2OC_{11}H_{23}$	Col	[338]
8.23	$(CH_2)_2OC_4H_9$	Cr 94 I	[27]
8.24	$CH_2SC_5H_{11}$	Cr 75 Col 138 I	[27]
8.25	$(CH_2)_2SC_3H_7$	Cr 86 Col 198 I	[27]
8.26	$(CH_2)_2SC_4H_9$	Cr 57 Col 202 I	[27]
8.27	$(CH_2)_3SC_3H_7$	Cr 98 I	[27]
8.28	$(CH_2)_2SC_5H_{11}$	Col$_r$ 203 I	[27]
8.29	$(CH_2)_5Br$	Cr 151 I	[339]
8.30	$(CH_2)_7Br$	Cr 78 I	[339]
8.31	$CH(Br)C_7H_{15}$	Cr 217 I	[339]
8.32	$CH(Cl)CH_2CH(CH_3)_2$	Cr (267 Col) 268 I	[343]
8.33	$CH(Cl)CH(CH_3)C_2H_5$	Cr (244 Col) 264.1 I	[343]
8.34	$(CH_2)_2CH(CH_3)$ $(CH_2)_2CH(CH_3)_2$	Col 112.1 I	[343]
8.35	$(CH_2)_7CH=CH_2$	Cr 55 Col$_r$ 88 I	[344]
8.36		Cr 50 Col$_h$ 83 N$_D$ 206 I	[341]
8.37	$(CH_2)_9Si(CH_3)_2OSi(CH_3)_3$	Col$_h$ 125 I	[344]
8.38		Cr 290 dec	[302]
8.39		Cr (200 Col) 207.5 I	[332]

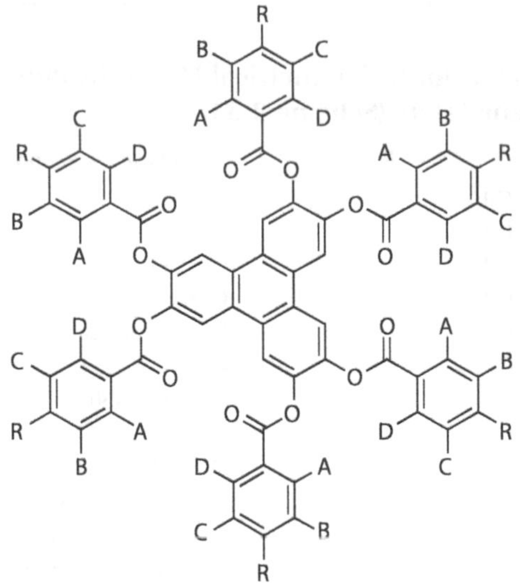

9

FIGURE 2.4 A general structural template for triphenylene hexabenzoates.

have been prepared by statistical approach involving the oxidative trimerization of a mixture of two different 1,2-dialkoxybenzene derivatives (Scheme 2.35). The reaction yields a mixture of products and often the mixture is extremely difficult to separate. The partial alkylation of hexaacetoxy-TP (Scheme 2.36) can yield a mixture of hexaalkoxy-TP, monoacetoxy-pentaalkoxy-TP, diacetoxy-tetraalkoxy-TP (different isomers), and other acetoxy-alkoxy-TP derivatives, which can be used to prepare unsymmetrical triphenylene derivatives. The amount of these products in the mixture depends on the reaction conditions, and some of the products can be isolated by column chromatography [364,365]. Currently, these routes are rarely used to prepare unsymmetrical derivatives. A well-defined but lengthy synthesis of unsymmetrically substituted TP derivatives was described by Wenz in 1985 (Scheme 2.37). The methodology involves the synthesis of a terphenyl having two different types of peripheral substitutions that can be cyclized chemically or photochemically [366].

The simplest and most straightforward route to prepare unsymmetrical hexaalkoxy-TPs is the oxidative coupling of a tetraalkoxybiphenyl with dialkoxybenzene derivatives. The incorporation of tetramethoxybiphenyl with veratrole to prepare hexamethoxy-TP was reported in 1965 by Musgrave [262]. The recent development of this rational synthesis (Scheme 2.38) allowed producing these unsymmetrical derivatives in large amounts with high purity. A variety of unsymmetrical triphenylene discotics have recently been prepared following this process [282–284,287,367–375].

Unsymmetrically substituted TP derivatives have been obtained from the oxidation of a mixture of 1,2-dialkoxybenzene and 3,3′,4,4′-tetraalkoxybiphenyl using $FeCl_3$, $MoCl_5$, or $VOCl_3$ oxidants (Scheme 2.38). Cross coupling reaction leading to unsymmetrical TPs dominates over the oxidative dimerization of tetraalkoxybiphenyl or trimerization of dialkoxybenzene. In view of this efficient cross coupling, it has been suggested [262,266] that tetraalkoxybiphenyl is probably the intermediate product during the oxidative trimerization of 1,2-dialkoxybenzene to hexaalkoxy-TP. However, any tetraalkoxybiphenyl in the oxidative trimerization of 1,2-dialkoxybenzene could not be isolated. The exact mechanism of the Scholl reaction is still not clear. The reaction is easy to perform and have been applied to a number of substrates. However, it should be noted that the reaction is not regiospecific. Thus, if the phenyl and biphenyl are not C_2 symmetric, a mixture of regioisomers will form. Moreover, the reaction works well only with activated (alkoxy-substituted) phenyl and

TABLE 2.39
Thermal Behavior of Triphenylene Hexabenzoates (Figure 2.4)

Structure	R	A	B	C	D	Phase Transition	Ref.
9.1	C_7H_{15}	H	H	H	H	Cr 210 I	[346]
						Cr 130 Col 210 I	[345]
9.2	C_8H_{17}	H	H	H	H	Cr 208 N 210 I	[346]
						Cr 179 N 192 I	[345]
						Cr 125 Col 222 I	[285]
9.3	C_9H_{19}	H	H	H	H	Cr 175 Col$_r$ 183 N 192 I	[346]
9.4	$C_{10}H_{21}$	H	H	H	H	Cr 185 Col$_r$ 189 I	[346]
9.5	OC_4H_9	H	H	H	H	Cr 257 N >300 I	[346]
9.6	OC_5H_{11}	H	H	H	H	Cr 224 N 298 I	[346]
9.7	OC_6H_{13}	H	H	H	H	Cr 186 Col$_t$ 193 N 274 I	[346]
9.8	OC_7H_{15}	H	H	H	H	Cr 168 N 253 I	[346]
						Cr 168 N 251.5 I	[323]
9.9	OC_8H_{17}	H	H	H	H	Cr 152 Col$_t$ 168 N 244 I	[346]
9.10	OC_9H_{19}	H	H	H	H	Cr 154 Col$_t$ 183 N 227 I	[346]
						Cr 180 N 222 I	[345]
9.11	$OC_{10}H_{21}$	H	H	H	H	Cr 142 Col$_t$ 191 N 212 I	[346]
						Cr 155 N 177 I	[345]
9.12	$OC_{11}H_{23}$	H	H	H	H	Cr 145 Col$_t$ 179 N 185 I	[346]
9.13	$OC_{12}H_{25}$	H	H	H	H	Cr 146 Col$_t$ 174 I	[346]
9.14	$O(CH_2)_3CH(CH_3)$ C_2H_5	H	H	H	H	Cr 192.5 N 246.5 I	[347]
9.15	$O(CH_2CH_2O)_2CH_3$	H	H	H	H	Cr 129.9 N 223.5 I	[359]
9.16	$O(CH_2CH_2O)_3CH_3$	H	H	H	H	Cr 25 N 147.4 I	[359]
9.17	$O(CH_2CH_2O)_4CH_3$	H	H	H	H	N 76 I	[359]
9.18	OC_8H_{17}	F	H	H	H	Cr 150 N 213 I	[361]
9.19	OC_8H_{17}	H	F	H	H	Cr 196 Col$_t$ 210 Col$_{r'}$ 237 Col$_{r''}$ 400 < Dec	[361]
9.20	OC_8H_{17}	F	H	H	F	Cr 116 M 176 I	[361]
9.21	OC_8H_{17}	H	F	F	H	Cr 140 Col$_h$ 400 < Dec	[361]
9.22	OC_6H_{13}	F	F	F	F	Cr 157 Col$_h$ 301 I	[360]
9.23	OC_8H_{17}	F	F	F	F	< 25 M 300 I	[285]
						Cr 133 Col$_h$ 308 I	[361]
9.24	$OC_{10}H_{21}$	F	F	F	F	Cr 109 Col$_h$ 302 I	[360]
9.25	$OC_{10}H_{21}$	CH_3	H	H	H	Cr 109 N 164 I	[351]
						Cr 107 N 162 I	[352]
9.26	$OC_{10}H_{21}$	C_2H_5	H	H	H	Cr 117 N 131 I	[352]
9.27	$OC_{10}H_{21}$	$CH(CH_3)_2$	H	H	H	Cr (70 N) 93 I	[352]
9.28	$OC_{10}H_{21}$	H	H	CH_3	H	? 129 N 170 I	[351]
						Cr 102 N 192 I	[352]
9.29	$OC_{10}H_{21}$	H	H	C_2H_5	H	Cr 129 N 206 I	[352]
9.30	$OC_{10}H_{21}$	H	H	$CH(CH_3)_2$	H	Cr 161 N 202 I	[352]
9.31	$OC_{10}H_{21}$	H	H	$C(CH_3)_3$	H	Cr 194 N 225 I	[352]
9.32	$OC_{10}H_{21}$	H	H	NO_2	H	Col >300 I	[354]
9.33	$OC_{10}H_{21}$	H	H	OCH_3	H	Col 265 I	[354]
9.34	$OC_{10}H_{21}$	H	H	$OC_{10}H_{21}$	H	Cr 79 I	[354]
9.35	$OC^*H(CH_3)C_6H_{13}$	H	H	H	H	Cr 161 Col 213 N* 216 BP 220 I	[354]
9.36	$OC^*H(CH_3)C_6H_{13}$	H	H	OCH_3	H	Cr 120 Col 237 I	[354]

(continued)

TABLE 2.39 (continued)
Thermal Behavior of Triphenylene Hexabenzoates (Figure 2.4)

Structure	R	A	B	C	D	Phase Transition	Ref.
9.37	OC*H(CH$_3$)C$_6$H$_{13}$	H	H	Cl	H	Col >300 I	[354]
9.38	OC$_6$H$_{13}$	H	CH$_3$	CH$_3$	H	Cr 150 Col$_h$ 210 N 243 I	[350]
9.39	OC$_8$H$_{17}$	H	CH$_3$	CH$_3$	H	Cr 170 Col$_h$ 195 N 215 I	[350]
9.40	OC$_{10}$H$_{21}$	H	CH$_3$	CH$_3$	H	Cr 157 Col$_h$ 167 N 182 I	[350]
9.41	OC$_{12}$H$_{25}$	H	CH$_3$	CH$_3$	H	Col$_h$ 143 N 151 I	[350]
9.42	OCH(CH$_3$) C$_6$H$_{13}$	H	CH$_3$	CH$_3$	H	Cr 125 Col$_h$ 156 N 183 I	[350]
9.43	OC$_6$H$_{13}$	CH$_3$	H	H	CH$_3$	Cr 170 N 196 I	[350]
9.44	OC$_8$H$_{17}$	CH$_3$	H	H	CH$_3$	Cr 155 N 170 I	[350]
9.45	OC$_{10}$H$_{21}$	CH$_3$	H	H	CH$_3$	Cr 108 N 134 I	[350]
9.46	OC$_{12}$H$_{25}$	CH$_3$	H	H	CH$_3$	Cr 88 N 99 I	[350]
9.47	OCH(CH$_3$) C$_6$H$_{13}$	CH$_3$	H	H	CH$_3$	Cr 161 I	[350]

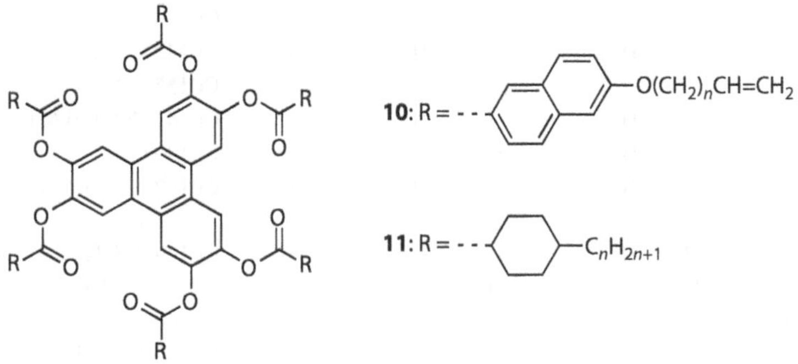

10: R = - - [naphthalene]—O(CH$_2$)$_n$CH=CH$_2$

11: R = - - [cyclohexane]—C$_n$H$_{2n+1}$

FIGURE 2.5　Molecular structures of triphenylene hexanaphthoates and cyclohexanoates.

TABLE 2.40
Transition Temperatures of Triphenylene Hexanaphthoates (Figure 2.5)

Structure	n	Phase Transition	Ref.
10.1	2	Col$_r$ 249.9 N 330 I	[362]
10.2	3	Col$_r$ 227.4 N 327.5 I	[362]
10.3	4	Col$_r$ 227.4 N 288.8 I	[362]
10.4	5	Col$_r$ 222 N 286.5 I	[362]
10.5	6	Col$_r$ 198.2 N 273.1 I	[362]
10.6	7	Col$_r$ 174.2 N 235.7 I	[362]
10.7	8	Col$_r$ 174.2 N 224.9 I	[362]
10.8	9	Col$_r$ 155.5 N 217.6 I	[362]
11.1	5	Cr 158 Col$_h$ 300 I	[349]
11.2	6	Cr 146 Col$_h$ 240 I	[349]
11.3	7	Cr 150 Col$_h$ 255 I	[349]
11.4	9	Cr 108 Col$_h$ 245 I	[349]
		Col >345 I	[285]

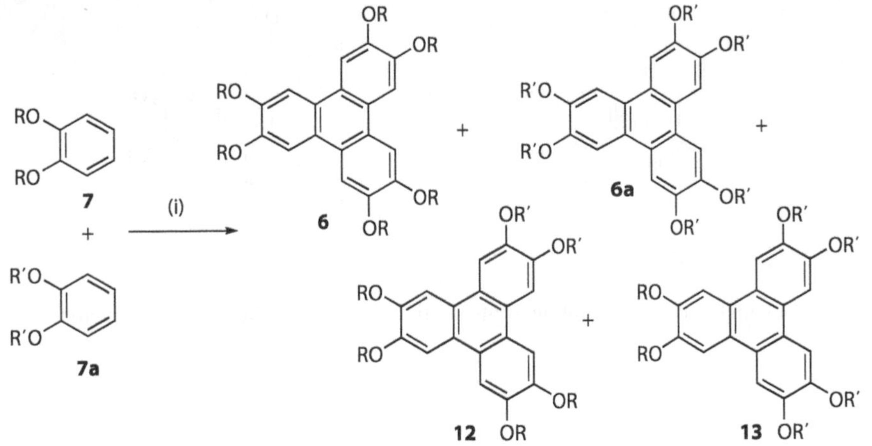

SCHEME 2.35 Statistical oxidative trimerization approach to the synthesis of unsymmetrical triphenylene derivatives: (i) chloranil or FeCl$_3$.

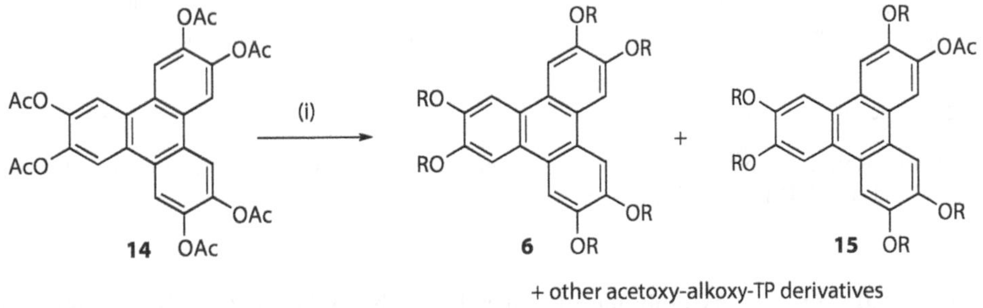

SCHEME 2.36 Synthesis of unsymmetrical triphenylenes via partial alkylation of hexaacetoxy-triphenylene: (i) RBr, K$_2$CO$_3$.

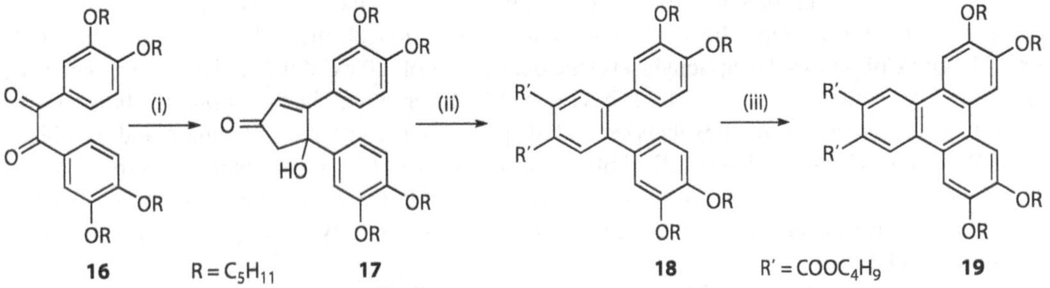

SCHEME 2.37 A rational synthetic route to unsymmetrical triphenylene derivatives: (i) CH$_3$COCH$_3$, EtOH, ^{t}BuOK, reflux; (ii) TsOH, C$_6$H$_5$Cl, C$_4$H$_9$CO$_2$C≡CCO$_2$C$_4$H$_9$; (iii) I$_2$, hv.

biphenyl. The biphenyls required for this reaction can be obtained via the classical Ullman reaction or via Suzuki-type reactions, while the 1,2-dialkoxybenzene can be conveniently prepared from catechol or guaiacol.

Triphenylene derivatives with any degree of substitution with full regiocontrol can be obtained with the help of organometallic chemistry [376–384]. An example is shown in Scheme 2.39. The

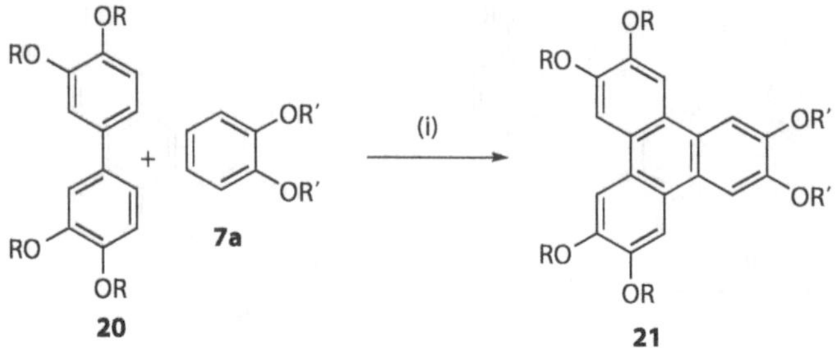

SCHEME 2.38 Phenyl–biphenyl coupling route to the synthesis of hexaalkoxytriphenylenes: (i) $FeCl_3$, CH_2Cl_2.

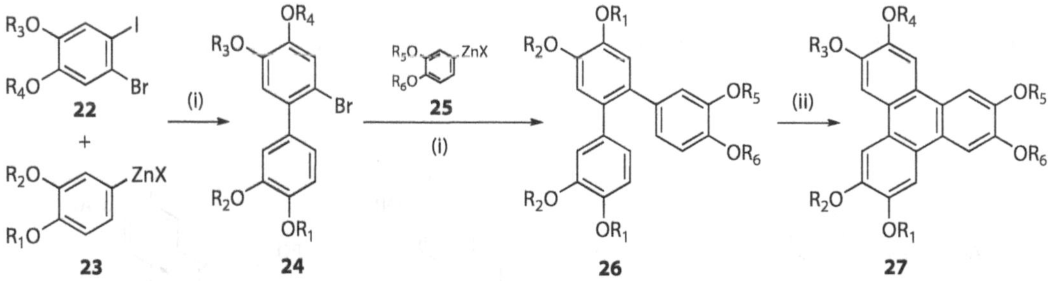

SCHEME 2.39 Rational synthetic route involving organometallic chemistry to unsymmetrical triphenylene derivatives: (i) Pd(0), THF, reflux; (ii) $FeCl_3$, CH_2Cl_2.

strategy essentially involves assembling a terphenyl, which can be oxidatively cyclized to triphenylene. Palladium-catalyzed Suzuki or Heck type of reactions are commonly used to prepare terphenyls.

Using the above-mentioned methodologies, a few unusual triphenylene discotics having three different types of peripheral substituents have been prepared, as shown in Scheme 2.40 [378]. The nature of peripheral substituents around the core has a great influence on the mesophase structure and its stability. For example, hexaesters of TP form Col_r and Col_h mesophases, TP hexabenzoates form N_D and Col phases, hexaphenylacetylene derivatives of TP exhibit N_D phases, symmetrically substituted hexaether derivatives of TP having 5–13 carbon atoms in the aliphatic chains form a single Col_h mesophase, while hexabutyloxy-TP displays a more ordered Col_p phase and hexahexyl-thio-TP exhibits a highly ordered helical phase, in addition to a Col_h phase. Numerous rod-like molecules with a variety of peripheral substitutions are known to exhibit various types of mesophases. However, probably because of synthesis problems, such a rich variety of peripheral substitutions are uncommon in DLCs.

The monobromination of 2,3,6,7-tetrakis(pentyloxy)-TP **30**, prepared by Suzuki, coupling of 2-iodo-3′,4,4′,5-tetrakis(pentyloxy)biphenyl **28** and phenylboronic acid followed by cyclization, yields 10-bromo-2,3,6,7-tetrakis(pentyloxy)-TP **31**. The nucleophilic aromatic displacement of the bromine with the potassium salt of pentanethiol, followed by bromination, yields 2-bromo-6,7,10,11-tetrakis(pentyloxy)-3-(pentylsulfanyl)-TP **33** having a bromo-, thioalkyl-, and alkoxy-substituted periphery of the TP nucleus. The reaction of **33** with copper(I) cyanide gives the cyano-TP derivative **34**, while the palladium-copper catalyzed alkynylation of **33** results in the synthesis of the substituted alkyne derivative **35**. The deprotected alkyne **36** can be converted to dimer **37** (see also Chapter 3). All the hexa-substituted triphenylenes **33–36** form hexagonal columnar phases, while the dimer **37** shows a discotic nematic phase.

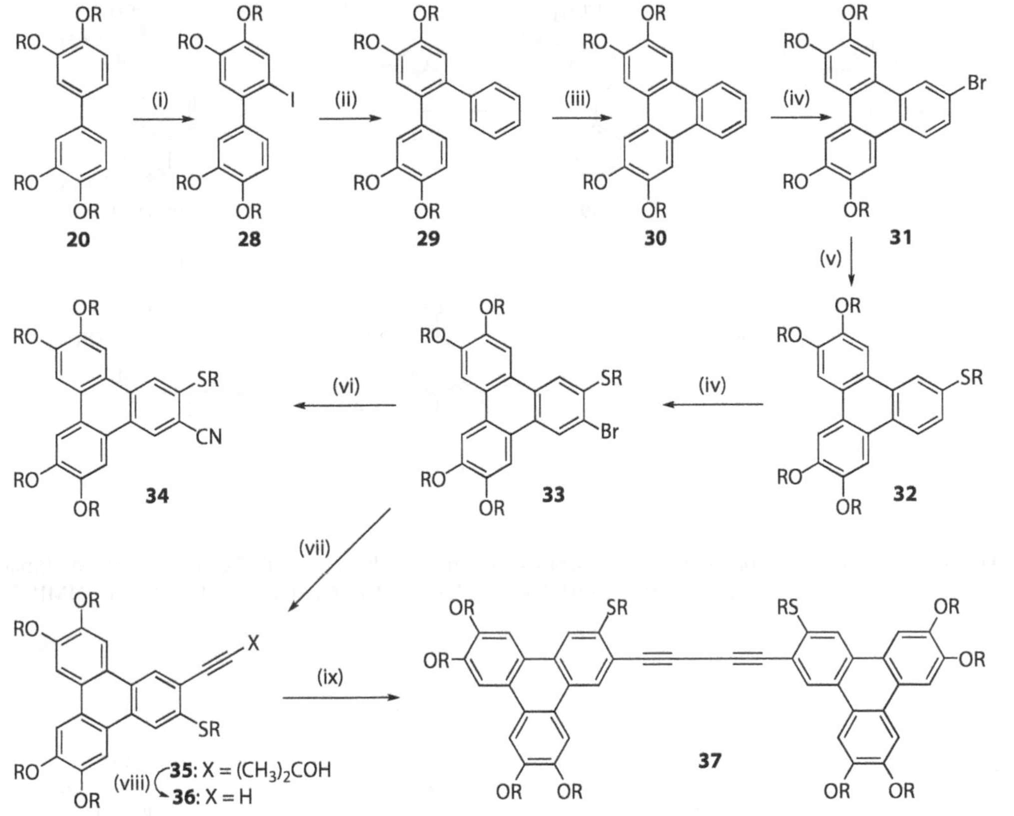

SCHEME 2.40 Rational synthetic route to triphenylene discotics having three different types of peripheral substituents: (i) iodine/iodic acid, CH_2Cl_2-AcOH, reflux; (ii) $C_6H_5B(OH)_2$, $[(C_6H_5)_3P]_4Pd$, Na_2CO_3, THF, reflux; (iii) CH_2Cl_2, $VOCl_3$; (iv) Br_2, CH_2Cl_2; (v) $C_5H_{11}SK$, NMP, $[(C_6H_5)_3P]_4Pd$; (vi) CuCN, NMP, reflux; (vii) 2-methyl-3-butyn-2-ol, CuI, $PdCl_2$ $(PPh_3)_2$, Et_3N, 60°C–70°C; (viii) KOH, toluene, reflux, 98%; (ix) $Cu(Ac)_2$, THF/C_5H_5N/CH_3OH, reflux.

Cammidge and Gopee applied the terphenyl as well as the phenyl–biphenyl coupling routes to prepare a number of alkoxy-thioalkyl discotics [382–384] to investigate the effect of sequential replacement of the oxygen linkers with sulfur linkers. The synthesis of these materials is shown in Schemes 2.41 and 2.42. The biphenyl and terphenyl can be prepared using Suzuki cross coupling. While the intramolecular cyclization of **45** provide the triphenylene **46** (Scheme 2.42), the phenyl–biphenyl coupling of **39** and 1,2-dimethoxybenzene yields the triphenylene **40** (Scheme 2.41). These triphenylenes can be brominated to generate dibromo-tetraalkoxy-TPs **47** and **42**, respectively. The nucleophilic replacement of bromine atoms with thiolates produces hexa-substituted triphenylenes **43** and **48** having two thioalkyl and four alkoxy groups. Similarly, triphenylenes having the thioalkoxy groups at other positions, **49**, **50**, and **51** (Scheme 2.42), can be prepared. The terphenyl route has also been utilized to prepare mixed alkoxy-alkyl-TPs **52** by replacing bromine atoms with alkylzinkiodide (Scheme 2.42).

A variety of unsymmetrical triphenylenes can be derived from various hydroxyl-alkoxy-TPs. The chemistry of these materials is presented in the following sections.

2.6.4 Hydroxy-Alkoxy-TPs

The hydroxy-functionalized triphenylenes are very valuable precursors for the synthesis of dimers, oligomers, polymers, networks, mixed tails, lower and higher degree substituted derivatives. Though

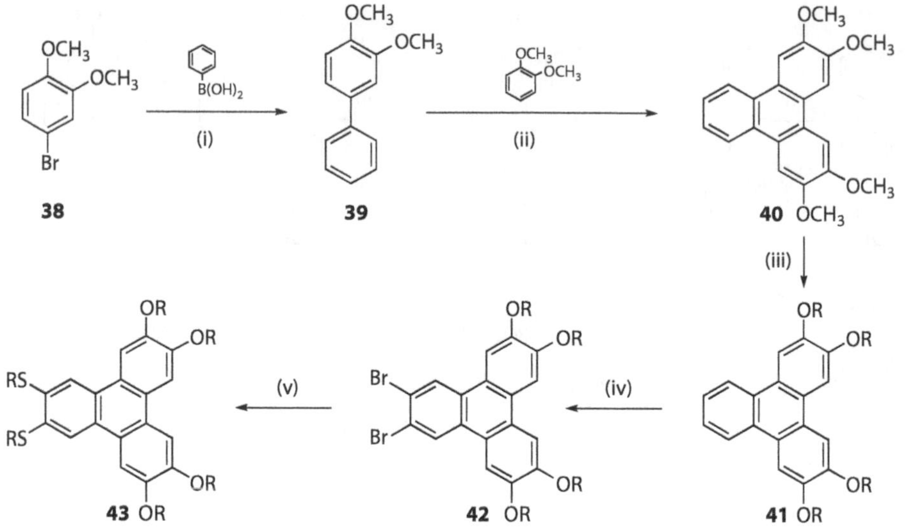

SCHEME 2.41 Synthesis of triphenylene discotics with mixed alkoxy-thioalkyl chains via phenyl-biphenyl coupling route: (i) Pd; (ii) FeCl$_3$; (iii) HBr/AcOH; RBr, K$_2$CO$_3$; (iv) Br$_2$, CH$_2$Cl$_2$; (v) RSH, ^{t}BuOK, NMP, 70°C.

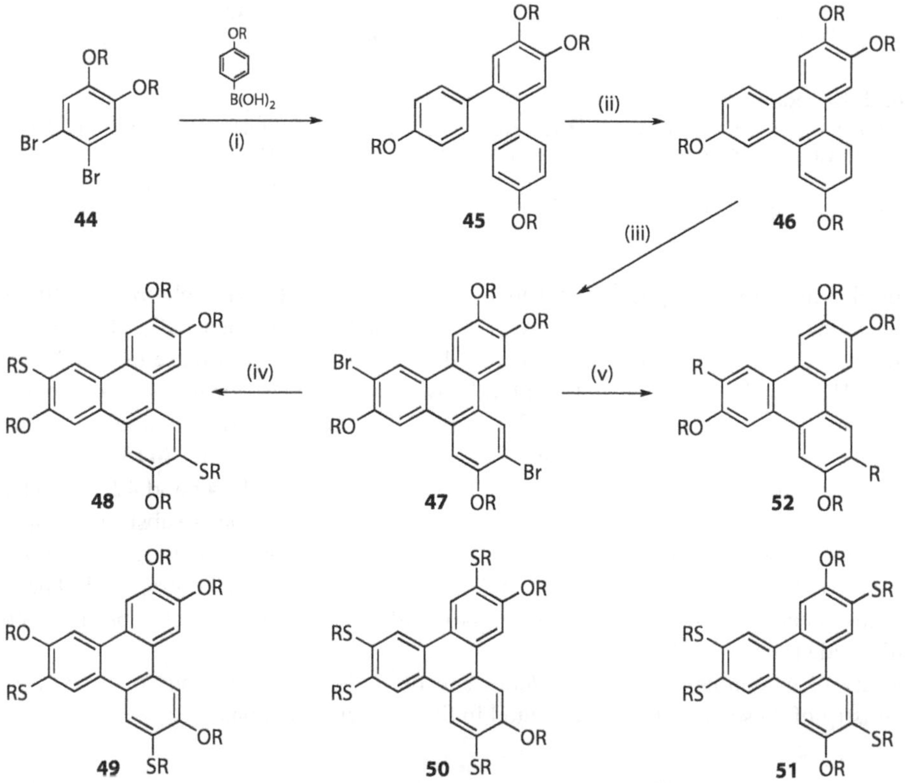

SCHEME 2.42 Synthesis of triphenylene discotics with mixed alkoxy-thioalkyl and alkoxy-alkyl chains via terphenyl route: (i) PdCl$_2$, Na$_2$CO$_3$, PPh$_3$, toluene/EtOH/H$_2$O reflux; (ii) I$_2$, hv; (iii) Br$_2$, CH$_2$Cl$_2$; (iv) RSH, ^{t}BuOK, NMP, 70°C; (v) RZnI, Pd(0), THF, reflux.

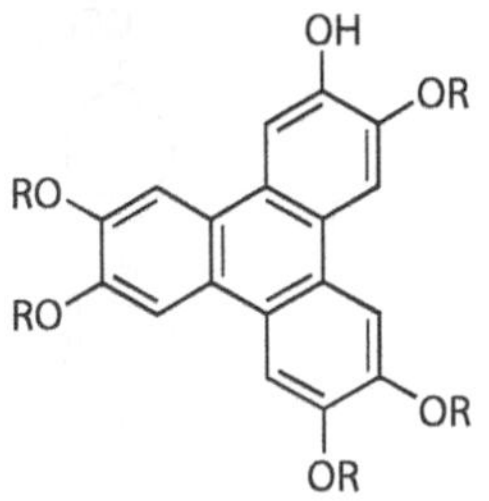

FIGURE 2.6 Molecular structures of different hydroxyl-alkoxy-triphenylenes.

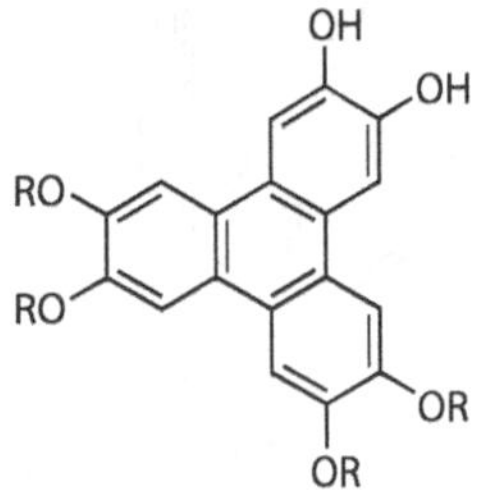

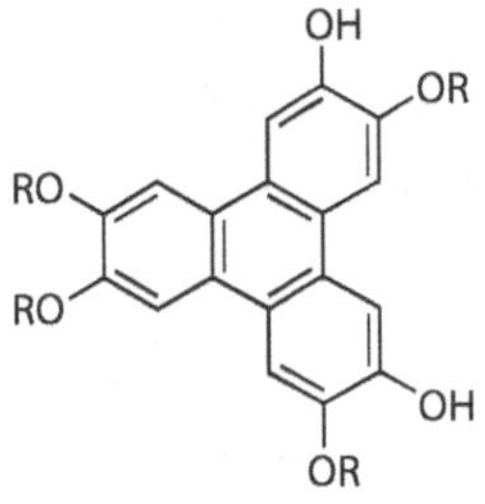

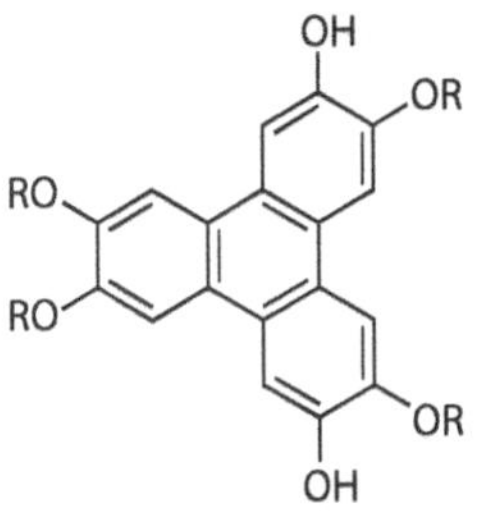

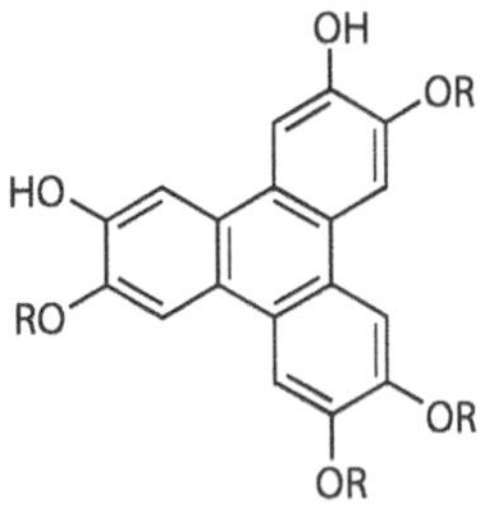

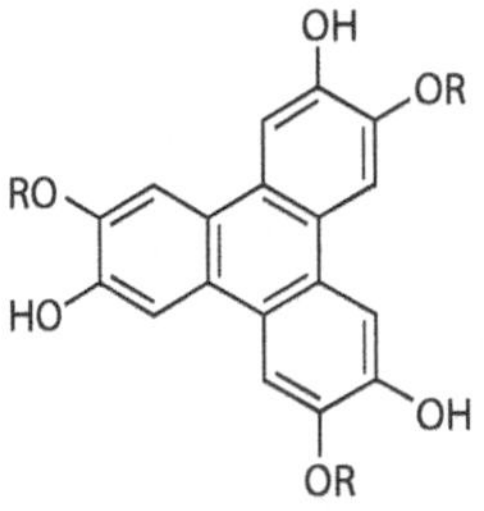

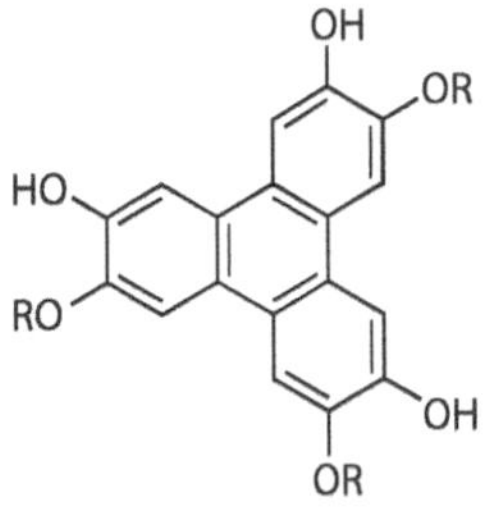

there are several possibilities of hydroxyl-alkoxy-TPs, seven different types of alkoxy-hydroxytriphenylene derivatives have been prepared using selective or non-selective chemical methods. The chemical structures of these derivatives are shown in Figure 2.6. These are monohydroxy-pentaalkoxy-TP, **53**; 2,3-dihydroxy-6,7,10,11-tetraalkoxy-TP, **54**; 2,6-dihydroxy-3,7,10,11-tetraalkoxy-TP, **55**; 2,7-dihydroxy-3,6,10,11-tetraalkoxy-TP, **56**; 2,11-dihydroxy-3,6,7,10-tetraalkoxy-TP, **57**; 2,6,10-trihydroxy-3,7,11-trialkoxy-TP, **58**; and 2,7,10-trihydroxy-3,6,11-trialkoxy-TP, **59**.

The synthesis of mono-functionalized triphenylenes can be achieved in different ways. One of the earliest reported methods [364] of monohydroxy-pentaalkoxy-TP synthesis involves the partial alkylation of hexaacetoxy-TP to monoacetyl-pentaalkoxy-TP in low yield (Scheme 2.36). This can be hydrolyzed to monohydroxy-pentaalkoxy-TP. Some other important methods are shown in Scheme 2.43. A non-selective cleavage of one of the alkoxy groups of hexaalkoxytriphenylene using a calculated amount of 9-Br-BBN gives a mixture of products containing unreacted hexaalkoxytriphenylene (26%), monohydroxytriphenylene (39%), and a minor amount of dihydroxytriphenylenes (10%). The desired product can be purified by column chromatography. An arylmethyl ether can be selectively cleaved in the presence of other arylalkyl ethers, and this strategy was utilized to develop a rational synthesis of various mono- and di hydroxyl-TP. A selective cleavage of the methyl ether of monomethoxypentaalkoxy-TP with lithium diphenylphosphide affords the monohydroxy-TP in high yield [374,385]. The monomethoxy-TP can be prepared by the so-called biphenyl route. The main drawback of this process is the poor yield of the biphenyl in the classical Ullman coupling reaction, and the use of highly sensitive and hazardous lithium diphenylphosphide. However, now better but expensive methods are available to prepare tetraalkoxybiphenyls [386]. Bushby and Lu have demonstrated the use of the isopropoxy masking group in a biphenyl–phenyl oxidative coupling route for the preparation of monohydroxy-TP [387]. The synthesis of monohydroxy-TP by directly coupling tetraalkoxybiphenyl and alkoxyphenol has also been reported. The reaction works well with guaiacol but with other long chain alkoxyphenols, generally, it gives poor and complicated results. This compound can also be obtained as a side product in the oxidative trimerization of dialkoxybenzene with various oxidants in the presence of an excess of H_2SO_4 [282,388]. Though it is a one-step process, the overall yield in the reaction

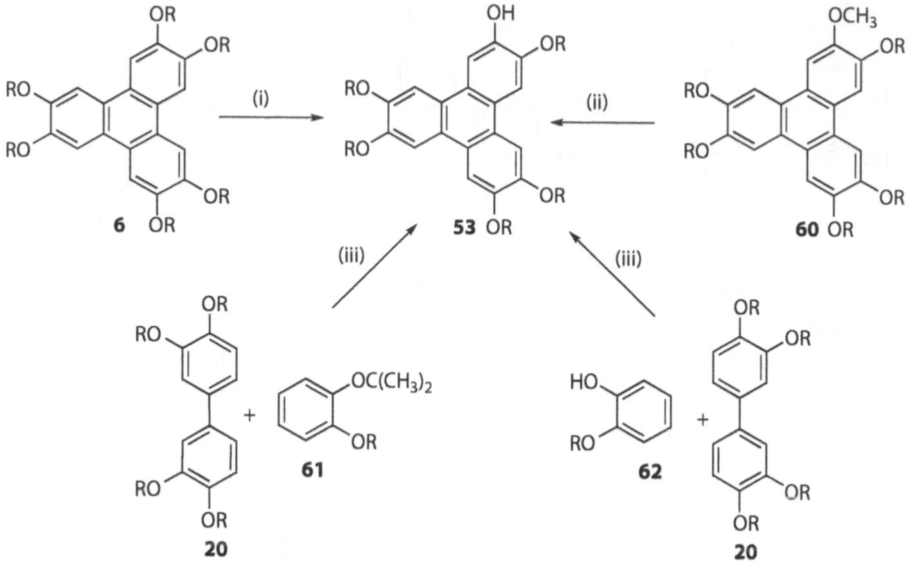

SCHEME 2.43 Synthetic routes to monohydroxy-pentaalkoxytriphenylenes: (i) bromocatecholborane (ii) lithium diphenylphosphide; (iii) FeCl$_3$, CH$_2$Cl$_2$.

does not go above about 20%. Bromocatecholborane, an alternative to 9-Br-BBN, has been found to give almost 70% of mono-functionalized triphenylene [389]. This reaction is highly efficient for the preparation of mono-, di-, and trifunctionalized triphenylene derivatives. An environmentally benign procedure for the preparation of monohydroxy-functionalized triphenylenes using simple ionic reagents such as pyridinium hydrochloride, pyridinium hydrobromide, N-methyl pyridinium iodide, N-ethyl pyridinium bromide, and 1-n-butyl-3-methyl imidazolium bromide have been reported to prepare various monohydroxypentaalkoxy-TPs [390]. This methodology avoids all types of toxic, volatile, and hazardous reagents. Developing green chemistry methodologies is one of the main themes of modern synthetic chemistry. In this context, the use of ionic liquids and microwave heating are powerful tools. Although this method is not the best, yield-wise, it is an economic and green synthesis method for the preparation of monohydroxy-pentaalkoxy-TP derivatives. No side reaction occurs in the reaction and the unreacted starting material can be isolated easily and can be recycled. However, it should be noted that the separation of the mono-hydroxy-pentaalkoxy-TP from unreacted hexaalkoxy-TP is tedious and needs highly efficient column chromatography. The use of aluminum oxide instead of silica gel column chromatography helps the purification.

The two hydroxy groups in a dihydroxy-tetraalkoxy-TP may be present in four different ways. These are at the 2,3-, 2,6-, 2,7-, and 3,6-positions [368,373]. The synthesis of dihydroxy-TP was achieved by statistical methods. The alkylation of hexaacetoxy-TP with alkyl halide provides a mixture of dihydroxy-TP in low yield [364]. Dihydroxytetraalkoxy-TP can be obtained by cleaving two alkoxy chains of a hexaalkoxy-TP using 9-Br-BBN, bromocatecholborane, and other reagents. However, in all cases, a mixture of various dihydroxy-TPs forms. It is generally very difficult to purify any pure product and, therefore, it is advisable to use selective methyl ether cleavage route to prepare these materials. The difunctionalized triphenylene of well-defined structure can be prepared by using the biphenyl route and a selective ether cleavage [373,374]. The multiple-step synthesis involves the preparation of different dimethoxytetraalkoxy-TP and finally the two methoxy groups were selectively cleaved by lithium diphenylphosphide to dihydroxy-TP [368]. As an example, the synthesis of 2,7-dihydroxy-3,6,8,9-tetraalkoxy-TP using this method is shown in Scheme 2.44. Similarly, all other dihydroxy-tetraalkoxy-TPs can be prepared. This rational

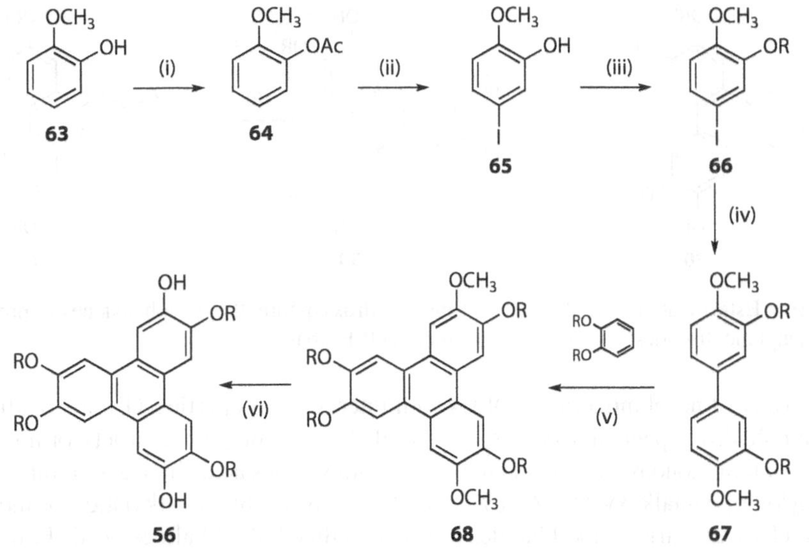

SCHEME 2.44 Synthesis of 2,7-dihydroxy-3,6,8,9-tetraalkoxy-TP: (i) CH₃COCl; (ii) I₂, HIO₃; (iii) RBr, base; (iv) Cu; (v) FeCl₃; (v) lithium diphenylphosphide.

synthesis is extremely important for the preparation of authentic samples, but the overall poor yield due to multiple steps, the use of highly sensitive, hazardous, and costly reagents makes this method uneconomical on a large scale.

Two isomeric trihydroxy-trialkoxy-TPs exist: the symmetrical 2,6,10-trihydroxy-3,7,11-trialkoxy-TP **58** and the non-symmetrical 2,7,10-trihydroxy-3,6,11-trialkoxy-TP **59** (Figure 2.6). Both symmetric **58** and unsymmetric **59** trihydroxy-trialkoxy-TPs are relatively easy to prepare and purify (Scheme 2.45). Thus, treatment of hexapentyloxy-TP (H5TP) with 3.6 equivalents of Cat-B-Br in dichloromethane at room temperature for 36 h gives exclusively two products: the symmetrical 2,6,10-trihydroxy-3,7,11- tris(pentyloxy)-TP **58** (61%) and the non-symmetrical 2,7,10-trihydroxy-3,6,11-tris(pentyloxy)-TP **59** (38%). The reagent, 9-Br-BBN, also gives similar result but it is more hazardous, sensitive, and expensive. None of the hydroxyl-alkoxy-TP is mesomorphic in nature.

2.6.5 DISCOTICS DERIVED FROM HYDROXY-ALKOXY-TPS

The hydroxy-alkoxy-TPs are valuable precursors for the synthesis of a variety of DLCs [208,209,213,391–460]. The monohydroxy-pentaalkoxy-TPs are extremely important not only to prepare discotic dimers (Chapter 3), oligomers (Chapter 4), and polymers (Chapter 5), but also to

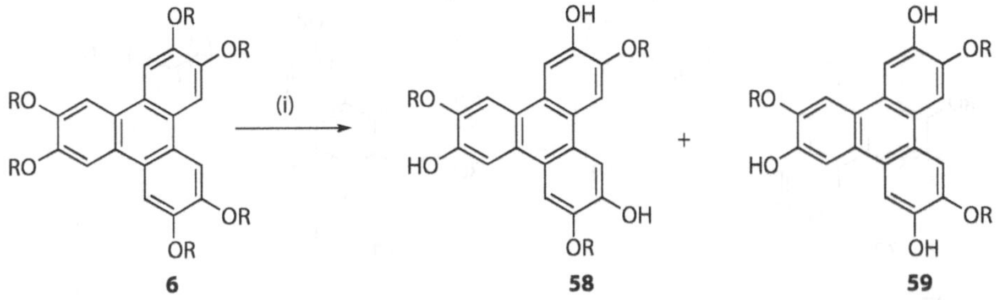

SCHEME 2.45 Synthesis of trihydroxy-trialkoxy-TPs: (i) bromocatecholborane.

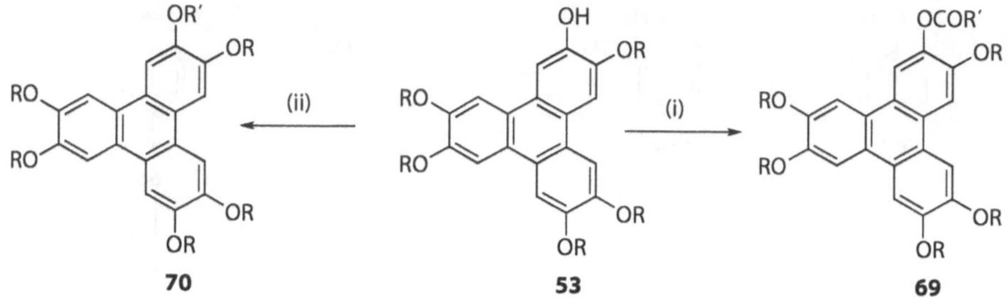

SCHEME 2.46 Esterification and alkylation of monohydroxy-pentaalkoxytriphenylenes to prepare unsymmetrical triphenylene discotics: (i) R'COCl, pyridine; (ii) R'Br, base.

generate a variety of novel monomeric DLCs with interesting properties. Given a readily available supply of monohydroxy-pentaalkoxy-TPs, it is not difficult to prepare a variety of novel discotics. The esterification of monohydroxy-pentaalkoxy-TP with various acid chlorides or anhydrides yields monoalkanoyloxy-pentaalkoxy-TPs, **69** (Scheme 2.46). Similarly, the alkylation of monohydroxy-pentaalkoxy-TPs with various alkyl halides or functionalized alkyl halides yields hexaalkoxy-TPs, **70**, having one alkyl chain different than other five (Scheme 2.46). Recently, microwaves dielectric heating has been applied to prepare unsymmetrical room temperature branched chain triphenylene discotics from monohydroxy-pentaalkoxy-TP [444]. It may be noted that such unsymmetrical TP derivatives can also be prepared via biphenyl–phenyl coupling route. However, in some cases it is difficult to apply oxidative coupling and hence the alkylation of monohydroxy-pentaalkoxy-TP is preferred. Similar to hexa-substituted-TPs, a variety of monofunctionalized-TPs with different terminal group, such as carbazole, anthraquinone, tetrathiafulvalene, ferrocene, etc., can be prepared from monohydroxy-pentaalkoxy-TP (Scheme 2.46).

The presence of a terminal bulky group generally destroys the mesomorphism. However, TP-tethered imidazolium salts **72** that were synthesized either by the quaternization of 1-methylimidazole with an ω-bromo-substituted TP **71** or by the quaternization of a TP-substituted imidazole **73** with methyl iodide (Scheme 2.47) [435] show columnar mesophase properties over a wide

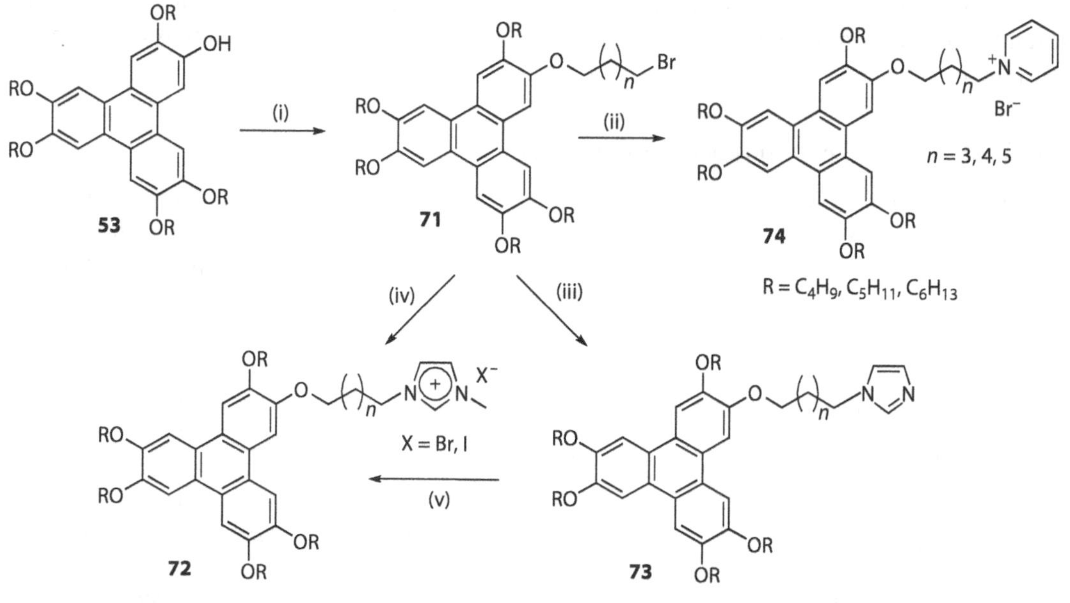

SCHEME 2.47 Synthesis of imidazolium and pyridinium-based DLCs: (i) Br(CH$_2$)$_n$Br, Cs$_2$CO$_3$; (ii) Py/toluene/reflux; (iii) imidazole/THF/NaH/reflux; (iv) 1-methylimidazole, toluene, reflux; (v) CH$_3$I.

temperature range. The iodide salt exhibited slightly lower isotropic transition temperature than the bromide salt. This could be due to the bulky nature of the counterion. Interestingly, the non-ionic imidazole-substituted TP derivative 73 was found to be non-liquid crystalline. Therefore, the mesomorphism in ionic salts is induced due to ionic self-assembly. Similarly, pyridinium salts 74 were obtained by reacting the bromo-substituted compounds 71 with pyridine (Scheme 2.47). This way, several members with different peripheral chains and spacer were prepared and characterized. Increasing the number of carbon atoms on the peripheral chains of the TP core stabilized the columnar phase, while increasing the spacer length connecting the TP unit with the pyridine moiety destabilized the mesophase [436].

In a novel approach, a series of functionalized triphenylene DLCs was synthesized by reductive removal of hydroxy group(s) followed by electrophilic aromatic substitution at the exposed β-site (Scheme 2.48) [393,397,398]. Thus a variety of functional groups can be introduced in the triphenylene nucleus. Such unsymmetrical and low degree-substituted triphenylenes can also be prepared using highly efficient but sophisticated organometallic chemistry. The reduction of the phenolic group of hydroxyalkoxy-TP was accomplished via activation with a tetrazole unit followed by catalytic hydrogenation to lead pentaalkoxy-TP, 75. Interestingly and somewhat surprisingly, the classical nitration of 75 occurs preferentially in the sterically hindered 1-position to yield 76 indicating

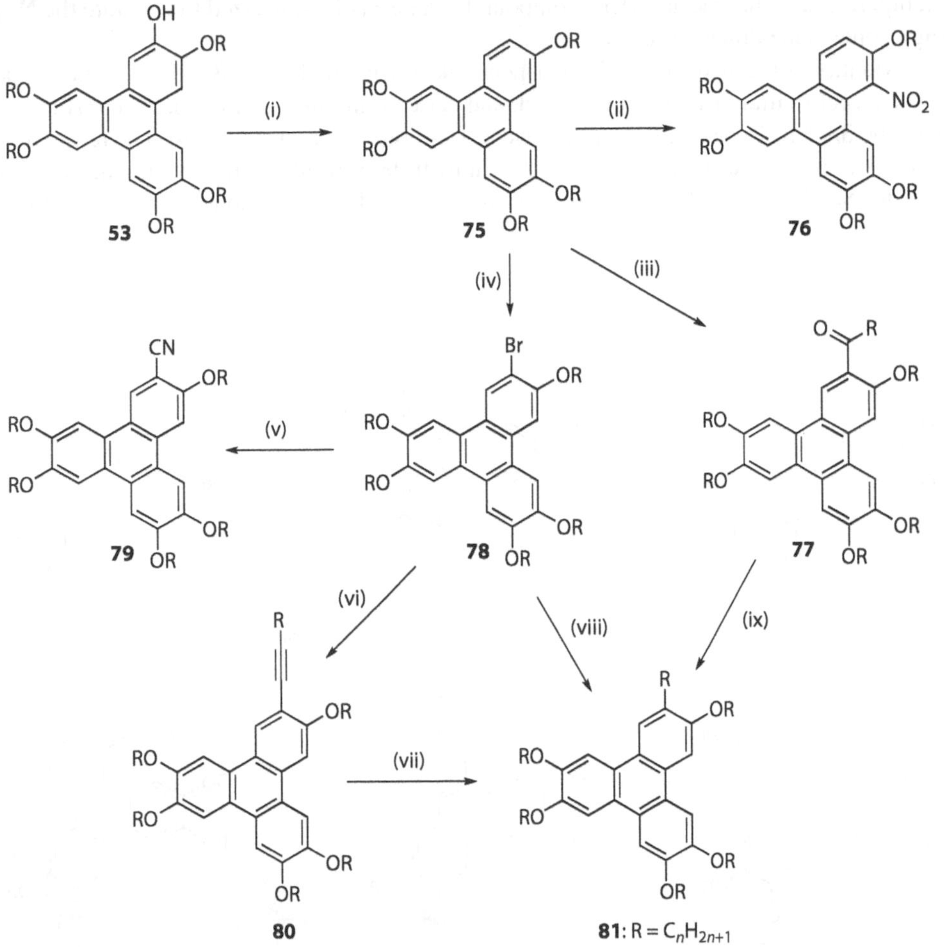

SCHEME 2.48 Discotics derived from mono-functionalized-TP: (i) 5-chloro-1-phenyl-tetrazole/K$_2$CO$_3$; Pd/C, EtOH/H$_2$; (ii) HNO$_3$; (iii) AlCl$_3$/RCOCl; (iv) Br$_2$/CH$_2$Cl$_2$; (v) CuCN/NMP; (vi) RC≡CH/Pd/Cu/PPh$_3$/ Et$_3$N; (vii) H$_2$/Pd/C; (viii) RZnCl, Pd(0); (ix) NH$_2$NH$_2$, KOH.

that electronic effects dominate in the nitration of alkoxytriphenylenes. Friedel–Crafts acylation of **75** with acetyl chloride and aluminum trichloride proceeds in high yield with acylation occurring in the 2-position (compound **77**). The bromination of the non-liquid crystalline pentaalkoxy-TP gives the liquid crystalline monobromo derivative **78**. It is interesting to note that the precursor molecules monomethoxy-TP, monohydroxy-TP, and pentaalkoxy-TP are not liquid crystalline. Dipolar interactions and/or extension of the core due to the bulky bromine atom may be the driving force for the induction of a mesophase. Compound **78** can also be prepared by a direct coupling of a tetraalkoxybiphenyl with 1-bromo-2-alkoxybenzene [288]. Several derivatives, such as cyano and alkynes, can be prepared by replacing the bromine atom. The reaction of **78** with copper cyanide gives the cyano-TP **79** in excellent yield. Palladium/copper-catalyzed alkynylation of **78** can be accomplished in excellent-to-quantitative yields with a variety of substituted alkynes, as demonstrated with compounds **80** (R = TMS; (CH$_3$)$_2$COH; C$_4$H$_9$). In addition, TPs **80a** (R = TMS) and **80b** (R = CH$_3$)$_2$COH) can be deprotected to yield free phenylacetylenes for further elaboration or potential polymerization. Compound **80c** (R = C$_4$H$_9$) on hydrogenation with catalytic palladium yields the alkyl/alkoxy-TP derivative **81**. This material can also be prepared directly from **78** by reacting with alkylzinc halide or from **77** by reducing the keto group. These LCs show a significantly broader range of mesogenicity relative to the parent compound 2,3,6,7,10,11-hexakis(pentyloxy)-TP. Moreover, some of the mesogens exhibit a more ordered mesophase relative to the hexagonal columnar phase at lower temperatures. The free acetylene compound can be easily dimerized to generate the N$_D$ phase forming triphenylene dimers (see Chapter 3).

The oxidation of 2-hydroxy-3,6,7,10,11-pentaalkoxytriphenylenes **53** with various oxidizing agents such as chromium trioxide, nitric acid, and ceric ammonium nitrate yields the ring-oxidized products, the 3,6,7,10,11-pentaalkoxytriphenylene-1,2-diones **82** (Scheme 2.49). The reductive acetylation of these o-quinones with zinc and acetic anhydride in triethylamine results in the formation of diacetates **83** (R′ = CH$_3$). The reductive esterification of the o-diquinone with various long chain

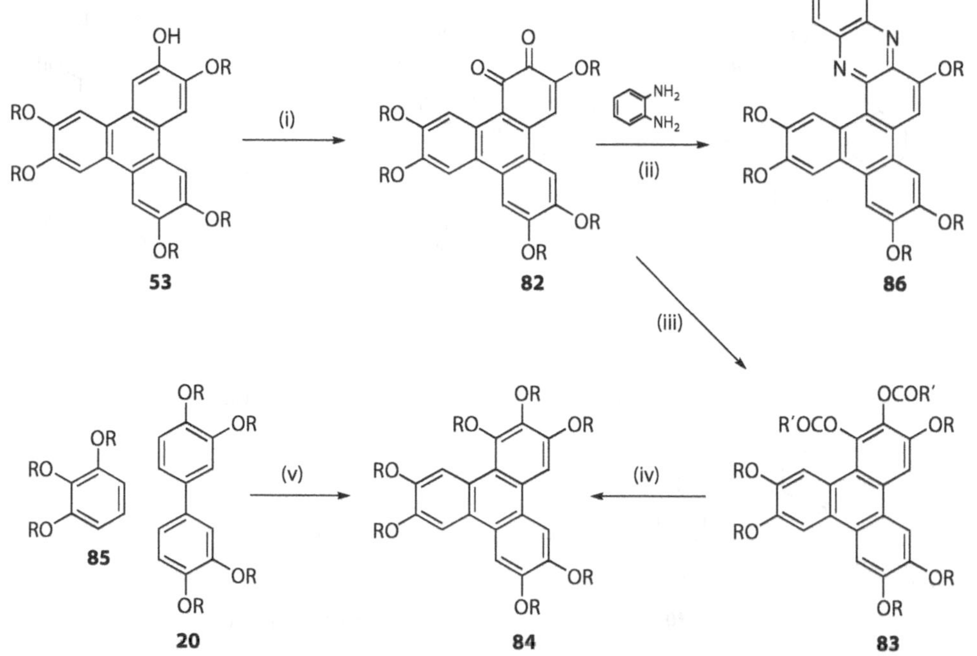

SCHEME 2.49 Synthesis of heptaalkoxy-TP and phenanthro[a]phenazine derivatives: (i) CAN, CH$_3$CN; (ii) C$_6$H$_6$-AcOH; (iii) Zn, RCOCl, reflux; (iv) DMSO, KOH, RBr; (v) FeCl$_3$, CH$_2$Cl$_2$.

acid chlorides yields discotic liquid crystalline 1,2-dialkanoyloxy-3,6,7,10,11-pentaalkoxy-TPs **83**. The diacetate **83** (R′=CH$_3$) can be directly alkylated with various alkyl halides to produce symmetrical (having all seven chains identical) or unsymmetrical (two peripheral chains are different) heptaalkoxy-TPs **84** in very high yield [407]. Hepta-substituted triphenylenes can also be prepared by a biphenyl–phenyl oxidative coupling route (Scheme 2.49). Thus, when tetraalkoxybiphenyls **20** were reacted with trialkoxybenzenes **85** under oxidative coupling conditions using either MoCl$_5$, VOCl$_3$, or FeCl$_3$, heptaalkoxy-TPs were formed in moderate yields. Similarly, unsymmetrical heptaalkoxy-TPs can also be prepared using tetraalkoxybiphenyl and trialkoxybenzene derivatives having different chains. It should be noted that the unsymmetrical heptaalkoxy-TPs prepared from these two routes are different. The o-quinone route gives an unsymmetrical heptaalkoxy-TP where two out of seven chains are different, while the biphenyl route gives a product in which three out of seven chains are different. Moreover, mixed esters–ethers **83** can be achieved only via the o-quinone route as the phenyl–biphenyl coupling usually works only with highly electron-rich precursors. The o-quinone can be easily coupled with o-phenylenediamines to generate phenazines. The reaction of o-quinone with an equimolar amount of 1,2-phenylenediamine in benzene-acetic acid at room temperature for 2 h furnished orange-red 2,3,6,7-tetrapentyloxy-phenanthro[a]phenazine **86** (Scheme 2.49) [410].

The monohydroxy-pentaalkoxy-TP can be used to couple triphenylene molecules with nanomaterials [455–460,209]. In the past two decades, the field of nanoscience and nanotechnology has made enormous growth because of its potential industrial applications. The properties of the materials change drastically when their dimensions scale down to nanometer length scale. For instance, the conductivity of bulk gold is very high (4.3×10^7 Ω^{-1} m^{-1}), while gold nanoparticles (GNPs) of 1–2 nm are only semiconducting (1.4×10^{-3} Ω^{-1} m^{-1}). The incorporation of nanoparticles in the supramolecular order of DLCs would provide materials that possess the properties of the nanoparticles as well as the processing, handling, and self-assembling properties of LCs and, therefore, is likely to lead to novel materials for many device applications.

GNPs are the most stable metal nanoparticles and they present fascinating aspects, such as their self-assembly, the behavior of individual particles, size-related electronic, magnetic, and optical properties, and their applications as a catalysis and in the bottom-up approach of nanotechnology. Monolayer-protected GNPs can be handled as simple organic materials and a variety of chemical reactions can be performed on functionalized GNPs. By providing metal nanoclusters with liquid crystalline properties, one could introduce the self-assembling ability and the reversible control of morphology of their aggregates by external stimuli, as well as the control of their properties by a simple and solvent-free method. The reversible control of the properties of metal nanocluster aggregates will cultivate their potential as functional nanomaterials.

The terminally thiol-functionalized triphenylene, namely, 6-{[3,6,7,10,11-pentakis(hexyloxy)-2-triphenylenyl]oxy}-1-hexanethiol, **87**, can be easily prepared from monohydroxy-TP [422]. Triphenylene-stabilized gold clusters **88** were prepared by reducing HAuCl$_4$ with NaBH$_4$ in the presence of terminally thiol-functionalized triphenylene (Scheme 2.50). These discotic-functionalized

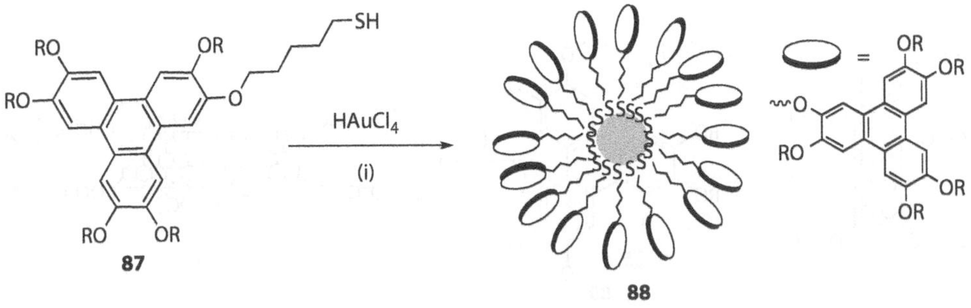

SCHEME 2.50 Synthesis of discotic-decorated gold nanoparticles: (i) NaBH$_4$.

gold nanoparticles were found to be highly soluble in common organic solvents like dichloromethane, chloroform, THF, etc. The virgin TP-GNPs were found to be non-liquid crystalline. However, their doping in the columnar phase forming triphenylene-based DLCs does not disturb the nature of the mesophase. Several binary mixtures of TP-GNPs and hexaalkoxy-TPs were prepared by sonicating the two components in dichloromethane followed by the removal of the solvent and drying in vacuum. All the composites were reported to be liquid crystalline in nature. A gradual decrease in the isotropic temperature upon increasing the amount of GNPs was observed. This is expected because of the insertion of the nanoparticles into the columnar matrix.

Similarly, monohydroxy-pentaalkoxy-TPs can also be coupled with single-wall carbon nanotubes [209,460]. CNTs, the one-dimensional carbon allotropes, are well ordered all-carbon hollow cylinders of graphite with a high aspect ratio. The combination of superlative mechanical, thermal, and electronic properties displayed by CNTs make them ideal for a wide range of applications, such as conductive and high-strength composites, catalyst supports in heterogeneous catalysis, energy-storage and energy-conversion devices, field emitters, transistors, sensors, gas storage media, tips for scanning probe microscopy, and molecular wires. There has been growing interest in the field of dispersion of CNTs in both thermotropic and lyotropic LCs. The insertion (dispersion) of CNTs in the supramolecular order of discotic liquid crystalline monomers and polymers, particularly those having a stable columnar phase at room temperature, may lead to novel materials with interesting properties that are useful for device applications. TP-functionalized single-wall carbon nanotubes (SWNTs) and commercially available octadecylamine (ODA)-functionalized SWNTs have been used to disperse in hexaalkoxy-TP discotics as well as in other monomeric and polymeric DLCs [209,460].

Carboxylic acid groups-terminated SWNTs are commercially available but can also be prepared and purified from "as prepared grade" SWNTs [460]. The reaction of these functionalized SWNTs with thionylchloride results in more reactive acid chloride groups functionalized nanotubes, which can be coupled with terminally hydroxyl-functionalized triphenylene **89** (Scheme 2.51). Like discotic-decorated GNPs, these discotic-functionalized SWNTs **90** are also highly soluble in common organic solvents. The formation of discotic-functionalized SWNTs was confirmed by infrared, ^{1}H NMR, ^{13}C NMR, thermogravimetric analysis (TGA), and scanning tunneling microscopy studies.

The triphenylene-capped SWNTs were found to be nonliquid-crystalline. The waxy solid melts at about 40°C to form the isotropic phase. Composites of functionalized-SWNTs and triphenylene-based discotic liquid crystals were prepared by sonicating the two components in dichloromethane followed by the removal of solvent and drying in a vacuum. Several compositions (by weight) of functionalized nanotubes in DLCs were prepared and analyzed for liquid crystalline properties by differential scanning calorimetry, polarizing optical microscopy, and x-ray diffractometry.

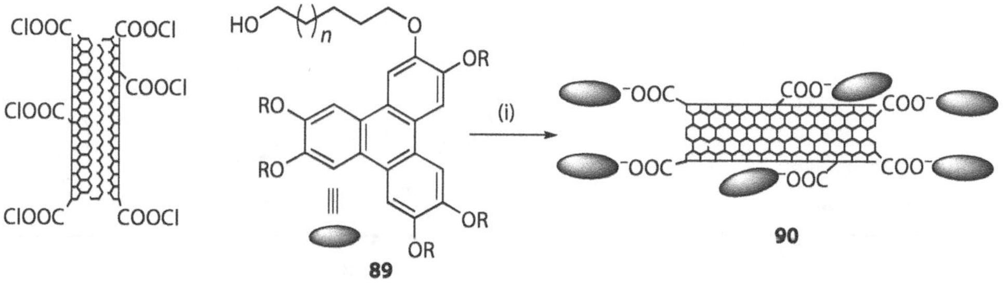

SCHEME 2.51   Synthesis of triphenylene-coupled single-wall carbon nanotubes: (i) pyridine, reflux.

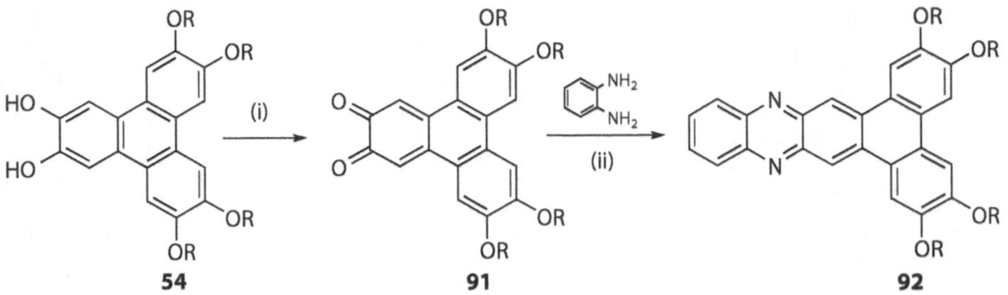

SCHEME 2.52 Synthesis of phenanthro[b]phenazine derivaties: (i) CAN, CH$_3$CN; (ii) C$_6$H$_6$/AcOH, reflux.

The insertion of either discotic-functionalized SWNTs or commercially available ODA-functionalized SWNTs does not affect the mesophase structure of the pure compounds, but brings down the isotropic transition temperatures. With an increase in the amount of CNTs, the isotropic transition temperature decreases in all of the composites. Commercial ODA-functionalized SWNTs can be dispersed in DLCs only in small amounts (~1%), while a large amount (10%) of discotic-functionalized SWNTs can be dispersed in a columnar matrix. X-ray studies indicate that SWNTs align in the hexagonal columnar phase along the director. The room temperature liquid crystalline nanocomposites with broad mesophase ranges and different electronic properties may be important for many device applications.

Similar to discotics derived from monohydroxy-pentaalkoxy-TP, a number of triphenylene derivatives can also be prepared from dihydroxy-tetraalkoxy-TPs. However, dihydroxy-tetraalkoxy-TPs are primarily interesting to prepare discotic trimers, oligomers, and main-chain polymers (Chapters 3 through 5). Because of synthetic difficulties in obtaining pure dihydroxy-tetraalkoxy-TP isomers, much work has not been done on these compounds. However, a few discotic monomers have been derived from these di-functional molecules.

The oxidation of 2,3-dihydroxy-6,7,10,11-tetrapentyloxy-TP **54** yields 6,7,10,11-tetrapentyloxy-TP-2,3-dione, **91** (Scheme 2.52). The reaction of this *o*-quinone with an equimolar amount of 1,2-phenylenediamine yields 2,3,6,7-tetrapentyloxy-phennanthro[b]phenazine **92**, which exhibited a metastable monotropic columnar mesophase [416]. The synthesis of these novel ring structures opens a route to prepare a variety of new materials using various alkyl and alkoxy-substituted 1,2-phenylenediamines. These unsymmetrical, colored, polar, heteroaromatic supramolecular architectures could be potential candidates for various devices, such as photovoltaic solar cells, light-emitting diodes, etc.

Both symmetric **58** and unsymmetric **59** trihydroxy-trialkoxy-TPs are relatively easy to prepare and purify (Section 2.6.2). Following the chemistry developed for the synthesis of TP derivatives derived from monohydroxy-pentaalkoxy-TPs (Scheme 2.44), a number of new materials from symmetric as well as from unsymmetric trihydroxy-trialkoxy-TPs can be prepared. Some of these discotics derived from the symmetric 2,6,10-trihydroxy-3,7,11-tris(alkoxy)-TPs are presented in Scheme 2.53. Similarly, a number of discotics have been derived from the non-symmetrical 2,7,10-trihydroxy-3,6,11-tris(alkoxy)-TPs.

2.6.6 Discotics Derived from Hexaalkoxy-TPs

2.6.6.1 Electrophilic Aromatic Substitution in Hexaalkoxy-TPs

Electrophilic aromatic substitution in unsubstituted triphenylene **1** is directed by steric and electronic effects. Substitution at β-position or 2-position is favored compared with α- or 1-position, presumably owing to a steric hindrance effect [240]. The bromination of triphenylene is reported to yield mainly 2,3,5,6,10,11-hexabromo-TP [461]; however, the electronic effect plays a major role

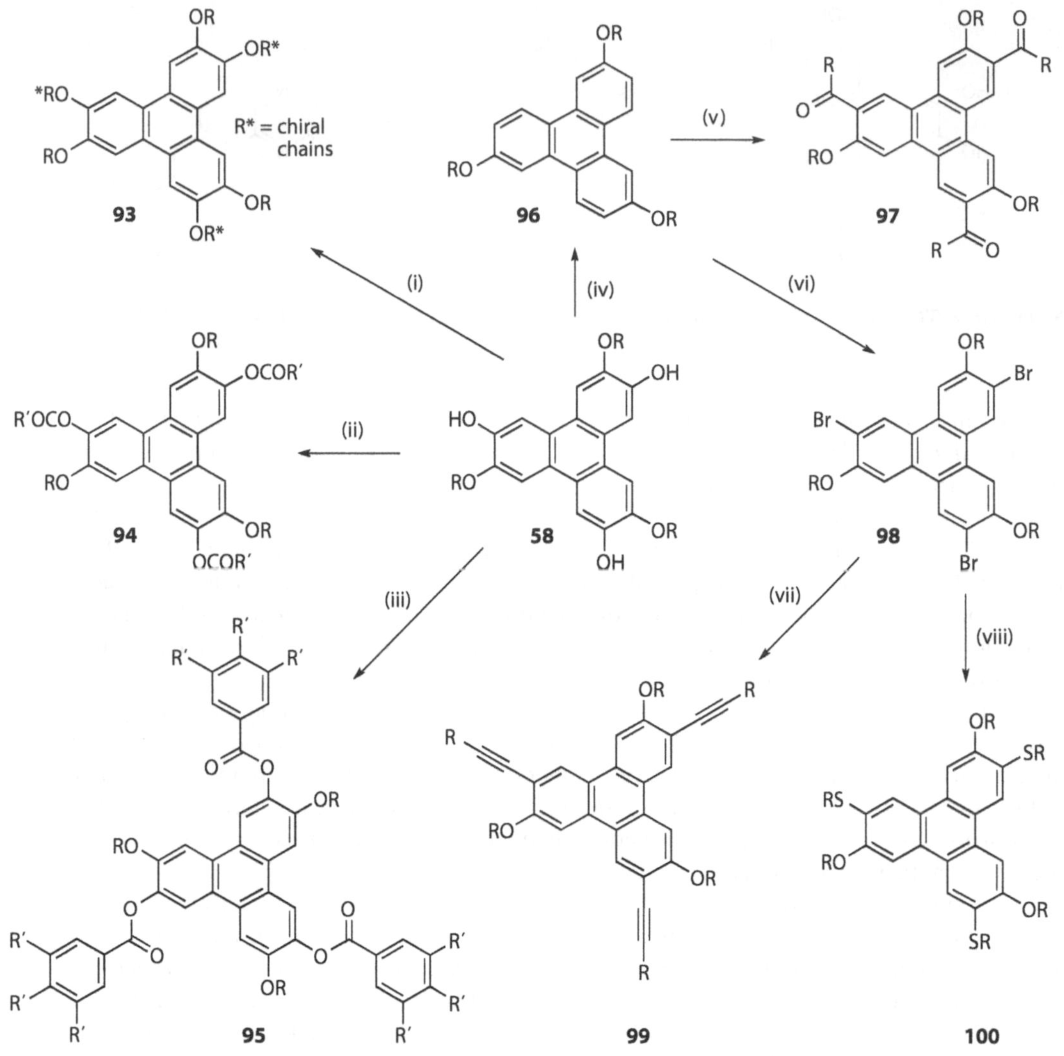

SCHEME 2.53 Discotics derived from trifunctionalized-TP: (i) R*COCl, pyridine; (ii) R'COCl, pyridine; (iii) substituted benzoyl chloride, pyridine; (iv) 5-chloro-1-phenyl-tetrazole/K$_2$CO$_3$; Pd/C, EtOH/H$_2$; (v) AlCl$_3$/ RCOCl; (vi) Br$_2$/CH$_2$Cl$_2$; (vii) RC≡CH/Pd/Cu/PPh$_3$/Et$_3$N; (viii) RSH, ^{t}BuOK, NMP, 70°C.

in the nitration of TP. The nitration of the parent hydrocarbon triphenylene is reported to yield a trinitro derivative, but under controlled conditions a mixture of 1-nitro- and 2-nitro-triphenylenes is formed [240]. In a pentaalkoxy-TP, where both α- and β-positions are available, classical electrophilic aromatic substitution reactions such as Friedel–Crafts acylation and halogenation give only β-substitution, nitration gives primarily the α-nitro product [398]. In hexaalkoxy-TP, only the α-positions, ("bay regions," 1-, 4-, 5-, 8-, 9-, and 12-positions) are free for further substitution. The nitration of 2,3,6,7,10,11-hexaalkoxy-TP provides the α-nitro product in very high yields [462–466]. Solvents play an important role in the nitration of hexaalkoxytriphenylenes. In a mixture of ether–acetic acid, almost exclusively the α-nitro product is formed. Even under exhaustive conditions, in this solvent system, only a small amount of trinitrohexaalkoxy-TP forms. Changing the solvent system from ether–acetic acid to dichloromethane–nitromethane imparts a dramatic effect and all three rings of triphenylene can be successively nitrated under very mild conditions

[465,466]. The trinitration proceeds with high regioselectivity to give exclusively one isomer having C_3 symmetry, that is, 1,5,9-trinitro-2,3,6,7,10,11-hexaalkoxy-TP. Alkoxynitro-TPs are valuable precursors to several other derivatives, such as amino, mono- and di-alkylamino, acylamino and azo derivatives, and thus a number of new triphenylene derivatives (Scheme 2.54) can be prepared [288,427,462–471].

Nitro-TPs **101**, **102**, and **103** were easily reduced to the corresponding amino derivatives **104**, **105**, and **106** with hydrazine and palladium, or with nickel chloride and sodium borohydride (Scheme 2.54). The reduction of mononitro-TP **101** also works well with tin and acetic acid, but this method gives very poor results with di- and tri-nitro derivatives. The amino derivative **104** was readily converted to an α, α-diazo compound **110** by the diazotization of the amine with nitrous

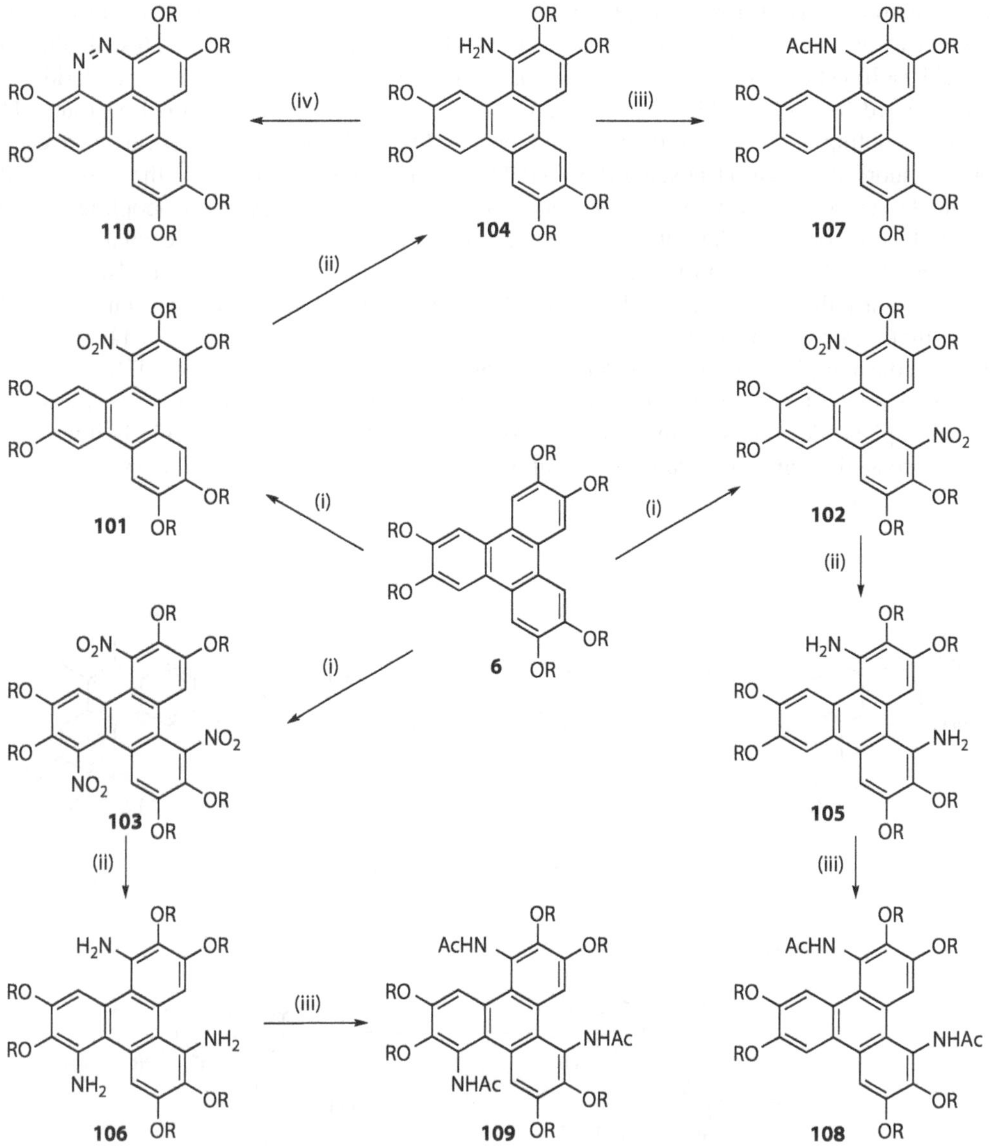

SCHEME 2.54 Synthesis of nitrotriphenylenes and other derivatives: (i) HNO_3/CH_3NO_2; (ii) NH_2NH_2/Pd or $NiCl_2/NaBH_4$; (iii) Ac_2O/Py; (iv) $NaNO_2/AcOH$.

acid followed by cyclization. The acylation of amines **104**, **105**, and **106** with acetic anhydride in pyridine converted them into their *N*-acyl derivatives **107**, **108**, and **109**. Many of these derivatives exhibited columnar mesophases.

Interestingly, the monohydroxypentaalkoxy-TP can also be nitrated at the α-position, but this reaction is complicated and often only the oxidized products, the 3,6,7,10,11-pentaalkoxytriphenylene-1,2-diones, are isolated. Under very careful conditions 1-nitro-2-hydroxy-3,6,7,10,11-pentaalkoxy-TPs can be prepared [466]. These double functionalized TP derivatives are extremely important precursors as the functional group such as nitro, amino, azo, etc., can be utilized to modify the electronic nature of the core and at the same time the hydroxy functional group may be converted to a polymerizable group.

The chlorination of 2,3,6,7,10,11-hexahexyloxy-TP (HAT6) with iodine monochloride gives a mixture of 1-chloro-, 1,8-dichloro-, and 1,5-dichloro-HAT6 (Scheme 2.55) [472,473]. Under exhaustive conditions, it gives 1,4,5,9-tetrachloro-HAT6 [473]. Praefcke et al. have also reported the synthesis of 1-chloro-HAT6 and 1-bromo-HAT6 by the chlorination of HAT6 with aluminum trichloride/sulfuryl chloride and bromination of HAT6 with bromine in carbon tetrachloride [474]. α-Halogenated hexaalkoxy-TPs can also be prepared via phenyl–biphenyl coupling route. Thus, oxidative coupling of 3,3′,4,4′-tetrahexyloxybiphenyl **20** and 1-fluoro-2,3-dihexyloxybenzene **113** affords 1-fluoro-2,3,6,7,10,11-hexahexyloxy-TP **114** in good yield [473]. While the α-bromo atom could not be replaced by alkylthiolates, it can be easily replaced with a cyano nucleophile [370]. The functionalization of the nucleus at the α-position is important not only to induce color, a molecular dipole, enhanced liquid crystalline properties, and a chemically reactive site but also to helically deform the normally planar triphenylene core [288,370,427,462–475]. This has recently been verified by single crystal x-ray analysis of 2,3,5,6,10,11-hexaethoxy-1-nitrotriphenylene [473].

Despite the non-planarity of the nucleus, many of the α-substituted hexaalkoxy-TPs exhibit, quite surprisingly, much larger mesophase ranges. Within an α-substituted halogen series, the stability of the mesophase decreases on increasing the size of the halogen atom. Particularly interesting is the α-fluorinated compound. Fluoro-substitution in the core of calamitic liquid crystals generally

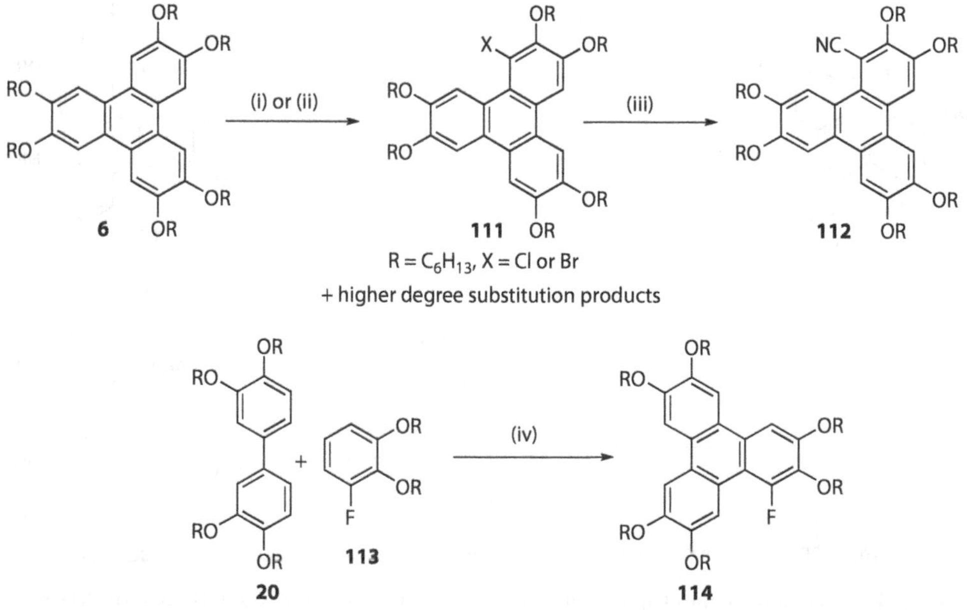

SCHEME 2.55 Synthesis of α-halogeno and α-cyano-substituted hexaalkoxy-TPs: (i) ICl/CH$_2$Cl$_2$; (ii) Br$_2$/CH$_2$Cl$_2$; (iii) CuCN, DMF; (iv) FeCl$_3$, CH$_2$Cl$_2$.

reduces the mesophase stability because of the increase in molecular breadth. On the other hand, fluoro-substitution in the core of discotic liquid crystals is in the plane of the core and, therefore, does not affect the molecular breadth, but fills the space around the core and enhances the intermolecular forces. Consequently, the columnar phase stability increases significantly. The replacement of halogen atoms by cyano groups, as in the case of β-substitution, significantly increases the stability of the columnar phase. Most of the α-nitrohexaalkoxy-TPs display wider mesophase ranges than the parent hexaalkoxy-TPs. The induction of a columnar phase in some non-liquid crystalline hexaalkoxy-TPs was also observed upon α-nitration [288]. While the dinitrohexabutyloxytriphenylene is not liquid crystalline, its two higher homologues are mesogenic. All the trinitro-hexaalkoxy-TPs are non-liquid crystalline [466].

2.6.6.2 Chromium-Arene Complex of Hexaalkoxy-TPs

Refluxing hexaalkoxy-TPs with $Cr(CO)_6$ in Bu_2O/THF for several hours resulted in the formation of chromium-arene complexes **115** (Scheme 2.56). In these complexes, with the exception of hexadecyloxy-TP, the chromium tricarbonyl moiety is exclusively attached to one terminal alkoxysubstituted aryl ring. In the case of hexadecyloxy-TP, an inseparable mixture of two chromium-arene complexes was formed. As expected, the self-assembly of triphenylene molecules was severely disturbed by the attachment of chromium tricarbonyl moiety. All the complexes, but the hexanonyloxy-TP chromium complex, were found to be non-liquid crystalline. The hexanonyloxy-TP chromium complex displays a nematic phase in between 37°C and 58°C [476].

2.6.7 THERMAL BEHAVIOR OF UNSYMMETRICAL TRIPHENYLENE DISCOTICS

Using the above-mentioned methodologies, a large number of unsymmetrical triphenylene derivatives have been prepared [363–492]. The thermal behavior of these discotics is given in 13 tables (Tables 2.41 through 2.53). Table 2.41 covers the thermal behavior of hexa-substituted triphenylenes having five identical alkoxy tails and one different substitution (Structure **116**, Figure 2.7). In Table 2.42, triphenylene derivatives in which the differing substituent is connected to the core via an ester linkage (Structure **117**, Figure 2.8) are collected. The thermal behavior of hexa-substituted triphenylenes having four identical alkoxy tails at the 6-, 7-, 10-, and 11-positions, and two identical substituents at 2- and 3-positions (Structure **118**, Figure 2.9) is listed in Table 2.43. In Table 2.44, while the four alkoxy chains are identical, the remaining two groups at the 2- and 3-positions are of a different nature (Structure **119**, Figure 2.9). Compounds in which two non-identical substituents are at the 2,7-, 2,11-, or at 2,6-positions (Structures **120**, **121**, and **122**, Figure 2.9) are summarized in Table 2.45. Compounds in which the three phenyl rings of the triphenylene core each have differing sizes of alkoxy chains (Structure **123**, Figure 2.9) are shown in Table 2.46. Symmetrical (Structure **124**, Figure 2.9) and unsymmetrical (Structure **125**, Figure 2.9) mixed tail hexa-substituted triphenylenes having three alkoxy chains and three other identical groups are listed in Tables 2.47

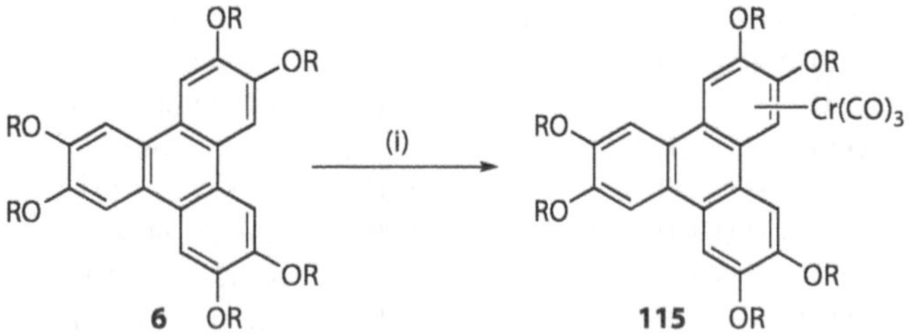

SCHEME 2.56 Preparation of hexaalkoxytriphenylene-chromium complex: (i) $Cr(CO)_6$/Bu_2O/THF, 156°C.

TABLE 2.41
Thermal Behavior of Monofunctionalized Triphenylene Pentaethers with the General Structure 116 (Figure 2.7)

Structure	X	R	Thermal Behavior	Ref.
116.1	C_6H_{13}	C_5H_{11}	Cr 81 I	[398]
116.2	OCH_3	C_5H_{11}	Cr 72 Col$_h$ 101 I	[374]
116.3	OCH_3	C_6H_{13}	Cr 56 Col 72 I	[394]
116.4	OC_2H_5	C_6H_{13}	Cr 58.7 Col$_h$ 81.3 I	[477]
			Cr 57.3 Col$_h$ 79.1 I	[287]
116.5	OC_3H_7	C_6H_{13}	Cr 62.3 Col$_h$ 87.9 I	[477]
			Cr 60.6 Col$_h$ 84.8 I	[287]
116.6	OC_4H_9	C_6H_{13}	Cr 61 Col$_h$ 98.4 I	[477]
			Cr 59.7 Col$_h$ 95.1 I	[287]
116.7	OC_5H_{11}	C_6H_{13}	Cr 59 Col$_h$ 100.8 I	[477]
			Cr 57.7 Col$_h$ 94.5 I	[287]
116.8	OC_7H_{15}	C_6H_{13}	Cr 59.6 Col$_h$ 97.2 I	[477]
			Cr 58.4 Col$_h$ 94.7 I	[287]
116.9	OC_8H_{17}	C_6H_{13}	Cr 54.8 Col$_h$ 91.3 I	[477]
			Cr 52 Col$_h$ 86 I	[287]
116.10	OC_9H_{19}	C_6H_{13}	Cr 49.7 Col$_h$ 82.4 I	[477]
116.11	$OC_{10}H_{21}$	C_6H_{13}	Cr 45 Col$_h$ 71.8 I	[477]
116.12	$OC_{11}H_{23}$	C_6H_{13}	Cr 46.8 Col$_h$ 62.9 I	[477]
116.13	$OC_{12}H_{25}$	C_6H_{13}	Cr 45.5 Col$_h$ 50.9 I	[477]
116.14	$OC_{13}H_{27}$	C_6H_{13}	Cr 41 I	[477]
116.15	$OC_{14}H_{29}$	C_6H_{13}	Cr 39.7 I	[477]
116.16	$OC_{16}H_{33}$	C_6H_{13}	Cr 38.4 I	[477]
116.17	$OC_{18}H_{37}$	C_6H_{13}	Cr 36.1 I	[477]
116.18	$OCH(CH_3)_2$	C_6H_{13}	Cr 54.6 I	[287]
116.19	$OCH(CH_3)C_4H_9$	C_5H_{11}	Cr (27 Col) 51 I	[365]
116.20	$O(CH_2)_3CH(CH_3)_2$	C_6H_{13}	Cr 61.5 Col$_h$ 105.3 I	[287]
116.21	$OCH_2CH(CH_2CH_3)_2$	C_6H_{13}	Cr 44.4 Col$_h$ 78.1 I	[287]
116.22	$OCH_2CH(CH_3)_2$	C_6H_{13}	Cr 59.4 Col$_h$ 85.4 I	[287]
116.23	$O(CH_2)_2CH(CH_3)_2$	C_6H_{13}	Cr 54.5 Col$_h$ 105.0 I	[287]
116.24	$OCH_2CH(C_2H_5)C_4H_9$	C_4H_9	Cr 71.6 Col$_h$ 95 I	[444]
116.25	$OCH_2CH(C_2H_5)C_4H_9$	C_5H_{11}	Cr 54.3 Col$_h$ 88.1 I	[444]
116.26	$OCH_2CH(C_2H_5)C_4H_9$	C_6H_{13}	Cr 53.3 Col$_h$ 74.5 I	[444]
116.27	$OCH_2CH(C_2H_5)C_4H_9$	C_7H_{15}	Cr 48.7 Col$_h$ 68.8 I	[444]
116.28	$OCH_2CH(C_2H_5)C_4H_9$	C_8H_{17}	Cr 51 Col$_h$ 58.5 I	[444]
116.29	$O(CH_2)_2CH(CH_3)(CH_2)_3CH(CH_3)_2$	C_4H_9	Cr 51.8 Col$_h$ 80.1 I	[444]
116.30	$O(CH_2)_2CH(CH_3)(CH_2)_3CH(CH_3)_2$	C_5H_{11}	Cr 39.7 Col$_h$ 88.6 I	[444]
116.31	$O(CH_2)_2CH(CH_3)(CH_2)_3CH(CH_3)_2$	C_6H_{13}	Cr 44 Col$_h$ 84.4 I	[444]
116.32	$O(CH_2)_2CH(CH_3)(CH_2)_3CH(CH_3)_2$	C_7H_{15}	Cr 42.7 Col$_h$ 86.3 I	[444]
116.33	$O(CH_2)_2CH(CH_3)(CH_2)_3CH(CH_3)_2$	C_8H_{17}	Cr 51.7 Col$_h$ 80.5 I	[444]
116.34	$O(CH_2)_{11}(CF_2)_7CF_3$	C_5H_{11}	Cr 55 I	[296]
116.35	OSO_2CF_2	C_5H_{11}	Cr 54 Col 85 Col$_h$ 188 I	[393]
116.36	$O(CH_2CH_2O)_2CH_3$	C_6H_{13}	Cr 55.4 Col$_h$ 69.3 I	[425]
116.37	$O(CH_2CH_2O)_2CH_3$	C_6H_{13}	Cr 57 Col$_h$ 72 I	[482]
116.38	$O(CH_2CH_2O)_3CH_3$	C_6H_{13}	Cr (22 Col) 43 I	[482]
116.39	$O(CH_2CH_2O)_4CH_3$	C_6H_{13}	Cr (22 Col) 38 I	[482]
116.40	$O(CH_2)_2CH=CH_2$	C_5H_{11}	Cr 55.5 Col$_h$ 126.5 I	[374]
116.41	$O(CH_2)_3CH=CH_2$	C_6H_{13}	Cr 56.9 Col$_h$ 99.2 I	[305]

TABLE 2.41 (continued)
Thermal Behavior of Monofunctionalized Triphenylene Pentaethers with the General Structure 116 (Figure 2.7)

Structure	X	R	Thermal Behavior	Ref.
116.42	$O(CH_2)_4CH=CH_2$	C_5H_{11}	Cr 60 Col$_h$ 114 I	[374]
116.43	$O(CH_2)_6CH=CH_2$	C_5H_{11}	Cr 51 Col$_h$ 100 I	[374]
116.44	$O(CH_2)_9CH=CH_2$	C_5H_{11}	Cr 32 Col$_h$ 55 I	[364]
116.45	$C\equiv CSiMe_3$	C_5H_{11}	g −50 Col$_h$ 184 I	[398]
116.46	$C\equiv CMe_2OH$	C_5H_{11}	Cr 58 Col$_h$ 180 I	[398]
116.47	$C\equiv CC_4H_9$	C_5H_{11}	Cr 44 Col$_h$ 157 I	[398]
116.48	$C\equiv CH$	C_5H_{11}	Col$_h$ 146 I	[398]
116.49	$COCH_3$	OC_5H_{11}	Cr 76 Col$_h$ 169 I	[398]
116.50	$OCH_2COOC_2H_5$	C_4H_9	Col$_p$ 128 Col$_h$ 149 I	[446]
116.51	$OCH_2COOC_2H_5$	C_5H_{11}	Cr 61 Col$_h$ 137 I	[446]
			Cr 65 Col$_h$ 140 I	[481]
116.52	$OCH_2COOC_2H_5$	C_6H_{13}	Cr 48 Col$_h$ 118 I	[446]
116.53	$OCH_2COOC_2H_5$	C_7H_{15}	Cr 34 Col$_h$ 112 I	[446]
116.54	$OCH_2COOC_2H_5$	C_8H_{17}	Cr 37 Col$_h$ 99 I	[446]
116.55	$OCH_2COOC_2H_5$	C_9H_{19}	Cr 36 Col$_h$ 96 I	[446]
116.56	$OCH_2COOC_4H_9$	C_4H_9	Cr 70 Col$_h$ 132 I	[446]
116.57	$OCH_2COOC_4H_9$	C_5H_{11}	Cr 52 Col$_h$ 132 I	[446]
116.58	$OCH_2COOC_4H_9$	C_6H_{13}	Cr 30 Col$_h$ 119 I	[446]
116.59	$OCH_2COOC_4H_9$	C_7H_{15}	Cr 28 Col$_h$ 114 I	[446]
116.60	$OCH_2COOC_4H_9$	C_8H_{17}	Cr 41 Col$_h$ 104 I	[446]
			Cr 41 Col$_h$ 102 I	[448]
116.61	$OCH_2COOC_4H_9$	C_9H_{19}	Cr 28 Col$_h$ 100 I	[446]
116.62	$O(CH_2)_6OCOCH_3$	C_5H_{11}	Cr 45 Col$_h$ 50 I	[481]
116.63	$O(CH_2)_2OCOCH=CH_2$	C_5H_{11}	Cr 65 X 68 Col$_h$ 97 I	[480]
116.64	$O(CH_2)_4OCOCH=CH_2$	C_5H_{11}	Cr 44 Col$_h$ 84 I	[480]
116.65	$O(CH_2)_6OCOCH=CH_2$	C_5H_{11}	Cr 46 Col$_h$ 54 I	[480]
116.66	$O(CH_2)_6OCOCH=CH_2$	C_6H_{13}	Cr 46 I	[369]
116.67	$O(CH_2)_6OCOC(CH_3)=CH_2$	C_6H_{13}	Cr 48 I	[369]
116.68	$O(CH_2)_{11}OCOC(CH_3)=CH_2$	C_6H_{13}	Cr 35 I	[375]
116.69	$O(CH_2CH_2O)_2COCH=CH_2$	C_6H_{13}	Cr 48 Col 64 I	[369]
116.70	$O(CH_2CH_2O)_2COC(CH_3)=CH_2$	C_6H_{13}	Cr 45 Col 56 I	[369]
116.71	$O(CH_2)_2OCOCH=CH_2$	C_4H_9	Cr 64 Col$_h$ 116 I	[480]
116.72	$O(CH_2)_3OCOCH=CH_2$	C_4H_9	Cr 83 X 89 Col$_h$ 102 I	[480]
116.73	$O(CH_2)_2OH$	C_5H_{11}	Cr 69 Col$_h$ 101 I	[481]
116.74	$O(CH_2)_3OH$	C_5H_{11}	Cr 55 Col$_h$ 106 I	[374]
116.75	$O(CH_2)_6OH$	C_5H_{11}	Cr 66 Col$_h$ 77 I	[481]
116.76	$O(CH_2)_6OH$	C_6H_{13}	Cr 62 Col 67 I	[369]
116.77	$O(CH_2)_2O(CH_2)_2OH$	C_6H_{13}	Cr 60 Col 66 Col 78 I	[369]
116.78	$OCH_2CONHC_4H_9$	C_4H_9	Cr 93 Col$_h$ 147 I	[446]
116.79	$OCH_2CONHC_4H_9$	C_5H_{11}	Cr 70 Col$_h$ 151 I	[446]
116.80	$OCH_2CONHC_4H_9$	C_6H_{13}	Cr 69 Col$_h$ 142 I	[446]
116.81	$OCH_2CONHC_4H_9$	C_7H_{15}	Cr 61 Col$_h$ 141 I	[446]
116.82	$OCH_2CONHC_4H_9$	C_8H_{17}	Cr 52 Col$_h$ 135 I	[446]
116.83	$OCH_2CONHC_4H_9$	C_9H_{19}	Cr 47 Col$_h$ 128 I	[446]
116.84	$O(CH_2)_5NHCONHC_6H_{13}$	C_6H_{13}	—	[434]
116.85	$O(CH_2)_4NHCOC_2H_5$	C_6H_{13}	Cr 112 I	[484]

(continued)

TABLE 2.41 (continued)
Thermal Behavior of Monofunctionalized Triphenylene Pentaethers
with the General Structure 116 (Figure 2.7)

Structure	X	R	Thermal Behavior	Ref.
116.86	$O(CH_2)_{10}NHCOC_7H_{15}$	C_6H_{13}	Cr 75 I	[484]
116.87	$O(CH_2)_4NHCONHC_2H_5$	C_6H_{13}	Cr 134 I	[484]
116.88	$O(CH_2)_6NHCONHC_6H_{13}$	C_6H_{13}	Cr 91 I	[484]
116.89	$O(CH_2)_{10}NHCONHC_6H_{13}$	C_6H_{13}	Cr 77 I	[484]
116.90	$O(CH_2)_3NHCSNHCH_3$	C_6H_{13}	Col_h 75 I	[484]
116.91	$O(CH_2)_3NHCSNHC_2H_5$	C_6H_{13}	Col_h 69 I	[484]
116.92	$O(CH_2)_4NHCSNHCH_3$	C_6H_{13}	Col_h 63 I	[484]
116.93	$O(CH_2)_4NHCSNHC_2H_5$	C_6H_{13}	Cr 85 I	[484]
116.94	$O(CH_2)_{10}NHCSNHC_2H_5$	C_6H_{13}	Cr 37 I	[484]
116.95	Br	C_5H_{11}	Col 19 Col_h 165 I	[393]
116.96	Br	C_6H_{13}	Cr 54 Col_h 142 I	[384]
116.97	$O(CH_2)_{10}Br$	C_4H_9	Cr 56.5 I	[478]
116.98	$O(CH_2)_{10}Br$	C_5H_{11}	Cr 25 I	[479]
116.99	$O(CH_2)_{10}Br$	C_6H_{13}	Cr 45 I	[433]
116.100	$O(CH_2)_{11}Br$	C_6H_{13}	Cr 46 I	[375]
116.101	CN	C_5H_{11}	Cr 51 Col 85 Col_h 226 I	[398]
116.102	CN	C_6H_{13}	Cr 93 Col 214 I	[370]
116.103	$O(CH_2)_5SH$	C_6H_{13}	Cr 94 Col 105 I	[409]
116.104	$O(CH_2)_{10}SH$	C_5H_{11}	—	[422]
116.105	$O(CH_2)_2O(CH_2)_2SH$	C_6H_{13}	Cr 75 Col 90.4 I	[403]
116.106	$O(CH_2)_2O(CH_2)_2SCOCH_3$	C_6H_{13}	Cr 60.4 Col 74.7 I	[403]
116.107	$(CH_2)_5Si(CH_3)_2OSi(CH_3)_3$	C_6H_{13}	Cr 43 Col_h 76.4 I	[305]
116.108	$(CH_2)_5Si(CH_3)_2OSi(CH_3)_2H$	C_6H_{13}	Cr 41.6 Col_h 74.5 I	[305]
116.109		C_5H_{11}	Ss 67 Col 101 I	[435]
116.110		C_5H_{11}	Ss 64 Col 83 I	[435]
116.111		C_6H_{13}	Cr 20.4 I	[431]
116.112		C_6H_{13}	Cr 22.2 I	[431]
116.113		C_6H_{13}	Cr 8.4 Col_h 39.7 I	[431]
116.114		C_6H_{13}	Cr 8.7 Col_h 44.3 I	[431]
116.115		C_5H_{11}	Cr 105.8 I	[414]
116.116		C_5H_{11}	Cr 106 I	[302]
116.117		C_5H_{11}	Cr 72 I	[302]
116.118		C_5H_{11}	g −6.9 I	[451]
116.119		C_5H_{11}	Cr 52-55 I	[303]
116.120		C_6H_{13}	Cr 53-57 I	[303]
116.121		C_5H_{11}	Cr 43 M 67 I	[419]
116.122		C_6H_{13}	Cr 52 M 61 I	[419]
116.123		C_7H_{15}	Cr 38 I	[419]
116.124		C_8H_{17}	Cr 40 M 53 I	[419]
116.125		C_9H_{19}	Cr 31 M 44 I	[419]
116.126		$C_{10}H_{21}$	Cr 49 M 63 I	[419]
116.127		C_6H_{13}	Cr 52-54 I	[304]
116.128		C_6H_{13}	Cr 42-44 I	[304]
116.129		C_5H_{11}	g 19.4 Col 56.3 I	[415]

TABLE 2.42

Thermal Behavior of Unsymmetrical Hexasubstituted Triphenylene Derivatives with the General Structure 117 Having One Ester and Five Alkoxy Substitutions (Figure 2.8)

Structure	X	R	Thermal Behavior	Ref.
117.1	CH_3	C_5H_{11}	Cr? Col$_h$ 161 I	[137]
117.2	CH_3	C_6H_{13}	Cr 66 Col$_h$ 136 I	[486]
117.3	CH_3	C_7H_{15}	Cr 50 Col$_h$ 117 I	[485]
117.4	C_4H_9	C_4H_9	g −58 Col$_h$ 177 I	[365]
117.5	C_4H_9	C_5H_{11}	Cr 44 Col$_h$ 177 I	[365]
117.6	$C(CH_3)_3$	C_5H_{11}	Col$_p$ 65 Col$_h$ 178 I	[404]
			g −43 Col$_p$ 65 Col$_h$ 178 I	[399]
117.7	C_5H_{11}	C_5H_{11}	Cr 47 Col$_h$ 178 I	[413]
117.8	C_5H_{11}	C_6H_{13}	Cr 49 Col$_h$ 154 I	[486]
117.9	C_8H_{17}	C_4H_9	Col$_h$ 126 I	[447]
117.10	C_8H_{17}	C_5H_{11}	Col$_h$ 139 I	[447]
117.11	C_8H_{17}	C_6H_{13}	Col$_h$ 137 I	[447]
117.12	C_8H_{17}	C_7H_{15}	Cr 6.9 Col$_h$ 141 I	[447]
117.13	C_8H_{17}	C_8H_{17}	Cr 23.2 Col$_h$ 138 I	[447]
117.14	C_8H_{17}	C_9H_{19}	Cr 13.6 Col$_h$ 141 I	[447]
117.15	$(CH_2)_4(CF_2)_3CF_3$	C_5H_{11}	Cr 38 Col$_h$ 98 I	[296]
117.16	$(CH_2)_2(CF_2)_5CF_3$	C_4H_9	Col$_1$ 102 Col$_2$ 146 Col$_3$ 156 I	[447]
117.17	$(CH_2)_2(CF_2)_5CF_3$	C_5H_{11}	Col$_1$ 40.4 Col$_2$ 72.9 Col$_3$ 166 I	[447]
117.18	$(CH_2)_2(CF_2)_5CF_3$	C_6H_{13}	Cr 39.4 Col$_h$ 167 I	[447]
117.19	$(CH_2)_2(CF_2)_5CF_3$	C_7H_{15}	Cr 34.3 Col$_h$ 178 I	[447]
117.20	$(CH_2)_2(CF_2)_5CF_3$	C_8H_{17}	Cr 40.1 Col$_h$ 178 I	[447]
117.21	$(CH_2)_2(CF_2)_5CF_3$	C_9H_{19}	Cr 38.3 Col$_h$ 176 I	[447]
117.22	$CH(Cl)C_3H_7$	C_4H_9	g −48 Col$_h$ 191 I	[365]
117.31	$CH(Cl)C_3H_7$	C_5H_{11}	Col$_h$ 191 I	[365]
117.24	$CH(Cl)CH_2CH(CH_3)_2$	C_4H_9	Col 200.7 I	[343]
117.25	$CH(Cl)CH_2CH(CH_3)_2$	C_5H_{11}	Col 191.1 I	[343]
117.26	$CH(Cl)CH(CH_3)C_2H_5$	C_5H_{11}	Col$_h$ 182 I	[365]
117.27	$C^*H(Cl)C^*H(CH_3)C_2H_5$	C_4H_9	Col 192.7 I	[343]
			g −46 Col$_h$ 182 I	[365]
117.28	$CH(Cl)CH(CH_3)C_3H_7$	C_5H_{11}	Col$_h$ 182 I	[365]
117.29	$C^*H(Cl)C^*H(CH_3)C_3H_7$	C_4H_9	g −41 Col$_h$ 182 I	[365]
117.30	$C^*H(Cl)CH_2CH(CH_3)_2$	C_5H_{11}	Col$_h$ 191.1 I	[343]
117.31	$C^*H(Cl)C^*H(CH_3)C_2H_5$	C_5H_{11}	-	[392]
117.32	$CH_2CH=CH_2$	C_4H_9	Col$_p$ 79 Col$_h$ 197 I	[408]
117.33	$CH_2CH_2CH=CH_2$	C_4H_9	g −43 Col$_p$ 78 Col$_h$ 188 I	[418]
			g −41 Col$_p$ 80 Col$_h$ 192 I	[408]
117.34	$(CH_2)_8CH=CH_2$	C_4H_9	g −43 Col$_h$ 86 I	[426]
117.35	$CH_2CH=CHC_6H_5$	C_6H_{13}	Cr 92 Col$_h$ 152 I	[455]
117.36	$CH_2C≡CC_6H_5$	C_4H_9	Col$_h$ 163 I	[455]
117.37	$CH_2C≡CC_6H_5$	C_5H_{11}	Col$_h$ 166 I	[455]
117.38	$CH_2C≡CC_6H_5$	C_6H_{13}	Cr 116 Col$_h$ 150 I	[455]
117.39	$CH_2C≡CC_6H_5$	C_7H_{15}	Cr 62 Col$_h$ 140 I	[455]
117.40	$CH_2C≡CC_6H_5$	C_8H_{17}	Col$_h$ 128 I	[455]
117.41	$CH_2C≡CC_6H_5$	C_9H_{19}	Col$_h$ 121 I	[455]
117.42		C_5H_{11}	g −43 Col$_h$ 190 I	[487]
117.43		C_5H_{11}	Col$_p$ 135 Col$_h$ 186 I	[404]
117.44		C_4H_9	Cr 116 Col$_p$ 156 Col$_h$ 205 I	[452]
117.45		C_5H_{11}	Col$_h$ 193 I	[413]
			g −30 Col$_h$ 193 I	[408]

(continued)

TABLE 2.42 (continued)
Thermal Behavior of Unsymmetrical Hexasubstituted Triphenylene Derivatives with the General Structure 117 Having One Ester and Five Alkoxy Substitutions (Figure 2.8)

Structure	X	R	Thermal Behavior	Ref.
117.46		C_6H_{13}	Col_h 170 I	[452]
117.47		C_7H_{15}	Col_h 160 I	[452]
117.48		C_8H_{17}	Col_h 147 I	[452]
117.49		C_9H_{19}	Col_h 144 I	[452]
117.50		C_5H_{11}	Cr 94 Col_h 135 I	[413]
117.51		C_6H_{13}	Cr 55 Col_h 147 I	[486]
117.52		C_6H_{13}	Cr < 0 Col_h 182 I	[486]
117.53		C_5H_{11}	g 0 Col_h 186 I	[487]
117.54		C_6H_{13}	Cr 59 Col_h 175 I	[486]
117.55		C_5H_{11}	Cr 203 Col_h? I	[413]
117.56		C_5H_{11}	Cr 74 Col_h 190 I	[413]
			g −5 Col_h 190 I	[408]
117.57		C_6H_{13}	Cr 59 Col_h 175 I	[486]
117.58		C_5H_{11}	g −11 Col_h 191 I	[408]
117.59		C_4H_9	Cr 178 Col_h 205 I	[452]
117.60		C_5H_{11}	Cr 131 Col_h 201 I	[413]
			Cr 143 Col_h 209 I	[452]
117.61		C_6H_{13}	Cr 110 Col_h 192 I	[486]
			Cr 118 Col_h 208 I	[452]
117.62		C_7H_{15}	Cr 90 Col_h 204 I	[452]
117.63		C_8H_{17}	Cr 66 Col_h 199 I	[452]
117.64		C_9H_{19}	Cr 61 Col_h 194 I	[452]
117.65		C_5H_{11}	Cr 103 Col_h 158 I	[413]
117.66		C_6H_{13}	Cr < 0 Col_h 121 I	[486]
117.67		C_5H_{11}	g 8 Col_h 180 I	[408]
117.68		C_4H_9	Cr 101 Col 115 I	[408]
117.69		C_5H_{11}	Cr 78 Col 120 I	[408]
117.70		C_6H_{13}	Cr 66 Col 100 I	[408]
117.71		C_7H_{15}	Cr 57 Col 101 I	[408]
117.72		C_8H_{17}	Cr 55 Col 94 I	[408]
117.73		C_9H_{19}	Cr 53 Col 89 I	[408]
117.74		C_5H_{11}	g −22 Col_h 201 I	[408]
117.75		C_5H_{11}	Cr 71 I	[441]
117.76		C_5H_{11}	g −10 Col_{ob} 82 I	[441]
117.77		C_5H_{11}	g −20 Col_{ob} 110 I	[441]
117.78		C_5H_{11}	Cr (168 Col_h) 199 I	[413]
117.79		C_5H_{11}	Cr 101 Col_h 170.7 I	[453]
117.80		C_6H_{13}	Cr 82.4 Col_h 150.9 I	[453]
117.81		C_5H_{11}	Col_h 155.5 I	[453]
117.82		C_5H_{11}	Cr (124 Col_h) 147 I	[413]
117.83	Y = H; n = 3	C_5H_{11}	Cr 77 I	[302]
117.84	Y = H; n = 5	C_5H_{11}	Cr 99 I	[302]
117.85	Y = H; n = 7	C_5H_{11}	Cr 78 I	[302]
117.86	Y = COC_4H_9; n = 5	C_5H_{11}	Cr 58 I	[429]
117.87	Y = COC_4H_9; n = 10	C_5H_{11}	Cr 63 I	[429]
117.88	Y = $COC_{14}H_{29}$; n = 4	C_5H_{11}	Cr 65 I	[429]
117.89	Y = COC_4H_9; n = 5	C_6H_{13}	Cr 65 I	[429]
117.90	Y = COC_4H_9; n = 5	C_7H_{15}	Cr 63 I	[429]
117.91	Y = C_6H_{13}; n = 5	C_4H_9	Cr 70 I	[429]

TABLE 2.43

Thermal Behavior of Unsymmetrical Hexasubstituted Triphenylene Derivatives with the General Structure 118 (Figure 2.9)

Structure	X	R	Thermal Behavior	Ref.
118.1	C_7H_{15}	C_6H_{13}	Cr 164 I	[383]
118.2	OCH_3	C_5H_{11}	Cr 80 I	[372]
118.3	OC_4H_9	C_6H_{13}	Cr 59 Col$_h$ 90 I	[372]
118.4	OC_4H_9	C_8H_{17}	Cr 60 Col$_h$ 67 I	[372]
118.5	OC_4H_9	$C_{10}H_{21}$	Cr 65 Col$_h$ 76 I	[372]
118.6	OC_5H_{11}	CH_3	Cr 80 I	[372]
118.7	OC_5H_{11}	C_9H_{19}	Cr 53 Col$_h$ 71 I	[363]
118.8	OC_5H_{11}	$C_{10}H_{21}$	Cr 61 Col$_h$ 63 I	[363]
118.9	OC_6H_{13}	CH_3	Cr 142.5 I	[372]
118.10	OC_6H_{13}	C_8H_{17}	Cr 46 Col$_h$ 84 I	[372]
118.11	OC_6H_{13}	$C_{10}H_{21}$	Cr 44 Col$_h$ 72 I	[372]
118.12	OC_8H_{17}	C_6H_{13}	Cr 47 Col$_h$ 84 I	[372]
118.13	OC_8H_{17}	$C_{10}H_{21}$	Cr 60 Col$_h$ 86 I	[372]
118.14	OC_9H_{19}	C_5H_{11}	Cr 54 Col$_h$ 74 I	[363]
118.15	$OC_{10}H_{21}$	CH_3	Cr 74 I	[372]
118.16	$OC_{10}H_{21}$	C_5H_{11}	Cr (56 Col$_h$) 57 I	[372]
118.17	$OC_{10}H_{21}$	C_6H_{13}	Cr 58 Col$_h$ 74 I	[372]
118.18	$OC_{10}H_{21}$	C_8H_{17}	Cr 48 Col$_h$ 73 I	[372]
118.19	$OC_{11}H_{23}$	C_5H_{11}	Cr 61 I	[372]
118.20	$OC_{12}H_{25}$	CH_3	Cr 125 I	[372]
118.21	$OC_{12}H_{25}$	C_6H_{13}	Cr 64 I	[372]
118.22	$OC_{12}H_{25}$	C_8H_{17}	Cr 51 Col$_h$ 61 I	[372]
118.23	$OC_{12}H_{25}$	$C_{10}H_{21}$	Cr 41 Col$_h$ 62 I	[372]
118.24	$O(CH_2)_2CH(CH_3)(CH_2)_3CH(CH_3)_2$	C_5H_{11}	Col$_h$ 63 I	[284]
118.25	$O(CH_2)_2OH$	C_5H_{11}	Cr 128 Col$_h$ 139 I	[373]
118.26	$O(CH_2)_6OH$	C_5H_{11}	Cr (35 LC) 66 I	[401]
118.27	$O(CH_2)_5SH$	C_6H_{13}	Cr 67 Col$_h$ 79 I	[409]
118.28	$O(CH_2)_3COOH$	C_6H_{13}	Cr (117 Col$_h$) 130 I	[488]
118.29	$O(CH_2)_4COOH$	C_6H_{13}	Cr 94 Col 102 Col$_h$ 105.3 I	[488]
118.30	$O(CH_2)_5COOH$	C_6H_{13}	Cr (78 Col$_h$)103.8 I	[488]
118.31	$O(CH_2)_6COOH$	C_6H_{13}	Cr (68 Col$_h$) 106.1 I	[488]
118.32	$O(CH_2)_7COOH$	C_6H_{13}	Cr 95.4 I	[488]
118.33	$O(CH_2)_{10}COOH$	C_6H_{13}	Cr 69.1 I	[488]
118.34	Br	C_6H_{13}	Cr 130 Col$_h$ 141 I	[383]
118.35	$SC6H13$	C_6H_{13}	Cr (81Col$_h$) 88 I	[383]
118.36	CN	C_8H_{17}	Cr 183.2 Col$_h$ 187.2 I	[443]
118.37	CN	$C_{10}H_{21}$	Cr 164.1 Col$_h$ 190.3 I	[443]
118.38	CN	$C_{12}H_{25}$	Cr 141.9 Col$_h$ 190.1 I	[443]
118.39	CN	$C_{14}H_{29}$	Cr 121.4 Col$_h$ 189.8 I	[443]

TABLE 2.44
Thermal Behavior of Unsymmetrical Hexasubstituted Triphenylene Derivatives with the General Structure 119 (Figure 2.9)

Structure	X	Y	R	Thermal Behavior	Ref.
119.1	OCH_3	OC_5H_{11}	C_4H_9	Cr 86.9 Col_h 110.4 I	[284]
119.2	OCH_3	$O(CH_2)_3OH$	C_6H_{13}	Cr 95 LC 116 I	[490]
119.3	$O(CH_2)_3OH$	$O(CH2)6CH=CH2$	C_5H_{11}	Cr (64 Col_h) 68 I	[374]
119.4	$C{\equiv}CH$	SC_5H_{11}	C_5H_{11}	Cr 57.7 Col_h 125.4 I	[378]
119.5	$C{\equiv}CMe_2OH$	SC_5H_{11}	C_5H_{11}	Cr 81.8 Col_h 167.8 I	[378]
119.6	Br	CN	C_6H_{13}	Cr 170 Col 190 I	[379]
119.7	Br	SC_5H_{11}	C_5H_{11}	Cr 75.6 Col_h 144.8 I	[378]
119.8	CN	SC_5H_{11}	C_5H_{11}	Cr 121 Col_h 193.8 I	[378]
119.9	OCH_3	**Structure 116.121**	C_5H_{11}	Cr 113 I	[303]
119.10	OCH_3	**Structure 116.121**	C_6H_{13}	Cr 89 Col 102 I	[303]
119.11	OCH_3	**Structure 116.127**	C_6H_{13}	Cr 42-44 I	[304]
119.12	OCH_3	**Structure 116.128**	C_6H_{13}	Cr 76-78 I	[304]

TABLE 2.45
Thermal Behavior of Unsymmetrical Triphenylene Derivatives with the General Structures 120, 121, and 122 (Figure 2.9)

Structure	X	R	Thermal Behavior	Ref.
120.1	C_7H_{15}	OC_6H_{13}	Cr 170 I	[383]
120.2	OC_6H_{13}	C_7H_{15}	Cr 157 I	[383]
120.3	OC_6H_{13}	SC_6H_{13}	Cr 54 Col_h 100.5 I	[383]
120.4	$O(CH_2CH_2O)_2CH_3$	OC_6H_{13}	Cr 44 I	[425]
120.5	$OCH_2COOC_2H_5$	OC_5H_{11}	Cr 81 Col 152.4 I	[448]
120.6	$OCH_2COOC_2H_5$	OC_6H_{13}	Cr 88.5 Col_h 129.2 I	[428]
120.7	$OCH_2COOC_4H_9$	OC_5H_{11}	Cr 40 Col 133.7 I	[448]
120.8	$OCH_2CONHC_4H_9$	OC_5H_{11}	Cr 167.7 Col 185.2 I	[448]
120.9	$O(CH_2)_6CH=CH_2$	OC_5H_{11}	Cr 55 Col_h 73 I	[491]
120.10	$O(CH_2)_3OCOCH=CH_2$	OC_5H_{11}	Cr 94 I	[396]
120.11	Br	OC_6H_{13}	Cr 40 Col_h 179 I	[383]
120.12	SC_6H_{13}	OC_6H_{13}	Cr 24 Col_h 102 I	[383]
120.13	**Structure 116.118**	OC_5H_{11}	g 24.7 N_D^* 51 I	[451]
120.14	**Structure 8.38**	OC_5H_{11}	Cr (90 Col_h)114 I	[302]
120.15	**Structure 6.41**	OC_5H_{11}	Cr (112 Col_h)147 I	[302]
121.1	C_7H_{15}	OC_6H_{13}	Cr 175 I	[383]
121.2	OC_6H_{13}	C_7H_{15}	Cr 155 I	[383]
121.3	OC_6H_{13}	SC_6H_{13}	Cr 72 Col_h 111 I	[383]
			Cr 54 Col_h 100 I	[387]
121.4	$O(CH_2)_2OH$	OC_5H_{11}	Cr 157 I	[401]
121.5	$O(CH_2)_3OH$	OC_5H_{11}	Cr 87 Col_1 99 Col_2 118 I	[373]
121.6	$O(CH_2)_6OH$	OC_5H_{11}	Cr (55 Col_h) 65 I	[401]
121.7	Br	OC_6H_{13}	Cr 63 Col_h 179 I	[383]
121.8	SC_6H_{13}	OC_6H_{13}	Cr 45 Col_h 123 I	[383]
122.1	$O(CH_2)_6OH$	OC_5H_{11}	Cr 49 Col_h 67 I	[401]

TABLE 2.46
Thermal Behavior of Triphenylene Derivatives with the General Structures 123 (Figure 2.9)

Structure	R_1	R_2	R_3	Thermal Behavior
123.1	CH_3	C_4H_9	$C_{12}H_{25}$	Cr 88 I
123.2	C_4H_9	C_6H_{13}	C_8H_{17}	Cr 48 Col_h 72 I
123.3	C_4H_9	C_8H_{17}	$C_{10}H_{21}$	Cr 36 Col_h 46 I
123.4	C_4H_9	C_8H_{17}	$C_{12}H_{25}$	Cr 52 I
123.5	C_6H_{13}	$C_{10}H_{21}$	$C_{12}H_{25}$	Cr 43 Col_h 51 I
123.6	$C_{10}H_{21}$	C_6H_{13}	C_8H_{17}	Cr 47 Col_h 75 I
123.7	$C_{12}H_{25}$	C_6H_{13}	C_8H_{17}	Cr 39 Col_h 75 I
123.8	$C_{12}H_{25}$	C_8H_{17}	$C_{10}H_{21}$	Cr 47 Col_h 65 I

Source: Data from Cross, S.J. et al., *Liq. Cryst.*, 25, 1, 1998.

TABLE 2.47
Thermal Behavior of Triphenylene Derivatives with the General Structures 124 (Figure 2.9)

Structure	R	X	Thermal Behavior	Ref.
124.1	C_5H_{11}	OCH_3	Cr 142 I	[421]
124.2	C_5H_{11}	OC_2H_5	Cr 108 I	[421]
124.3	C_8H_{17}	OC_2H_5	Cr 65 I	[421]
124.4	C_5H_{11}	OC_3H_7	Cr 68 Col_h 105 I	[421]
124.5	C_7H_{15}	OC_3H_7	Cr 66 I	[421]
124.6	C_5H_{11}	OC_4H_9	Cr 55 Col_h 129 I	[421]
124.7	C_6H_{13}	OC_4H_9	Cr 58 Col_h 96 I	[421]
124.8	C_5H_{11}	OC_6H_{13}	Cr 48 Col_h 106 I	[421]
124.9	C_5H_{11}	OC_7H_{15}	Cr 58 Col_h 91 I	[421]
124.10	C_5H_{11}	OC_8H_{17}	Cr 46 Col_h 67 I	[421]
124.11	C_5H_{11}	OC_9H_{19}	Cr 44 Col_h 59 I	[421]
124.12	C_6H_{13}	$OC_{10}H_{21}$	Cr 43 Col 55 I	[437]
124.13	C_6H_{13}	$O(CH_2)_3CF_3$	Cr 108 Col 142 I	[437]
124.14	C_5H_{11}	$OCOCH_3$	Cr 185 I	[411]
124.15	C_5H_{11}	$OCOC_3H_7$	Cr 132.7 Col_h 253.9 I	[411]
124.16	C_5H_{11}	$OCOC_4H_9$	Cr 130.6 Col_h 244.6 I	[411]
124.17	C_5H_{11}	$OCOC_5H_{11}$	Cr 109.9 Col_h 218.4 I	[411]
124.18	C_5H_{11}	$OCOC_7H_{15}$	Cr 61.8 Col_h 168.3 I	[411]
124.19	C_5H_{11}	$OCOC_8H_{17}$	Cr 31.5 I	[411]
124.20	C_5H_{11}	$OCOCH(Cl)CH(CH_3)C_2H_5$	Cr 200.9 I	[343]
124.21	C_6H_{13}	$OCOC_6H_2(OC_{12}H_{25})_3$	Cr −11 Col 47 I	[442]
124.22	C_4H_9	$OCH_2COOC_4H_9$	Cr 15.5 Col 138.7 I	[448]
124.23	C_5H_{11}	$OCH_2COOC_2H_5$	Cr 93 Col 151 I	[438]
124.24	C_5H_{11}	$OCH_2COOC_4H_9$	Cr 50 Col 142 I	[448]
124.25	C_6H_{13}	$OCH_2COOC_2H_5$	Cr 80 Col 131 I	[438]
124.26	C_6H_{13}	$OCH_2COOC_4H_9$	Col 138 I	[448]
124.27	C_7H_{15}	$OCH_2COOC_2H_5$	Cr 74 Col 117 I	[438]
124.28	C_7H_{15}	$OCH_2COOC_4H_9$	Cr −1 Col 135 I	[448]

(*continued*)

TABLE 2.47 (continued)
Thermal Behavior of Triphenylene Derivatives with the General Structures 124 (Figure 2.9)

Structure	R	X	Thermal Behavior	Ref.
124.29	C_8H_{17}	$OCH_2COOC_2H_5$	Cr 55 Col 108 I	[438]
124.30	C_8H_{17}	$OCH_2COOC_4H_9$	Col 129 I	[448]
124.31	C_4H_9	$OCH_2CONHC_4H_9$	Cr 173 I	[448]
124.32	C_5H_{11}	$OCH_2CONHC_4H_9$	Cr 100 Col$_p$ 163 Col$_h$ 196 I	[438]
124.33	C_6H_{13}	$OCH_2CONHC_4H_9$	Col$_p$ 139 Col 199 I	[448]
			Cr 99 Col$_p$ 143 Col$_h$ 204 I	[438]
124.34	C_7H_{15}	$OCH_2CONHC_4H_9$	Col$_p$ 117 Col 206 I	[448]
			Cr 72 Col$_p$ 111 Col$_h$ 205 I	[438]
124.35	C_8H_{17}	$OCH_2CONHC_4H_9$	Col$_p$ 122 Col 204 I	[448]
			Cr 98 Col$_p$ 119 Col$_h$ 204 I	[438]
124.36	C_6H_{13}	$O(CH_2)_3CH=CH_2$	Cr 64.8 Col$_h$ 95.8 I	[305]
124.37	C_6H_{13}	$O(CH_2)_5Si(CH_3)_2O\,Si(CH_3)_3$	Cr 31.2 Col$_h$ 86 I	[305]
124.38	C_5H_{11}	$COCH_3$	Col 194 Col$_h$ 250 I	[397]
124.39	C_5H_{11}	COC_5H_{11}	Col$_h$ 274 I	[397]
124.40	C_5H_{11}	Br	Cr 187 Col$_h$ 216 I	[397]
124.41	C_5H_{11}	SC_5H_{11}	Cr 27 Col$_h$ 127 I	[397]
124.42	C_5H_{11}	$C\equiv CC_4H_9$	Cr 17 Col 33 Col$_h$ 165 I	[397]
124.43	C_5H_{11}	$C\equiv CCMe_2OH$	Cr 118 Col$_h$ 250 dec	[397]
124.44	C_5H_{11}	$C\equiv CSiMe_3$	Cr 213 Col 236 I	[397]
124.45	C_5H_{11}	$O(CH_2)_5COO$-cholesterol	g 30.2 N$_{D*}$ 67.3 I	[451]
124.46	C_5H_{11}	$O(CH_2)_{10}COO$-cholesterol	g 25.7 N$_{D*}$ 53 I	[451]
124.47	C_7H_{15}	$O(CH_2)_5COO$-cholesterol	g 28.9 N$_{D*}$ 61.2 I	[451]
124.48	C_9H_{19}	$O(CH_2)_5COO$-cholesterol	g 21.4 N$_{D*}$ 51.4 I	[451]
124.49	$C_{11}H_{23}$	$O(CH_2)_5COO$-cholesterol	g 16.5 N$_{D*}$ 51.2 I	[451]
124.50	$(CH_2)_5SH$	OC_6H_{13}	Col$_h$ 44.1 I	[409]

TABLE 2.48
Thermal Behavior of Unsymmetrical Triphenylene Derivatives with the General Structures 125 (Figure 2.9)

Structure	R	X	Thermal Behavior	Ref.
125.1	C_6H_{13}	OC_3H_7	Cr 61 Col 69 I	[437]
125.2	C_6H_{13}	OC_4H_9	Cr 57 Col 98 I	[437]
125.3	C_6H_{13}	OC_5H_{11}	Cr 45 Col 102 I	[437]
125.4	C_6H_{13}	OC_7H_{15}	Cr 47 Col 89 I	[437]
125.5	C_6H_{13}	OC_8H_{17}	Cr 44 Col 79 I	[437]
125.6	C_6H_{13}	OC_9H_{19}	Col 66 I	[437]
125.7	C_6H_{13}	$OC_{10}H_{21}$	Col 51 I	[437]
125.8	C_6H_{13}	$OC_{12}H_{25}$	Col 48 I	[437]
125.9	C_6H_{13}	$OC_{14}H_{29}$	Cr 46 I	[437]
125.10	$C_{10}H_{21}$	OCH_3	LC (42.5 Col) 68 I	[471]
125.11	C_5H_{11}	$OCOCH_3$	Cr 146.5 I	[411]
125.12	C_5H_{11}	$OCOC_3H_7$	Cr 61.7 Col$_h$ 230.2 I	[411]
125.13	C_5H_{11}	$OCOC_4H_9$	Cr 82.5 Col$_h$ 227.6 I	[411]

TABLE 2.48 (continued)
Thermal Behavior of Unsymmetrical Triphenylene Derivatives with the General Structures 125 (Figure 2.9)

Structure	R	X	Thermal Behavior	Ref.
125.14	C_5H_{11}	$OCOC_5H_{11}$	Cr 69.2 Col_h 203.3 I	[411]
125.15	C_5H_{11}	$OCOC_7H_{15}$	Cr 61.9 Col_h 155.3 I	[411]
125.16	C_5H_{11}	$OCOC_8H_{17}$	Cr 56.5 Col_h 126.7 I	[411]
125.17	C_6H_{13}	$OCOC_6H_2(OC_{12}H_{25})_3$	Cr 28 Col 109 I	[442]
125.18	C_5H_{11}	$O(CH_2)_3OH$	Cr 86 Col 96 Col_h 110 I	[401]
125.19	C_6H_{13}	$O(CH_2)_3CH=CH_2$	Cr 47.7 Col_h 96.4 I	[305]
125.20	C_6H_{13}	$O(CH_2)_3CF_3$	Cr 100 Col 146 I	[437]
125.21	C_6H_{13}	$O(CH_2CH_2O)_2CH_3$	Cr 36 I	[425]
125.22	C_6H_{13}	$O(CH_2)_5Si(CH_3)_2O\,Si(CH_3)_3$	Cr 2.5 Col_h 83.3 I	[305]
125.23	C_4H_9	$OCH_2COOC_4H_9$	Col_h 128.7 I	[448]
125.24	C_5H_{11}	$OCH_2COOC_2H_5$	Cr 64 Col 157 I	[438]
125.25	C_5H_{11}	$OCH_2COOC_4H_9$	Col_h 139 I	[438]
125.26	C_6H_{13}	$OCH_2COOC_2H_5$	Cr 50 Col 133 I	[438]
125.27	C_6H_{13}	$OCH_2COOC_4H_9$	Col_h 137 I	[438]
125.28	C_7H_{15}	$OCH_2COOC_2H_5$	Cr 61 Col 125 I	[438]
125.29	C_7H_{15}	$OCH_2COOC_4H_9$	Col_h 136 I	[438]
125.30	C_8H_{17}	$OCH_2COOC_2H_5$	Cr 58 Col 104 I	[438]
125.31	C_8H_{17}	$OCH_2COOC_4H_9$	Col_h 131 I	[438]
125.32	C_4H_9	$OCH_2CONHC_4H_9$	Cr 162 I	[448]
125.33	C_5H_{11}	$OCH_2CONHC_4H_9$	Col_p 144 Col 179 I	[448]
			Cr 105 Col_p 140 Col_h 178 I	[438]
125.34	C_6H_{13}	$OCH_2CONHC_4H_9$	Col_p 146 Col 188 I	[448]
			Cr 100 Col_p 143 Col_h 188 I	[438]
125.35	C_7H_{15}	$OCH_2CONHC_4H_9$	Col_p 138 Col 187 I	[448]
			Cr 91 Col_p 132 Col_h 184 I	[438]
125.36	C_8H_{17}	$OCH_2CONHC_4H_9$	Col_p 143 Col 192 I	[448]
			Cr 98 Col_p 142 Col_h 193 I	[438]
125.37	C_5H_{11}	OCH_2COO-cholesterol	Cr 192.5 I	[451]
125.38	C_5H_{11}	$O(CH_2)_5COO$-cholesterol	g 40 N_D^* 79.5 I	[451]
125.39	C_5H_{11}	$O(CH_2)_{10}COO$-cholesterol	g 26.2 N_D^* 57.9 I	[451]
125.40	C_7H_{15}	$O(CH_2)_5COO$-cholesterol	g 29.5 N_D^* 75.5 I	[451]
125.41	C_9H_{19}	$O(CH_2)_5COO$-cholesterol	g 19.2 N_D^* 64.3 I	[451]
125.42	$C_{11}H_{23}$	$O(CH_2)_5COO$-cholesterol	g 13.7 N_D^* 58.4 I	[451]
125.43	C_5H_{11}	Br	Cr 104 Col_h 195 I	[397]
125.44	C_5H_{11}	$C\equiv CSiMe_3$	Cr 117 Col_h 217 I	[397]

and 2.48, respectively. Table 2.49 shows the thermal behavior of heptaalkoxy-triphenylenes **126** (Figure 2.9), while the thermal behavior of 1-nitro-hexaalkoxy-TPs, **127** (Figure 2.9), is summarized in Table 2.50. The thermal data of other hepta-substituted triphenylenes are collected in Table 2.51. The octa-substituted triphenylenes **129–131** (Figure 2.9) are listed in Table 2.52. The thermal behavior of miscellaneous unsymmetrical triphenylenes with structure **132** (Figure 2.9) is summarized in Table 2.53.

As expected, the thermal behavior is largely dependent on the nature of the peripheral substituents and generalization is difficult. Some common observations are as follows. Unsymmetrical derivatives generally show reduced transition temperatures compared to corresponding

TABLE 2.49
Thermal Behavior of Heptaalkoxy-Triphenylenes with the General Structures 126 (Figure 2.9)

Structure	R	R_1	R_2	Thermal Behavior
126.1	C_3H_7	C_3H_7	C_3H_7	Cr 92.3 I
126.2	C_3H_7	C_3H_7	$C_{12}H_{25}$	Cr 50.3 I
126.3	C_4H_9	C_4H_9	C_3H_7	Cr 74.3 I
126.4	C_4H_9	C_4H_9	C_4H_9	Cr 65.7 Col 70.1 I
126.5	C_4H_9	C_4H_9	C_5H_{11}	Cr 58.7 Col$_h$ 69.3 I
126.6	C_4H_9	C_4H_9	C_6H_{13}	Cr 56.2 Col$_h$ 59.1 I
126.7	C_4H_9	C_4H_9	C_7H_{15}	Col$_h$ 47.6 I
126.8	C_5H_{11}	C_5H_{11}	C_4H_9	Cr 47.9 Col$_h$ 70.5 I
126.9	C_5H_{11}	C_5H_{11}	C_5H_{11}	Cr 45.7 Col$_h$ 75.0 I
126.10	C_5H_{11}	C_5H_{11}	C_6H_{13}	Cr 47.6 Col$_h$ 68.6 I
126.11	C_6H_{13}	C_6H_{13}	C_6H_{13}	Cr 44.6 Col$_h$ 69.5 I
126.12	C_7H_{15}	C_7H_{15}	C_7H_{15}	Cr 46.6 Col$_h$ 71.8 I
126.13	C_8H_{17}	C_8H_{17}	C_8H_{17}	Cr 44.3 Col$_h$ 68.7 I
126.14	C_8H_{17}	C_8H_{17}	$C_{10}H_{21}$	Col$_h$ 63.9 I
126.15	C_3H_7	C_3H_7	$COCH_3$	Cr 152.0 I
126.16	C_4H_9	C_4H_9	$COCH_3$	Cr 122.3 I
126.17	C_4H_9	C_4H_9	COC_3H_7	Col$_x$ 97.4 Col$_h$ 121.1 I
126.18	C_5H_{11}	C_5H_{11}	$COCH_3$	Cr 88.1 Col$_h$ 105.7 I
126.19	C_5H_{11}	C_5H_{11}	COC_4H_9	Col$_h$ 129.1 I
126.20	C_6H_{13}	C_6H_{13}	$COCH_3$	Cr 85.6 Col$_h$ 90.0 I
126.21	C_7H_{15}	C_7H_{15}	$COCH_3$	Cr 72.9 Col$_h$ 90.6 I
126.22	C_8H_{17}	C_8H_{17}	$COCH_3$	Cr 57.8 Col$_h$ 88.7 I
126.23	C_8H_{17}	C_8H_{17}	COC_7H_{15}	Cr 55.5 Col$_h$ 121.7 I
126.24	C_4H_9	C_5H_{11}	C_5H_{11}	Cr 60.6 Col$_h$ 73.5 I

Source: Data from Kumar, S. et al., *J. Mater. Chem.*, 10, 2483, 2000.

symmetrical derivatives. However, the melting and clearing temperatures depend upon the nature of unsymmetrical substitutions. In most cases, the maximum clearing temperature is found when all six alkoxy chains have an equal number of carbon atoms. Compounds with shorter and longer alkyl chains connected to the core alternatively also show clearing temperatures lower than that of shorter-chain hexaalkoxy-TPs. The interdigitation of chains is reported in such molecules [420].

In contrast to hexafluorinated chains, the use of a single long fluorinated chain in compound **116.34** markedly reduces the melting temperature and the compound is non-mesomorphic. However, the columnar phase is retained when a short fluorinated chain is attached via an ester group. The compound **117.15** possesses reduced stability when compared with the parent hexaalkoxy-TP. However, the mesophase range improves because not only the isotropic temperature decreases, but the melting temperature also decreases significantly due to the unsymmetrical nature of the system [296].

The replacement of one of the alkoxy chains of a hexaalkoxy-TP by an ester group results in a large effect on the nature and stability of the mesophase. Usually these monoalkanoyloxy-pentaalkoxy-TPs (Table 2.42) show a broader mesophase range than the parent hexaalkoxy or hexaalkanoyloxy-TP. When the ester group is bulky in nature, such as adamantanoate or pivaloate, a suppression of the crystallization and a stabilization of the mesophase were observed. In addition, these

TABLE 2.50
Thermal Behavior of α-Nitro-Triphenylene Derivatives 127
(Figure 2.9)

Structure	R	X	Phase Transition	Ref.
127.1	C_2H_5	C_2H_5	Cr 179 I	[288]
127.2	C_3H_7	C_3H_7	Cr 131.6 Col 147.5 I	[288]
127.3	C_4H_9	C_4H_9	Cr 61.2 Col$_h$ 141.3 I	[465]
			Cr 53 Col 143.3 I	[288]
127.4	C_5H_{11}	C_5H_{11}	Col$_h$ 141.4 I	[466]
			Cr < −20 Col 141.4 I	[288]
127.5	C_6H_{13}	C_6H_{13}	Cr 42.9 Col$_h$ 137.3 I	[474]
			Cr < −20 Col 137 I	[288]
			Col$_h$ 136 I	[384]
127.6	C_7H_{15}	C_7H_{15}	Col$_h$ 129.8 I	[474]
			Cr < −20 Col 129.3 I	[288]
127.7	C_8H_{17}	C_8H_{17}	Cr < −20 Col 128.4 I	[288]
127.8	C_9H_{19}	C_9H_{19}	Cr 0 Col 127 I	[288]
127.9	$C_{10}H_{21}$	$C_{10}H_{21}$	Cr 7.1 Col 118.4 I	[288]
127.10	$C_{11}H_{23}$	$C_{11}H_{23}$	Cr 37.3 Col 116 I	[288]
127.11	$C_{12}H_{25}$	$C_{12}H_{25}$	Cr 39.8 Col 107.7 I	[288]
127.12	$C_{14}H_{29}$	$C_{14}H_{29}$	Cr 52 Col 97 I	[288]
127.13	$C_{16}H_{33}$	$C_{16}H_{33}$	Cr 63.9 Col 86.1 I	[288]
127.14	C_4H_9	H	Cr 106 dec	[466]
127.15	C_5H_{11}	H	Cr 105 dec	[466]
127.16	C_4H_9	$COCH_3$	Cr 155 I	[466]
127.17	C_5H_{11}	$COCH_3$	Cr 129.8 Col$_h$ 139 I	[466]
127.18	C_4H_9	CH_2CH_2OH	Col 127.8 Col$_h$ 144.4 I	[466]
127.19	C_5H_{11}	CH_2CH_2OH	Cr 95.6 Col$_h$ 130.3 I	[466]
127.20	C_5H_{11}	$(CH_2CH_2O)_2COCH=CH_2$	Cr 68 Col$_h$ 141 I	[427]

compounds also exhibit a more ordered plastic columnar phase at lower temperature. Monoesters with a strongly polar group generally show higher melting points; an increase in the size of the ester group often decreases the mesophase range.

Replacing one of the alkoxy chains by a polar group such as Br or CN in a hexaalkoxytriphenylene greatly stabilizes the columnar phase, and its range is enhanced. The presence of a more ordered mesophase and the enhancement of fluorescence were also reported in some of these polar monofunctionalized triphenylene discotics. The replacement of a vicinal alkoxy chain by another Br atom increases the melting point significantly, and consequently the columnar phase range decreases. On the other hand, the replacement of the alkoxy chain at other positions gives compounds such as 2,4-dibromo- and 2,11-dibromo-tetraalkoxytriphenylene with better mesophase ranges. Both symmetrical and unsymmetrical tribromotrialkoxytriphenylenes show very high melting and clearing temperatures. The replacement of β-halogen by the more polar cyano group increases the stability of columnar phase significantly. This effect is also observed when only one of the two halogens is converted to a cyano group. The β-halogen(s) can be displaced by thioalkyl group(s) to give a series of mixed alkoxythioalkyl-TPs. The mesophase behavior of these compounds is dependent on the position and relative number of substituents.

The mesophase range of mixed tail symmetrical and unsymmetrical triphenylenes having three alkoxy and three alkanoyloxy substitutions was reported to be much higher when compared with

TABLE 2.51
Thermal Behavior of Hepta-Substituted Triphenylene Derivatives 128 (Figure 2.9)

Structure	R	X	Thermal Behavior	Ref.
128.1	C_4H_9	NH_2	Cr 102 I	[465]
128.2	C_4H_9	$NHCOCH_3$	Cr 132.9 Col_h 172.1 I	[465]
			Cr 144 Col_h 170 I	[468]
128.3	C_4H_9	Cl	Cr 55 Col 96 Col_h 112.2 I	[284]
128.4	C_6H_{13}	CH_3	Cr 60 I	[384]
128.5	C_6H_{13}	NH_2	Cr 53.6 Col_h 69.	[474]
			Cr 35 Col_h 77 I	[463]
128.6	C_6H_{13}	$NHCOCH_3$	Cr 99 Col_h 162 I	[463]
128.7	C_6H_{13}	$NHCOC_2H_5$	Cr 104 Col 167 I	[462]
128.8	C_6H_{13}	$NHCOC_3H_7$	Cr 98 Col 176 I	[462]
128.9	C_6H_{13}	$NHCOC_4H_9$	Cr 103 Col 186 I	[462]
128.10	C_6H_{13}	$NHCOC_5H_{11}$	Cr 96 Col 186 I	[462]
128.11	C_6H_{13}	$NHCOC_6H_{13}$	Cr 90 Col 191 I	[462]
128.12	C_6H_{13}	$NHCOC_7H_{15}$	Cr 90 Col 191 I	[462]
128.13	C_6H_{13}	$NHCOC_8H_{17}$	Cr 84 Col 192 I	[462]
128.14	C_6H_{13}	$NHCOC_9H_{19}$	Cr 78 Col 190 I	[462]
128.15	C_6H_{13}	$NHCOC_{12}H_{25}$	Cr 68 Col 181 I	[462]
128.16	C_6H_{13}	Cl	Cr 31.1 Col_h 96.1 I	[474]
			Cr 37 Col_h 98 I	[384]
128.17	C_6H_{13}	Br	Cr 37 Col_h 83 I	[473]
128.18	C_6H_{13}	F	Cr 39 Col_h 116 I	[473]
128.19	C_6H_{13}	N_3	Cr 70 Col_h 118 I	[384]
128.20	C_7H_{15}	NH_2	Cr 47 LC 75 I	[473]
128.21	C_7H_{15}	NHCO-2,3,5-triiodobenzene	Cr 43 LC 57 I	[473]
128.22	C_7H_{15}	NHCO-3,4,5-trihydroxybenzene	Cr 39 LC 45 I	[473]
128.23	C_7H_{15}	NHCO-4-nitrobenzene	Cr 5 LC 42 I	[473]

TABLE 2.52
Thermal Behavior of Octasubstituted Triphenylene Derivatives 129, 130, and 131 (Figure 2.9)

Structure	R	X	Phase Transition	Ref.
129.1	C_6H_{13}	F	Col 121.2 I	[492]
130.1	C_4H_9	NO_2	Cr 109 I	[466]
130.2	C_5H_{11}	NO_2	Cr 60.6 Col_h 96 I	[466]
131.1	C_7H_{15}	NO_2	Col_h 80–85 I	[466]
131.2	C_6H_{13}	CN	Col_h 236.9 I	[370]

hexaether or hexaester derivatives of triphenylene. Again, symmetrical mixed ether–ester derivatives are thermally more stable than unsymmetrical alkoxy-alkanoyloxy-TP derivatives. While mixed alkoxy, alkoxy-thioalkyl, alkoxy-alkanoyloxy, alkoxy-alkenyl, alkoxy-ketoalkyl compounds are generally liquid crystalline, mixed alkoxy-alkyl derivatives of triphenylene do not show mesomorphism.

TABLE 2.53
Thermal Behavior of Unsymmetrical Triphenylene Derivatives with the Structure 132 (Figure 2.9)

Structure	A	B	C	D	R	Phase Transition	Ref.
132.1	H	OC_6H_{13}	H	OC_6H_{13}	C_6H_{13}	Cr 70 (Col$_h$ 60) I	[462]
132.2	OC_6H_{13}	H	H	OC_6H_{13}	C_6H_{13}	Cr 83 I	[384]
132.3	H	OC_6H_{13}	H	H	C_6H_{13}	Cr 75 I	[384]
132.4	OCH_3	OCH_3	OCH_3	NO_2	C_6H_{13}	Cr 58 I	[462]
132.5	OC_6H_{13}	H	OC_6H_{13}	NO_2	C_6H_{13}	Cr 63 Col 89 I	[462]

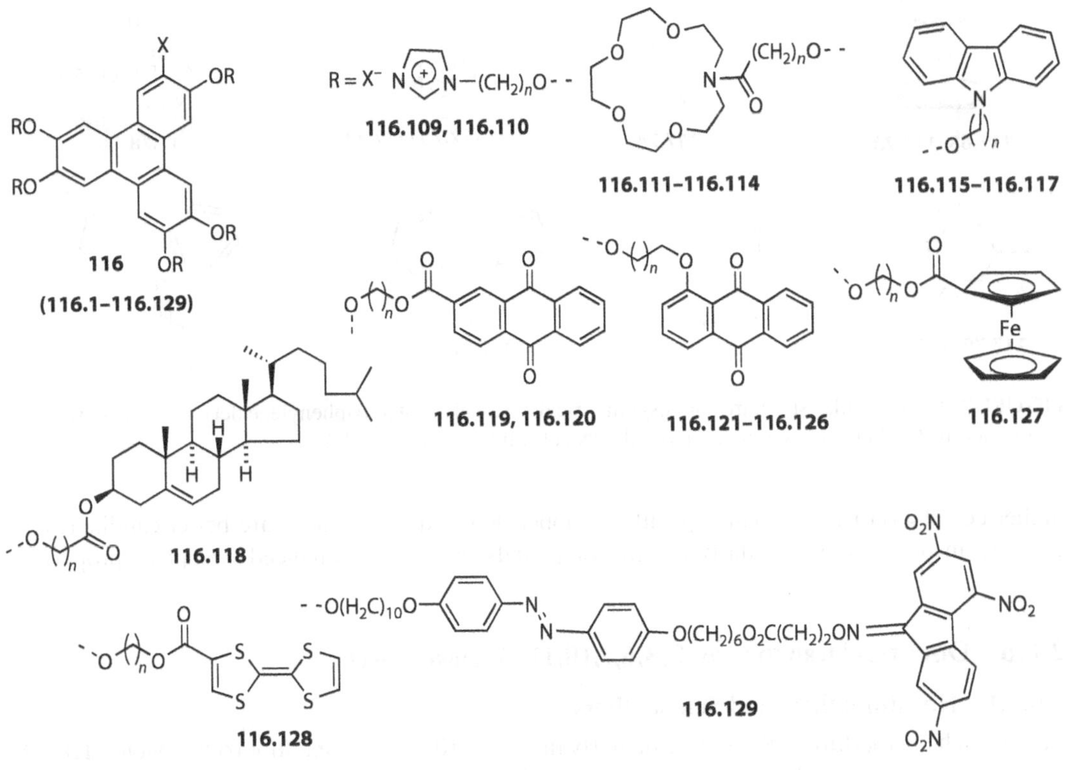

FIGURE 2.7 Molecular structures of monofunctionalized triphenylene pentaethers (Table 2.41).

α-Alkoxy substitution reduces both the melting and clearing temperatures compared with the hexaalkoxy-TPs. This may be due to the presence of the additional alkoxy chain and the steric hindrance caused by it. In the heptaalkoxy-TP series, the melting and clearing temperatures do not change significantly. In all the compounds, mesophases supercool to room temperature. No members of the series show any sign of crystallization upon storing at room temperature for long periods. Mixed ether–ester hepta-substituted triphenylenes show relatively higher isotropization temperatures. The presence of an unidentified high ordered mesophase in the lower homologue of this series was also reported. These mixed ether–ester derivatives have smaller core–core separations and

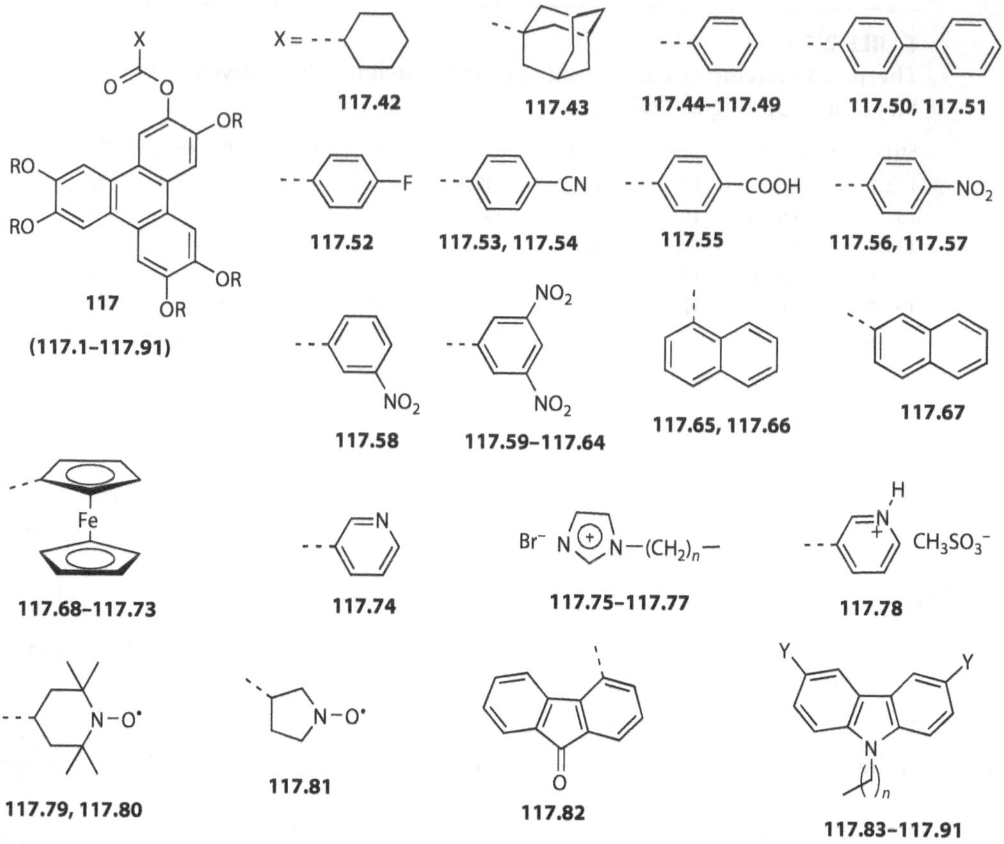

FIGURE 2.8 Molecular structures of unsymmetrical hexa-substituted triphenylene derivatives with the general structure **117** having one ester and five alkoxy substitutions (Table 2.42).

higher correlation lengths than heptaalkoxytriphenylenes and, therefore, are better candidates for charge transport studies. Similarly, α-nitro compounds also exhibit enhanced mesophase properties.

2.6.8 DISCOTICS DERIVED FROM 2,3,6,7,10,11-HEXABROMO-TP

2.6.8.1 Hexathioethers and Selenoethers

Interest in hexaalkylthio-TPs, and particularly in 2,3,6,7,10,11-hexahexylthiotriphenylene (HHTT) [493–500], arises from the fact that this material displays a unique helical columnar phase having very high photoconductivity. HHTT forms a self-organized helical columnar phase at low temperatures with nearly crystalline order, in addition to the normal columnar phase. The photo-induced charge carrier mobilities up to 0.1 cm² V s⁻¹ were achieved in the H phase. With the exception of organic single crystals, these were the highest electronic mobility values reported until recently. However, now much higher mobilities have been achieved in some DLCs. The synthesis of hexaalkylthio-TPs involves the thiolate anion substitution of 2,3,6,7,10,11-hexabromo-TP with an excess of sodium alkylthiolate in a polar aprotic solvent DMEU at 100°C for 1–2 h in 40%–55% yield (Scheme 2.57). An improvement in this process was reported by generating the thiolates with sodium hydride in dry ether, instead of sodium ethoxide in ethanol, and subsequently heating with hexabromo-TP in DMEU at 70°C for 30 min. This enhanced the yield to 78%–84% and the product was less contaminated with side products [498]. The poor yield and contamination could

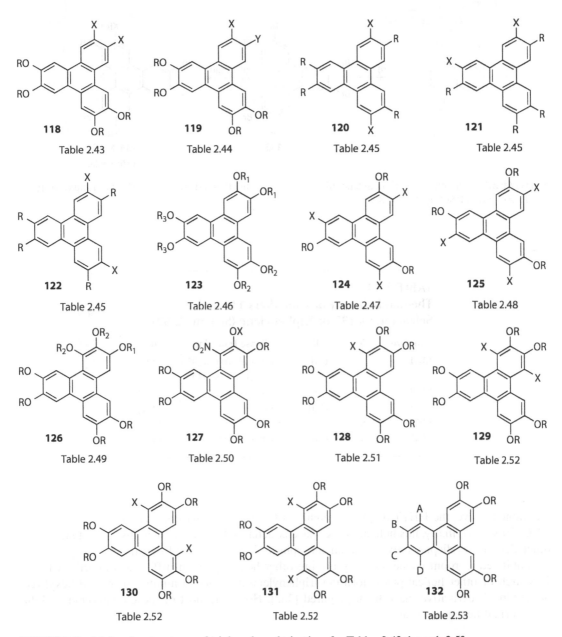

FIGURE 2.9 Molecular structures of triphenylene derivatives for Tables 2.43 through 2.53.

be due to the nucleophilic dealkylation reactions, where the initially formed hexaarylalkyl sulfide may be attacked by excess thiolate anion to produce dialkyl sulfide and thiophenolate. In order to re-alkylate these exposed thiophenolate groups, the reaction mixture was worked up with the appropriate 1-bromo- or 1-iodoalkane. This modification not only improved the yield substantially (~95%) but also afforded very high purity (~99.9%). Thus, thiolates were generated with potassium-t-butoxide in NMP and subsequently heated with hexabromo-TP at 70°C for 25 min in the same solvent, followed by quenching the reaction with an appropriate 1-bromoalkane to afford the desired products [500]. Similarly, selenoethers of triphenylene can also be prepared. The mesophase behavior of these thioethers and selenoethers is summarized in Table 2.54. As

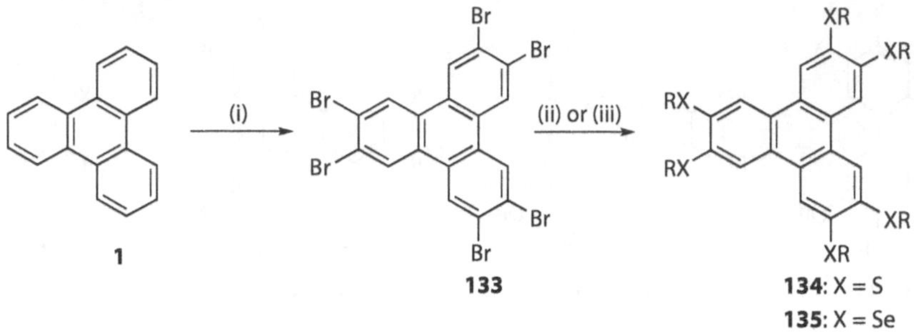

SCHEME 2.57 Synthesis of triphenylene thioethers and selenoethers: (i) Br$_2$, Fe, C$_6$H$_5$NO$_2$, reflux; (ii) RSK, NMP, 70°C; (iii) RSeNa, HMPA, 80°C.

TABLE 2.54
Thermal Behavior of Thioethers 134 and
Selenoethers 135 of Triphenylene (Scheme 2.57)

Structure	X	R	Phase Transition	Ref.
134.1	S	C$_6$H$_{13}$	Cr 62 Col$_h$ 70 Col$_h$ 93 I	[494]
			Cr 62 H 70 Col$_h$ 93 I	[497]
134.2	S	C$_8$H$_{17}$	Cr 55 Col$_h$ 87 I	[491]
134.3	S	C$_{10}$H$_{21}$	Cr 67 Col$_h$ 70 I	[491]
135.1	Se	C$_6$H$_{13}$	Cr 40 Col 72 I	[496]
135.2	Se	C$_7$H$_{15}$	Cr 40 Col 70 I	[496]

---

mentioned above, the HHTT displays a highly ordered helical phase in the temperature range of 62°C–70°C. Originally, this helical phase was characterized as a hexagonal disordered phase. All other derivatives display only columnar mesophase. It may be noted that the helical phase can be stabilized to room temperature by mixing other homologues in different proportions [500]. In search of other helical phase forming triphenylene discotics, a number of mixed hexyloxy-hexylthio-TP derivatives have been prepared [382]. However, most of these compounds exhibit only normal hexagonal phase.

2.6.8.2 Hexaalkynyltriphenylenes

Multiynes are interesting materials as they display discotic nematic phases. Palladium catalyzed cross-coupling of hexabromo-TP and *p*-alkyl-substituted phenylacetylene easily yields liquid crystalline hexaalkynyl derivatives of triphenylene (Scheme 2.58). Only two compounds in this series have been prepared [75,154]. Both exhibit a single enantiotropic nematic phase. Increasing the peripheral alkyl chain length decreases the isotropic temperature significantly. Thus compound **136.1** with R=C$_5$H$_{11}$ melts at 157°C and goes to the isotropic phase at 237°C, while the peripheral heptyl chain derivative **136.2** (R=C$_7$H$_{15}$) exhibits these transitions at 122°C and 176°C. Interestingly, straight-chain alkylacetylene derivatives, that is, without the benzene rings, do not exhibit any mesomorphism. The reduction of these alkylacetylenes can generate hexaalkyl-TPs, but these materials are more conveniently prepared from triphenylene hexatriflate via Pd-catalyzed cross coupling with alkylzink iodides [383].

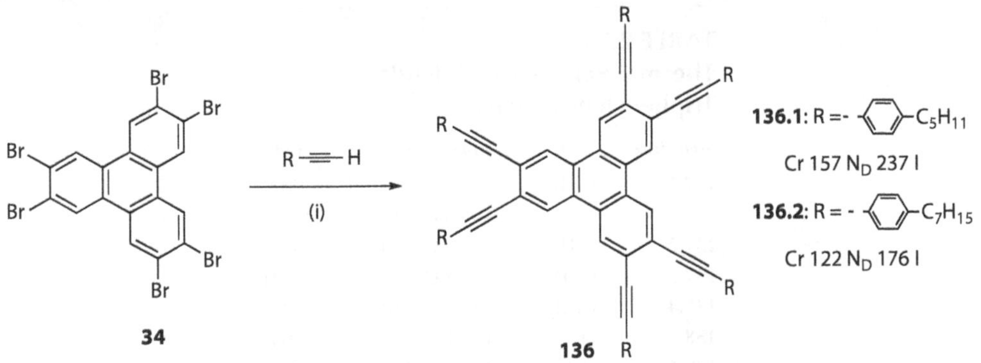

SCHEME 2.58 Synthesis of hexaalkynyltriphenylenes: (i) Pd(PPh$_3$)$_2$Cl$_2$/Cu/PPh$_3$/Et$_3$N, reflux.

2.6.8.3 Hexaphenyltriphenylenes

The hexaphenyltriphenylenes **137–140** are conveniently obtained via palladium-catalyzed cross-coupling between alkoxyphenylboronic acid and 2,3,6,7,10,11-hexabromo-TP (Scheme 2.59). To investigate the effect of variation in the position of an alkoxy chain in the phenyl ring, *ortho*-, *meta*-, and *para*-substituted derivatives were prepared and studied for mesomorphic properties [227,501–505].

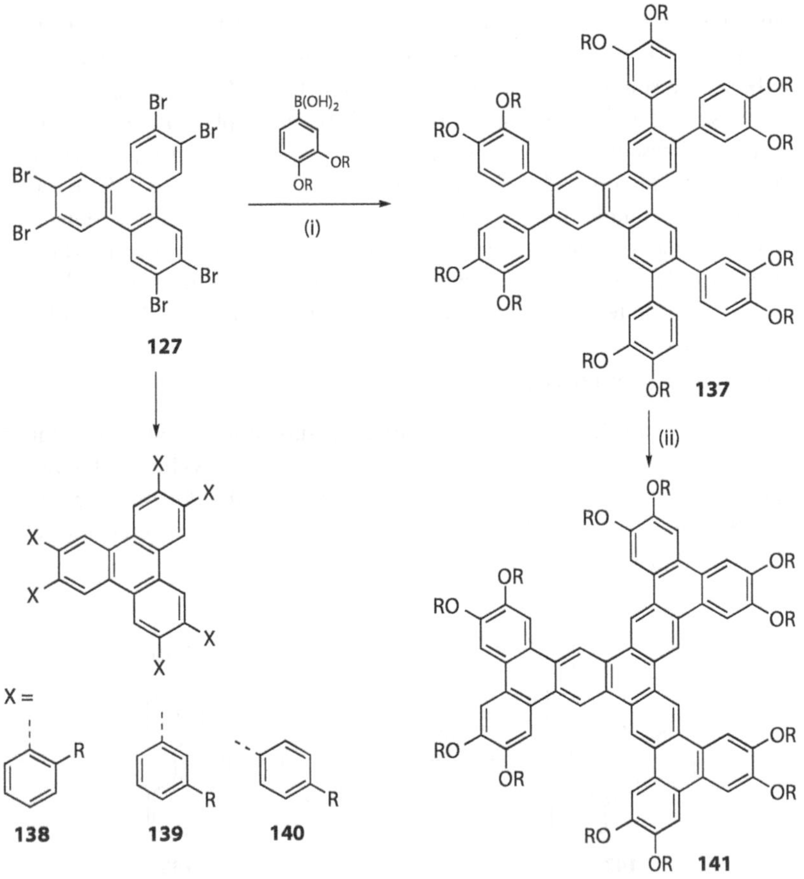

SCHEME 2.59 Synthesis of extended triphenylenes: (i) Pd(0), toluene/EtOH/H$_2$O, reflux; (ii) FeCl$_3$, CH$_2$Cl$_2$.

TABLE 2.55
Thermal Behavior of Extended Triphenylenes (Scheme 2.58)

Structure	R	Phase Transition	Ref.
137.1	C_6H_{13}	Cr 111 Col 126 I	[502]
		Cr 65 Col 135 I	[501]
137.2	C_8H_{17}	Cr 85 Col 104 I	[502]
137.3	$C_{10}H_{21}$	Cr 74 Col 103 I	[502]
137.4	$C_{12}H_{25}$	Cr 47 Col 101 I	[502]
138.1	OC_6H_{13}	Cr 81.4 I	[505]
139.1	OC_6H_{13}	Col$_h$ 111 I	[505]
140.1	OC_6H_{13}	Cr 153 I	[503]
140.2	C_9H_{19}	Cr 59 I	[503]
140.3	$OC_{11}H_{23}$	Cr 66 I	[503]
140.4	$C_{12}H_{23}$	Cr 36 I	[503]
141.1	C_6H_{13}	Cr 180 Col 430 I	[502]
141.2	C_8H_{17}	Cr 150 Col 370 I	[502]
141.3	$C_{10}H_{21}$	Cr 122 Col 322 I	[502]
141.4	$C_{12}H_{25}$	Cr 104 Col 306 I	[502]

In polyphenylated triphenylenes, a single *para*-alkoxy chain per phenyl ring is insufficient to induce mesomorphism (Table 2.55). This may be because of the unfilled space around the large core. Dodecaalkoxy-substituted polyphenylated triphenylenes **137** are liquid crystalline. Stable columnar mesophases are realized when an alkoxy substituent is placed in the meta-position but not in the ortho-position. These large core liquid crystalline as well as non-liquid crystalline triphenylenes have recently been mixed with several hexaalkoxytriphenylenes to enhance their mesophase range. It is reported that complementary polytropic interactions (CPI) are responsible for the stabilizing effect. The cyclodehydrogenation of dodecaalkoxy-substituted polyphenylated triphenylenes **137** using ferric chloride produced triphenylene trimer **141**. These tristriphenylenes exhibit much higher melting and clearing temperatures than uncyclized triphenylenes **137** (Table 2.55).

2.6.9 TRISUBSTITUTED TRIPHENYLENE DISCOTICS

Usually triphenylenes with less than six peripheral substitutions are not liquid crystalline. Exceptionally, a few alkyl esters of triphenylene 2,6,10-tricarboxylic acid have recently been claimed to be liquid crystalline [227]. These were prepared by the dehydrating cyclotrimerization of 4-methylcyclohexanone with $ZrCl_2$, as shown in Scheme 2.60. The thermal behavior of these derivatives is presented in Table 2.56.

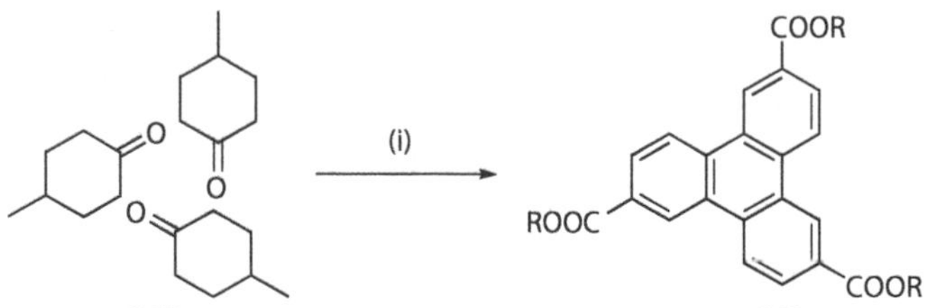

SCHEME 2.60 Synthesis of triphenylene triesters: (i) $ZrCl_2$; Pd; $Na_2Cr_2O_7$; $SOCl_2$; ROH.

TABLE 2.56
Thermal Behavior of Trisubstituted
Triphenylene Derivatives (Scheme 2.60)

Structure	R	Phase Transition
143.1	CH_3	Cr 258 I
143.2	C_2H_5	Cr (135 Col) 161 I
143.3	C_3H_7	Cr (82 N_D) 137 I
143.4	C_4H_9	Cr (59 N_D) 86 I
143.5	C_5H_{11}	Cr 111 I
143.6	C_6H_{13}	Cr 101 I
143.7	C_8H_{17}	Cr 77 I
143.8	$CH_2CH(C_2H_5)C_4H_9$	Cr 111 Col 124 I

Source: Data from Hassheider, T. et al., *Angew. Chem. Int. Ed.*, 40, 2060, 2001.

2.6.10 PHYSICAL STUDIES

Triphenylene-based DLCs have been extensively studied for various physical properties such as structural analysis [506–526], molecular ordering and dynamics [527–560], alignment [561–597] and self-assembling behavior on surfaces [598–626,1500], electro-optical properties [627–666], charge and energy migration [667–672, see Chapter 6 also], charge-transfer complexation [673–682], high-pressure studies [683–688], theoretical [689–724], and other studies. A detailed discussion of these properties is beyond the scope of this chapter.

2.7 PERYLENE CORE

The perylene nucleus **1** consists of two naphthalene molecules linked together at the 1- and 8-positions. Excellent electronic absorption and fluorescence emission properties of this material have received attention of scientists working in various areas of materials science. Depending on the substituent position and the degree of derivatization, perylene derivatives can fluoresce anywhere from yellow to red region. Therefore, perylene-based compounds are proficient materials for various applications such as organic solar cells [725,726], field effect transistors [727,728], light emitting diodes [729,730], etc.

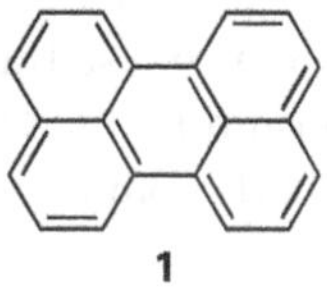

1

In order to combine the excellent electroluminescence behavior of perylene derivatives with efficient charge transport and processing properties of columnar liquid crystals, a number of liquid crystalline derivatives of perylene have been prepared. Perylene-based liquid crystalline materials can be categorized according to the type of functional group connecting the central perylene core to peripheral substituents. Accordingly, these are of two types: ester based and imide based.

2.7.1 3,4,9,10-TETRA-(N-ALKOXYCARBONYL)-PERYLENES

Various synthetic methods to prepare the tetracarboxylate of perylene have been reported in the literature [213,227,228,731–738]. Refluxing a mixture of dimethyl sulfate and

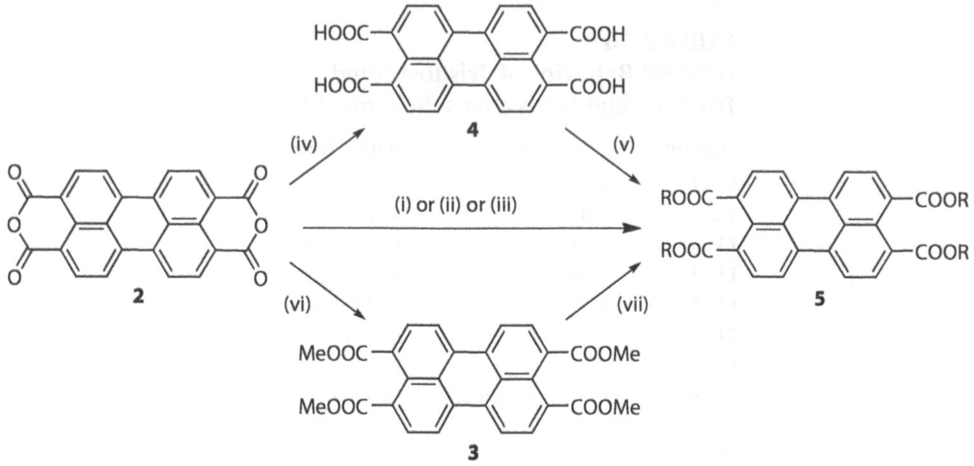

SCHEME 2.61 Synthesis of perylene tetraesters: (i) ROH/RBr (1:1), K_2CO_3, 7 days; (ii) ROH/RBr (1:1), DBU, CH_3CN, 12–16 h; (iii) (a) KOH, H_2O, 70°C, (b) HCl, (c) $(C_8H_{17})_4$NBr [TOAB], RBr, reflux, 2 h; (iv) aq KOH, reflux, aq H_2SO_4 or HCl; (v) $(C_8H_{17})_4$NBr [TOAB], RBr, reflux, 2 h; (vi) $(CH_3)_2SO_4$, aq NaOH, reflux, 24 h, aq HCl; (vii) ROH, RONa.

perylene-3,4,9,10-tetracarboxylic acid dianhydride **2** in aqueous NaOH for 24 h followed by neutralization with aq HCl furnished 3,4,9,10-tetrakis(carbimethoxy)perylene **3** (Scheme 2.61) [731]. The transesterification of perylenetetracarboxylate tetramethyl ester **3** with alkanol in the presence of sodium alkanoate gave 3,4,9,10-tetra-(n-alkoxycarbonyl)-perylene **5** in good yield [732]. These tetraester **5** can also be prepared by reacting perylene dianhydride **2** with equimolar amount of alkyl bromide or iodide and alkanol in the presence of a base. Vigorous heating of perylene-3,4,9,10-tetracarboxylic acid dianhydride **2** with K_2CO_3 in a 1:1 mixture of alkyl halide and alkanol for 2–7 days afforded 3,4,9,10-tetra-(n-alkoxycarbonyl)-perylene **5** in a single step [227,733,734]. Another attractive alternative of this time-consuming reaction was the replacement of potassium carbonate as base by 1,8-diazabicyclo[5.4.0]undec-7-ene (DBU) [735]. Efficient product formation was achieved by heating to reflux a mixture of perylene-3,4,9,10-tetracarboxylic acid dianhydride **2**, alkyl bromide, alkanol [1:1 molar ratio], and DBU in acetonitrile for overnight. Another route [736–738] goes via hydrolysis of perylene-3,4,9,10-tetracarboxylic acid dianhydride **2** by refluxing it in aqueous potassium hydroxide followed by acidification to pH 8–9 with aqueous dilute H_2SO_4 [736] or HCl [737,738] to provide perylene-3,4,9,10-tetracarboxylic acid **4**. The use of tetraoctylammonium bromide (TOAB) [737,738] as phase transfer catalyst is a modified way of esterification to avoid complicated and time-consuming mode of conventional transesterification [732]. Refluxing vigorously a mixture of **2** and alkyl bromide in water and TOAB for about 2 h produces tetraalkylester **5** in good yield [737,738].

Though the tetraalkyl esters of perylene were reported previously [732], the liquid crystalline property of these materials was first identified by Bennings et al. in 2000 [227,733]. The thermal behavior of alkyl esters of perylene-3,4,9,10-tetracarboxylic acid **5** is given in Table 2.57. All compounds, but very short and very long peripheral chains, are liquid crystalline and exhibit hexagonal columnar phases. A regular decrease in melting as well as isotropic temperatures was observed for all the liquid crystalline derivatives on increasing the peripheral chain length. The use of branched chains instead of normal alkyl chains suppresses the melting point significantly; compound **5i** shows a very broad mesophase, which is stable at room temperature. Lattice parameter measurement of some of the perylene tetraesters revealed the presence of matching columnar arrangement in crystal as well as in liquid crystalline phase for lower homologues **5b–5g**, but the same was not true in the

TABLE 2.57
Thermal Behavior of Discotic Perylene Tetraalkyl Esters

Structure	R	Phase Transition	Ref.
5a	CH_3	Cr 322 I	[227]
5b	C_2H_5	Cr 244 Col_h 313 I	[733]
5c	C_3H_7	Cr 193 Col_h 287 I	[733]
5d	C_4H_9	Cr 161 Col_h 242 I	[733]
		Cr 164.3 Col_h 242.6 I	[738]
5e	C_5H_{11}	Cr 114 Col_h 203 I	[733]
		Cr 114 Col_h 210 I	[734]
5f	C_6H_{13}	Cr 72 Col_h 177 I	[733]
		Cr 85.7 Col_h 187.3 I	[738]
5g	C_7H_{15}	Cr 63 Col_h 152 I	[733]
		Cr 63.8 Col_h 137.1 I	[738]
5h	C_8H_{17}	Cr 62 Col_h 132 I	[733]
5i	$CH_2CH(CH_2CH_3)C_4H_9$	Col_h 240 I	[227]
5j	C_9H_{19}	Cr_1 44 Cr_2 60 Col_h 100 I	[733]
5k	$C_{10}H_{21}$	Cr 51 I	[733]
		Cr 69.8 I	[738]
5l	$C_{12}H_{25}$	Cr 81.2 I	[738]

case of longer chain homologues **5h**, **5j**. Tetraesters **5b** and **5e** are solid at room temperature in their pure form, but a mixture of these two compounds in equimolar amount showed room temperature glassy columnar liquid crystalline phase.

2.7.2 Perylene Bisimides

Perylene tetracarboxylic acid bisimides commonly known as perylene bisimides (PBIs) are readily available, inexpensive, and thermally very stable materials. They are well known in the dye and pigment industry for a long time. More recently their applications in organic electronics have been realized [739–741]. PBIs are among the best n-type organic semiconductors known to date. The self-assembly of these derivatives has their relevance as efficient component in solar cells [725,742,743], organic light emitting diodes (OLED) [744–746], and organic thin film transistors (OTFT) [728,747–749]. During the last two decades, a number of PBI derivatives have been synthesized and explored for excellent electronic and liquid crystalline properties [750–763].

The existence of mesomorphic behavior in PBIs was first realized by Cormier and Gregg [750,751]. These materials can be easily prepared by condensing perylene-3,4,9,10-tetracarboxylic dianhydride **2** with suitable primary amines in the presence of pyridine at 100°C–125°C (Scheme 2.62) [750,751]. Currently, the most popular method for the synthesis of PBI derivatives involves zinc acetate catalyzed condensation of appropriate alkyl or aryl amine with perylene-3,4,9,10-tetracarboxylic acid dianhydride **2**.

A number of PBI derivatives **6** with normal alkyl, oligoethylene, and other branched chains have been prepared. Preference has been given to polyoxyethylene chains over normal alkyl chains in order to increase conformational flexibility, reduce transition temperature, and hence to stimulate mesomorphic behavior [750,751]. Some of these were reported to display liquid crystalline properties, but the exact nature of the mesophase has not been revealed. As the shape of such molecules is more or less rod-like, most likely they form smectic phases. In order to induce columnar phases,

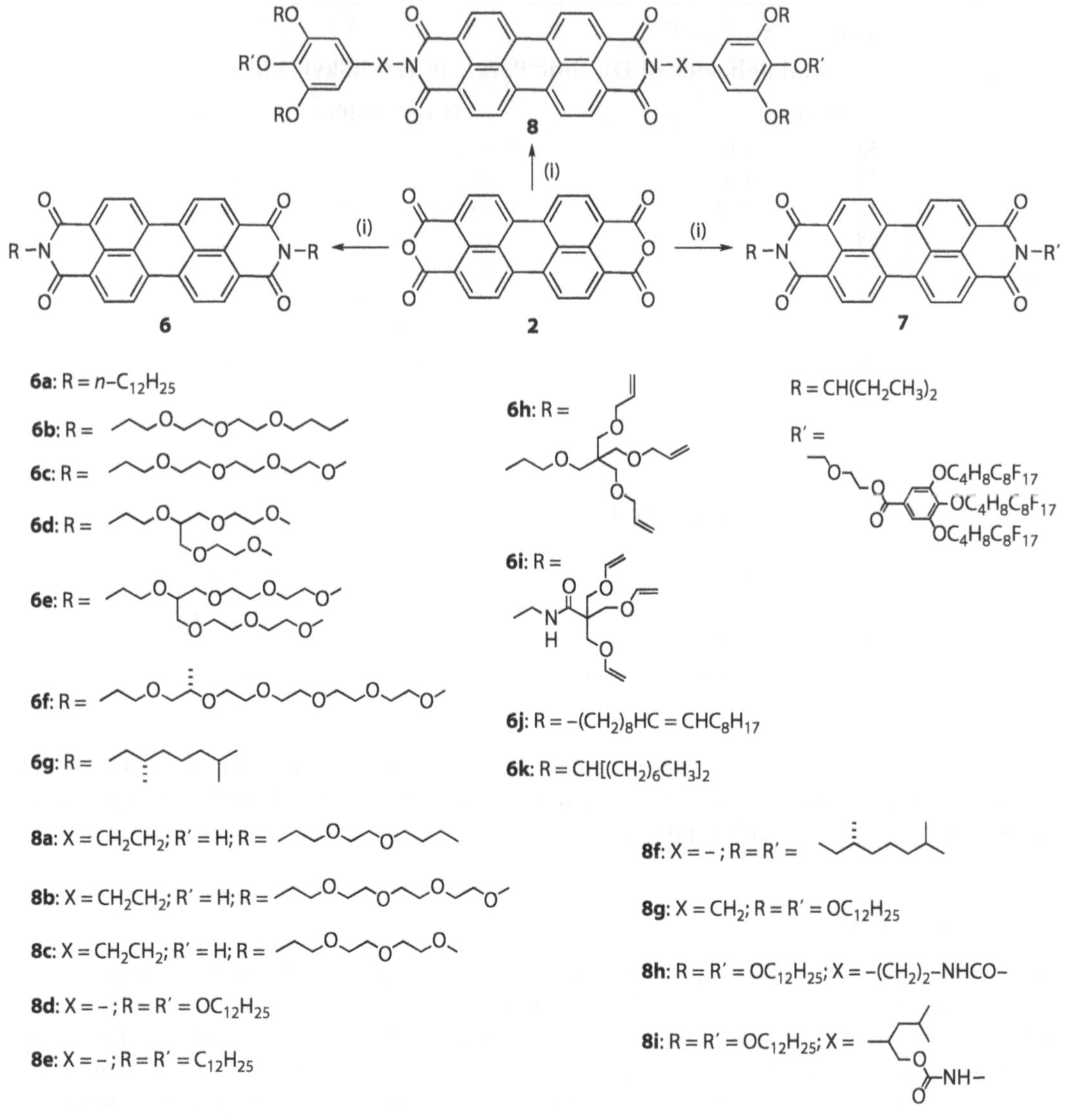

6a: R = n-C$_{12}$H$_{25}$

6b: R =

6c: R =

6d: R =

6e: R =

6f: R =

6g: R =

6h: R =

6i: R =

6j: R = $-(CH_2)_8HC = CHC_8H_{17}$

6k: R = $CH[(CH_2)_6CH_3]_2$

R = $CH(CH_2CH_3)_2$

R′ =

OC$_4$H$_8$C$_8$F$_{17}$
OC$_4$H$_8$C$_8$F$_{17}$
OC$_4$H$_8$C$_8$F$_{17}$

8a: X = CH$_2$CH$_2$; R′ = H; R =

8b: X = CH$_2$CH$_2$; R′ = H; R =

8c: X = CH$_2$CH$_2$; R′ = H; R =

8d: X = – ; R = R′ = OC$_{12}$H$_{25}$

8e: X = – ; R = R′ = C$_{12}$H$_{25}$

8f: X = – ; R = R′ =

8g: X = CH$_2$; R = R′ = OC$_{12}$H$_{25}$

8h: R = R′ = OC$_{12}$H$_{25}$; X = $-(CH_2)_2$–NHCO–

8i: R = R′ = OC$_{12}$H$_{25}$; X =

SCHEME 2.62 Synthesis of perylene bisimides: (i) (a) RNH$_2$ or ArNH$_2$, Zn(OAc)$_2$, quinoline or imidazole, 130°C–180°C, 2–3 h or (b) RNH$_2$, pyridine, 100°C–125°C.

bulky groups were attached on the imide N atoms directly or via other groups (Scheme 2.62) as well as in the bay positions (Scheme 2.63).

The chlorination of the dianhydride **2** was reported in a patent literature [764] to produce the tetrachlorinated product **13** (Scheme 2.63). The 1,7-dibrominated bisimide **9** was achieved by treating a mixture of **2**, iodine, sulfuric acid and this was followed by drop-wise addition of bromine at 80°C–85°C over a period of 10 h [765]. These halogenated compounds can be converted to liquid crystalline imides **10**, **16** by treating with amines. Further, halogen substituents in the bay regions are reactive toward substitution with nucleophiles. In a typical reaction, halogenated bisanhydrides were reacted with phenol in the presence of potassium carbonate and NMP at 80°C to afford phenoxy-substituted bisanhydrides **11**, **14**, which can be condensed with various amines to produce liquid crystalline imides **12**, **15**. Some of these exhibit discotic liquid crystalline properties. The thermal behavior of these PBI derivatives is summarized in Tables 2.58 and 2.59.

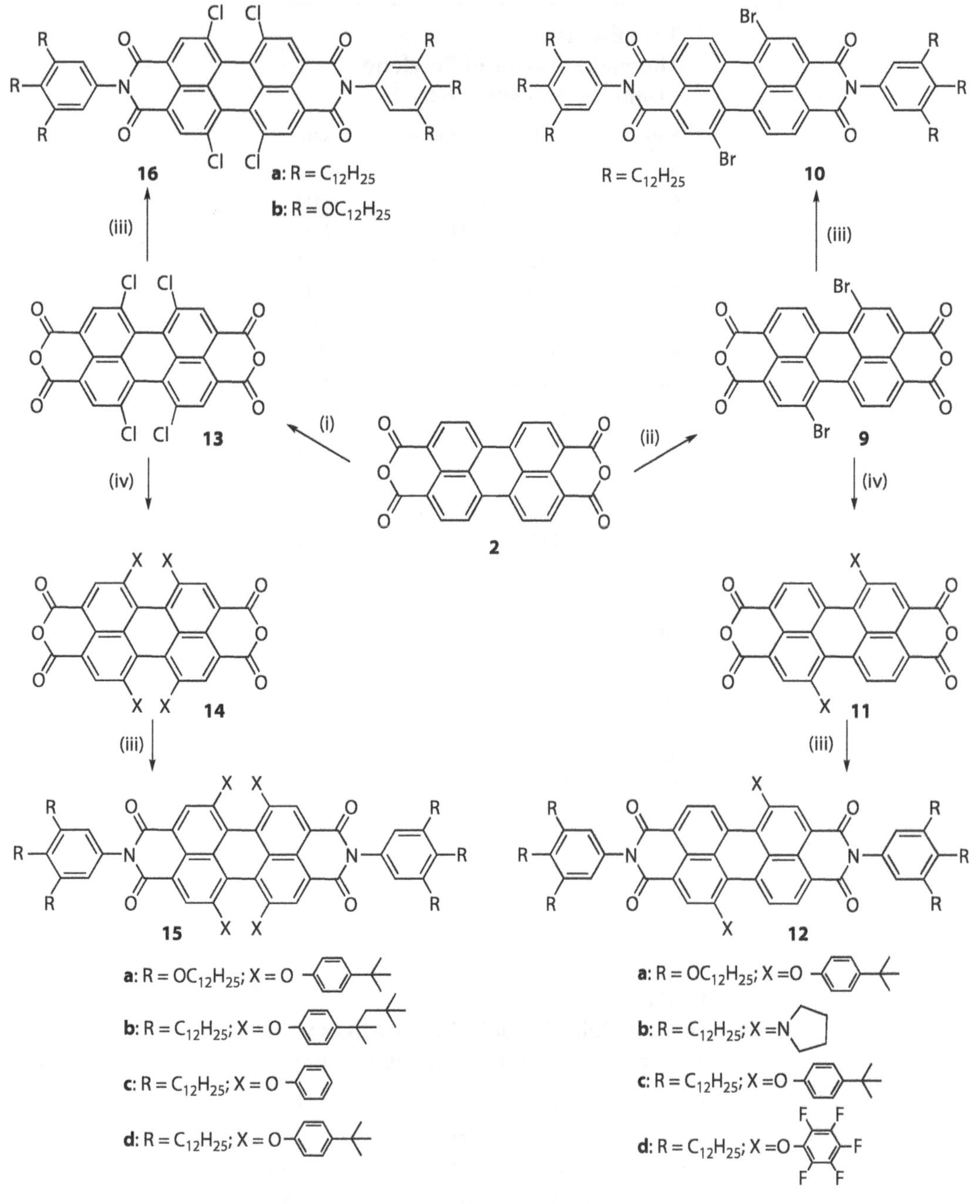

SCHEME 2.63 Synthesis of bay-positions substituted perylene bisimides: (i) sulfuryl chloride, $C_6H_5NO_2$; (ii) Br_2, I_2, H_2SO_4, 85°C, 10 h; (iii) $ArNH_2$, $Zn(OAc)_2$, quinoline; (iv) ArOH, NMP, K_2CO_3, 80°C.

Compound **6a** with normal alkyl chains does not show mesomorphism. The presence of polyethyleneoxy chains induces mesomorphism; however, the shape of these materials is more calamitic than discotic. Columnar phases can be generated by filling the space around the core with the use of di- or trisubstituted phenyl group. Bulky substitution at one side of the molecule is sufficient to induce columnar phase in compound **7** as the space around the core is filled due to the antiparallel supramolecular packing. Non-symmetric derivative **7** showed two enantiotropic columnar rectangular mesophases. Compound **8b** with the highest number of oxygen atoms in side chains

TABLE 2.58
Thermal Behavior of Perylene Bisimides (Scheme 2.62)

Structure	Phase Transition	Ref.
6a	Non-LC	[751]
6b	Cr 73 M 289 I	[751]
6c	Cr 122 M 269 I	[751]
6d	Cr 122 M 167 I	[751]
6e	Cr −43 M 55 I	[751]
6f	Cr (185 M) 192 I	[753]
6g	Cr 366 I	[753]
6h	Non-LC	[751]
6i	Non-LC	[751]
6j	Cr 178 Col_r 292 I	[756]
6k	Cr 130 I	[757]
7	Col_r 137 $Col_{r'}$ 200 I	[760]
8a	M 136 I	[751]
8b	M 119 I	[751]
8c	M 169 I	[751]
8d	Col_h 373 I	[752]
8e	Cr 85 Col 300 I	[762]
8f	Col_h 349 I	[763]
8g	Cr −46 Col_h 219 I	[753]
8h	Cr 126 Cr (plastic) 235 I	[758]
8i	Non-LC	[753]

TABLE 2.59
Thermal Behavior of PBI Derivatives with Bay Region Substituents (Scheme 2.63)

Structure	R	Phase Transition	Ref.
10	$C_{12}H_{25}$	Col_h 160 I	[761]
12a	$OC_{12}H_{25}$	Col_h 283 I	[752]
12b	$C_{12}H_{25}$	Non-LC	[761]
12c	$C_{12}H_{25}$	Col_h 130 I	[761]
12d	$C_{12}H_{25}$	Non-LC	[761]
15a	$OC_{12}H_{25}$	Col_h 346 I	[752]
15b	$OC_{12}H_{25}$	Col_h 283 I	[752]
15c	$OC_{12}H_{25}$	Cr 243 I	[752]
15d	$C_{12}H_{25}$	Col_h 285 I	[761]
16a	$OC_{12}H_{25}$	Cr 86 Col 214 I	[754]
16b	$C_{12}H_{25}$	Col_r 110 I	[761]

was in liquid crystalline state at ambient temperature. This compound also exhibits the lowest clearing point as well as enthalpy change corresponding to isotropization as compared to the other two derivatives **8a** and **8c**. The substitution of methylene group for one oxygen atom (compare **8a** vs. **8b**) and the elimination of ethoxy units (compare **8c** with **8b**) from the side chains tend to increase the isotropization temperature and consequently widening of mesophase range. The PBI derivative **8f** with six branched chiral alkyl chains [763] exhibits higher values of enthalpy and entropy change corresponding to the isotropization process. This indicates a higher order of columnar mesophase in **8f** as compared to its normal dodecyloxy and dodecyl chain counterparts **8d** and **8e**. Compound **8g** is similar to **8d**, but having additional methylene spacer between imide nitrogen and 3,4,5-tridodecyoxyphenyl group. The smaller methylene spacer reduces the rigidity as well as co-planarity of perylene ring and benzene ring. Consequently, the isotropization temperature of **8g** was much lower compared to the isotropic temperature of **8d**. The replacement of methylene spacer with long spacer containing amide linkage as in compound **8h** results in loss of mesomorphism. Compound **8h** exhibits only crystal–crystal polymorphism. This could be because of strong intermolecular interactions due to hydrogen bonding. The PBI derivative **8i** was glassy and goes to the isotropic state at 220°C.

Wurthner and coworkers [752] reported a series of PBI derivatives **8d, 12a, 15a–c** in which the nitrogen atoms of imide ring were linked to mesogenic 3,4,5-tridodecyoxyphenyl group without using any spacer and the hydrogen atoms at bay positions (1, 6, 7, 12) were replaced by zero (**8d**), two (**12a**), and four (**15a–c**) phenoxy substituents. Out of those, four compounds **8d, 12a, 15a, 15b** were found to exhibit hexagonal columnar phases. Compound **15c** did not display any mesomorphism and showed layered crystalline structure. The presence of non-alkylated phenoxy groups at bay positions resulted in extra lateral π–π interactions and therefore caused stacking of π systems in a second dimension. This phenomenon destroyed mesomorphism in **15c**. Whereas for compounds **12a, 15a–b**, containing phenoxy unit with bulkier alkyl chain in the bay positions, additional packing of π system is restricted by dodecyloxy chains on the periphery and *tert*-alkyl groups on the phenoxy substituents. Consequently, these derivatives showed hexagonal columnar phase over a wide temperature range. Increase in the bulkiness of alkyl group at phenoxy substituent (compare **15b** vs. **15a**) caused a decrease in isotropization temperature.

Halogen substituents at the bay area impart significant effect on mesomorphic behavior. Thus, the clearing temperature of tetrachloro-substituted PBI **16a** was about 159°C lower than that of **8d**. Contrary to **8d**, this compound was crystalline at room temperature in its pristine state. Similar behavior was observed in case of **8e, 16b**, and **10**. While **8e** cleared at around 300°C, its halogenated derivatives **16b** and **10** have isotropic temperatures of 110°C and 160°C, respectively.

In PBI derivatives **8e, 10, 12b–d, 15d, 16b** having dodecyl chains instead of dodecyloxy peripheral chains, the introduction of two (**12c**) or four (**15d**) *p*-tertiaryalkylphenoxy substituents in the bay region decreases isotropization temperature as compared to the unsubstituted compound **8e**. On comparing clearing points of compounds of this series **8e, 10, 12b–d, 15d, 16b** with those of structurally related compounds **8d, 12a, 15a–b** discussed above, it is also notable that the isotropic temperature decreases on replacing dodecyloxy with dodecyl peripheral chain length.

Perylene derivatives of the type **19** (2,9-dialkyl-5,6,11,12-tetrahydrocoronene-5,6,11,12-tetracarboxylic acid bisphenyl- or alkylimide) were investigated for their mesomorphic behavior by Mullen and coworkers [766–768]. The synthesis of these compounds is presented in Scheme 2.64. The cycloaddition of dienophiles to perylene under normal conditions results only monoaddition product. A low yield of diaddition product could be achieved only under drastic conditions. This is because of less solubility and reduced reactivity of the enophile produced after monoaddition reaction. Reasonable yield of **19** was possible by treating 3,10-dialkyl perylene **17** with excess amount of 4-*n*-alkyl-3,5-dioxatriazole **18** in xylene and refluxing the reaction mixture. Small portions of **18** were added to the above reaction mixture until complete disappearance of yellow starting material as well as red monoaddition product.

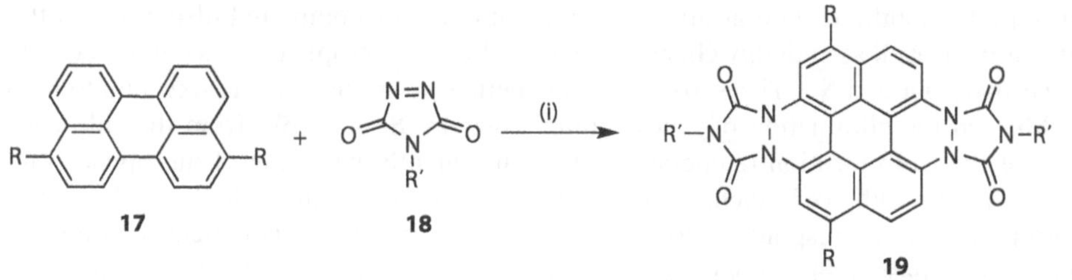

SCHEME 2.64 Synthesis of 2,9-dialkyl-5,6,11,12-tetrahydrocoronene-5,6,11,12-tetracarboxylic acid bisalkylimide derivatives: (i) xylene, reflux.

TABLE 2.60
Thermal Behavior of 2,9-Dialkyl-5,6,11,12-Tetrahydrocoronene-5,6,11,12-Tetracarboxylic Acid Bisalkylimide Derivatives (Scheme 2.64)

Structure	R	R′	Phase Transition	Ref.
19a	C_5H_{11}	C_5H_{11}	Cr 27 Col 319 I	[766]
19b	C_5H_{11}	$C_{17}H_{35}$	Cr 85 Col$_h$ 202 I	[766]
19c	$C_{11}H_{23}$	$C_{12}H_{25}$	Cr −20 Col$_h$ 373 I	[766]
19d	$C_{12}H_{25}$	$C_{11}H_{23}$	Cr −20 Col$_h$ 373 I	[766]
19e	$C_{12}H_{25}$	$C_{17}H_{35}$	Cr 15 Col$_h$ 316 I	[766]
19f	$C_{18}H_{37}$	C_5H_{11}	Cr 69 Col 261 I	[766]
19g	$C_{18}H_{37}$	$C_{17}H_{35}$	Cr 45 Col$_h$ 305 I	[766]
			Cr 82 Col$_h$ 303 I	[767]
19h	H	$C_{17}H_{35}$	Cr 69 Sm 341 I	[767]

The thermal behavior of these derivatives **19** is summarized in Table 2.60. All the derivatives display very broad mesophase. Deviation from the symmetry, that is, difference between the length of alkyl chains (R ≠ R′) leads to decrease in isotropization temperature (compare **19b** vs. **19a**; **19e** vs. **19d**; **19f** vs. **19g**). The disubstituted derivative **19h** having only peripheral alkyl chains connected to hetrocyclic triazole ring exhibits, as expected, a layered structure.

Thermally as well as photochemically stable blue perylene imide derivatives **21** were prepared by Muller et al. [769]. Diels–Alder reaction of *N*-alkylated perylene monoimide **20** with *N*-pentadecyl-1,2,4-triazoline-3,5-dione **18′** in boiling xylene yields the T-shaped mesogens **21** (Scheme 2.65). The low reactivity of **20** and poor stability of **18′** demand addition of large excess (10 equivalent) of **18′** in several portions to a solution of **20** in refluxing *m*-xylene to achieve **21** in good yield.

All the three derivatives show columnar phases with very high isotropization temperatures (Scheme 2.65). High-temperature mesophase of **21a** and **21b** was proved to be ordered hexagonal columnar with an intracolumnar distance of 3.4 Å and intercolumnar distance of 27.4 Å. Low-temperature phase of **21a** exhibits non-uniformly organized hexagonal order. The exact morphology of the low-temperature columnar mesophase of **21b** could not be deduced successfully because of more complex x-ray patterns. The replacement of normal peripheral alkyl chain by more rigid 2,6-diisopropylphenyl group in **21c** imparts a striking influence on thermal behavior.

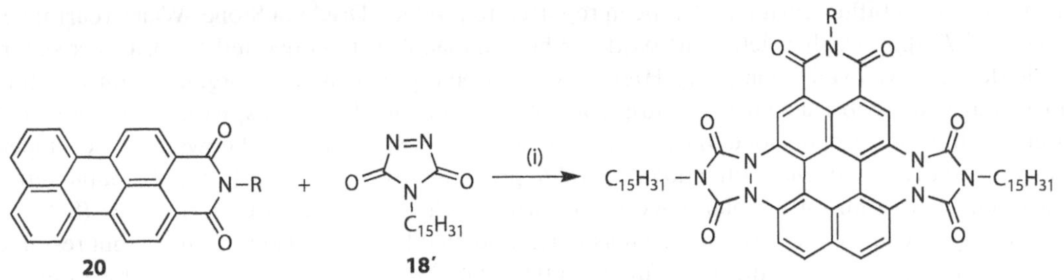

21a: R = C_8H_{17} (Cr 11 Col$_h$ 211 Col$_h$ > 340 I)
21a: R = $C_{18}H_{37}$ (Cr 36 Col$_h$ 186 Col$_h$ > 340 I)
21a: R = 2,6-diisopropylphenyl (Cr 293 Col 315 I)

SCHEME 2.65 Synthesis of T-shaped perylene mesogens: (i) *m*-Xylene, reflux.

This compound cleared at 315°C. An unidentified mesophase was found for this compound in the temperature range of 293°C–315°C.

The related discotic compounds benzo[g,h,i]perylenes **22** and **23** are also known to exhibit columnar phases [770,771]. The oxidative Diels–Alder reaction of perylene-3,4,9,10-tetracarboxylic diimide with maleic anhydride yields benzo[g,h,i]perylenehexacarboxylic 4,5,10,11-diimide-1,2-anhydride, which can be converted to liquid crystalline diimide–diesters **22**. Similarly, the Diels–Alder reaction of commercial perylene-3,9- and 3,10-dicarboxylic acid diisobutyl ester mixture with an acrylate under oxidative conditions affords liquid crystalline triesters **23**.

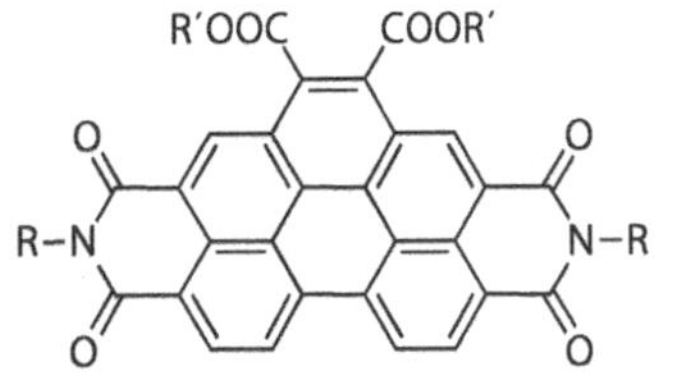

22a: R = R′ = 2-ethylhexyl; (Cr 197 Col 325 I)
22b: R = 4-heptyl, R′ = 3-octyl; (Col 192 I)

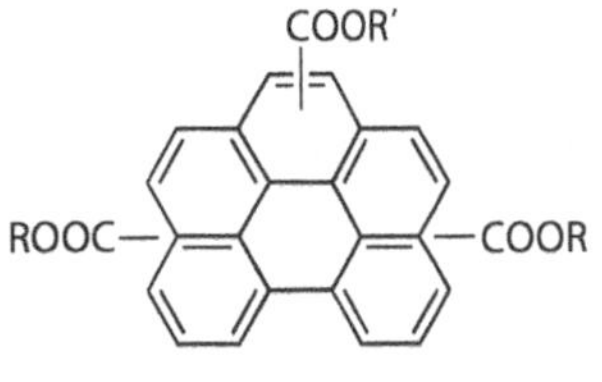

23a: R = isobutyl, R′ = cyclohexyl; (Col 134 I)
23b: R = R′ = methyl; (Cr 113 Col 209 I)
23c: R = R′ = ethyl; (Cr 61 Col 187 I)

Compound **22a** with four 2-ethylhexyl chains melts at 197°C and clears at 325°C. On the other hand, the use of 3-octyl ester chains with 4-heptyl imide group produces a room temperature liquid crystalline material **22b** that goes to the isotropic phase at 192°C. The benzopyrene triester **23a** with isobutyl and cyclohexyl chains (as a mixture of four isomers) also exhibits a hexagonal columnar phase at ambient temperature and clears at 134°C. Corresponding short chain triester **23b** (methyl ester) and **23c** (ethyl ester) melts at 113°C and 61°C, respectively, and clears at 209°C and 187°C, respectively. These discotics have been reported to be excellent electron-transporting materials [770,771].

2.8 DIBENZO[G,P]CHRYSENE CORE

Dibenzo[g,p]chrysene (DBC) derivatives are attractive because of their notable fluorescent properties such as good quantum yields, small Stoke shifts, and long-lived excited states [772]. The parent polycyclic aromatic hydrocarbon can be obtained easily from fluorenone [773,774].

The pyrolysis of bifluorenylidene has been reported to produce DBC via Stone–Wales rearrangement [775,776]. Tetraphenylethylene oxide or benzopinacol were converted to DBC via superacidic dehydrative cyclization [777]. Highly efficient blue-green emissive organic light-emitting diodes have been fabricated using bis(diphenylamino)chrysene derivatives, which were prepared from 2,7-dibromo-9-fluorenone via sequential reactions [778]. Swager and coworkers developed a novel oxidative acetylene cyclization method to prepare DBC derivatives [772]. A double intramolecular cyclization of bis(biaryl)acetylenes with $SbCl_5$ produces DBC derivatives (Scheme 2.66) in high yields. Other oxidants such as $FeCl_3$ and $NOSbF_6$ have also been used, but resulted in lower yield. A facile synthesis of functionalized DBC derivatives was reported by Li et al., which involves the iodination of bis(biaryl)acetylene with iodine monochloride followed by the Mizoroki–Heck coupling reaction [779].

Though a simple MM2 energy-minimized structure of DBC appeared planar [780], x-ray crystal structural analysis has revealed its significant twisted nature [772,781]. The possibility of mesophase formation by DBC upon peripheral alkyl chains substitution was realized in 2002 by Kumar and coworkers [780]. The strategy adopted to prepare octaalkoxy-DBC is outlined in Scheme 2.67 [780]. A palladium-catalyzed annulation reaction [782] of diphenylacetylenes **1** with 2-iodobiaryls **2**, followed by oxidative coupling using $FeCl_3$ yielded octaalkoxy-DBC **4**. The intermediate compound **3** can also be prepared by the photochemical cyclization of tetraphenylethene **6**, which was prepared from benzophenone derivative **5** via McMurry coupling [783–785]. Two octaalkoxy-DBC

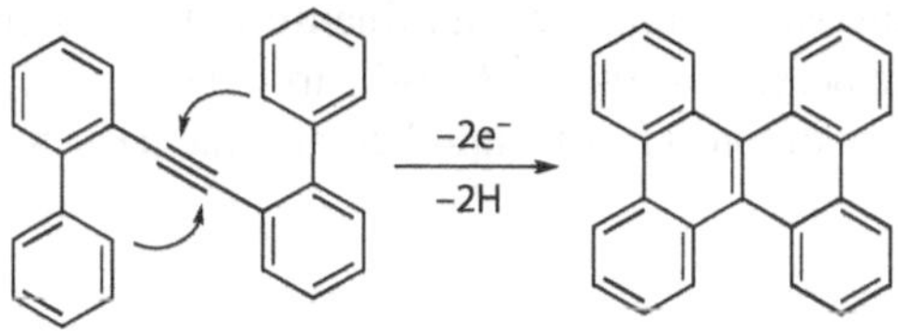

SCHEME 2.66   Building of DBC skeleton via double intramolecular cyclization.

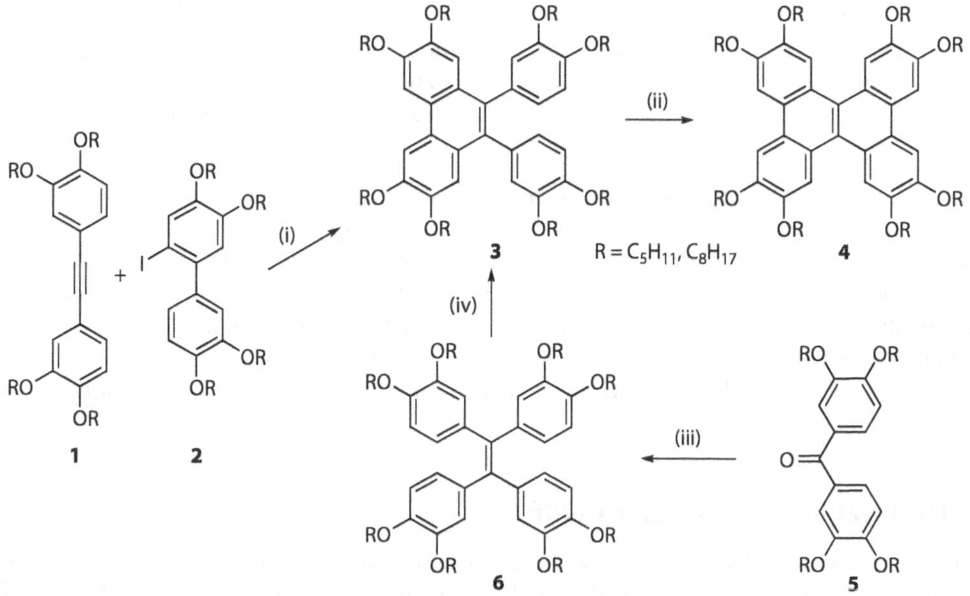

SCHEME 2.67   Synthesis of octaalkoxy-DBC derivatives: (i) $Pd(OAc)_2$, THF; (ii) $FeCl_3$, DCM; (iii) $TiCl_4$, Zn, pyridine; (iv) toluene, $h\nu$.

derivatives were prepared. Both octapentyloxy- and octaoctyloxy-DBC derivatives did not display mesomorphism in the virgin state; however, their charge-transfer (CT) complexes with trinitrofluorenone (TNF) exhibited columnar phases [780].

Enantiotropic DLCs could be obtained by substituting the peripheral positions by dialkoxybenzoate groups [784]. The synthetic route followed to prepare liquid crystalline derivative **16** is shown in Scheme 2.68 [772,784]. Compound **16** melted into hexagonal columnar phase at 117.8°C and cleared at 230.3°C. Charge mobility in the columnar phase of **5** was measured by time-of-flight method. It displayed a hole mobility up to 9.3×10^{-6} cm^2 V^{-1} s^{-1} for fields from 2.4×10^5 to 4.0×10^5 V cm^{-1} and electron mobility up to 3.9×10^{-6} cm^2 V^{-1} s^{-1} for fields from 3.2×10^5 to 4.4×10^5 V cm^{-1}.

Interestingly, the intermediates tetraphenylethylene **3** and tetraphenylphenanthrene **6** when substituted with trialkoxyphenyl units also exhibit columnar phases [783,785]. Synthetic routes to prepare these derivatives are shown in Scheme 2.69. The key precursor **18** can be easily prepared from 4,4-dimethoxybenzophenone **17** via McMurry reaction followed by demethylation using BBr$_3$. The esterification of ethene **18** with benzoic acids **19** using DCC-DMAP furnished the tetraphenylethenes **20** in good yield [783]. The thermal behavior of these discotics is summarized in Table 2.61. Trialkoxyphenyl substitution is required to fill the space around the core to induce mesomorphism.

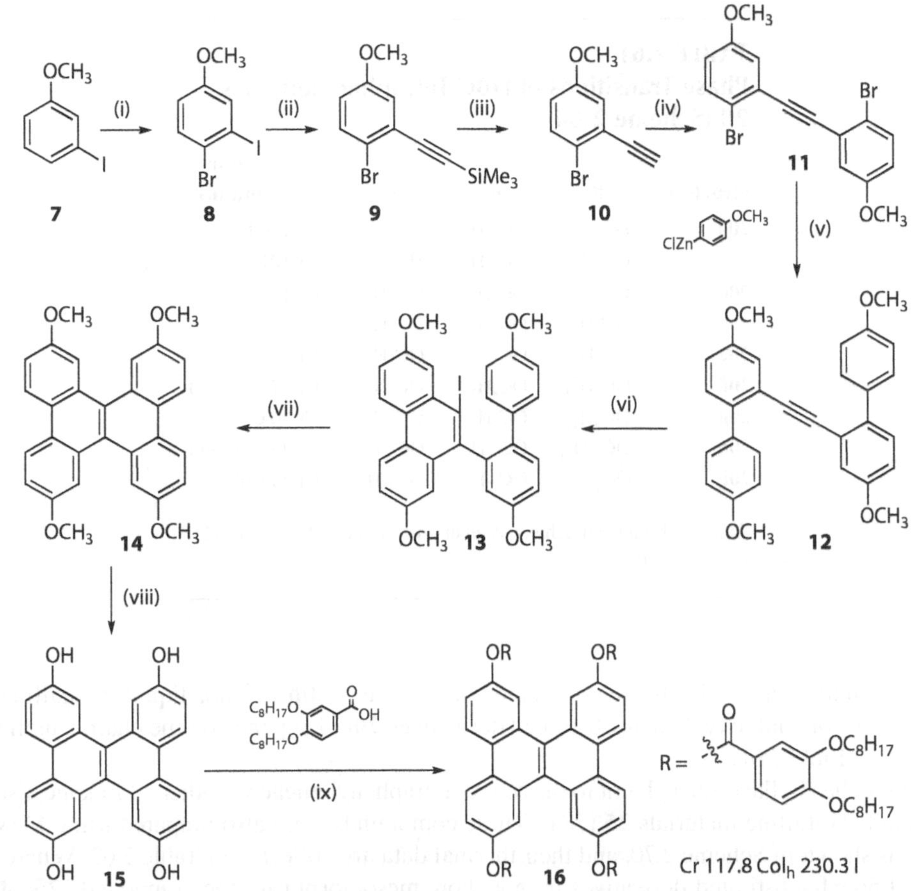

SCHEME 2.68 Synthesis of DBC derivatives via sequential ICl-induced cyclization and Mizoroki–Heck coupling: (i) NBS, DMF/DCM; (ii) TMSA, PdCl$_2$(PPh$_3$)$_2$, CuI, Et$_3$N; (iii) K$_2$CO$_3$, THF/MeOH; (iv) **6**, PdCl$_2$(PPh$_3$)$_2$, CuI, Et$_3$N; (v) PdCl$_2$(PPh$_3$)$_2$, Et$_2$O/THF; (vi) ICl, DCM; (vii) PdCl$_2$(PPh$_3$)$_2$, DMA/NaOAc; (viii) BBr$_3$, DCM; (ix) DCC, DMAP, DCM.

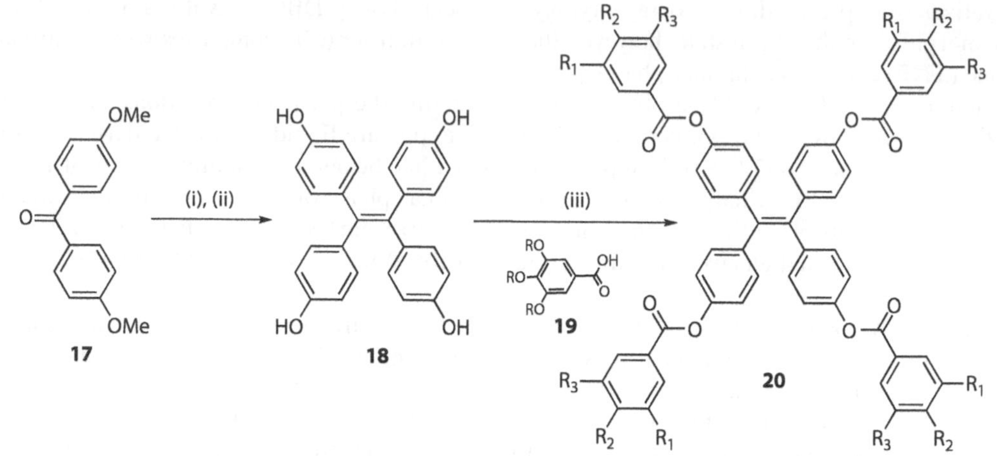

SCHEME 2.69 Synthesis of tetraphenylethylene derivatives **20**: (i) TiCl$_4$, Zn, pyridine; (ii) BBr$_3$, CH$_2$Cl$_2$; (iii) DCC, DMAP, CH$_2$Cl$_2$.

TABLE 2.61
Phase Transitions of DBC Tetraphenylethenes 20 (Scheme 2.64)

Structure	R$_1$	R$_2$	R$_3$	Transition Temperature
20a	H	OC$_{10}$H$_{21}$	H	Cr 139 I
20b	OC$_{10}$H$_{21}$	OC$_{10}$H$_{21}$	H	Cr 101 I
20c	OC$_5$H$_{11}$	OC$_5$H$_{11}$	OC$_5$H$_{11}$	Cr 126 I
20d	OC$_6$H$_{13}$	OC$_6$H$_{13}$	OC$_6$H$_{13}$	Cr 55 I
20e	OC$_7$H$_{15}$	OC$_7$H$_{15}$	OC$_7$H$_{15}$	Cr 37 Col$_h$ 62 I
20f	OC$_8$H$_{17}$	OC$_8$H$_{17}$	OC$_8$H$_{17}$	Cr 47 Col$_h$ 64 I
20g	OC$_9$H$_{19}$	OC$_9$H$_{19}$	OC$_9$H$_{19}$	Cr 40 Col$_h$ 69 I
20h	OC$_{10}$H$_{21}$	OC$_{10}$H$_{21}$	OC$_{10}$H$_{21}$	Cr 29 Col$_h$ 69 I
20i	OC$_{12}$H$_{25}$	OC$_{12}$H$_{25}$	OC$_{12}$H$_{25}$	Cr 34 Col$_h$ 58 I

Source: Data from Schultz, A. et al., *Adv. Funct. Mater.*, 11, 441, 2001.

Mono- and dialkoxyphenyl-substituted compounds **20a** and **20b** are not liquid crystalline. The phase behavior of trialkoxyphenyl-substituted derivatives **20c–i** depends on the chain length of the alkyl groups (Table 2.61).

Attaching the trialkoxyphenyl unit to the central tetraphenylethene via ethereal linkage also generates liquid crystalline materials **25** [784]. These compounds were also prepared using McMurry reaction, as shown in Scheme 2.70, and their thermal data are collected in Table 2.62. Whereas the octyl- and nonyl-substituted derivatives do not show mesomorphism, the compounds **25c–f** with longer chains exhibit monotropic columnar phases. Compound **25f** does not display any mesophase transition in DSC, but columnar phase textures appeared under polarizing optical microscopy on rapid cooling from the isotropic phase.

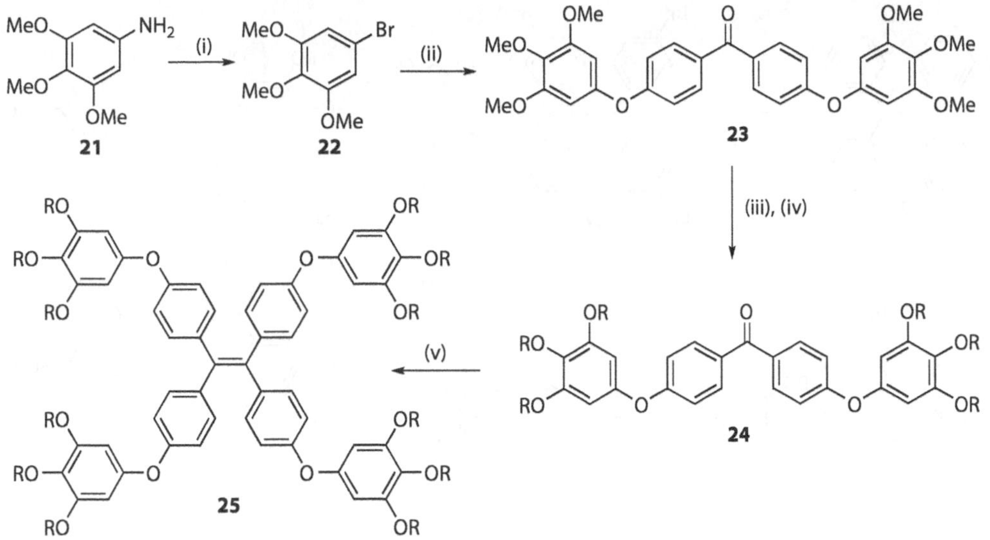

SCHEME 2.70 Synthesis of DBC tetraphenylethylene derivatives **25**: (i) (a) NaNO$_2$, H$_2$SO$_4$, H$_2$O, 10°C; (b) CuBr, H$_2$O, reflux; (ii) CuO, pyridine, K$_2$CO$_3$, 140°C; (iii) BBr$_3$, CH$_2$Cl$_2$, −60°C; (iv) K$_2$CO$_3$, DMF, RBr, 80°C; (v) TiCl$_4$, Zn, pyridine.

TABLE 2.62
Phase Transitions of DBC
Tetraphenylethenes 25 and 29
(Schemes 2.65 and 2.66)

Structure	R	Transition Temperature
25a	C$_8$H$_{17}$	Cr 54 I
25b	C$_9$H$_{19}$	Cr 43 I
25c	C$_{10}$H$_{21}$	Cr (4 Col$_h$) 38 I
25d	C$_{11}$H$_{23}$	Cr (11 Col$_h$) 42 I
25e	C$_{12}$H$_{25}$	Cr (17 Col$_h$) 52 I
25f	C$_{16}$H$_{33}$	Cr 43 I
29a	C$_{10}$H$_{21}$	Cr 65 I
29b	C$_{12}$H$_{25}$	Cr 49 Col$_h$ 65 I

Source: Data from Schultz, A. et al., *Eur. J. Org. Chem.*, 2829, 2003.

The trialkoxyphenyl groups were directly linked to the central tetraphenylethene unit to generate compounds **29**. Two synthetic routes have been used to prepare these materials, as shown in Scheme 2.71 [785]. The classical bromination of tetraphenylethene **26** yields tetrakis(p-bromophenyl)ethene **27**, which can be coupled with trialkoxyphenyl boronic acid **28** to generate tetraphenylethene derivatives **29**. Alternatively, compound 4,4-dibromobenzophenone **30** can be coupled with boronic acid **28** to afford compound **31** and the final McMurry coupling of **31** yields the tetraphenylethene **29**. The biphenyl derivative **29b** exhibits a columnar phase in the range of 49°C–65°C but its lower homologue **29a** does not show any mesomorphism (Table 2.62).

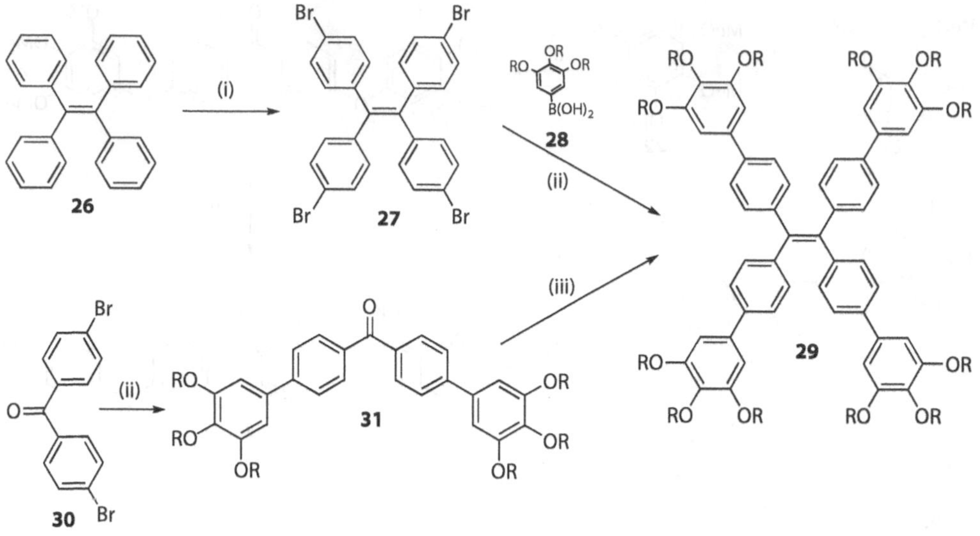

SCHEME 2.71 Synthesis of tetraphenylethylene derivatives **29**: (i) Br$_2$; (ii) Pd(PPh$_3$)$_4$, DME, H$_2$O, 100°C; (iii) TiCl$_4$, Zn, pyridine.

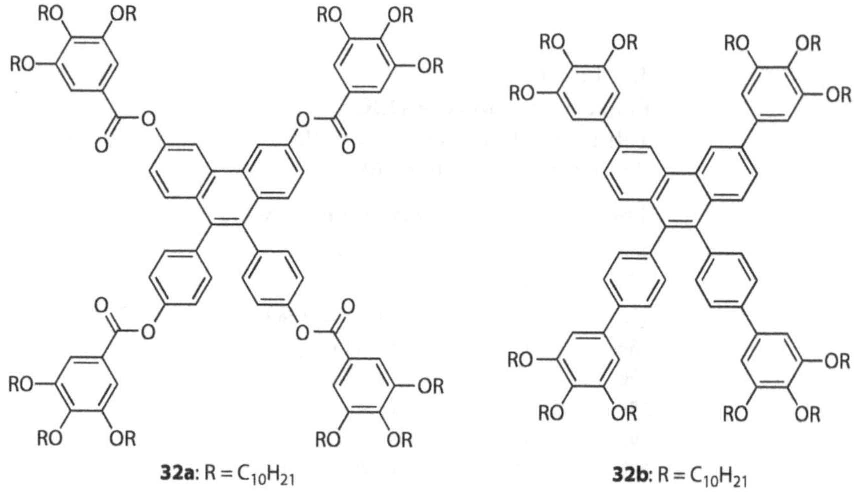

32a: R = C$_{10}$H$_{21}$ **32b**: R = C$_{10}$H$_{21}$

FIGURE 2.10 Molecular structure of discotic liquid crystalline tetraphenylethene and tetraphenylphenanthrene derivatives.

The photocyclization of tetraphenylethene **20h** yields liquid crystalline phenanthrene derivative **32a**. Compound **32a** melts at 110°C to form a hexagonal columnar phase that clears at 121°C. On the other hand, compound **32b** (Figure 2.10) prepared from the photocyclization of tetraphenylethene **29a** does not show any mesomorphism [785].

2.9 DIBENZO[*FG,OP*]NAPHTHACENE CORE

Dibenzo[*fg,op*]naphthacene (DBN) consists of 24 carbon atoms bound together in 6 adjoining C$_6$-membered rings to form the planar disk-shaped aromatic molecule. This C$_{24}$ hexacycle, which is also known as dibenzopyrene (DBP), has a diameter of about 10.0 Å. The DBN core appeared in the liquid crystal field recently [340], but became the focus of attention because of the ferroelectric switching of columnar mesophase of some of its derivatives [340,786]. The ferroelectricity in tilted chiral smectic liquid crystalline phases was discovered by Meyer et al. [787]. After the discovery of columnar mesophases in disk-shaped molecule, it was predicted that they could be ferroelectric if the disk-like molecules are chiral and tilted with regards to the axis of column [788]. Experimentally, this suggestion was verified by Bock and Helfrich in 1992 [340]. They demonstrated the existence of ferroelectricity in the columnar liquid crystals of chiral DBN derivatives.

Before discussing the chemistry of DBN discotics, it is worth pointing out an error in the identification of these materials. The synthesis of octa-substituted DBN was reported by Musgrave and Webster [263,789,790]. The oxidation of 3,3',4,4'-tetramethoxybiphenyl 1 by chloranil in 70% sulfuric acid yielded an over-oxidized quinone 2 as the sole isolable product (Scheme 2.72). The quinone was formed by repeated Scholl reactions followed by oxidative demethylation of the resultant octamethoxy compound. It furnished a leucoacetate $C_{24}H_6(OMe)_6(OAc)_2$, m.p. 317°C–317.5°C and a leucomethyl ether 3, $C_{24}H_6(OMe)_8$, m.p. 222.5°C–223.5°C on reductive acetylation and methylation, respectively. Based on NMR studies on DBN derivative prepared from partially deuterated starting material, the structure of this diketone was assigned as 2,5,6,9,12,13-hexamethoxy-1,8-quinone 2. Musgrave and Webster also reported that oxidation of veratrole under similar reaction conditions yields 2,3,6,7,10,11-hexamethoxytriphenylene as the major product and a diketone 5 (7%) as the minor product, which furnish a leucoacetate $C_{24}H_6(OMe)_6(OAc)_2$, m.p. 275°C–277.5°C and a leucomethyl ether 6, $C_{24}H_6(OMe)_8$, m.p. 283°C–284°C on reductive acetylation and methylation, respectively. The minor product was characterized as 2,5,6,9,12,13-hexamethoxy-1,10-quinone 5 (Scheme 2.72).

The octamethoxy-DBN has recently been prepared via another route, as shown in Scheme 2.73 [791]. The cyclization of octamethoxy tetraphenyl 11 was expected to yield the octamethoxy-DBN 3 reported by Musgrave; however, single-crystal x-ray diffraction analysis revealed that the product formed was 1,2,5,6,9,10,12,13-octamethoxy-DBN, 6, via the carbon skeleton rearrangement. The m.p. and spectral data of this product were matched with Musgrave's oxidative product. Therefore, it was concluded that Musgrave's oxidative dimerization of tetramethoxybiphenyl actually produced 6, not 3. On the basis of these facts, it is deduced that all the octa-substituted DBN-based DLCs prepared following Musgrave's method should be named as 1,2,5,6,9,10,12,13-octa-substituted-DBN

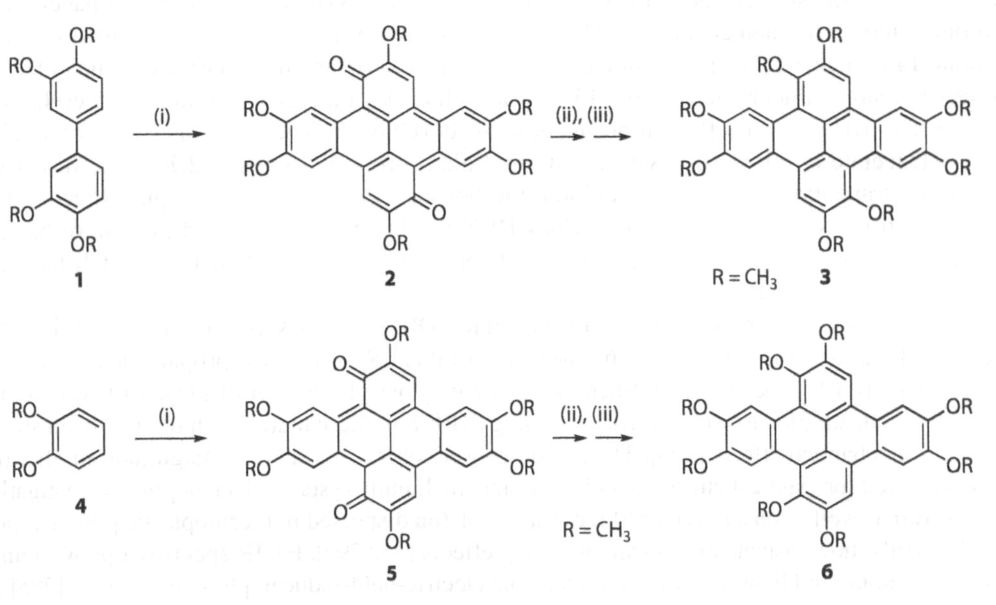

SCHEME 2.72 Musgrave's synthesis of octaalkoxy-DBN derivatives: (i) chloranil, H_2SO_4; (ii) Zn, Ac_2O, Et_3N; (iii) RBr, DMSO, KOH.

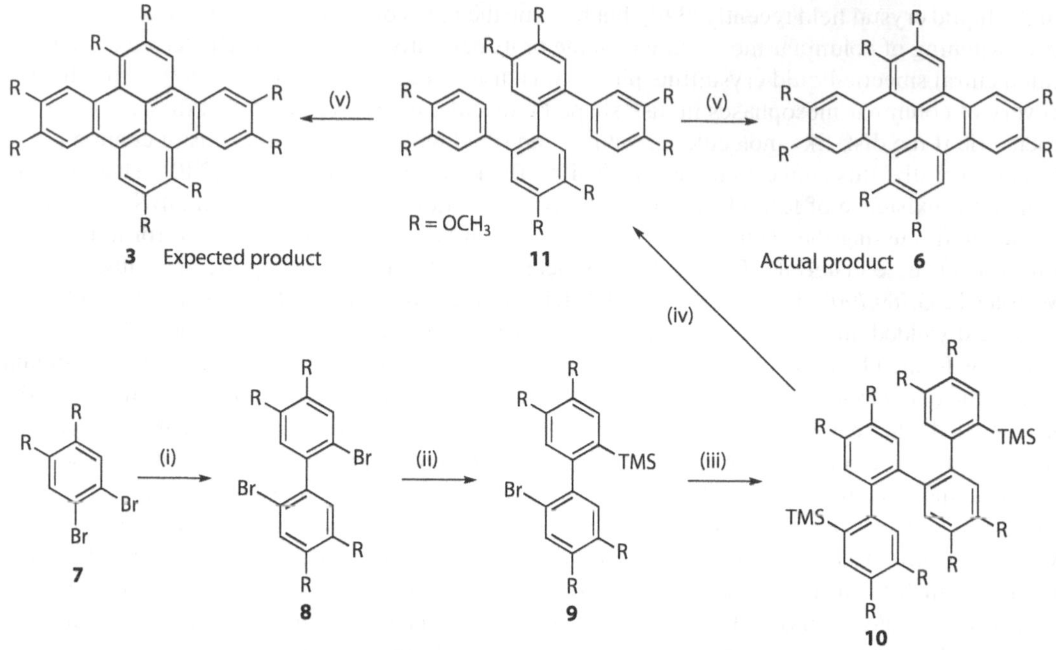

SCHEME 2.73 Synthesis of octaalkoxy-DBN derivatives: (i) *n*-BuLi, THF; (ii) *n*-BuLi, THF, TMSCl; (iii) Ni(COD)$_2$, COD, bipy, DMF/toluene, 80°C, (8) 40%; (iv) HBF$_4$, CH$_2$Cl$_2$, 0°C; (v) PIFA, BF$_3$.Et$_2$O, CH$_2$Cl$_2$.

with molecular structure **6** rather than 1,2,5,6,8,9,12,13-octa-substituted-DBN with molecular structure **3**. However, a single-crystal x-ray analysis of the DBN derivatives prepared by Musgrave's method is required to confirm it.

All the octasubstituted-DBN discotic liquid crystals were synthesized following the method of Musgrave and Webster, which involved the oxidation of 3,3′,4,4′-tetramethoxybiphenyl **1** by chloranil in 70% sulfuric acid to 2,5,6,8,12,13-hexamethoxydibenzo[*fg,op*]naphthacene-1,10-quinone **5** (corrected nomenclature). The cleavage of methoxy groups yields hexahydroxy-1,10-quinone **14** (Scheme 2.74). The quinone can be reduced to octahydroxy-DBN **15**. The octaesters **16** were prepared either by acetylating **15** with acid chloride or directly by reductive acetylation of **14** with zinc/pyridine or triethylamine and desired acid chloride. The octaethers **17** were obtained from octaacetate by alkylating with the desired alkyl bromide (Scheme 2.74). The direct oxidation of tetrapentyloxybiphenyl by chloranil to hexapentyloxydibenzo[*fg,op*]naphthacene-1,10-quinone and its conversion to octapentyloxy-DBN by the above-mentioned procedure has also been reported by the Ringsdorf group [374]. Other oxidizing reagents such as VOCl$_3$ have also been used to prepare DBN quinone [284].

The thermal behavior of all the octa-substituted-DBN discotics is presented in Table 2.63. Lactic acid-based chiral esters, 1,2,5,6,9,10,12,13-octakis-((*S*)-2-hexyloxypropanoyloxy)-DBN **16a** and 1,2,5,6,9,10,12,13-octakis-((*S*)-2-heptyloxypropanoyloxy)-DBN **16b** displayed tilted columnar mesophases. In addition to ferroelectric switching [340,786], these materials have also been studied for electromechanical effects [793]. The responses were observed orders of magnitude smaller than those observed for classical chiral ferroelectric smectic liquid crystals. Electrooptical investigations on pure **16b** as well as on a polymer film consisted of **16a** dispersed in thermoplastic polymer, poly-(vinylbutyral) show bistable and linear switching effects [342,794]. FT-IR spectroscopy with unpolarized and polarized light was used to understand electric-field-induced phase transitions [795] and orientational order and reorientation dynamics [796].

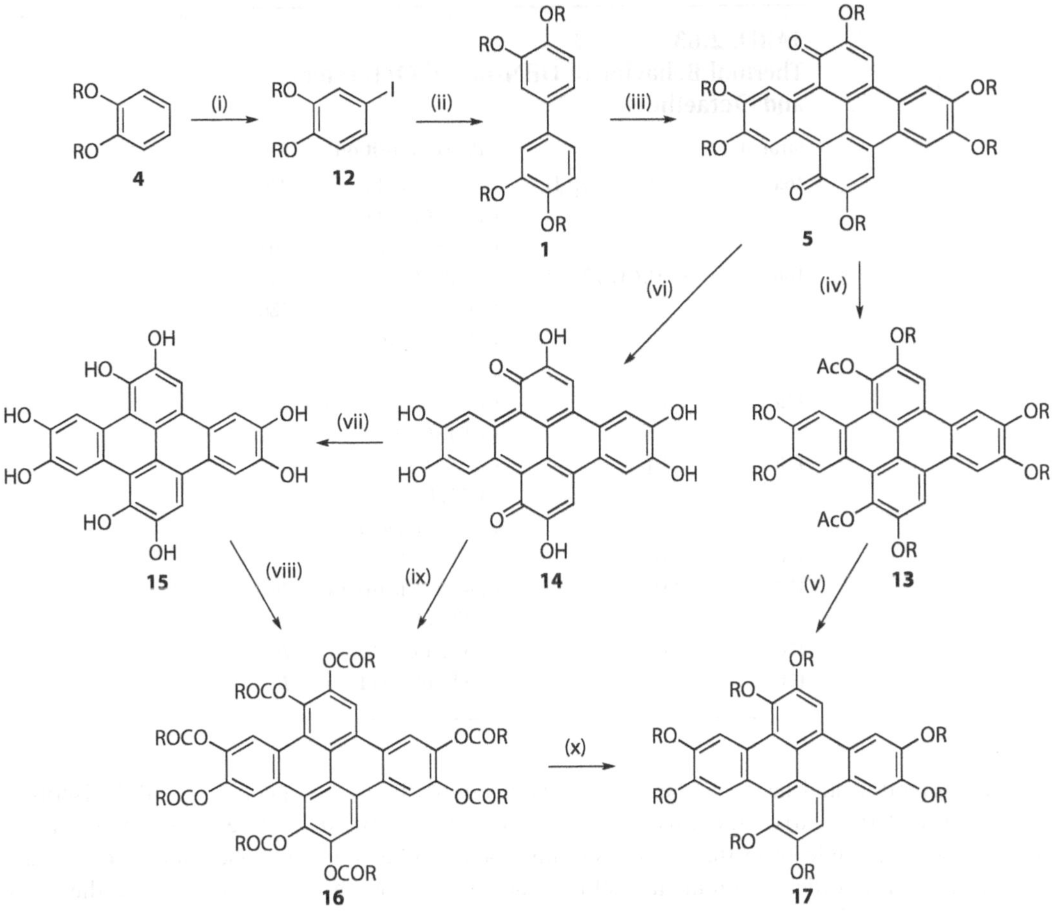

SCHEME 2.74 Synthesis of DBN-based discotic octaesters and octaethers: (i) HIO_4, I_2, CH_3COOH; (ii) Cu; (iii) chloranil, H_2SO_4; (iv) Zn, Ac_2O, pyridine; (v) DMSO, KOH, RBr; (vi) BBr_3; (vii) $Na_2S_2O_4$; (viii) RCOCl; (ix) Zn, pyridine, RCOCl; (x) RBr, base.

All the octaalkoxy-DBN discotics **17** exhibited a single hexagonal columnar phase (Table 2.63). The lowest homologue exhibited an unusual thermal behavior. The crystalline material **17a** melts at 70°C to columnar phase, which transforms to the isotropic phase at 83.6°C. The sample solidifies on annealing in mesophase for some time to give another solid polymorph, which displayed only a monotropic columnar phase [792]. The octapentyloxy derivative was reported to be liquid crystalline at ambient temperature with a glass transition at low temperature. All other derivatives were reported to display typical enantiotropic columnar mesophases. The addition of the electron acceptor TNF enhanced the mesophase stability significantly with a maximum stability for the equimolar donor–acceptor mixtures. The formation of a lyotropic nematic columnar phase was observed on addition of alkanes to the 1:1 charge transfer complexes [792]. Octaalkoxy-DBN derivatives displayed intense absorption at 231, 296, and 313 nm and an intense fluorescence peak at 402 nm. The peripheral side-chain length does not have any influence on the absorption and fluorescence spectra of both solution and thin film, but the profile of these spectra greatly depends on the molecular rearrangement [797,798].

As described in Section 2.6, triphenylene discotics have been well recognized as a new class of organic semiconductors. Being a larger core (~10 Å) than triphenylene (~7 Å), a greater degree

TABLE 2.63
Thermal Behavior of DBN-Based Octaesters and Octaethers

Structure	R	Phase Transition	Ref.
16a	C*H(CH$_3$)OC$_6$H$_{13}$	Cr 84 Col 124 I	340
		Cr ~80 Col ~118 I	786
		Cr 88.8 Col$_t$ 129.5 I	794
16b	C*H(CH$_3$)OC$_7$H$_{15}$	Col$_t$ 126 I	342
		Col ~115 I	786
		G 70 Col 120 I	793
		Cr 70 Col 126 I	796
17a	C$_4$H$_9$	Cr(A) 70 Col$_h$ 83.6 I	792
		Cr(B) 86.5 I	792
17b	C$_5$H$_{11}$	Col$_h$ 93.5 I	792
		g (-35) Col$_h$ 98 I	374
		g (-35) Col$_h$ 96 I	797
17c	C$_6$H$_{13}$	Cr 55.2 Col$_h$ 89.6 I	792
17d	C$_7$H$_{15}$	Cr 43.3 Col$_h$ 104.3 I	792
		Cr 47 Col$_h$ 107 I	797
17e	C$_8$H$_{17}$	Cr 38.3 Col$_h$ 103.7 I	792
17f	C$_{10}$H$_{21}$	Cr 42 Col$_h$ 105 I	797

of π–π interaction and hence higher charge carrier mobility was expected in DBN discotics. However, the charge carrier mobility in the octasubstituted DBN derivatives was realized to be one order of magnitude lower than in the columnar phase of hexaalkoxytriphenylenes. This could be because of less ordered columnar packing due to the steric hindrance caused by the "way region" alkoxy chains, which resulted in the lower charge carrier mobility. It was anticipated that the removal of these way-region alkoxy chains would give a better core–core interaction and, therefore, high charge-carrier mobility. This leads to the development of a novel, versatile, and regiospecific synthesis of variable degree substituted dibenzo[*fg,op*]naphthacene derivatives [799,800]. For the synthesis of various hexaalkoxy-DBN derivatives **24**, the strategy outlined in Scheme 2.75 was envisaged. The synthesis is based on the preparation of key intermediate tetraphenyl **23** starting from 3-nitrobromobenzene **18**. The 2,2′-dibromo-4,4′diaminobiphenyl **19** was prepared by the reductive dimerization of **18** followed by benzidine rearrangement. Compound **19** was converted to bisphenol **20** via Sandmeyer reaction. The alkylation of diphenol **20** with appropriate 1-bromoalkane resulted in 2,2′-dibromo-4,4′dialkoxybiphenyls **21**. The tetraphenyls **23** were prepared either by coupling **21** with dialkoxyboronic acids or by first converting the dibromobiphenyls **21** to diboronic acids **22** followed by Suzuki coupling with dialkoxyiodobenzenes. The photocyclodehydrogenation of tetraphenyls **23** in the presence of an excess of iodine furnished the desired 2,5,6,9,12,13-hexaalkoxy-DBN derivatives **24**. The cyclization can also be achieved by oxidative coupling using FeCl$_3$ as oxidizing agent. All the hexaalkoxy-DBN derivatives exhibit a single mesophase, which was identified by x-ray diffractometry and optical microscopy as hexagonal columnar phase (Table 2.64).

All the hexa-substituted derivatives were found to be thermally more stable (having higher isotropic temperature) than the octasubstituted dibenzonaphthacenes. The lower melting and clearing temperatures of octaalkoxy-DBN could be due to the presence of two extra alkoxy chains and the steric hindrance caused by these chains. Replacing two *n*-butyloxy tails in compound **24a** by branched chains, both melting and clearing temperatures were lowered. The resulting compound **24f** exhibits about 71°C lower isotropic temperature. The lower melting and clearing temperatures

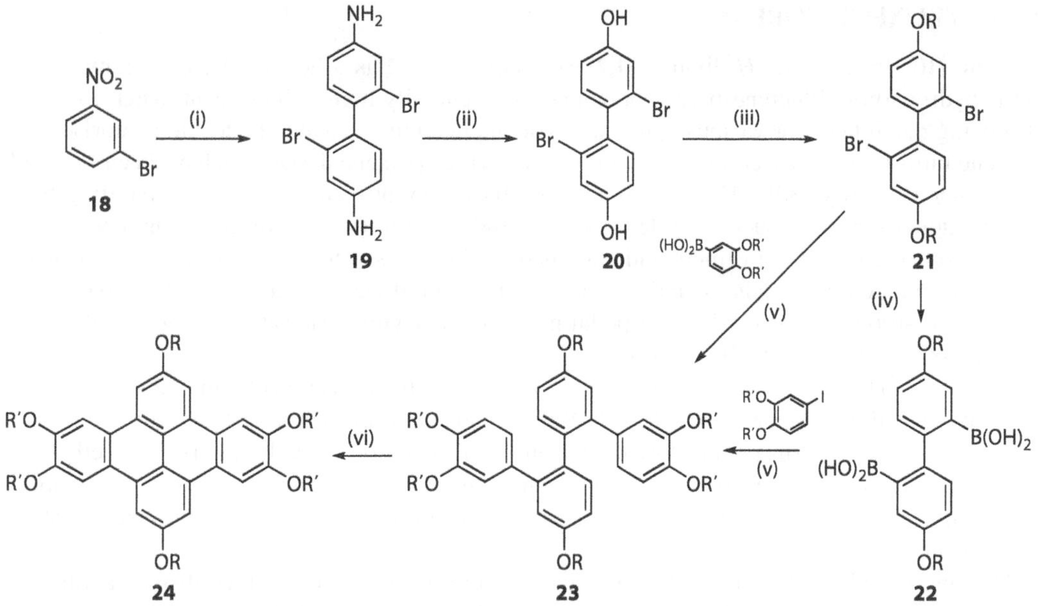

SCHEME 2.75 Synthesis of hexaalkoxynaphthacene derivatives: (i) Zn/NaOH; HCl; (ii) NaNO$_2$/HCl; (iii) DMSO, KOH, RBr; (vi) *n*-BuLi, B(OMe)$_3$; HCl; (v) Pd(PPh$_3$)$_4$, Na$_2$CO$_3$; (vi) FeCl$_3$ or I$_2$, *h*v.

TABLE 2.64
Thermal Behavior of Hexaalkoxy-DBN Discotics

Structure	R	R′	Phase Transition
24a	C$_4$H$_9$	C$_4$H$_9$	Cr 147.7 Col$_h$ 236.9 I
24b	C$_5$H$_{11}$	C$_5$H$_{11}$	Cr 120.5 Col$_h$ 223.1 I
24c	C$_6$H$_{13}$	C$_6$H$_{13}$	Cr 143.0 Col$_h$ 188.0 I
24d	C$_7$H$_{15}$	C$_7$H$_{15}$	Cr 143.4 Col$_h$ 179.7 I
24e	C$_8$H$_{17}$	C$_8$H$_{17}$	Cr 134.0 Col$_h$ 161.0 I
24f	(CH$_2$)$_2$CH(CH$_3$) (CH$_2$)$_3$CH(CH$_3$)$_2$	C$_4$H$_9$	Cr 118.7 Col$_h$ 165.5 I

Source: Data from Kumar, S. et al., *J. Mater. Chem.*, 12, 1335, 2002.

---

of the compound **24f** could be due to the unsymmetrical nature (by virtue of non-identical peripheral chains) and the presence of two branched chains in the molecule.

X-ray studies have been carried out on octapentyloxy- and hexapentyloxy-DBN derivatives under similar conditions. It was observed that the disk diameter remains the same ~21 Å for both octapentyloxy-DBN and hexapentyloxy-DBN irrespective of the number of alkoxy chain substitution, that is, whether it is 8 or 6. Also the disk diameter remains constant with varying temperature for both the compounds. The core–core distance shows strong temperature dependence for octapentyloxy-DBN, but has a smaller variation for hexapentyloxy-DBN. The value of the core–core distance was small for **24b** with six alkoxy substitution than for the compound with eight alkoxy substitution. Further, the core–core correlation length was higher for the compound with six alkoxy substitution (**24b**) than for the compound with eight alkoxy substitutions. Thus, x-ray results confirmed that hexa-substituted-DBN discotics are more ordered than octasubstituted derivatives [800].

2.10 TRUXENE CORE

Truxene (10, 15-dihydro-$5H$-diindeno[1,2-a;1'2'-c]fluorene) **1** is a heptacyclic C_3 symmetric aromatic hydrocarbon. Fluorene-based compounds are generally highly fluorescent materials and are promising candidates for various opto-electronic applications. Truxene molecule comprises three fluorene units that share a common central benzene ring. Truxene derivatives have been known for more than a century [801,802] but only recently have they attracted attention as building blocks for functional materials such as fullerenes [803,804]. During the past decade, a number of truxene-derived monodisperse dimers and star-shaped oligomers were prepared and investigated as electroluminescent materials, organic semiconductors, nonlinear optical materials, photochromophores, hole-transport materials, C_3 tripodal materials in asymmetric catalysis and chiral recognition, fluorescent probes, etc. [805–816].

Truxene can be easily synthesized by the acid-catalyzed trimerization of 1-indanone **2** or 3-phenylpropionic acid **4** (Scheme 2.76) [817,818]. Strong acids or Lewis acids, such as hydrochloric acid, sulfuric acid, phosphorus pentachloride, phosphorus pentaoxide, polyphosphoric acid, ethyl polyphosphate, etc., have generally been employed to achieve this trimerization. In dilute solutions of acid, mainly the dimeric intermediate **3** forms, which can be converted to **1** upon treatment with phosphorus pentaoxide.

Discotic liquid crystals derived from truxene core are interesting as they display a variety of phases and, therefore, have been studied extensively [589,819–842]. Truxene discotics with six peripheral chains were synthesized by the self-condensation of dimethoxy indanone (5,6-dimethoxy-2,3-dihydro-1H-indane-1-one) **5** in polyphosphate ethyl ester (PPE) at 160°C (Scheme 2.77). At lower temperature (80°C), only dimeric product **7** could be isolated. It is worth mentioning that the parent hydrocarbon was obtained in quantitative yield at this reaction temperature. The yield of hexamethoxytruxene **6** depends on the quality of PPE and is usually between 50% and 90%. Generally a mixture of dimer **7** and trimer **6** forms in the reaction, but the product can be easily isolated simply by washing the mixture with solvents as the trimer was not soluble in common organic solvents. Hexamethoxytruxene **6** was demethylated yielding the key intermediate, 2,3,7,8,12,13-hexahydroxytruxene **8**. Hexaesters of truxene **9** were obtained by the esterification of hexahydroxytruxene with the required acid chloride in pyridine [829,832]. The addition of a catalytic amount of DMAP improved the yield significantly [835]. Similarly, hexabenzoates **10** were prepared using substituted benzoyl chlorides [829].

Aryl ethers are generally prepared conveniently by reacting phenol with alkyl halides under basic condition. However, the Na and K salts of hexahydroxytruxene were reported to be very unstable and, therefore, the desired alkyl chains were attached prior to trimerization step (Scheme 2.78). Dihydroxy phenyl propionic acid **11** was alkylated with alkyl halide using KOH in ethanol. The resultant 3-(3,4-dialkoxyphenyl)propionic acids **12** were cyclized to indanones **13** in PPA. A solution

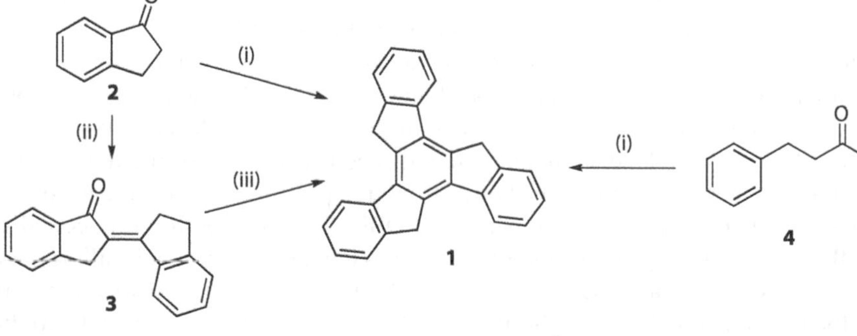

SCHEME 2.76 Synthetic routes to truxene nucleus: (i) strong acid, 80°C–160°C; (ii) dilute acid; (iii) P_2O_5.

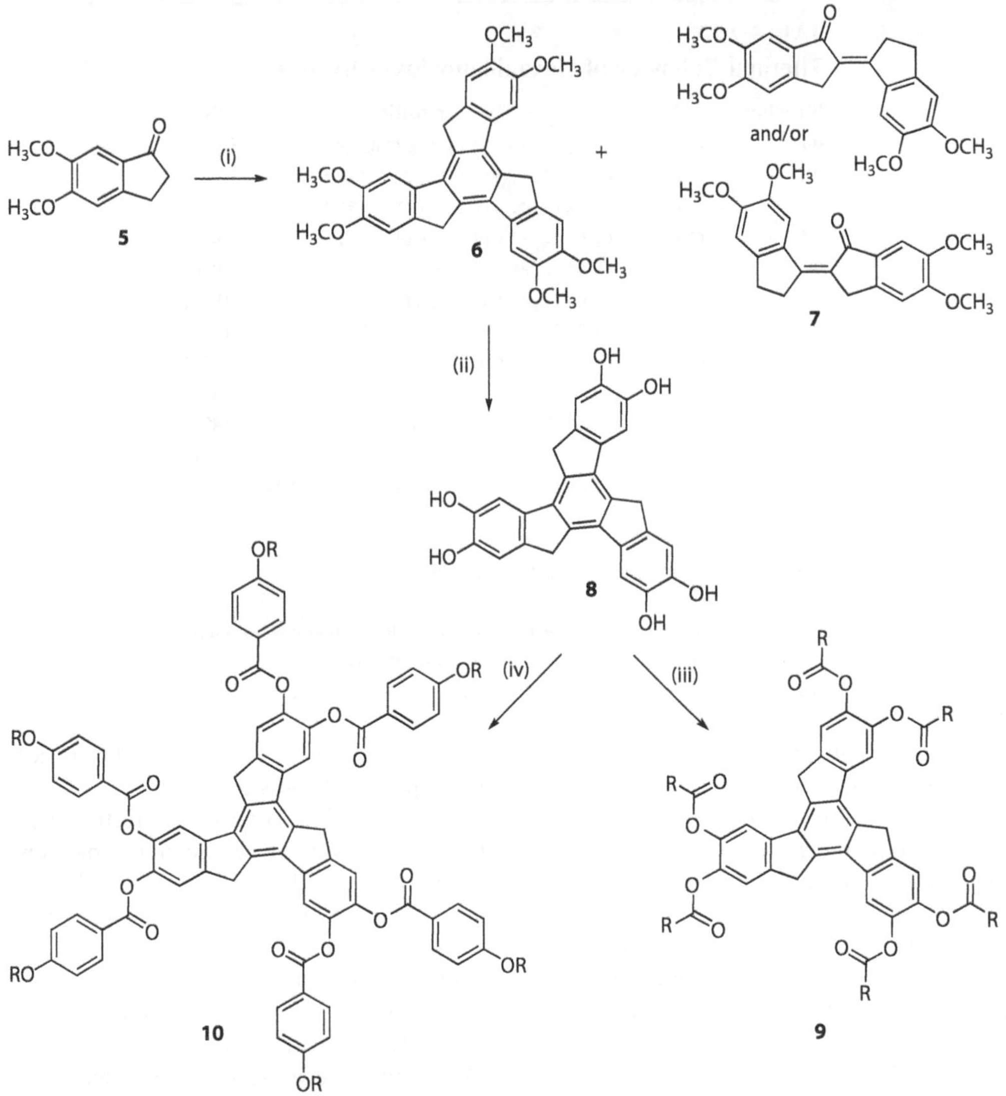

SCHEME 2.77 Synthesis of truxene hexaesters and benzoates: (i) PPE, 130°C; (ii) Py-HCl, 218°C or BBr₃-(CH₃)₂S, ClCH₂CH₂Cl, reflux; (iii) RCOCl, Py, DMAP; (iv) ROC₆H₄COCl, Py, DMAP.

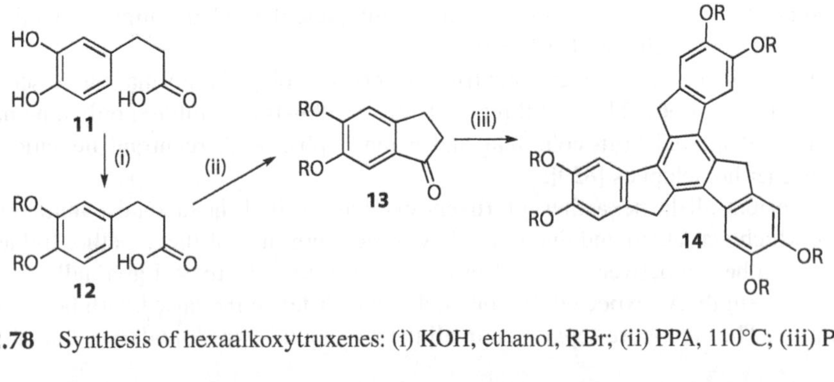

SCHEME 2.78 Synthesis of hexaalkoxytruxenes: (i) KOH, ethanol, RBr; (ii) PPA, 110°C; (iii) PPA, 140°C.

TABLE 2.65
Thermal Behavior of Hexaalkanoyloxytruxenes

Structure	R	Phase Transition	Ref.
9a	C_6H_{13}	Cr 112 (96 N_D) Col$_r$ 138 Col$_h$ 280 I	[820]
9b	C_7H_{15}	Cr 98 (85 N_D) Col$_r$ 140 Col$_h$ 280 I	[820]
9c	C_8H_{17}	Cr 88 (87 N_D) Col$_r$ 141 Col$_h$ 280 I	[820]
9d	C_9H_{19}	Cr 68 N_D 85 Col$_r$ 138 Col$_h$ 280 I	[820]
9e	$C_{10}H_{21}$	Cr 62 N_D 89 Col$_r$ 118 Col$_h$ 250 I	[820]
9f	$C_{11}H_{23}$	Cr 64 N_D 83.5 Col$_r$ 130 Col$_h$ 250 I	[820]
9g	$C_{12}H_{25}$	Cr 57 N_D 84 Col$_r$ 107 Col$_h$ 249 I	[820]
9h	$C_{13}H_{27}$	Cr 58 N_D 83 Col$_h$ 235 I	[820]
		Cr 61 ^aN_D 84 Col$_r$ 112 Col$_h$ 241 I	[823]
9i	$C_{14}H_{29}$	Cr 64 ^bN_D 82 Col$_r$ 95 Col$_h$ 221 I	[821]
		Cr 84 $^cCol_r$ 95 Col$_h$ 201 I	[821]
9j	$C_{15}H_{31}$	Cr 69 (62 Col$_h$) N_D 84 Col$_r$ 95 Col$_h$ 210 I	[829]
9k	$C_{17}H_{31}$	Cr 58 Col 67 N_D 82 I 89 Col 183 I	[833]

[a] Exhibits Col$_h$ phase on cooling to 56°C.

[b] Exhibits Col$_h$ phase on cooling to 58°C.

[c] On slow (1°C min^{-1}) heating and cooling, exhibits N_D phase followed by Col$_h$ phase on cooling to 58°C and 53°C, respectively.

of 5,6-dialkoxy-1-indanones when heated at 140°C in PPA produced desired truxene hexaethers **14** in low yield [828]. Thus a series of nine homologues was prepared and studied.

Hexaesters of truxene **9** exhibited complex polymorphism (Table 2.65). All derivatives showed nematic and columnar phases. While higher chain-length esters displayed enantiotropic nematic phase, it was monotropic in nature in the shorter chain esters. It should be noted that all these compounds displayed an inverted nematic-columnar phase sequence. The normal sequence of phase transitions in discotic liquid crystals is Cr-Col-N_D-I, while in these materials the less ordered nematic phase was observed after the crystalline phase but below the viscous columnar phase during heating cycle. The highly purified hexapentadecanoate of truxene **9i** was reported to exhibit a reentrant hexagonal columnar phase at lower temperature [821]. A reentrant isotropic phase was also reported in the highest homologue of this series [833]. A mixture of 87% hexatetradecanoyloxytruxene **9h** and 13% of hexa-(4-dodecyloxy-benzoyloxy)truxene **10g** was also reported to display a reentrant isotropic phase with Cr 67 N_D 112 I 129 Col$_h$ 214 I phase sequence. An equal mixture of octanoyloxy-, decanoyloxy-, and tetradecanoyloxy-truxene was reported to exhibit only a nematic phase with phase sequence Cr 55 N_D 90 I [826]. A binary mixture (15:85) of **10b** and **9d** also exhibited only a nematic phase. The two columnar phases of **9d** disappeared in the mixture [829]. It should be noted that pure as well as binary mixtures of truxene esters and benzoates showed signature of decomposition with time at elevated temperature, and degraded samples sometimes did not show reproducible phase behavior [832,835].

The lower homologues of hexabenzoate truxene series displayed a normal phase sequence, that is, Cr-Col-N_D-I (Table 2.66). The first three derivatives (**10a–10c**) exhibited only a nematic phase, other derivatives displayed both columnar and nematic phases. A reentrant nematic phase was observed in higher homologues [829].

Unlike hexaesters, all the hexaethers of truxene exhibited a single hexagonal columnar mesophase (Table 2.67), which was attributed due to the low steric hindrance of the six ether linkages, which allows a strong cohesion between cores. Isotropic temperature decreased gradually on increasing peripheral chain length. As expected, the intercolumnar distance increased with both chain length and temperature. The core–core separation in all the compounds was found to be about 3.6 Å.

TABLE 2.66
Thermal Behavior of Hexa(alkoxybenzoyloxy)truxenes

Structure	R	Phase Transition	Ref.
10a	C_6H_{13}	Cr 238 N_D >290 I	[829]
10b	C_7H_{15}	Cr 190 N_D >290 I	[829]
10c	C_8H_{17}	Cr 148 N_D >290 I	[829]
10d	C_9H_{19}	Cr 126 Col_r 131 N_D >290 I	[829]
10e	$C_{10}H_{21}$	Cr 106 Col_r 131 N_D >290 I	[829]
10f	$C_{11}H_{23}$	Cr 90 Col_r 137 N_D 171 Col_r 284 N_D 297 I	[822]
10g	$C_{12}H_{25}$	Cr 90 (79 N_D) 179 Col_r 260 N_D >290 I	[829]
10h	$C_{13}H_{27}$	Cr 80.5 (75 N_D) 139 Col_r 223 Col_h >290 I	[829]
10i	$C_{14}H_{29}$	Cr 78 (72 N_D) 121 Col_r 225 Col_h >290 I	[829]

TABLE 2.67
Thermal Behavior of Hexaalkoxytruxenes

Structure	R	Phase Transition
14a	C_6H_{13}	Cr 79 Col_h >300 I
14b	C_7H_{15}	Cr 70 Col_h >300 I
14c	C_8H_{17}	Cr 86 Col_h >300 I
14d	C_9H_{19}	Cr 67 Col_h 274 I
14e	$C_{10}H_{21}$	Cr 67 Col_h 260 I
14f	$C_{11}H_{23}$	Cr 64 Col_h 220 dec
14g	$C_{12}H_{25}$	Cr 73 Col_h 230 I
14h	$C_{13}H_{27}$	Cr 59 Col_h 220 dec
14i	$C_{14}H_{29}$	Cr 75 Col_h 205 I

Source: Data from Foucher, P. et al., *Mol. Cryst. Liq. Cryst.*, 108, 219, 1984.

2.11 DECACYCLENE CORE

Decacyclene, a putative fullerene precursor, is a symmetric, large polycyclic aromatic molecule present in the carbonaceous mesophase. It can be synthesized via several methods such as metal-catalyzed cyclotrimerization of acenaphthalene [843–845], by dehydrogenating cyclotrimerization of acenaphthalene with elemental sulfur at high temperature [846], or via pyrolysis of acenaphthalene [847]. It is produced abundantly as one of the main products when acenaphthalene pitch is pyrolyzed at temperatures between 320°C and 470°C [848,849]. On further heating, it polymerizes, forming the carbonaceous mesophase. Recently, decacyclene has been carbonized in a glass tube or a molten salt at various temperatures to investigate the influence of contained pentagonal rings on the change of carbon structures with heat treatment [850]. Conducting carbon nanofibers can be prepared by vapor deposition of decacyclene on a flat substrate followed by ion-beam irradiation [851,852].

It may be recalled that since the discovery of the carbonaceous mesophase in the early 1960s [853–856], there has been a continuous search for a well-defined, chemically stable system that could serve as a model for this unstable phase, and these efforts have resulted in the discovery of discotic liquid crystals in 1977 [1]. The carbonaceous mesophase is an intermediate phase appeared at temperatures of 400°C–500°C during the process of coke and graphite manufacture via carbonization of organic precursors [857,858]. The quality of final graphitic products, e.g., carbon fibers, critically depends upon the structural features of this chemically unstable, lamellar mesophase [859].

The common DLC has been characterized as a flat molecule comprised of a rigid polycyclic aromatic core surrounded by four to nine aliphatic side chains [860]. It was a common belief that "three elongated chains would insufficiently fill the space around the core and, therefore, could not allow the existence of columnar mesophase" [861]. It was anticipated that increasing the size of polycyclic aromatic core and keeping the minimal constraint around the core to avoid crystallization, one could achieve liquid crystallinity regardless of the number, nature, and size of the side chains [862]. These mesophases could model the carbonaceous mesophase more closely than other known discotic systems.

Accordingly, trialkanoyldecacyclene-based discotic mesogens **2** were designed and synthesized by a direct Friedel–Crafts acylation of commercially available parent hydrocarbon (Scheme 2.79) [862]. The reaction proceeds with very high regioselectivity and only the C_3 symmetric 1,7,13-trialkanoyldecacyclene could be isolated. Only two LC derivatives, triheptanoyloxy-, **2a**, and trioctanoyloxy-decacyclene, **2b**, have been reported [862]. Both compounds show a high tendency to aggregate, even at high dilution, as evident by NMR studies. Their behavior is presented in Table 2.68. Hysteresis is observed on cooling. X-ray diffraction studies reveal a two-dimensional square lattice in the M1 phase of both the compounds. The M2 phase of **2a** was characterized as a columnar rectangular phase. However, the M2 phase of **2b** could not be fully characterized as this

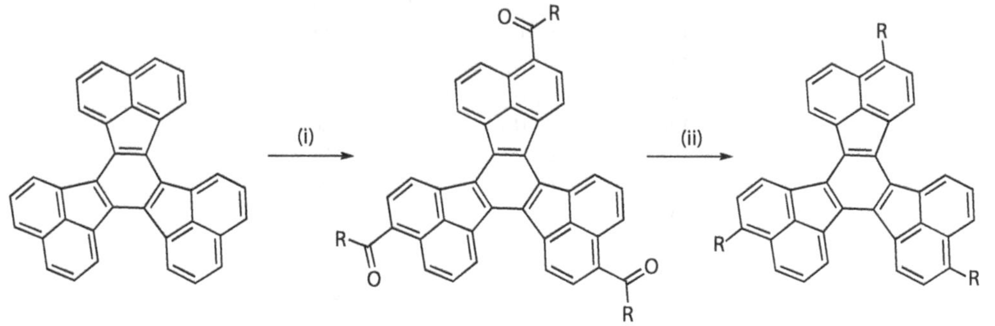

2a: R = C_6H_{13}
2b: R = C_7H_{15}

3a: R = C_7H_{15}
3b: R = C_8H_{16}

SCHEME 2.79 Synthesis of decacyclene derivatives: (i) RCOCl, AlCl$_3$, ClCH$_2$CH$_2$Cl, reflux, 38%; (ii) NH$_2$NH$_2$.H$_2$O, KOH, reflux, 50%.

TABLE 2.68
Transition Temperatures of Decacyclene Derivatives

Structure	Phase Transition
2a	Cr 92.8 Col$_t$ 115 Col$_r$ 262 I
2b	Cr 98 Col$_t$ 108.5 Col 240.5 I
3a	Cr 118 I
3b	Cr 120 I

Source: Data from Keinan, E. et al., *Adv. Mater.*, 3, 251, 1991.

displays only one reflection. The deoxygenated hydrocarbons **3a** and **3b,** obtained by reduction of the keto groups, were found to be non-mesomorphic (Table 2.68). Both compounds show double melting behavior. 1,7,13-triheptyldecacyclene **3a** first melts at 90°C to give an isotropic liquid that crystallizes immediately into fine needles and finally melts at 118°C. Similarly, 1,7,13-trioctylde-cacyclene **3b** first melts at 92°C to give an isotropic liquid that crystallizes at 95.7°C to another isomorph, which finally melts at 120°C.

Decacyclene molecule itself has been extensively studied for various physical properties. Electrochemical studies [863,864] revealed its multiple redox states because of the orbital degeneracy. A variety of decacyclene-based metal sandwich complexes have recently been prepared and studied [865–868]. Decacyclene has been used as a dopant to improve the lifetime of organic electroluminescent device [869]. Acharya et al. studied spectroscopic properties of decacyclene Langmuir–Blodgett (LB) film mixed with stearic acid [870]. Such decacyclene LB film have been used to fabricate light-emitting devices [871]. Longer alkyl chains substituted 1,7,13-trialkanoylde-cacyclene do not display liquid crystalline properties, but their molecular self-assembly observed by scanning tunneling microscope was reported by Li et al. [872]. The liquid crystalline 1,7,13-tri-heptyldecacyclene **3a** has recently been studied for its electrochemical properties and a photovoltaic device using **3a** as an electron transport material has been prepared [873].

Evidently, decacyclene-based DLCs have huge potential. More LC derivatives of this core, particularly low temperature or room-temperature DLCs, are required to investigate intriguing properties of these materials and to find their applications in industry.

2.12   HEXABENZOCORONENE CORE

Hexa-*peri*-hexabenzocoronene (HBC) is one of the largest and highly symmetrical *all-benzenoid* polycyclic aromatic hydrocarbons that acts as the core fragment of discotic liquid crystals [6,9,257,874–880]. HBC derivatives are the discotic materials for which many of the proposed application potentials of discotic liquid crystals have been successfully demonstrated such as in field effect transistors, photovoltaic solar cells, etc. [742,881–884]. The parent HBC and its mesomorphic and non-mesomorphic derivatives have been studied extensively for their various physical properties. The physical property studies include order and dynamics of molecules in the columnar phase studied by NMR, macroscopic alignment of processable derivatives either from solution or from melt, the relationship between supramolecular order and charge carrier mobility, self-assembly of molecules on flat surfaces studied by STM at the air–water interface or solid–liquid interface, etc. Moreover, they act as precursors for functional carbon nanomaterials [291,885–924]. Owing to their excellent performance in electro-optic devices, theoretical treatments of their temperature-driven change of the mesophase structure and the associated change of charge carrier mobilities in columnar phase of HBCs have been carried out, which would provide better understanding and designing molecules with desired properties [925–928]. Using atomistic molecular dynamics simulations, the solid and liquid crystalline columnar discotic phases formed by alkyl-substituted HBC mesogens have been studied. Correlations between the molecular structure, packing, and dynamic properties of these materials have been established. Theoretical studies also rationalize the difference in charge carrier mobilities in the solid herringbone and hexagonal mesophase. The charge carrier mobilities predicted by theoretical treatments agree with the values from pulse-radiolysis time-resolved microwave conductivity (PRTR-MC) experiments (see Chapter 6).

The parent hydrocarbon, **1,** is known in the chemical literature since the late 1950s [929–931]. The parent hydrocarbon was first synthesized by Clar (Scheme 2.80) [929,930], the pioneer of polycyclic aromatic hydrocarbons (PAHs). Hexa-*peri*-hexabenzocoronene contains 42 carbon atoms and 13 benzene rings so it can be considered as a nano-graphene. Moreover, due to its D_{6h}-symmetry and large conjugated π-system (three times the size of triphenylene), HBC has been considered as "super benzene," where each peripheral benzene ring is equivalent to sp^2 carbon of benzene.

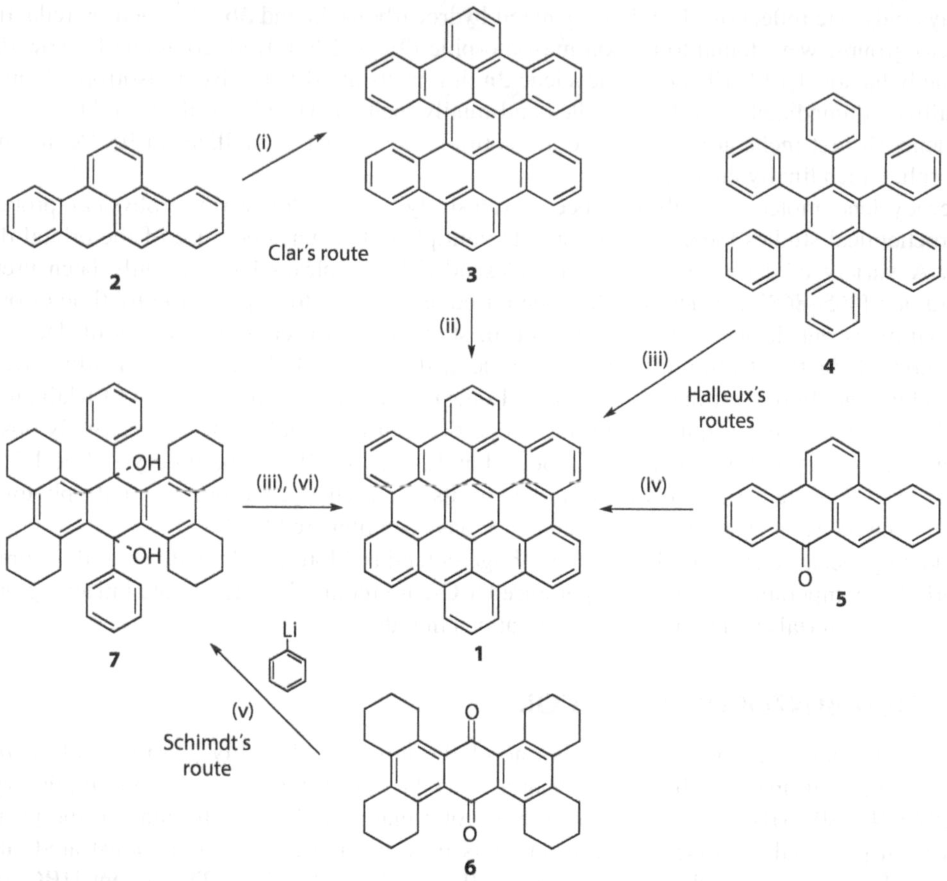

SCHEME 2.80 Synthetic routes to HBC hydrocarbon: (i) Br_2, C_6H_6 153°C; (ii) 481°C; (iii) $AlCl_3$/NaCl, 120°C; (iv) $Zn/ZnCl_2$, 330°C; (v) Et_2O/C_6H_6; (vi) Cu, 400°C.

Accordingly, the well-known substitution pattern of benzene can be extended to HBC derivatives. However, the analogous scheme suffers from complications when substituents are placed in the bay positions of the core. The remarkable thermal and chemical stability of HBC is evident by its existence and survival in the interstellar space [932,933].

The synthesis optimization of parent HBC **1** has taken a long time since the first synthesis by Clar and coworkers in 1958 [929]. Later, Halleux et al. [934] and Schmidt et al. [933] (Scheme 2.80) reported alternative methods toward the parent HBC. However, the original synthesis as well as the alternative methods yield HBC in very low quantities and involve complicated work-up. More recently, the progress of modern synthetic methods has allowed the synthesis of HBC and its functional derivatives under mild conditions in high yields. Mullen and associates have developed an efficient way to prepare the parent HBC, its derivatives, and related PAH structures by Scholl-type intramolecular oxidative cyclodehydrogenation of branched oligophenylenes with Cu(II) salts such as $CuCl_2$ and $Cu(OTf)_2$ catalyzed by Lewis acid $AlCl_3$ [6,9,257,874–880]. Subsequently, the use of an additional oxidant was avoided by using another Lewis acid $FeCl_3$ since iron(III) chloride acts both as a fairly strong Lewis acid and a mild oxidizing agent, and possesses an oxidation potential sufficient for the C–C bond formation. The general synthetic protocol of HBC and its sixfold symmetric derivatives is shown in Scheme 2.81 [935]. The synthesis of HBC starts with commercially available 4-alkyl aniline **8**, which is transformed into 4-alkyliodobenzene **9** via the Sandmayer reaction; the 4-alkyl iodobenzene **9** is coupled with trimethylsilylacetylene to yield compound **10**. After removal

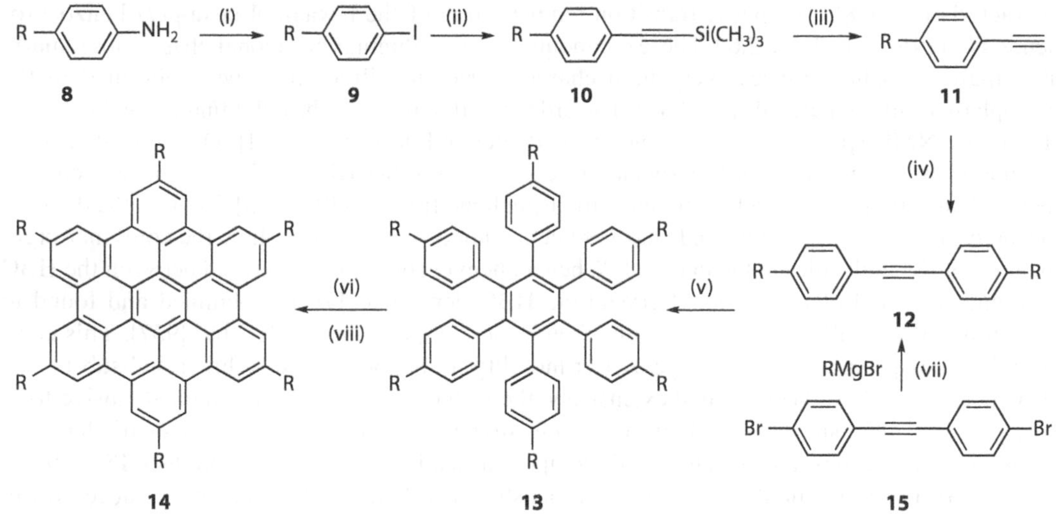

SCHEME 2.81 Synthesis of HBC discotics: (i) $C_5H_{11}NO_2$, KI; (ii) TMSA, $[PPh_3]_2PdCl_2$, PPh_3, CuI, piperidine; (iii) KF, DMF; (iv) **9**, $[PPh_3]_4Pd$, CuI, piperidine; (v) $Co_2(CO)_8$; (vi) $AlCl_3$, $Cu(CF_3SO_3)_2$, CS_2; (vii) $[PdCl_2(dppf)]$, THF; (viii) $FeCl_3/CH_3NO_2$, CH_2Cl_2.

of the trimethylsilyl group, the 4-alkylphenylacetylene **11** is coupled with 4-alkyliodo benzene **9** by Sonogashira coupling, yielding the 4,4′-di-*n*-alkyltolane **12**, which is the crucial building block of HBC derivatives. Then the tolane is cyclotrimerized to **13** under the catalytic action of cobalt octa-carbonyl $Co_2(CO)_8$. The key step for the synthesis of compound **14** is the oxidative cyclodehydroge-nation (intramolecular Scholl reaction) of hexaalkylphenyl benzenes **13** in the presence of $AlCl_3$ and $Cu(CF_3SO_3)_2$. While the synthetic protocol for HBC materials in Scheme 2.81 is simple, flexible, and high yielding, some limitations have been encountered. First the synthesis requires multiple steps to occur with hexaphenyl benzene precursors containing the desired functional groups (R). Carrying these substituents through several steps can sometimes lead to low over all yield. Second, the identity of R is limited by the tolerance of the cyclotrimerization catalyst. Third, the R is limited by its compatibility with the final oxidation step. For example, the cyclodehydrogenation of alkoxy-substituted hexaphenylbenzenes resulted in ether cleavage [936]. Finally, as shown in Scheme 2.81, a four-step synthetic sequence was utilized to prepare alkyl substituted 4,4′-diphenylacetylenes. Two disadvantages of this scheme are that two of the four steps are catalyzed by expensive palladium catalysts and the introduction of different substituents cannot be accomplished on the level of the 4,4′-diphenylacetylene, but must be introduced right at the beginning of the reaction sequence. The desired diphenylacetylene **12** may be obtained from 4,4′-dibromodiphenylacetylene **15**, which is a much faster, cheaper, and high-yielding one-step process (Scheme 2.81) [937]. Attempts to func-tionalize **15** under common Grignard coupling conditions using Ni(II) or Pd(0) were unsatisfactory. However, by using $PdCl_2(dppf)$ catalyst under Kumuda coupling conditions, both aryl and alkyl magnesium bromides can be coupled with **15** to substituted diphenylacetylenes. This procedure is highly attractive when the group to be attached is either expensive or must be prepared by multistep synthesis. Then the cyclotrimerization, followed by intramolecular oxidative dehydrogenation by ferric chloride ($FeCl_3$), provides the desired HBC derivative.

HBC as a core fragment of discotic liquid crystals was introduced in the year 1996 by Mullen and coworkers [935]. They reported the synthesis and mesophase characterization of several symmetrical hexaalkylsubstituted hexa-peri-hexabenzocoronenes (Scheme 2.81). It is interesting to note that prior to this report, the authors have studied the self-assembly and supramolecular organization of one of the derivatives on highly oriented pyrolytic graphite by scanning tunnel-ing microscopy and scanning tunneling spectroscopy, which exhibited diode-like current–voltage characteristics [938]. The phase transition temperatures of the hexaalkyl hexa-*peri*-benzocoronenes **14** are listed in Table 2.69. These compounds show columnar hexagonal phase with remarkably high mesophase ranges. Very high charge carrier mobilities have been obtained in the mesophase of these materials [939,940]. The order and dynamics in the columnar phase have been studied by NMR spectroscopy with one of the deuterated derivatives [941]. Owing to their high thermal stability and strong self-association, recently two other HBC derivatives have been synthesized and studied as model compounds for asphaltene fractions [942,943]. Though the detailed mesomorphism in these compounds has not been studied, x-ray studies show the columnar organization of the molecules in the material. When a phenyl group was introduced between the HBC core and the alkyl chains, a liquid crystalline HBC derivative **14i** was obtained and found to exhibit liquid crystalline behavior starting from room temperature (Table 2.69) [937]. This compound also displays very high charge carrier mobility in the mesophase. While alkyl-substituted HBC derivatives have been studied extensively, their alkoxy counterparts remained elusive for a long period of time since the oxidative cyclodehydrogenation step resulted in the dealkylation of alkoxy-substituted hexaphenylbenzene **17** compound, leading to a quinone product **18** (Scheme 2.82) [936]. Recently, an alternative method, as shown in Scheme 2.83, has been successful in obtaining the hexaalkoxy-substituted HBC **24** derivatives [944]. However, their mesomorphic behavior is yet to be reported.

TABLE 2.69
Phase Transition Temperatures of Hexaalkylsubstituted HBC Derivatives

Structure	R	Phase Transition	Ref.
14a	C_4H_9	Not observed	[943]
14b	C_6H_{13}	< 30 Cr > 600 I	[942]
14c	C_9H_{19}	< 30 Cr > 600 I	[942]
14d	$C_{10}H_{21}$	Cr 124 Col$_h$ > 400 I	[940]
14e	$C_{12}H_{25}$	Cr 60 Col$_h$ 399 I	[935]
14f	$CD_2C_{11}H_{23}$	—	[941]
14g	$C_{14}H_{29}$	Cr 62 Col$_h$ 350 I	[935]
14h	$C_{16}H_{33}$	Cr 65 Col$_h$ 323 I	[935]
14i	$PhC_{12}H_{25}$	Col$_1$ 18 Col$_2$ 83 Col$_h$ > 400 I	[940]

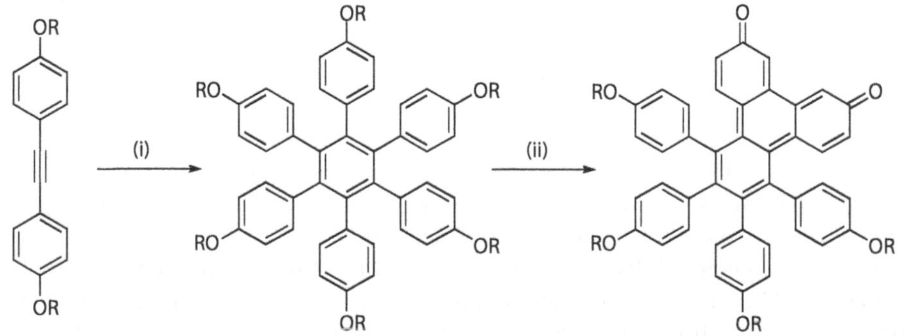

SCHEME 2.82 Attempted synthesis of hexaalkoxy-HBC discotics: (i) [Co$_2$(CO)$_8$], dioxane; (ii) FeCl$_3$, MeNO$_2$, CH$_2$Cl$_2$.

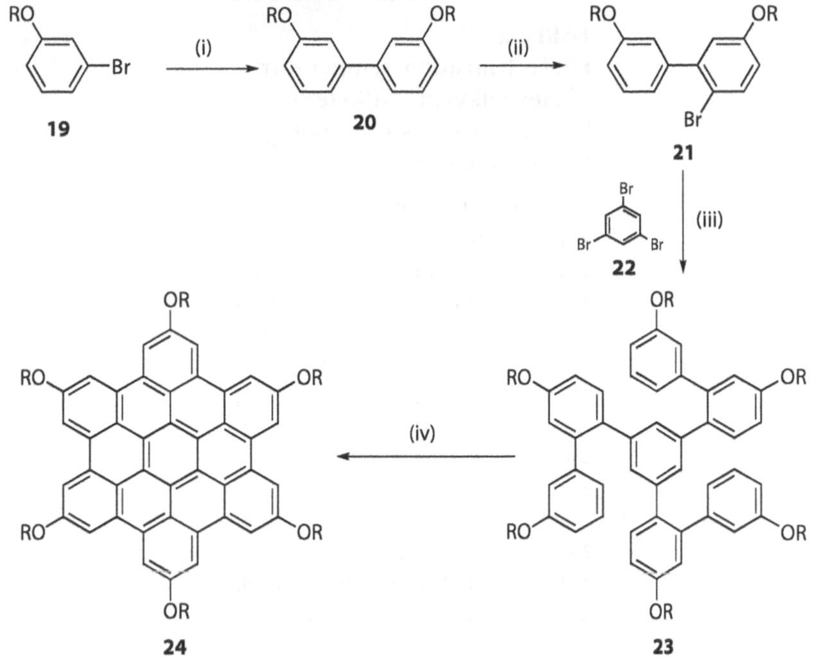

SCHEME 2.83 Synthesis of hexaalkoxy-HBC discotics: (i) THF, –78°C, BuLi, B(OMe)$_3$, (ii) toluene/EtOH/H$_2$O/Na$_2$CO$_3$/Pd(PPh$_3$)$_4$; (iii) NBS, MeCN; (iv) FeCl$_3$, CH$_3$NO$_2$, CH$_2$Cl$_2$.

Symmetrical hexa-substituted HBCs bearing branched alkyl **25** and alkylether **26** chains have been synthesized by following the synthetic procedure given in Scheme 2.81. The chemical structures of the synthesized compounds are shown in Figure 2.11 [292,945–949]. The mesophase behavior of these materials is shown in Table 2.70. The isotropic transition temperatures of some of these compounds are remarkably low, and a stable isotropic liquid phase could be attained, which is in contrast to their *n*-alkyl chain counterparts [948]. This is probably because of the stereoheterogeniety of the racemic branched chains and the steric bulk of the side chains, which

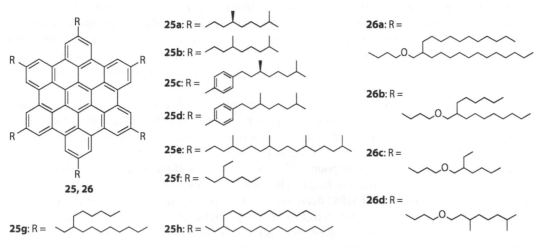

FIGURE 2.11 Chemical structure of symmetrical hexa-substituted HBCs bearing branched alkyl and alkylether chains.

TABLE 2.70
Phase Transition Temperatures
of Hexaalkyl and Alkylether
HBC Derivatives Containing
Branched Chains

Structure	Phase Transition	Ref.
25a	Cr 96 Col$_h$ 430 I	[945]
25b	Cr 81 Col$_h$ 420 I	[945]
25c	Col$_p$ > 500 I	[946]
25d	Not observed	[946]
25e	Cr −36 Col$_h$ 231 I	[947]
25f	Cr 97 Col$_h$ 420 I	[292]
25g	Cr 24 Col$_h$ 93 I	[292]
25h	Cr 46 I	[292]
26a	Col −15 Col$_p$ 42 I	[948]
26b	Col$_p$ −1 Col$_h$ 162 I	[948]
26c	Col$_p$ 55 Col$_h$ 420 I	[948]
26d	Col$_p$ 0 Col$_h$ 370 I	[948]

reduces the strong π–π interaction between the discotic cores. When the thermal behavior of an optically pure compound and its corresponding racemic HBC derivatives were compared, it has been observed that the optically pure compound possesses a stable columnar phase as compared to its racemic derivative, indicating the effect of stereohomogeneity on the mesophase stability [945]. Similarly, the self-assembly of optically active and racemic HBC derivatives at the solid–liquid interface also display distinctly different orientations as revealed by submolecularly resolved scanning tunneling microscopy images on highly orientated pyrolytic graphite surfaces. Thus, the self-assembling processes are governed by the interplay of intramolecular, intermolecular, as well as interfacial interactions [946]. The room temperature liquid crystalline derivative 25e exhibits spontaneous homeotropic alignment between two electrodes, which has allowed to study and evaluate its optical and optoelectronic properties [947]. Thermal properties and self-assembly of HBC derivatives with dove-tailed alkyl chains of various lengths have been investigated. It has been observed that the introduction of ether linkages within the side chains enhances the affinity of the discotic molecules toward polar surfaces thereby resulting in spontaneous alignment of these materials between two surfaces or on one surface [948]. The enhanced solubility, lower isotropization temperature, and spontaneous alignment of these branched chain derivatives between electrodes make them suitable candidates for application in photovoltaic applications.

A hexafluoro-substituted hexa-peri-hexabenzocoronene 27 (Figure 2.12) has been synthesized, which acts as an active material for n-type semiconductors [950]. A field effect transistor was fabricated by using this material and showed n-type performance. The electron-withdrawing effect of the fluorine substituents changed the polarity from p-type to n-type. Moreover, the attempted synthesis of this compound 27 from 4,4′-difluorodiphenylacetylene failed at the cobalt-catalyzed cyclotrimerization stage. Therefore, the alternative method developed for the synthesis of unsymmetrical HBC derivatives [945], that is, the Diels–Alder reaction with 4,4′-difluorodiphenylacetylene and tetrakis(4-fluorophenyl)cyclopentadienone, was adopted and found to be successful. Though the synthesis of hexaalkoxy-substituted HBCs suffers from dealkylation in the final step and results in quinone structures, a liquid crystalline hexaalkoxy-substituted HBC derivative 28 (Figure 2.12) has been prepared that contains fluorine atoms on

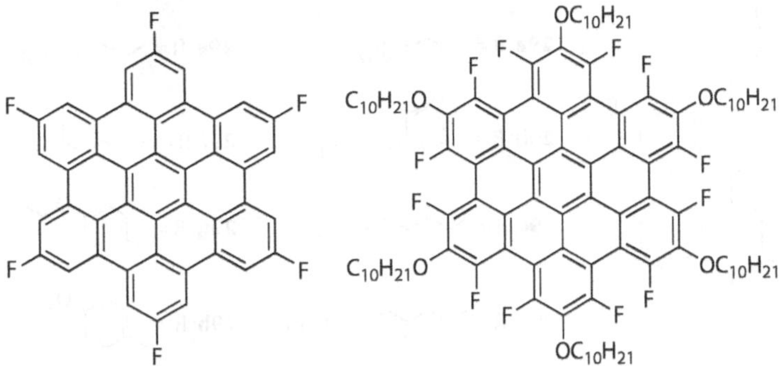

27

28   Cr 39 Col$_h$ 310 I

FIGURE 2.12   Chemical structure of symmetrical fluoro-substituted HBC derivatives.

the discotic core [951]. Probably the electron-withdrawing effects of the core-substituted fluorine atoms are reduced by the alkoxy groups and hence the cobalt(0)-catalyzed cyclotrimerization step proceeds smoothly, and subsequently the fluorine substituents help in obtaining the hexaalkoxy-substituted HBC derivative in the intramolecular cyclodehydrogenation step instead of the previously observed quinone structure. This compound possesses a columnar hexagonal mesophase with a wide range of temperatures (Table 2.71). The charge carrier mobility values measured for this compound by pulse radiolysis time-resolved microwave conductivity technique are comparable to that of other HBC materials.

Jenny et al. have prepared various partially perfluorinated alkyl chain-substituted liquid crystalline HBC derivatives **29**. The chemical structures of the compounds are shown in Figure 2.13, and their thermal phase behavior is collected in Table 2.71 [952–955]. Powder x-ray diffraction and differential scanning calorimetry of these perfluoroalkylated HBCs proved liquid crystalline properties whose transition temperatures, mesophase stability, and nature depend on the detailed structure of the side chains, that is, the ratio between aliphatic and perfluorinated parts. The insertion of phenyl spacer between the aromatic core and the lateral chains increases the LC transition temperature and interestingly switches the mesophase structure from columnar to smectic. Since the rotation of the discotic cores are hindered or reduced around the molecular axis parallel to the layer normal in the smectic phase, the biaxiality of the smectic phase is claimed in these compounds. Furthermore, the phase-transition temperatures observed for such compounds were found to be much higher than for their purely aliphatic analogue. This observation has been attributed to

TABLE 2.71
Phase Transition Temperatures of Perfluoro HBC Derivatives

Structure	R	Phase Transition	Ref.
29a	$(CH_2)_2C_8F_{17}$	Cr 180 Col$_r$ ~300 dec	[952]
29b	$(PhCH_2)_2C_8F_{17}$	Cr 236 SmA ~300 dec	[952]
29c	$(CH_2)_4C_6F_{13}$	Cr 194.1 Col$_h$ >250 I	[955]
29d	$(CH_2)_6C_6F_{13}$	Cr 109 Col$_h$ ~300 dec	[952]
29e	$(CH_2)_4C_8F_{17}$	Cr 120 Col$_h$ ~300 dec	[952]
29f	$(CH_2)_3CH(CH_2C_6F_{13})_2$	Amorphous material	[955]
29g	$PhO(CH_2)_4C_8F_{17}$	Cr 222 SmA ~300 dec	[952]
29h	$PhO(CH_2)_6C_6F_{13}$	Cr ~200 Col$_h$ ~300 dec	[952]

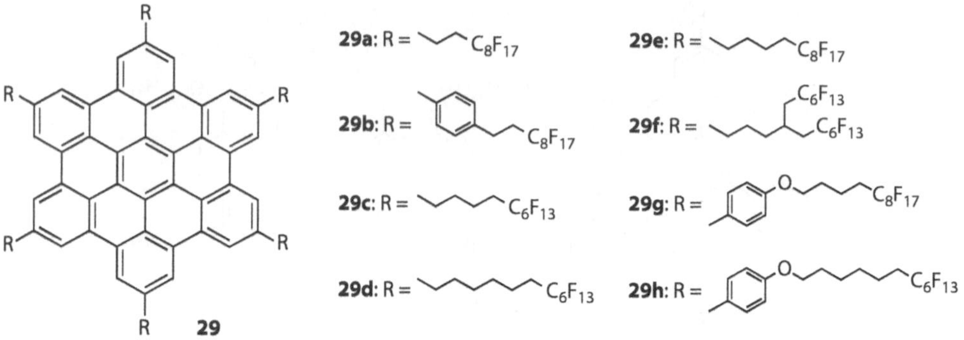

FIGURE 2.13 Chemical structure of symmetrical HBC derivatives with partially perflorinated alkyl chains.

the formation of much longer columnar structures, which compensates for the reduced lateral inter-actions due to perfluorinated chains. It should be noted that the attempted synthesis of HBC deriva-tives with directly attached perfluoroalkylated chains onto the HBC core prevents the formation of the target molecule at the last step of the synthesis, that is, planarization by cyclodehydrogenation, the key step in obtaining HBC derivatives. Neither of the two known conditions for this reaction, either the mild $FeCl_3/CH_3NO_2$ combination or the harsh $AlCl_3/Cu(OTf)_2$ mixture, is effective. This is most likely due to the strong electron-withdrawing nature of the perfluoroalkyl chains. Hence an aliphatic spacer is inserted between the hexaphenyl benzene core and perfluoroalkyl chain to reduce the electron-withdrawing effect that furnishes the desired compounds.

HBC derivatives carrying substituents in different regioisomeric patterns, unsymmetrically substituted with different groups and mono- and di-functionalized HBCs were synthesized by an alternative route (Scheme 2.84) [945,956]. The unsymmetrical and/or functionalized hexa-phenylbenzene precursor **34** is prepared via the [4+2] Diels–Alder cycloaddition of a suitably substituted diphenyl acetylene **33** and 2,3,4,5-tetrarylcyclopenta-2,4-dien-1-one **32**. The cyclopen-tadienone is synthesized via double Knovenagel condensation between a 4,4′-substituted benzyl **30** and 1,3-diarylacetone **31**, as shown in Scheme 2.84. Finally, the hexaphenylbenzene deriva-tives **34** further underwent oxidative cyclodehydrogenation with Iron(III) chloride to give fused HBC derivatives **35**. The present route is a versatile way to obtain both unsymmetrical and sym-metrical derivatives.

The synthesis and mesomorphic properties of mono-functionalized HBC derivatives **36–40** with branched alkyl chains have been reported. They have been synthesized according to the above-described synthetic route (Scheme 2.84). The chemical structures of these compounds are shown in Figure 2.14, and their phase transition temperatures are collected in Table 2.72 [957–960]. The dipole functionalized derivatives **36** are found to destabilize the columnar crystalline phase [957]. The HBC-pyrene dyad **37** exhibits more ordered columnar phase with dramatically lowered isotro-pization temperature [958]. At solid–liquid interface, this dyad displays two-dimensional crystal-line monolayer with a uniform nanoscale segregation of the large and small aromatic systems as revealed by STM. Several covalently linked dyads **38** and multiads **39** of electron-rich HBC and electron-deficient anthraquinone have been synthesized and their mesomorphism has been studied [959]. These compounds are expected to provide side-by-side percolation pathways for holes and electrons in solar cells, that is, they can act as model compounds for "molecular double cables." HBC derivatives carrying amido or ureido groups **40** capable of hydrogen bonding adjacent to the aromatic cores have been synthesized to study the effect of intracolumnar hydrogen bonds on the self-assembly behavior of HBCs in solution and in the bulk state [960]. In the bulk state, the typi-cal columnar supramolecular arrangement of HBCs was substantially stabilized. Additionally, the combined effect of the hydrogen bonds and π-stacking of the aromatic moieties led to the formation of fluorescent organogels.

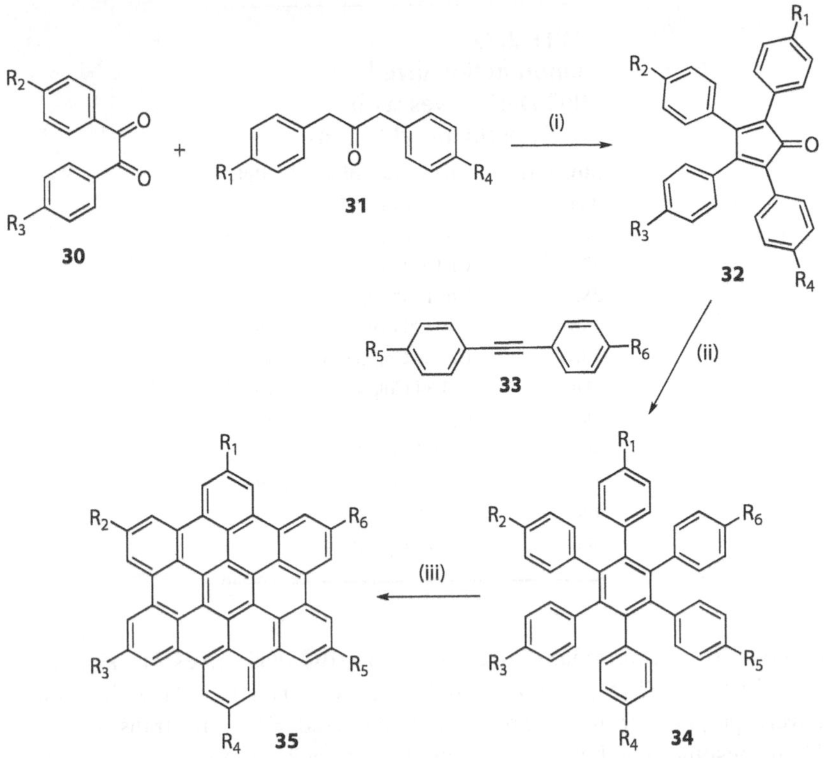

SCHEME 2.84 Synthesis of unsymmetrically substituted HBC derivatives: (i) Bu$_4$NOH, MeOH, *t*-BuOH or KOH, EtOH, reflux; (ii) Ph$_2$O, 260°C; (iii) FeCl$_3$, CH$_2$Cl$_2$, MeNO$_2$.

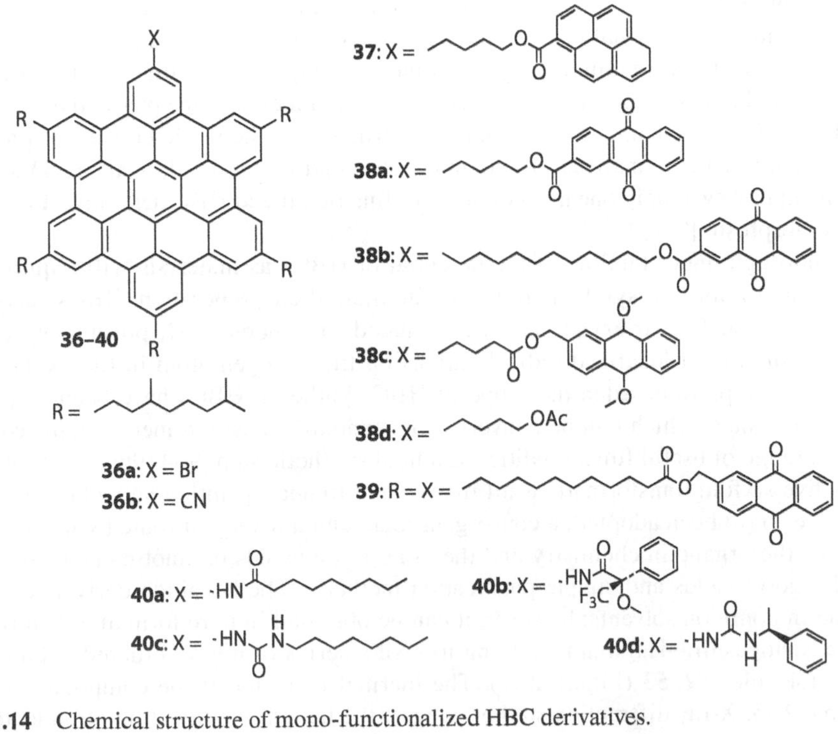

FIGURE 2.14 Chemical structure of mono-functionalized HBC derivatives.

TABLE 2.72
Monofunctionalized
HBC Derivatives with
3,7-Dimethyloctyl Chains

Structure	Phase Transition	Ref.
36a	G −35 Col$_h$ >170 I	[957]
36b	G −39 Col$_h$ >170 I	[957]
37	Cr 105 Col$_h$ 136 I	[958]
38a	Col 136 Col$_h$ 280 I	[959]
38b	Col 100 Col$_h$ 120 I	[959]
38c	Col 69 Col$_h$ 325 I	[959]
38d	Col 84 Col$_h$ 437 I	[959]
39	Cr 162 I	[959]
40a	−100 Col$_h$ >250 I	[960]
40b	−100 Col$_h$ >250 I	[960]
40c	−100 Col$_h$ >250 I	[960]
40d	−100 Col$_h$ >250 I	[960]

Several mono-, di-, and terminally functionalized HBC derivatives 41–45 have been synthesized, as shown in Figure 2.15 [957,961–963]. The HBC derivatives 41 with cyano, ether, ester, amino, etc., were prepared from the bromo-functionalized HBCs by transition-metal catalyzed chemistry. Their mesomorphic behavior as well as their packing in two and three dimensions was studied and found that the bulk structure in the mesophase is insensitive to changes of the substitution pattern. Mono- and bisfunctionalized derivatives are accessible via the Sonogashira coupling of bromo-substituted HBCs and functionalized alkynes. Terminally functionalized hexa-substituted derivatives 45 (Figure 2.16) are obtained using the cobalt octacarbonyl catalyzed cyclotrimerization of suitably substituted diphenyl acetylenes followed by an oxidative cyclodehydrogenation with ferric chloride. The thermal phase behavior of the compounds is summarized in Tables 2.73 and 2.74. HBC derivatives containing acryloyl or methacryloyl functionalities at the terminal positions of the alkyl chains 45 thermally polymerize in the liquid crystalline phase; thereby a network is obtained in which the columnar hexagonal superstructure of the liquid crystalline phase is preserved [963]. Para-bisfunctionalized bromo and corresponding amino derivatives 43 also exhibit mesomorphism; however, only one meta dimethoxy functionalized HBC derivative 44 is known to exhibit mesomorphism [957].

A more thorough understanding of the potential of HBCs as materials will require a library of derivatives and improved synthetic routes to facilitate their preparation. HBCs carrying electroactive moieties such as triarylamines are excluded in Scheme 2.81, presumably due to the preferential formation of localized radical cations on the nitrogen atom in the oxidative cyclodehydrogenation step. To broaden the scope of HBC synthesis, HBCs have been prepared with versatile reactive sites, which can be converted by standard transition metal catalyzed coupling reactions to a range of useful functionalities as a final synthetic step [964–966]. The total synthesis involves five sixfold transformations all in excellent to near quantitative yields. Two different synthetic routes have been adopted, a convergent route and a divergent route (Scheme 2.85). One route involves the organotin chemistry and the other route avoids organotin chemistry. Both the routes render good yields and simple purification methods. The hexaiodo derivative 51 is sparingly soluble in common solvents; however, it can be obtained in pure form and then functionalized via the Sonogashira–Hagihara coupling to give a series of highly ordered columnar liquid crystalline molecules 52, 53 (Figure 2.17). The thermal behavior of the compounds is summarized in Table 2.75. X-ray diffraction studies suggest the helical organization of molecules in the

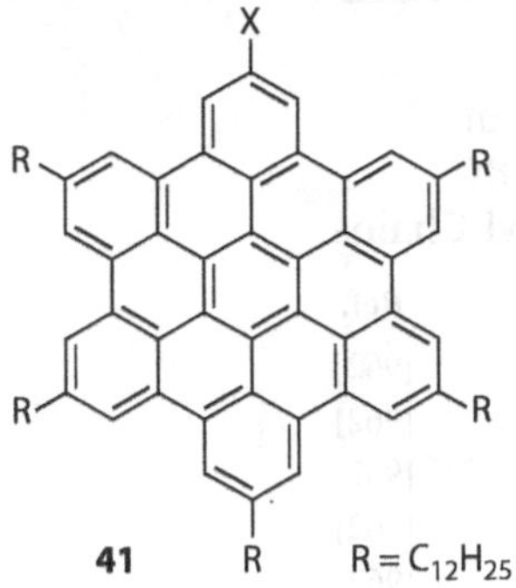

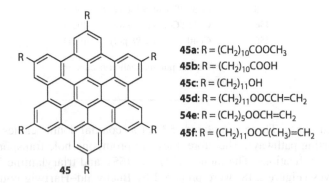

41   R   R = C$_{12}$H$_{25}$

41a: X = Br
41b: X = CN
41c: X = COOCH$_3$
41d: X = ⬡—O
41e: X = NH-C$_{11}$H$_{23}$
41f: X = ⬡N—

41g: X = (CH$_2$)$_{11}$OH
41h: X = (CH$_2$)$_{10}$COOH
41i: X = (CH$_2$)$_{11}$OOCCH=CH$_2$
41j: X = CC(CH$_2$)$_8$COOCH$_3$
41k: X = (CH$_2$)$_{10}$COOCH$_3$
41l: X = CCH

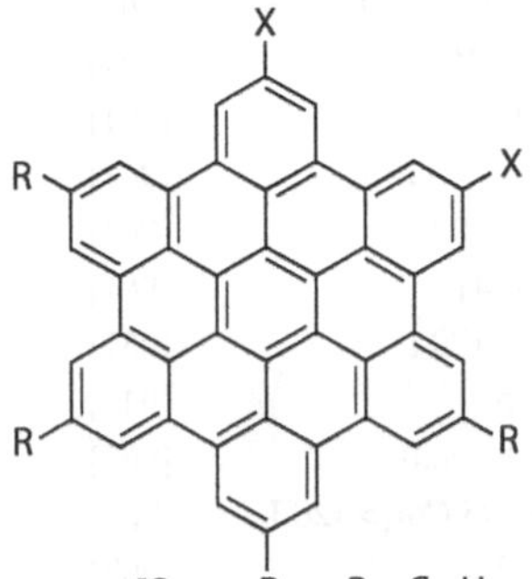

42   R   R = C$_{12}$H$_{25}$

42a: X = H
42b: X = Br
42c: X = NH-C$_{11}$H$_{23}$
42d: X = COOCH$_3$
42e: X = CC(CH$_2$)$_8$COOCH$_3$
42f: X = (CH$_2$)$_{10}$COOCH$_3$
42g: X = (CH$_2$)$_{11}$OH
42h: X = (CH$_2$)$_{10}$COOH

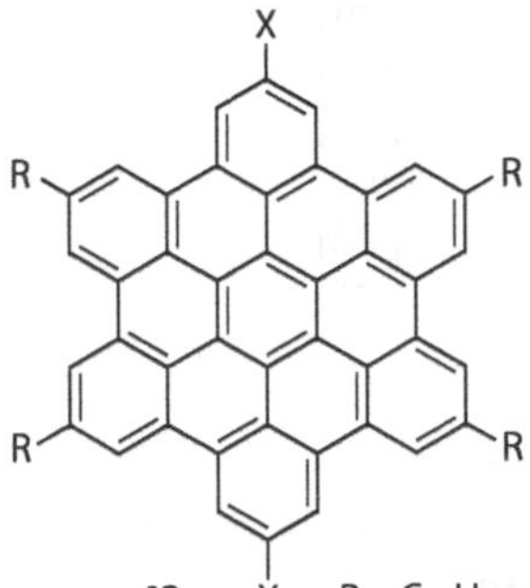

43   X   R = C$_{12}$H$_{25}$
43a: X = Br   **43b**: X = NH-C$_{11}$H$_{23}$

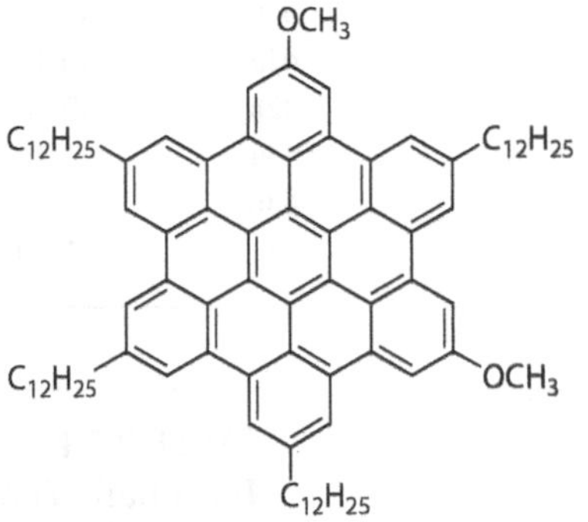

44

FIGURE 2.15 Chemical structure of some unsymmetrically substituted HBC derivatives.

45   R

45a: R = (CH$_2$)$_{10}$COOCH$_3$
45b: R = (CH$_2$)$_{10}$COOH
45c: R = (CH$_2$)$_{11}$OH
45d: R = (CH$_2$)$_{11}$OOCCH=CH$_2$
54e: R = (CH$_2$)$_5$OOCH=CH$_2$
45f: R = (CH$_2$)$_{11}$OOC(CH$_3$)=CH$_2$

FIGURE 2.16 Chemical structure of terminally functionalized hexa-substituted HBC derivatives.

TABLE 2.73
Phase Transition Temperatures of Mono- and Difunctionalized HBC Derivatives Containing C$_{12}$ Alkyl Chains

Structure	Phase Transition	Ref.
41a	Cr 74.1 Col$_h$ > 420 I	[962]
41b	Cr 77.3 LC > 420 I	[962]
41c	Cr 82.6 Col$_h$ > 420	[962]
41d	Cr 95.9 LC 404 I	[962]
41e	Cr 89.4 Col$_h$ 410 I	[962]
41f	Cr 101.7 Col$_h$ > 420 I	[962]
41g	Cr 106 Col 410 I	[961]
41h	Cr 108 Col$_h$	[961]
41i	Cr 75–100 Col > 140 poly	[961]
41j	Cr 69 Col$_h$ 390 I	[961]
41k	Cr 100 Col$_h$ 413 I	[961]
41l	Cr 32 Col$_h$ > 170 I	[957]
42a	Col$_p$ 146 Col$_h$	[943]
42b	Cr 59.3 Col$_h$ > 420 I	[962]
42c	Cr 89.7 LC 95 Col$_h$ > 420 I	[962]
42d	Cr 83 Col$_h$ 380 I	[962]
42e	Cr 56 Col	[961]
42f	Cr 73 Col	[961]
42g	Cr 115 Col	[961]
42h	Cr 108 Col	[961]
43a	Cr 104.3 Col$_h$ > 420 I	[962]
43b	Cr 95.6 Col$_r$ 115 Col$_h$ 360 I	[962]
44	Cr 109 Col$_h$ > 170 I	[957]

TABLE 2.74
Terminally Functionalized and Polymerizable HBC Derivatives

Structure	Phase Transition	Ref.
45a	Cr 79 Col > 370 I	[961]
45b	Cr 100–140 Col	[961]
45c	Cr 158 Col	[961]
45d	Cr 80–97 Col$_h$ > 100 poly	[961]
45e	Cr 112 Col$_h$ > 150 poly	[963]
45f	Cr 40–80 Col$_h$ > 98 poly	[961]

Note: Poly = thermal polymerization occurs on heating.

columnar hexagonal phase. These compounds **53** act as coaxial double cables owing to two different hole transporting pathways. Therefore, they are promising hole transporting materials for organic electronics applications. The mono-, (**54**), di-, (**55**), and triarylamine 2-2 functionalized (**56**) HBC derivatives (Figure 2.18) were prepared by Buchwald–Hartwig coupling reactions of arylamines with corresponding iodo HBC derivatives [966].

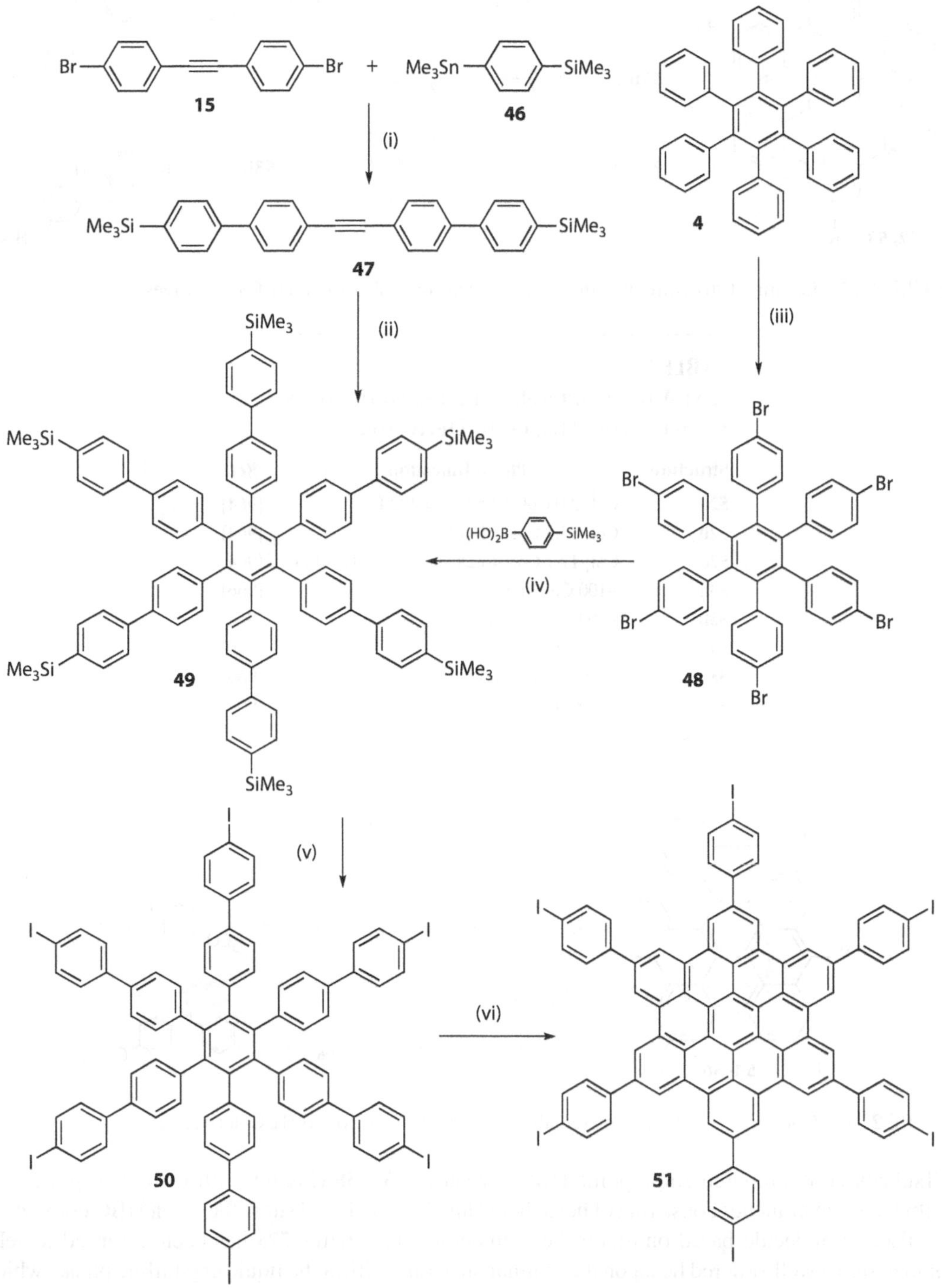

SCHEME 2.85 Synthesis of symmetrical HBC derivatives with versatile reactive sites which can be converted to a range of useful functionalities: (i) [Pd(PPh₃)₄], toluene 125°C; (ii) [Co₂(CO)₈], dioxane 125°C; (iii) bromine, −20°C; (iv) [Pd(PPh₃)₄], K₂CO₃ (aq) 95°C; (v) ICl, CHCl₃; (vi) FeCl₃/nitromethane, CH₂Cl₂/CS₂.

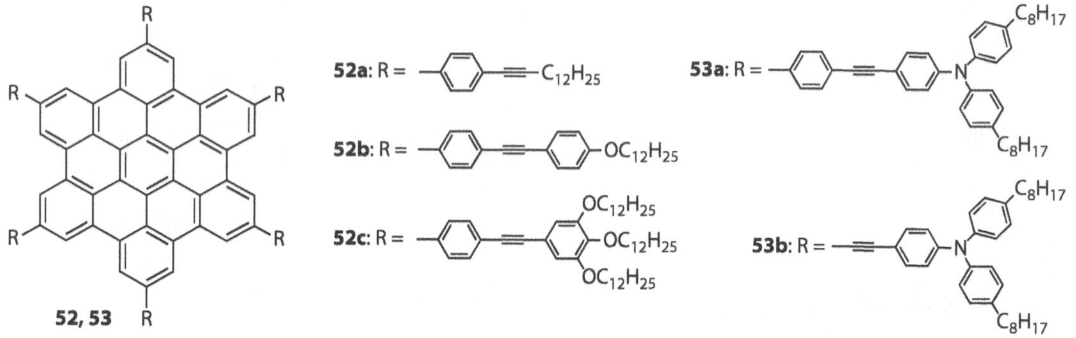

FIGURE 2.17 Chemical structure of symmetrical arylamine-substituted HBC derivatives.

TABLE 2.75
Aryl Amine and Related HBC Derivatives
Derived from Hexaiodo Derivative

Structure	Phase Transition	Ref.
52a	Col_1 2.0 Col_2 19.8 Col_h > 400 I	[964]
52b	Col_h 85.5 Col_h > 400 I	[964]
52c	Col_1 17.7 Col_2 48.2 Col_h 74.4 Col > 400 I	[964]
53a	−100 Col > 400 I	[966]
53b	−100 Col > 400 I	[966]
54	Cr 162 Col_h > 400 I	[966]
55	Cr 217 Col_h > 400 I	[966]
56	Cr 387 I	[966]

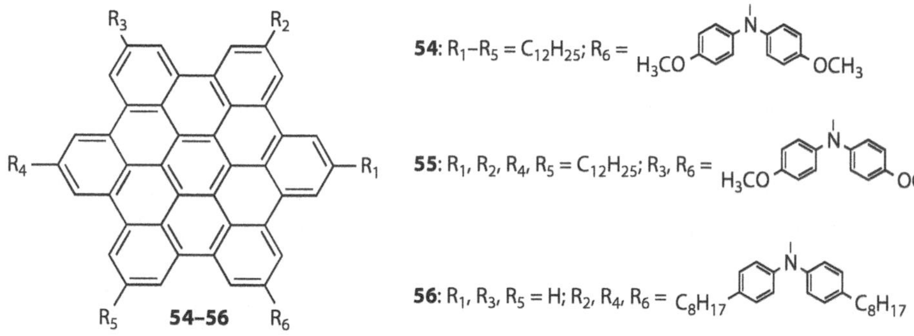

FIGURE 2.18 Chemical structure of mixed alkyl-arylamine-substituted HBC derivatives.

Tsukruk et al. prepared amphiphilic HBC derivatives **57, 58** (Figure 2.19) with hydrophilic oligoethylene side branches consisting of hexaphenyl and hexabiphenyl substituents on HBC core [967]. The discotic molecule based on dibranched oligoether side chains **57a** has been observed to self-organize into a well-ordered hexagonal columnar structure within the liquid crystalline phase, which possesses high thermal stability and a wide temperature range. The molecule based on dibranched oligoethers having biphenyl spacer **58** shows liquid crystalline phase at room temperature without any sign of crystallization. The liquid crystalline state is retained at elevated temperature up to decomposition temperature. However, the discotic molecule based on tetrabranched oligoether **57b** shows only an isotropic phase at room temperature. This is most likely because of the higher conformation entropy of bulkier peripheral chains and because of the fact that the volume fraction

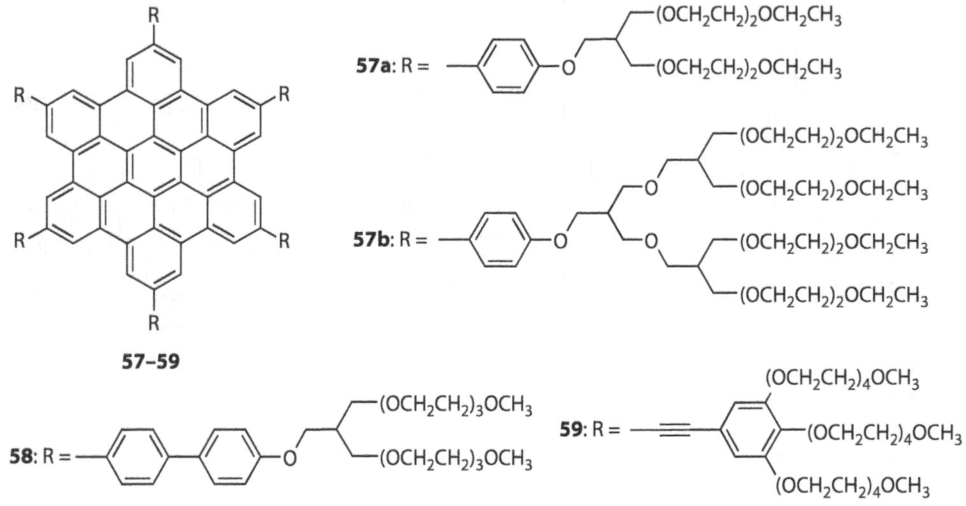

FIGURE 2.19 Chemical structure of amphiphilic HBC derivatives.

of the rigid core became small, preventing their effective segregation into ordered columns. These HBC derivatives are packed in a face-on arrangement at the air–water interface and on solid surfaces. Recently, a water-soluble HBC **59** (Figure 2.19) was prepared and shown to undergo a ordered columnar self-assembly either in water solution or in bulk and, therefore, served as a template for the fabrication of porous silica with aligned nanochannels [968]. This compound also displays a thermotropic columnar liquid crystalline phase at room temperature with lower isotropization temperature as compared to the above dibranched oligoether-substituted derivatives. The phase transition temperatures of these oligoethylene-containing HBC derivatives are collected in Table 2.76.

The controlled catalytic hydrogenation of aromatic hydrocarbons, which is very unpredictable and uncontrollable, has been achieved with HBCs. Exposure of HBCs to moderate hydrogen pressure in the presence of palladium on activated carbon as catalyst results in the quantitatively regiospecific conversion of suitably substituted HBCs to the first peralkylated coronenes **60** [969]. Some of the derivatives **60c** and **60e** exhibit liquid crystalline phase behavior just like the precursor HBCs from which they have been derived. The chemical structures of these compounds are depicted in Figure 2.20, and their thermal data are collected in Table 2.77. These materials also exhibit high charge carrier mobilities and lifetimes within the bulk material. On one hand, HBCs' existence around the stars manifests its remarkable stability while its susceptibility to catalytic hydrogenation under moderate pressure despite being an *all-benzenoid* polycyclic hydrocarbon makes it very intriguing.

A lightly substituted unsymmetrical HBC derivative **61** containing only three alkyl chains on one of the phenyl rings has been synthesized and characterized (Figure 2.20) [970]. Despite being

TABLE 2.76
Phase Transition Temperatures of Amphiphilic and Related HBCs with Ethylene Glycol Chains

Structure	Phase Transition	Ref.
57a	$Col_h > 350$ dec	[967]
57b	Liquid	[967]
58	$Col_h > 350$ dec	[967]
59	Col_h 150 I	[968]

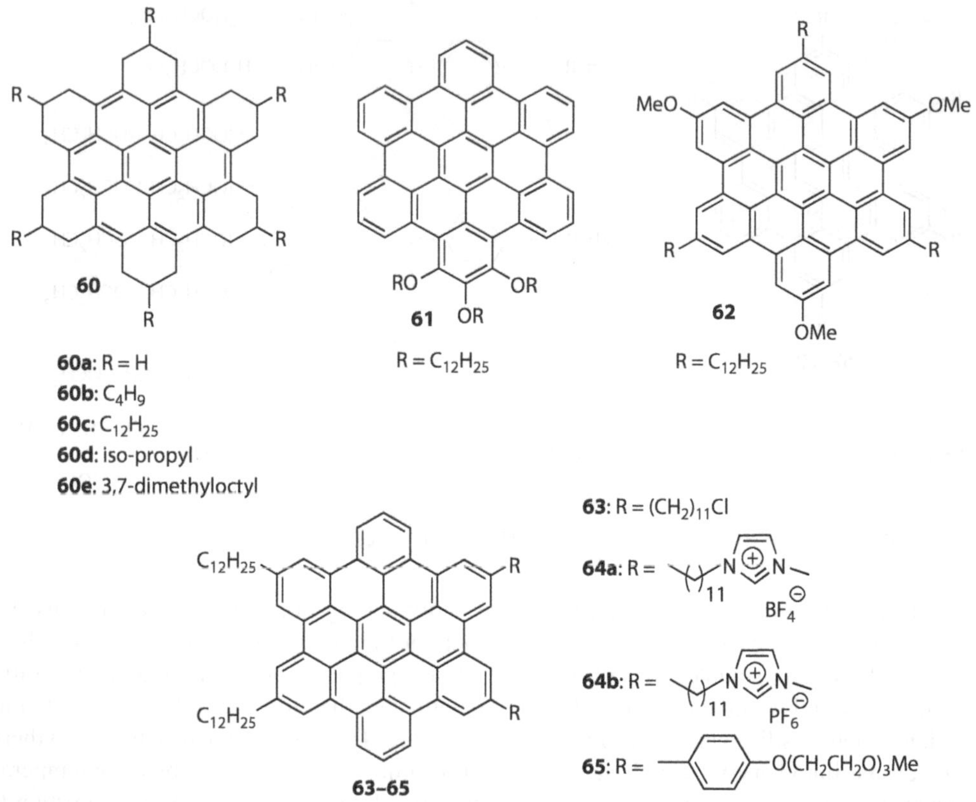

60

60a: R = H
60b: C$_4$H$_9$
60c: C$_{12}$H$_{25}$
60d: iso-propyl
60e: 3,7-dimethyloctyl

61

R = C$_{12}$H$_{25}$

62

R = C$_{12}$H$_{25}$

63–65

63: R = (CH$_2$)$_{11}$Cl

64a: R = ... BF$_4^\ominus$

64b: R = ... PF$_6^\ominus$

65: R = —⬡—O(CH$_2$CH$_2$O)$_3$Me

FIGURE 2.20   Chemical structure of some hydrogenated and other unconventional HBC derivatives.

TABLE 2.77
Phase Transition Temperatures of Some Miscellaneous HBC Discotics (Figure 2.20)

Structure	Phase Transition	Ref.
60c	Col$_h$ 113 I	[969]
60e	Col$_t$ 110 I	[969]
61	Col 48 Col$_r$ 190 I	[970]
62	Cr 145 Col$_h$	[971]
63	Cr 190 Col$_h$	[972]

"unwrapped," it forms a stable columnar liquid crystalline phase with a practically accessible isotropization temperature (Table 2.77), which is in contrast to the alkyl-substituted derivatives, which usually have a very high isotropization temperature and even their thermal decomposition precedes their isotropization. Moreover, this alkoxy-substituted HBC could be successfully obtained via the ferric chloride–mediated intramolecular Scholl reaction without dealkylation, and the compound is soluble in common organic solvents. The phase behavior of the compound is shown in Table 2.77.

Another C$_3$ symmetric and trimethoxy-functionalized liquid crystalline HBC derivative **62** is synthesized (Figure 2.20) and its mesomorphism has been investigated (Table 2.77) [971]. The compound exhibits helical organization in the bulk state and forms exceptionally long fibers in the solution. The facile transformation of the methoxy groups of this compound into hydroxyl groups is anticipated to afford novel alkoxy-substituted HBC derivatives. The liquid crystalline ω-chloro-substituted gemini-shaped HBC derivative **63** acts as a precursor for amphiphilic imidazolium-based HBC derivatives **64** [972]. Though the imidazolium-based ionic HBC derivatives do not form any LC phase, they are shown to form fibrous aggregates upon solution drop casting onto solid substrates. Recently, Aida and coworkers have intensively and systematically investigated such gemini-shaped molecular architectures of HBC **65** containing alkyl or alkylphenyl hydrophobic side chains on one side and hydrophilic triethyleneglycol (TEG) side chains on the other side [973–978]. These amphiphilic HBC derivatives form tubular assemblies in solution in contrast to the above-mentioned fibrous aggregates. The hierarchical nanotubular structures are composed of helically coiled bilayer tapes. Each tape consists of π-stacked HBC units, where the inner and outer HBC layers are connected by the interdigitation of paraffinic side chains.

2.12.1 HEXA-*CATA*-HEXABENZOCORONENE

The hexa-*cata*-hexabenzocoronene **68**, a homonym of hexa-peri-hexabenzocoronene **1**, and its derivatives **72**, **73** have recently been synthesized (Scheme 2.86) [979,980]. The first synthesis of hexa-*cata*-hexabenzocoronene was reported by Clar et al. in 1965 [931]. They prepared the tetraol compound **67** from **66** and found that condensation of **67** with copper powder at 400°C provided the HBC **68** in about 2% yield after column chromatography. In contrast to the insoluble parent hexa-*peri*-hexabenzocoronene **1**, the hexa-*cata*-hexabenzocoronene **68** is soluble in common organic solvents and has a melting point of 516°C. Recently, Nuckolls et al. have reported a new synthesis of hexa-*cata*-hexabenzocoronene and its derivatives **72** and **73** by photocyclization and a combination of photocyclization and ferric chloride–mediated cyclodehydrogenation of the bisolefins **71** [980]. This good yielding method allows large-scale preparation of HBC.

As disclosed by single-crystal analysis, due to the steric congestion in its proximal carbon atoms, the aromatic core was distorted away from planarity and the three intersecting pentacene subunits

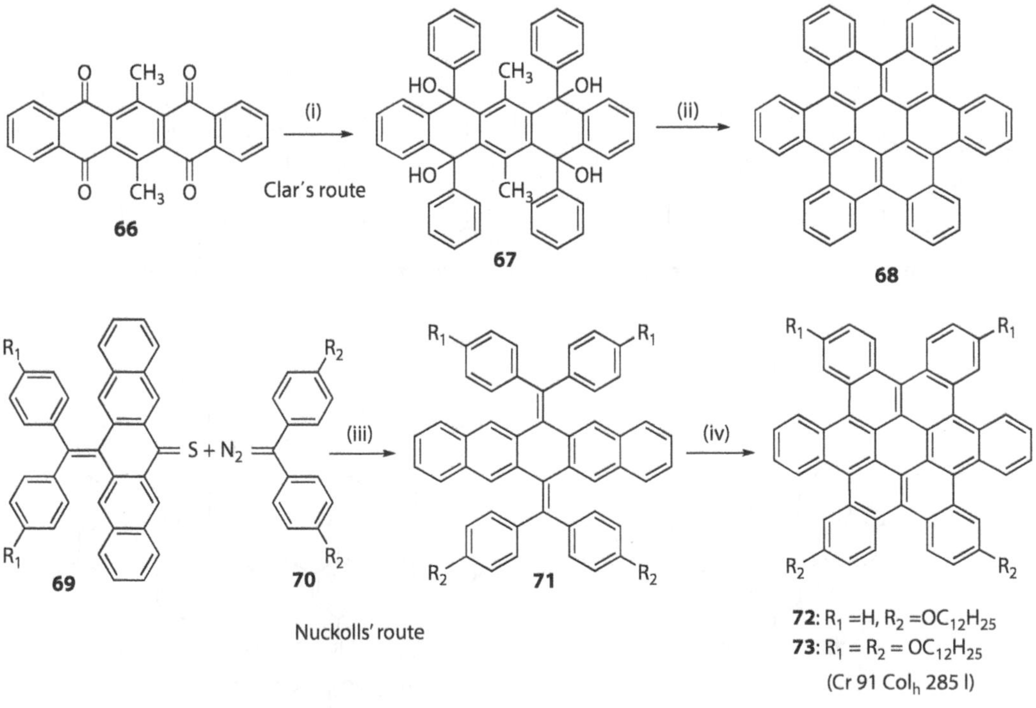

SCHEME 2.86 Synthesis of hexa-*cata*-hexabenzocoronene derivatives: (i) C_6H_6, PhLi; (ii) Cu, 400°C; (iii) THF, PPh$_3$; (iv) $h\nu$, I$_2$, propylene oxide.

contorted into a zigzag conformation. After the attachment of flexible alkyl chains, the non-planar compound **73** formed an ordered columnar liquid crystalline phase between 91°C and 285°C [979]. The relatively low isotropization temperature of this compound as compared to its planar counterparts is attributed to the non-planarity of the central discotic core, resulting in reduced stacking in the columnar phase. These columnar materials have good electrical properties in thin film transistors made with LC films, self-assembled monolayers, and nanostructured cables [981–984]. A thin-film transistor constructed from carbon nanotubes and HBC bilayer shows field effect mobility increased by a factor of 6 as compared to the HBC itself. Photoconductivity in these contorted HBC liquid crystals is found to be exclusively one-dimensional. Spectroscopic measurements and density functional theory support the existence of two π-systems attributed to a low-energy radialene-core and higher energy out-of-plane alkoxyphenyl cladding. Unlike the material with four alkoxy side chains **73**, the derivative with eight alkoxy side chains or two alkoxy chains does not show any mesophase property [983].

2.12.2 LARGER DISCOTIC CORES (GRAPHENES)

The $FeCl_3$ or $Cu(OTf)_2$-AlCl$_3$-mediated oxidative cyclodehydrogenation of branched hexaphenyl-benzene derivative was applied to the synthesis of giant graphene molecules with different sizes and shapes **74–85** (Figures 2.21 and 2.22) [943,985–991]. Increasing the size of the core of discotic materials is predicted to improve the order of columnar superstructures and phase widths due to the large overlaps of π-surfaces and thus enhance their mesophase stability and charge carrier mobility. Accordingly, the synthesis of alkyl-chain-substituted discotics with cores related to HBC **74–79** (Figure 2.21) and significantly larger than HBC **80–85** (Figure 2.22) has been accomplished. Appropriate branched oligophenylenes were first prepared by Diels–Alder reactions and then subjected to oxidative cyclodehydrogenation to give planar graphene disks [943,991]. The thermal phase transition temperatures of these larger discotic cores are summarized in Table 2.78. The number

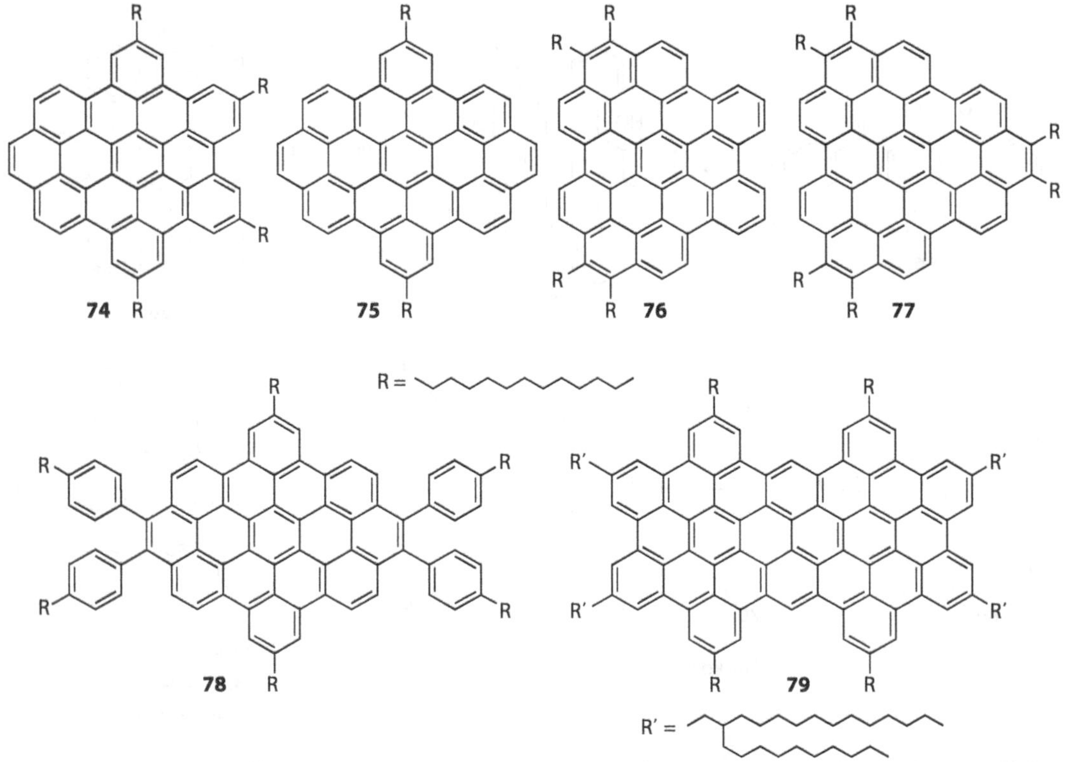

FIGURE 2.21   Chemical structure of some graphene molecules explored for mesomorphism.

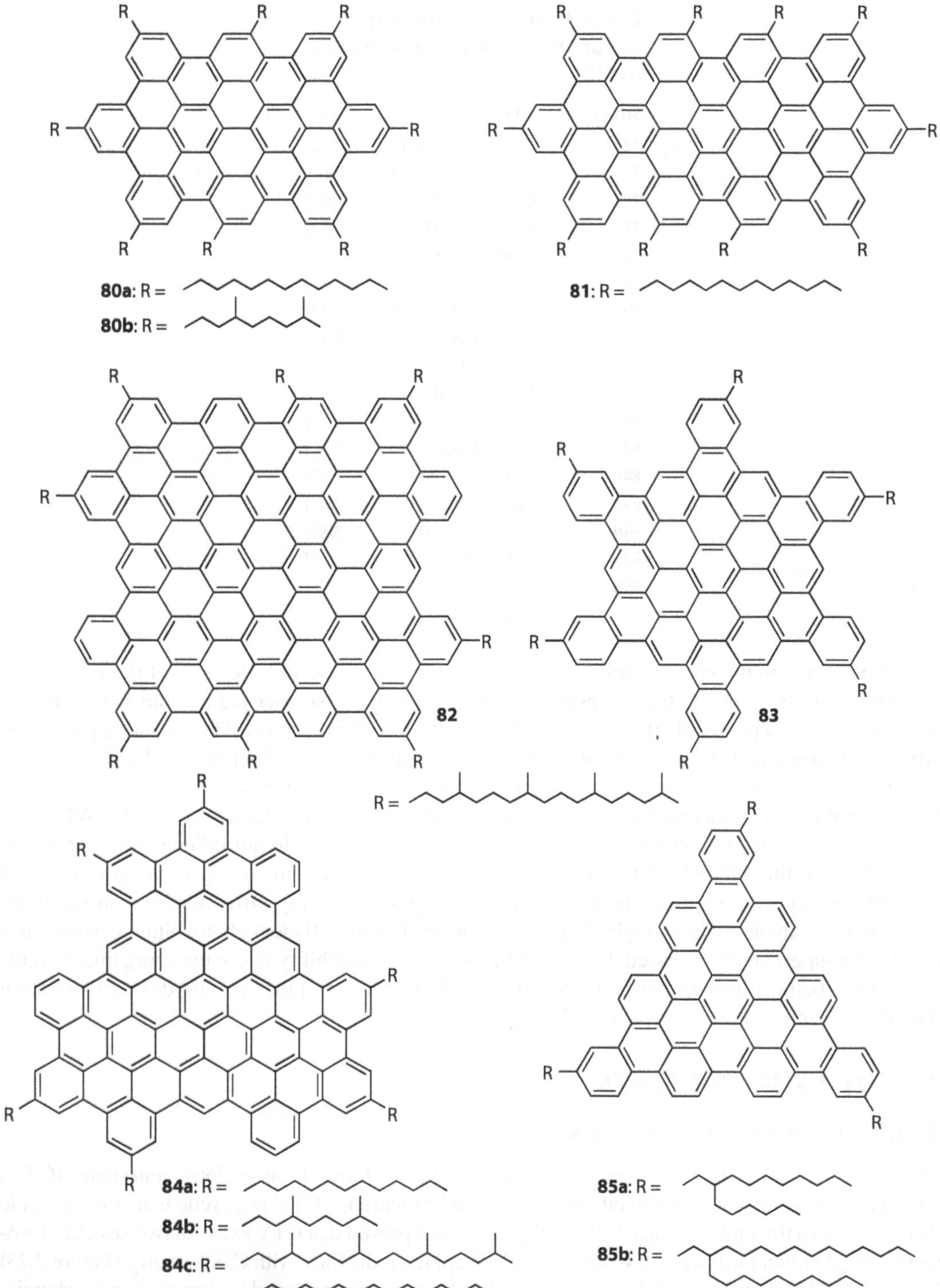

FIGURE 2.22 Chemical structure of some discotic graphene molecules significantly larger than HBC.

TABLE 2.78
Phase Transition Temperatures
of Larger Discotic Cores Related
to HBC

Structure	Phase Transition	Ref.
74	Cr 148 Col$_h$ 400 I	[986]
75	Cr 226 Col$_h$ > 500 I	[986]
76	Cr 173 Col$_c$ 210 I	[986]
77	Cr 48 Col$_h$ > 500 I	[987]
78	−150 Col > 400 I	[985]
79	Col	[989]
80a	Cr 104 Col$_h$ > 450 I	[943]
80b	Cr 109 Col$_h$	[943]
81	Cr 87 Col$_h$ > 450 I	[943]
82	Not observed	[943]
83	Not observed	[943]
84a	Col 38 Col$_h$ > 550 I	[987]
84b	−100 Col > 250 I	[987]
84c	−100 Col > 250 I	[987]
84d	−100 Col > 250 I	[987]
85a	Col$_h$ > 500 I	[988]
85b	Col 45 Col$_h$ > 500 I	[988]

of carbon atoms in the cores varies from 44 through 60, 78, 96 to 132. Because of the increase of π-conjugation, the color of the compounds is shifted from yellow over orange and red to purple. These compounds possess better optical and charge transport properties. The extended polycyclic aromatic hydrocarbons possessing armchair to zigzag peripheries have been realized and it has been found that the variation of nature of periphery, symmetry, size, and shape has a distinct impact upon the electronic properties and the organization into columnar superstructures [985,986]. While the number and the substitution patterns of attached peripheral chains do not influence the electronic properties, the thermal behavior and supramolecular organization are strongly influenced, which has been elucidated with differential scanning calorimetry and x-ray diffractometry on mechanically aligned samples. The triangle-shaped discotic **85** displays efficient photovoltaic performance [988]. The superphenalene-based discotics **84** possess good solubility in common organic solvents and exhibit better film-forming properties that are favorable for optoelectronic device fabrication with these macrodiscotic materials [987].

2.13　MACROCYCLIC CORES

2.13.1　TRIBENZOCYCLONONATRIENE CORE

The derivatives of tribenzocyclononatriene core are often referred to as cyclotriveratrylenes (CTV) after their parent molecule, hexamethoxy-tribenzocyclononatriene. CTV is a cyclic trimer of veratrole that possesses a threefold symmetry. Recently, it has been proved that CTV exists in two structural isomers: a rigid crown form with C$_3$ symmetry and a flexible saddle form with C$_1$ symmetry (Figure 2.23) [992–994]. CTV is an archetypal cyclophane scaffold commonly employed in supramolecular chemistry and has been extensively studied for its ability to complex with a variety of molecules [995–1009].

As CTV adopts a shallow cone-shaped conformation, it has been suggested that the liquid crystalline compounds derived from this nucleus may be termed as "pyramidic" [1010]. A crystal structure indeed showed the head-to-tail or ferroelectric order within the column [998,1011]; however,

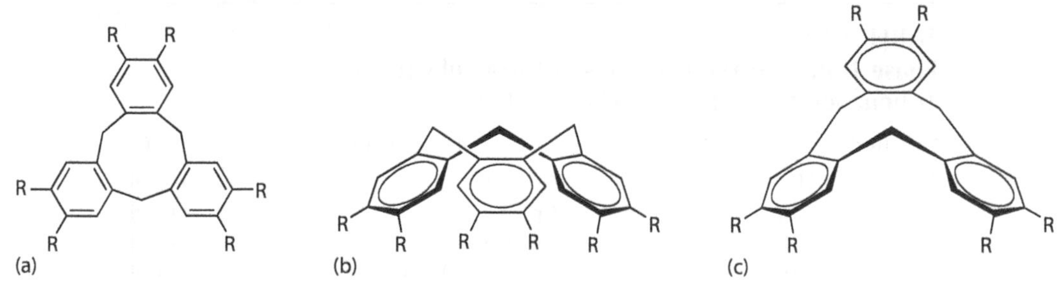

FIGURE 2.23 Molecular structure of (a) tribenzocyclononatriene, (b) its crown form, and (c) its saddle form.

it is not necessary that the same order persists in the liquid crystalline mesophase. At elevated temperature, the cone inverts very easily. The CTV nucleus is conveniently assembled via the acid-catalyzed condensation of veratrole **1** and formaldehyde. The reaction has been known for about a century [1012], but only in 1965 the products of this reaction (Scheme 2.87), the tribenzocyclononatriene and the tetrabenzocyclododecatetraene, could be identified [1013,1014]. The acid-catalyzed oligomerization of 3,4-dimethoxybenzyl alcohol **3** and chloromethylation of the diphenylmethane **4** derivative also reported to yield CTV [1013–1015]. The condensation of appropriate benzyl alcohol in an ionic liquid with only a catalytic amount of H_3PO_4 is reported to afford CTV in excellent yield [1016]. Simply heating the benzyl alcohol with catalytic amount of H_3PO_4 in the absence of any solvent also produces CTV, albeit in low yield. Discotic liquid crystalline hexa-substituted CTV derivatives [72,1010,1011,1017–1021] were prepared by the demethylation of **5** followed by the alkylation or acylation of the hexahydroxy product **6** with long alkyl chain alkyl halide or acid chloride. The thermal behavior of these derivatives is summarized in Table 2.79. As can be seen from

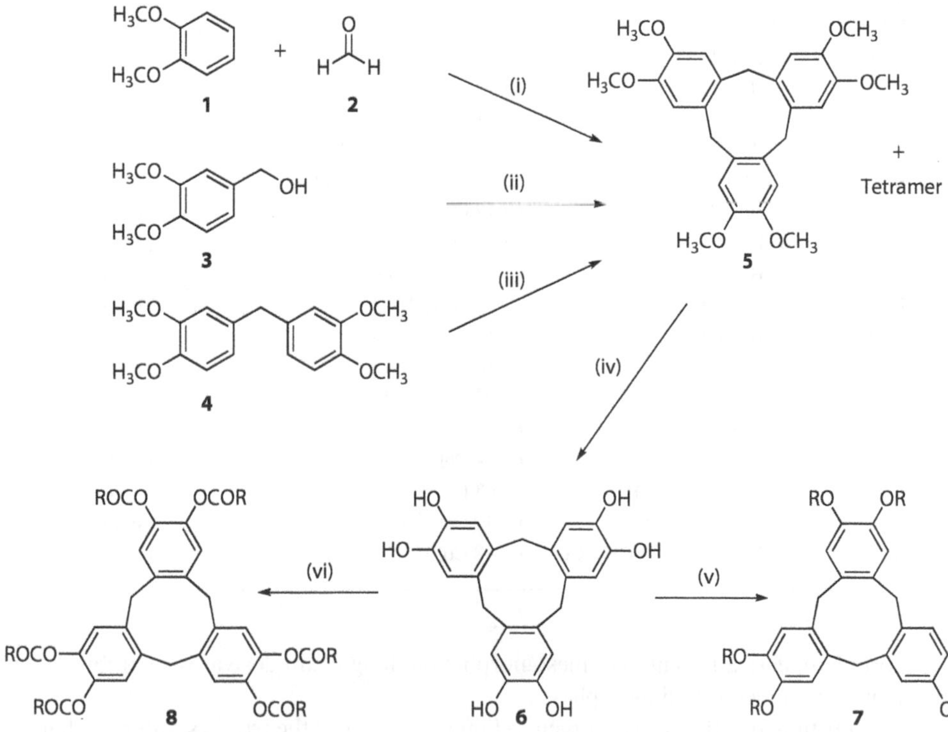

SCHEME 2.87 Synthesis of hexa-substituted CTV discotics: (i) AcOH, H_2SO_4, 100°C; (ii) H_2SO_4, 0°C; (iii) CH_2O, HCl; (iv) BBr_3, $CHCl_3$; (v) RBr, K_2CO_3, DMF; (vi) RCOCl.

TABLE 2.79
Phase Transition Temperatures of Hexa-Substituted
Tribenzocyclononatriene (Scheme 2.87)

Structure	R	Phase Transition	Ref.
7.1	CH_3	Cr 232 I	[1010]
7.2	C_4H_9	Cr 135.6 I	[1010]
7.3	C_5H_{11}	Cr 103.8 I	[1010]
7.4	C_6H_{13}	Cr 40.9 Col 92.2 I	[1010]
7.5	C_7H_{15}	Cr 25.0 Col 79.9 I	[1010]
7.6	C_8H_{17}	Cr 24.9 Col_h 71.5 I	[1010]
7.7	C_9H_{19}	Cr 18.7 Col_h 66.1 I	[1010]
7.8	$C_{10}H_{21}$	Cr 25.5 Col_h 63.2 I	[1010]
7.9	$C_{11}H_{23}$	Cr 34.8 Col 44.2 Col 62.0 I	[1010]
7.10	$C_{12}H_{25}$	Cr 48.3 Col_t 61.6 I	[1010]
8.1	C_3H_7	Cr 197 I	[1011]
8.2	C_7H_{15}	Cr 5.2 Col 153 I	[1010]
		Cr 150 I	[1011]
8.3	C_8H_{17}	Cr 23.9 Col 152.6 I	[1010]
8.4	C_9H_{19}	Cr 32.7 Col 146.2 I	[1010]
		Cr 50 Col 144.5 I	[1011]
8.5	$C_{10}H_{21}$	Cr 31.5 Col 38.6 Col 131.6 Col 140.8 I	[1010]
8.6	$C_{11}H_{23}$	Cr 58.1 Col 118.8 Col 140.6 I	[1010]
		Cr 55 Col 115.5 Col 138 I	[1011]
8.7	$C_{12}H_{25}$	Cr 67.4 Col 99.5 Col 139.2 I	[1010]
8.8	$C_{13}H_{27}$	Cr 73.4 Col 81.4 Col 136.2 I	[1010]
8.9	$C_{14}H_{29}$	Cr 80.5 Col 134.6 I	[1010]
8.10	$C_{15}H_{31}$	Cr 80 Col 129 I	[1011]
8.11	$C_7H_{15}C_6H_4$	Cr 33 Col 57 Col 149 I	[1017]
8.12	$C_8H_{17}C_6H_4$	Cr 12 Col 21 Col 157 I	[1017]
8.13	$C_{10}H_{21}C_6H_4$	Col 43 Col 162 I	[1017]
8.14	$C_{10}H_{21}OC_6H_4$	Cr 51 Col 100 Col 190 I	[1017]
8.15	$C_{12}H_{25}OC_6H_4$	Cr < 25 Col_1 75 Col_2 188 I	[1011]
8.16	$C_4H_9OCH_2$	Cr 175 I	[1020]
8.17	$C_5H_{11}OCH_2$	Cr 161 I	[1020]
8.18	$C_6H_{13}OCH_2$	Cr 140 Col_{ob} 156 I	[1020]
8.19	$C_7H_{15}OCH_2$	Cr 130 Col_h 154 I	[1020]
8.20	$C_8H_{17}OCH_2$	Cr 123 Col_h 160 I	[1020]
8.21	$C_{10}H_{21}OCH_2$	Cr 109 Col_h 165 I	[1020]
8.22	$C_{12}H_{25}OCH_2$	Cr 93 Col_h 164 I	[1020]
8.23	$C_5H_{11}CH(CH_3)OCH_2$	Col_t 188 I	[1020]
8.24	$C_8H_{17}SCH_2$	Cr 59 Col_h 138 I	[1020]
8.25	$C_5H_{11}O(CH_2)_2CO$	Cr 63 I	[1020]
8.26	$C_9H_{19}O(CH_2)_2CO$	Cr 20 Col 79 I	[1020]
8.27	$C_5H_{11}CH(CH_3)O(CH_2)_2CO$	Cr −8 Col 105 I	[1020]

this table, short chain derivatives are nonmesomorphic but long chain derivatives exhibit columnar as well as many other unidentified mesophases.

CTVs when substituted with nine substituents (Figure 2.24) loses the trifold symmetry that it possessed before; thus even in the absence of asymmetric carbon atom, they show structural chirality. Therefore, an as-synthesized compound contains racemic mixtures of optical isomers. When CTVs

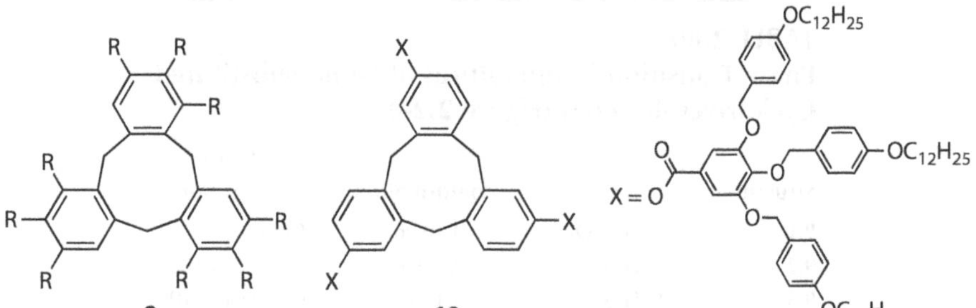

FIGURE 2.24   Molecular structure of CTV discotics with nine peripheral chains.

are hexa-substituted, the crown form is the only stable conformer. When the *ortho* positions of benzene groups are substituted with bulky substituents, the crown conformer is destabilized owing to steric hindrance due to the adjacent ring. Despite the steric hindrance of the *ortho* substituents, the more stable species at room temperature is crown form, due to its lower enthalpy. At higher temperatures, it converts to saddle form, which has an open structure and is highly flexible. This increases its entropy and thus the saddle form dominates over the crown form at higher temperatures [1022,1023].

The two conformations, crown and saddle form, can be achieved synthetically by low-temperature and high-temperature methods, respectively. The low-temperature method yields only compounds with crown conformation while the high-temperature method yields a mixture of both crown and saddle conformers, but saddle conformers dominate. The slow rate of inter-conversion at room temperature made it possible to quantitatively separate the two conformers by room temperature column chromatography and to purify them separately [1022].

The crown and saddle forms showed very interesting mesophase behavior; the saddle conformers from $n = 4$ to 14 and crown conformers from $n = 5$ to 14 are all mesogenic (Table 2.80), showing a single columnar hexagonal mesophase. The clearing temperatures of the crown isomers are higher than those for the corresponding saddle. The as-synthesized saddle mesogens were obtained in wax form apparently as supercooled mesophase. They did not exhibit clear melting transitions; the first endothermic transition corresponds to clearing temperature of the saddle form. In the isotropic liquid, equilibrium between the crown and saddle form is attained. At this temperature, the isomerization half time is of the order of a few seconds, so that as soon as the isotropic liquid is formed, part of the saddle molecules convert to crown. This crown form immediately crystallizes out of the system to form the more stable crown mesophase, leading to further saddle-crown inter-conversion, until eventually the entire sample is transformed into the crown mesophase [1022].

A trisubstituted, but having nine peripheral chains, compound **10** was prepared in racemic and optically active forms (Figure 2.24) [861]. Both (±)-**10** and (+)-**10** were obtained in liquid crystalline state at room temperature and display a hexagonal columnar mesophase that clears at about 150°C. The optically active compound (+)-**10** racemizes within 20 min at 100°C.

2.13.2   Tetrabenzocyclododecatetraene Core

The tetrabenzocyclododecatetraene (CTTV) molecule has fourfold symmetry and like CTV, it is also nonplanar. CTTV derivatives undergo rapid interconversion between two symmetry-related sofa conformations. As mentioned in the previous section, CTTV is formed as a minor product during the cyclotrimerization of 3,4-dimethoxybenzyl alcohol to prepare CTV. The minor product is isolated and explored for the preparation of CTTV derivatives. Later, Percec, and coworkers realized that CTTV can be made the major product by varying the reaction conditions [1024]. The liquid crystalline derivatives of CTTV **13, 14** are obtained by symmetrically substituting eight side chains

TABLE 2.80
Phase Transition Temperatures of Nona-Substituted Cyclotriveratrylenes (Figure 2.24)

Structure	R	Conformation	Phase Transition
9.1	$C_5H_{11}CO_2$	Crown	Cr? Col 164 I
9.2	$C_6H_{13}CO_2$	Crown	Cr 25 Col 162 I
9.3	$C_7H_{15}CO_2$	Crown	Cr 19 Col 170 I
9.4	$C_8H_{17}CO_2$	Crown	Cr 45 Col 164 I
9.5	$C_9H_{19}CO_2$	Crown	Cr 39 Col 160 I
9.6	$C_{10}H_{21}CO_2$	Crown	Cr 43 Col 160 I
9.7	$C_{11}H_{23}CO_2$	Crown	Cr 44 Col 164 I
9.8	$C_{12}H_{25}CO_2$	Crown	Cr 60 Col 159 I
9.9	$C_{13}H_{27}CO_2$	Crown	Cr 65 Col 160 I
9.10	$C_{14}H_{29}CO_2$	Crown	Cr 71 Col 156 I
9.11	$C_4H_9CO_2$	Saddle	Col 148 I
9.12	$C_5H_{11}CO_2$	Saddle	Col 135 I
9.13	$C_6H_{13}CO_2$	Saddle	Col 135 I
9.14	$C_7H_{15}CO_2$	Saddle	Col 127 I
9.15	$C_8H_{17}CO_2$	Saddle	Col 137 I
9.16	$C_9H_{19}CO_2$	Saddle	Col 115 I
9.17	$C_{10}H_{21}CO_2$	Saddle	Col 125 I
9.18	$C_{11}H_{23}CO_2$	Saddle	Col 128 I
9.19	$C_{12}H_{25}CO_2$	Saddle	Col 126 I
9.20	$C_{13}H_{27}CO_2$	Saddle	Col 126 I
9.21	$C_{14}H_{29}CO_2$	Saddle	Col 126 I

Source: Data from Zimmermann, H. et al., *J. Am. Chem. Soc.*, 124, 15286, 2002.

via ether or ester linkage (Scheme 2.88) [1025–1032]. The octaethers can also be prepared via direct cyclotetramerization of dialkoxybenzyl alcohol **15**; however, the mesophase behavior of the products obtained via two routes differs significantly. It is inferred that the product obtained by the cyclotetramerization of dialkoxybenzyl alcohol possesses some impurities other than CTTV, which is miscible in mesophase a leading to, decrease in melting temperature compared to pristine compound [1028]. The results are in accordance with the well-studied chemistry of hexaalkoxytriphenylene discotics (Section 2.6), where the cyclotrimerization of dialkoxybenzene results in the partial dealkylation of the product.

The expansion of CTV ring to give the CTTV core conserves the mesogenic character of the compounds but has noticeable consequence on the nature of mesophases formed. These changes involve the suppression of polymorphic mesomorphism and increase of the melting as well as clearing temperatures, suggesting the higher order in the CTTV mesophase than the mesophases of CTV. The CTTV derivatives do not appear to have a polar structure.

Alkoxy derivatives with chain length shorter than five carbons did not show any mesophase but the derivatives with alkyl chains between 5 carbon atoms to 16 carbon atoms showed columnar mesophases (Table 2.81). The higher homologues exhibit two columnar mesophases with the second mesophase appearing during the cooling run [1028]. Alkanoyloxy derivatives showed a columnar rectangular mesophase [1029]. Ethyleneoxy derivatives showed a columnar mesophase and both

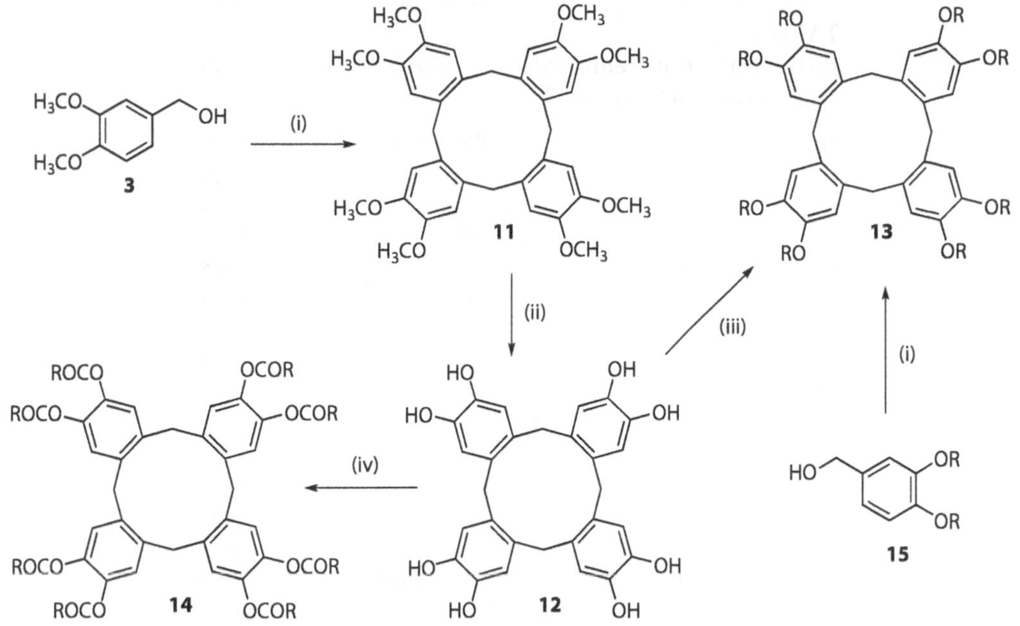

SCHEME 2.88 Synthetic routes to CTTV discotics: (i) TFA, CH$_2$Cl$_2$, 0°C; (ii) BBr$_3$, CHCl$_3$, 0°C; (iii) RBr, K$_2$CO$_3$, EtOH; (iv) RCOCl.

melting and clearing temperatures decrease with increase in ethyleneoxy chains. These derivatives also display lyotropic phases when mixed with water [1026].

The formation of smectic and columnar phases by CTV and CTTV derivatives incorporating calamitic units (Figure 2.25) has been realized [1033–1035]. Most of these materials behave as calamitic liquid crystals and display smectic phases, but a few derivatives exhibit columnar phases. However, the columnar mesophases observed for these oligomers do not result from their ability to adopt a discotic shape, but result from a steric frustration caused by the different space filling of the central cone-like core and the rod-like mesogenic groups [1035]. As these molecules belong to the calamitic group, their detailed discussion is not presented here.

2.13.3 METACYCLOPHANE

Like CTV, metacyclophanes also have a rigid cone-shaped or crown-shaped conformation and may lead to ferroelectric columnar phases, provided that the molecules organize themselves in head-to-tail fashion. A number of compounds consisting of a bowl-shaped metacyclophane core surrounded by 12 flexible alkyl chains have been prepared [1036–1043]. These materials are generally prepared by the acid-catalyzed condensation of pyrogallol **18** with 1,1-diethoxyethane **19**, followed by esterification of the resulting macrocyclic tetramer **20** with the appropriate alkanoyl chloride (Scheme 2.89). The ring methyl group (R) adopt an endo orientation. If the R groups are bulkier than methyl, the columnar packing disturbs to lead nonmesomorphic materials. Eight substituents are not enough to observe mesophase; only the presence of twelve alkyl chains enables the molecule to cover homogeneously the peripheral space surrounding the central core. Mesomorphism is observed only for alkanoyloxy derivative; their alkoxy and benzyloxy derivatives do not show any mesophase [1037]. Branching in the side chains also leads to the disappearance of mesophase. A columnar mesophase is obtained for chain lengths $12 \leq n \leq 17$. The crystal to mesophase transition temperatures decrease with increasing chain length, reaching a minimum for $n = 13$; then they

TABLE 2.81
Phase Transition Temperatures of Cyclotetraveratrylene Derivatives (Scheme 2.88)

Structure	R	Phase Transition	Ref.
13.1	C_4H_9	Cr 222.6 I	[1028]
13.2	C_5H_{11}	Cr 181.7 Col 190.4 I	[1028]
		Cr 177 Col 186.4 I	[1029]
13.3	C_6H_{13}	Cr 161.6 Col 172.4 I	[1028]
		Cr 160.1 Col 170 I	[1029]
13.4	C_7H_{15}	Cr 139.1 Col 161.6 I	[1028]
		Cr 136.0 Col 159.9 I	[1029]
13.5	C_8H_{17}	Cr 137.4 Col 154.4 I	[1028]
		Cr 137.2 Col 153.2 I	[1029]
13.6	C_9H_{19}	Cr 113.6 Col 147.9 I	[1028]
		Cr 109.4 Col 147.9 I	[1029]
13.7	$C_{10}H_{21}$	Cr 112.9 Col 140.3 I	[1028]
		Cr 111.9 Col 136.6 Col 142 I	[1029]
13.8	$C_{11}H_{23}$	Cr 100.5 Col 136.9 I	[1028]
		Cr 98 Col 128.5 Col 137.2 I	[1029]
13.9	$C_{12}H_{25}$	Cr 96.6 Col 131.2 I	[1028]
		Cr 94.4 Col 121.5 Col 133.2 I	[1029]
13.10	$C_{13}H_{27}$	Cr 92.9 Col 126.6 I	[1028]
		Cr 91.5 Col 112.6 Col 127.9 I	[1029]
13.11	$C_{14}H_{29}$	Cr 97.4 Col 120.2 I	[1028]
		Cr 96.1 Col 106.4 Col 123.1 I	[1029]
13.12	$C_{15}H_{31}$	Cr 95.7 Col 117.7 I	[1028]
		Cr 92.7 Col 102.1 Col 119.1 I	[1029]
13.13	$C_{16}H_{33}$	Cr 97.8 Col 99.6 Col 115.4 I	[1029]
13.14	$CH_3OCH_2CH_2$	Col 81.9 M 211 I	[1026]
13.15	$CH_3(OCH_2CH_2)_2$	Col 29.9 M 109.2 I	[1026]
13.16	$CH_3(OCH_2CH_2)_3$	Col 22 M 63 I	[1026]
14.1	$C_{11}H_{23}$	Cr 82 Col$_r$ 246 I	[1025]
		Cr 72 Col$_r$ 246 I	[1029]
14.2	$C_{13}H_{27}$	Cr 90 Col$_r$ 237 I	[1025]
		Cr 90 Col$_r$ 236 I	[1029]
14.3	$C_{14}H_{29}$	Cr 92 Col$_r$ 230 I	[1029]
14.4	$C_{15}H_{31}$	Cr 95 Col$_r$ 223 I	[1029]
14.5	$C_{10}H_{21}C_6H_4$	Col 96 M 151 I	[1025]

increase (Table 2.82). A similar but less pronounced effect is observed for clearing temperatures, where the minimum is observed at $n = 15$. Compared with other columnar liquid crystals, this class of compounds needs longer alkyl chains to form mesogenic properties [1037]. A monotropic mesophase was obtained for those compounds having R = H instead of the methyl group. They showed rectangular columnar phase, contrary to hexagonal columnar phase showed by structurally related methyl-substituted derivatives [1040].

Two liquid crystalline tungsten-oxocalix[4]arene were reported by Xu and Swager [1042]. The tungsten-oxo group caps the metacyclophane by coordinating to four phenolic oxygen atoms to give rise to mesomorphic compounds **21.9** and **21.10** (Scheme 2.89). Both **21.9** and **21.10** exhibit columnar phases that are stable over a wide temperature range. These materials display host–guest

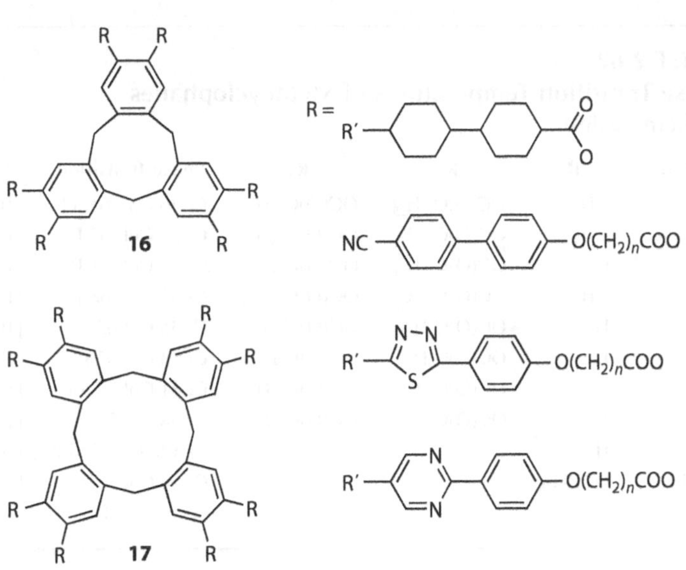

FIGURE 2.25 Molecular structure of CTV and CTTV derivatives incorporating calamitic units.

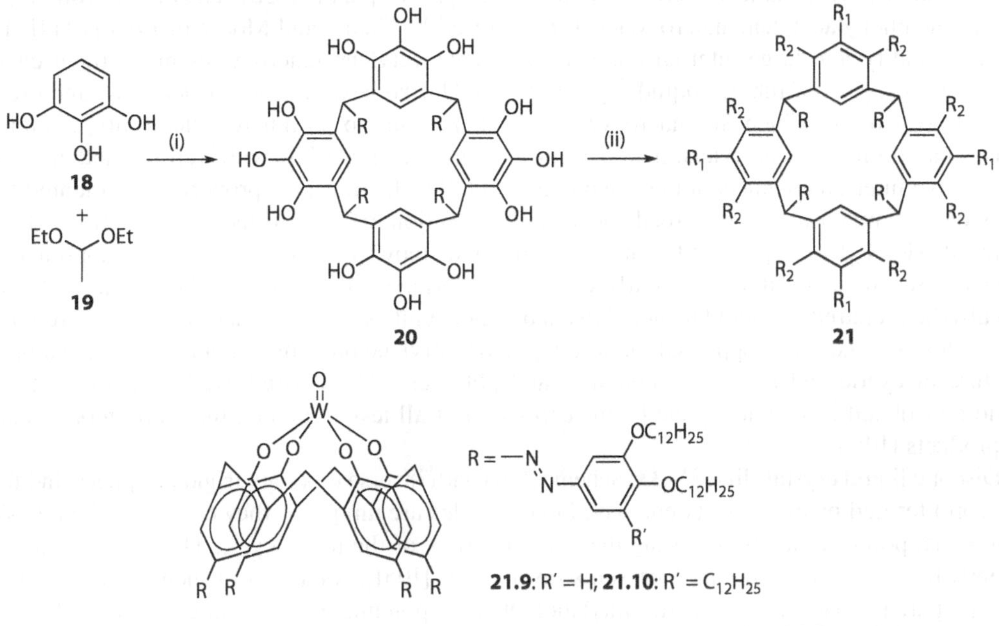

SCHEME 2.89 Synthesis of metacyclophane discotics: (i) HCl, EtOH, 0°C; (ii) RCOCl, 180°C.

interaction with Lewis bases. On complexing, the mesophase behavior of the mesogens changes drastically. The DMF-**21.9** host–guest complex is not mesomorphic and cleared directly to isotropic liquid at low temperature.

2.13.4 Phenylacetylene Macrocycles

The phenylacetylene macrocycles, recently described as shape-persistent macrocycles, are interesting because of their potential to self-organize into supramolecular nanochannels that could be

TABLE 2.82
Phase Transition Temperatures of Metacyclophanes
(Scheme 2.89)

Structure	R	R_1	R_2	Phase Transition	Ref.
21.1	CH_3	$OC(O)C_{12}H_{25}$	$OC(O)C_{12}H_{25}$	Cr 46 Col_h 65.5 I	[1037]
21.2	CH_3	$OC(O)C_{13}H_{27}$	$OC(O)C_{13}H_{27}$	Cr 31 Col_h 67 I	[1037]
21.3	CH_3	$OC(O)C_{15}H_{31}$	$OC(O)C_{15}H_{31}$	Cr 48 Col_h 61 I	[1037]
21.4	CH_3	$OC(O)C_{17}H_{35}$	$OC(O)C_{17}H_{35}$	Cr 58 Col_h 68 I	[1037]
21.5	H	$OC(O)C_8H_{17}$	$OC(O)C_8H_{17}$	Cr 35 Col 63 I	[1040]
21.6	H	$OC(O)C_9H_{19}$	$OC(O)C_9H_{19}$	Cr 33 Col 65 I	[1040]
21.7	H	$OC(O)C_{10}H_{21}$	$OC(O)C_{10}H_{21}$	Cr 31 Col 52.5 I	[1040]
21.8	H	$OC(O)C_{11}H_{23}$	$OC(O)C_{11}H_{23}$	Cr 38 Col 54 I	[1040]
21.9	H	—	—	Cr 136 Col_h 320 I	[1042]
21.10	$OC_{12}H_{25}$	—	—	Cr 54 Col_{h1} 77 Col_{h2} 267 I	[1042]

processed into macroscopically aligned nanotubules [6,8]. A number of cyclic acetylenes were reported to display nematic as well as columnar mesophases [1044–1055]. The first discotic liquid crystalline phenylacetylene macrocycles were reported by Zhang and Moore in 1994 [1044]. The relative rigidity and large internal diameter of phenylacetylene macrocycles make them candidates for mesogens of tubular liquid-crystal phases. Unlike hexacyclenes, azacrowns, and crown ether derivatives in which the macrocycle can collapse due to its inherent flexibility, columnar phases based on phenylacetylene macrocycles may produce materials with well-defined and non-collapsable internal channels. It may be possible to tailor the transport properties of oriented thin films of such materials over a broad spectrum by functionalization to the endo positions. These materials ($R=R_1$) were prepared using a well-defined stepwise, repetitive synthetic approach, as shown in Scheme 2.90 [1044]. Similarly, compounds **28** having different peripheral chains ($R \neq R_1$) can also be prepared. It should be noted that a one-pot synthesis of such macrocycles was reported about 35 years ago. The copper salt of *m*-iodophenylacetylene on homocondensation gave a broad mixture of cyclic and noncyclic oligomers and polymers. The desired cyclic hexamer ($R=R_1$) could be isolated in very low yield by the extraction of all less rigid and, therefore, more soluble by-products [1056].

Discotic liquid crystals like **31–33** (Scheme 2.91) with inverse topology (rigid periphery and flexible core) formed by cyclic acetylene-based macrocycles are unique as they carry less flexible side chains that point to the inside, filling the internal cavity of the macrocycle. This architecture has recently been confirmed by single-crystal x-ray analysis [1051]. A number of such nanosized DLCs were prepared by Hoger et al. via the alkylation of corresponding phenols using classical alkylation or under Mitsunobu reaction conditions. Diyne **31** is obtained by intermolecular Glaser coupling of the half disk (bis-acetylene) **30**, which in turn can be prepared following the methodologies described above.

The hexaesters of phenylacetylene macrocycles **28** ($R=R_1=COOC_nH_{2n+1}$), owing to their strong π–π interactions, do not exhibit any nematic phase; however, the mixed ether–ester derivatives **28** ($R=OC_nH_{2n+1}$, $R_1=OCOC_nH_{2n+1}$) display a monotropic nematic phase followed by a columnar phase (Table 2.83). On the other hand, the hexaethers **28** ($R=R_1=OC_nH_{2n+1}$) and hexaalkanoates **28** ($R=R_1=OCOC_nH_{2n+1}$) exhibit stable discotic nematic phases over a wide range of temperature and display schlieren textures. The shape-persistent macrocycles (**31, 32**) have been reported to exhibit exclusively the discotic nematic phase. Apparently, the flexible alkyl chains should be placed in the

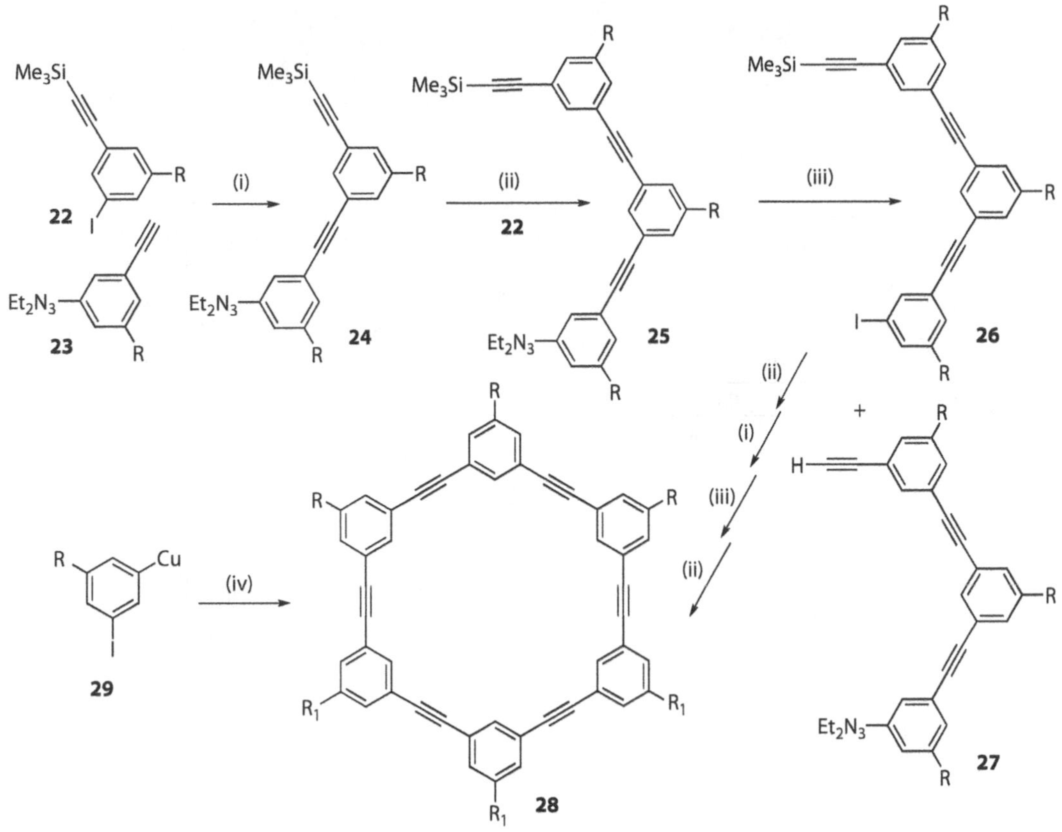

SCHEME 2.90 Synthesis of phenylacetylene macrocycles: (i) Pd(0); (ii) K$_2$CO$_3$, MeOH; (iii) MeI; (iv) pyridine, reflux.

adaptable positions of the macrocycles so that they can fill their interior by their own alkyl chains. If the alkyl chains are positioned at the extra-annular non-adaptable positions, then their back folding is unfavorable for enthalpic and entropic reasons and no mesomorphism is observed. In contrast, when the alkyl chains are positioned at the intra-annular adaptable positions, they fill the interior and exhibit discotic nematic mesophases. Replacing phenyl groups by the polycyclic aromatic backbone as in **33** decreases the size of the cavity but it still exhibits a nematic phase with intra-annular flexible alkyl chains [1051].

Linking three dialkoxybenzene units via acetylene groups leads to the formation of hexaalkoxy-cyclotriyne derivatives **38**. These materials are structurally similar to well-studied hexaalkoxytriphenylene discotics. Triangular *ortho*-phenylene ethylene cyclic trimers are one of the most compact shape-persistent macrocycles. The synthesis is shown in Scheme 2.92. The iodination of 1,2-dialkoxybenzene **34** with iodine in the presence of catalytic amount of mercuric acetate yields 4,5-bis(alkoxy)-1,2-diiodobenzene **35**. The palladium-catalyzed Sonogashira coupling of **35** with trimethylsilylacetylene generates 4,5-bis(alkoxyphenyl)-1-iodo-2-[(trimethylsilyl)ethynyl]benzene **36**, which can be deprotected with potassium fluoride to free acetylene **37**. The cyclotrimerization of **37** under modified Stephens–Castro coupling reaction furnishes a mixture of tris(4,5-dialkoxyphenyl)cyclotriynes **38** and tetrakis(4,5-dialkoxyphenyl)cyclotetraynes **39**. The mixture can be separated by column chromatography to obtain pure materials. Decyl and tetradecyl chains were used to induce mesomorphism but none of the compounds display liquid crystallinity [1047].

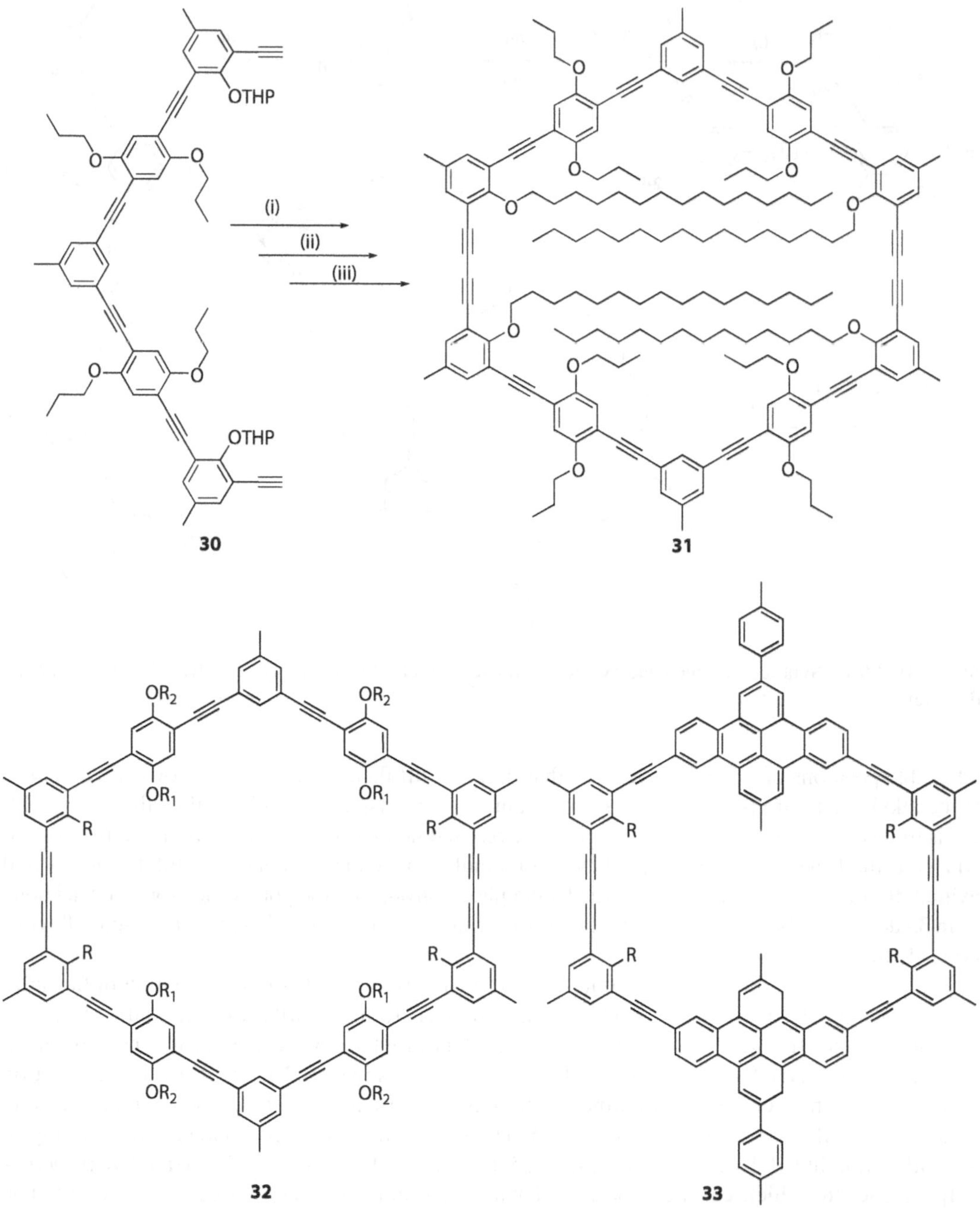

SCHEME 2.91   Synthesis of shape-persistent macrocycles: (i) Cu(I), (ii) H$^+$, (iii) RBr.

TABLE 2.83
Phase Transition Temperatures of Phenylacetylenes

Structure	R	R_1	R_2	Transition Temperature	Ref.
28.1	OC_7H_{15}	OC_7H_{15}		I 192 N 168 Cr	[1044]
28.2	$OCOC_7H_{15}$	$OCOC_7H_{15}$		I 241 N 121 Cr	[1044]
28.3	$COOC_6H_{13}$	OC_6H_{13}		I 202 N 130 Col_h 103 Cr	[1044]
32.1	H	$C_{18}H_{37}$	C_3H_7	Cr 185 N 207 I	[1048]
32.2	$OC_{18}H_{37}$	C_3H_7	C_3H_7	Cr 134 N 159 I	[1050]
32.3	$OC_{14}H_{29}$	C_3H_7	C_3H_7	Cr 148 N 222 I	[1051]
32.4	$OC_{10}H_{21}$	C_3H_7	C_3H_7	Cr 247 N? dec	[1051]
32.5	$O(CH_2)_2CH(CH_3)$ $(CH_2)_3CH(CH_3)_2$	C_3H_7	C_3H_7	Cr 266 N? dec	[1051]
33.1	$OC_{18}H_{37}$			Cr 226 N? dec	[1051]
33.2	$OC_{22}H_{45}$			Cr 218 N? dec	[1051]
44.1				Col_r 81 I	[1053]
44.2				Col_h 55 I	[1053]
44.3				N_D 54 I	[1053]

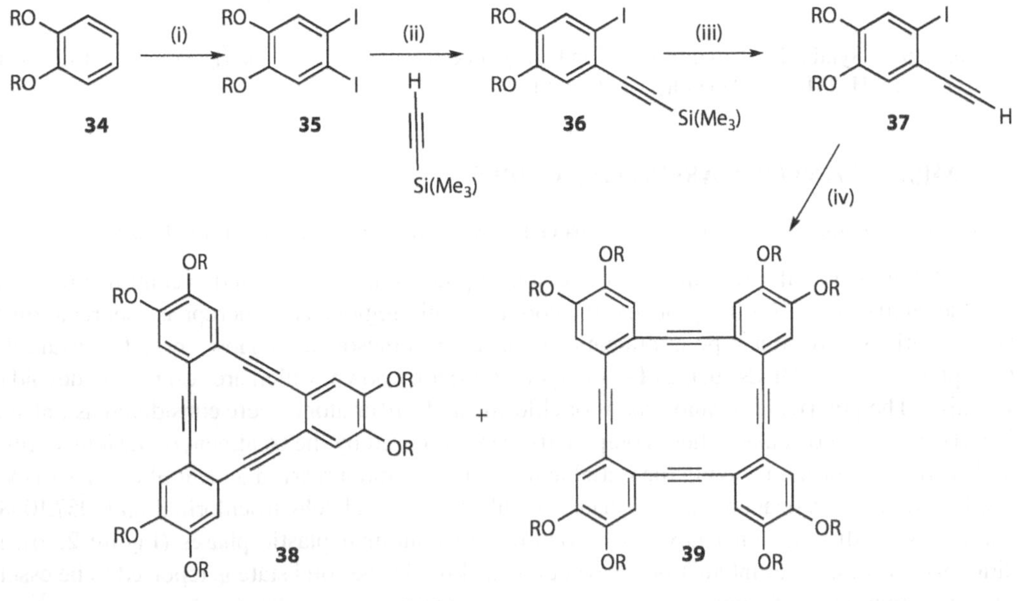

SCHEME 2.92 Synthesis of hexaalkoxy-cyclotriyne derivatives: (i) I_2, $Hg(OAc)_2$; (ii) $Pd(PhCN)_2Cl_2$, CuI, Ph_3P, (i-Pr)$_2$NH, 70°C; (iii) KF, THF, CH_3OH; (iv) t-BuOK, CuCl, pyridine, reflux.

Surprisingly, corresponding trialkoxy derivatives **44** with branched alkoxy- and/or triethylene glycol side chains are reported to show columnar mesophases [1053]. The synthesis of these symmetrical and unsymmetrical *ortho*-phenylene ethylene macrocycles is presented in Scheme 2.93. Compounds **44** are reported to be liquid crystalline at ambient temperature. X-ray studies indicate the presence of a rectangular columnar phase in **44.1**, a hexagonal columnar phase in **44.2**, and a nematic discotic phase in compound **44.3**.

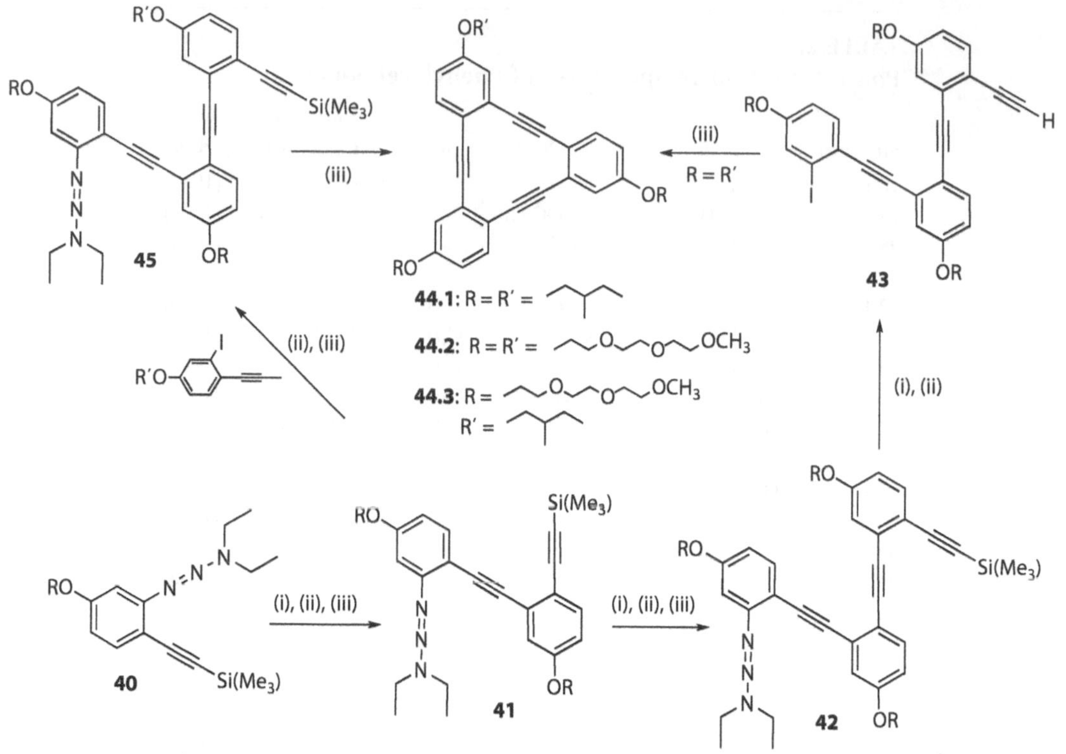

SCHEME 2.93 Synthesis of symmetrical and unsymmetrical trialkoxy-cyclotriynes: (i) CH$_3$I, 110°C; (ii) K$_2$CO$_3$, CH$_3$OH, THF; (iii) Pd(PPh$_3$)$_2$Cl$_2$, CuI, Et$_3$N, 45°C.

2.14 MISCELLANEOUS AROMATIC CORES

2.14.1 Indene and Pseudoazulene: Discotics without Flexible Aliphatic Chains

In general, liquid crystals are composed of flexible aliphatic chains connected to central core where alkyl side chains act as a soft region and the core as a rigid region. The microphase segregation of polar aromatic part from non-polar aliphatic chains creates mesophases. However, a few examples of Col$_h$ phase forming DLCs such as **1–11** (Figure 2.26) are known, which are completely devoid of side chains. The polarizable cyano groups or chlorine and sulfur atoms were considered as unusual soft parts in these molecules. These compounds were prepared by the treatment of various oximes with disulfur dichloride involving the initial abnormal Beckmann rearrangement of oximes to cyanides, followed by cyclization and/or exhaustive chlorination and dehydrochlorination [1057,1058]. These materials display either hexagonal columnar or columnar plastic phases (Figure 2.26). In addition to intermolecular interactions, molecular stacking in the solid state is reported to be essential for the appearance of columnar phases in these compounds.

2.14.2 Benzo[b]triphenylene Core

Hexaalkoxybenzo[b]triphenylene derivatives can be prepared, as shown in Scheme 2.94. The key compound in the synthesis is the 2,3-dialkoxy-6,7-dibromonaphthalene derivative **15**. The 2,3-dialkoxy-6,7-dibromonaphthalene cannot be prepared directly via the bromination of dialkoxynaphthalene **12** as electrophilic aromatic substitution occurs primarily at 1- and 4-positions and therefore, it is necessary to block these reactive positions. Treatment of compound **13** with *n*-butyllithium followed by methyl iodide affords compound **14**, which can now be brominated

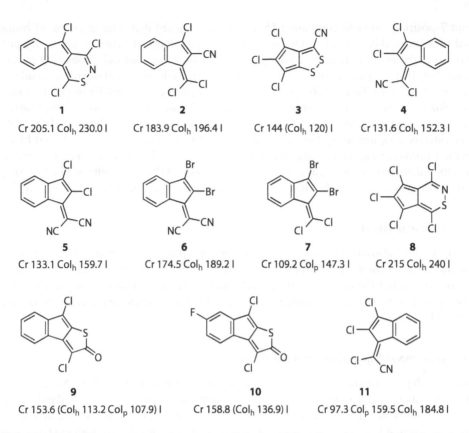

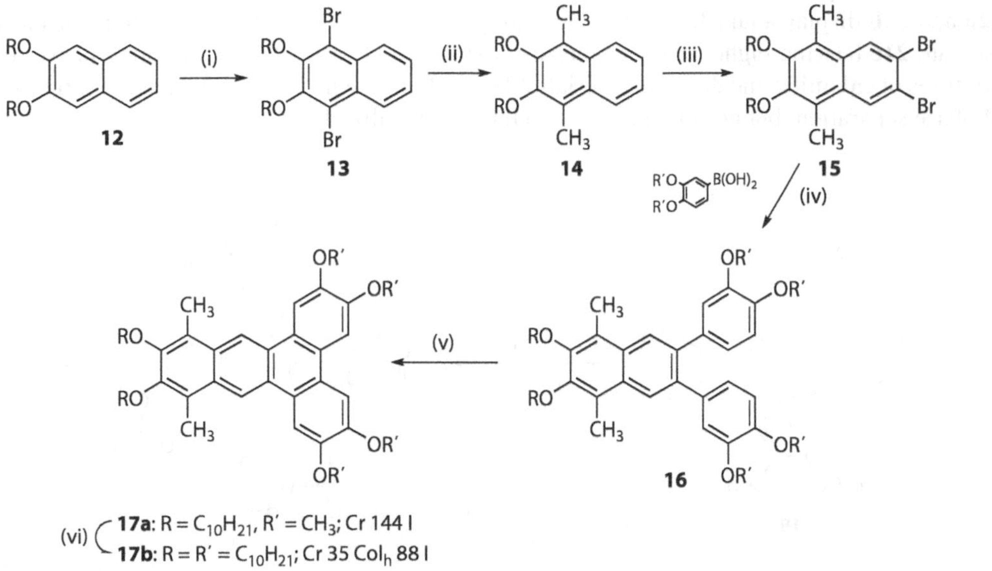

1
Cr 205.1 Col$_h$ 230.0 I
2
Cr 183.9 Col$_h$ 196.4 I
3
Cr 144 (Col$_h$ 120) I
4
Cr 131.6 Col$_h$ 152.3 I
5
Cr 133.1 Col$_h$ 159.7 I
6
Cr 174.5 Col$_h$ 189.2 I
7
Cr 109.2 Col$_p$ 147.3 I
8
Cr 215 Col$_h$ 240 I
9
Cr 153.6 (Col$_h$ 113.2 Col$_p$ 107.9) I
10
Cr 158.8 (Col$_h$ 136.9) I
11
Cr 97.3 Col$_p$ 159.5 Col$_h$ 184.8 I

FIGURE 2.26 Molecular structures and mesophase behavior of discotics without peripheral alkyl chains.

12
13
14
15
16
(i)
(ii)
(iii)
(iv)
(v)
(vi)
17a: R = $C_{10}H_{21}$, R′ = CH_3; Cr 144 I
17b: R = R′ = $C_{10}H_{21}$; Cr 35 Col$_h$ 88 I

SCHEME 2.94 Synthesis of hexaalkoxybenzo[b]triphenylene derivatives: (i) Br_2, ACOH; (ii) *n*-BuLi, −78°C; CH_3I; (iii) Br_2, CH_2Cl_2; $NaBH_4$, DMSO, 50°C; (iv) $Pd(Ph_3)_4$, Na_2CO_3; (v) $FeCl_3$, CH_2Cl_2; (vi) BBr_3, CH_2Cl_2; 1-bromodecane, K_2CO_3, DMF, 110°C.

at 6- and 7-positions to yield compound **15**. It should be noted that a large excess of bromine is required to achieve dibromination. Use of only two equivalents of bromine primarily produces monobromo derivative. However, with large excess of bromine, not only does ring substitution take place, but the benzylic positions also get brominated. Reductive work-up of the crude product removes bromine from the benzylic positions and thus pure compound **15** in good yield can be isolated. Suzuki coupling of the dibromo compound **15** with 3,4-dimethoxyphenylboronic acid yields the diphenylnaphthalene derivative **16**, which can be converted to the benzo[b]triphenylene **17a** via oxidative coupling using FeCl$_3$. The demethylation of **17a** with BBr$_3$ followed by realkylation with 1-bromodecane yields compound **17b**. Compound **17a** with four methoxy groups and only two long decyloxy chains is, as expected, not liquid crystalline. It melts at 144°C. Compound **17b** with six decyloxy chains exhibits a hexagonal columnar phase between 35°C and 88°C [1059].

2.14.3   TETRAPHENYLENES

A number of ester-substituted tetraphenylenes **19** were prepared from octamethoxytetraphenylene by demethylation followed by esterification, as shown in Scheme 2.95 [1060,1061]. All the members of the series, but the first one with pentyl chain, exhibit columnar mesophases, which are either hexagonal or rectangular in nature. The thermal behavior of these materials is summarized in Table 2.84.

2.14.4   TETRABENZO[A,C,H,J]ANTHRACENE CORE

Tetrabenzo[a,c,h,j]anthracene is a board-like polycyclic aromatic hydrocarbon. Several tetra- and octaalkoxy-tetrabenzo[a,c,h,j]anthracene derivatives were designed to envisage calamitic, discotic, and biaxial phases [1062]. The synthetic route used to prepare these materials is shown in Scheme 2.96. Commercially available 1,2,4,5-tetrabromobenzene **20** can be coupled with 4-alkoxyboronic acid or 3,4-dialkoxyboronic acid to yield corresponding tetraphenyls **21**. The oxidative cyclization of **21** using Lewis acid affords the tetrabenzo[a,c,h,j]anthracene derivatives **22**.

The thermal behavior of these materials is summarized in Table 2.85. The disk-shaped octa-substituted compound **22c** exhibits a columnar mesophase, whereas the tetra-substituted compounds **22a** and **22b** display a lamellar phase. Interestingly, the non-planar tetra-substituted benzene intermediate **21c** bearing eight alkoxy chains also exhibits a hexagonal columnar mesophase. Binary mixtures of calamitic and discotic materials **22a** and **22c** were mesogenic and did not show any sign of phase separation, but any biaxial phase could not be realized.

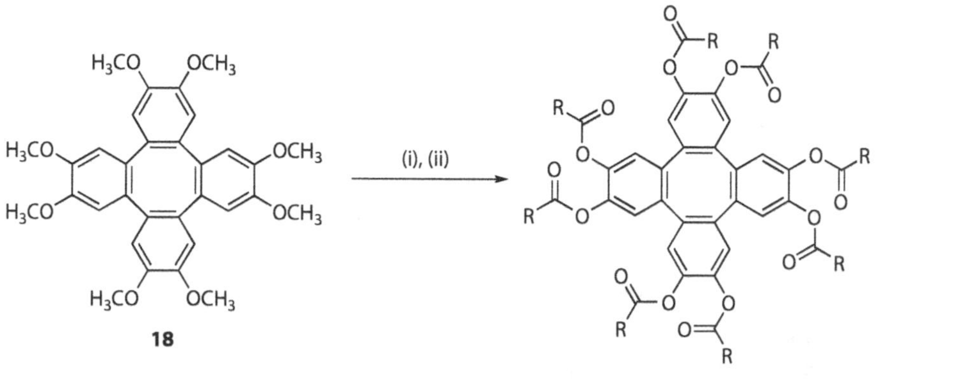

19a–k: R = C$_n$H$_{2n+1}$, n = 5–15; **19l**: R = ⟨◯⟩ OC$_{10}$H$_{21}$

SCHEME 2.95   Synthesis of tetraphenylene discotics: (i) BBr$_3$, CH$_2$Cl$_2$; (ii) RCOCl, DMAP, pyridine, CH$_2$Cl$_2$.

TABLE 2.84
Transition Temperatures of Tetraphenylene Octaesters

Structure	R	Phase Transition
19a	C_5H_{11}	Cr 112 I
19b	C_6H_{13}	Cr 45 Col 75 I
19c	C_7H_{15}	Cr 67 Col 88 I
19d	C_8H_{17}	Cr 52 Col 89 I
19e	C_9H_{19}	Cr 58 Col 93 I
19f	$C_{10}H_{21}$	Cr 59 Col 93 I
19g	$C_{11}H_{23}$	Cr 63 Col 95 I
19h	$C_{12}H_{25}$	Cr 67 Col 93 I
19i	$C_{13}H_{27}$	Cr 77 Col 89 I
19j	$C_{14}H_{29}$	Cr 75 Col 93 I
19k	$C_{15}H_{31}$	Cr 80 Col 88 I
19l		Cr 109 Col 135 I

Source: Data from Wuckert, E. et al., *Liq. Cryst.*, 33, 103, 2006.

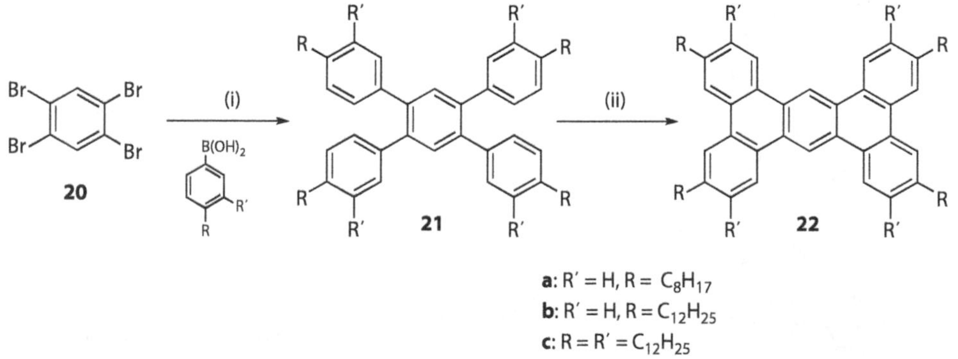

a: R′ = H, R = C_8H_{17}
b: R′ = H, R = $C_{12}H_{25}$
c: R = R′ = $C_{12}H_{25}$

SCHEME 2.96 Synthesis of tetrabenzo[a,c,h,j]anthracene discotics: (i) Pd(Ph₃)₄, Na₂CO₃; (ii) FeCl₃, CH₃NO₂.

TABLE 2.85
Transition Temperatures of Tetrabenzo[a,c,h,j] anthracene Discotics

Structure	R	R′	Phase Transition
21c	OC_8H_{17}	OC_8H_{17}	I 74.4 Col$_h$ −29.9 g
22a	OC_8H_{17}	H	I 179.1 Cryst B-Cr
22b	$OC_{12}H_{25}$	H	I 145.2 Cryst B-Cr
22c	$OC_{12}H_{25}$	$OC_{12}H_{25}$	I 251.8 Col$_h$ 154.6 Cr

Source: Data from Artal, M.C. et al., *J. Mater. Chem.*, 11, 2801, 2001.

2.14.5 HELICENE DISCOTICS

Alkoxy derivatives of helicene and heterohelicene are very interesting as these are derived from a chiral core. Self-assembly of these helical cores attached with aliphatic chains produces hexagonal arrayed columns not only from the melt but also in solution [1063–1071]. These structures exhibit interesting and some unprecedented optical properties. The synthetic method applied for the synthesis of a helicene molecule is depicted in Scheme 2.97. Though helicinebisquinones assemble into hexagonally arrayed columns, all the helicinebisquinone derivatives are not true discotic liquid crystals. Liquid crystalline derivatives are obtained via the partial reduction of helicinebisquinone followed by appropriate substitution of diols. Compound **30** is liquid crystalline at room temperature and clears at 85°C. Reduction of both quinones results in non-mesomorphic materials. Thus, it appears that in such structures, the quinone part is essential for stacking to occur. Liquid crystalline helicinebisquinones are realized by changing the type and number of side chains. Thus, compounds **31–33** display mesomorphism. In most of these compounds, the packing is reported to be hexagonal columnar [1071].

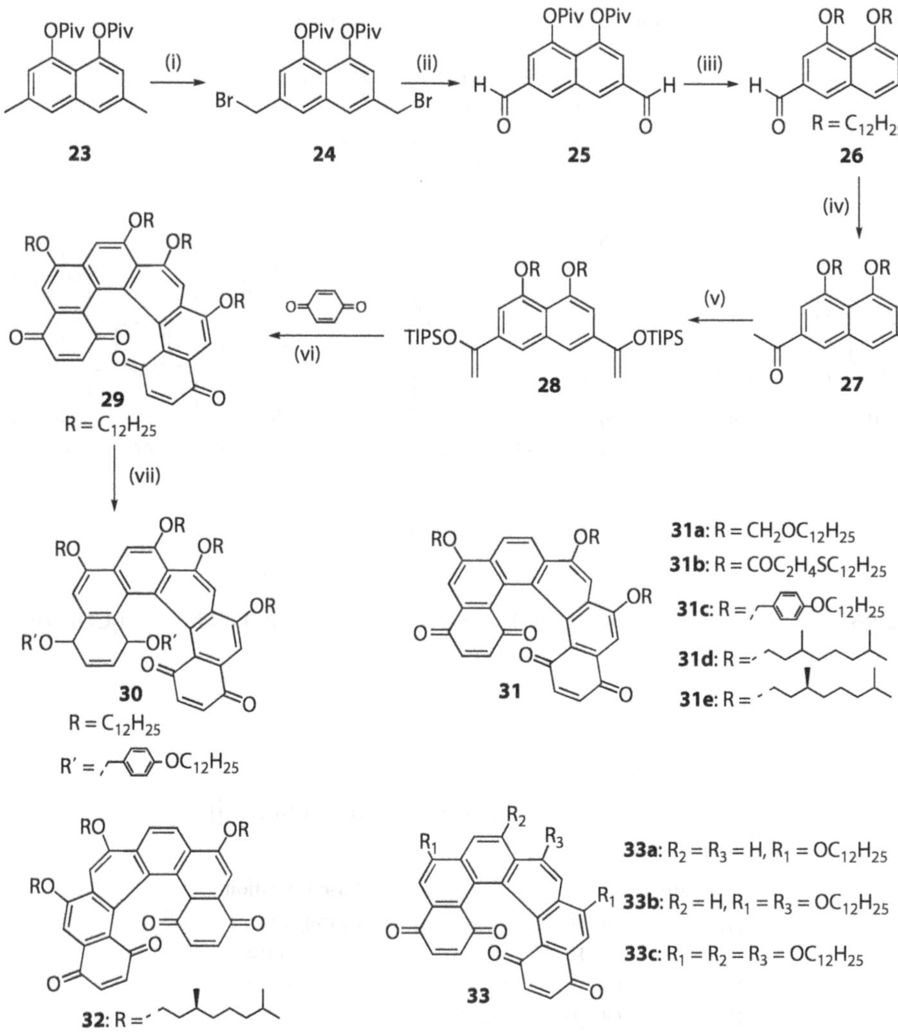

SCHEME 2.97 Synthesis of a helicene molecule: (i) NBS, AIBN, hv; (ii) (Bu$_4$N)$_2$Cr$_2$O$_7$; (iii) (a) KOH, (b) RI, K$_2$CO$_3$; (iv) MeLi; (v) TIPSOTf, Et$_3$N; (vi) toluene, reflux; Ac$_2$O, Et$_3$N; RI, K$_2$CO$_3$; (vii) Na$_2$S$_2$O$_4$, BnOCH$_2$COCl, Et$_3$N.

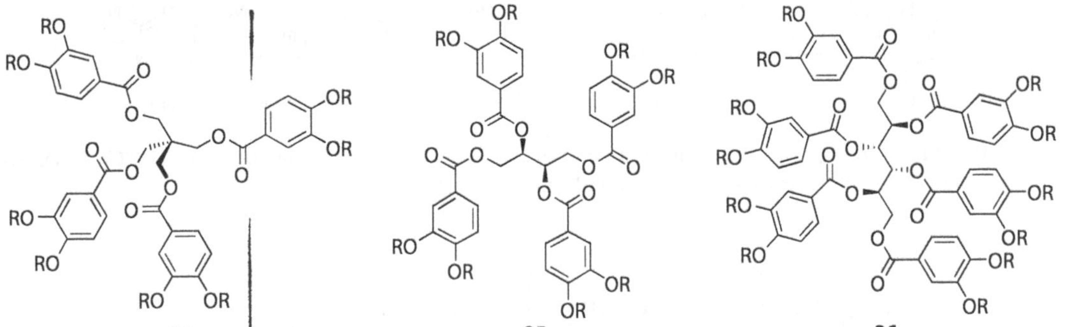

FIGURE 2.27 Molecular structures of some columnar phase forming liquid crystals derived from tetrahedral and other low aspect ratio organic materials.

2.14.6 TETRAHEDRAL AND OTHER LOW ASPECT RATIO ORGANIC MATERIALS

Some aliphatic alcohols like pentaerythritol, D-threitol and D-mannitol, when coupled with dialkoxybenzoic acids form columnar phases [1072,1073]. The benzoylation of these commercially available alcohols with appropriately substituted benzoyl chlorides produces liquid crystalline peracetylated pentaerythritol **8**, D-threitol **9**, and D-mannitol **10** derivatives (Figure 2.27). The tetrabenzoates of pentaerythritol **8** are interesting DLCs as they are derived from a tetrahedral dot-like molecule. However, it should be noted that the real core is not the central tetrahedral carbon atom but probably the part structure consisting of the carboxy groups, the aromatic rings, and oxygen atoms of the ether linkage acts as the core. The columnar phase formation is essentially driven by the microsegregation of their polar central cores from the non-polar flexible chains. Several such compounds have been prepared and they exhibit columnar phases; however, as these star-shaped dendrimers are not derived from true discotic cores, their synthesis and thermal properties are not presented here.

2.15 TRIAZINE CORE

1,3,5-Triazine derivatives have been known in chemistry literature for more than 150 years [1074]. A variety of triazine derivatives have been prepared due to their widespread applications in analytical chemistry, medicinal chemistry, electrochemistry, agricultural chemistry, and combinatorial chemistry [1075–1077]. Hydrogen-bonded melamine-cyanuric acid networks are well known in supramolecular chemistry [1078,1079]. 1,3,5-Triazine can be easily functionalized to give a wide range of new products useful as pesticides, herbicides, explosive, dyestuffs, surface active agents, optical bleachers, reagent in organic synthesis, nonlinear optical materials, catalysts, etc. 1,3,5-Triazine derivatives can be easily prepared from inexpensive, readily available 2,4,6-trichloro-1,3,5-triazine, commonly known as cyanuric chloride. Practically each chloride atom of cyanuric chloride can be substituted step-wise by any nucleophile like, amines, alcohols, thiols, etc. Mono-substitution is generally carried out at low temperature (≤0°C) due to the exothermic nature of the first substitution. The resultant mono-functionalized dichloro-triazine can be easily converted to unsymmetrical derivatives or main-chain polymers. The second substitution can be achieved at room temperature and finally the third substitution is carried out at elevated temperature (Scheme 2.98).

In the field of liquid crystals, initially some polymeric and monomeric calamitic liquid crystals were prepared from cyanuric chloride following the versatile synthetic pathway described above [1080,1081]. In 1984, LeBarny et al. proposed that the trisubstituted-*s*-triazine nucleus is favorable to produce discotic liquid crystals [1082]. To realize this idea, Latterman and Hocker prepared a number of tris(dialkylamino)-*s*-triazines **1** and 2,4-bis(dialkylamino)-6-chloro-*s*-triazine **2** (Figure 2.28) [1083]. However, these materials did not show mesomorphism. Paleos

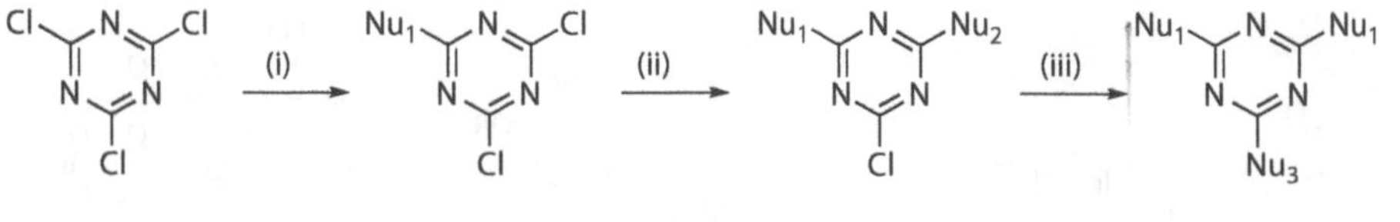

SCHEME 2.98 Mono-, di-, and tri-substitution of cyanuric chloride: (i) first nucleophile, (≥0°C), base; (ii) second nucleophile, r.t., base; (iii) third nucleophile, 60°C–200°C, base.

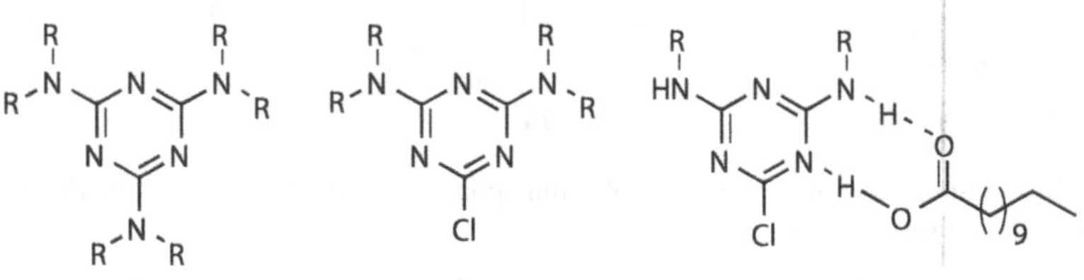

FIGURE 2.28 Chemical structures of tris(dialkylamino)-s-triazines **1**; and 2,4-bis(dialkylamino)-6-chloro-s-triazine **2**; and its H-bonded complex with lauric acid, **3**.

and coworkers reported that the hydrochloride salt of tris(didodecylamino)-s-triazine as well as the hydrogen-bonded complex **3** (Figure 2.28) of disubstituted derivative 2,4-bis(dodecylamino)-6-chloro-s-triazine with lauric acid displayed liquid crystalline properties, but the exact nature of mesophases formed by these systems were not revealed [1084]. Later, a number of other triazines tethered with three calamitic moieties were prepared [1082,1085–1089]. They displayed nematic and smectic phases. Huang and coworkers prepared 2,4,6-tris[4-(4-alkoxyanilideneamino)phenoxy]-1,3,5-triazine and reported the discotic liquid crystalline nature of these molecules without giving any phase behavior [1090]. It was deduced that the attachment of three normal calamitic moieties around the triazine nucleus does not form star-shaped discotic liquid crystals, but the peripheral calamitic units align parallel to give a more rod-like extended shape. Discotic shape may, however, be achieved in some cases by increasing molecular interactions and filling the space around the central core. DLCs derived from 1,3,5-triazine core are mainly of three types: (1) in which peripheral arms are attached to the core via a heteroatom, (2) in which peripheral arms are attached directly to triazine nucleus via carbon–carbon bond formation, and (3) DLCs derived from triazinetrione nucleus.

The reaction of cyanuric chloride with mono-, di-, and tri-substituted anilines under basic conditions formed 2,4,6-triarylamino-1,3,5-triazines (Scheme 2.99) [10,1091–1098]. IR, NMR, and x-ray diffraction studies have shown that 1,3,5-triazines with three and six peripheral chains are susceptible to forming H-bonded dimers, while triazine having nine peripheral chains stay in monomeric form due to steric crowding (Figure 2.29). The thermal behavior of 2,4,6-triarylamino-1,3,5-triazines is presented in Table 2.86. Triarylaminotriazines with only three alkyl chains **5a** and **5b** were reported to be non-mesogenic [1096]. Compound **5c** bearing three decyloxy chains around the triazine nucleus showed a monotropic smectic phase [1096] while its higher homologue with dodecyloxy chains **5d** was reported to be non-liquid crystalline [1098]. However, smectic as well as columnar phases can be induced upon doping with electron acceptors 2,4,7-trinitrofluorene-9-one (**6**) and 2,4,7-trinitrofluorene-9-ylidene malodinitrile (**7**) (Figure 2.30) [1093]. Surprisingly, the induction of smectic liquid crystalline structures through donor–acceptor interactions in columnar phase forming triazines **5e** and **5f** was observed [1094]. The thermal behavior of these charge-transfer complexes is given in Table 2.87.

H-bonded complexes of 2,4,6-triarylamino-1,3,5-triazines with various carboxylic acids have been extensively studied not only to understand the supramolecular organization of these heterodimeric complexes and their mesomorphic behavior [10,1092,1093,1095–1098] but also to induce

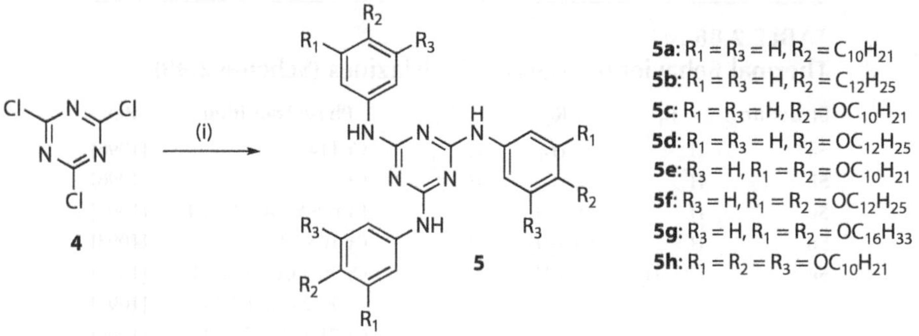

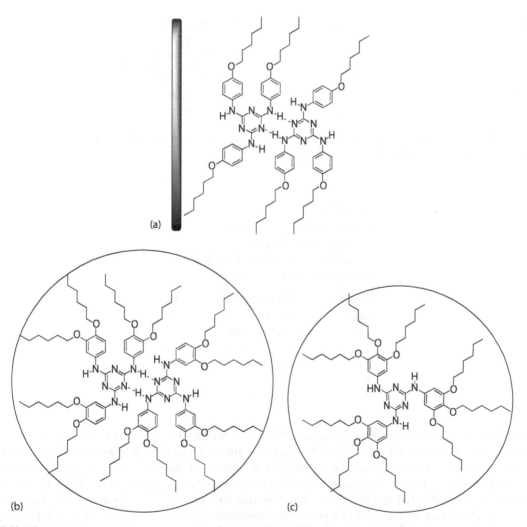

5a: $R_1 = R_3 = H$, $R_2 = C_{10}H_{21}$
5b: $R_1 = R_3 = H$, $R_2 = C_{12}H_{25}$
5c: $R_1 = R_3 = H$, $R_2 = OC_{10}H_{21}$
5d: $R_1 = R_3 = H$, $R_2 = OC_{12}H_{25}$
5e: $R_3 = H$, $R_1 = R_2 = OC_{10}H_{21}$
5f: $R_3 = H$, $R_1 = R_2 = OC_{12}H_{25}$
5g: $R_3 = H$, $R_1 = R_2 = OC_{16}H_{33}$
5h: $R_1 = R_2 = R_3 = OC_{10}H_{21}$

SCHEME 2.99 Synthesis of 2,4,6-triarylamino-1,3,5-triazines: (i) mono-, di-, or tri-substituted aniline, K_2CO_3.

FIGURE 2.29 Proposed dimeric form of (a) trisubstituted triazine (rod-shaped), (b) hexa-substituted triazine (disk-shaped), and (c) monomeric form of nonasubstituted triazine (disk-shaped).

TABLE 2.86
Thermal Behavior of Triarylaminotriazines (Scheme 2.99)

Structure	R$_1$	R$_2$	R$_3$	Phase Transition	Ref.
5a	H	C$_{10}$H$_{21}$	H	Cr 114.6 I	[1096]
5b	H	C$_{12}$H$_{25}$	H	Cr 106.7 I	[1096]
5c	H	OC$_{10}$H$_{21}$	H	Cr (66 SmA) 105.1 I	[1096]
5d	H	OC$_{12}$H$_{25}$	H	Cr 105.2 I	[1098]
5e	OC$_{10}$H$_{21}$	OC$_{10}$H$_{21}$	H	Cr 70.4 Col$_h$ 154.6 I	[1091]
				Cr 73.2 Col$_h$ 89.8 I	[1095]
				Cr 71.1 Col$_h$ 86.6 I	[1096]
5f	OC$_{12}$H$_{25}$	OC$_{12}$H$_{25}$	H	Cr 54.4 Col$_h$ 90.8 I	[1091]
				Cr 52.7 Col$_h$ 88.4 I	[1097]
5g	OC$_{16}$H$_{33}$	OC$_{16}$H$_{33}$	H	Cr 64.8 Col$_h$ 76.9 I	[1091]
5h	OC$_{10}$H$_{21}$	OC$_{10}$H$_{21}$	OC$_{10}$H$_{21}$	Col$_h$ 34.0 I	[1096]

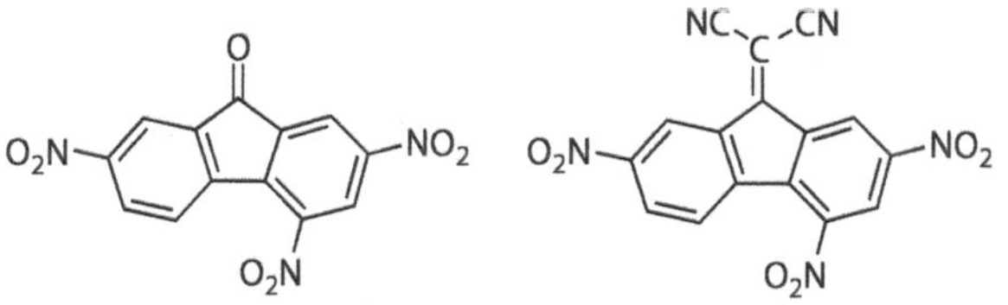

FIGURE 2.30 Chemical structures of electron acceptors 2,4,7-trinitrofluorene-9-one (**6**) and 2,4,7-trinitro-fluorene-9-ylidene malodinitrile (**7**).

TABLE 2.87
Phase Transition Temperatures of Equimolar Mixtures of the Triarylaminotriazines 5 with the Electron Acceptors 6 and 7

Complex	Phase Transition	Ref.
5d/6	Cr 135.1 Col$_r$ 154.9 I	[1093]
5d/7	Cr 98.2 SmA 168.4 I	[1093]
5e/6	Cr 68.1 SmA 110.4 I	[1094]
5f/7	Cr 68.6 SmA 115.0 I	[1094]
5f/7	Cr 93.7 Col$_r$ 147.2 I	[1094]

chirality into the system using chiral acids [1097]. The carboxylic acids used to prepare H-bonded complexes are shown in Figure 2.31. The binary mixtures of triarylaminotriazines and carboxylic acids were prepared by mixing and evaporating a solution of equimolar amount of two components followed by heating the materials to the isotropic phase. Thermal data of all the H-bonded complexes are collected in Table 2.88. On considering the data in Table 2.88 and comparing them with the data given in Table 2.86 for pure triarylaminotriazines, it can be deduced that, depending upon the nature of acid, intermolecular H-bonding may induce mesomorphism in non-mesogenic triazines; may change the nature of existing mesophase and, increase or decrease mesophase

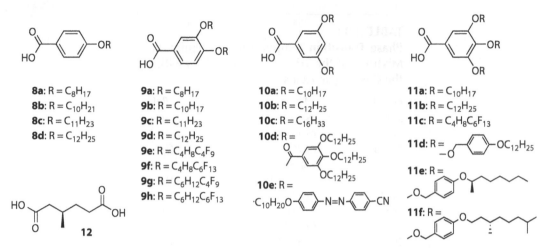

FIGURE 2.31 Chemical structures of carboxylic acids used for H-bonding with triarylaminotriazines.

stability. The association of carboxylic acid with the triazines frustrates the tendency of the aromatic carboxylic acid to form self-dimers.

A series of H-bonded complexes consisting of a mono-substituted melamine **13a** (Figure 2.32) and various polycatener benzoic acids was reported by Sierra and coworkers [1099]. The thermal data of these complexes are presented in Table 2.89. The pure 2,4-diamino-6-dodecylamino-1,3,5-triazine **13a** exhibited a monotropic smectic phase. However, its H-bonded complexes with bulky aromatic acids exhibited columnar phases. It may be noted that polycatener benzoic acids having three substituted phenyl units display columnar phases, while benzoic acid having only three long aliphatic chains is not liquid crystalline, and the H-bonded complex formed by this acid with triazine **13a** was also not liquid crystalline.

Mono-substituted triazine with a chiral chain was prepared to study chirality transfer from the molecular block to the columnar phase [1099]. The chiral triazine **13b** was not liquid crystalline (Table 2.89); however, when complexed with benzoic acid **11d**, it displayed a columnar phase that was stable at room temperature. Circular dichroism (CD) studies were carried out to investigate the transfer of the molecular chirality from the triazine to the columnar mesophase. The transfer of the molecular chirality was also achieved by taking chiral benzoic acids **11e** and **11f**. Optical activity was detected for all the H-bonded complexes formed by either chiral triazine or chiral acid. Thus, these complexes displayed a chiral columnar organization. The chirality transfer to the columnar organization from the chiral molecular building blocks occurs independently of the number of chiral centers. Similar H-bonded complexes were formed from V-shaped acids **14** [1100]. The thermal data of these complexes obtained during the cooling scan are given in Table 2.89.

An H-bonded complex of monosubstituted melamine **13a** with carboxylic acid containing azobenzene and cyano groups **10e** was prepared to realize an optical and electric field responsive supramolecular architecture [1101]. The complex was reported to display three endothermic transitions at 59°C, 108°C, and 122°C. However, from the DSC scan, it appears that the transition at 108°C could be due to crystals to mesophase transformation followed by mesophase to isotropic transition at 122°C. On cooling, the isotropic phase to mesophase transition occurred at 91.4°C. The nature of the mesophase was reported to be hexagonal columnar with lattice parameter $a = 3.98$ nm.

A di-substituted triazine **15** having photoswitchable azobenzene arms and its complexes with N-dodecylcyanuric acid **16** and tridodecyloxyphenyl-substituted barbituric acid **17** were reported by Yagai et al. (Figure 2.33) [1102]. The thermal behavior of these complexes is given in Table 2.89. Three molecules each of triazine and cyanuric acid aggregate to form a rosette **15$_3$.16$_3$** (Figure 2.33). The H-bonded aggregate was shown to stack into extended columnar aggregates. The formation of

TABLE 2.88
Phase Transition Temperatures of Equimolar
Mixtures of the Triarylaminotriazines with
the Carboxylic Acids

Complex	Phase Transition	Ref.
5a/8b	Cr 77.3 I	[1096]
5a/9b	Cr 80.9 I	[1096]
5a/10d	Cr 77.7 I	[1096]
5a/11a	Col$_h$ 50.3 I	[1096]
5a/11b	Col$_h$ 51.8 I	[1096]
5a/11d	Cr 79.5 Col$_h$ 95.3 I	[1096]
5b/8b	Cr 77.0 I	[1096]
5b/9b	Cr 80.4 I	[1096]
5b/10d	Cr 56.8 I	[1096]
5b/11a	Col$_h$ 50.9 I	[1096]
5b/11b	Col$_h$ 70.3 I	[1096]
5b/11d	Cr 86.3 Col$_h$ 89.6 I	[1096]
5c/8b	Cr 60.2 I	[1096]
5c/9b	Cr 78.1 I	[1096]
5c/10d	Cr 55.2 Col$_h$ 67.7 I	[1096]
5c/11a	Col$_h$ 64.3 I	[1096]
5c/11b	Col$_h$ 77.0 I	[1096]
5c/11d	Cr 84.8 Col$_h$ 118.2 I	[1096]
5d/8a	Cr 35.9 Col$_h$ 80.4 I	[1093]
5b/11c	Cr 68.0 Col$_h$ 131.0 I	[1098]
5d/9f	Cr 68.0 Col$_h$ 131.0 I	[1098]
5e/8a	Cr 34.9 Col$_h$ 79.8 I	[1095]
5e/8b	Col$_h$ 57.3 I	[1096]
5e/8c	Cr 33.1 Col$_h$ 78.9 I	[1095]
5e/8d	Cr 33.1 Col$_h$ 79.1 I	[1095]
5e/9a	Cr 70.5 Col$_r$ 94.1 I	[1095]
5e/9b	Cr 46.1 Col$_h$ 75.6 I	[1095]
5e/9c	Cr 62.2 Col$_r$ 88.7 I	[1095]
5e/9d	Cr 80.4 Col$_r$ 98.2 I	[1095]
5e/9f	Col$_h$ 114.0 I	[1098]
5e/10a	Cr 32.4 Col$_h$ 68.4 I	[1092]
5e/10b	Cr 34.5 Col$_h$ 75.1 I	[1092]
5e/10c	Cr 18.7 Col$_h$ 73.6 I	[1092]
5e/10d	Cr 62.5 Col$_h$ 66.6 I	[1096]
5e/11a	Col$_h$ 71.7 I	[1096]
5e/11b	Col$_h$ 82.2 I	[1096]
5e/11c	Col$_h$ 124.4 I	[1098]
5e/11d	Cr 111.5 Col$_h$ 132.8 I	[1096]
5e/12	Col$_r$ 51.2 I	[1098]
5f/8a	Cr 34.7 Col$_h$ 80.9 I	[1095]
5f/8c	Cr 33.7 Col$_h$ 80.8 I	[1095]
5f/8d	Cr 36.3 Col$_h$ 83.8 I	[1095]
5h/8b	Cr 29.8 Col$_r$ 70.2 I	[1096]
5h/9b	Cr 81.6 I	[1096]
5h/11a	Col$_r$ 33.9 I	[1096]
5h/11b	Col$_r$ 32.4 I	[1096]

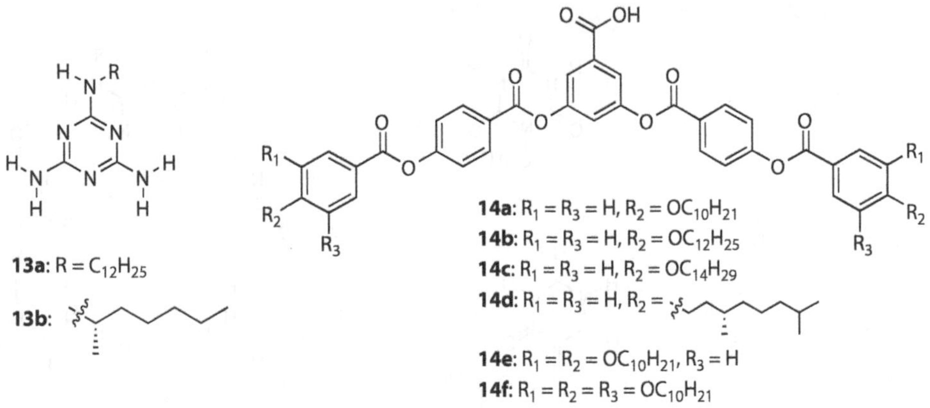

FIGURE 2.32 Chemical structures of mono-substituted melamines and V-shaped acids used for H-bonding.

TABLE 2.89
Phase Transition Temperatures of Mono- and Di-Substituted Melamines and Derived H-Bonded Complexes

Structure	Phase Transition	Ref.
13a	Cr (38.6 SmA) 102.7 I	[1099]
13b	g 15.0 I	[1099]
13a/11b	Cr 24.1 I	[1099]
13a/11d	Cr 48.5 Col$_h$ 127.7 I	[1099]
13a/11e	Cr 61.9 Col$_h$ 77.3 I	[1099]
13a/11f	Col$_h$ 40.1 Col$_h$ 118.2 I	[1099]
13b/11b	Cr 33.0 I	[1099]
13b/11d	Col$_h$ 104.0 I	[1099]
13a/14a	I 98.4 Col$_r$	[1100]
13a/14b	I 101.0 Col$_r$	[1100]
13a/14c	I 93.7 Col$_r$	[1100]
13a/14d	I 74.5 Col$_r$	[1100]
13a/14e	I 95.0 Col$_h$	[1100]
13a/14f	I 63.3 Col$_h$	[1100]
13a/10e	I 91.4 Col$_h$	[1101]
15/16	Cr 124 Col$_r$ 130 Col$_r$ 230 dec	[1102]
15/17	Cr 129 Col$_r$ 189 Col$_r$ 215 dec	[1102]

columnar aggregates and further hierarchically organized fibrous assembly could be controlled by photoswitching the peripheral azobenzenes [1102].

Intramolecularly H-bonded complexes of triazine **18** have recently been prepared, as shown in Scheme 2.100. The thermal data of these melamines are collected in Table 2.90. The exact nature of the columnar phase formed by these materials has not been revealed [1103].

A triazine **19** fused with three carbazole units was prepared by Bai et al. [1104]. It was prepared by coupling 3,6-didecanoyloxycarbazole with cyanuric chloride, as shown in Scheme 2.101. The carbazole trimer was reported to exhibit a columnar phase between 108°C and 166°C. A supercooling of about 30°C in the phase transition was observed during the cooling cycle.

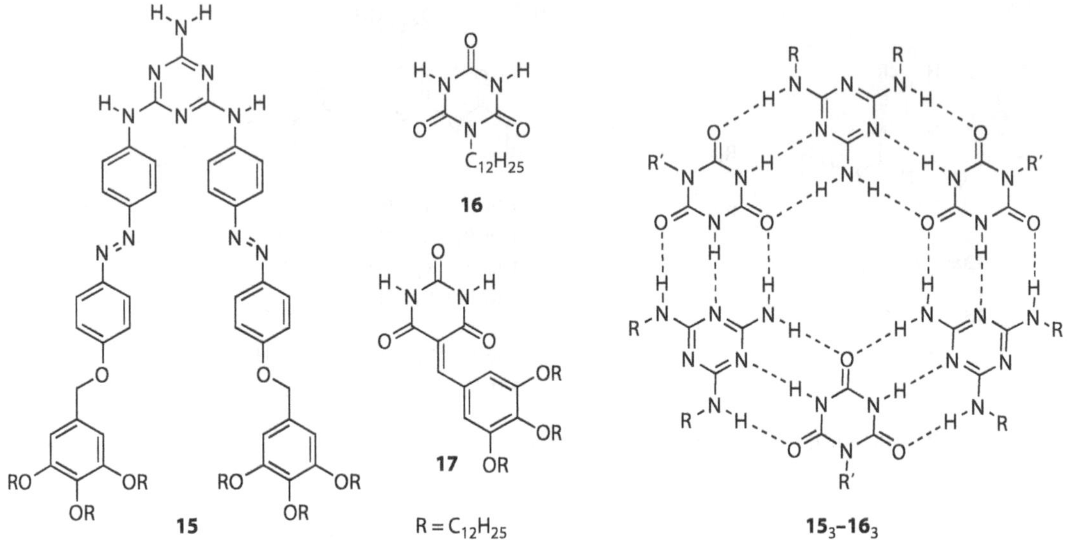

FIGURE 2.33 Molecular structure of a di-substituted melamine and its aggregation with a mono-substituted cyanuric acid.

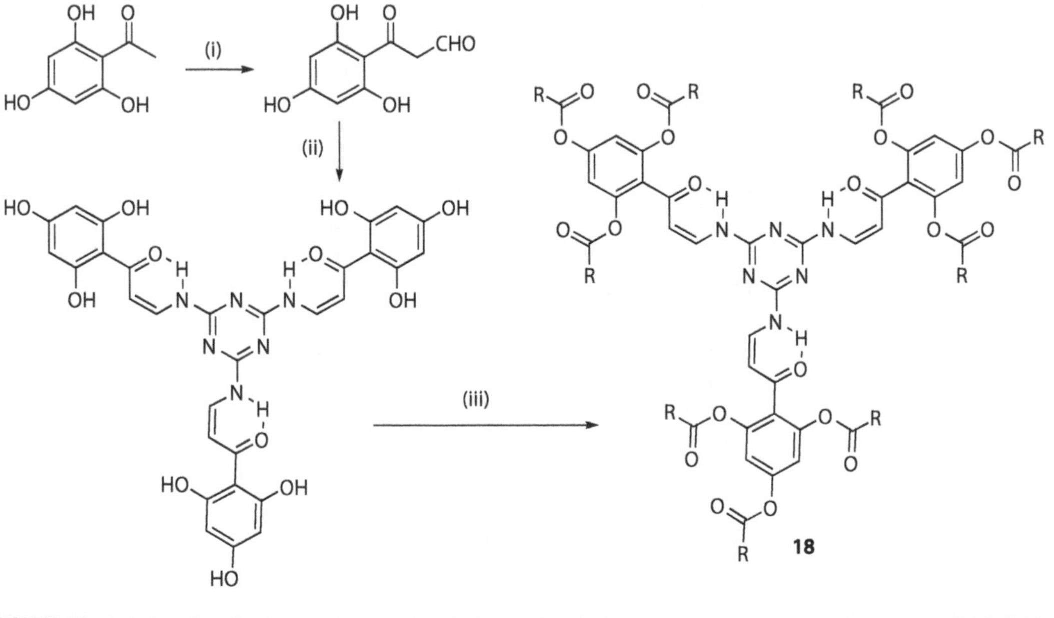

SCHEME 2.100 Synthesis of intramolecularly H-bonded triazine complexes: (i) Na, $HCOOC_2H_5$; (ii) melamine, CH_3OH-DMF, CH_3COOH; (iii) RCOCl.

Chang et al. prepared a number of 1,3,5-triazine derivatives **20** having diacetylene groups, as shown in Scheme 2.102 [1105,1106]. Only the diphenylacetylene derivative with secondary amino linkage and peripheral octyloxy chains **20c** displayed a columnar mesophase (Table 2.91). Corresponding ether-linked triazine **20d** did not show any mesomorphism. Triazines **20e**, **20f**, and **20g** were not liquid crystalline, but their charge-transfer complexes with different mole ratios of TNF exhibited columnar phases. Equimolar complexes showed the highest thermal

TABLE 2.90
Thermal Data for
Intramolecularly H-Bonded
Triazines (Scheme 2.100)

Structure	R	Phase Transition
18a	C_8H_{17}	Col 25.0 I
18b	C_9H_{19}	Col 54.0 I
18c	$C_{10}H_{21}$	Cr (29 Col) 33.0 I
18d	$C_{11}H_{23}$	Cr (25 Col) 28.0 I
18e	$C_{12}H_{25}$	Cr (35 Col) 48.3 I
18f	$C_{14}H_{29}$	Cr (47 Col) 52.6 I
18g	$C_{16}H_{33}$	Cr (54 Col) 62.3 I

Source: Data from Akopova, O.B. et al., *Rus. J. Gen. Chem.*, 77, 103, 2007.

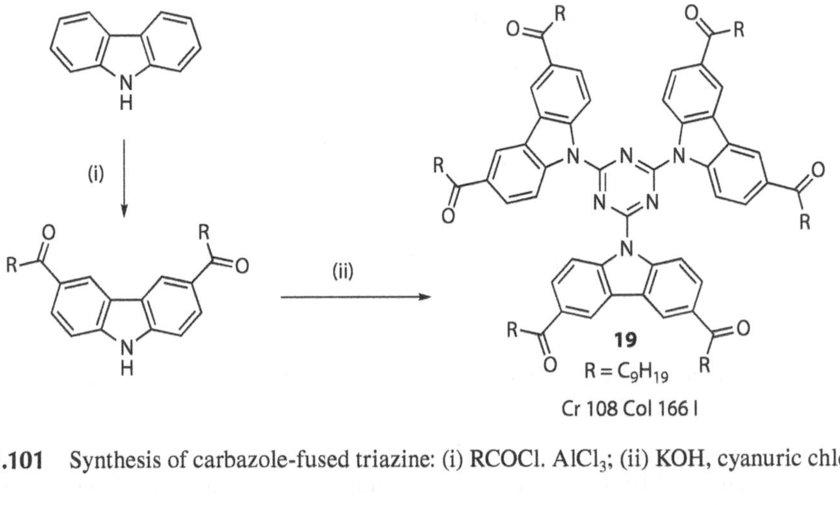

SCHEME 2.101 Synthesis of carbazole-fused triazine: (i) RCOCl. AlCl₃; (ii) KOH, cyanuric chloride.

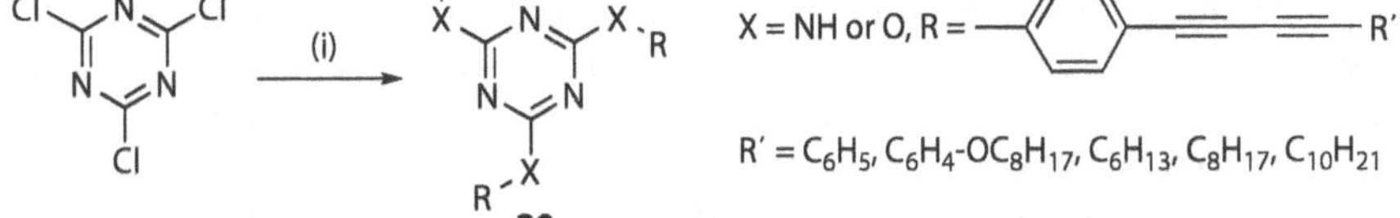

SCHEME 2.102 Synthesis of 1,3,5-triazine derivatives with diacetylene groups: (i) diacetylene-substituted aniline or phenol, K_2CO_3.

stability. The photopolymerization of **20c** or **20f**:TNF complex in mesophase produced oligomers with ordered structure.

Analogous to dialkoxyphenyl-substituted melamines, a few 2,4,6-tris(aryloxy)-1,3,5-triazines have also been prepared [1090,1107]. There are two major routes to prepare triaryloxytriazines. The nucliophilic displacement of chlorine atoms of cyanuric chloride by phenolates easily furnished

TABLE 2.91

Thermal Data for 1,3,5-Triazine Derivatives with Diacetylene Groups (Scheme 2.102)

Structure	X	R'	Phase Transition	Ref.
20a	NH	C_6H_5	Cr 250 poly	[1105]
20b	O	C_6H_5	Cr 280 poly	[1105]
20c	NH	C_6H_5-OC_8H_{17}	Cr 197 Col 250 poly	[1105]
20d	O	C_6H_5-OC_8H_{17}	Cr 228 I	[1105]
20e	NH	C_6H_{13}	—	[1106]
20f	NH	C_8H_{17}	Cr 64 I	[1106]
20g	NH	$C_{10}H_{21}$	Cr 77 I	[1106]
20e/TNF			Col 126 I	[1106]
20f/TNF			Col 134 I	[1106]
20g/TNF			Col 140 I	[1106]

Note: Poly, polymer (solid state polymerization occurred).

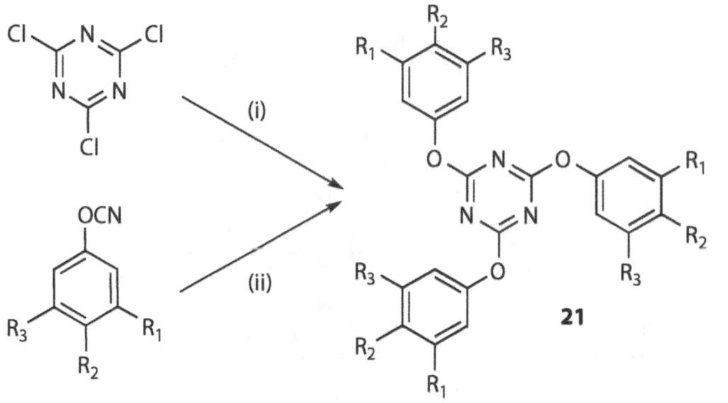

21a: $R_1 = R_2 = OC_4H_9$, $R_3 = H$
21b: $R_1 = R_2 = OC_8H_{17}$, $R_3 = H$
21c: $R_1 = R_2 = OC_{12}H_{25}$, $R_3 = H$
21d: $R_1 = R_3 = OCOCH_3$, $R_2 = H$
21e: $R_1 = R_3 = OCOC_4H_9$, $R_2 = H$

SCHEME 2.103   Synthesis of 2,4,6-triaryloxy-1,3,5-triazines: (i) 3,4-di- or 3,5-di-substituted phenol, K_2CO_3; (ii) heat.

aryloxytriazines (Scheme 2.103). The reaction of cyanuric chloride and phenol at elevated temperature in the absence of solvent and base has also been reported to produce aryloxytriazines in high yield [1108]. Mormann and Zimmermann recommended the use of silylated phenol to react with cyanuric chloride in the presence of a catalytic amount of acid or base [1082]. Trimethylchlorosilane formed in the reaction can be easily removed and, therefore, the purification of the product is convenient. Alternatively, aryloxytriazines can be prepared by the cyclotrimerization of an aromatic cyanate (Scheme 2.103).

The 3,4-dialkoxyphenoxyl triazines **21b** and **21c** with long-chain substituents were reported to exhibit mesomorphism at low temperature (**21b**: Cr 61 LC 88 I; **21c**: Cr 45 LC 81 I) [1107]. However, the exact nature of the mesophase was not reported. The lower homologue with butyloxy chains was reported to be non-liquid crystalline (**21a**: Cr 92 I). Compound **21d** displayed very broad melting between 167°C and 184°C, and the possibility of its liquid crystallinity was expressed, but no conclusion could be deduced. Compound **21e** did not show any mesomorphism [1089]. Many other calamitic triaryloxytriazines have also been prepared [1082,1085–1087,1089,1090]. Concrete evidences in favor of a columnar phase formation by triaryloxytriazines are not available. Filling the space around the central core by taking appropriate tri-substituted phenol instead of di-substituted may generate triaryloxytriazine discotics.

2,4,6-Triphenyl-substituted-1,3,5-triazines in which a phenyl ring is attached to the central triazine nucleus directly via carbon–carbon bond can be easily synthesized by trimerizing a substituted benzonitrile (Scheme 2.104). Treatment of 4-bromobenzonitrile or 4-hydroxybenzonitrile **22** with trifluoromethanesulfonic acid in dry chloroform at room temperature yielded the cyclotrimerized product **23** in high yield. Sonogashira coupling of this tribromide with mono- or dialkoxy-substituted phenylacetylenes produced compounds **24** in good yield [1109,1110]. Being a large π-conjugated system, all 2,4,6-Triphenyl-1,3,5-triazine derivatives produce green-blue photoluminescence with high quantum yields. The UV-vis absorption peak of **24** shifted to a longer wavelength ($\lambda_{max} = 365$) from that of 1,3,5-triazine ($\lambda_{max} = 272$) due to the extended π-conjugation. In the dialkoxyphenylacetylene-substituted series, only two members with decyloxy and undecyloxy chains displayed columnar mesomorphism (Table 2.92) [1109]. Other lower and higher homologues were non-liquid crystalline. On the other hand, all the monoalkoxyphenylacetylene-substituted triazines with peripheral tails of more than 10 carbons exhibited mesomorphism (Table 2.92). The nature of mesophases in these derivatives was established only from POM studies and, therefore, it is necessary to confirm the columnar phase formation in these derivatives by x-ray studies.

Treatment of triazine tribromide **23** with lithium bis(trimethylsilyl)amide in the presence of a palladium(0) catalyst generated 2,4,6-tris(4-aminophenyl)-1,3,5-triazine **25** (Scheme 2.104).

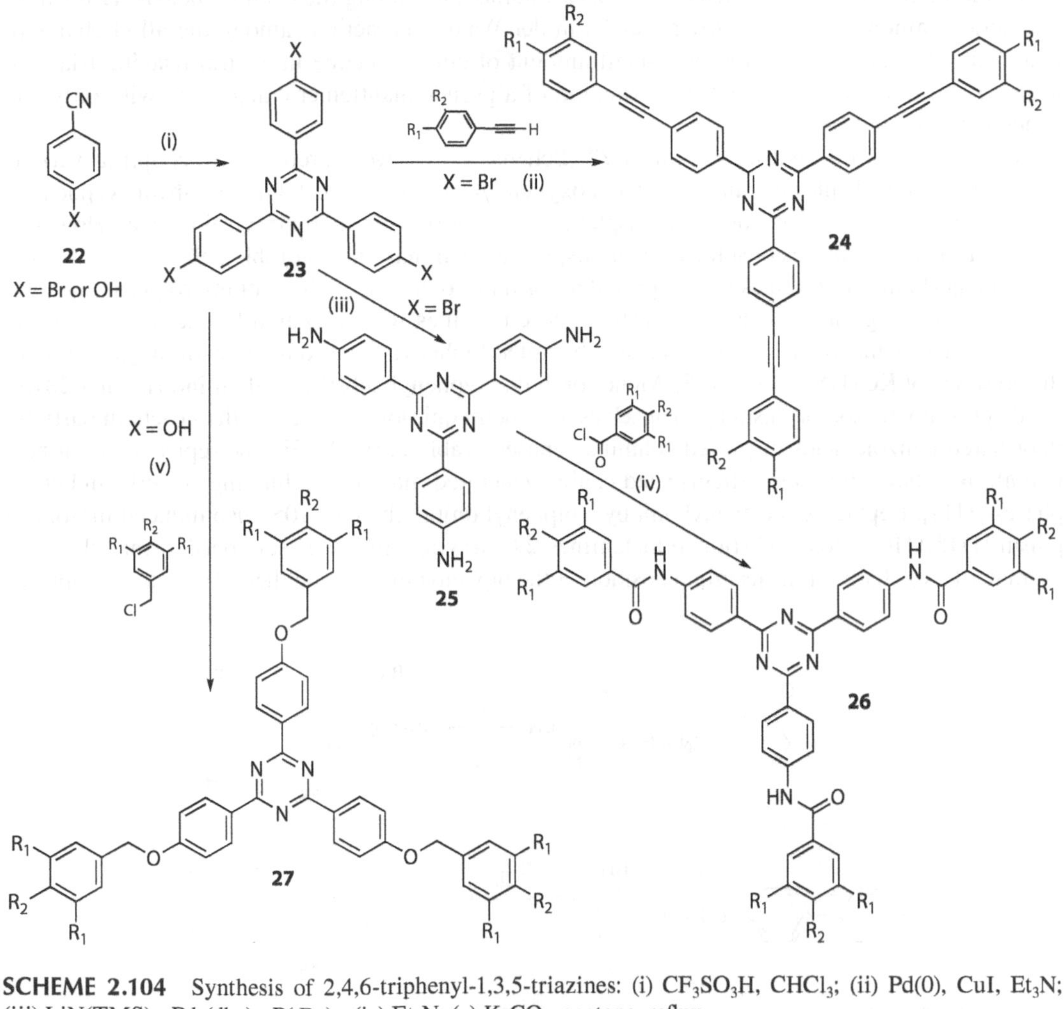

SCHEME 2.104 Synthesis of 2,4,6-triphenyl-1,3,5-triazines: (i) CF$_3$SO$_3$H, CHCl$_3$; (ii) Pd(0), CuI, Et$_3$N; (iii) LiN(TMS)$_2$, Pd$_2$(dba)$_3$, P(tBu)$_3$; (iv) Et$_3$N; (v) K$_2$CO$_3$, acetone, reflux.

TABLE 2.92

Phase Transition Temperatures of Phenylacetylene-Substituted 2,4,6-Triphenyl-1,3,5-Triazines (Scheme 2.104)

Structure	R^1	R^2	Phase Transition	Ref.
24a	$OC_{10}H_{21}$	$OC_{10}H_{21}$	Cr 85.6 Col$_h$ 99.4 I	[1109]
24b	$OC_{11}H_{23}$	$OC_{11}H_{23}$	Cr 92.0 Col$_h$ 102.7 I	[1109]
24c	$OC_{10}H_{21}$	H	Cr 57.2 Sm 66.9 Col$_h$ 141.1 I	[1110]
24d	$OC_{12}H_{25}$	H	Cr 116.3 Sm 124.7 Col$_h$ 129.3 I	[1110]
24e	$OC_{14}H_{29}$	H	Cr 66.4 Col$_h$ 104.8 I	[1110]
24f	$OC_{16}H_{33}$	H	Cr 105.1 Col$_h$ 116.2 I	[1110]
27a/TNF	OC_6H_{13}	OC_6H_{13}	Ss 56.2 Col 77.6 I	[1112]
27b/TNF	$OC_{12}H_{25}$	$OC_{12}H_{25}$	Ss 50.6 Col 101.5 I	[1112]

Compound **25** was easily converted into triazine triamides **26** by condensation with various aroyl chlorides [1111]. Several achiral and chiral derivatives of **26** were prepared but their mesomorphic properties in bulk were not studied. However, these triazine triamides self-assembled to form columnar helical aggregates through π–π interactions among the central nucleus, H-bonding interactions among the amide groups, and van der Waals interactions among the alkyl chains in nonpolar solvents. The addition of a small amount of chiral triazine in achiral triazine triamide leads to chiral amplification and the formation of a pseudoenantiomeric aggregate with only one handedness of the helix.

A number of triphenyltriazine ethers **27** (Scheme 2.104) were prepared by coupling various 3,5-dialkoxybenzyl chlorides and 3,4,5-trialkoxybenzyl chlorides with 2,4,6-tris(4-hydroxyphenyl)-1,3,5-triazine [1112]. None of them displayed mesomorphism in their virgin state. However, charge-transfer complexes of some of them displayed columnar phases (Table 2.92).

Some aryldiaminotriazines were reported to form mesophases by virtue of micro phase segregation or H-bonding interactions [1113–1116]. These triazines were prepared by the reaction of the di-substituted benzonitrile or mono- or di-substituted biphenyl-4-carbonitrile with dicyanamide in the presence of KOH (Scheme 2.105). Mono- or dialkoxyphenyl-substituted diaminotriazines **28** did not display any mesomorphism by themselves, but their equimolar adducts with two-chain partially fluorinated benzoic acids displayed columnar phases (Table 2.93) [1113]. The replacement of normal alkoxy chains by semiperfluorinated chains generated materials exhibiting smectic and cubic phases [1114]. Replacing the phenyl unit by a biphenyl unit (Scheme 2.105) also induced mesomorphism [1115,1116]. Mono-substituted melamines **29** carrying either one semiperfluorinated or two normal alkoxy chains at the terminal position of the biphenyl moiety and their H-bonded complexes

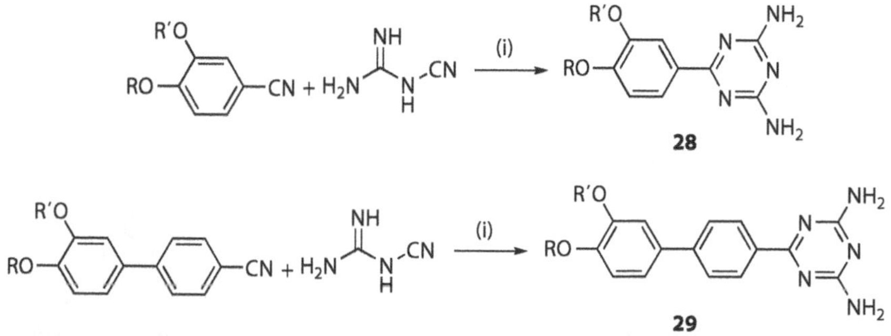

SCHEME 2.105 Synthesis of aryldiaminotriazines: (i) KOH, 2-methoxyethanol.

TABLE 2.93
Phase Transition Temperatures of Aryldiaminotriazines
and Their H-Bonded Complexes (Scheme 2.105)

Structure	R	R′	Phase Transition	Ref.
28a	$C_{12}H_{25}$	H	Cr 157.7 I	[1113]
28b	$C_{10}H_{21}$	$C_{10}H_{21}$	Cr 111.9 I	[1113]
28c	$C_4H_8C_4F_9$	H	Cr (167.0 SmA) 187.7 I	[1114]
28d	$C_6H_{13}C_4F_9$	H	Cr (151.5 SmA) 176 I	[1114]
28e	$C_4H_8C_6F_{13}$	H	Cr 182.1 SmA 200.9 I	[1114]
28f	$C_6H_{12}C_6F_{13}$	H	Cr 180.1 SmA 192.8 I	[1114]
28g	$C_4H_8C_4F_9$	$C_4H_8C_4F_9$	Cr 82.7 Cub 140.4 I	[1114]
28a/9d			Cr 116.4 I	[1113]
28a/9e			Cr 111.1 Col$_r$ 129.9 I	[1113]
28a/9f			Cr 128.6 Col$_r$ 154.6 I	[1113]
28a/9g			Cr 103.3 Col$_r$ 130.3 I	[1113]
28a/9h			Cr 106.7 Col$_r$ 132.7 I	[1113]
28b/9d			Cr 77.3 I	[1113]
28b/9e			Cr 71.3 Col$_{ob}$ 118.6 I	[1113]
28b/9f			Cr 82.3 Col$_{ob}$ 128.6 I	[1113]
28b/9g			Cr 60.1 Col$_{ob}$ 117.0 I	[1113]
28b/9h			Cr 64.5 Col$_{ob}$ 116.2 I	[1113]
29a	$C_{12}H_{25}$	H	Cr 255.0 I	[1115]
29b	C_6H_{13}	C_6H_{13}	Cr 135 I	[1115]
29c	C_8H_{17}	C_8H_{17}	Cr 110 Col$_h$ 143 I	[1115]
29d	$C_{10}H_{21}$	$C_{10}H_{21}$	Cr 108 Col$_h$ 154 I	[1115]
29e	$C_{12}H_{25}$	$C_{12}H_{25}$	Cr 106 Col$_h$ 158 I	[1115]
29f	$C_4H_8C_4F_9$	H	Cr 230.4 SmA 281.2 I	[1116]
29f/9f			Cr 180.4 Col$_r$ 219.6 I	[1116]
29e/9f			Cr 105.1 Col$_r$ 146.1 I	[1116]

with benzoic acids having two or three partially fluorinated chains were prepared by Vlad-Bubulak et al. [1116]. The biphenyl-substituted diaminotriazine **29a** having a single alkoxy chain did not display mesomorphism (Table 2.93) [1115]. The corresponding semiperfluorinated chain-substituted derivative **29f** exhibited a smectic phase [1116]. The dialkoxybiphenyl-substituted diaminotriazines displayed columnar phases if the length of peripheral chains was sufficiently large. Similar to the work of Yagai et al. [1102], H-bonded hexameric rosette formation in these compounds was presented [1116]. Complexes formed by these triazines with various carboxylic acids were also reported to exhibit columnar phases (Table 2.93).

2,4,6-Triphenyl-1,3,5-triazines can also be obtained from cyanuric chloride via the Suzuki coupling with a substituted phenylboronic acid (Scheme 2.106). Both **30a** and **30b** exhibited enantiotropic columnar phases (**30a**: Cr 35.6 Col$_h$ 145.3 I; **30b**: Cr −15.2 Col$_r$ 55.8 I) [1117]. Circular dichroism spectroscopy was used to understand the effect of the chiral alkyl chain substitution on the mesophase. A thin film of **30b** exhibited strong circular dichroism with a negative exciton splitting centered around 318 nm, which suggested that the discotic molecules were stacked along the column and tilted with respect to column axis to form a left-handed helix within the column. The optical activity of **30b** was not due to the presence of a chiral center in the chain but due to the helical arrangement of the disks induced by the chiral alkyl chains. The intramolecular H-bonded discotic triazine **31** melts at 111°C to columnar mesophase, which clears at 173°C [1118]. The reaction of cyanuric fluoride with trialkoxyphenyl acetalide generated the

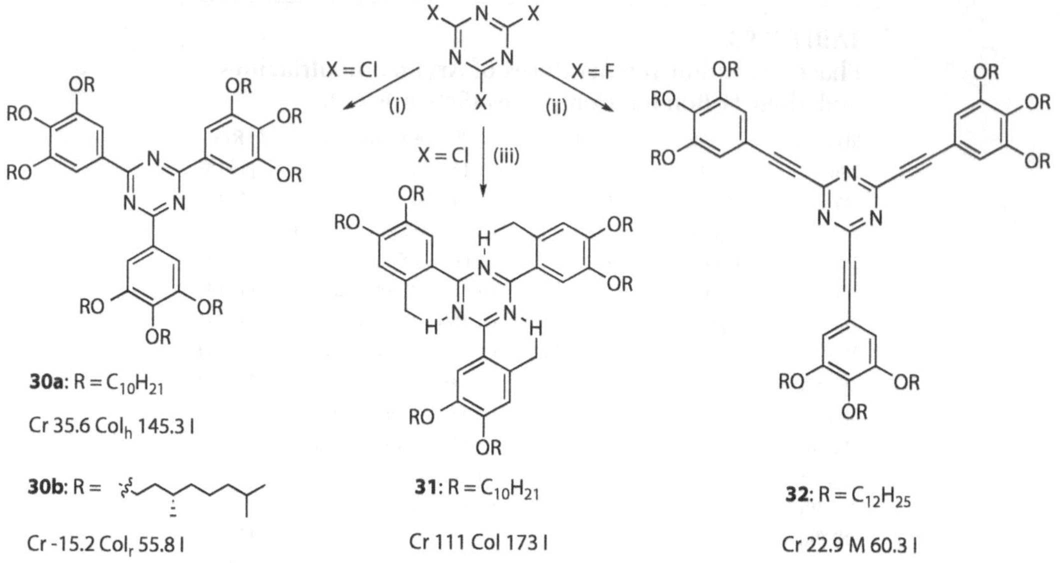

30a: R = C$_{10}$H$_{21}$

Cr 35.6 Col$_h$ 145.3 I

30b: R =

Cr -15.2 Col$_r$ 55.8 I

31: R = C$_{10}$H$_{21}$

Cr 111 Col 173 I

32: R = C$_{12}$H$_{25}$

Cr 22.9 M 60.3 I

SCHEME 2.106 Synthesis of 2,4,6-triphenyl-1,3,5-triazines: (i) trialkoxyphenylboronic acid, Na$_2$CO$_3$, Pd(PPh$_3$)$_4$; (ii) trialkoxyphenylacetylene, n-BuLi, THF, −78°C to 20°C; (iii) 2-hydroxy-4,5-didecyloxyphenyl-boronic acid, Na$_2$CO$_3$, Pd(PPh$_3$)$_4$.

discotic molecule **32** (Scheme 2.106), which displayed a mesophase between 22.9°C and 60.3°C. It is expected to exhibit a columnar phase but the exact nature of the mesophase could not be deduced even by x-ray studies [1119].

Meier and coworkers synthesized a number of 1,3,5-triazines with stilbenoid arms [134,1120–1122]. The alkaline condensation of 2,4,6-trimethyl-1,3,5-triazine and substituted benzaldehyde yield 2,4,6-tristyryl-1,3,5-triazines (Scheme 2.107). When substituted with an appropriate number and length of alkoxy chains, these molecules display columnar phases. Compounds with only three peripheral chains (**33a–g**) do not display mesomorphism (Table 2.94). At least six (**33h**) or better nine long alkoxy chains (**33i–n**) are required to provide mesomorphic properties. 2,4,6-tristyryl-1,3,5-triazines with dialkylamino groups at the periphery do not display mesomorphism [1121]. Increasing the length of peripheral chains causes a steady decrease of the isotropic temperature. The lateral branching of stilbenoid arms generated star-shaped triazines **34** (Figure 2.34) displaying nematic phases [1120]. Compound **34a** exhibited a nematic phase between 95°C and 107°C

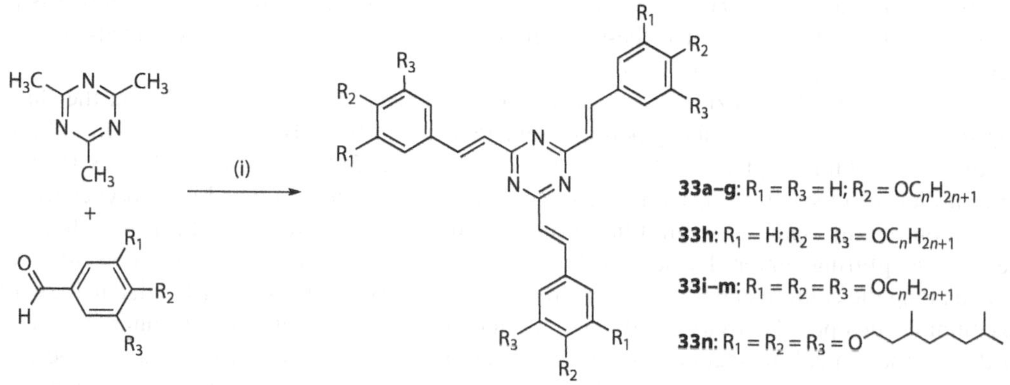

33a–g: R$_1$ = R$_3$ = H; R$_2$ = OC$_n$H$_{2n+1}$

33h: R$_1$ = H; R$_2$ = R$_3$ = OC$_n$H$_{2n+1}$

33i–m: R$_1$ = R$_2$ = R$_3$ = OC$_n$H$_{2n+1}$

33n: R$_1$ = R$_2$ = R$_3$ = O

SCHEME 2.107 Synthesis of 2,4,6-tristyryl-1,3,5-triazines: (i) KOH, CH$_3$OH.

TABLE 2.94
Phase Transition Temperatures of 2,4,6-Tristyryl-1,3,5-Triazines (Scheme 2.107)

Structure	R_1	R_2	R_3	Phase Transition
33a	H	OCH_3	H	Cr 228 I
33b	H	OC_3H_7	H	Cr 99 I
33c	H	OC_5H_{11}	H	Cr 109 I
33d	H	OC_6H_{13}	H	Cr 96 I
33e	H	OC_7H_{15}	H	Cr 85 I
33f	H	OC_8H_{17}	H	Cr 76 I
33g	H	$OC_{12}H_{25}$	H	Cr 63 I
33h	H	OC_8H_{17}	OC_8H_{17}	Cr 75 Col_h 82 I
33i	OC_6H_{13}	OC_6H_{13}	OC_6H_{13}	Cr 68.4 Col_h 112.0 I
33j	OC_8H_{17}	OC_8H_{17}	OC_8H_{17}	Cr 42.6 Col_h 101.8 I
33k	$OC_{10}H_{21}$	$OC_{10}H_{21}$	$OC_{10}H_{21}$	Cr −23.2 Col_h 90.1 I
33l	$OC_{12}H_{25}$	$OC_{12}H_{25}$	$OC_{12}H_{25}$	Cr 8.7 Col_h 90.1 I
33m	$OC_{16}H_{33}$	$OC_{16}H_{33}$	$OC_{16}H_{33}$	Cr 44.9 Col 50.3 Col_h 80.3 I
33n				Col_h 45.3 I

Source: Data from Holst, H.C. et al., *Tetrahedron*, 60, 6765, 2004.

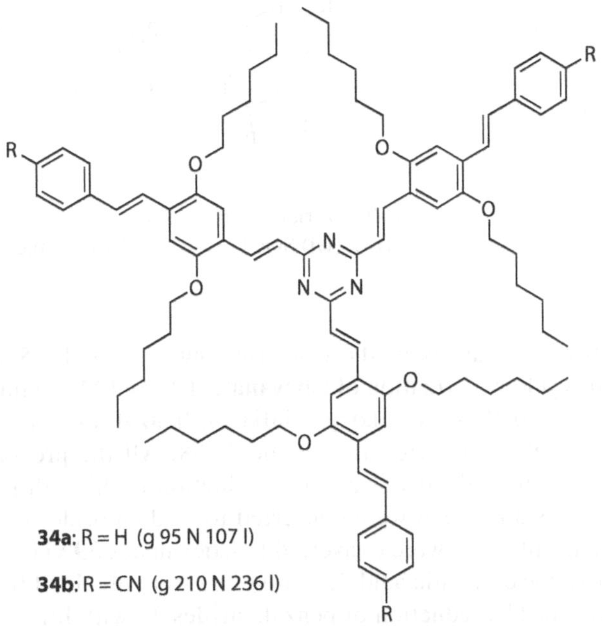

34a: R = H (g 95 N 107 I)

34b: R = CN (g 210 N 236 I)

FIGURE 2.34 Structure of lateral chain-substituted 2,4,6-tristyryl-1,3,5-triazines.

(T_g 95 N_D 107 I). Compound **34b** with terminal cyano group exhibited much higher transition temperatures compared to unsubstituted triazine **34a** due to its acceptor–donor–acceptor character. A nematic phase is obtained at $T_g = 210°C$ and cleared at 236°C. The irradiation of these stilbenoid triazines in solution or in the mesophase causes photocyclodimerization and finally generates cross-linked oligomers and polymers.

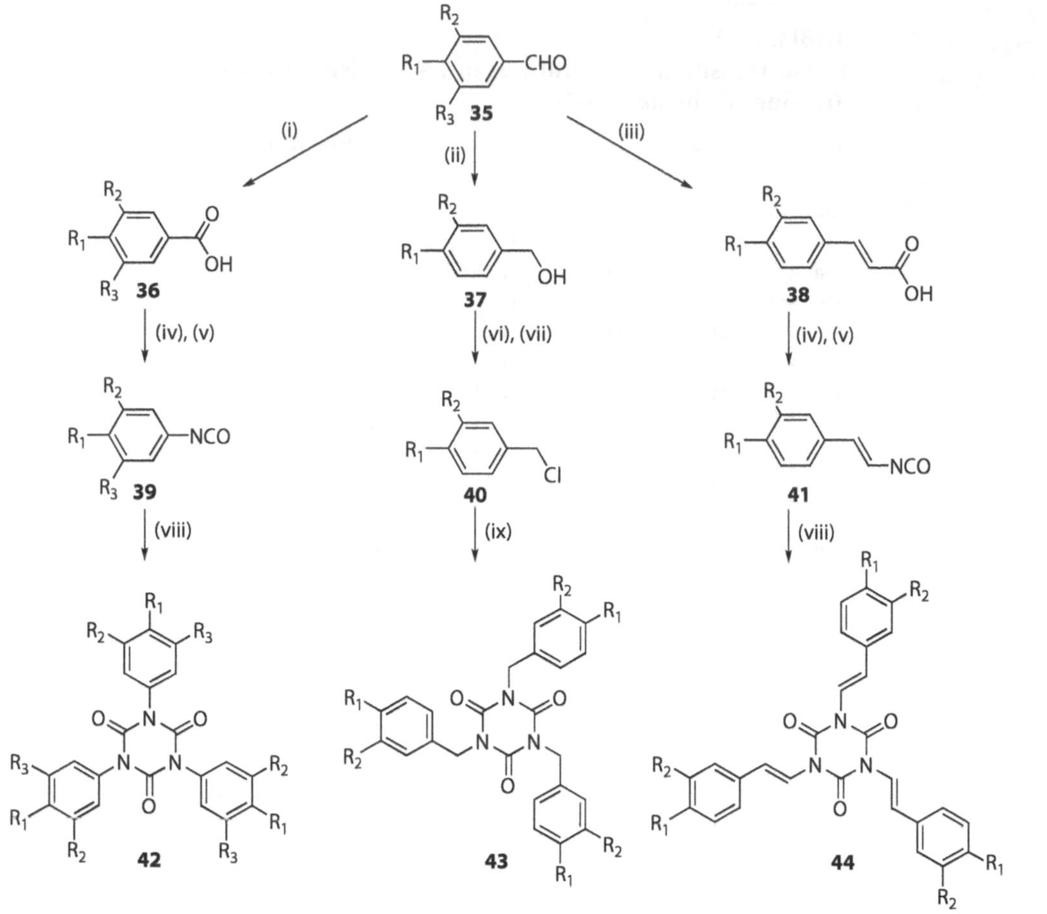

SCHEME 2.108 Synthesis of 1,3,5-triazine-2,4,6-trione derivatives: (i) KMNO$_4$; (ii) LAH; (iii) malonic acid/piperidine; (iv) SOCl$_2$; (v) NaN$_3$; (vi) HMDS; (vii) SOCl$_2$; (viii) CsF/crown ether; (ix) KOCN/DMF.

The liquid crystalline derivatives of the isomeric ring system, 1,3,5-triazine-2,4,6-trione, can be prepared by the cyclotrimerization of isocyanate. Irle and Mormann prepared a number of isocyanurates with monoalkoxy, dialkoxy, trialkoxy, benzyl, and styryl arms [1123]. The synthesis of these materials is depicted in Scheme 2.108. All the precursor molecules were obtained from substituted benzaldehydes **35**. The oxidation of **35** with potassium permanganate yielded benzoic acids **36**, which were converted to acid chlorides by refluxing in thionyl chloride. The crude acid chlorides were converted to isocyanates **39** via Curtius reaction using sodium azide. Similarly, the cinnamic acid derivatives **41** were obtained from substituted benzaldehydes (Scheme 2.108). The reduction of benzaldehydes **35** with lithium aluminum hydride yielded the benzyl alcohols **37**, which were converted to benzyl chlorides **40** via silyl ethers. The nucleophilic substitution of chlorine by potassium cyanate followed by in situ trimerization yielded the isocyanurates **43**, while the trimerization of **39** and **41** was achieved using CsF and dicyclohexyl-18-crown-6.

The thermal data of all isocyanurates are presented in Table 2.95. Enantiotropic columnar mesophases were observed only in the series of tris(3,4-dialkoxystyryl)isocyanurates **44** with eight or more carbon atoms in the chain. Compound **42e** displayed a monotropic mesophase.

TABLE 2.95

Phase Transition Temperatures of 2,4,6-Tristyryl-1,3,5-Triazines (Scheme 2.108)

Structure	R_1	R_2	R_3	Phase Transition
42a	H	H	H	Cr 280 I
42b	$OC_{10}H_{21}$	H	H	Cr 103 I
42c	OC_4H_9	OC_4H_9	H	Cr 123 I
42d	$OC_{10}H_{21}$	$OC_{10}H_{21}$	H	Cr 98 I
42e	$OC_{12}H_{25}$	$OC_{12}H_{25}$	H	Cr (85 Col$_h$) 110 I
42f	OC_4H_9	OC_4H_9	OC_4H_9	Cr 262 I
43a	H	H	H	Cr 158 I
43b	$OC_{10}H_{21}$	H	H	Cr 99 I
43c	$OC_{10}H_{21}$	$OC_{10}H_{21}$	H	Cr 87 I
43d	$OC_{12}H_{25}$	$OC_{12}H_{25}$	H	Cr 81 I
44a	H	H	H	Cr 231 I
44b	$OC_{10}H_{21}$	H	H	Cr 128 I
44c	OC_4H_9	OC_4H_9	H	Cr 131 I
44d	OC_8H_{17}	OC_8H_{17}	H	Cr 74 Col$_h$ 125 I
44e	$OC_{10}H_{21}$	$OC_{10}H_{21}$	H	Cr 90 Col$_h$ 133 I
44f	$OC_{12}H_{25}$	$OC_{12}H_{25}$	H	Cr 83 Col$_h$ 130 I

Source: Data from Irle, C. and Mormann, W., *Liq. Cryst.*, 21, 295, 1996.

2.16 PHENAZINES

Polycyclic aromatic nitrogen heterocycles play an important role not only in biological science but also in materials science. A large number of pharmaceuticals, pesticides, insecticides, dye-stuffs, solvents, polymers, and organic electronic materials are based on nitrogen heterocycles. Phenazine derivatives are well known for their antibacterial and insecticidal properties. Numerous phenazine-based dyes are used in the printing industry. The charge generating, transporting, and photophysical properties of various phenazine derivatives have been well documented [1124]. One of the simplest and common methods to prepare phenzine derivatives is condensing a 1,2-diamino precursor with a 1,2-dione unit (Scheme 2.109). The 1,2-diamines are generally prepared by the reduction of dinitro compounds. The 1,2-diketones can be prepared by the oxidation of tolanes or phenols. They can also be prepared by the Friedel–Crafts acylation of substituted benzenes with oxalyl chloride.

The shape of phenazine nucleus is a middle ground between rod and disk and, therefore, it is an interesting core system to generate liquid crystals. Discotic liquid crystals have not been realized from simple phenazine molecule; however, the core has been extended to give a variety of molecules in which mesomorphism can be induced by the proper engineering of alkyl chains around the core.

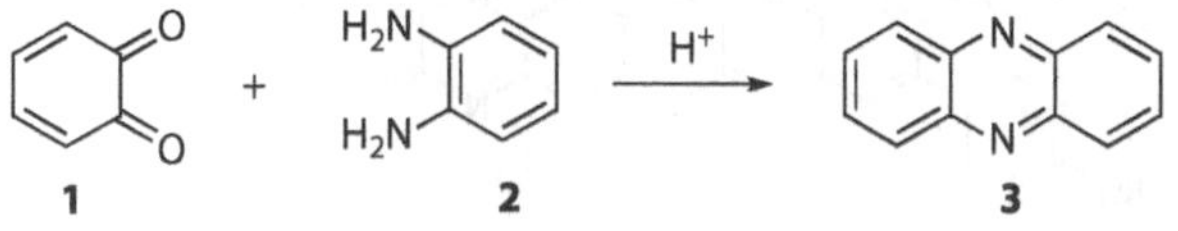

SCHEME 2.109 Synthesis of phenazine by condensing a 1,2-diamino derivative with a 1,2-dione unit.

Kumar and coworkers condensed 3,6,7,10,11-pentaalkoxy-triphenylene-1,2-dione and isomeric 6,7,10,11-tetrapentyloxy-triphenylene-2,3-dione (see Section 2.6) with 1,2-phenylenediamine in benzene-acetic acid to generate orange-red liquid crystalline 2,3,6,7-tetrapentyloxy-phennanthro[a]phenazine **4** and 2,3,6,7-tetrapentyloxy-phennanthro[b]pheazine **5** in high yield [410,416]. While phennanthro[a]phenazine derivatives exhibit enantiotropic columnar mesophase over a wide temperature range, phennanthro[b]phenazine **5** exhibited a metastable monotropic columnar mesophase.

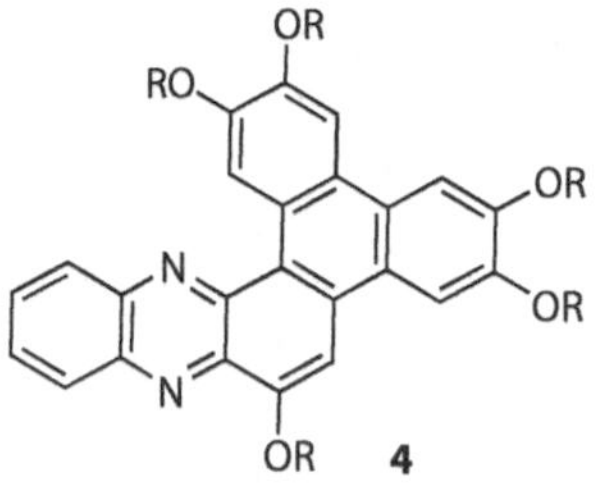

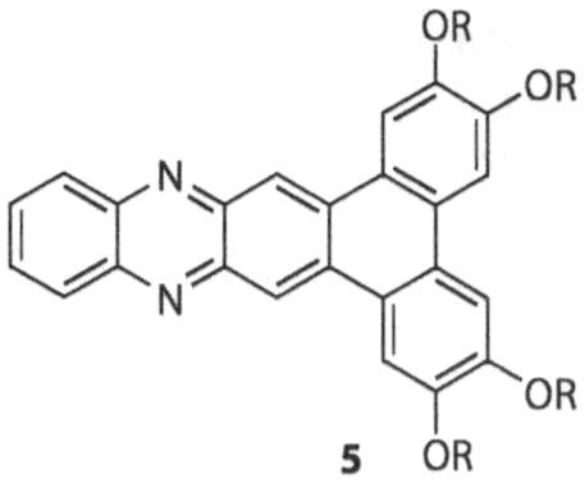

2.16.1 BISPHENAZINES

The fused bisphenazines **8** are synthesized, as shown in the Scheme 2.110. The condensation of two molecules of 1,2-bisalkoxy-4,5-diaminobenzene **7** with 2,7-di-*tert*-butyl-pyrene-4,5,9,10-tetraone

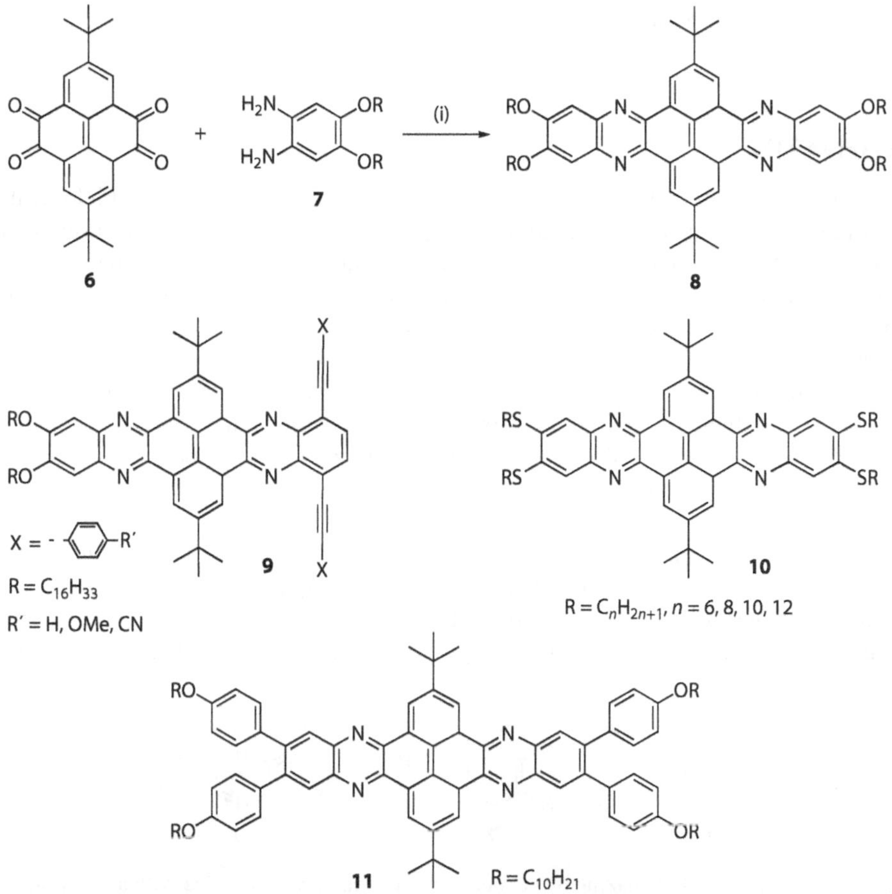

SCHEME 2.110   Synthesis of bisphenazine derivatives: (i) benzene, reflux.

TABLE 2.96
Phase Behavior of the Bisphenazines 8 (Scheme 2.110)

Structure	R	Thermal Behavior
8.1	$CH_2CH(Et)(CH_2)_3CH_3$	Cr >350 I
8.2	$(CH_2)_2CH(CH_3)(CH_2)_3CH(CH_3)_2$	Cr 316 I
8.3	$C_{10}H_{21}$	Cr 213 M 268 I
8.4	$C_{11}H_{23}$	Cr 210 M 234 I
8.5	$C_{14}H_{29}$	Cr 202 I

Source: Data from Hu, J. et al., *Chem. Mater.*, 16, 4912, 2004.

6 yields bisphenazines **8** [1125]. Five members in the series, three with normal alkyl chains and two having branched alkyl chains, were prepared. The *n*-alkyl substituted compounds **8.1** and **8.2** (Table 2.96) showed mesomorphism. Both melting and clearing temperatures decrease with increase in chain length, but the clearing temperature decreases more drastically than melting the temperature. The higher homologue **8.3** with normal tetradecyl chains and branched chain derivatives did not show any mesophase. A number of other bisphenazines such as T-shaped molecule **9**, peripheral thioether-substituted bisphenazines **10**, and acene-type molecular ribbons **11** containing 2–6 pyrazine units and up to 16 rectilinearly arranged fused aromatic rings have been synthesized and studied for various properties [1126–1129]. However, these materials are either non-liquid crystalline or their mesomorphic properties are not revealed.

2.16.2 DIBENZOPHENAZINES

Dibenzophenazines **14** can be conveniently prepared by coupling appropriately substituted 1,2 diamines **13** with 2,3,6,7-tetra-alkoxy phenanthrene-9,10-dione **12** (Scheme 2.111). A variety of molecules, as shown in Scheme 2.111, has been prepared to understand the structure–property relationship [1130–1134]. Williams et al. studied the effect of hydrogen bonding on mesophase behavior with a series of dibenzophenazine carboxylic acids and their methyl ester analogues. The acid derivatives form dimeric structure **15** by the hydrogen bonding between two monomers. These macrodiscotics are liquid crystalline, displaying columnar phases over broad temperature range (Table 2.97). The acid derivatives exhibit a much richer variety of liquid crystalline phases in contrast to ester derivatives, which show only hexagonal columnar phase. In ester series, with increase in chain length, both melting and clearing temperature decrease (Table 2.97), but the change is drastic for clearing temperature and this leads to a narrowing of the mesophase. In contrast, while the phase transitions for acids are also shifted to lower temperatures as chain length increases, the range over which the compound show mesophase remains constant. However, it is observed that the temperature range of less ordered nematic phase becomes broader with an increase in chain length. The clearing temperatures of the acid derivatives are 60°C–90°C higher than those of corresponding esters. The acid derivatives were also found to supercool from the columnar phase to room temperature while maintaining liquid crystalline ordering for prolonged periods. The esters are approximately disk shaped and can assemble into columns that can in turn pack in hexagonal lattice. On the other hand, the H-bonded acid dimers **15** are elliptical and the resulting columns are also expected to be elliptical and to pack in a rectangular rather than a hexagonal lattice [1130].

Williams and coworkers have also studied a structural feature that influences the self-assembly of mesogens. The effect of core size on mesophase behavior has been studied by considering three compounds with slightly different structures. These comprise a series in which core size increases progressively from four to six rings. The four ring (dibenzoquinoxaline, see Section 2.16.3) and five

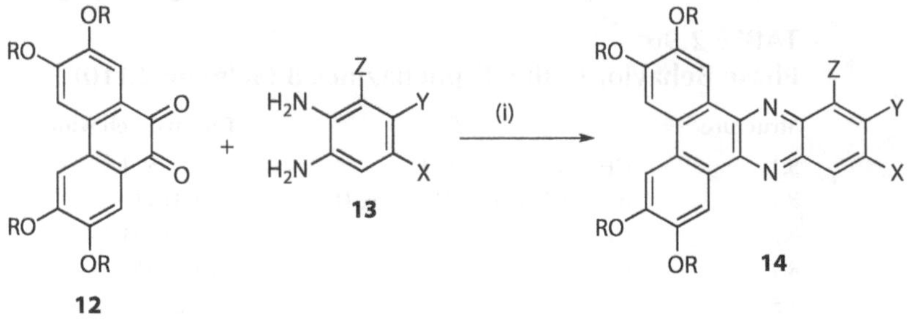

14.1–14.5: X = COOH, Y = Z = H, R = C_nH_{2n+1}, n = 6, 8, 10, 12, 14

14.6–15.9: X = COOCH$_3$, Z = Y = H, R = C_nH_{2n+1}, n = 6, 8, 10, 12

14.10–14.16: X = F, Cl, CN, NO$_2$, CH$_3$, OCH$_3$;

Z = Y = H, R = C_nH_{2n+1}, n = 6, 14

14.17–14.20: Z = H, X = Y = H, Cl, CH$_3$, OCH$_3$, R = C_nH_{2n+1}, n = 6

14.21–14.28: X = Y = H, Z = COOH, COOCH$_3$, NO$_2$, R = C_nH_{2n+1}, n = 6, 10, 14

14.29–14.33: Z = H, X = Y = OC_nH_{2n+1}, R = C_nH_{2n+1}, n = 4, 6, 8, 10, 12

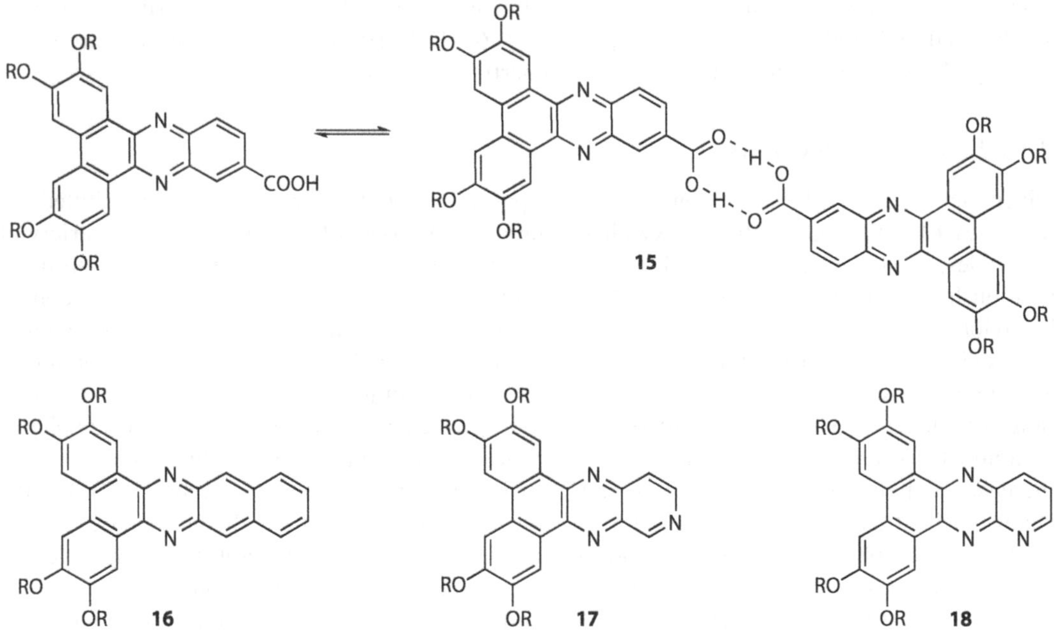

SCHEME 2.111 Synthesis of dibenzophenazines: (i) acetic acid, reflux.

ring systems did not show any mesophase but the naphthalene derivative **16** is liquid crystalline. This implies that the large core size is favorable to form a columnar phase. This effect can be understood in terms of the interactions between neighboring molecules within a columnar phase. However, it should be noted that appropriately substituted four- and five-ring systems do exhibit mesomorphism. Therefore, the formation of mesophases not only depends on π–π interactions but also on other forces like dipolar interactions and filling the space around the core. The type of substituent influences the type of mesophase formed. There is a striking correlation between the tendency of the molecules to form columnar phase and the electron-withdrawing or electron-donating ability of

TABLE 2.97
Phase Behavior of the Dibenzophenazines (Scheme 2.111)

Structure	R	X	Y	Z	Thermal Behavior	Ref.
14.1	C_6H_{13}	COOH	H	H	Cr 173 Col$_h$ 263 I	[1130,1133]
14.2	C_8H_{17}	COOH	H	H	Cr 162 Col$_r$ 192 Col$_r$ 227 N 232 I	[1130]
14.3	$C_{10}H_{21}$	COOH	H	H	Cr 122 Col$_r$ 196 N 209 I	[1130,1133]
14.4	$C_{12}H_{25}$	COOH	H	H	Cr 122 Col$_r$ 196 N 209 I	[1130]
14.5	$C_{14}H_{29}$	COOH	H	H	Cr 47 Col$_r$ 154 N 193 I	[1133]
14.6	C_6H_{13}	COOCH$_3$	H	H	Cr 111 Col$_h$ 200 I	[1130]
14.7	C_8H_{17}	COOCH$_3$	H	H	Cr 108 Col$_h$ 162 I	[1130]
14.8	$C_{10}H_{21}$	COOCH$_3$	H	H	Cr 98 Col$_h$ 137 I	[1130]
14.9	$C_{12}H_{25}$	COOCH$_3$	H	H	Cr 77 Col$_h$ 106 I	[1130]
14.10	C_6H_{13}	F	H	H	Cr (154 Col$_h$)161 I	[1131]
14.11	C_6H_{13}	Cl	H	H	Cr 137 Col$_h$ 188 I	[1131]
14.12	C_6H_{13}	CN	H	H	Cr 101 Col$_h$ 208 I	[1131]
14.13	C_6H_{13}	NO$_2$	H	H	Cr 131 Col$_h$ 229 I	[1131,1133]
14.14	$C_{14}H_{29}$	NO$_2$	H	H	Cr 107 Col$_h$ 136 I	[1133]
14.15	C_6H_{13}	CH$_3$	H	H	Cr 125 I	[1131]
14.16	C_6H_{13}	OCH$_3$	H	H	Cr 149 I	[1131]
14.17	C_6H_{13}	H	H	H	Cr 170 I	[1131]
14.18	C_6H_{13}	Cl	Cl	H	Cr 142 Col$_h$ 238 I	
14.19	C_6H_{13}	CH$_3$	CH$_3$	H	Cr 155 I	[1131]
14.20	C_6H_{13}	OCH$_3$	OCH$_3$	H	Cr 171 I	[1131]
14.21	C_6H_{13}	H	H	COOH	Cr 88 Col$_h$ 260 I	[1133]
14.22	$C_{10}H_{21}$	H	H	COOH	Cr 107 Col$_h$ 242 I	[1133]
14.23	$C_{14}H_{29}$	H	H	COOH	Cr 85 Col$_h$ 210 I	[1133]
14.24	C_6H_{13}	H	H	COOCH$_3$	Cr 158 I	[1133]
14.25	$C_{10}H_{21}$	H	H	COOCH$_3$	Cr 127 I	[1133]
14.26	$C_{14}H_{29}$	H	H	COOCH$_3$	Cr 110 I	[1133]
14.27	C_6H_{13}	H	H	NO$_2$	Cr 132 Col$_h$ 175 I	[1133]
14.28	$C_{14}H_{29}$	H	H	NO$_2$	Cr (109 Col$_h$) 115 I	[1133]
14.29	C_4H_9	OC$_4$H$_9$	OC$_4$H$_9$	H	Cr 138 Col$_h$ 204 I	[1134]
14.30	C_6H_{13}	OC$_6$H$_{13}$	OC$_6$H$_{13}$	H	Cr 70 Col$_h$ 159 I	[1134]
14.31	C_8H_{17}	OC$_8$H$_{17}$	OC$_8$H$_{17}$	H	Cr 65 Col 70 Col$_h$ 148 I	[1134]
14.32	$C_{10}H_{21}$	OC$_{10}$H$_{21}$	OC$_{10}$H$_{21}$	H	Cr 55 Col 59 Col$_h$ 130 I	[1134]
14.33	$C_{12}H_{25}$	OC$_{12}$H$_{25}$	OC$_{12}$H$_{25}$	H	Cr 57 Col$_h$ 111 I	[1134]
16	C_6H_{13}	—	—	—	Cr 123 Col$_h$ 172 I	[1131]
17.1	C_6H_{13}	—	—	—	Cr 155 Col$_h$ 179 I	[1132]
17.2	$C_{10}H_{21}$	—	—	—	Cr 123 Col$_h$ 135 I	[1132]
18.1	C_6H_{13}	—	—	—	Cr 145 I	[1132]
18.2	$C_{10}H_{21}$	—	—	—	Cr 113 I	[1132]

the functional groups attached to the core. Compounds with electron-withdrawing groups formed columnar phases, whereas those with relatively electron-donating groups were nonmesogenic. The compound substituted with a weakly electron withdrawing fluoro substituent forms only a metastable monotropic phase. The nitro derivative, which is the most electron-deficient mesogen, showed the highest clearing temperature in the series. The addition of electron-withdrawing groups minimizes the repulsive interactions between adjacent aromatic π-systems and favors the π-stacking. Dipole–dipole interactions also contribute to the formation of columnar phase in molecules substituted with electron-withdrawing groups. X-ray analysis shows anti-ferroelectric stacking of these mesogens

[1131]. While phenazines with small electron-donating groups do not show mesomorphism, long alkoxy chain-substituted derivatives exhibit columnar phases due to effective space filling around the core [1134]. Ong et al. synthesized five hexaalkoxy-substituted dibenzophenazines **14.29–14.32** and studied their mesophase behavior. All the derivatives showed columnar mesophase. While the C_4, C_6, and C_{12} derivatives showed only Col_h phase, the octyloxy- and decyloxy-derivatives possess an additional low temperature columnar phase. Melting temperatures of the hexa-substituted derivatives are much lower compared to tetra-substituted derivatives with electron-withdrawing groups [1134].

The nitrogen analogues of dibenzophenazine **17** and **18** were studied for their mesophase behavior as well as for the mesophase behavior of their hydrogen-bonded complexes with various acid derivatives of dibenzophenazine [1132]. The position of pyridyl nitrogen atom has a large effect on phase behavior as compounds **18.1** and **18.2** melt directly to isotropic liquid while the corresponding compounds **17.1** and **17.2** exhibit a columnar hexagonal mesophase over a broad temperature range. Stoichiometric binary mixtures of acid **14** with **17** behave as a single component and showed mesophase behavior distinct from those of their two individual constituents. The binary mixture of **17.1–14.1** formed a columnar hexagonal phase over an extremely broad temperature range similar to the behavior of **17.1**. On the other hand, the binary mixtures of acids **14** with dibenzophenazine derivatives **18** having nitrogen atom in the side did not behave as a single component. They formed different domains showing different textures. The binary mixture of **18.1** and **14.1** showed liquid crystalline phase and isotropic liquid phase coexisted for more than 50°C. Compound **18** possess a very different structure on complexing with acid compared to H-bonded complex with isomeric compound **17**. The overall shape of the **18–14** complexes is probably incompatible for the formation of liquid crystalline phases.

The acid derivatives **14.21–14.23** do not form H-bonded dimers since the hydroxyl hydrogen is intramolecularly hydrogen bonded to pyrazinyl nitrogen atom. Compounds **14.21–14.23** show columnar hexagonal mesophase. The corresponding ester derivatives **14.24–14.26** melt directly to isotropic liquid without passing through the mesomorphic state. Moving a nitro group from the top of a molecule in **14.13** to the side in **14.27** leads to dramatic decrease in clearing temperature while the melting temperatures remain unchanged. Thus hexagonal columnar phase of **14.13** has a narrow range compared to **14.27**. Overall, it may be concluded that changing the position of the group from the top to the side of the phenyl ring led to dramatic differences in the mesomorphic behavior.

2.16.3 DIBENZOQUINOXALINE

The dibenzoquinoxaline derivatives are structurally related to dibenzophenazine discotics and are prepared in a similar fashion. Thus, the condensation of 1,2-diketone **12** with diaminomaleonitrile **19** furnished the dibenzoquinoxaline **20** (Scheme 2.112). Ohta et al. reported the discotic liquid crystalline nature of dibenzoquinoxaline **20** bearing four peripheral decyloxy chains [1135]. Williams and coworkers compared the mesophase behavior of the symmetrical and unsymmetrical

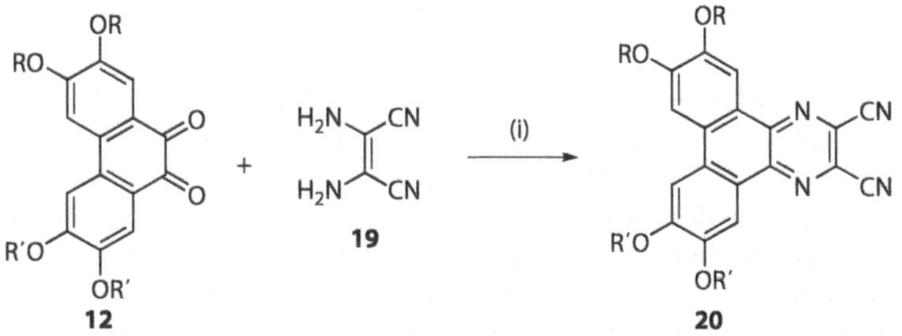

SCHEME 2.112 Synthesis of dibenzoquinoxaline discotics: (i) acetic acid, reflux.

TABLE 2.98
Phase Behavior of Dibenzoquinoxalines 20
(Scheme 2.112)

Structure	R	R′	Thermal Behavior	Ref.
20.1	C_6H_{13}	C_6H_{13}	Cr 72.4 Col_h 256.1 I	[1137]
20.2	C_8H_{17}	C_8H_{17}	Cr 85.7 Col_h 252.1 I	[1137]
20.3	$C_{10}H_{21}$	$C_{10}H_{21}$	Cr 71.6 Col_h 254.2 I	[1137]
			Cr 58 Col_h 258 I	[1135]
20.4	$C_{10}H_{21}$	CH_3	Cr 264 I	[1137]
20.5	$C_{10}H_{21}$	C_6H_{13}	Cr 37.8 Col_h 215.6 I	[1137]
20.6	$C_{10}H_{21}$	C_8H_{17}	Cr 56.2 Col_h 231.2 I	[1137]

substituted dibenzoquinoxalines [1136,1137]. The thermal behavior of these materials is summarized in Table 2.98. Removing strongly electron-withdrawing cyano groups makes the compound nonmesomorphic [1136]. The unsymmetrical peripheral substitution also imparts drastic influence on phase behavior [1137]. While the symmetrical derivatives show columnar phase over similar temperature ranges as unsymmetrical derivatives, the melting and clearing temperatures of unsymmetrical mesogens are appreciably lower than symmetrical derivatives.

2.17 HEXAAZATRIPHENYLENE CORE

1,4,5,8,9,12-Hexaazatriphenylene (HAT) is a heteroaromatic polycyclic discotic core that behaves as an electron-deficient scaffold. Praefcke and coworkers systematically investigated the hexaalkyl and hexaalkoxy derivatives of hexaazatriphenylene after the discovery of mesomorphism in triphenylene derivatives [1138]. However, none of these compounds were found to be liquid crystalline in nature. The synthesis of the HAT derivatives involves one-step cyclocondensation of benzene hexamine **1** and various symmetrical 1,2-diketones **2**, as shown in Scheme 2.113. It is interesting to note that the hexa-substituted derivatives of HAT are devoid of aromatic protons. Moreover, Praefcke and coworkers could not isolate the heterocyclic parent aromatic core **3a** in pure form. Subsequently, however, Rogers reported the improved synthesis of the parent core **3a** by the condensation of glyoxal **2a** with freshly prepared hexaaminebenzene **1**, as shown in Scheme 2.114 [1139]. Later in the same year, Czarnik and coworkers reported the synthesis of hydrogen-free hexaazatriphenylenehexacarbonitrile **7**, which acts as a potential precursor for the preparation of various liquid crystalline derivatives of HAT [1140]. Furthermore, they also synthesized functionalized HAT derivatives such as hexaamides **8**, hexamethoxycarbonyls **9**, and hexaacids **10**, as shown in Scheme 2.115 [1141–1143], some of which later also served as important precursors for the preparation of novel liquid crystalline derivatives. In addition, the derivatives of HAT have received considerable attention because of their easy synthetic accessibility, diversity in peripheral functionality,

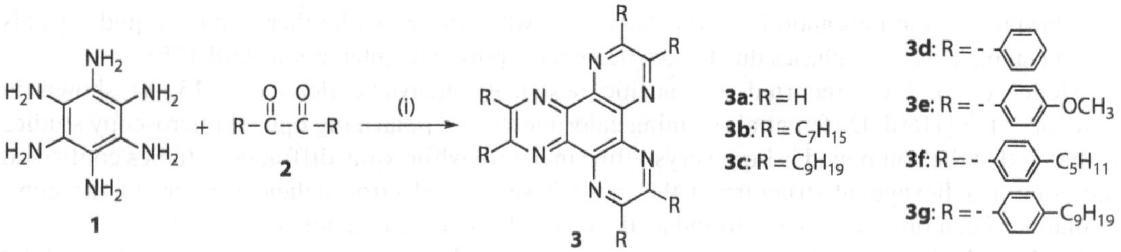

SCHEME 2.113 Synthesis of hexaazatriphenylene derivatives **3**: (i) AcOH.

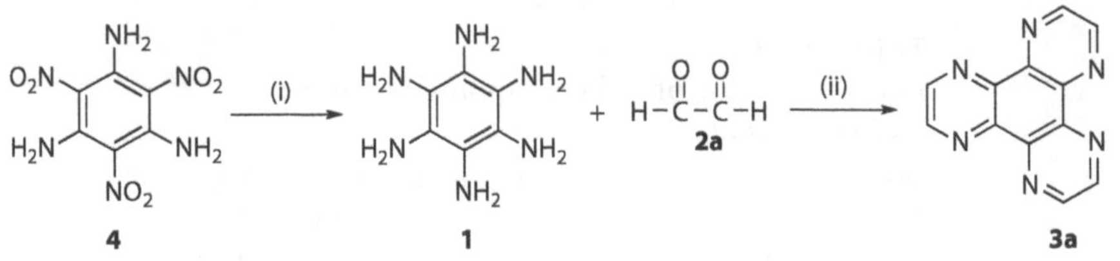

SCHEME 2.114 Synthesis of unsubstituted HAT: (i) MeOH, Na/Liq. NH$_3$; (ii) AcOH.

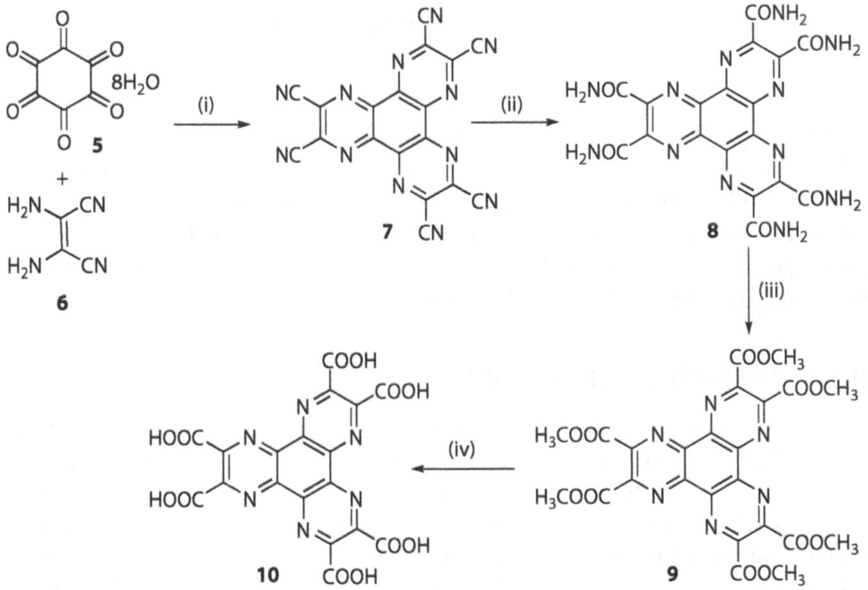

SCHEME 2.115 Synthesis of hexaazatriphenylenehexacarbonitrile **7**: (i) AcOH; (ii) Conc. H$_2$SO$_4$; (iii) MeOH, H$_2$SO$_4$; (iv) Et$_3$N/H$_2$O.

electron deficiency, coordination properties, and π-complexation ability. Many HAT derivatives have been exploited in a variety of applications such as self-assembled organogels, n-type semiconductors, magnetic materials, fluorescent dyes, octupolar nonlinear chromophores, etc. [1144–1152]. Moreover, the coordination chemistry of HAT in the context of multinuclear complexes is very rich and a variety of metal complexes have been prepared and studied [1153–1158].

The first liquid crystalline derivatives of HAT were reported by Bushby and coworkers [501,503,1159]. They prepared several HAT derivatives **12** with different peripheral chains, as shown in Scheme 2.116. Compound **12a** with 12 hexyloxy substituents displays mesomorphism. It exhibits two columnar phases; the high temperature phase was characterized as a columnar rectangular phase. Non-mesomorphic HAT derivatives when mixed with other discotic liquid crystals exhibit stable columnar phases due to complementary polytopic interactions [501,1159].

Meijer et al. have reported a dendritic hexaazatriphenylene derivative **13**, as shown in Scheme 2.117 [1160]. Differential scanning calorimetry and polarizing optical microscopy studies revealed that the compound is liquid crystalline in nature while x-ray diffraction studies confirmed the columnar hexagonal structure of the mesophase. The electron-deficient nature of the compound has been observed by photo-induced electron transfer with a donor polymer.

HATs with alternative donor and acceptor substituents have been prepared and found to exhibit columnar mesomorphism [1161]. The synthesis of the symmetrical difunctional HATs **14** has

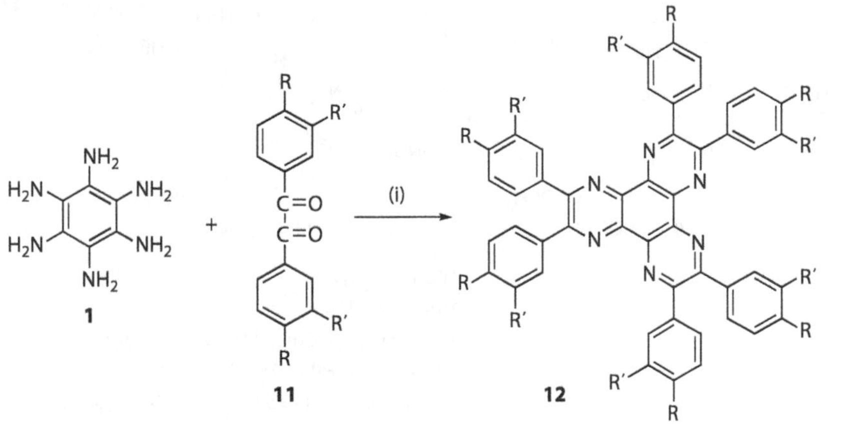

12a: R = R′ = OC_6H_{13}; Cr 147 Col_1 150 Col_2 194 I

SCHEME 2.116 Synthesis of liquid crystalline HAT derivatives **12**: (i) AcOH.

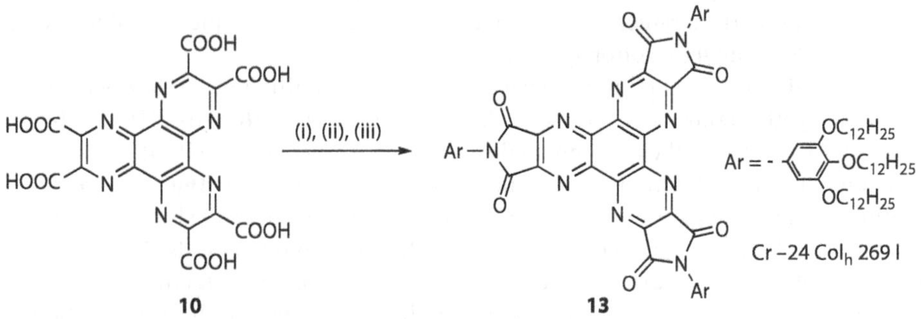

Ar = - $OC_{12}H_{25}$ $OC_{12}H_{25}$ $OC_{12}H_{25}$

Cr −24 Col_h 269 I

SCHEME 2.117 Synthesis of dendritic hexaazatriphenylene derivative **13**: (i) Ac_2O; (ii) Ar-NH_2; (iii) Ac_2O, TFA.

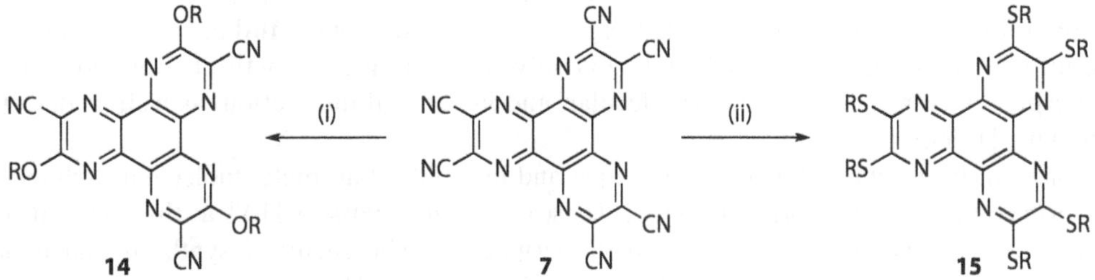

14a: R = CH_3; Cr > 288 I
14b: R = C_6H_{13}; Cr 141 Col_h 148.2 I

15a–d: R = C_nH_{2n+1}, n = 6, 8, 10, 12
15e: R = $CH_2CH(Et)CH_2CH_2CH_2CH_3$

SCHEME 2.118 Synthesis of HAT derivatives with alternative donor and acceptor substituents **14** and hexa-thioethers of HAT **15**: (i) ROH, CH_3CN; (ii) RSH, K_2CO_3, DMF.

been achieved by the regioselective displacement of hexaazatriphenylenehexacarbonitrile **7** with alkoxy groups, as shown in Scheme 2.118. Compound **14b** exhibits enantiotropic columnar liquid crystalline behavior as observed by polarizing optical microscopy and differential scanning calorimetry. X-ray diffraction study confirms the ordered columnar hexagonal mesophase structure in this compound.

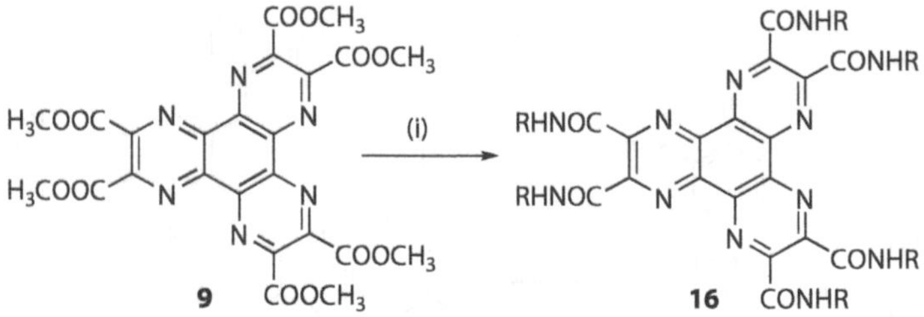

16a: R = C$_6$H$_{13}$
16b: R = C$_{10}$H$_{21}$
16c: R = C$_{12}$H$_{25}$; Col$_{h1}$ 47 Col$_{h2}$ >250 dec.
16d: R = CH$_2$CH(Et)CH$_2$CH$_2$CH$_2$CH$_3$;
g 80 Col$_{ob1}$ 119.5 Col$_{ob2}$ 175.3 Col$_{ob3}$ > 250 dec.

SCHEME 2.119   Synthesis of hexacarboxamido-HAT derivatives **16**: (i) RNH$_2$.

A series of hexathioethers **15** of HAT has been prepared as shown in Scheme 2.118 to investigate the liquid crystalline properties [1162]. Surprisingly, none of the synthesized compounds exhibit liquid crystalline properties; rather some of the derivatives exhibit multiple crystal to crystal transitions before transforming into isotropic liquid.

Hydrogen bond–enforced columnar hexagonal liquid crystalline phase has been achieved by designing and synthesizing hexacarboxamido-HAT derivatives **16** [1163,1164]. The synthesis and chemical structures of these compounds are shown in Scheme 2.119 along with the phase transition temperatures of liquid crystalline materials. Compounds **16a** and **16b** decompose upon heating to higher temperatures. Remarkably, the compound **16c** exhibits smallest intracolumnar inter-disk distance (3.18 Å) ever found in columnar liquid crystals. Moreover, the intracolumnar correlation length varies between 120 and 180 Å, thereby extending over 40–55 disks. Owing to the small intracolumnar distance and high correlation length, it was anticipated that these materials should display high charge carrier mobility. The charge carrier mobility in this compound was measured by pulse-radiolysis time-resolved microwave conductivity and found to vary from 0.04 to 0.08 cm^2 V^{-1} s^{-1}. Another hexaamide derivative of HAT **16d** containing branched alkyl chain is found to display multiple columnar oblique phases before its decomposition at higher temperatures. The self-organization of compound **16d** at surfaces was found to be influenced by the type of substrate and solvent, leading to growth of films governed by the interplay of intramolecular, intermolecular, and interfacial interaction, as well as dewetting phenomena [1165,1166].

Ishi-i et al. have reported a series of unique and unusual all-aromatic molecular architectures **18** to form columnar mesomorphism [1167]. These compounds possess HAT as the central discotic core and the core is surrounded by aromatic peripheries. The chemical synthesis and molecular structures of these compounds are shown in Scheme 2.120. The phase transition temperatures of the compounds are collected in Table 2.99. The hexagonal symmetry of the columnar phases exhibited by the liquid crystalline derivatives has been characterized by x-ray diffraction studies. Furthermore, in certain solvents, some of the compounds form organogels in which one-dimensional aggregates consisting of HAT molecules are self-assembled into three-dimensional network structures.

2.17.1 HEXAAZATRINAPHTHYLENE

5,6,11,12,17,18-Hexaazatrinaphthylene **20a** (HATNA) is a C_3 symmetric heterocyclic aromatic core that also acts as a core fragment for discotic liquid crystals (Scheme 2.121). The molecule

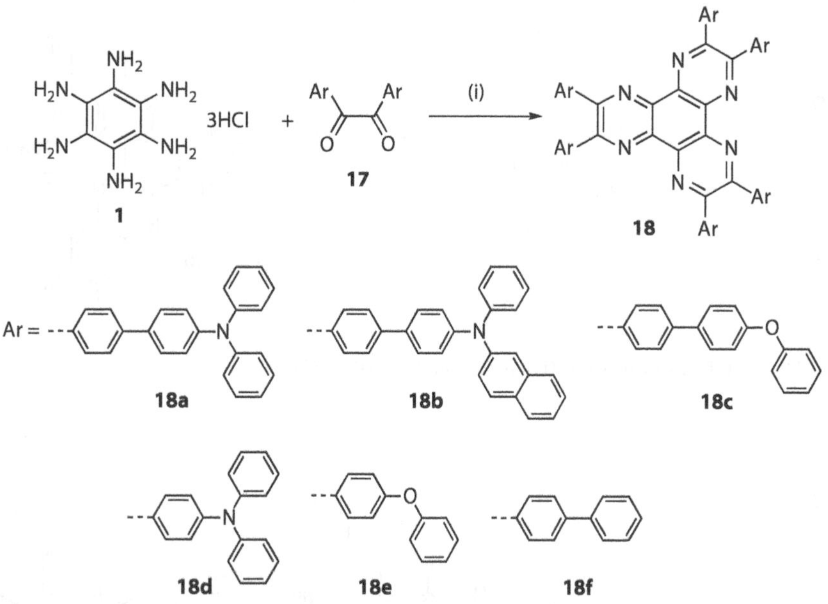

SCHEME 2.120 Synthesis of all-aromatic HAT derivatives **18**: (i) AcOH.

actually belongs to the phenazine group (diquinoxalino[2,3-*a*; 2′,3′-*c*]phenazine); however, because of its structural similarity with HAT, it is grouped here. The presence of six nitrogen atoms in the aromatic core is anticipated to significantly increase the first reduction potential, thereby facilitating electron injection and high electron mobility. The electron-deficient nature of the core has encouraged the design and synthesis of its liquid crystalline as well as non-mesomorphic derivatives that can act as electron-transporting materials in organic semiconductor devices. Various derivatives of HATNA have been studied as model compounds to determine the crystal structure, ionization potential, and electron affinity in addition to carrier mobilities [1168,1169]. The controlled self-assembly and reorientation of HATNAs on gold electrodes has been probed by electrochemical scanning tunneling microscopy (ECSTM) [1170,1171]. Nanoporous network, nanorods, and nanowires have been obtained by the self-assembly of HATNA. Moreover, this heteroaromatic core acts as an excellent scaffold for the preparation and study of trinuclear complexes [1172–1176].

TABLE 2.99
Phase Transition Temperatures of All-Aromatic Hexaazatriphenylene Derivatives (Scheme 2.120)

Structure	Thermal Behavior
18a	Cr 343 Col$_h$ 385 I
18b	Cr 278 Col$_h$ 398 I
18c	Cr 357 Col$_h$ 464 I
18d	Cr 380 I
18e	Cr 290 I
18f	Cr 484 I

Source: Data from Ishi-i, T. et al., *Langmuir*, 21, 1261, 2005.

The parent compound **20a** and 2,3,8,9,14,15-hexakismethyl-5,6,11,12,17,18-hexaazatrinaphthylene **20b** (Scheme 2.121) have been known in the chemical literature well before the discovery of discotic liquid crystals [1177]. However, the first discotic liquid crystalline derivatives of HATNA were introduced at the beginning of the twenty-first century by Geerts and Lehmann [1178,1179]. They designed and synthesized a series of hexathioethers of HATNA **23** and studied their mesophase behavior. The synthesis of the compounds is shown in Scheme 2.122, which involves the condensation of cyclohexane-1,2,3,4,5,6-hexaone octahydrate with 4,5-dichlorobenzene-1,2-diamine to give **22** and subsequent nucleophilic substitution of chlorine atoms by appropriate alkyl thiolates to furnish **23**. The thermal phase behavior of these compounds is summarized in Table 2.100. The thermotropic behavior is strongly dependent on the thioalkyl chain length. For these compounds, no clearing temperatures could be obtained below

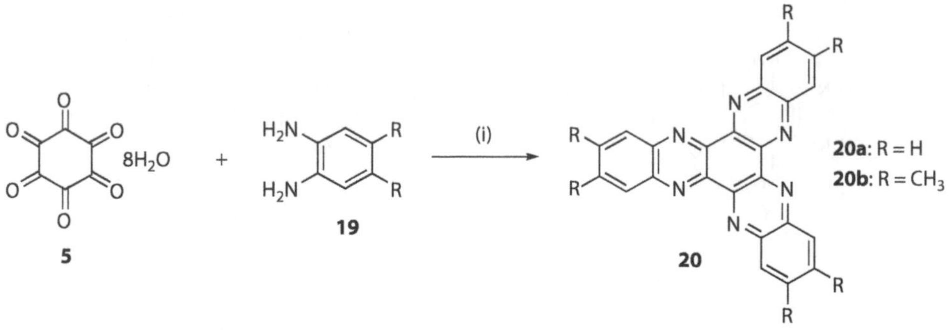

SCHEME 2.121 Synthesis of hexaazatrinaphthylene **20**: (i) AcOH.

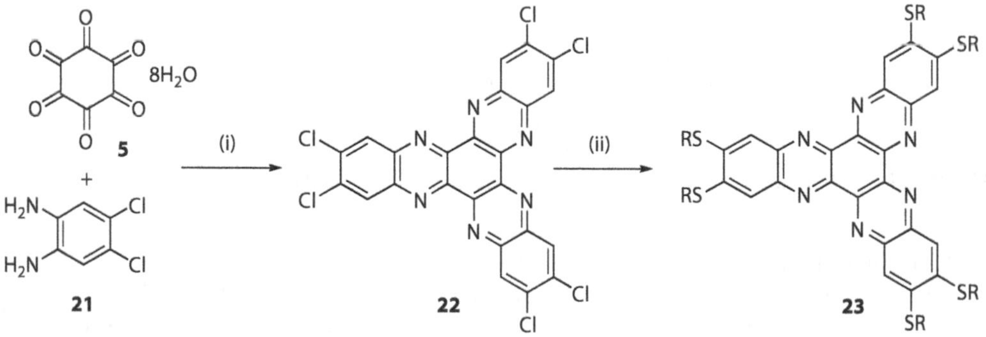

23a–d: R = C$_n$H$_{2n+1}$, n = 6, 8, 10, 12

23e: R = CH$_2$CH(Et)CH$_2$CH$_2$CH$_2$CH$_3$

SCHEME 2.122 Synthesis of hexathioethers of HATNA **23**: (i) AcOH; (ii) RSH, K$_2$CO$_3$/DMF.

TABLE 2.100
Phase Transition Temperatures of Hexathioether
Derivatives of Hexaazatrinaphthylene (Scheme 2.122)

Structure	Thermal Behavior	Ref.
23a	Cr 205 Col$_h$ > 250 dec	[1179]
23b	Cr 175 Col$_h$ > 250 dec	[1179]
23c	Cr 113 Col$_{r1}$ 134 Col$_{r2}$ 181 Col$_{r3}$ 222 Col$_h$ > 250 dec	[1179]
23d	Cr 90 Col$_r$ 149 Col$_h$ > 250 dec	[1179]
23e	Cr > 250 dec	[1162]

---

their decomposition. The branched chain derivative **23e** exhibits multiple crystalline phases prior to decomposition and fails to exhibit any liquid crystalline phase [1162]. Compounds **23a** and **23b** exhibit columnar hexagonal phase while compounds **23c** and **23d** exhibit both columnar rectangular and columnar hexagonal phases. However, the appearance of multiple columnar rectangular phases in compound **23c** is remarkable. The supramolecular organization of the columnar phases of these discotic compounds has been revealed by x-ray diffraction studies on both powder and oriented samples. The electron-deficient nature of these derivatives has been studied by cyclic voltametry technique, which demonstrates three successive reduction potentials corresponding to the formation of radical anions. Charge carrier mobilities of the thioethers have been measured with the help of pulse-radiolysis time-resolved microwave conductivity technique. Carrier mobility value has been found to be $0.3\,cm^2\,V^{-1}\,s^{-1}$ in compound **23c**. Moreover, it is observed that the charge carrier mobilities strongly depend on the nature of the side chains and varies from 0.02 to $0.32\,cm^2\,V^{-1}\,s^{-1}$. By space charge-limited current and time of flight experiments it has been found that the carrier mobilities are predominantly electron-mobilities [1180]. The blend of electron transporting HATNA **23d** with a hole-transporting polymer material has been studied to understand the charge recombination and charge separation. They form films with good intermixing of the two materials and good interfacial contact surface area; however, due to the strong localization of excitons at the interface, it results in low yield of charge separation [1181,1182].

The hexaalkoxy derivatives **28** of HATNA have been reported to exhibit columnar mesomorphism. The synthesis of these compounds involves the condensation of 1,2-dialkoxy-4,5-diaminobenzene with freshly prepared cyclohexane-1,2,3,4,5,6-hexaone, as shown in Scheme 2.123 [1183,1184]. The transition temperatures of these compounds are summarized in Table 2.101. Unlike their thioether counterparts, hexaethers exhibit clear mesophase to isotropic transitions, as has been observed by polarized light microscopy and differential scanning calorimetry. These compounds exhibit a very wide columnar mesophase range with rectangular and/or hexagonal symmetry; however, the lower members fail to be liquid crystalline. The introduction of six electron-donating side chains could not override the electron-deficiency of the core, which has been confirmed by measuring the reduction potential of these compounds by cyclic voltametry studies.

Triesters of HATNA **31** have been prepared and found to exhibit mesomorphism [1185]. Surprisingly, some of these compounds exhibit highly ordered columnar phases at room temperature, and interestingly these materials also form a high-temperature nematic liquid crystalline

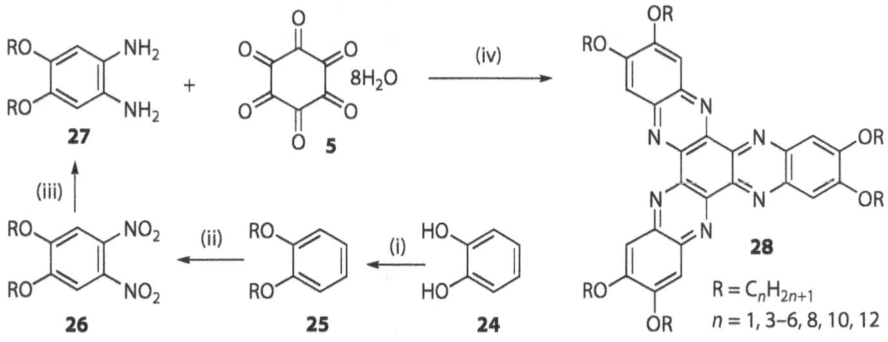

SCHEME 2.123 Synthesis of hexaalkoxy derivatives of HATNA **28**: (i) RBr, K_2CO_3, acetone; (ii) HNO_3, AcOH, CH_2Cl_2; (iii) H_2, Pd/C, EtOH; (iv) H^+, EtOH.

TABLE 2.101
Phase Transition Temperatures of Hexaethers of Hexaazatrinaphthylene (Scheme 2.123)

Structure	Thermal Behavior	Ref.
28e	Cr 187.1 Col$_r$ 230.3 I	[1183]
28f	Cr 114.5 Col$_r$ 176 Col$_h$ 224.4 I	[1184]
28g	Cr 86 Col$_h$ 214.6 I	[1184]
28h	Cr 94.2 Col$_h$ 206.1	[1184]

phase, as shown in Table 2.102. These compounds can be obtained by threefold condensation of 3,4-diaminobenzoic acid with hexaketocyclohexane, as shown in Scheme 2.124. In principle, two regio-isomers can be formed in the reaction, while, interestingly, it has been confirmed by spectral and analytical techniques that the less symmetric 2,8,15-trisubstituted isomer is exclusively obtained.

TABLE 2.102
Phase Transition Temperatures of Triester Derivatives of Hexaazatrinaphthylene (Scheme 2.124)

Structure	Thermal Behavior
31a	Cr (290 Col 325 N) 340 I
31b	Cr (229 Col 258 N) 265 I
31c	Cr (183Col 208 N) 242 I
31d	Cr (177 N) 195 I
31e	Cr (132 N) 158 I
31f	Cr 121 I
31g	Cr 119 I
31h	Cr 115 I
31i	Cr 280 I
31j	Cr (224 N) 300 I
31k	Cr (147 N) 172 I
31l	Col 180 N 221 I
31m	Cr (154 Col 192 N) 249 I
31n	Cr (179 Col 194 N) 230 I
31o	Col 130 N 145 I

Source: Data from Bock, H. et al., *Chem. Phys. Chem.*, 532, 2002.

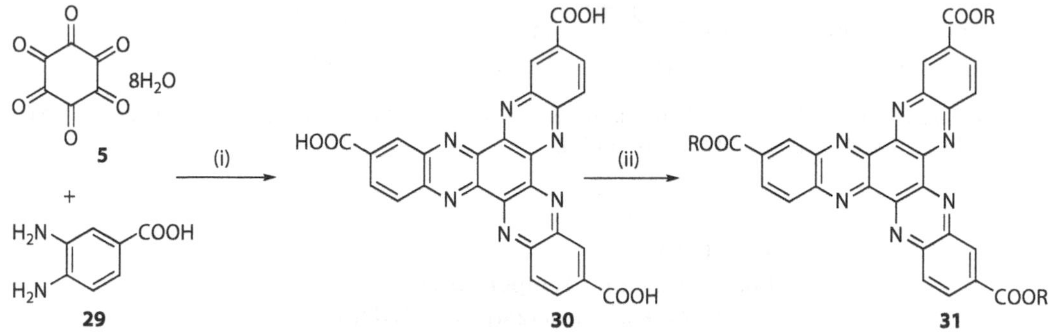

31a–h: R = C_nH_{2n+1}, n = 2–9
31i: CH(Et)CH₂CH₃
31j: CH₂CH(CH₃)CH₃
31k: CH₂CH(Et)CH₂CH₃
31l: CH₂CH₂OCH₂CH₃
31m: CH₂CH₂OCH(CH₃)CH₃
31n: CH₂CH₂CH(OCH₃)CH₃
31o: CH(CH₂OC₂H₅)₂

SCHEME 2.124 Synthesis of triesters of HATNA **31**: (i) AcOH; (ii) SOCl₂, ROH.

2.18 HETEROTRUXENES

2.18.1 OXATRUXENE

Oxatruxene (benzo [1,2-b: 3,4-b′: 5,6-b″]trisbenzofuran) core is structurally similar to that of truxene. Like truxene this is also a heptacyclic C_3 symmetric aromatic hydrocarbon. The three methylene bridges of truxene are replaced by oxygen atoms to obtain trioxatruxene. The first information on trioxatruxene-based mesomorphic materials appeared in 1982 [1186]. Trioxatruxene can be prepared in a similar way to that of truxene, as shown in Scheme 2.125 [1187]. 3,4-dimethoxyphenol 1 was treated with ethyl bromoacetate in the presence of potassium carbonate. This was followed by the hydrolysis of the product with ethanolic potassium hydroxide to afford 3,4-dimethoxyphenoxyacetic acid 2. The polyphosphoric acid (PPA)-catalyzed cyclization of 2 followed by the polyphosphate ethyl ester (PPE)-mediated self-condensation of resulting 3,4-dimethoxy-2H-benzofuran-3-one 3 gave hexamethoxytrioxatruxene 4. The demethylation of 4 followed by the in situ esterification of hexahydroxytrioxatruxene with alkanoyl chloride in pyridine afforded the desired hexa-n-alkanoyloxytrioxatruxene 5.

Like their truxene counterparts, the hexaesters of trioxatruxene 5 exhibited complex polymorphism (Table 2.103). It should be noted that all these compounds displayed an inverted nematic-columnar phase sequence. All derivatives showed enantiotropic rectangular columnar and monotropic nematic phases. Higher chain length esters (5f–h) displayed an enantiotropic hexagonal columnar phase at higher temperature in addition to Col$_r$ and monotropic nematic phases at lower temperature. It is interesting to note that the exchange of the three methylenes of the truxene by oxygen atom leads to the appearance of a new oblique columnar phase. Surprisingly, not much work has been done on this C_3 symmetric tripodal molecule. It would be interesting to synthesize various hexaethers, benzoates, etc., of oxatruxene core to study structure–property relationships.

2.18.2 THIATRUXENE

Trithiatruxene (benzo [1,2-b: 3,4-b′: 5,6-b″]trisbenzothiophene) core can be achieved by the heteroatomic substitution of the three methylene groups of the truxene core by sulfur atoms. The

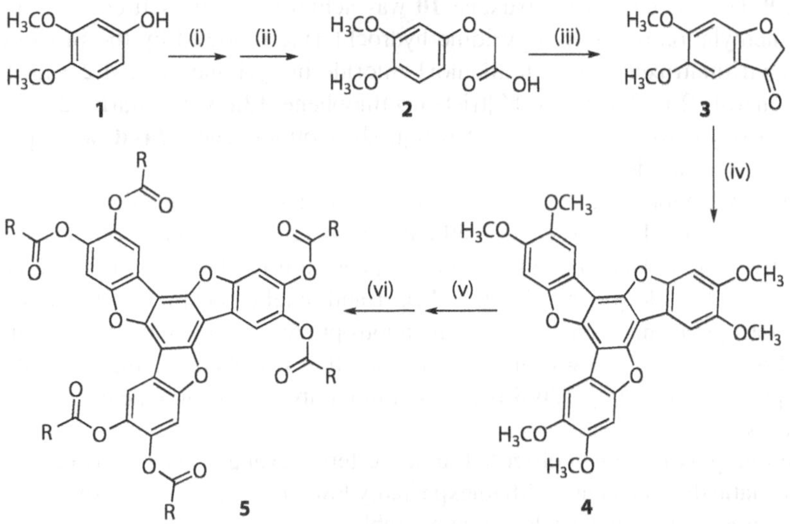

SCHEME 2.125 Synthesis of trioxatruxene hexaesters: (i) BrCH$_2$COOEt, K$_2$CO$_3$, acetone, reflux; (ii) KOH, EtOH, reflux; (iii) PPA, 110°C; (iv) PPE, 140°C; (v) Py-HCl, 240°C; (vi) RCOCl, pyridine.

TABLE 2.103
Thermal Behavior of
Hexaalkanoyloxytrioxatruxene Derivatives
(Scheme 2.125)

Structure	R	Phase Transition
5a	C_7H_{15}	Cr 96 (89 N_D) Col$_r$ 194 I
5b	C_8H_{17}	Cr 90 (75 N_D) Col$_r$ 197 I
5c	C_9H_{19}	Cr 82 (67.5 N_D) Col$_r$ 192 I
5d	$C_{10}H_{21}$	Cr 76 (61.5 N_D) Col$_r$ 194 I
5e	$C_{11}H_{23}$	Cr 78 (58 N_D 64 Col$_{ob}$) Col$_r$ 184 I
5f	$C_{12}H_{25}$	Cr 78 (59 N_D 64 Col$_{ob}$) Col$_r$ 172 Col$_h$ 177 I
5g	$C_{13}H_{27}$	Cr 74 (58 N_D 71 Col$_{ob}$) Col$_r$ 158 Col$_h$ 166 I
5h	$C_{15}H_{31}$	Cr 80 (60 N_D) Col$_{ob}$ 84 Col$_r$ 138 Col$_h$ 152 I

Source: Data from Destrade, C. et al., *Mol. Cryst. Liq. Cryst.*, 114, 139, 1984.

rich polymorphism observed in truxene and trioxatruxene derivatives along with the reentrant nematic phase found in truxene derivatives provided motivation to study the effect of replacement of methylene by sulfur atom on the mesomorphic properties. Moreover, the presence of sulfur atom is expected to increase the electron donor property of the resultant nucleus [1188,1189]. The trithiatruxene nucleus can be assembled following the methodology developed for the synthesis of truxene and trioxatruxene, as shown in Scheme 2.126. Compound **6** (4-aminoveratrol) was treated with 4 M HCl, and to this cold suspension (0°C) was added cold solution of NaNO$_2$ in water. This diazonium salt solution was slowly added to warm [50°C] solution of potassium ethyl xanthate in water. The xanthate was now refluxed with KOH in ethanol. The resulting slurry was dissolved in water and acidified with 15N H$_2$SO$_4$ to afford 3,4-dimethoxythiophenol **7**. The reaction of **7** with ethylbromoacetate in the presence of NaOEt in EtOH solution, followed by refluxing in ethanolic KOH and acidification with cold solution of 3M HCl furnished 3,4-dimethoxyphenylthioacetic acid **8**. Polyphosphoric acid (PPA)–mediated ring closure of **8** gives 5,6-dimethoxy-2H-benzothiophene-3-one, **9**. Hexamethoxytrithiatruxene **10** was achieved by the self-condensation of **9** with PPE. The demethylation of **10** with pyridine hydrochloride followed by the in situ esterification of hexahydroxytrithiatruxene **11** with alkanoyl chloride in pyridine yielded 2,3,7,8,12,13-hexa-*n*-alkanoyloxybenzo-[1,2-b: 3,4-b': 5,6-b'']trisbenzothiophene **12a–i**. Similarly, 2,3,7,8,12,13-hexa-*n*-alkoxybenzoyloxybenzo-[1,2-b: 3,4-b': 5,6-b']trisbenzothiophene **13a–d** were prepared using substituted benzoyl chlorides.

The thermal behavior of all the hexa-*n*-alkanoyloxytrithiatruxene discotics is presented in Table 2.104. All the alkyl hexaesters of trithiatruxenes **12a–i** exhibited complex polymorphism. All the derivatives showed a rectangular columnar phase followed by a hexagonal columnar phase. While the higher chain length alkyl esters **12h–i** and smaller chain length alkyl esters **12b–c** displayed monotropic nematic phase, it was enantiotropic for intermediate peripheral chain length compounds, **12d–g**. This phase was absent in the smallest peripheral chain length derivative **12a**. The nematic phase exhibited by **12b–i** is inverted in nature, that is, it appeared below the ordered columnar phases.

Unlike the rich polymorphism observed in hexaesters, hexabenzoates of trithiatruxenes **13** displayed only nematic discotic phase with unexpectedly low Cr-N_D transition enthalpy. Isotropic transition temperatures of these materials are comparable to those of hexa(alkoxybenzoyloxy)truxenes having the same peripheral chain length.

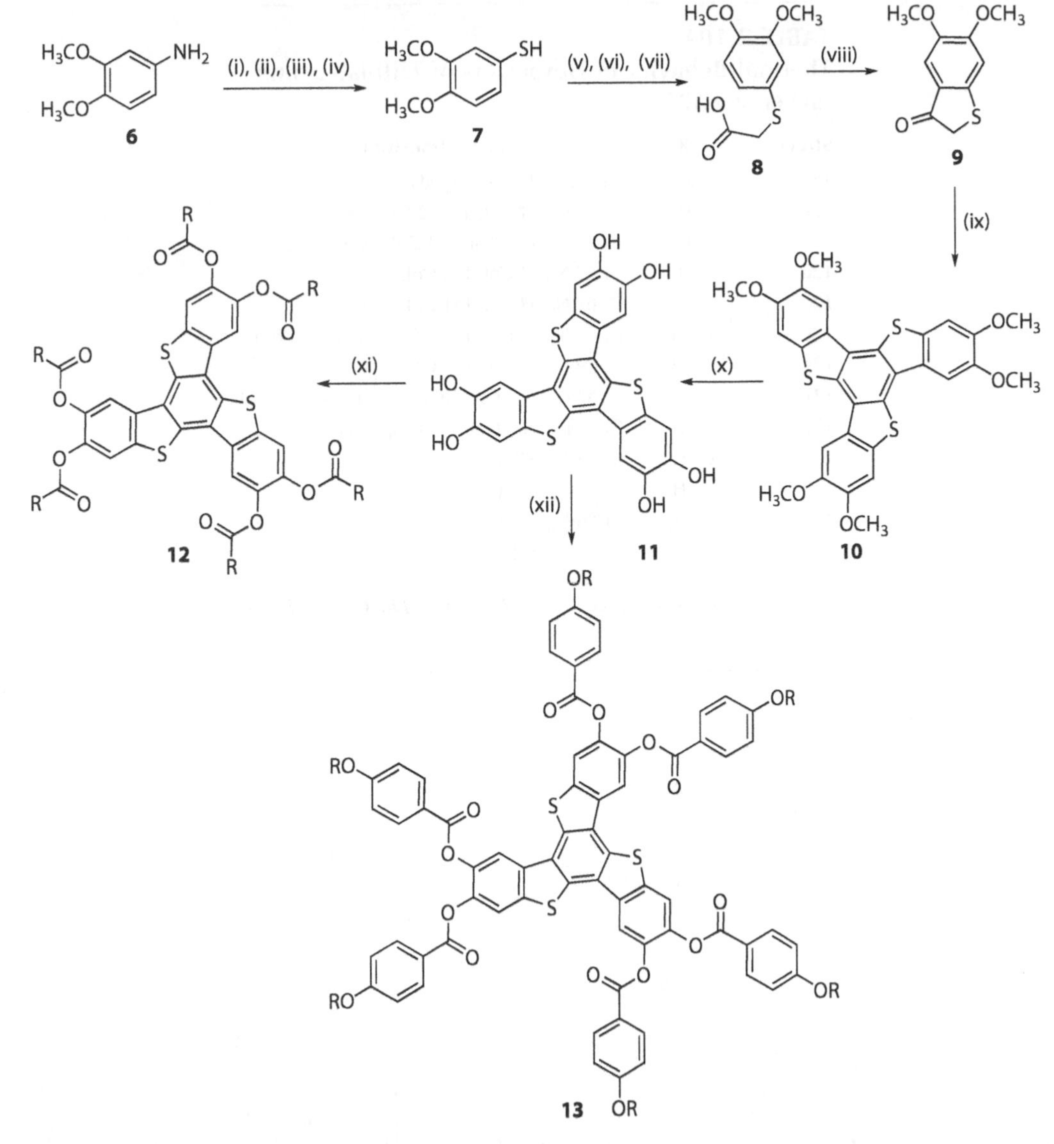

SCHEME 2.126 Synthesis of trithiatruxene hexaesters and benzoates: (i) NaNO$_2$, HCl, 0°C; (ii) KSCSOEt, H$_2$O, 50°C; (iii) KOH, EtOH, reflux; (iv) H$_2$SO$_4$, H$_2$O; (v) BrCH$_2$COOEt, NaOEt, EtOH, reflux; (vi) KOH, EtOH, reflux; (vii) HCl, H$_2$O; (viii) PPA, 110°C; (ix) PPE, 130°C; (x) Py-HCl, 240°C; (xi) RCOCl, Py; (xii) ROC$_6$H$_4$COCl, Py.

2.18.3 TRIINDOLE

Similar to truxene, oxatruxene, and thiatruxene cores, triindole (10,15-dihydro-5H-diindolo[3,2-a:3′,2′-c]carbazole) core is a heptacyclic C$_3$ symmetric aromatic hydrocarbon in which the central benzene ring is shared by three carbazole units. Thus, the heteroatom in this core is nitrogen. Multi-step routes [1190,1191] for the synthesis of triindole derivatives have been realized earlier; however, these methods are complicated and low yielding. Recently, a straightforward method was developed by Robinson et al., which involves the oxidative trimerization of indole using bromine [1192]. The reaction of indole **14** in acetonitrile with three equivalents of bromine provided indole trimer **15** (Scheme 2.127). The hexaalkynyl derivatives **17** were designed to produce liquid crystalline materials

TABLE 2.104
Thermal Behavior of Hexaesters of Trithiatruxene
(Scheme 2.126)

Structure	R	Phase Transition
12a	C_6H_{13}	Cr 133 Col$_r$ 218 Col$_h$ 241 I
12b	C_7H_{15}	Cr 103 (92 N$_D$) Col$_r$ 212 Col$_h$ 236 I
12c	C_8H_{17}	Cr 90 (82 N$_D$) Col$_r$ 173 Col$_r$ 191Col$_h$ 229 I
12d	C_9H_{19}	Cr 87 N$_D$ 93 Col$_r$ 185 Col$_h$ 210 I
12e	$C_{10}H_{21}$	Cr 62 N$_D$ 98 Col$_r$ 151 Col$_h$ 193 I
12f	$C_{11}H_{23}$	Cr 64 N$_D$ 93 Col$_r$ 145.5 Col$_r$ 149 Col$_h$ 179 I
12g	$C_{12}H_{25}$	Cr 79 N$_D$ 87 Col$_r$ 136 Col$_h$ 191 I
12h	$C_{13}H_{27}$	Cr 82 (72 N$_D$) Col$_r$ 131.5 Col$_r$ 134 Col$_h$ 179 I
12i	$C_{14}H_{29}$	Cr 88 (81 N$_D$) Col$_r$ 121 Col$_r$ 125 Col$_h$ 178 I
13a	$C_{10}H_{21}$	Cr 83 N$_D$ >300 I
13b	$C_{12}H_{25}$	Cr 81 N$_D$ 295 I
13c	$C_{13}H_{27}$	Cr 86 N$_D$ 280 I
13d	$C_{16}H_{33}$	Cr 96 N$_D$ 241 I

Source: Data from Cayuela, R. et al., *Mol. Cryst. Liq. Cryst.*, 177, 81, 1989.

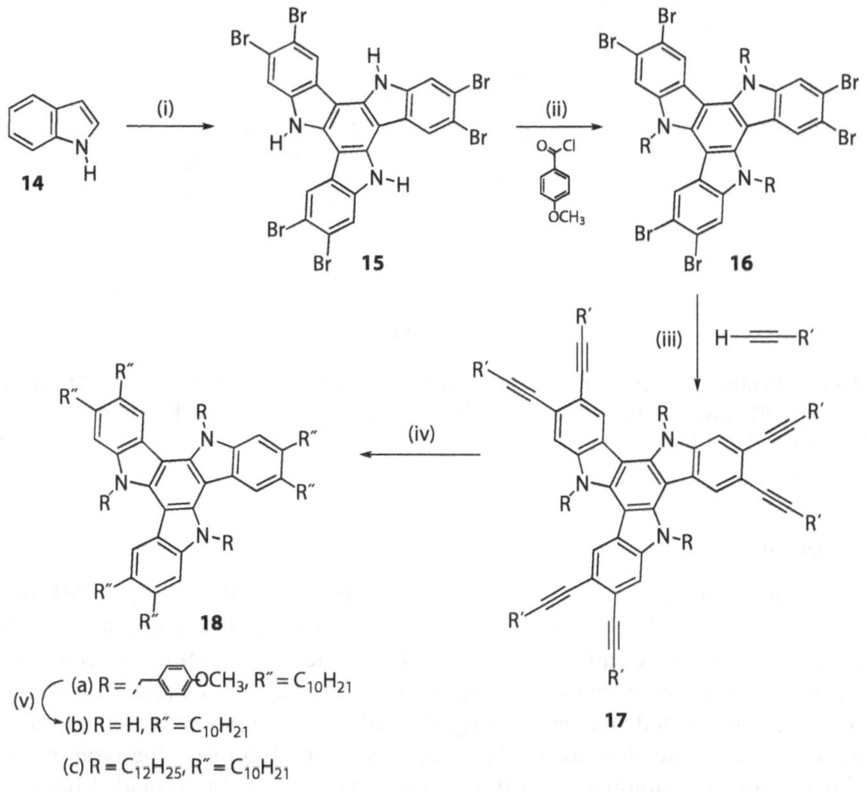

SCHEME 2.127 Synthesis of triindole derivatives: (i) Br$_2$, CH$_3$CN; (ii) NaH, THF; (iii) PdCl$_2$(dppf), CuI, Et$_3$N, THF, 110°C; (iv) HCOONH$_4$, Pd/C, EtOAc; (v) AlCl$_3$, CH$_2$Cl$_2$.

TABLE 2.105
Thermal Behavior of Triindole
Derivatives (Scheme 2.127)

Structure	Phase Transition	Ref.
18a	Cr 84.7 Col$_h$ 151.5 I	[1193,1194]
18b	Cr 76 Col$_h$ 154.8 I	[1193,1194]
18c	Cr 38.9 Col$_h$ 67.3 I	[1194]

as hexaalkynyl derivatives of benzene and triphenylene are well known to show discotic nematic phases. The amino groups of **15** were first protected with the benzyl group to avoid reaction complications during the Sonogashira coupling. The hexaalkynyl derivatives **17** were found to be non-mesomorphic. The reduction of the triple bonds yields liquid crystalline hexaalkyl derivative **18a**, which can be converted to another liquid crystalline derivative **18b** via a deprotection of the benzyl group [1193,1194]. The transition temperatures can be significantly lowered by the introduction of three additional alkyl chains at the nitrogen of trisindole (Table 2.105). A number of phenylethynyl triindoles have been reported to self-associate in solution through arene–arene interactions [1195].

2.19 TRICYCLOQUINAZOLINE CORE

The parent polyaza-polycyclic hydrocarbon tricycloquinazoline (TCQ) **1** is a potent carcinogen and has been extensively studied by biologists [1196–1198]. The high carcinogenicity of TCQ is probably due to its ability to intercalate into DNA [1199,1200] and this reflects its strong tendency to stacking and aggregation, as may also be suggested by its high melting point (323°C) [1201] and its crystal structure [1202]. However, the general observation that polysubstitution destroys the carcinogenic activity [1203,1204] suggests that peripherally substituted TCQ derivatives can be handled safely with adequate precautions. TCQ is readily formed in pyrolytic reactions of a number of anthranilic acid derivatives, for example, methyl anthranilate [1205]. In view of the wide occurrence of the latter in plant materials and the ease of TCQ production from them by combustion, it was considered essential to investigate its carcinogenic activity extensively.

From a materials science point of view, the TCQ molecule is very attractive for a variety of reasons. It possesses a trigonal symmetry, its derivatives are colored, it shows extraordinary thermal and chemical stability, it sublimes without decomposition under atmospheric pressure at very high temperatures, it tolerates strong oxidants such as chromic anhydride in concentrated sulfuric acid, it is highly resistant to biological oxidation, it does not couple with diazotized arylamines, and it exhibits low ionization potential and interesting spectroscopic and electronic properties [1206–1210]. TCQ forms charge-transfer complexes with DNA and acts as an electron acceptor [1200]. Due to its electron-deficient nature, it can be doped with a reducing agent and can be used as an *n*-type organic semiconductor.

TCQ was synthesized by Cooper and Partridge during investigations on cyclic amidines [1211]. The cyclotrimerization of *o*-aminobenzaldehyde or *o*-aminonitrile or anthranil can easily yield TCQ nucleus [1211–1213]. The trimerization of a disubstituted anthranil in the presence of ammonium acetate in sulfolane-acetic acid has been used to prepare various DLCs [268,293,422,1214–1222]. In general, this trimerization is a poor-yielding process. The addition of ammonium acetate periodically during the reaction improves the yield marginally.

Two series of TCQ discotics were prepared. The first series was concerned with the 2,3,7,8,12,13-hexathioalkyl-TCQ derivatives with alkyl side-chain length varying from 3 to 18 carbon atoms [422,1214–1217]. The synthesis of these derivatives is depicted in Scheme 2.128. The nitration of 3,4-dichlorotoluene, **2**, yielded 3,4-dichloro-2-nitrotoluene, **3**, which on oxidation with chromium trioxide in acetic anhydride afforded 3,4-dichloro-2-nitro-α-α-diacetoxytoluene, **4**. The hydrolysis of **4** furnished 3,4-dichloro-2-nitrotbenzaldehyde, **5**. Compound **5** can also be prepared directly from **3** in a single step using ceric ammonium nitrate. The partial reduction of **5** yielded anthranil **6**, which can be trimerized to hexachloro-TCQ, **7**. Hexabromo-TCQ can also be prepared in the same manner but in lower yield. Treatment of hexachloro- or hexabromo-TCQ with alkylthiolate in DMF or NMP at 100°C afforded hexathioethers of TCQ, **8**. It should be noted that although both hexachloro- and hexabromo-TCQ are insoluble in either DMF or NMP, they react readily with alkylthiolates and give rise to soluble products. 2,6,10-trithioalkyl-TCQ derivatives were also prepared following the same route starting from 4-chloro- or 4-bromotoluene [1214]. The mesophase behavior of hexathioalkyl-TCQ discotics is summarized in Table 2.106. All derivatives form a single hexagonal columnar mesophase over a wide temperature range. While the melting transition of all the members of the series was found to be irreversible, the clearing transition shows highly reversible behavior. X-ray diffraction studies carried out on some members confirm the presence of a hexagonal columnar phase with a core–core distance of about 3.4–3.5 Å, which is typical of columnar discotics having aromatic cores. The corresponding trialkylthio-TCQ derivatives were found to be non-liquid crystalline [1214].

The second series was based on 2,3,7,8,12,13-hexaalkoxy-TCQ derivatives [268,293,1218–1220]. The nucleophilic substitution of hexahalo-TCQ with alkoxide could produce TCQ hexaethers; however, this reaction with hexachloro-TCQ failed to produce hexaalkoxy-TCQ derivatives. Therefore, these derivatives were prepared from hexahydroxy-TCQ, as shown in Scheme 2.129.

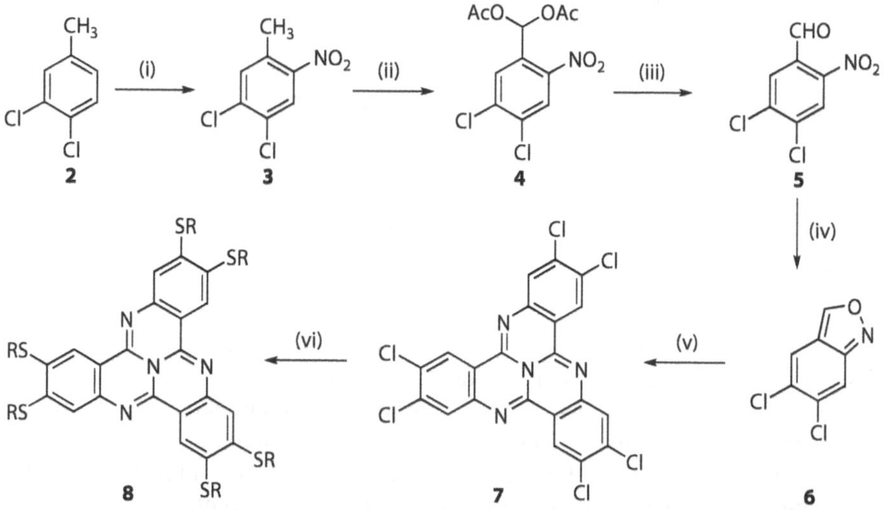

SCHEME 2.128 Synthesis of hexaalkylthio-TCQ discotics: (i) AcOH, H_2SO_4, HNO_3, 0°C–5°C; (ii) Ac_2O, H_2SO_4, CrO_3, 0°C–5°C; (iii) EtOH, H_2O, HCl, reflux; (iv) AcOH, Sn; (v) sulfolane, AcOH, NH_4OAc, 150°C, 7 h; (vi) RSK, NMP or DMF, 100°C.

TABLE 2.106
Thermal Behavior of TCQ-Hexathioethers

Structure	R	Phase Transition	Ref.
8a	C_3H_7	Cr 159.7 Col_h 274.3 I	[1214]
8b	C_4H_9	Cr 155.5 Col_h 240.9 I	[1214]
8c	C_5H_{11}	Cr 113.7 Col_h 223.2 I	[1214]
8d	C_6H_{13}	Cr 94.8 Col_h 213.5 I	[1214]
		Cr 78.0 Col_h 213.0 I	[1215]
8e	C_7H_{15}	Cr 69.9 Col_h 211.7 I	[1216]
8f	C_8H_{17}	Cr 80.7 Col_h 207.4 I	[1214]
8g	$C_{12}H_{25}$	Cr 53.9 Col_h 170.3 I	[1214]
8h	$C_{16}H_{33}$	Cr 56.5 Col_h 152.0 I	[1214]
8i	$C_{18}H_{37}$	Cr 92.2 Col_h 183.7 I	[1214]

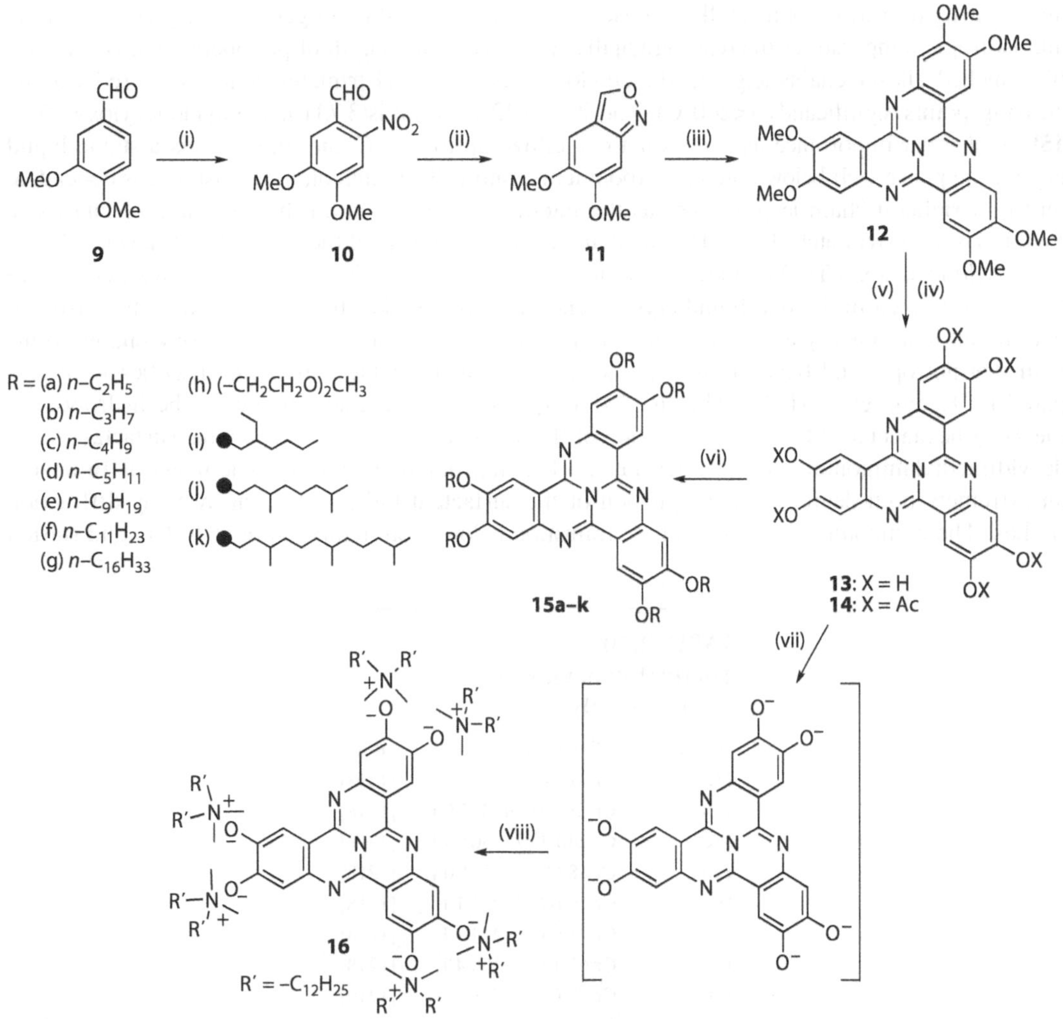

$R =$ (a) n-C_2H_5 (h) (–$CH_2CH_2O)_2CH_3$
(b) n-C_3H_7
(c) n-C_4H_9 (i) ●
(d) n-C_5H_{11}
(e) n-C_9H_{19} (j) ●
(f) n-$C_{11}H_{23}$
(g) n-$C_{16}H_{33}$ (k) ●

$R' = $ –$C_{12}H_{25}$

SCHEME 2.129 Synthesis of hexaalkoxy-TCQ discotics: (i) HNO_3; (ii) AcOH, Sn; (iii) sulfolane, AcOH, NH_4OAc, reflux; (iv) pyridine, HCl, 220°C; (v) pyridine, Ac_2O; (vi) KOH, DMSO, RBr; (vii) NaOH; (viii) surfactant.

Veratraldehyde, **9**, was nitrated to 2-nitroveratraldehyde, **10**, at low temperature in high yield. The partial reduction of **10** using tin foil and acetic acid produced 5,6-dimethoxyanthranil, **11**, which was successfully trimerized in the presence of ammonium acetate in refluxing sulfolane-acetic acid to produce hexamethoxy-TCQ, **12**, in low yield. The demethylation of all six methoxy groups of **12** was carried out in molten pyridinium hydrochloride at elevated temperatures, affording hexahydroxy-TCQ, **13**, in about **50%** yield. The hexaalkylation of **13** was achieved in low to moderate yields using an excess of the appropriate *n*-bromoalkane in KOH/DMSO. The yield of final product **15** depends upon the purity of hexaphenol **13**, which is sensitive to air oxidation and, therefore, a stored hexa-hydroxy-TCQ generally resulted in poor yield of **15**. The in situ conversion of hexaphenol **13** into its hexaacetate **14** followed by the direct alkylation of hexaacetate improved the yield of hexaalkoxy-TCQ significantly [268,293]. Efforts have also been made to introduce the desired alkoxy chains prior to the trimerization step. Thus, the trimerization of 5,6-bis[2-(2-methoxyethoxy)ethoxy]anthra-nil produced 2,3,7,8,12,13-hexakis[2-(2-methoxyethoxy)ethoxy]-TCQ **15h** in 18% yield [1219].

The thermal behavior of hexaalkoxy-TCQ discotics is summarized in Table 2.107. Similar to hexaalkylthio series, all hexaalkoxy-TCQ discotics also display a single hexagonal columnar meso-phase over a wide temperature range. However, unlike the hexaalkylthio series, both the melting and clearing transitions of hexaalkoxy series exhibited reversibility. In general, the phase transition and isotropic temperatures decrease gradually with increasing length of peripheral chains. The use of branched alkoxy chains (e.g., 3,7-dimethyloctyloxy and 3,7,11-trimethyldodecyloxy) reduces the clearing points significantly (<200°C). The 2,3,7,8,12,13-hexakis(3,7,11-trimethyldodecyloxy)-TCQ **15k** is of great importance as it does not crystallize up to −50°C and, thus, forms a stable liquid crystalline phase well below and above room temperature. The intercolumnar distance is dependent on the peripheral chain length, whereas the interdisk distance within the discotic molecules was found to be approximately 3.4 Å. The small angle x-ray scattering of the 2,3,7,8,12,13-hexakis(3,7,11-trimethyldodecyloxy)-TCQ displays a core to core distance of 3.29 Å, one of the lowest core–core separation known in discotic liquid crystals and, therefore, makes the molecule very attractive for one-dimensional energy and charge migration. The TCQ discotic with ethyleneoxy chains forms both thermotropic and lyotropic mesophases [1219]. Water soluble ionic TCQ discotics were pre-pared by Kadam et al. [1221]. The charged TCQ core was generated in situ by the hydrolysis of the TCQ-hexaacetate **14**. The complexation of the TCQ hexaanion with the double surfactant dido-decyldimethylammonium bromide generated a lamellar liquid crystalline ionic material **16** having an extremely high degree of interdigitation of the surfactant tails, as evident by x-ray diffraction studies. The compound displayed a phase transition below room temperature at 10.1°C and started

TABLE 2.107
Thermal Behavior of
TCQ-Hexaethers

Structure	Phase Transition	Ref.
15a	Cr 315 I	[1218]
15b	Cr 286.0 Col$_h$ 305.6 I	[1218]
15c	Cr 240.2 Col$_h$ 305.9 I	[1218]
15d	Cr 187.8 Col$_h$ 301.0 I	[1218]
15e	Cr 79.6 Col$_h$ 237.1 I	[1218]
15f	Cr 71.9 Col$_h$ 214.7 I	[1218]
15g	Cr 67.1 Col$_h$ 166.4 I	[1218]
15h	Cr 77 Col$_h$ 233 I	[1219]
15I	Cr 152.8 I	[293]
15j	Cr 117.9 Col$_h$ 190.8 I	[293]
15k	Col$_h$ 143.1 I	[293]

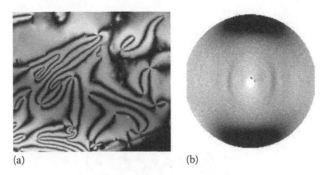

FIGURE 1.20 Typical (a) polarizing optical microscopic Schlieren texture and (b) x-ray diffraction pattern of an aligned sample of discotic nematic LCs.

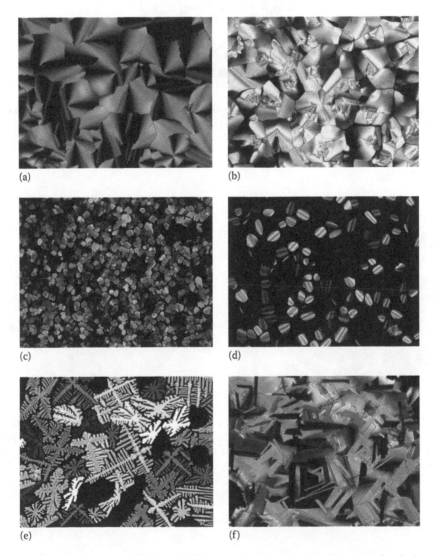

FIGURE 1.26 Optical photomicrographs of different columnar phases under crossed polarizers: (a) focal conic, (b) pseudo focal conic, (c) mosaic, (d) texture with rectilinear defects of Col_h phases, (e) dehydritic texture, and (f) texture of helical (H) phase with crystalline order.

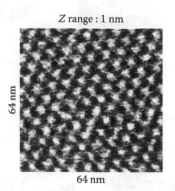

Z range : 1 nm

64 nm

64 nm

FIGURE 1.47 STM image of a triphenylene discotic on HOPG surface. (Reproduced from Gupta, S.K. et al., *J. Phys. Chem. B*, 113, 12887, 2009. With permission from ACS.)

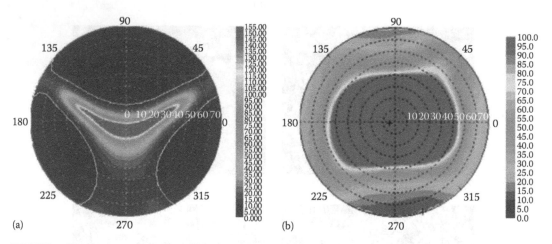

(a)

(b)

FIGURE 6.4 Measured iso-CR plots for TN-LCDs without (a) and with (b) the discotic optical compensation films. Clearly, there is a remarkable widening of the viewing angle characteristics of the TN-LCD with the negative optical compensation film. (Reproduced from Mori, H., *J. Display. Tech.*, 1, 179, 2005. With permission. Copyright @ 2005 IEEE.)

decomposition at 125°C. The corresponding single-tail derivative could not be prepared in pure form while the corresponding longer chain double-tailed derivatives did not display mesomorphism [1221].

All the TCQ derivatives were highly fluorescent absorbing at about 246, 282, 322, 400, 422, 450, and 482 nm with very intense emission at 522, 563, 609, and 662 nm [1218]. A significant shift in the absorption and fluorescence spectra was realized in different TCQ derivatives [1214,1216,1218,1219]. Cyclic voltametry and NMR studies on 2,3,7,8,12,13-hexakis(2-(2′-methoxyethoxy)ethoxy-TCQ indicated its aggregation in solutions [1220]. Due to the low reduction potential of TCQ discotics, they can be doped with electron donor and, accordingly, compound **8d** was doped with 6 mol % of potassium metal [1215]. The doped system displayed conductivity 2.9×10^{-5} S m^{-1} in the columnar phase with an anisotropy ($\sigma_{\parallel}/\sigma_{\perp}$) of 518. In order to enhance the level of potassium doping, TCQ discotic with polar peripheral chains **15h** was prepared, which displayed substantially higher conductivity upon 10 mol % potassium doping [1219]. Time-of-flight charge transport studies on **15h** exhibited electron mobility as $\sim 10^{-4}$ cm^2 V^{-1} s^{-1} at 150°C. The mobilities for the positive and negative carriers in room-temperature discotic TCQ derivative **15k** were estimated to be 2×10^{-2} and 4×10^{-2} cm^2 V^{-1} s^{-1}, respectively, at room temperature (20°C) [1222]. TCQ-thioethers have been shown to self-assemble face-on from dilute solution onto gold surfaces. These self-assembled monolayers were characterized by scanning tunneling microscopy, infrared spectroscopy, quartz crystal microbalance, and atomic force microscopy [422,1217].

A novel homologous series of covalently linked disk–rod oligomers containing six cyanobiphenyl-based rod-shaped moieties attached radially to a central tricycloquinazoline core via flexible alkyl spacers (Figure 2.35) was reported by Bisoyi et al. [330]. These materials were prepared by alkylating TCQ hexaacetate with ω-bromo-substituted alkoxycyanobiphenyls using potassium hydroxide in dimethylsulfoxide. A nanophase segregated layered phase (SmA) with alternating calamitic and discotic layers was observed in these materials. Solvent crystallized compounds show phase transitions without any characteristic texture for enantiotropic compounds **17b–d** on heating; however, on cooling from the isotropic liquid, both enantiotropic and monotropic compounds exhibit fan-shaped textures characteristic of smectic A phase. Interestingly, compound **17a** exhibits an optically isotropic mesophase below the smectic phase. The thermal behavior of these oligomers is summarized in Table 2.108.

The crystal to mesophase and the mesophase to isotropic transition temperatures decrease, as anticipated, with an increase in the length of the methylene spacer connecting the discotic core to the calamitic peripheries. With increase in the spacer length, the compounds show a tendency to vitrify below the mesophase instead of crystallization. However, remarkable is the fact that the transition enthalpies of these compounds associated with the mesophase to isotropic transition are very high and are of the order of *ca.* 20 kJ mol^{-1}. Such high values of transition enthalpies suggest

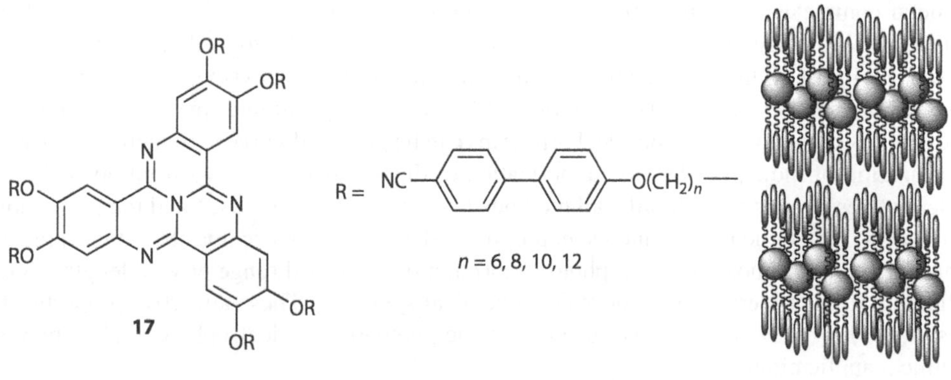

FIGURE 2.35 The proposed model for the nanophase segregated structure in the mesophase of the oligomers, wherein the disks and rods segregate into different sub-layers.

TABLE 2.108
Phase Transition Temperatures of
TCQ-Cyanobiphenyl Hybrids

Structure	n	Phase Transition
17a	6	Cr (173.4 SmA 169.5 Cub) 188.2 I
17b	8	Cr 131.0 SmA 161.7 I
17c	10	Cr 109.4 SmA 155.6 I
17d	12	Cr 93.5 SmA 147.4 I

Source: Data from Bisoyi, H.K. et al., *Chem. Commun.*, 7003, 2009.

the highly ordered organization of the mesogens in the mesophase. The suppression of the columnar and discotic nematic phase and the emergence of the lamellar phase in these disk–rod hybrid compounds suggest that the central discotic core acts like a connecting unit to the calamitic peripheral units. The substitution of six calamitic units on the central discotic core has sufficiently distorted the gross molecular shape to exhibit the calamitic mesophase. Probably because of strong recognition among the calamitic cyanobiphenyl units in the mesophase, the molecular shape is no longer circular. It has been distorted to a non-circular shape by the uneven distribution of peripheral substituents around it. The bigger core size and high propensity of TCQ core to stack up and aggregate also plays a role in exhibiting lamellar phase than a fluid nematic phase by intermolecular recognition. Besides geometric elements like molecular shape, various intra- and intermolecular interactions play a significant role in determining the supramolecular organization in disk–rod hybrid oligomers. On the basis of x-ray and microscopy data, a structure for the mesophase of these compounds (Figure 2.35) is proposed. In this SmA structure, the rods and disks microphase separate into alternating layers.

2.20 PORPHYRIN CORE

Porphyrins are biochemically important, medically useful, and synthetically interesting compounds. Porphyrins are natural products and play a major role in life-sustaining biochemical reactions. These are a ubiquitous class of compounds with many important biological representatives, including hemes, chlorophylls, myoglobins, cytochromes, catalases, peroxidases, and several others. They are described as "pigments of life" because of their involvement in various vital processes in the living systems [1223]. Their aromatic character, inner chelating pocket, and varying peripheral carbon chains have allowed scientists to discover new and unique chemical reactions. Porphyrins and their metal complexes have also stirred interdisciplinary interest due to their intriguing physical, chemical, and biological properties [1224–1227]. Because of their biological and materials science importance, almost all the metals have been incorporated in the porphyrin nucleus (central cavity). The porphyrin macrocycle consists of four pyrrole rings joined by four inter-pyrrolic methine bridges to give a highly conjugated macrocycle. Porphyrins can be prepared via several methods, such as the tetramerization of monopyrroles, the dimerization of dipyrromethanes, and from open chain pyrrolic derivatives. However, generally, all methods give only moderate yields and the purification of the product is often tedious. The interest in porphyrin derivatives for organic electronic applications originates from their photo-stability, photo-absorption over a broad range of wavelengths, interesting photophysical properties, and convenient chemical synthesis. These attractive properties make them suitable for many applications, such as organic photovoltaic, electrophotographic, photoelectrochemical applications, etc.

Mother Nature uses the principle of "order and mobility" for various specific functions. Obviously, it is of great fundamental importance to incorporate porphyrins into self-organizing supramolecular liquid crystalline systems. Porphyrin derivatives form both thermotropic as well as lyotropic liquid crystals. Among the lyotropic liquid crystalline compounds, porphyrins belong to the category of chromonic metallomesogens [1228]. The porphyrin complexes **1** of copper(II), cobalt(II), zinc(II), nickel(II), etc., form lyomesophases at very low dilution and in basic conditions [1229–1232]. Both hexagonal as well as nematic columnar phases appear at different concentrations. The stability of the mesophases results from π–π interactions between successive macrocycles with the possible contributions from hydrogen bonding between the sulfonyl moieties in polar solvents. So it implies that the number and position of the polar substituents have an impact on the mesomorphism. Some of the thermotropic discotic mesogens based on porphyrin also exhibit lyomesophases, that is, they behave as amphotropic liquid crystals.

X = H or Na
M = 2H, Ni, Cu, Co, Zn

A large number of compounds based on porphyrins are known to display thermotropic mesophases. About 300 discotic liquid crystalline derivatives derived from porphyrin nucleus have already been reported. These may be broadly divided into the following categories: β-substituted porphyrin derivatives, *meso*-substituted porphyrin derivatives, and miscellaneous porphyrin derivatives.

2.20.1 β-SUBSTITUTED PORPHYRIN DERIVATIVES

The thermotropic mesomorphism of porphyrins was first reported in 1980 by Goodby et al. [1233]. The uro-porphyrin I octa-*n*-dodecyl ester **2** was prepared from commercially available uro-porphyrin I dihydrochloride by acid-catalyzed esterification. It exhibited a monotropic columnar mesophase over a very narrow range of temperature, that is, 0.1°C.

Cr 107 I (I 96.8 Col 96.7 Cr)

The work was followed by Gregg et al., who synthesized a series of octaesters and octaethers of porphyrin [1234,1235]. The porphyrin octaesters and their metal complexes were synthesized using the methodology shown in Scheme 2.130. The Knorr method of pyrrole synthesis [1236] was

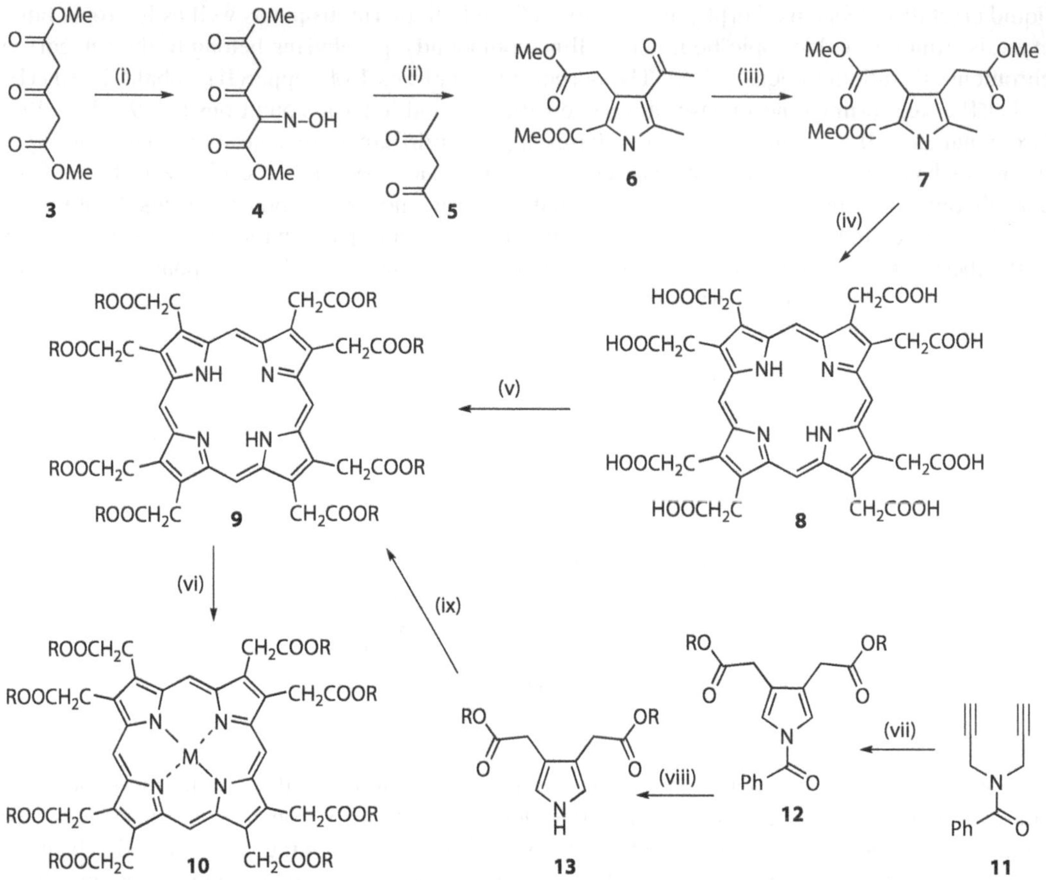

SCHEME 2.130 Synthesis of porphyrin octaesters: (i) NaNO$_2$/H$_2$O; (ii) Zn/AcOH; (iii) Tl(NO$_3$)$_3$/MeOH, HClO$_4$; (iv) Pb(OAc)$_4$, AcOH; 20% HCl (reflux); O$_2$; (v) ROH/H$_2$SO$_4$; (vi) MCl$_2$/M(OAc)$_2$, DMF; (vii) Pd/C/ KI/ROH, CO/O$_2$/DMA; (viii) NEt$_3$/MeOH; (ix) (CH$_2$O)$_n$, BF$_3$·Et$_2$O, DDQ.

applied to prepare the key precursor **6**. The reaction between 2,4 pentanedione **5** and the oxime **4** (generated in situ from the β-keto ester **3**), yields pyrrole **6**. Pyrrole trimethylester **7** was prepared by treating **6** with thalium nitrate followed by oxidation with lead tetraacetate. Porphyrin molecule **8** was obtained by refluxing **7** with 20% hydrochloric acid. The porphyrin octaesters of different chain length **9** can be achieved by esterifying the octaacid **8** with appropriate alcohols. Metallated derivatives **10** were prepared by heating the metal-free compound **9** either with the metal chloride or metal acetate.

A simplified method to produce the octaesters has recently been reported by Paganuzzi et al. [1237]. The long alkyl chain-substituted pyrrole diesters **13** can be easily prepared from dipropargylamine **11** (Scheme 2.130). The palladium catalyzed oxidative alkoxycarbonylation of **11** with simultaneous cyclization produces the protected pyrroles **12**, which can be easily deprotected upon treatment with triethylamine in methanol. The free pyrrole on condensation with paraformaldehyde in the presence of BF$_3$.Et$_2$O generates porphyrins. This method works well for a wide spectrum of *n*-alkyl chains in the range of 3–14 carbon atoms. A number of β-substituted porphyrins have been prepared and studied for various physical properties [1234–1251].

The thermal behavior of all the metal-free porphyrin octaesters **9** and their metal complexes is presented in Table 2.109. In general, metal-containing derivatives, with the exception of very large size metal atoms like lead, are more stable than metal-free porphyrin octaesters. Because

TABLE 2.109
Thermal Behavior of Metal-Free Porphyrin
Octaesters 9 and Their Metal Complexes 10
(Scheme 2.130)

Structure	R	M	Thermal Behavior	Ref.
9.1	C_3H_7	—	Cr 249 I	[1237]
9.2	C_3H_6Cl	—	Cr 250 Col 262 Col 275 I	[1238]
9.3	C_4H_9	—	Cr 178Col 222 I	[1234]
9.4	C_4H_8Cl	—	Cr 180 Col 200 I	[1238]
9.5	C_5H_{11}	—	Cr 135 Col 200 I	[1238]
9.6	C_6H_{13}	—	Cr 61 Col 136 Col 232 I	[1234]
9.7	$C_6H_{12}Cl$	—	Cr 146 I	[1238]
9.8	C_8H_{17}	—	Cr 96 Col 99 Col 166 I	[1234]
9.9	$C_{10}H_{21}$	—	Cr 110 Col$_r$ 159 I	[1237]
9.10	$C_{11}H_{23}$	—	Cr 106 Col$_r$ 166 I	[1237]
9.11	$C_{12}H_{25}$	—	Cr 116 Col$_r$ 160 I	[1237]
9.12	$C_{14}H_{29}$	—	Cr 117 Col$_r$ 140 I	[1237]
10.1	C_3H_7	Zn	Cr 297 I	[1237]
10.2	C_4H_9	Zn	Cr 184 Col 273 I	[1234]
10.3	C_4H_8Cl	Cu	Cr 155Col 242 I	[1238]
10.4	C_6H_{13}	Zn	Cr 61 Col 136 Col 232 I	[1234]
10.5	$C_6H_{12}Cl$	Cu	Cr 155 Col 178 I	[1238]
10.6	C_8H_{17}	Zn	Cr 91 Col 101 Col 208 I	[1234]
10.7	$C_{10}H_{21}$	Pt	Cr 111 Col$_r$ 210 I	[1237]
10.8	$C_{10}H_{21}$	Pd	Cr 101 Col$_r$ 203 I	[1237]
10.9	$C_{10}H_{21}$	Zn	Cr 110 Col$_r$ 194 I	[1237]
10.10	$C_{10}H_{21}$	Cu	Cr 101 Col$_r$ 191 I	[1237]
10.11	$C_{10}H_{21}$	Ni	Cr 104 Col$_r$ 168 I	[1237]
10.12	$C_{10}H_{21}$	Cd	Cr 111 Col$_r$ 177 I	[1239]
10.13	$C_{11}H_{23}$	Zn	Cr 99 Col$_r$ 186 I	[1237]
10.14	$C_{12}H_{25}$	Zn	Cr 113 Col$_r$ 189 I	[1237]
10.15	$C_{14}H_{29}$	Zn	Cr 114 Col$_r$ 180 I	[1237]

of the large size, lead ion does not fit in the porphyrin cavity and protrudes out of the plane. This distortion of planarity strongly disturbs the π–π stacking and consequently destroys mesomorphism. All the other derivatives are planar and exhibit columnar phases. The first member of the series with a propyl chain and its zinc complex are not liquid crystalline. However, the columnar phase can be induced by terminal heavy atom substitution. Thus, compound **9b** exhibits a columnar mesophase [1238]. The nature of the central metal ion affects the clearing temperatures by affecting the strength of π–π interactions [1239]. The influence is of both electrostatic and geometric in nature.

Gregg et al. prepared a series of porphyrin-based octaether discotic liquid crystals [1235] by modifying the synthetic methodology used for the synthesis of octaesters. The synthetic route is outlined in Scheme 2.131. The pyrrole trimethyl ester **7** on selective hydrolysis using sodium methoxide in methanol yields the diacid monoester **14**, which on reduction by BH$_3$/THF followed by tosylation resulted in the ditosylated derivative **16**. The ditosylate derivative on heating with the appropriate alcohol forms the corresponding diether **17**, which on oxidation with lead tetraacetate led to compound **18**. Pyrrole **18** was converted to porphyrin **19** in one pot synthesis using KOH to hydrolyze the ester, hydrobromic acid for decarboxylation and cyclization, and chloranil

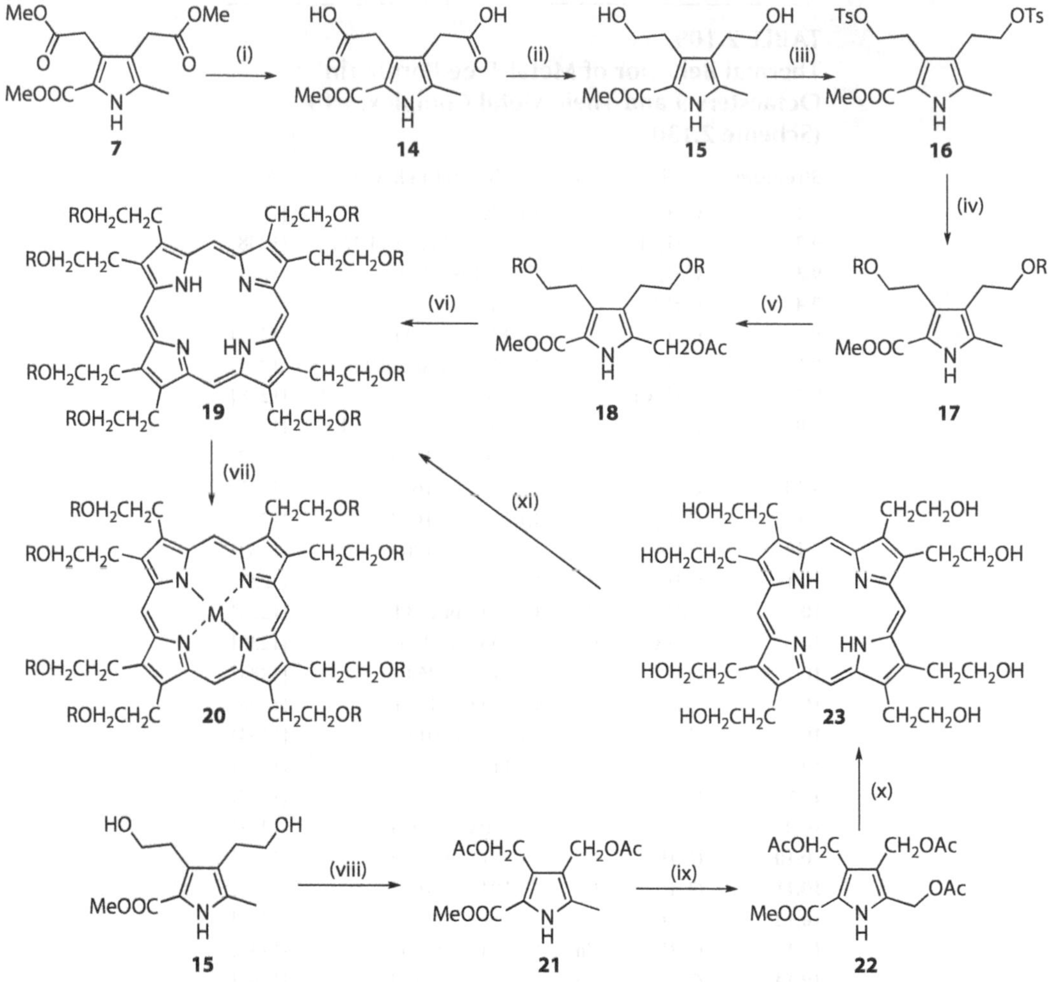

SCHEME 2.131 Synthesis of octaethanol porphyrin and its octaethers: (i) NaOMe/MeOH; (ii) BH₃/THF; (iii) TsCl/pyridine; (iv) ROH/toluene; (v) Pb(OAc)₄; (vi) (a) KOH/EtOH; (b) HBr/EtOH; (c) chloranil; (vii) M(OAc)₂, CH₂Cl₂/MeOH, reflux; (viii) AcCl/NEt₃; (ix) Pb(OAc)₄; (x) (a) KOH/EtOH; (b) HBr/EtOH; (c) O₂; (xi) NaH/DMSO, RX.

for the oxidation of the macrocycle to the porphyrin discotics. Alternatively, these porphyrin ethers can also be prepared from the porphyrin octaethanol **23**, which in turn can be generated from pyrrole **15**.

The thermal behavior of all the metal-free and metallated porphyrin octaethers **19** and **20** is summarized in Table 2.110. All the octaethers exhibit lower isotropic temperatures than the corresponding octaesters. This may be attributed to the absence of polar ester groups. None of the free bases, but the octyl derivative, displays mesomorphism. On the other hand, all the metallated derivatives display columnar phases. Within the octyl series, the clearing points follow the sequence $Zn > Cd > Cu > Pd > 2H$, but the mesophase width is in the following order: $Zn > Cu > Pd > Cd > 2H$. In comparison, the octaesters with decyl periphery show the following thermal stability: $Pt > Pd > Zn > Cu > Cd > Ni > 2H$ and the mesophase range in the following order: $Pd > Pt > Cu > Zn > Cd > Ni > 2H$. Some theoretical models have been proposed to account for the role of metals on the mesophase stability of these materials [1239,1240]; however, it is difficult to rationalize the effects of metal size and electronic nature on mesomorphism.

TABLE 2.110
Thermal Behavior of Porphyrin Octaethers 19 and 20 (Scheme 2.131)

Structure	R	M	Thermal Behavior
19.1	C_4H_9	2H	Cr 154 I
19.2	C_6H_{13}	2H	Cr 111 I
19.3	C_8H_{17}	2H	Cr 84 Col 89 I
19.4	$C_{10}H_{21}$	2H	Cr 69 I
20.1	C_4H_9	Zn	Cr 159 Col 164 I
20.2	C_6H_{13}	Zn	Cr 114 Col 181 I
20.3	C_8H_{17}	Zn	Cr 107 Col 162 I
20.4	C_8H_{17}	Cu	Cr 84 Col 132 I
20.5	C_8H_{17}	Pd	Cr 89 Col 123 I
20.6	C_8H_{17}	Cd	Cr 103 Col 136 I
20.7	$C_{10}H_{21}$	Zn	Cr 86 Col 142 I

Source: Data from Gregg, B.A. et al., *J. Am. Chem. Soc.*, 111, 3024, 1989.

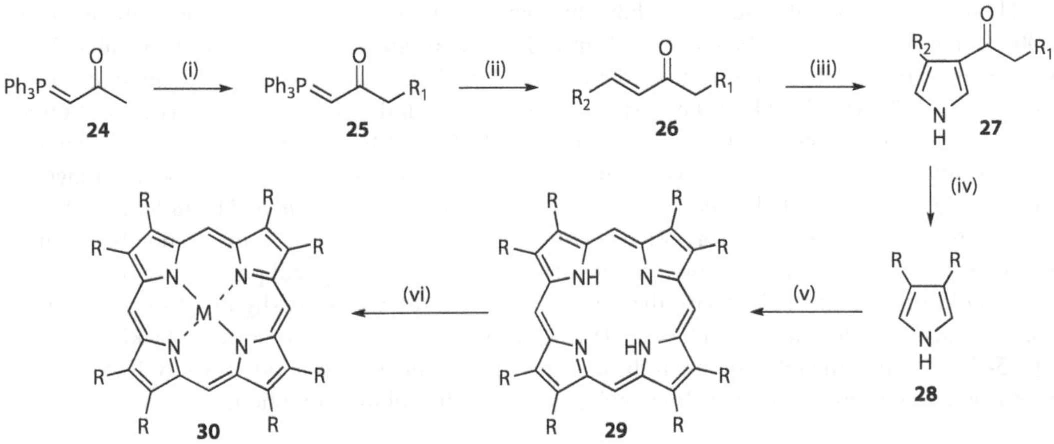

SCHEME 2.132 Synthesis of octaalkylporphyrins: (i) (a) *n*-BuLi, THF, (b) R_1-Br; (ii) R_2-CHO, DCM; (iii) tosylmethyl isocyanide, NaH, Et_2O, DMSO; (iv) $LiAlH_4$, Et_2O; (v) HCHO, HBr, EtOH, O_2; (vi) $M(OAc)_2$, MeOH, $CHCl_3$.

Recently, Shearman et al. synthesized a number of liquid crystalline octaalkyl porphyrins [1241], as shown in Scheme 2.132. The thermal behavior of these porphyrins is summarized in Table 2.111. Like other porphyrin discotics, the melting and the clearing points of these materials are also a function of chain length, and the metallation increases the clearing points of all the derivatives. The metal-free porphyrins are non-liquid crystalline whereas their Zn complexes display rectangular columnar phases.

The unsymmetrical β-substituted porphyrins **36**, **39** can be easily prepared from hemin **31**. Hemin is a naturally occurring, biologically active, iron-containing asymmetrically substituted porphyrin. Liquid crystalline derivatives of this molecule can be conveniently synthesized via appropriate chemical transformations (Scheme 2.133) [1252]. The synthesis of **36** basically involves two reactions: first, a Heck-type coupling between the two vinyl groups of the hemin and the iodo-phenyl

TABLE 2.111
Thermal Behavior of Octaalkylporphyrins 29
and 30 (Scheme 2.132)

Structure	R	M	Thermal Behavior
29.1	C_8H_{17}	2H	Cr 137 I
29.2	$C_{10}H_{20}$	2H	Cr 115.4 I
29.3	$C_{12}H_{25}$	2H	Cr 106.6 I
30.1	C_8H_{17}	Zn	Cr 172.5 Col$_r$ 180.8 Col$_r$ 220.4 I
30.2	$C_{10}H_{20}$	Zn	Cr 145.6 Col$_r$ 182.3 I
30.3	$C_{12}H_{25}$	Zn	Cr 124.5 Col$_r$ 152 I

Source: Data from Shearman, G.C. et al., *J. Mater. Chem.*, 19, 598, 2009.

derivatives and second, the esterification of the two propeonic groups of the hemin. On the other hand, the synthesis of the tetraester **38** involves the oxidation of porphyrin dimethylester to diformyl product either directly or via bis-glycol followed by Wittig reaction with methyl(triphenylphosphoranylidene)acetate and saponification to yield the tetraacid. This acid can be esterified with the desired alkoxy-substituted phenol to yield asymmetrically β-substituted porphyrin-based DLCs.

These hemin-derived discotics exhibit mesomorphism from ambient temperature to nearly 300°C. The mesomorphic behavior of hemin-derived discotics is summarized in Table 2.112. Compound **36.1** exhibits an enantiotropic discotic lamellar mesophase in the temperature range of –2°C to 70°C [1252], while the porphyrins with 3,5-didodecyloxyphenyl acrylic ester groups **38** display columnar phases [1253–1255]. The introduction of benzoate groups and the change in the substitution pattern of the phenyl groups provide the mesogens with wider mesophase range and higher order. As expected, the alkyl chain length affects the melting point and metallation enhances the thermal stability of the mesophase due to the enhancement of the π–π interaction between the neighboring units of the porphyrins. Among the two columnar hexagonal phases that are shown by these molecules, in the Col$_{h1}$ phase the supramolecular arrangement is slightly different from the usual hexagonal columnar mesophase. The Col$_{h2}$ phase exhibits the normal π–π stacking distance of 3.5–3.6 Å, while the intracolumnar distance in the Col$_{h1}$ phase was approximately 10.5 Å, which is attributed to the association of three molecules along the column direction.

2.20.2 *Meso*-Substituted Porphyrins

The *meso*-substituted porphyrins, though not naturally occurring, are widely preferred candidates in various fields such as biomimetic models, material chemistry, photodynamic therapy, catalysis, electron transfer, etc. [1256]. The synthesis of these derivatives is much simpler compared to their β-substituted counterparts. The first *meso* derivative synthesized was tetramethyl porphyrin by Rothemund et al. in 1935, by the condensation of acetaldehyde and pyrrole [1257]. It was found that along with the mesoporphyrin, some amount of side product was also formed, which was characterized as the chlorine (a porphyrin-related macrocycle largely aromatic but not aromatic through the entire circumference of the ring). However, chlorins can be easily oxidized to form the corresponding porphyrins. In the 1960s, a new method was devised by Alder et al. who synthesized tetraphenyl derivative by the condensation reaction of benzaldehyde and pyrrole in the presence of acidic solvents under refluxing conditions. The yield was much higher compared to the tetramethyl derivative synthesized by Rothenmund et al. This method also involves the interference of chlorins, which was oxidized by the use of DDQ. The methodology was modified slightly by Longo et al. and is now commonly known as the Alder–Longo method [1258]. Refluxing pyrrole and benzaldehyde

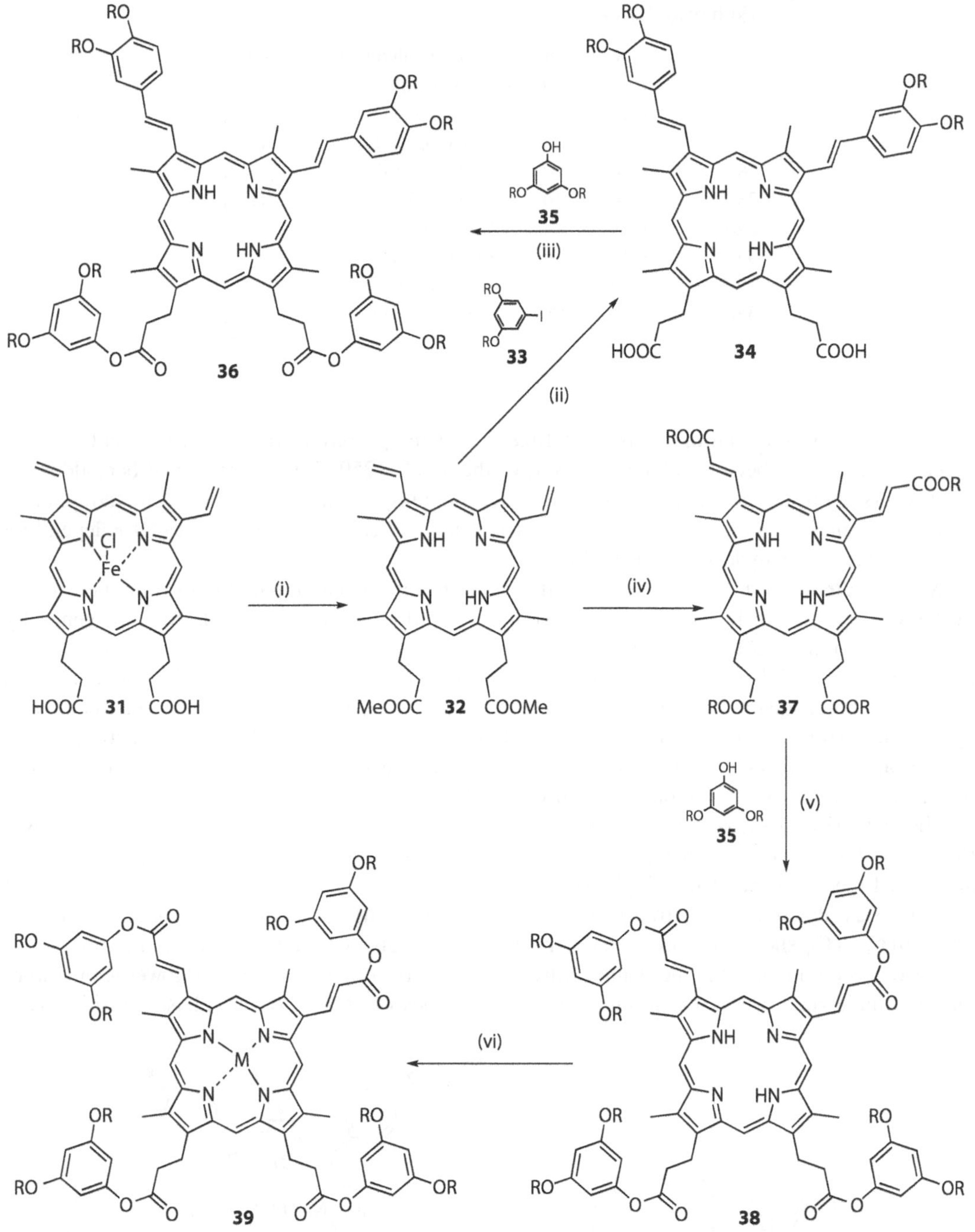

SCHEME 2.133 Synthesis of hemin-based unsymmetrical porphyrin derivatives: (i) FeSO$_4$·7H$_2$O, HCl(g), MeOH, CH$_2$Cl$_2$, R.T, 85%; (ii) (a) DMF, Pd catalyst, NaOAc, (b) DME/KOH; (iii) DCCI/DMAP, CH$_2$Cl$_2$; (iv) (a) OsO$_4$, N-methylmorpholine, NaHSO$_3$, (b) HIO$_4$, (c) Zn(OAc)$_2$, (d) Ph$_3$P=CHCOOR; (v) (a) KOH/DME/H$_2$O (b) SOCl$_2$ (c) Et$_3$N, CH$_2$Cl$_2$, reflux.; (vi) M(OAc)$_2$, DMF, 115°C.

TABLE 2.112
Thermal Behavior of Hemin-Derived Discotics
(Scheme 2.133)

Structure	R	M	Thermal Behavior	Ref.
36.1	$C_{12}H_{25}$	2H	Cr −2.5 D_L 70 I	[1252]
38.1	C_4H_9	2H	g 13 Col_h 193 I	[1255]
38.2	C_8H_{17}	2H	Cr −10 Col_{h1} 50 Col_{h2} 168	[1255]
38.3	$C_{12}H_{25}$	2H	Cr −33 Col_h 51 Col_h 139 I	[1253]
38.4	$C_{16}H_{33}$	2H	Cr 22 Col_{h1} 53 Col_{h2} 108 I	[1255]
39.1	C_4H_9	Zn	g 11 Col_h 303 dec	[1255]
39.2	C_8H_{17}	Zn	g −5 Col_h 244 I	[1255]
39.3	$C_{12}H_{25}$	Zn	G −25 Col_h 192 I	[1255]
39.4	$C_{12}H_{25}$	Cu	Cr −31 Col_h 174 I	[1255]
39.5	$C_{16}H_{33}$	Zn	Cr 23 Col_h 153 I	[1255]

in propeonic acid for 30 min yields crystalline tetraphenylporphyrin in about 20% yield. Several other methods have been developed to improve the yield [1259,1260]. Pyrrole and benzaldehyde react reversibly in the presence of acid catalyst at room temperature to form the cyclic tetraphenyl-porphyrinogen that can be oxidized to porphyrin. The yield can be achieved up to 50% depending on the choice of aldehyde and acid [1259].

A large number of *meso*-tetra-substituted porphyrins have been prepared to explore their meso-morphic properties [6,1261–1315]. While simple tetraalkyl-substituted porphyrins are generally non-mesomorphic [1261,1262], many tetra-(4-alkylphenyl), tetra-(4-alkoxyphenyl), tetra-(3,4-dial-kylphenyl), tetra-(3,4-dialkoxyphenyl), tetra-(4-carboxyphenyl), tetra-(alkanoyloxyphenyl), etc., display mesomorphism. These materials are generally prepared by reacting appropriate benzalde-hyde with pyrrole using standard reaction conditions (Scheme 2.134). *Meso*-tetra-substituted por-phyrins are of 15 different structural types, as shown in Figure 2.36. The thermal behavior of these materials is summarized in Tables 2.113 through 2.119.

The tetrakis(4-alkylphenyl)porphyrins **43** generally exhibit discotic lamellar mesophases [6,1263–1278]. The thermal behavior of these materials is summarized in Table 2.113. Shimizu et al. studied the influence of the alkyl chains and metal ion on the mesophase behavior. The pentyl derivatives exhibited the direct transition to isotropic liquid while derivatives with chain length rang-ing from C_6 to C_{10} show single mesophase; the undecyl or higher derivatives form two mesophases. It was also observed that the higher temperature mesophase is of high viscosity and lower temperature ones resemble rigid crystalline forms. The entropy values of crystal to mesophase transition vary

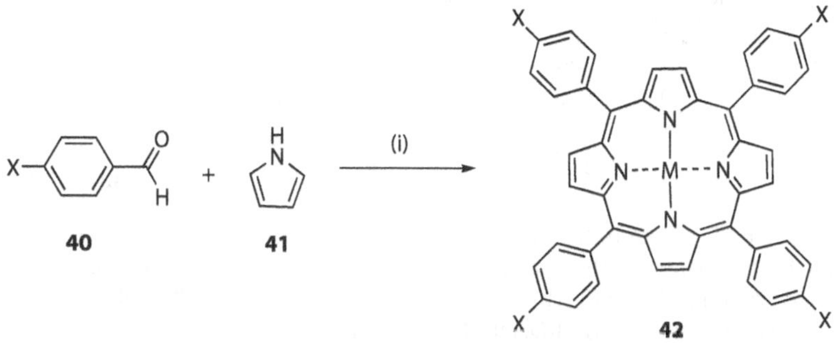

SCHEME 2.134 Synthesis of *meso*-tetra(4-alkylphenyl)porphyrins: (i) propionic acid, reflux; DDQ; $M(OAc)_2$, DMF.

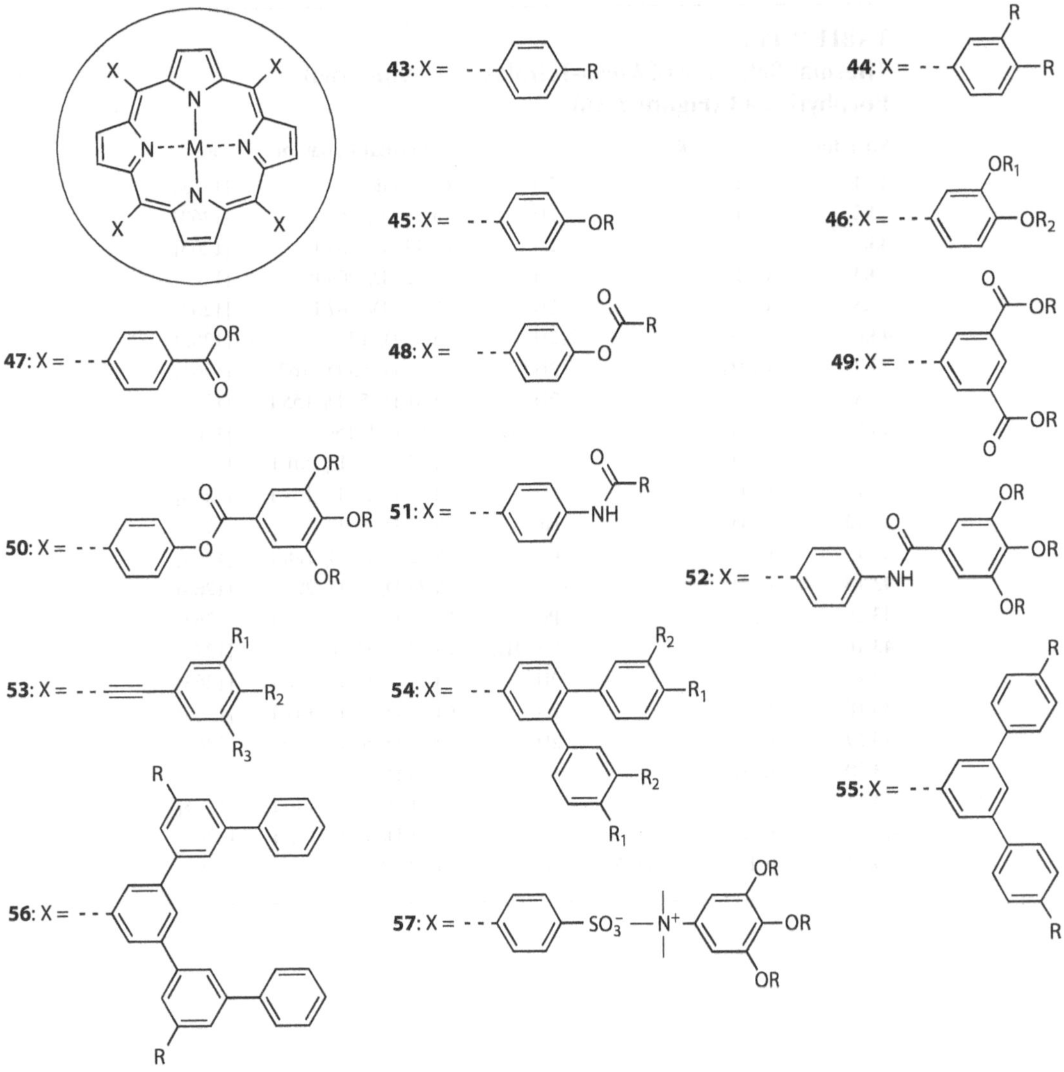

FIGURE 2.36 General structures for *meso*-substituted porphyrins.

with chain length of the alkyl derivatives, but the entropy values of mesophase to isotropic transition are approximately the same for all homologues. The metal ions have greater influence on the enhancement of the thermal stability of mesophase in the order Zn > Cu > Pd > Co > 2H > Ni. Sugino et al. synthesized silicon metal complexes of the tetradodecylphenyl porphyrin with axial hydroxyl ligand and it was found to exhibit lamellar mesophase. The axial hydroxyl groups are connected by weak hydrogen bonding in the columnar structure. Nagata et al. synthesized molybdenum complex of tetradodecyl phenyl porphyrin with chlorine and oxo groups as axial ligands, which also exhibited lamellar mesophase. The molybdenum complex, which has two different groups in the axial coordination, showed lamellar transition over a broad temperature range. The complexes of molybdenum with oxygen and hydroxyl groups as the axial ligands had been found to stabilize the mesophase by dipole–dipole interactions. The VO complex showed two lamellar phases. The mesomorphism in complexes with axial coordination depends on the molecular geometry, intermolecular interaction, and steric hindrance. The hydroxo-aluminum complex forms hexagonal columnar phase, but the phase behavior changes once it is subjected to heating above the clearing point, which is due to the formation of μ-oxo

TABLE 2.113
Thermal Behavior of *Meso*-Tetrakis(4-Alkylphenyl)
Porphyrins 43 (Figure 2.36)

Structure	R_1	M	Thermal Behavior	Ref.
43.1	C_5H_{11}	2H	Cr 313 dec	[1263]
43.2	C_6H_{13}	2H	Cr 190 D_L 263 I	[1263]
43.3	C_7H_{15}	2H	Cr 145 D_L 237 I	[1263]
43.4	C_8H_{17}	2H	Cr 126 D_L 209 I	[1263]
43.5	C_9H_{19}	2H	Cr 107 D_L 187 I	[1263]
43.6	$C_{10}H_{21}$	2H	Cr 52 D_L 173 I	[1263]
43.7	$C_{11}H_{23}$	2H	Cr 23 D_L 55 $D_{L'}$ 162 I	[1263]
43.8	$C_{12}H_{25}$	2H	Cr 31 D_L 52 D_L 155 I	[1263]
43.9	$C_{12}H_{25}$	Al-OH	Cr 84 Col_h 150 I	[1267]
43.10	$C_{12}H_{25}$	MoOCl	Cr 37 Cr 58 D_L 201 I	[1266]
43.11	$C_{12}H_{25}$	Co	Cr 28 D_L 50 D_L 161 I	[1263]
43.12	$C_{12}H_{25}$	Ni	Cr 44 D_L 129 I	[1263]
43.13	$C_{12}H_{25}$	Cu	Cr 32 D_L 56 D_L 188 I	[1263]
43.14	$C_{12}H_{25}$	Zn	Cr 37 D_L 52 D_L 220 I	[1263]
43.15	$C_{12}H_{25}$	Pd	Cr 30 D_L 60 D_L 186 I	[1263]
43.16	$C_{12}H_{25}$	$Si(OH)_2$	Cr 84 M 211 I	[1278]
43.17	$C_{13}H_{27}$	2H	Cr 37 D_L 60 D_L 148 I	[1263]
43.18	$C_{14}H_{29}$	2H	Cr 52 D_L 57 D_L 141 I	[1263]
43.19	$C_{15}H_{31}$	2H	Cr 56 D_L 66 $D_{L'}$ 135 I	[1263]
43.20	$C_{15}H_{31}$	Ni	Cr 68 D_L 122 I	[1263]
43.21	$C_{15}H_{31}$	VO	Cr 51 D_L 82 D_L 140 I	[1274]
43.22	$C_{16}H_{33}$	2H	Cr 65 D_L 71 D_L 129 I	[1263]
43.23	$(CH_2)_5O(C_2H_4O)_3Me$	2H	Cr 70–75 I	[1268]

TABLE 2.114
Thermal Behavior of *Meso*-Tetrakis(3,4-Dialkoxyphenyl)
Porphyrins 46 (Figure 2.36)

Structure	R_1	R_2	M	Thermal Behavior	Ref.
46.1	$C_{10}H_{21}$	CH_3	Yb-OH	Cr −10.6 Col_h 42.3 I	[1284]
46.2	C_2H_5	$C_{12}H_{25}$	2H	Cr −53.5 Col_h 64 I	[1286]
46.3	C_2H_5	$C_{12}H_{25}$	Yb-OH	Cr −33.8 Col_h 71.9 I	[1286]
46.4	C_2H_5	$C_{12}H_{25}$	Er-OH	Cr −43.7 Col_h 101 I	[1286]
46.5	C_2H_5	$C_{12}H_{25}$	Dy-OH	Cr 8.5 Col_h 15.4 I	[1286]
46.6	C_2H_5	$C_{12}H_{25}$	Tb-OH	Cr −32.8 Col_h 6.9 I	[1286]
46.7	C_2H_5	$C_{14}H_{29}$	2H	Cr 6.7 Col 36.1 Col 69.6 I	[1286]
46.8	C_2H_5	$C_{14}H_{29}$	Yb-OH	Cr −4.1 Col 37 Col 43.6 I	[1286]
46.9	C_2H_5	$C_{14}H_{29}$	Er-OH	Cr −3.5 Col 29.9 Col 50.3 I	[1286]
46.10	C_2H_5	$C_{14}H_{29}$	Tb-OH	Cr −2.8 Col 49.5 I	[1286]
46.11	C_2H_5	$C_{16}H_{33}$	2H	Cr −0.3 Col 22.2 Col 37.2 Col 65.6 I	[1286]

TABLE 2.115
Thermal Behavior of Esters of Porphyrin 48 (Figure 2.36)

Structure	R	M	Thermal Behavior	Ref.
48.1	C_7H_{15}	2H	Cr 90.6 Col 133.9	[1288]
48.2	C_7H_{15}	Zn	Cr 100.8 Col 190.8 I	[1288]
48.3	C_9H_{19}	2H	Cr 24.0 Col 48.4 Col 124.7 I	[1288]
48.4	C_9H_{19}	Zn	Cr −9.0 Col 35.5 Col 166 I	[1288]
48.5	C_9H_{19}	Ni	Cr 35.6 Col 159.7 I	[1289]
48.6	$C_{11}H_{23}$	2H	Cr 26 Col 134.5 I	[1288]
48.7	$C_{11}H_{23}$	Zn	Cr −36.4 Col −19.3 Col 10.8 Col 39.8 I	[1288]
48.8	$C_{11}H_{23}$	Co	Cr −17.0 Col 93.4 Col 156.1 I	[1289]
48.9	$C_{11}H_{23}$	Ni	Cr 40.5 Col 106.5 I	[1289]
48.10	$C_{13}H_{27}$	Co	Cr 28.1 Col 47.6 Col 130.6 I	[1289]
48.11	$C_{13}H_{27}$	Ni	Cr −6.1 Col 34.3 Col 61.2 Col 101.4 I	[1289]
48.12	$C_{15}H_{31}$	Co	Cr 4.4 Col 46.9 Col 70.2 Col 116.6 I	[1289]
48.13	$C_{15}H_{31}$	Ni	Cr 30.9 Col 75.1 I	[1289]
48.14	$C_{17}H_{35}$	Co	Cr 10 Col 48.8 Col 113.8 I	[1289]
48.15	$C_{17}H_{35}$	Ni	Cr −22.5 Col 52.8 Col 94.6 I	[1289]

TABLE 2.116
Thermal Behavior of *Meso*-Tetrakis (3,5-Dialkoxycarbonylphenyl) Porphyrins 49 (Figure 2.36)

Structure	R	M	Thermal Behavior
49.1	C_8H_{17}	2H	Cr 132 I
49.2	$C_{10}H_{22}$	2H	Cr −3 Col_h 121 I
49.3	$C_{12}H_{25}$	2H	Cr 17 Col_h 114 I
49.4	$C_{12}H_{25}$	VO	Cr 13 Col_h 233 I
49.5	$C_{14}H_{29}$	2H	Cr 36 Col_h 113 I
49.6	$C_{16}H_{33}$	2H	Cr 49 Col_h 103 I
49.7	$C_{18}H_{37}$	2H	Cr 63 Col_h 106 I
49.8	$C_{20}H_{41}$	2H	Cr 70 Col_h 103 I
49.9	$C_{22}H_{45}$	2H	Cr 78 Col_h 101 I

Source: Data from Patel, B.R. and Suslick, K.S., *J. Am. Chem. Soc.*, 120, 11802, 1998.

dimer, which shows several transitions apart from the two lamellar ones. The charge transfer complex of Mn porphyrin complex with tetracyanoquinodimethane was found to display mesomorphic behavior. Increasing the chain length does not change the mesomorphic behavior drastically, but replacement of the alkyl chains by oligo(ethyleneoxide) chains destroys the mesomorphism.

Ohta et al. synthesized tetrakis(3,4-dialkylphenyl)porphyrin derivatives **44** to study the effects of additional peripheral chains on mesomorphism [1279]. Three derivatives bearing octyl, dodecyl, and octadecyl peripheral chains were prepared. While octyl and dodecyl derivatives were found to be isotropic liquid at room temperature, tetrakis(3,4-dioctadecyl)porphyrin and its Ni and Cu complexes display complex thermal behavior. The presence of two liquid crystalline phases and two unidentified phases were reported. One of the mesophases was identified as discotic lamellar

TABLE 2.117
Thermal Behavior of Tetrakis(3,4,5-Trialkoxybenzoatephenyl) Porphyrins 50 (Figure 2.36)

Structure	R	M	Thermal Behavior	Ref.
50.1	C_7H_{15}	2H	Col_r 169.1 I	[1291]
50.2	C_8H_{17}	2H	I 166.8 Cr	[1292]
50.3	$C_{10}H_{21}$	2H	I 162.3 Cr	[1292]
50.4	$C_{12}H_{25}$	2H	Cr −33 Col_r 146.5 I	[1291]
			I 139.7 Col_h 137 Col_h −50 Cr	[1292]
50.5	$C_{12}H_{25}$	Zn	Cr −28 Col_r 116.5 I	[1291]
50.6	$C_{12}H_{25}$	Ni	Cr 8.6 Col_r 136.7 I	[1291]
50.7	$C_{14}H_{29}$	2H	I 135.4 Col_h 115.7 Col_h −14 Cr	[1292]
			I 130.7 Col_h 104.9 Col_h 2 Cr	[1291]
50.8	$C_{16}H_{33}$	2H	Cr 17.5 Col_r 97.7 Col 109.5 I	[1291]
			I 121 Col_h 96.7 Col_h −9.7 Cr	[1292]
50.9	$C_{14}F_9H_{20}$	2H	I 120.4 Col_h 120.2 Col_h < 0	[1295]

TABLE 2.118
Thermal Behavior of Mesomorphic Amide Derivatives of Porphyrin 51, 52 (Figure 2.36)

Structure	R	M	Thermal Behavior	Ref.
51.1	$C_{11}H_{23}$	2H	Cr 182 Col 270 I	[1299]
51.2	$C_{13}H_{27}$	2H	Cr 163 Col 177 Col 274 I	[1299]
51.3	$C_{15}H_{31}$	2H	Cr 121 Col 156 Col 281 I	[1299]
51.4	$C_{17}H_{35}$	2H	Cr 115 Col 143 Col 315 I	[1299]
52.1	$C_{12}H_{25}$	2H	Cr −26 Col_r 225.7 I	[1296]
52.2	$C_{12}H_{25}$	Zn	Cr −29.9 Col_r 187.6 Col 203.1 I	[1296]
52.3	$C_{16}H_{33}$	2H	Cr 19.3 Col_r 189.3 I	[1296]
53.1	$R_1 = C_{10}H_{21}$ $R_2 = R_3 = H$	2H	Cr 149 M Col_h 182 I	[1297]
53.2	$R_1 = R_3 = C_{10}H_{21}$ $R_2 = H$	2H	Cr 107 M 114 Col_h 182 I	[1297]
53.3	$R_1 = R_2 = C_{10}H_{21}$ $R_3 = H$	2H	Cr 108 M >260 dec	[1297]

TABLE 2.119
Thermal Behavior of Mesomorphic Dendritic (54,55,56) and Amphiphilic (57) Porphyrins (Figure 2.36)

Structure	R	M	Thermal Behavior	Ref.
54.1	$OC_{12}H_{25}$	2H	Cr −35 Col_h 59 I	[1302]
54.2	$OC_{12}H_{25}$	Cu	Cr −37 Col_h 57 I	[1302]
55.1	$C_{12}H_{25}$		g −50 X 43 I	[1304]
56.1	$C_{12}H_{25}$		g −54 X 39 Col_r 110 I	[1304]
57.1	C_8H_{17}		Col_h 170 dec	[1305]
57.2	$C_{12}H_{25}$		Col_h 170 dec	[1305]

phase. These phases were observed only by very slow relaxation at a certain temperature or by rapid cooling of the isotropic phase. Despite the presence of eight peripheral chains, these molecules do not show any columnar phase.

The tetrakis(4-alkoxyphenyl)porphyrin derivatives **45** do not show mesomorphism. Some of these derivatives were claimed [1280,1281] to be liquid crystalline but they are ill characterized and it is unlikely that they display mesomorphism [1262,1282]. The non-mesomorphic behavior of the tetrakis(4-alkoxyphenyl)porphyrin was attributed to the stronger intermolecular force between the cores because the electron pumping ability of alkoxy groups increases the electron density of the tetraphenylporphyrin core. It is also well known that these alkoxy substituents can adopt planar chain extended conformation more readily than alkyl substituents. As a result, the isotropization of the compounds precedes their mesophase. However, the introduction of one more alkoxy chain per phenyl ring leads to mesophase formation probably because of the weak intermolecular interaction due to steric hindrance.

The tetrakis(3,4-dialkoxyphenyl)porphyrins **46** have been reported to show the mesophase [1283–1286]. Ohta and coworkers prepared a series of tetrakis(3,4-dialkoxyphenyl)porphyrins bearing butyloxy to octadecyloxy chains [1283]. All these derivatives exhibit complicated phase behavior and they show mesomorphism under unusual conditions (rapid cooling from the isotropic phase) and, therefore, are not true liquid crystalline materials. The lanthanide complexes of porphyrins displaying liquid crystalline behavior was reported in 2002 [1284]. The coordination numbers of the lanthanide ions are more compared to that of the transition elements, which in turn as an impact on the geometries of liquid crystals. Since lanthanide ions are known for their luminescence properties, the incorporation of these ions in the liquid crystal assemblies will lead to luminescent mesomorphic compounds that may find applications in the design of emissive devices [1285]. Yu et al. studied the mesomorphism of analogues of 5, 10, 15, 20-tetra [(4-alkoxy-3-ethyloxy)phenyl]porphyrin hydroxy lanthanide complexes and found the existence of hexagonal columnar mesophase [1286]. The mesomorphism of these materials is summarized in Table 2.114.

The tetrakis(4-*n*-alkoxycarbonylphenyl)porphyrins **47** with dodecyl, hexadecyl, and octadecyl chains, and their Zn and Cu complexes did not show any mesomorphism [1287]. However, esters derived from tetrakis(4-hydroxyphenylporphyrins), that is, tetrakis(4-*n*-alkanoyloxyphenyl)porphyrins **48** exhibit columnar phases [1288,1289]. Their mesophase behavior is summarized in Table 2.115. The formation of more than one columnar phase is reported in many derivatives of this series. The nature of these columnar phases was reported to be hexagonal, but a distinction between different columnar phases has not been made.

Patel et al. reported the mesomorphism of 5,10,15,20-tetrakis(3,5-dialkoxycarbonylphenyl) porphyrins **49** where the alkyl groups act as protecting pockets on both faces of the macrocycle [1290]. The phase behavior of these compounds is summarized in Table 2.116. The homologue corresponding to the octyl chain is not mesomorphic whereas decyl and dodecyl derivatives display hexagonal columnar mesophase at room temperature. These materials retain their mesomorphism in the presence of axial ligation, which was proved by observing the mesomorphic behavior of the VO complex of the dodecyl derivative [1290].

The tetrakis (3,4,5-trialkoxybenzoate) phenyl porphyrins **50** are observed to exhibit columnar mesomorphism [1291]. The flat rigid core of these molecules promotes the formation of highly aligned columnar mesophase and, therefore, excellent candidates for device applications [1292,1293]. The fullerene adduct of these compounds also shows mesomorphism and photovoltaic behavior [1294]. It has been demonstrated by Zhou et al. that partial perflourination in the large π-conjugated discotic molecules leads to highly ordered nanostructure, which enhances the tendency toward homeotropic alignment [1295]. The mesomorphic behavior of these compounds is summarized in Table 2.117.

The condensation of 4-nitrobenzaldehyde with pyrrole yields tetrakis (4-nitroporphyrin), which can be reduced to tetraamino derivative. The amide derivatives **51, 52** are prepared by reacting it with various acid chlorides [1296–1300]. Among the *meso*-tetra(*p*-alkylamidophenyl) porphyrin series, only free bases with long side chains (>12 carbon atoms) exhibit mesomorphism.

Lower homologues as well as metal complexes of higher homologues are not mesomorphic. The electrochemical and luminescence behavior of lauroylimidophenyl-porphyrin and their corresponding metal complexes have been studied. All these derivatives are reported to be good p-type semiconductors [1300]. It has been observed that as the chain length increases, the span of the mesophase range increases, whereas the melting temperature decreases. The mesophases are lamellar in nature. The reason for the non-mesomorphic behavior of metal complexes could be due to the amide N-chains coordinating with the metal ions, which lead to multiporphyrin aggregation. The *meso*-tetra[4-(3,4,5-n-trialkoxybenzoylamino)phenyl]porphyrins **52** and their corresponding zinc complexes exhibit columnar phases over a wide range of temperatures [1296]. The thermal behavior of mesogenic porphyrin amide derivatives **51, 52** is presented in Table 2.118.

Milgrom et al. attached 4-alkoxyphenyl-, 3,4-dialkoxyphenyl-, and 3,5-dialkoxyphenyl-groups to the porphyrin nucleus via ethynyl linkages [1301]. The 5,10,15,20 tetraarylethynylporphyrinato zinc(II) complexes **53** display columnar mesophases. The phase transition temperatures observed for these compounds are presented in Table 2.118.

The substitution of *ortho*-terphenyl moieties in all the four *meso*-positions of porphyrins **54** induces the overall macrocycle to adopt a disk-like structure [1302]. The terphenyl aldehyde required to react with pyrrole to generate these porphyrins is prepared following the method of Wenz [1303]. The derivative bearing 16 dodecyloxy chains exhibits a complicated phase behavior, but its corresponding copper complex displays rather simpler phase transition behavior. It displays a columnar phase at −37°C, which clears at 57°C. The free base exhibits this transition at 59°C (Table 2.119).

Kimura et al. reported the mesomorphism of two series of phenylene-based dendritic porphyrins **55, 56**. Dendrimer **56**, where the phenylene moiety acts as a mesogenic architecture and their packing results in columnar aggregates, exhibits mesomorphism. The transition temperatures are collected in Table 2.119. Dendrimer **55** did not display any mesogenic properties [1304]. The inclusion of C_{60} in the dendritic structure affects the mesophase structure. It also resulted in the quenching of the fluorescence. The electro reduction potential studies indicate the formation of stable complex between the C_{60} and dendrimer **56**.

Ionic self-assembled complexes provide an easy way to produce liquid crystalline materials and thin films that are stable over a wide temperature range. This flexibility leads to the production of various soft materials with optical, magnetic, and electronic properties. This ability was tested using ammonium amphiphiles with tetrakis(4-sulfonatophenyl)porphyrin **57**. Complexes with C_8 ammonium as well as C_{12} ammonium amphiphiles showed the columnar mesophase (Table 2.119) [1305]. The infrared studies infer that the amide subunit in the ammonium framework is engaged in supramolecular hydrogen-bonded network that stabilizes the thermotropic mesophase.

Surprisingly, a number of porphyrins and their lanthanide complexes having only a single alkoxy or alkanoyloxy peripheral chain **58** (Figure 2.37) are reported to display hexagonal columnar phases [1306–1308]. The columnar phase formation in these materials could be due to the supramolecular assemblies of porphyrin molecules. The thermal behavior of these porphyrins is summarized in Table 2.120. Multiple columnar phases are reported in some of these derivatives; however, more studies are required to understand the columnar mesomorphism in such materials.

Extending the porphyrins along one axis through the incorporation of substituents in 5 and 15 positions transforms the discotic porphyrins into calamatic molecules and accordingly they show calamatic nematic and smectic phases at elevated temperatures. This was experimentally shown by Bruce et al. [1309–1313]. Ohta and coworkers also synthesized three series of di-substituted porphyrins and studied their mesomorphism. It was found that the substituents around the porphyrin core play a vital role in deciding the nature of the mesophases [1302].

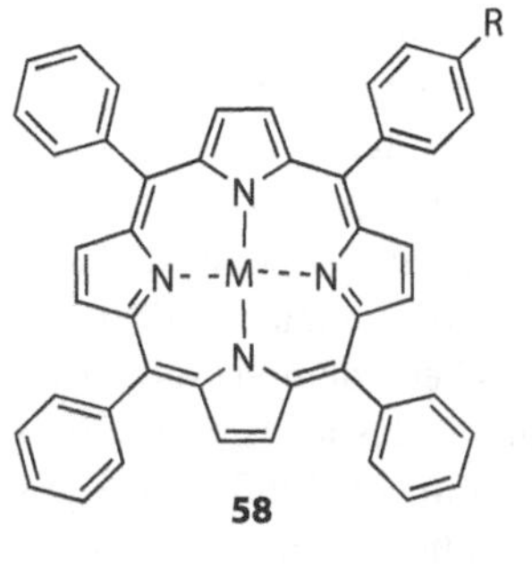

58

FIGURE 2.37 Structure of columnor phase forming monosubstituted porphyrin.

TABLE 2.120
Thermal Behavior of Porphyrins 58 Possessing Only a Single Peripheral Chain (Figure 2.37)

Structure	R	M	Thermal Behavior	Ref.
58.1	$OCOC_9H_{19}$	2H	Cr 35 Col 179 I	[1306]
58.2	$OCOC_9H_{19}$	Ho-OH	Cr 42 Col 55.3 I	[1306]
58.3	$OCOC_{11}H_{23}$	2H	Cr 63.3 Col 142 I	[1306]
58.4	$OCOC_{11}H_{23}$	Yb-OH	Cr 20 Col 55.3 I	[1306]
58.5	$OCOC_{11}H_{23}$	Er-OH	Cr −77 Col 28.9 I	[1306]
58.6	$OCOC_{11}H_{23}$	Ho-OH	Cr 4.4 Col 37.9 I	[1306]
58.7	$OC_{12}H_{25}$	2H	Cr 31.7 Col −15.5 I	[1307]
58.8	$OC_{12}H_{25}$	Er-OH	Cr −33.6 Col −18.1 Col 4.9 I	[1307]
58.9	$OC_{12}H_{25}$	Dy-OH	Cr −26.1 Col 10.7 I	[1307]
58.10	$OC_{12}H_{25}$	Tb-OH	Cr −25.3 Col 20.3 I	[1307]
58.11	$OC_{14}H_{29}$	2H	Cr −9.7 Col −1.7 Col −4.3 Col 20.5 I	[1307]
58.12	$OC_{14}H_{29}$	Yb-OH	Cr −7.5 Col 7.1 Col 19.3 Col 25.1 I	[1307]
58.13	$OC_{14}H_{29}$	Er-OH	Cr −3.7 Col 6.9 Col 16.9 I	[1307]
58.14	$OC_{14}H_{29}$	Dy-OH	Cr −11.1 Col 18.3 Col 26.5 I	[1307]
58.15	$OC_{14}H_{29}$	Tb-OH	Cr −8.3 Col 3.5 Col 15.1 Col 32.5 I	[1307]
58.16	$OC_{16}H_{33}$	2H	Cr −4.5 Col 4.2 Col 17.3 Col 31 I	[1307]
58.17	$OC_{16}H_{33}$	Yb-OH	Cr 16 Col 35.4 I	[1307]
58.18	$OC_{16}H_{33}$	Er-OH	Cr 14.9 Col 18.6 Col 30.8 I	[1307]
58.19	$OC_{16}H_{33}$	Tb-OH	Cr 12.4 Col 28.8 Col 38.2 I	[1307]

2.20.3 Miscellaneous Porphyrin Derivatives

The mesomorphic derivatives of porphyrins belonging to this class include derivatives that are substituted in both β- and *meso*-positions, expanded porphyrins, porphycenes, etc. These substituted porphyrins are gaining importance due to their material applications with extended large π-system with high thermal stability. The choice of substituents at the periphery of the porphyrin moiety plays a vital role in altering the property of the macrocycle. The control of intermolecular interactions in the solid state introduces porphyrins with the desired functional properties. The steric crowding around the periphery of porphyrins induces the nonplanarity to the core. This nonplanarity influences the properties of discotic derivatives of the porphyrins.

Bhyrappa et al. synthesized a series of β-tetrabrominated tetraalkylporphyrins [1314]. The synthesis was accomplished using methods reported in the literature, which involve the synthesis of tetraalkoxyporphyrins **45** by the method proposed by Patel et al. [1290]. Compound **45** on reaction with NBS followed by metallation under standard conditions results tetrabromoderivative **59** (Scheme 2.135). The mesomorphic temperature ranges of these derivatives were reduced when compared to their nonbrominated counterparts. The mesomorphic behaviors of these derivatives are summarized in Table 2.121. However, it may be noted that none of these materials were fluid in the claimed mesophase temperature range. Moreover, the mesomorphism of starting materials used in this study is also controversial.

Recently, expanded porphyrins were reported to display columnar mesophases [1315]. Expanded porphyrins are the oligopyrolic macrocycles that have ring sizes larger than porphyrins with either more than four heterocycle subunits or a greater number of π electrons within their peripheries. These hydrazinophyrins **61** were synthesized by the acid-catalyzed condensation of 3,4-dialkoxy-2,5-diformylpyrroles with one equivalent of hydrazine (Scheme 2.136). Although the 2 + 2 condensation is also possible and has been realized in other cases, Sessler et al. obtained

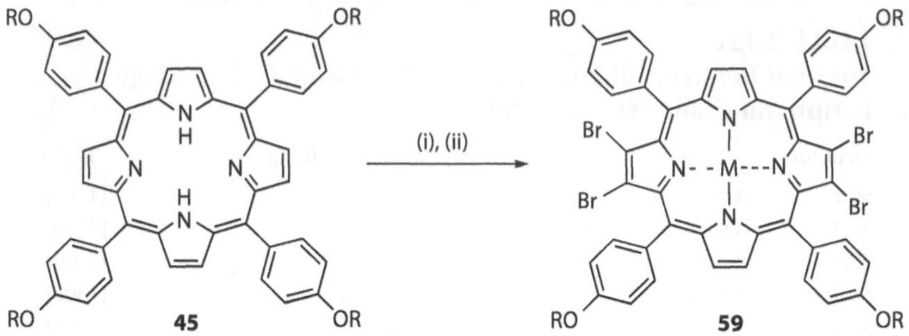

SCHEME 2.135 Synthesis of β-tetrabrominated-tetraalkoxyporphyrins **59**: (i) NBS/CHCl$_3$, (ii) M(OAc)$_2$, DMF, 115°C.

TABLE 2.121
Thermal Phase Behavior of Brominated Porphyrin Derivatives (Scheme 2.135)

Structure	R	M	Thermal Behavior	Ref.
59.1	C$_4$H$_9$	2H	Cr 197 I	[1314]
59.2	C$_6$H$_{13}$	2H	Cr 56 Col 204 I	[1314]
59.3	C$_8$H$_{17}$	2H	Cr 3 Col 188 I	[1314]
59.4	C$_{10}$H$_{21}$	2H	Cr −4 Col 101 I	[1314]
59.5	C$_{12}$H$_{25}$	2H	Cr 12 Col 92 I	[1314]
59.6	C$_{14}$H$_{29}$	2H	Cr13 Col 88 I	[1314]
59.7	C$_{16}$H$_{33}$	2H	Cr 56 Col 90 I	[1314]
59.8	C$_{18}$H$_{37}$	2H	Cr 68 Col 88 I	[1314]
59.9	C$_4$H$_9$	Cu	Cr 230 I	[1314]
59.10	C$_6$H$_{13}$	Cu	Cr 1 Col 227 I	[1314]
59.11	C$_8$H$_{17}$	Cu	Cr 19 Col 216 I	[1314]
59.12	C$_{10}$H$_{21}$	Cu	Cr 15 Col 140 I	[1314]
59.13	C$_{12}$H$_{25}$	Cu	Cr 12 Col 146 I	[1314]
59.14	C$_{14}$H$_{29}$	Cu	Cr 22 Col 137 I	[1314]
59.15	C$_{16}$H$_{33}$	Cu	Cr 38 Col 112 I	[1314]
59.16	C$_{18}$H$_{37}$	Cu	Cr 59 Col 102 I	[1314]

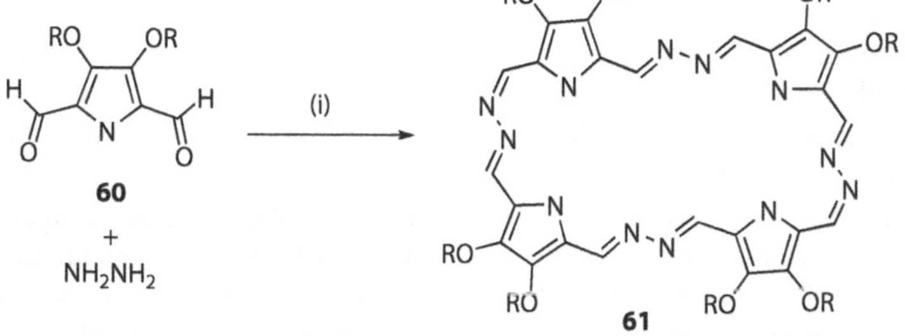

Cr ? Col 120–160 I/dec

SCHEME 2.136 Synthesis of expanded hydrazinophyrins **61**: (i) NH$_2$NH$_2$, HCl, MeOH.

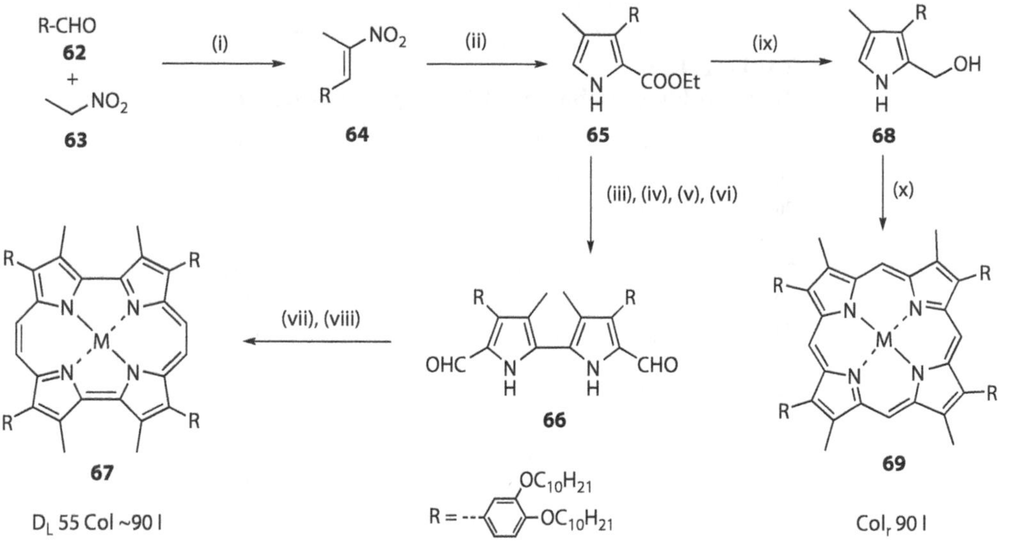

SCHEME 2.137 Synthesis of porphycenes **67** and corresponding porphyrins **69**: (i) AcONH$_4$; (ii) CNCH$_2$COOEt, THF/iPr-OH, DBU; (iii) I$_2$, NaI, NaHCO$_3$, DCE, H$_2$O; (iv) Boc$_2$O, DMAP, CH$_2$Cl$_2$; Cu, toluene; 180°C, vacuum; (v) KOH, glycol, 200°C; (vi) POCl$_3$, DMF/DCE; AcONa, H$_2$O; (vii) TiCl$_4$, Zn, CuCl, THF; (viii) M(OAc)$_2$, CHCl$_3$, MeOH; (ix) LiAlH$_4$, THF; (x) silica gel, CH$_2$Cl$_2$; DDQ; M(OAc)$_2$, CHCl$_3$, MeOH.

only the 4+4 product in 30%–60% yield [1315]. The hydrazine units in the macrocycle are all in *trans* configuration. Three derivatives bearing hexyloxy, decyloxy, and tetradecyloxy chains were prepared to investigate their mesomorphic properties. All compounds started decomposing above 120°C on heating and reached to complete isotropic/decomposed state by 160°C. Also, none of them exhibit a clear melting point. The formation of the columnar phase was deduced only on the basis of microscopy. More studies are required to confirm these findings.

Porphycene is a macrocyclic system isomeric with porphyrin and is also known as [18] porphyrin (2.0.2.0). The optical properties of porphycenes are drastically different than porphyrins and, therefore, are interesting materials for many device applications. The porphycene nucleus was originally synthesized by Vogel and coworkers [1316]. Discotic liquid crystalline porphycenes **67** were synthesized by Sessler and coworkers, as shown in Scheme 2.137 [1317]. Corresponding porphyrins **69** (as a mixture of different isomers) were also prepared to compare the thermal behavior. While four peripheral chains are not sufficient to induce mesomorphism, the Zn porphycene with peripheral 3,4-didecyloxyphenyl group exhibits a lamellar columnar mesophase. It shows mesomorphism at room temperature and goes to the isotropic phase at about 90°C. In contrast, the corresponding porphyrin **69** displays a rectangular columnar mesophase at room temperature, which also clears at 90°C.

Qi and Liu prepared a large number of benzoporphyrins **70** and their metal complexes [1318,1319]. All these *meso*-tetraalkyl-tetrabenzoporphyrin derivatives exhibit hexagonal columnar mesophase at room temperature. Their thermal behavior is summarized in Table 2.122.

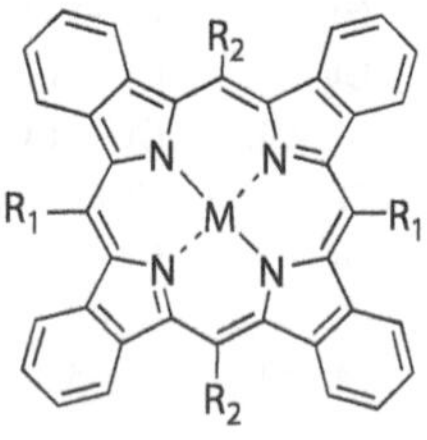

70

TABLE 2.122
Thermal Phase Behavior of Benzoporphyrins 70

Structure	R	M	Thermal Behavior	Ref.
70.1	C_6H_{13}	Zn	Cr −20.8 Col$_h$ 73.4 I	[1318]
70.2	C_6H_{13}	2H	Cr −42.6 Col$_h$ 74.2 I	[1318]
70.3	C_6H_{13}	Co-Cl	Cr −42. 5 Col$_h$ 75.2 I	[1318]
70.4	C_6H_{13}	Yb-OH	Cr −44.1 Col$_h$ 72.5 I	[1318]
70.5	C_6H_{13}	Ni	Cr −42 Col$_h$ 75 I	[1319]
70.6	C_6H_{13}	Cu	Cr −45 Col$_h$ 72	[1319]
70.7	C_6H_{13}	Mn-Cl	Cr −44 Col$_h$ 76 I	[1319]
70.8	C_8H_{17}	2H	Cr −50.6 Col$_h$ 26.5 I	[1318]
70.9	C_8H_{17}	Zn	Cr −51.8 Col$_h$ 24.4 I	[1318]
70.10	C_8H_{17}	Co-Cl	Cr −50.2 Col$_h$ 25.6 I	[1318]
70.11	C_8H_{17}	Yb-OH	Cr −50 Col$_h$ 19.2 I	[1318]
70.12	C_8H_{17}	Ni	Cr −50 Col$_h$ 20 I	[1319]
70.13	C_8H_{17}	Cu	Cr −48 Col$_h$ 28 I	[1319]
70.14	C_8H_{17}	Mn-Cl	Cr −51 Col$_h$ 30 I	[1319]
70.15	$C_{10}H_{21}$	2H	Cr −44.9 Col$_h$ 55.6 I	[1318]
70.16	$C_{10}H_{21}$	Zn	Cr −20.6 Col$_h$ 56.4 I	[1318]
70.17	$C_{10}H_{21}$	Co-Cl	Cr −46.5 Col$_h$ 55.5 I	[1318]
70.18	$C_{10}H_{21}$	Yb-OH	Cr −46.5 Col$_h$ 54.8 I	[1318]
70.19	$C_{10}H_{21}$	Ni	Cr −44 Col$_h$ 58 I	[1319]
70.20	$C_{10}H_{21}$	Cu	Cr −42 Col$_h$ 58 I	[1319]
70.21	$C_{10}H_{21}$	Mn-Cl	Cr −45 Col$_h$ 55 I	[1319]
70.22	$C_{12}H_{25}$	2H	Cr −46.6 Col$_h$ 52.6 I	[1318]
70.23	$C_{12}H_{25}$	Zn	Cr −20.5 Col$_h$ 50.4 I	[1318]
70.24	$C_{12}H_{25}$	Co-Cl	Cr −48.9 Col$_h$ 50.7 I	[1318]
70.25	$C_{12}H_{25}$	Yb-OH	Cr −51.5 Col$_h$ 51.1 I	[1318]
70.26	$C_{12}H_{25}$	Ni	Cr −46 Col$_h$ 55 I	[1319]
70.27	$C_{12}H_{25}$	Cu	Cr −45 Col$_h$ 55 I	[1319]
70.28	$C_{12}H_{25}$	Mn-Cl	Cr −48 Col$_h$ 53 I	[1319]
70.29	$C_{14}H_{29}$	2H	Cr −49.9 Col$_h$ 51.3 I	[1318]
70.30	$C_{14}H_{29}$	Zn	Cr −20.5 Col$_h$ 50.8 I	[1318]
70.31	$C_{14}H_{29}$	Co-Cl	Cr −50.8 Col$_h$ 51.6 I	[1318]
70.32	$C_{14}H_{29}$	Yb-OH	Cr −51.4 Col$_h$ 50 I	[1318]
70.33	$C_{14}H_{29}$	Ni	Cr −48 Col$_h$ 51 I	[1319]
70.34	$C_{14}H_{29}$	Cu	Cr −48 Col$_h$ 49 I	[1319]
70.35	$C_{14}H_{29}$	Mn-Cl	Cr −50 Col$_h$ 50 I	[1319]
70.36	$C_{16}H_{33}$	2H	Cr −52.3 Col$_h$ 22.3 I	[1318]
70.37	$C_{16}H_{33}$	Zn	Cr −21 Col$_h$ 48.8 I	[1318]
70.38	$C_{16}H_{33}$	Co-Cl	Cr −52.2 Col$_h$ 17.5 I	[1318]
70.39	$C_{16}H_{33}$	Yb-OH	Cr −52.3 Col$_h$ 19.1 I	[1318]
70.40	$C_{16}H_{33}$	Ni	Cr −51 Col$_h$ 35 I	[1319]
70.41	$C_{16}H_{33}$	Cu	Cr −51 Col$_h$ 34 I	[1319]
70.42	$C_{16}H_{33}$	Mn-Cl	Cr −51 Col$_h$ 48 I	[1319]
70.43	$C_{18}H_{37}$	2H	Cr −52 Col$_h$ 20 I	[1319]
70.44	$C_{18}H_{37}$	Ni	Cr −52 Col$_h$ 30 I	[1319]
70.45	$C_{18}H_{37}$	Cu	Cr −52 Col$_h$ 31 I	[1319]
70.46	$C_{18}H_{37}$	Mn-Cl	Cr −52 Col$_h$ 45 I	[1319]

2.21 PORPHYRAZINE CORE

Porphyrazines are macrocyclic compounds where the methine bridges of the porphyrin cavity are replaced by aza bridges. They are otherwise known as tetraazaporphyrins. Along with the phthalocyanines and porphyrins, they also play a vital role in commercial applications such as photovoltaic cells, photoconductors and semiconductors, gas sensors, photodynamic therapy, etc. They exhibit excellent chemical, thermal, and photochemical stability.

The first liquid crystalline derivative of porphyrazine was reported in 1990 by Doppelt et al. who studied the mesomorphism of various metal complexes of octakis(octylthio)tetraazaporphyrins displaying hexagonal columnar mesophases [1320]. The mesomorphic nature of the copper complex of the same ligand was confirmed by Morelli et al. [1321]. The octathioether derivatives **4** show great promise in the field of materials science because the presence of sulfur atom near the delocalized core of porphyrazine moiety maximizes the intra-stack interactions and also the intra- and intercolumnar electron motions by their strong polarizing power. The synthesis of octathioether derivatives allows easier incorporation of various alkyl homologues with high yields. The synthesis of octathioethers involves the preparation of 1,2-dicyano-1,2-bis(alkylthio) ethylene **2** followed by the reaction with magnesium propoxide and trifluroacetic acid to lead the magnesium derivative of the octasubstituted tetraazaporphyrin **3**. The demetallation of the **3** with trifluoroacetic acid followed by remetallation with other metals yields porphyrazines **4** (Scheme 2.138) [1322,1323]. The thermal behavior of these octathioether porphyrazines **4** is summarized in Table 2.123. Among the free base mesomorphic compounds, homologues corresponding to butyl and hexyl chains undergo two solid–solid phase transitions before the formation of mesophase. Higher homologues of the series do not display mesomorphism. All the metal complexes are mesomorphic in nature, and the incorporation of metal atoms stabilizes the mesophase. Among the metal complexes, cobalt and copper have the largest stability. Interestingly, these materials can be spread at the air–water surface to form monolayer and multilayer Langmuir–Blodgett (LB) films [1324]. The electronic and vibrational spectra of some of the octathioether porphyrazines in solid mesophase as well as in isotropic liquid suggest that the π–π inter macrocycle interactions are influenced by the axial interaction of the metal ions with the sulfur atom of the adjacent macrocycles [1325]. The mesomorphism of porphyrazine octathioethers with oligo(oxyethylene) side chains is similar to their alkyl counterparts [1326]. For mesomorphism in these compounds, the nature of metal ion is very important and all the metal complexes except Zn were mesomorphic with hexagonal columnar phase.

Belviso et al. synthesized a series of alkenyl(sulfanyl)porphyrazines and studied their mesomorphic behavior [1327,1328]. The phase transition temperature data of these derivatives are also collected in Table 2.123. Increasing the chain stiffness by introducing the terminal double bond reduces the thermal range in which the liquid crystallinity is observed compared to the corresponding saturated chains. This effect is pronounced in the case of free base, Cu and Ni complexes, and weaker in case of cobalt porphyrazines. This is due to the axial Co–S interaction operating within the columns in the liquid crystalline phase.

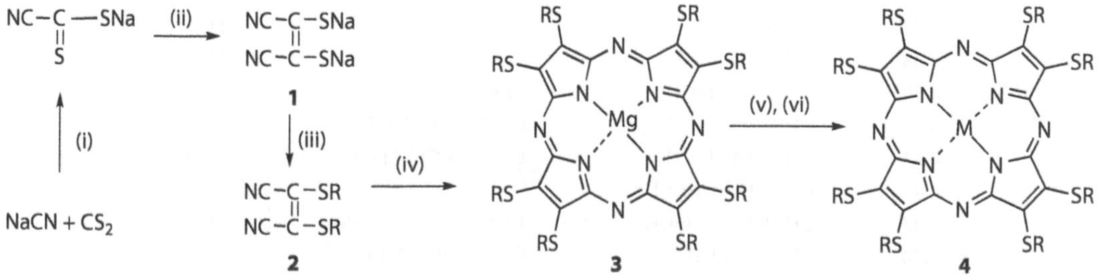

SCHEME 2.138 Synthesis of octathioethers of porphyrazines **4**: (i) DMF, 0°C, 30 min; (ii) H$_2$O, 12 h; (iii) RBr, MeOH, 0°C, 24 h; (iv) Mg(OPr)$_2$, PrOH, 100°C, 12 h; (v) CF$_3$COOH; (vi) M(OAc)$_2$, EtOH, Br-CH$_2$CH$_2$OH, 20 min.

TABLE 2.123
Thermal Behavior of Octakis(octylthio)tetraazaporphyrins 4
(Scheme 2.138)

Structure	R	M	Thermal Behavior	Ref.
4.1	C_4H_9	2H	Cr 113.1 Col 121.9	[1323]
4.2	C_4H_9	Co	Cr 96 Col 271 I	[1323]
4.3	C_4H_9	Ni	Cr 116.9 Col 183.5 I	[1323]
4.4	C_4H_9	Cu	Cr 111.7 Col 222.8 I	[1323]
4.5	C_4H_9	Zn	Cr 53.6 Col 208 I	[1323]
4.6	C_6H_{13}	2H	Cr 77.6 Col 92.2 I	[1323]
4.7	C_6H_{13}	Co	Cr 45.2 Col 243.6 I	[1323]
4.8	C_6H_{13}	Ni	Cr 75.3 Col 154.3 I	[1323]
4.9	C_6H_{13}	Cu	Cr 80.3 Col 189.7 I	[1323]
4.10	C_6H_{13}	Zn	Cr 46 Col 167.7 I	[1323]
4.11	C_7H_{15}	2H	Cr 75 I	[1323]
4.12	C_7H_{15}	Cu	Cr 70.2 Col_h 169.5 I	[1323]
4.13	C_8H_{17}	2H	Cr 81.5 I	[1323]
4.14	C_8H_{17}	Co	Cr 55 Col_h 208 I	[1323]
4.15	C_8H_{17}	Ni	Cr 67.6 Col_h 118.8 I	[1323]
4.16	C_8H_{17}	Cu	Cr 67.6 Col_h 151.7 I	[1323]
4.17	C_8H_{17}	Zn	Cr 40.8 Col_h 119.6 I	[1323]
4.18	C_9H_{19}	2H	Cr 78.6 I	[1323]
4.19	C_9H_{19}	Cu	Cr 61.8 Col_h 132.9 I	[1323]
4.20	$C_{10}H_{21}$	2H	Cr 80.7 I	[1323]
4.21	$C_{10}H_{21}$	Co	Cr 62.1 Col_h 176.8 I	[1323]
4.22	$C_{10}H_{21}$	Ni	Cr 75.8 Col_h 90.3 I	[1323]
4.23	$C_{10}H_{21}$	Cu	Cr 68.7 Col_h 121.7 I	[1323]
4.24	$C_{10}H_{21}$	Zn	Cr 53.7 Col_h 87.3 I	[1323]
4.25	$C_{12}H_{25}$	2H	Cr 85.4 I	[1323]
4.26	$C_{12}H_{25}$	Co	Cr 73 Col_h 153.2 I	[1323]
4.27	$C_{12}H_{25}$	Cu	Cr 83.2 Col_h 102.3 I	[1323]
4.28	$(CH_2CH_2O)_2CH_3$	2H	Cr 55 I	[1326]
4.29	$(CH_2CH_2O)_2CH_3$	Co	Cr −5.5 Col 170.4 I	[1326]
4.30	$(CH_2CH_2O)_2CH_3$	Ni	Cr 44.2 Col_h 115.3 I	[1326]
4.31	$(CH_2CH_2O)_2CH_3$	Cu	Cr 61 Col_h 152 I	[1326]
4.32	$(CH_2CH_2O)_2CH_3$	Zn	Tg −51	[1326]
4.33	$(CH_2CH_2O)_3CH_3$	2H	Tg −75.8	[1326]
4.34	$(CH_2CH_2O)_3CH_3$	Co	g −76.5 Col 64.2 I	[1326]
4.35	$(CH_2CH_2O)_3CH_3$	Ni	Cr −33.5 Col 10.3 Col_h 40.2 I	[1326]
4.36	$(CH_2CH_2O)_3CH_3$	Cu	Cr −16 Col_h 56.2 I	[1326]
4.37	$(CH_2CH_2O)_3CH_3$	Zn	Tg −63.8	[1326]
4.38	$CH_2CH_2CH=CH_2$	2H	Cr 125.8 N_D 154.7 I	[1327]
4.39	$CH_2CH_2CH=CH_2$	Co	Cr 64.9 Col 237.2 I	[1327]
4.40	$CH_2CH_2CH=CH_2$	Cu	Cr 98.2 Col 220 I	[1327]
4.41	$CH_2CH_2CH=CH_2$	Ni	Cr 160.4 Col 217.8 I	[1327]
4.42	$CH_2(CH_2)_2CH=CH_2$	2H	Cr 73.6 N_D 107.2 I	[1327]
4.43	$CH_2(CH_2)_2CH=CH_2$	Co	Cr 52.8 Col 246.2 dec	[1327]
4.44	$CH_2(CH_2)_2CH=CH_2$	Cu	Cr 72 Col 212.4 I	[1327]
4.45	$CH_2(CH_2)_2CH=CH_2$	Ni	Cr 61.5 Col 174.8 I	[1327]
4.46	$CH_2(CH_2)_3CH=CH_2$	2H	Cr 76.5 N_D	[1327]

TABLE 2.123 (continued)
Thermal Behavior of Octakis(octylthio)tetraazaporphyrins 4
(Scheme 2.138)

Structure	R	M	Thermal Behavior	Ref.
4.47	$CH_2(CH_2)_3CH=CH_2$	Co	Cr 63 Col 234.8 I	[1327]
4.48	$CH_2(CH_2)_3CH=CH_2$	Cu	Cr 67.5 Col 178.8 I	[1327]
4.49	$CH_2(CH_2)_3CH=CH_2$	Ni	Cr 62.5 Col 136.3 I	[1327]
4.50	$CH_2(CH_2)_4CH=CH_2$	2H	Cr 71.3 N_D	[1327]
4.51	$CH_2(CH_2)_4CH=CH_2$	Co	Cr 39.2 Col 193.6 I	[1327]
4.52	$CH_2(CH_2)_4CH=CH_2$	Cu	Cr 58.3 Col 131.1 I	[1327]
4.53	$CH_2(CH_2)_4CH=CH_2$	Ni	Cr 67.2 Col 89.7 I	[1327]

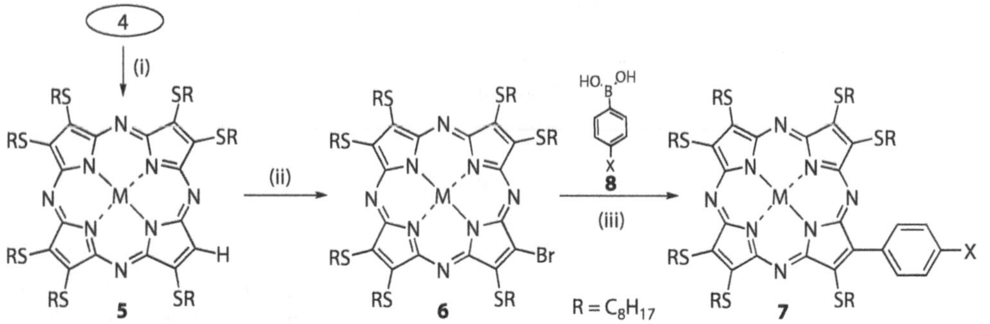

SCHEME 2.139 Synthesis of unsymmetrically substituted porphyrazines **7**: (i) $CrCl_2$, TCB, n-BuOH, 190°C, 7 h; (ii) NBS, room temp, $CHCl_3$, 15 min; (iii) **8**, $Pd(PPh_3)_4$, toluene, DMF, K_2CO_3.

Unsymmetrical porphyrazines **5, 6, 7** in which one of the thioether side chains is replaced by either a hydrogen atom or a bromine atom or an aryl group are also able to display mesophase behavior [1329]. The synthesis and chemical structures of these compounds are depicted in Scheme 2.139 and their thermal data are collected in Table 2.124. It is to be noted that the unsymmetrical derivatives have been obtained from their corresponding symmetrical counterparts. The Ni complex of the aryl derivatives shows great promise in the development of nonlinear optical devices. Both the bromo derivative **6** [1330] and its corresponding Ni complex exhibit columnar hexagonal phase during the heating cycle and columnar rectangular mesophase during the cooling cycle. The nickel complex of **5** also shows similar behavior. Phenyl-substituted porphyrazine **7.1** is nonmesomorphic but its Ni complex exhibits columnar mesophase. Compound **7.2** is liquid crystalline at ambient temperature and goes to the isotropic phase at about 43°C; on cooling, the columnar phase appears at 37°C that crystallizes at low temperature. Compounds **7.3** and **7.4** display similar behavior. The substitution of the phenyl ring with the NMe_2 group destroys mesomorphism in compound **7.5**.

Ohta et al. synthesized a series of tetrapyrazinoporphyrazine **12** complexes (Scheme 2.140) that form rectangular columnar structure, and their mesomorphic behavior is very similar to the corresponding phthalocyanine homologues [1331]. These moieties can be used as electron acceptors in mixed columnar liquid crystals. The thermal phase behavior of these compounds is summarized in Table 2.125. Both free bases and metal complexes exhibit multiple crystal transitions before passing to mesophases, and thermal decomposition precedes their isotropization temperature. It has been found that the axially substituted tetrapyrazinoporphyrazinatoindium chloride shows an optical limiting behavior at 532 nm [1332].

Ohta and coworkers also reported the mesomorphism of octakis(4-alkoxyphenyl)- and octakis(3,4-dialkoxyphenyl)tetrapyrazinoporphyrazinatometal(II) derivatives **15** [1333]. These were synthesized

TABLE 2.124
Thermal Behavior of Unsymmetrically Substituted Porphyrazines 5, 6, 7 (Scheme 2.139)

Structure	X	M	Thermal Behavior
5.1	—	2H	Cr 51.7 Col$_h$ 99.7 I
5.2	—	Ni	M 51.7 Col$_h$ 92.5 I
6.1	—	2H	Cr 37.3 Col$_h$ 113.7 I
6.2	—	Ni	M 47.4 Col$_h$ 120 I
7.1	H	2H	Cr 46.2 I
7.2	H	Ni	Col 43.4 I
7.3	Ph	2H	Col 77.2 I
7.4	Ph	Ni	Col 62.5 I
7.5	NMe$_2$	Ni	Cr 77.1 I

Source: Data from Belviso, S. et al., *Mol. Cryst. Liq. Cryst.*, 481, 56, 2008.

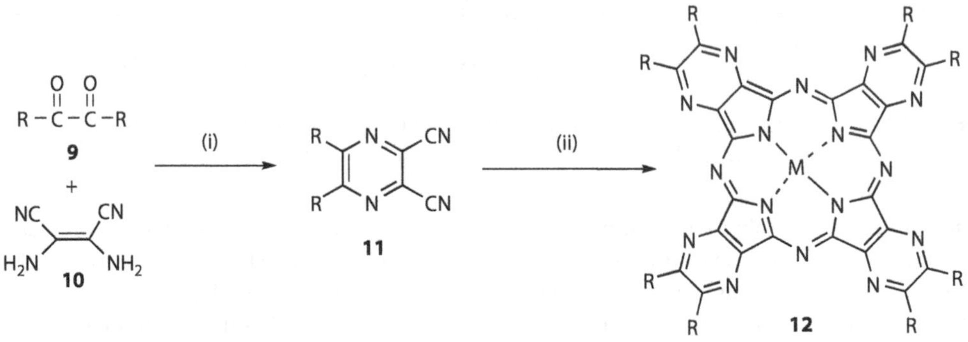

SCHEME 2.140 Synthesis of tetrapyrazinoporphyrazines **12**: (i) AcOH, (ii) MX$_2$, DBU, alcohol.

TABLE 2.125
Thermal Behavior of Tetrapyrazino Porphyrazine 12 Complexes (Scheme 2.140)

Structure	R	M	Thermal Behavior
12.1	C$_{12}$H$_{25}$	2H	Cr 118 Col$_h$ 238 dec
12.2	C$_{12}$H$_{25}$	Cu	Cr 114 Col$_r$ 288 dec
12.3	C$_{12}$H$_{25}$	Ni	Cr 118 Col$_r$ 264 dec
12.4	C$_{12}$H$_{25}$	Co	Cr 74 Col$_r$ 255 dec
12.5	CH$_2$CH(C$_2$H$_5$)C$_4$H$_9$	2H	Cr 74.8 M$_1$ 193 M$_2$ 219 I
12.6	CH$_2$CH(C$_2$H$_5$)C$_4$H$_9$	Cu	Cr 90 M 232 I

Source: Data from Ohta, K. et al., *Mol. Cryst. Liq. Cryst.*, 196, 13, 1991.

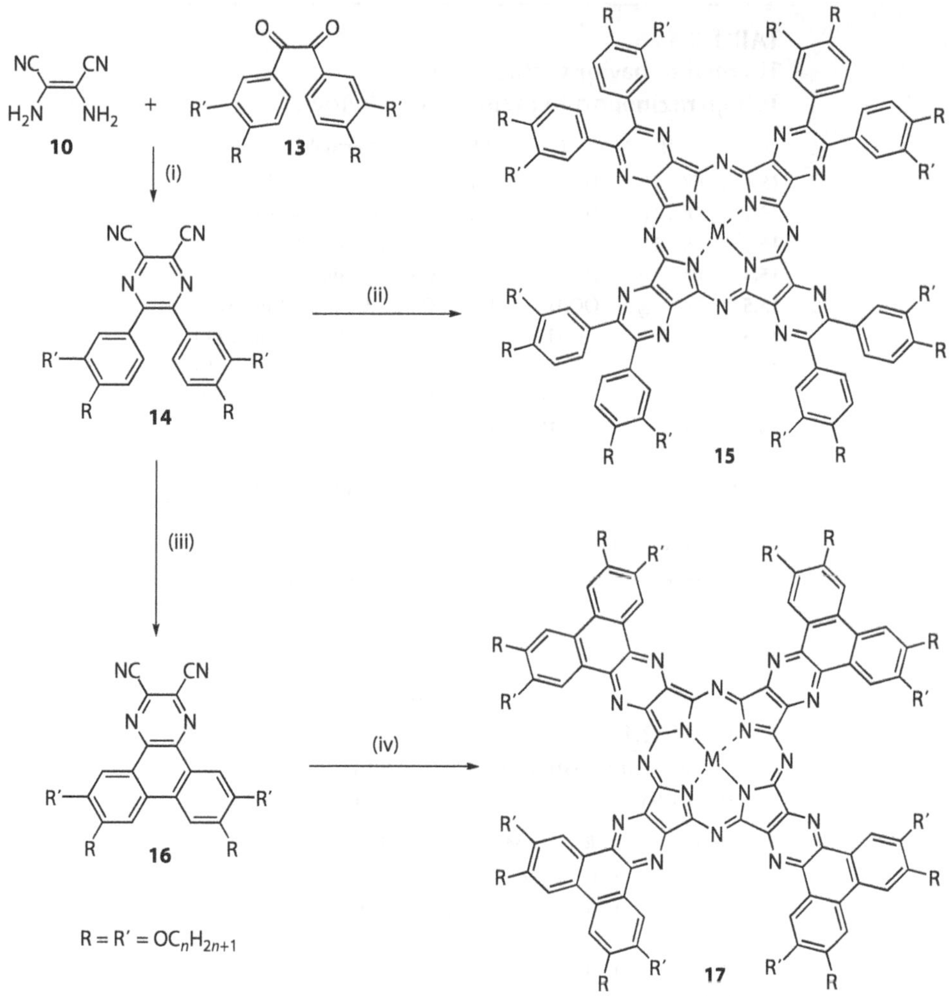

SCHEME 2.141 Synthesis of octaphenyl tetrapyrazino porphyrazines **16** and triphenylenoporphyrazines **17**: (i) AcOH; (ii) DBU/n-C$_5$H$_{11}$OH, MX$_2$; (iii) VOF$_3$, BF$_3$/Et$_2$O, CH$_2$Cl$_2$, room temperature, 45 min; (iv) MCl$_2$, DBU, 2-methylbutan-2-ol, 72 h.

by the common method of porphyrazine synthesis with the appropriate substitution, as shown in Scheme 2.141. The thermal behavior of these derivatives is described in Table 2.126. Metal complexes of the p-alkoxyphenyl substituted derivatives **15** display multiple mesophase with tetragonal, rectangular, and hexagonal symmetry, and the thermal decomposition of these materials precedes their isotropization. Metal complexes of 3,4-dialkoxyphenyl-substituted derivatives **15** also exhibit columnar mesomorphism of comparable stability. The lower members of the series exhibit multiple columnar rectangular phases while the higher members display both columnar rectangular and hexagonal phases. It is interesting to note that the higher symmetry hexagonal phase appears below the lower symmetry rectangular columnar phase in the higher members.

Some of the core-extended porphyrazines such as triphenylenoporphyrazines **17** display mesomorphism. The synthetic pathway involves oxidative aryl–aryl coupling of **14** with VOF$_3$ in the presence of borontriflouride diethyl ether to afford **16**, which can be tetramerized under standard reaction conditions to yield triphenylenoporphyrazines **17** (Scheme 2.141). The phase transition temperatures of these compounds are produced in Table 2.127. These derivatives show uniform homeotropic alignment at room temperature over a macroscopic area, and they display tetragonal

TABLE 2.126
Thermal Behavior Octaphenyl
Tetrapyrazinoporphyrazines 15 (Scheme 2.141)

St	R	R'	M	Thermal Behavior
15.1	$OC_{10}H_{21}$	H	Cu	Col_t 149 Col_h 230 dec
15.2	$OC_{12}H_{25}$	H	Cu	Col_t 109 Col_r 230 dec
15.3	$OC_{10}H_{21}$	H	Ni	Col_t 150 Col_h 210 dec
15.4	$OC_{12}H_{25}$	H	Ni	Col_t 109 Col_r 230 dec
15.5	OC_8H_{17}	OC_8H_{17}	Cu	Cr −29 Col_{r1} 247 Col_{r2} 323 dec
15.6	OC_8H_{17}	OC_8H_{17}	Ni	Cr −41 Col_{r1} 226 Col_{r2} 324 I
15.7	$OC_{10}H_{21}$	$OC_{10}H_{21}$	Cu	Cr −13 Col_{r1} 221 Col_{r2} 281 I
15.8	$OC_{10}H_{21}$	$OC_{10}H_{21}$	Ni	Col_h 41 Col_{r1} 214 Col_{r2} 269 I
15.9	$OC_{12}H_{25}$	$OC_{12}H_{25}$	Cu	Col 14 Col_h 44 Col_{r1} 194 Col_{r2} 241 I 320 dec
15.10	$OC_{12}H_{25}$	$OC_{12}H_{25}$	Ni	Col_h 40 Col_r 206 I 320 dec

Source: Data from Ohta, K. et al., *J. Mater. Chem.*, 9, 2313, 1999.

TABLE 2.127
Phase Transition Temperatures of
Triphenylenoporphyrazines (Scheme 2.141)

Structure	R	M	Thermal Behavior	Ref.
17.1	C_8H_{17}	Cu	Col_t 340 dec	[1334]
17.2	$C_{10}H_{21}$	Cu	Cr −100 Col_t 300 I	[1135]
			Col_t 340 dec	[1334]
17.3	$C_{10}H_{21}$	Ni	Cr −100 Col_t 300 I	[1135]
17.4	$C_{12}H_{25}$	Cu	Col_t 350 dec	[1334]
17.5	$C_{14}H_{29}$	Cu	Col_t 350 dec	[1334]

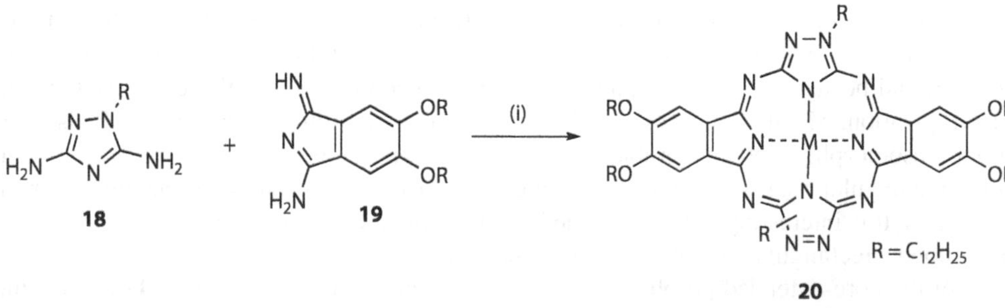

20.1: M = 2H; Cr 39 Col_h 165 I
20.2: M = Cu; Cr 40.5 Col_h 313.3 I
20.3: M = Ni; Cr 37.1 Col_h 303.8 I

SCHEME 2.142 Synthesis of triazolehemiporphyrazines: (i) 2-ethoxyethanol, reflux.

columnar structure [1135,1134]. Some of the derivatives pass into stable isotropic phase while others decompose before their isotropization.

A few hexa-substituted hemiporphyrazine discotics were prepared by Fernandez et al., as shown in Scheme 2.142 [1335]. Though the shape of the molecule is elliptical, it displays hexagonal columnar phase. The hexadodecyloxy-substituted macrocycles melt at relatively low temperature to liquid crystalline phase which was stable up to 300°C (Scheme 2.142). The corresponding dioctylamino-carbonilmethoxy derivatives were found to be non-liquid crystalline.

2.22 PHTHALOCYANINE CORE

Phthalocyanine (Pc), **1**, is a two-dimensional symmetrical aromatic heterocyclic compound composed of four iminoisoindoline units with a cavity of sufficient size to accommodate various metal ions. In other words, phthalocyanines (Pcs) are macrocyclic compounds having an alternative nitrogen atom–carbon atom ring structure and act as a tetradentate ligand. Moreover, owing to their close resemblance to porphyrins, Pcs are otherwise known as tetrabenzo-tetraazaporphyrins [6,1336–1340]. Pcs and their metallo-derivatives have recently attracted increasing interest not only for the preparation of dyes and pigments but also as building blocks for the construction of new molecular materials for electronics and optoelectronics. The versatility, architectural flexibility, non-toxicity, and ease of processing make them eligible candidates for use in electronics. Interestingly, Pcs can act both as electron acceptor and electron donor with respect to other moieties in donor–acceptor ensembles. In addition to acting as components of molecular electronics, Pcs find biological and biomedical applications. Recently, Pc-conjugated quantum dots have been explored as materials for the treatment of cancer by photodynamic therapy and in quantum computing [1341–1343]. The chemical flexibility of Pcs allows the preparation of a large variety of related structures and consequently, the tailoring of the physical, electronic, and optical properties, as well as the improvement of processability.

1

The peripheral substitution of Pcs with hydrocarbon chains not only enhances their solubility and permits facile deposition onto substrates but also induces liquid crystallinity [4,9,879,1344,1345]. Pcs can be self-assembled into columnar structure to form lyotropic and thermotropic liquid crystalline phases and highly ordered thin films. Mesogenic Pcs offer the possibility of combining the excellent optoelectronic properties of Pcs with the orientational control of conventional liquid crystal systems and the facile fabrication of ordered thin films in the liquid crystalline state. Since the initial discovery [1346], many examples of mesogenic Pcs have been prepared with variations in the nature, number, length, and position of the flexible side chains. In addition, the linking group attaching the side chains to the Pc core and the central metal ion all have substantial influence on the structure and thermal stability of the resultant mesophase. Pcs enjoy the luxury of incorporating about 70 different metal ions into the cavity and thereby optimize the electronic and optical properties [4,9,879,1344,1345]. Moreover, the central metal atoms can carry additional axial ligands. Depending on the size and oxidation state of the metal ions, a library of Pcs with different ability to self-assembly and improved physical properties can be designed and synthesized. It is interesting to note that lyotropic mesomorphism in aqueous solution of the sodium salt of Pc tetracarboxylic acid was reported in 1979 [1347] before the discovery of thermotropic mesomorphism in these materials in 1982 [1346].

Mesomorphic Pcs have been extensively studied for their various useful physical properties, including charge migration, crystal structure, electron spin resonance, laser-induced triplet excitons and energy migration, luminescence, electric and magnetic properties, nonlinear optical properties, etc. [1348–1369]. Owing to their large core size, Pc derivatives aggregate in solution to form supramolecular fibers that can be transferred to substrates and, therefore, Langmuir monolayers and Langmuir–Blodgett films of amphiphilic Pcs have been extensively studied. Self-assembled monolayers of Pcs from solutions have been visualized by scanning tunneling microscopy on inert surfaces. Moreover, macroscopic alignments (homeotropic and homogeneous) of mesomorphic Pc thin films have been achieved on/between various substrates (see Chapter 1). In an effort to comprehend the excellent electronic, optical, and bulk self-assembly properties of Pc derivatives, various theoretical studies have been undertaken, including molecular dynamic studies [1370,1371]. Discotic liquid crystalline phthalocyanines can be categorized into the following groups: (1) peripherally octaalkoxymethyl-substituted Pcs; (2) peripherally octaalkoxy-substituted Pcs; (3) peripherally octaalkyl-substituted Pcs; (4) peripherally octathioalkyl-substituted Pcs; (5) peripherally octathiaalkylmethyl-substituted Pcs; (6) peripheral octaalkoxyphenyl- and alkoxyphenoxy-substituted Pcs; (7) octaalkyl esters of Pc; (8) non-peripherally substituted octaalkyl and octaalkoxymethyl Pcs; (9) non-symmetrical octa-, hepta-, hexa-, and penta-substituted Pcs; (10) unsymmetrical nonperipheral Pcs; (11) tetraalkoxy-substituted Pcs; (12) tetrathiaalkyl and tetraalkylthiamethyl-substituted Pcs; (13) tetraesters of phthalocyanine; (14) crown-ether-substituted Pcs; (15) core-extended macrodiscotic Pcs; (16) subphthalocyanines; and (17) miscellaneous compounds structurally related to Pcs.

2.22.1 Octaalkoxymethyl-Substituted Phthalocyanines

In 1982, Piechocki et al. synthesized octa(dodecyloxymethyl)phthalocyanine, which was the first phthalocyanine-based discotic liquid crystal [1346]. The synthesis is shown in Scheme 2.143. *o*-Xylene **2** is treated with Br$_2$ to obtain dibromo derivative of xylene, **3**. The bromination of the methyl groups of **3** is accomplished by treating with NBS (N-bromosuccinimide) to afford the tetrabrominated

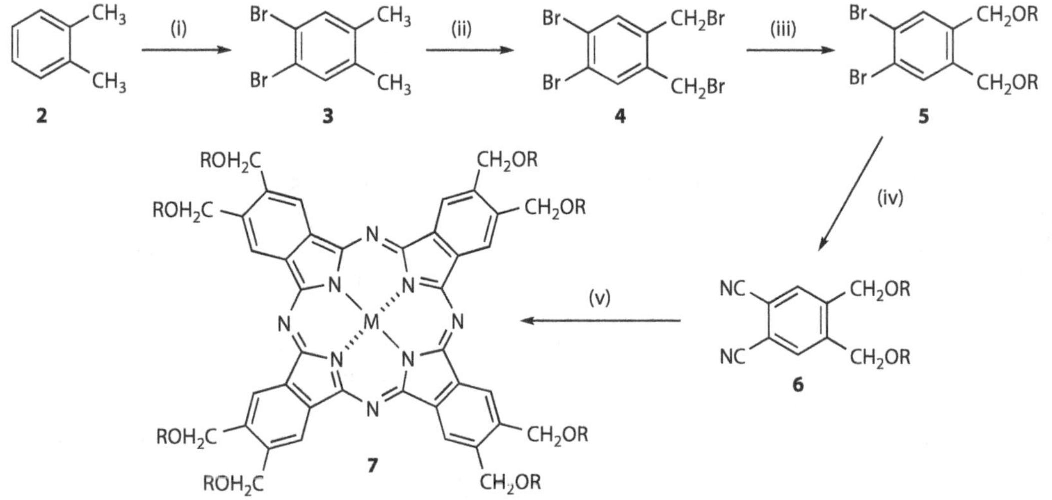

SCHEME 2.143 Synthesis of alkoxymethyl-substituted phthalocyanines: (i) Br$_2$; (ii) NBS; (iii) ROH, K$_2$CO$_3$; (iv) CuCN, DMF; (v) DMAE; metal salt.

compound **4**. Treatment of **4** with appropriate alkoxide displaces the Br attached to methylene group and furnished **5**, which can be converted to phthalonitrile **6** by reacting with Cu(I)CN in DMF. The cyclization of phthalonitrile **6** to phthalocyanine **7** is achieved by refluxing in 2-dimethylaminoethanol. The octaalkoxymethyl-substituted phthalocyanine with simple alkyl chain and with no central metal ion or small central metal ions show columnar hexagonal mesophase [1346,1354,1372–1384]. The thermal phase behavior of these compounds is collected in Table 2.128. The incorporation of small metal ions stabilizes the discotic mesophase. As expected, lengthening of alkyl chains leads to the destabilization of the mesophase. Octaalkoxymethyl-Pcs exhibited wide mesophase behavior when coordinated with a wide variety of divalent metal ions (M=Cu, Zn, Pb, Ni, Sn, Co). Complexation to metal ion increased the stability of the mesophase due to long range stacking order. The transition temperatures of the mesophase are dependent on the central metal ion to which it was coordinated. Small-angle x-ray diffraction patterns indicate a two-dimensional hexagonal order with an almost constant lattice parameter (intercolumnar distance) of 31 Å regardless of the central metal atom.

Copper or zinc insertion in Pc forms stable planar complexes; thus they have higher melting temperatures than metal-free derivatives [1345,1346,1375,1378]. Nickel and cobalt complexes show similar mesophase ranges as metal-free derivatives [1345,1379]. Manganese derivatives also show a hexagonal columnar mesophase [1378]. Alkoxymethyl derivatives with $n=8$, 12, 18 have been studied with lead as a central metal atom and have shown mesophase at very low temperature compared to their other analogues [1377]. This is because lead(II) ions, owing to their larger size, cannot be accommodated in the phthalocyanine cavity and form out-of-plane complexes. Tin (II) compound

TABLE 2.128
Phase Transition Temperatures of Octaalkoxymethyl Substituted Pcs (Scheme 2.143)

Structure	M	R	Transition Temperature	Ref.
7.1	2H	C_8H_{17}	Cr 66 Col$_h$ 322 I	[1345]
7.2	2H	$C_{12}H_{25}$	Cr 79 M 260 I	[1373]
7.3	2H	$C_{13}H_{27}$	Cr 80 Col$_h$	[1372]
7.4	2H	$C_{14}H_{29}$	Cr 83 Col$_h$ 234 I	[1375]
7.5	Cu	C_8H_{17}	Cr 82 Col$_h$ >300 dec	[1345]
7.6	Cu	$C_{12}H_{25}$	Cr 53 Col$_h$ >300 I	[1378]
7.7	Cu	$C_{14}H_{29}$	Cr 80 Col$_h$ 280 I	[1375]
7.8	Pb	C_8H_{17}	Cr −45 Col$_h$ 155 I	[1377]
7.9	Pb	$C_{12}H_{25}$	Cr −12 Col$_h$ 125 I	[1377]
7.10	Pb	$C_{18}H_{37}$	Cr 46 Col 60 I	[1377]
7.11	Zn	C_8H_{17}	Cr 88 Col$_h$ >300 dec	[1345]
7.12	Zn	$C_{12}H_{25}$	Cr 78 Col$_h$ >300 I	[1378]
7.13	Ni	C_8H_{17}	Cr 67.5 Col$_h$ >300 I	[1379]
7.14	Ni	$C_{12}H_{25}$	Cr 58 Col$_h$ >300 I	[1345]
7.15	Co	C_8H_{17}	Cr 72 Col$_h$ >300 I	[1379]
7.16	Co	$C_{12}H_{25}$	Cr 71 Col$_h$ >300 I	[1345]
7.17	NaCo(CN)$_2$	C_8H_{17}	Cr 73 Col$_h$ >300 I	[1379]
7.18	Mn	$C_{12}H_{25}$	Cr 44 Col$_h$ 280 I	[1378]
7.19	Si(OH)$_2$	$C_{12}H_{25}$	Cr −7 Col$_h$ >300 I	[1380]
7.20	Sn(OH)$_2$	$C_{12}H_{25}$	Cr 59 Col$_r$ 95 M 114 I	[1381]
7.21	2H	$CH_2CH(CH_3)O(CH_2)_{11}CH_3$	Cr 23 M 158 I	[1384]
7.22	Cu	$CH_2CH(CH_3)O(CH_2)_{11}CH_3$	Cr 29 M 191 I	[1384]
7.23	2H	$C_2H_4SC_{12}H_{25}$	Cr 52 Col$_h$ 247 I	[1382]
7.24	Cu	$C_2H_4SC_{12}H_{25}$	Cr 70 Col$_h$ 255 I	[1382]

in the presence of air rapidly transforms into dihydroxy Sn(IV) compound, which shows rectangular columnar mesophase and another mesophase at higher temperature. Eventually, in the isotropic liquid, the compound undergoes polymerization, which can be attributed to loss of water [1381].

The effect of decreasing side-chain length has been studied for cobalt(II) complexes, and for chain lengths smaller than $n = 8$ in $CH_2OC_nH_{2n+1}$, no mesophase was observed [1379]. The alkoxymethyl substitution to the phthalocyanine ring orients itself at an angle of 35° to phthalocyanine planes, which leads to a slightly disordered arrangement of phthalocyanine rings in columns. Some branched alkoxymethyl-substituted derivatives have also been synthesized and studied, which exhibit mesomorphism [1382].

2.22.2 Octaalkoxy-Substituted Phthalocyanines

Octaalkoxy-substituted phthalocyanines are the most widely studied phthalocyanine discotics. The most convenient and common method to prepare these derivatives is via the cyclization of 1,2-dicyano-4,5-bis(alkoxy)benzene **8** (Scheme 2.144). In order to improve the yield, the cyclization reaction has also been carried out using 5,6-dialkoxy-1,3-diiminoisoindoline **9**, which can be obtained from 1,2-dicyano-4,5-bis(alkoxy)benzene by the reaction of ammonia. However, generally, not much improvement in the overall yield was realized. The dialkoxyphthalonitrile can be easily obtained from catechol via alkylation, bromination, and cyanation reactions. The cyclization is commonly carried out in a high boiling solvent in the presence of a non-nucleophilic strong base. Metal salt can be added in the same reaction to obtain the metallated phthalocyanine **11**. Alternatively, free phthalocyanine **10** can be first prepared and then reacted with appropriate metal salt. However, the isolation of metal-free phthalocyanine is often tedious and low yielding and, therefore, it is convenient to prepare metallated phthalocyanine directly unless otherwise the metal-free phthalocyanine is required.

The steric and electronic effects of the groups that link alkyl chain to phthalocyanine core strongly influence the transition temperature and phase structures of the mesophase. Molecular modeling calculations suggest that the relatively small alkoxy linking groups allows the side chains to sit comfortably in the plane of Pc ring since there is no steric hindrance in these compounds unlike their alkyl or alkoxymethyl-substituted counterparts in which there is considerable steric crowding, leading to a tilting of side chains at an angle of 35° to the plane of the Pc ring. The alkoxy linking groups favor the untilted hexagonal columnar mesophase with a molecular stacking distance of about 3.4 Å [1385].

The mesophase data of the alkoxy-substituted derivatives are collected in Table 2.129. Octaalkoxy-Pcs with normal chains form stable ordered hexagonal columnar phases [1338,1345,1385–1403]. The

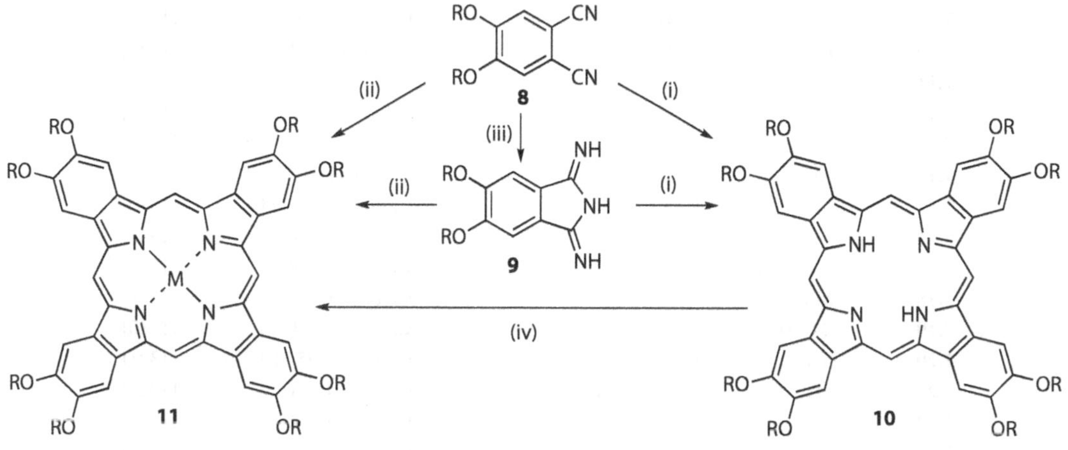

SCHEME 2.144 Synthesis of alkoxyl-substituted phthalocyanines: (i) reflux in a high boiling solvent; (ii) reflux in a high boiling solvent with metal salt; (iii) NH₃, MeOH, NaOMe; (iv) reflux in EtOH with metal salt.

TABLE 2.129
Phase Transition Temperatures of Octalkoxy-Substituted Phthalocyanines (Scheme 2.144)

Structure	M	R	Transition Temperature	Ref.
11.1	2H	C_6H_{13}	Cr 119 Col_h	[1385]
11.2	2H	C_7H_{15}	Cr 104 Col_h	[1385]
11.3	2H	C_8H_{17}	Cr 94 Col_h	[1385]
11.4	2H	C_9H_{19}	Cr 101 Col_h	[1385]
11.5	2H	$C_{10}H_{21}$	Cr 94 Col_h 345 I	[1385]
11.6	2H	$C_{11}H_{23}$	Cr 83 Col_h 334 I	[1385]
11.7	2H	$C_{12}H_{25}$	Cr 83 Col_h 309 I	[1385]
11.8	2H	$C_{18}H_{37}$	Cr 98 Col_h 247 I	[1338]
11.9	2H	$CH_2CH(Et)C_4H_9$	Cr 170 Col_t 223 N_D 270 I	[1394]
11.10	Cu	C_6H_{13}	Cr 120 Col_h	[1385]
11.11	Cu	C_7H_{15}	Cr 110 Col_h	[1385]
11.12	Cu	C_8H_{17}	Cr 112 Col_h	[1385]
11.13	Cu	C_9H_{19}	Cr 106 Col_h	[1385]
11.14	Cu	$C_{10}H_{21}$	Cr 104 Col_h	[1385]
11.15	Cu	$C_{11}H_{23}$	Cr 92 Col_h	[1385]
11.16	Cu	$C_{12}H_{25}$	Cr 95 Col_h	[1385]
11.17	Cu	$CH_2CH(Et)C_4H_9$	Cr 204 Col_t 242 N_D 290 I	[1401]
11.18	Co	C_4H_9	Cr 218 Col_h	[1386]
11.19	Co	C_5H_{11}	Cr 133 Col_h	[1386]
11.20	Co	C_6H_{13}	Cr 126 Col_h	[1386]
11.21	Co	C_7H_{15}	Cr 116 Col_h	[1386]
11.22	Co	C_8H_{17}	Cr 123 Col_h	[1386]
11.23	Co	C_9H_{19}	Cr 112 Col_h	[1386]
11.24	Co	$C_{10}H_{21}$	Cr 111 Col_h	[1386]
11.25	Co	$C_{11}H_{23}$	Cr 102 Col_h	[1386]
11.26	Co	$C_{12}H_{25}$	Cr 101 Col_h	[1386]
11.27	Co	$C_{13}H_{27}$	Cr 91 Col_h	[1386]
11.28	Co	$C_{14}H_{29}$	Cr 95 Col_h 328 I	[1386]
11.29	Co	$C_{15}H_{31}$	Cr 91 Col_h 305 I	[1386]
11.30	Co	$C_{16}H_{33}$	Cr 92 Col_h 306 I	[1386]
11.31	Co	$C_{17}H_{35}$	Cr 96 Col_h 292 I	[1386]
11.32	Co	$C_{18}H_{37}$	Cr 64 Col_h 266 I	[1386]
11.33	VO	C_8H_{17}	Cr 63 Col_r 250 dec	[1402]
11.34	VO	$C_{12}H_{25}$	Cr 51 Col_r 250 dec	[1402]
11.35	VO	$C_{16}H_{33}$	Cr 60 Col_r 250 dec	[1402]
11.36	$Si(OH)_2$	C_2H_5	Non-LC	[1389]
11.37	$Si(OH)_2$	C_4H_9	Cr 111.4 Col_h	[1389]
11.38	$Si(OH)_2$	C_6H_{13}	Cr 71 Col_h	[1389]
11.39	$Si(OH)_2$	C_8H_{17}	Cr 59.2 Col_h 84 I	[1389]
11.40	$Si(OH)_2$	$C_{10}H_{21}$	Cr 58.2 Col_h 87.3 I	[1389]
11.41	$Si(OH)_2$	$C_{12}H_{25}$	Cr 58.8 Col_h 93.6 I	[1389]
11.42	$Si(OH)_2$	$C_{14}H_{29}$	Cr 55.4 Col_h 107.2 I	[1389]
11.43	$Si(OH)_2$	$C_{16}H_{33}$	Cr 56.2 Col_h 89.3 I	[1389]
11.44	$Si(OH)_2$	$C_{18}H_{37}$	Cr 56.7 Col_h 86.6 I	[1389]
11.45	Zn	C_8H_{17}	Cr 135 Col_h >300 I	[1345]
11.46	Zn	$C_{12}H_{25}$	Cr 99 Col_h 375 I	[1390]

(*continued*)

TABLE 2.129 (continued)
Phase Transition Temperatures of Octalkoxy-Substituted
Phthalocyanines (Scheme 2.144)

Structure	M	R	Transition Temperature	Ref.
11.47	Zn	$(CH_2)_2CH(CH_3)$ $[(CH_2)_3CH(CH_3)]_3CH_3$	Cr -79 Col$_h$ 121 I	[1391]
11.48	Ni	C_8H_{17}	Cr 110 Col$_h$ > 300 I	[1387]
11.49	Ni	$C_{10}H_{21}$	Cr 87 Col$_h$ > 300 I	[1387]
11.50	Ni	$C_{12}H_{25}$	Cr 97 Col$_h$ 254 I	[1390]
11.51	Pb	$CH_2CH(Et)C_4H_9$	Col$_t$ 79 M > 240 I	[1401]
11.52	Pt	$C_{10}H_{21}$	Cr 64 Col$_h$ > 300 I	[1387]
11.53	Pt	$C_{12}H_{25}$	Cr 77 M > 350 I	[1388]
11.54	Pt	$CH_2CH(Et)C_4H_9$	Col$_{ob}$ 205 I	[1401]
11.55	Pd	$C_{10}H_{21}$	Cr 90 M	[1387]
11.56	2H	$(CH_2)_{10}OH$	Cr 120 Col$_h$ 254 I	[1395]
11.57	Cu	$(CH_2)_{10}OH$	Cr 134 Col$_h$	[1395]
11.58	Zn	$(CH_2)_{10}OH$	Cr 139 Col$_h$	[1395]
11.59	Co	$(CH_2)_{10}OH$	Cr 112 Col$_h$	[1395]
11.60	Ni	$(CH_2)_{10}OH$	Cr 66 Col$_h$	[1395]
11.61	Pb	$(CH_2)_{10}OH$	Cr 110 Col$_h$	[1395]
11.62	2H	$(CH_2)_{10}OC(O)CH_3$	Cr 85 Col$_h$ 235 I	[1395]
11.63	Cu	$(CH_2)_{10}OC(O)CH_3$	Cr 75 Col$_h$	[1395]
11.64	Zn	$(CH_2)_{10}OC(O)CH_3$	Cr 82 Col$_h$	[1395]
11.65	Co	$(CH_2)_{10}OC(O)CH_3$	Cr 82 Col$_h$	[1395]
11.66	Ni	$(CH_2)_{10}OC(O)CH_3$	Cr 63 Col$_h$	[1395]
11.67	2H	$(CH_2)_{10}OC(O)CHCH_2$	Cr 85 Col$_h$	[1395]
11.68	Cu	$(CH_2)_{10}OC(O)CHCH_2$	Cr 70 Col$_h$	[1395]
11.69	Zn	$(CH_2)_{10}OC(O)CHCH_2$	Cr 77 Col$_h$	[1395]
11.70	Co	$(CH_2)_{10}OC(O)CHCH_2$	Cr 75 Col$_h$	[1395]
11.71	Ni	$(CH_2)_{10}OC(O)CHCH_2$	Cr 73 Col$_h$	[1395]
11.72	2H	$(CH_2)_{10}OCOC(CH_3)CH_2$	Cr 54 Col$_h$	[1395]
11.73	Cu	$(CH_2)_{10}OCOC(CH_3)CH_2$	Cr 51 Col$_h$	[1395]
11.74	Zn	$(CH_2)_{10}OCOC(CH_3)CH_2$	Cr 80 Col$_h$	[1395]
11.75	Co	$(CH_2)_{10}OCOC(CH_3)CH_2$	Cr 85 Col$_h$	[1395]
11.76	Ni	$(CH_2)_{10}OCOC(CH_3)CH_2$	Cr 50 Col$_h$	[1395]
11.77	Cu	$C_{10}H_{20}(OEt)_3OCH_3$	Cr 136 Col$_h$ 220 I	[1396]
11.78	2H	$(EtO)_3Me$	Col$_h$ 253 I	[1397]
11.79	2H	$(CH_2)_{11}N^+H(CH_3)_2$ OAc$^-$	Cr 73 M 89 Col$_h$ 270 I	[1398]
11.80	Cu	$(CH_2)_2OCH_2Ph$	Cr 63 Col$_h$	[1399]
11.81	2H	$CH_2C(O)N(C_8H_{17})_2$	Col$_h$ 91 I	[1400]
11.82	Cu	$CH_2C(O)N(C_8H_{17})_2$	Col$_h$ 95 I	[1400]
11.83	Ni	$CH_2C(O)N(C_8H_{17})_2$	Col$_h$ 81 I	[1400]
11.84	2H	$CH_2CH(CH_3)$ $(CH_2)_3C^*H(CH_3)(CH_2)_3CH_3$	(R,S) Cr 70 Col$_h$ 295 I	[1392]
11.85	2H	$CH_2CH(CH_3)$ $(CH_2)_3C^*H(CH_3)(CH_2)_3CH_3$	(S) Cr 14 Col$_h$ 111 Col$_r$ 295 I	[1392]
11.86	2H	$CH_2CH_2(CH(CH_3)$ $(CH_2)_3)_2CH(CH_3)_2$	Col$_h$ 34 Col$_h$ 185 I	[1393]

isotropic temperature of these derivatives is generally very high (>300°C) and often they decompose before reaching to the isotropic phase. As observed commonly in DLCs, the melting points decrease with increasing chain length, although there are some deviations for lower homologues which show odd–even effects. Melting points are marginally higher for the copper derivatives; this can be due to the stabilization of the mesophase by small metal ions [1385]. The cobalt complexes of Pc show a columnar hexagonal mesophase and display the general trend of decreasing clearing temperature upon increasing chain length [1386]. The phthalocyanine complexed to silicon, which has two axial hydroxyl substitutions, shows columnar mesophase, with low clearing temperature. This can be inferred due to a destabilization of the ordered stacking of phthalocyanine cores by axially substituted hydroxyl groups [1389].

The phthalocyaninatoplatinum(II) complex substituted with eight dodecyloxy groups forms columnar hexagonal mesophase with an intracolumnar distance of 3.29 Å and a hexagonal lattice constant of 34.3 Å. The unusually short Pt....Pt distance was attributed to attractive overlap of $d_z{}^2$ platinum orbitals [1387,1388]. Branched phthalocyanines like octa(2-ethylhexyloxy)phthalocyanine with M = 2H, Cu, Pb, Pt form columnar tetragonal mesophase at lower temperature; on further heating, a nematic mesophase is formed in some of these compounds [1391–1394,1401,1403]. Phthalocyanine complex with VO as central group has oxo functionality, which protrudes out of phthalocyanine plane and shows Col_r mesophase at low temperature [1402]. Many terminally functionalized Pc derivatives, which can act as precursors for polymers and networks, have been synthesized and studied. They generally exhibit disordered columnar hexagonal phase [1395,1398]. The derivatives with amide group having two pendant alkyl chains show room temperature Col_h mesophase, which clears to the isotropic liquid at less than 100°C [1400].

2.22.3 Octaalkyl-Substituted Phthalocyanines

Peripherally substituted octaalkyl phthalocyanine can be synthesized by alkylating o-dichloro benzene 12 with a Grignard reagent in the presence of a nickel catalyst to give dialkylbenzenes 13. Bromination with molecular bromine followed by reaction with copper cyanide in DMF yields the dialkyl phthalonitrile 15. Finally, the phthalonitrile is cyclized to phthalocyanine 16 by refluxing in 2-dimethylaminoethanol (Scheme 2.145) [1404].

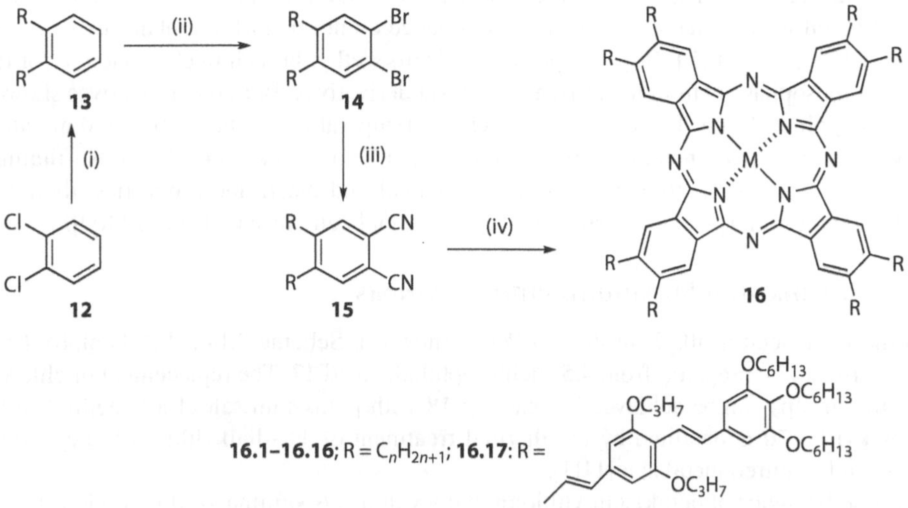

16.1–16.16; R = C_nH_{2n+1}; **16.17**: R =

SCHEME 2.145 Synthesis of alkyl-substituted phthalocyanines: (i) RMgBr, Ni catalyst; (ii) Br$_2$; (iii) CuCN, DMF; (iv) DMAE.

TABLE 2.130
Phase Transition Temperatures of Octaalkyl-Substituted Pcs
(Scheme 2.145)

Structure	M	R	Transition Temperature	Ref.
16.1	2H	C_5H_{11}	Cr 323 Col$_h$ 379 I	[1405]
16.2	2H	C_6H_{13}	Cr 250 Col$_h$ 363 I	[1405]
16.3	2H	C_8H_{17}	Cr 186 Col$_h$ 325 I	[1405]
16.4	2H	$C_{10}H_{21}$	Cr 163 Col$_h$ 282 I	[1405]
16.5	2H	$C_{12}H_{25}$	Cr 120 Col$_h$ 252 I	[1372]
16.6	2H	$C_{16}H_{33}$	Cr 108 Col$_r$ 170 Col$_h$ 196 I	[1406]
16.7	2H	$CH_2CH(C_2H_5)C_4H_9$	Cr 267 I	[1405]
			Col$_{r1}$ 63.2 Col$_{r2}$ 269 I	[1394]
16.8	Cu	C_5H_{11}	Cr 342 Col$_h$	[1405]
16.9	Cu	C_6H_{13}	Cr 259 Col$_h$	[1405]
16.10	Cu	C_8H_{17}	Cr 180 Col$_h$	[1405]
16.11	Cu	$C_{10}H_{21}$	Cr 169 Col$_h$ 351 I	[1405]
16.12	Ni	C_6H_{13}	Cr 260 Col$_h$ 412 I	[1405]
16.13	Ni	C_7H_{15}	Cr 203 M	[1387]
16.14	Ni	C_8H_{17}	Cr 190 Col$_h$ 373 I	[1405]
16.15	Ni	$C_{10}H_{21}$	Cr 168 Col$_h$ 333 I	[1405]
16.16	Ni	$CH_2CH(C_2H_5)C_4H_9$	Cr 293 I	[1405]
16.17	Zn	(a)	Cr 16 Col$_h$ 220 I	[1407]

The alkyl-substituted Pcs are less explored than their alkoxy and alkoxymethyl counterparts. Octaalkyl-substituted Pc with alkyl chain length equal to or longer than five carbon atoms exhibit mesophase, characterized by hexagonal columnar arrangement for both metal-free and metal-substituted derivatives [1373,1387,1394,1405–1407]. The mesophase behavior of these derivatives is summarized in Table 2.130. The metal-free hexadecyl derivative shows columnar rectangular mesophase at low temperature, which changes to columnar hexagonal phase at higher temperature [1406]. Copper- and nickel-substituted Pcs showed high melting and clearing temperatures compared to metal-free alkyl Pc, showing the stabilization of the mesophase by metal atoms. Copper and nickel metal ions fit well into the cavity and hence do not disturb the planarity of the macrocyclic core [1373,1387,1405,1406]. The presence of bulky ethyl branch in close vicinity of the core destroys the mesophase both in metal-free and nickel derivatives. Both the derivatives showed very high melting points [1405]. The melting and clearing temperatures of metal-free and metallated Pc derivatives decrease with increasing the chain length [1374,1387,1405,1406]. An oxo-titanium(IV) derivative of Pc has been studied for nonlinear optical and electronic properties; though it was reported to be mesomorphic, its thermal behavior has not been presented [1408,1409].

2.22.4 OCTATHIOALKYL-SUBSTITUTED PHTHALOCYANINES

The synthesis of octathioalkyl-substituted Pcs is shown in Scheme 2.146. 1,2-dichloro-4,5-dicyanobenzene **18** can be prepared from 4,5-dichlorophthalic acid **17**. The replacement of chloro atoms by alkylthio groups can be achieved by reacting **18** with potassium salt of alkanethiol in DMSO. Octathioalkyl-Pc **20** is obtained by the thermal treatment of 1,2-dialkylthio-4,5-dicyanobenzene with DBU and required metal salt [1411].

The phase behavior of octakis(alkylthio)phthalocyanines is summarized in Table 2.131. Metal-free derivatives have a very high clearing temperature (about 300°C) and show columnar hexagonal mesophase. The shorter chain derivatives decompose at about 340°C. While the clearing

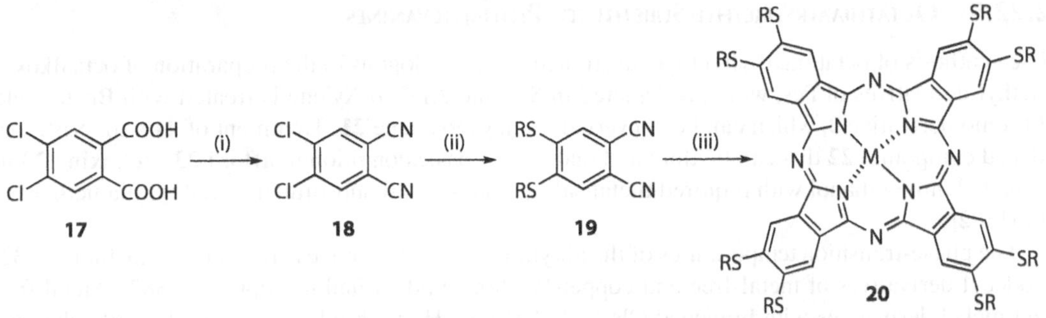

SCHEME 2.146 Synthesis of thioalkyl-substituted phthalocyanines: (i) Ac_2O; $HCONH_2$, reflux; NH_3; $SOCl_2$; (ii) RSH, K_2CO_3/DMSO; (iii) DBU, metal salt, $C_5H_{11}OH$.

TABLE 2.131
Phase Transition Temperatures of Octathiaalkyl-Substituted Pcs (Scheme 2.146)

Structure	M	R	Transition Temperature	Ref.
20.1	2H	C_8H_{17}	Cr 68.2 Col$_h$ 340 dec	[1410]
20.2	2H	$C_{10}H_{21}$	Cr 77 Col$_h$ 340 dec	[1410]
20.3	2H	$C_{12}H_{25}$	Cr 88 Col$_h$ 286.1 I	[1410]
20.4	2H	$C_{16}H_{33}$	Cr 94.8 Col$_h$ 228.4 I	[1410]
20.5	Cu	C_8H_{17}	Cr 76.8 Col$_h$ 350 dec	[1410]
20.6	Cu	$C_{10}H_{21}$	Cr 53.9 Col$_h$ 290 dec	[1410]
20.7	Cu	$C_{12}H_{25}$	Cr 77.5 Col$_h$ 322.8 X 340 dec	[1410]
20.8	Cu	$C_{16}H_{33}$	Cr 97.8 Col$_h$ 261.7 I	[1410]
20.9	Cu	$C_{18}H_{37}$	Cr 52 Col	[1411]
20.10	Ni	C_6H_{13}	Cr 36 Col$_h$ 300 dec	[1412]
20.11	Ni	$C_{12}H_{25}$	Cr 20 Col$_h$ 275 dec	[1412]
20.12	Co	$C_{12}H_{25}$	Cr 52 Col 300 dec	[1411]
20.13	VO	$C_{18}H_{37}$	Cr 56 Col 295 I	[1413]
20.14	Zn	C_8H_{17}	Cr 76 Col$_h$	[1415]
20.15	Zn	$C_{16}H_{33}$	Cr 89 Col$_h$ 260 I	[1414]
20.16	TiO	$C_{18}H_{37}$	Cr 59 Col$_h$ 68 Col$_h$ 292 I	[1416]
20.17	2H	$CH(OC_{12}H_{25})_2$	Cr 37 Col$_r$	[1417]
20.18	Ni	$CH(OC_{12}H_{25})_2$	Cr 38 Col$_r$	[1417]
20.19	Pb	$CH(OC_{12}H_{25})_2$	Col$_h$ −9.62 I	[1418]
20.20	Ni	$CH(CH_2O(CH_2CH_2O)_2C_2H_5)_2$	Liquid	[1419]
20.21	2H	$CH(CH_2O(CH_2CH_2O)_2C_2H_5)_2$	Liquid	[1419]

temperatures decrease with an increase in side-chain length, the melting temperatures increase. The copper complexes also exhibit hexagonal columnar mesophase. The longer chain derivatives clear into isotropic liquid at higher temperatures, but the shorter chain analogues decompose at higher temperature [1410,1411]. The nickel complexes show mesophase near room temperature and clear at about 300°C [1412]. Vanadyl, zinc, and oxo-titanium derivatives also show hexagonal columnar mesophase at elevated temperatures [1413–1416]. While metal-free and metallated octakis(13,17-dioxanonacosane-15-sulfanyl)phthalocyanines exhibit broad columnar mesophase [1417,1418], similar octa-poly(oxyethylene)-substituted metal-free and Ni phthalocyanines are non-mesomorphic [1419].

2.22.5 OCTATHIAALKYLMETHYL-SUBSTITUTED PHTHALOCYANINES

The synthesis of octathiaalkylmethyl-substituted Pcs is analogous to the preparation of octaalkoxymethyl derivatives of Pcs, which is depicted in Scheme 2.147. *o*-Xylene is treated with Br$_2$ to yield dibromo derivative **2**, which can be converted to dicyanoxylene **21**. Treatment of **21** with NBS furnished compound **22** that can be thiolated under mild basic condition to afford **23**. Refluxing **23** in dimethylaminoethanol with required metal salt provides metal-substituted octaalkylthiamethyl-Pc, **24** [1382].

The phase-transition temperatures of the alkylthiamethyl derivatives are collected in Table 2.132. Dodecyl derivatives of metal-free and copper-Pc show unidentified mesophase [1382]. Metal-free and nickel derivatives with branched CH$_2$SCH(CH$_2$OC$_{12}$H$_{25}$)$_2$ peripheral chains showed columnar hexagonal mesophase at low temperatures, stable down to room temperature. On the other hand, the lead derivative of the same ligand exhibits hexagonal columnar mesophase at very low temperature that clears to isotropic liquid at −1.91°C [1418,1420].

2.22.6 PERIPHERAL OCTAALKOXYPHENYL- AND ALKOXYPHENOXY-SUBSTITUTED PHTHALOCYANINES

Liquid crystalline Pcs with extended π-conjugation by alkoxyphenyl or alkoxyphenoxy groups have been designed and synthesized to investigate the effect of π-conjugation and bulky groups on phase structure and stability. Accordingly, Pcs peripherally substituted with alkoxyphenyl groups that are linked either directly, **28**, or through oxygen atom **31** to the Pc core have been synthesized. The phenyl rings carry one to three alkoxy groups. The synthesis of octakis(2,3-dialkoxyphenyl) phthalocyanine **28** has been outlined in Scheme 2.148. 3,3′,4,4′-tetra-n-alkyloxybenzil **25** was

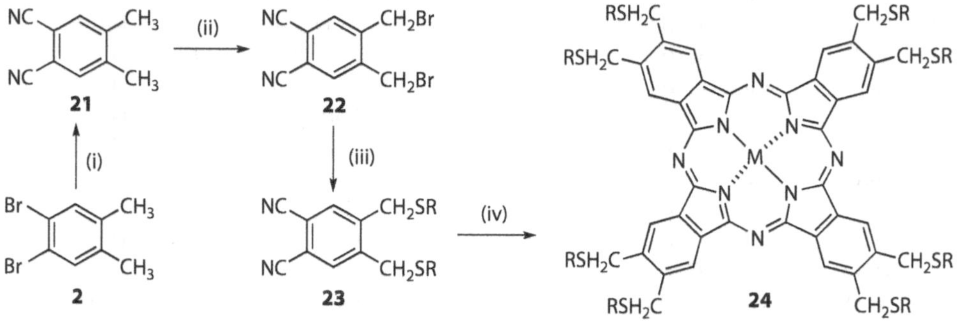

SCHEME 2.147 Synthesis of thioalkylmethyl-substituted phthalocyanines: (i) CuCN, DMF; (ii) NBS; (iii) RSH, DMSO, K$_2$CO$_3$; (iv) DMAE, metal salt.

TABLE 2.132
Phase Transition Temperatures of Octathioalkylmethyl-Substituted Pcs (Scheme 2.147)

Structure	M	R	Transition Temperature	Ref.
24.1	2H	C$_{12}$H$_{25}$	Cr 95 M 267 I	[1382]
24.2	Cu	C$_{12}$H$_{25}$	Cr 108 M 304 I	[1382]
24.3	Ni	CH(CH$_2$OC$_{12}$H$_{25}$)$_2$	Cr −7.56 Col$_h$ 28.87 I	[1420]
24.4	2H	CH(CH$_2$OC$_{12}$H$_{25}$)$_2$	Cr −9.39 Col$_h$ 23.55 I	[1420]
24.5	Pb	CH(CH$_2$OC$_{12}$H$_{25}$)$_2$	Col$_h$ −1.91 I	[1418]

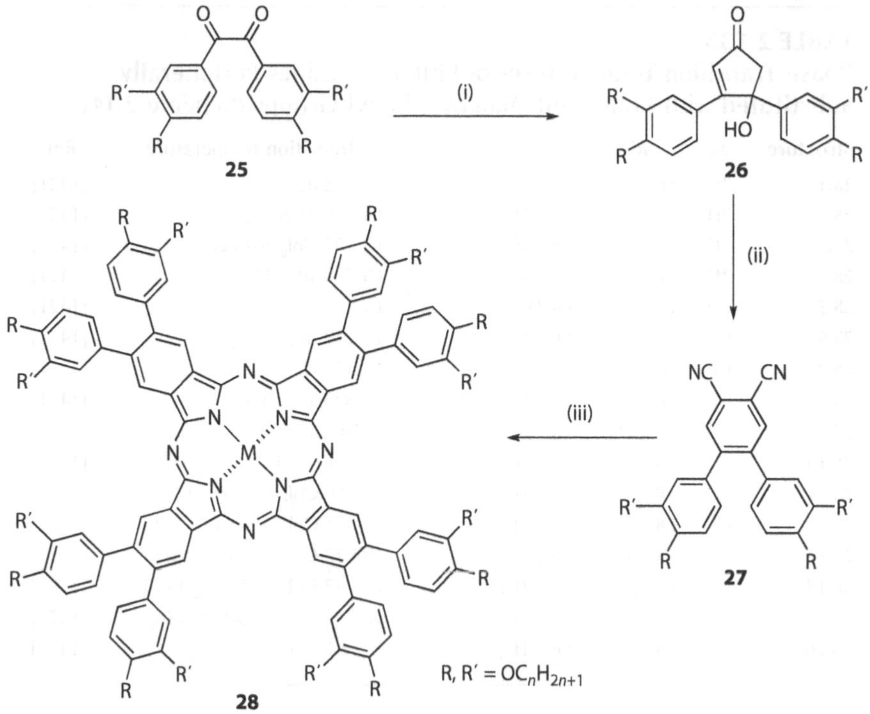

SCHEME 2.148 Synthesis of alkoxyphenyl-substituted phthalocyanines: (i) potassium *t*-butanolate, acetone, ethanol; (ii) *p*-toluenesulfonic acid, chlorobenzene; (iii) DBU, *n*-pentanol, metal salt.

treated with acetone and potassium *t*-butanoate to give 3,4-bis(3,4-dialkyloxyphenyl)-4-hydroxy-2-cyclopenten-1-one **26**. This product on reaction with dicyanoacetylene gives 3,3′,4,4′-tetradecyloxy-*o*-terphenyl-4′,5′-dicarbonitrile **27**, which on cycloteramerization with DBU in n-pentanol affords octaphenyl-substituted phthalocyanines **28**. The phase-transition temperatures of these compounds are collected in Table 2.133.

The metal-free Pcs substituted with *p*-alkoxy-substituted phenyl groups **28** (R′=H) show hexagonal columnar mesophase. The melting temperature decreases with increase in peripheral chain length. The alkyl chain with less than 10 carbon atoms did not show any mesophase; branched alkyl chain (2-ethylhexyl) derivative also did not show any mesophase. All these compounds decompose above 250°C. While metal-free dodecyloxy and octadecyloxy derivatives exhibit hexagonal columnar phase, corresponding Cu complexes display rectangular columnar phase. The change in the two-dimensional lattice could be due to the tendency of the copper ion to coordinate axially with the *meso*-nitrogen atom of neighboring phthalocyanine molecule. The Cu-Pc substituted with alkoxy phenyl groups where alkyl chain length is shorter than 10 carbon atoms did not show mesophase [1421].

The 3,4-dialkoxyphenyl-substituted Pcs **28.9–28.16** (R′=R=OC$_n$H$_{2n+1}$) were observed to show multiple mesophases. The metal-free lower homologues showed columnar mesophases but the exact structures of all the phases could not be identified; however, one of the high-temperature mesophases was identified as columnar tetragonal in the case of C$_{10}$ derivative, and rectangular columnar in the case of C$_{12}$ derivative. The copper complexes showed multiple columnar rectangular mesophases (Col$_{r1}$ and Col$_{r2}$). Nickel derivatives also showed columnar structure but the symmetry could not be identified. Ni complexes showed lower melting and clearing temperatures compared to copper complexes. All the dialkoxyphenyl-substituted Pcs **28.9–28.16** cleared to isotropic liquid

TABLE 2.133

Phase Transition Temperatures of Phthalocyanines Peripherally Substituted with Mono and Dialkoxyphenyl Groups (Scheme 2.148)

Structure	M	R′	R	Transition Temperature	Ref.
28.1	2H	H	OC_8H_{17}	Cr 285 dec	[1421]
28.2	2H	H	$OC_{10}H_{27}$	Cr 229 M 267 dec	[1421]
28.3	2H	H	$OC_{12}H_{25}$	Cr 192 Col_h 261 dec	[1421]
28.4	2H	H	$OC_{18}H_{37}$	Cr 78 Col_h 253 dec	[1421]
28.5	2H	H	$OCH_2CH(Et)C_4H_9$	Cr 275 dec	[1421]
28.6	Cu	H	$OC_{10}H_{21}$	Cr 222 dec	[1421]
28.7	Cu	H	$OC_{12}H_{25}$	Cr 120 Col_r 227 dec	[1421]
28.8	Cu	H	$OC_{18}H_{37}$	Cr 82 Col_r 220 dec	[1421]
28.9	2H	$OC_{10}H_{21}$	$OC_{10}H_{21}$	X 56 Col_1 106 Col_2 184 I	[1422]
28.10	2H	$OC_{11}H_{23}$	$OC_{11}H_{23}$	X 66 Col 190 I	[1422]
28.11	2H	$OC_{12}H_{25}$	$OC_{12}H_{25}$	Col 66 Col_{r1} 185.6 Col 187.2 I	[1422]
28.12	Cu	$OC_{10}H_{21}$	$OC_{10}H_{21}$	Col 75 Col_{r1} 196 Col_{r2} 219 I	[1422]
28.13	Cu	$OC_{11}H_{23}$	$OC_{11}H_{23}$	Col 72 Col_{r1} 185 Col_{r2} 204 I	[1422]
28.14	Cu	$OC_{12}H_{25}$	$OC_{12}H_{25}$	Col 72 Col_{r1} 192 Col_{r2} 197 I	[1422]
28.15	Ni	$OC_{10}H_{21}$	$OC_{10}H_{21}$	X 59 Col_1 100 Col_2 175 I	[1422]
28.16	Ni	$OC_{11}H_{23}$	$OC_{11}H_{23}$	X 58 Col 195 I	[1422]

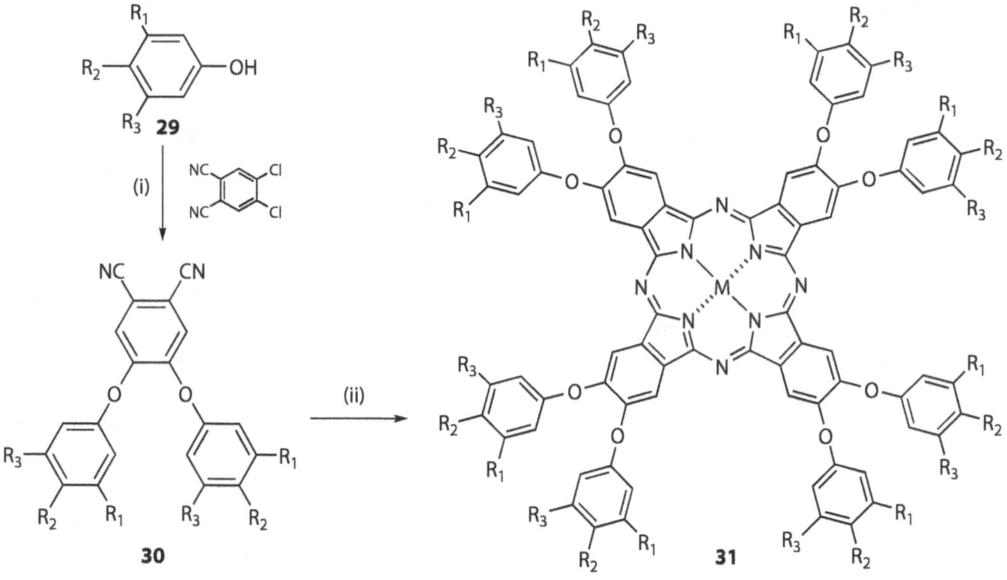

SCHEME 2.149 Synthesis of alkoxyphenyloxy-substituted phthalocyanines: (i) K_2CO_3/DMSO; (ii) DBU/metal salt/n-hexanol.

at around 200°C, unlike the mono-substituted derivatives **28.1–28.8**, which decomposed at higher temperatures [1422].

Pcs peripherally substituted with alkoxyphenyl groups that are linked to the Pc core through oxygen atom are synthesized, as shown in the Scheme 2.149. The key precursor **30** can be prepared by coupling mono-, di-, or trialkoxyphenol **29** with 4,5-dichlorophthalonitrile in mild basic conditions. The 4,5-bis[(alkoxy)phenoxy]-1,2-dicyanobenzene, **30**, on cyclotetramerization under

TABLE 2.134

Phase Transition Temperatures of Phthalocyanine Peripherally Substituted with Mono-, Di-, Trialkoxyphenoxy Groups (Scheme 2.149)

Structure	M	R_1	R_2	R_3	Transition Temperature	Ref.
31.1	Cu	H	$OC_{10}H_{21}$	H	Cr 122.6 Col$_h$ 362.2 I	[1423]
31.2	Cu	H	$OC_{12}H_{25}$	H	Cr 116.2 Col$_h$ 338.3 I	[1423]
31.3	Cu	H	$OC_{14}H_{29}$	H	Cr 113.9 Col$_h$ 292.3 I	[1423]
31.4	Cu	OC_9H_{19}	OC_9H_{19}	H	Col$_h$ 94.1 Col$_{r1}$ 128.8 Col$_{r2}$ 150.8 Col$_{r3}$ 217 I	[1424]
31.5	Cu	$OC_{10}H_{21}$	$OC_{10}H_{21}$	H	Col$_h$ 104.4 Col$_{r1}$ 131.2 Col$_{r2}$ 142.4 Col$_{r3}$ 202.1 I	[1423]
31.6	Cu	$OC_{11}H_{23}$	$OC_{11}H_{23}$	H	X 77.1 Col$_h$ 106.3 Col$_{r1}$ 136.2 Col$_{r2}$ 175.7 Col$_t$ 201.5 I	[1424]
31.7	Cu	$OC_{12}H_{25}$	$OC_{12}H_{25}$	H	X 30 Col$_h$ 106 Col$_{r1}$ 132.1 Col$_{r2}$ 149 Col$_t$ 195.3 I	[1423]
31.8	Cu	$OC_{13}H_{27}$	$OC_{13}H_{27}$	H	X 20 Col$_h$ 104.7 Col$_r$ 125.5 Col$_t$ 183.6 I	[1424]
31.9	Cu	$OC_{14}H_{29}$	$OC_{14}H_{29}$	H	X 20 Col$_h$ 111.2 Col$_r$ 125.5 Col$_t$ 180.9 I	[1423]
31.10	Cu	$OC_{10}H_{21}$	$OC_{10}H_{21}$	$OC_{10}H_{21}$	X 37.6 Col$_r$ 71.2 Y 81.2 cub 176.9 I	[1423]
31.11	Cu	$OC_{12}H_{25}$	$OC_{12}H_{25}$	$OC_{12}H_{25}$	X 27.6 Col$_r$ 56 Y 75.1 cub 159.6 I	[1423]
31.12	Cu	$OC_{14}H_{29}$	$OC_{14}H_{29}$	$OC_{14}H_{29}$	Cr 36.8 Col$_r$ 45.9 Y 59.6 cub 144.8 I	[1423]

usual reaction conditions (refluxing with DBU and metal salt in *n*-pentanol) gives octakis(*p*-alkoxyphenoxy)phthalocyaninato metal complex **31** [1423]. The phase transition temperatures of the compounds are collected in Table 2.134.

p-Alkoxyphenyloxy-substituted Pcs **31.1–31.3** ($R_1 = R_3 = H$; $R_2 = OC_nH_{2n+1}$) having copper as central metal show hexagonal columnar mesophase and follow the common trend, that is, the melting and clearing temperatures decrease with increase in alkyl chain length. The change was drastic for clearing temperatures, while the melting points decrease only marginally. All three *p*-alkoxyphenyloxy-substituted compounds cleared into isotropic liquid at higher temperature unlike the phenyl derivatives, which decomposed at higher temperatures [1423].

3,4-Dialkoxyphenyloxy-substituted Pcs **31.4–31.9** ($R_1 = R_2 = OC_nH_{2n+1}$; $R_3 = H$) with copper as central metal show complex mesophase behavior with hexagonal, rectangular, and tetragonal orders. Each of the derivatives shows a Col$_h$ mesophase at lower temperatures and at least one Col$_r$ mesophase at higher temperatures. Three different Col$_r$ mesophases, Col$_{r1}$, Col$_{r2}$, and Col$_{r3}$ for C_9 and C_{10} derivatives; two Col$_r$ mesophases, Col$_{r2}$ and Col$_{r3}$, for C_{11} and C_{12}; and one mesophase, Col$_{r2}$, for C_{13} and C_{14} were observed. A Col$_{tet}$ mesophase appears over their Col$_r$ mesophases for C_{11}–C_{14} homologues. All derivatives cleared to isotropic liquid at around 200°C. The C_{11}–C_{14} derivatives also show a low temperature unidentified mesophase denoted as X in Table 2.134 [1424].

The 3,4,5-trialkoxyphenyloxy-substituted Pcs **31.10–31.12** ($R_1 = R_2 = R_3 = OC_nH_{2n+1}$) with copper as central metal atom show Col$_r$ mesophase at lower temperature; on further heating, it transformed into an optically isotropic phase having no fluidity. The phase could not be identified from the photomicrograph observation, and thus was denoted as an unidentified Y phase, which, on slight heating formed cubic mesophase before going to the isotropic phase [1423].

2.22.7 Octaalkyl Esters of Phthalocyanine

Octaester **35** and octaalkanoate **36** derivatives of Pc also exhibit columnar mesomorphism. Phthalocyanine octacarboxylic acid **34** is prepared from benzene tetracarbonitrile (tetracyanobenzene) **32**. Refluxing the tetracarbonitrile with propyl lithium in propanol yields the octacyano-Pc **33** as shown in Scheme 2.150. The hydrolysis of phthalocyanine octacarbonitrile with potassium hydroxide in triethylene glycol yields octaacid, which is esterified with appropriate alkanol [1425].

Metal-free derivatives of Pc octasubstituted with alkoxycarbonyl group bearing alkyl chains of different lengths showed a very wide temperature range hexagonal columnar mesophase

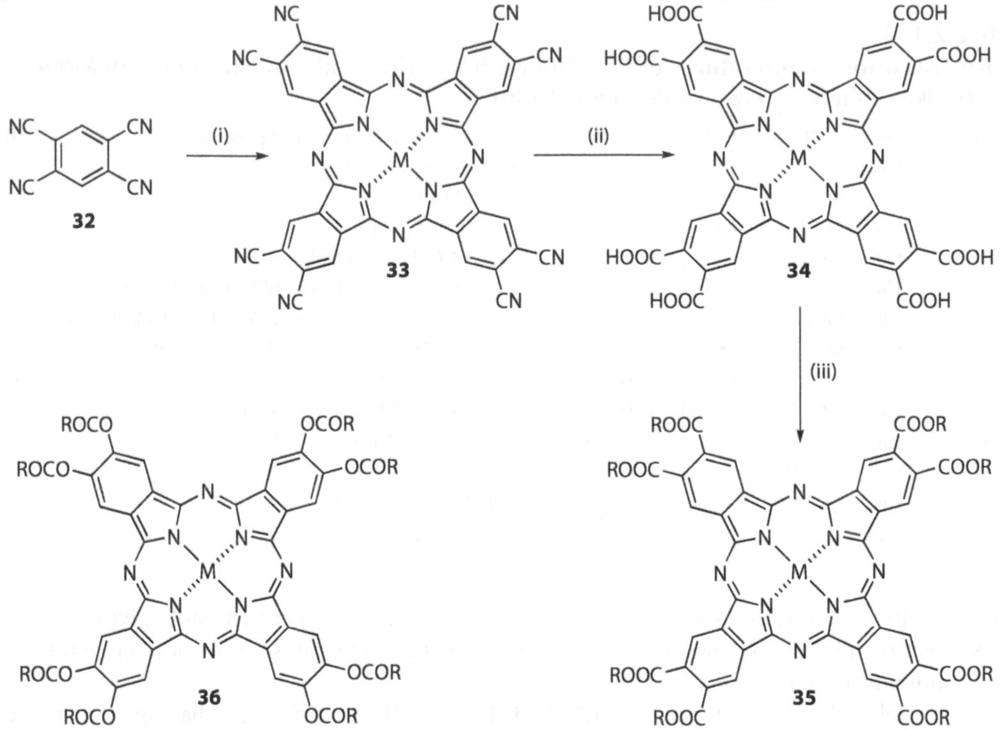

SCHEME 2.150 Synthesis of phthalocyanine octaesters: (i) PrOLi/PrOH; (ii) KOH/triethylene glycol; (iii) RBr/DBU/CH₃CN.

TABLE 2.135
Phase Transition Temperatures of Phthalocyanine Esters (Scheme 2.150)

Structure	M	R	Transition Temperature	Ref.
35.1	2H	C_5H_{11}	<−60 Col$_h$ >300 I	[1425]
35.2	2H	C_6H_{13}	<−60 Col$_h$ >300 I	[1425]
35.3	2H	C_7H_{15}	<−60 Col$_h$ >300 I	[1425]
35.4	2H	C_8H_{17}	<−60 Col$_h$ >300 I	[1425]
35.5	2H	C_9H_{19}	<−60 Col$_h$ >300 I	[1425]
35.6	2H	$C_{10}H_{21}$	Cr < −16.8 Col$_h$ >300 I	[1425]
35.7	2H	$C_{11}H_{23}$	Cr −0.7 Col$_h$ >300 I	[1425]
35.8	2H	$C_{12}H_{25}$	Cr 12.1 Col$_h$ >300 I	[1425]
35.9	Ge(OH)₂	$C_{12}H_{25}$	Cr −20.1 Col$_h$ 170.9 poly	[1426]
35.10	Zn	$CH_2CH(C_4H_9)$ $(CH_2)_5CH_3$	Col $_{r1}$ 44 Col$_{r2}$ 250 I	[1427]
36.1	2H	$C_{11}H_{23}$	Cr 58 Col$_h$ 303 I	[1382]

(Table 2.135). Increase in chain length of the octaesters brought about an increase in melting temperature [1425]. The extraordinarily large temperature range over which the mesophase is stable shows the stabilizing effect of alkoxycarbonyl side chains. When Pc was synthesized with Ge(OH)₂ as central metal atom, it showed hexagonal columnar mesophase at lower temperature and forms polygermoxane at elevated temperature. Because of polycondensation, no clearing point could be observed for this compound [1426]. The wide range of temperature over which mesophase is stable is ascribed again to the stabilization of mesophase by alkoxycarbonyl side chains. Branched derivative with zinc as central metal atom showed two columnar rectangular mesophases [1427].

Esters can also be prepared by condensing phenol with appropriate acid or acid derivatives. They are termed as alkanoates. Thus, the esterification of octahydroxy-Pc with dodecanoyl chloride yields phthalocyanine 2,3,9,10,16,17,23,24-octa-n-dodecanoates **36.1**. Like the esters **35**, octaalkanoates **36** showed columnar hexagonal mesophase but the melting temperature increases significantly, demonstrating the effect of orientation of linking groups connecting the central core to the peripheral chains [1382].

2.22.8 NON-PERIPHERALLY SUBSTITUTED OCTAALKYL AND OCTAALKOXYMETHYL PHTHALOCYANINES

While non-peripherally substituted octaalkoxy-Pcs do not exhibit any mesomorphism, their alkyl and alkoxymethyl counterparts display mesomorphism [1350,1387,1428–1438]. The synthesis of liquid crystalline non-peripherally substituted octaalkyl Pcs is shown in Scheme 2.151. 3,6-Dialkylphthalonitrile **40** is prepared from 2,5-dialkylfuran **38**, which is synthesized by lithiation and alkylation of furan. The Diels–Alder reaction of 2,5-dialkylfuran with fumaronitrile furnishes **39**. The reaction is reversible and requires several weeks at low temperature to reach the equilibrium. The use of excess fumaronitrile to push the reaction toward adduct formation causes problems as the product cannot be isolated by conventional purification methods and is used as such to prepare phthalonitrile **40** by reacting the adduct **39** with lithium bis(trimethylsilyl)amide (a non-nucleophilic base). Unreacted fumaronitrile reacts with the base and creates problems in obtaining pure material. It is also generally difficult to isolate pure phthalonitrile **40** and, therefore, in most cases was used as prepared for conversion into the phthalocyanine **41** [1428].

Structurally, the nonperipheral positions are more demanding than the peripheral substituted ones, and allow stronger interaction between the macrocycle and the core chain linking group. The strongest steric interactions for nonperipheral side chains are those between chains situated on neighboring benzo-moieties. These interactions are likely to result in the displacement of some of the side chains from the plane of the Pc ring. This results in lower co-facial attractions of the aromatic cores and to the induction of mesophase with less ordered stacking at low temperatures. Increasing the chain length of the octaalkyl derivatives **41** results in depression of both melting and clearing temperatures. The phase behavior of these materials is summarized in Table 2.136. Several members of the series are polymorphic and exhibit one or two additional mesophases with hexagonal (Col_h) or rectangular (Col_r) symmetry at low temperatures. The free base compounds with side chains shorter than C_6 did not show any mesophase. Zinc, copper, and nickel complex of Pc showed columnar hexagonal mesophase. Cobalt complexes also showed unidentified columnar mesophase [1345,1428,1431,1436].

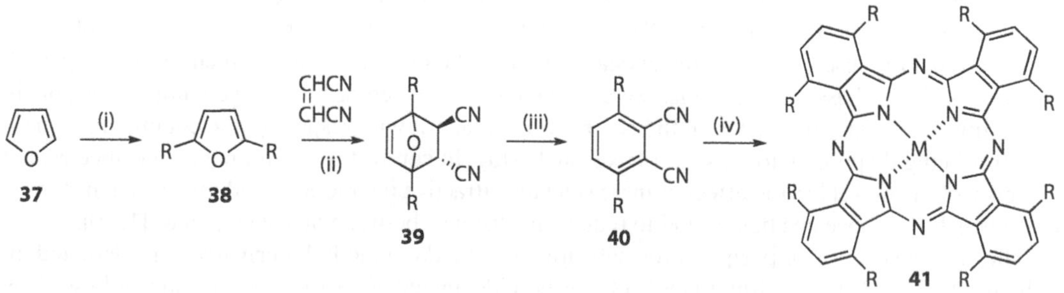

SCHEME 2.151 Synthesis of non-peripherally alkyl-substituted phthalocyanines: (i) BuLi; RBr; (ii) LiN(SiMe$_3$)$_2$, THF; (iii) H$_2$O; (iv) LiOC$_5$H$_{11}$/C$_5$H$_{11}$OH; metal salt.

TABLE 2.136
Phase Transition Temperatures of Non-Peripherally
Alkyl-Substituted Pcs (Scheme 2.151)

Structure	M	R	Transition Temperature	Ref.
41.1	2H	C_4H_9	Cr 230 I	[1428]
41.2	2H	C_5H_{11}	Cr 218 I	[1428]
41.3	2H	C_6H_{13}	Cr 161 Col$_h$ 171 I	[1428]
41.4	2H	C_7H_{15}	Cr 113 Col$_h$ 145 Col$_{h'}$ 163 I	[1428]
41.5	2H	C_8H_{17}	Cr 85 Col$_r$ 101 Col$_h$ 152 I	[1428]
41.6	2H	C_9H_{19}	Cr 103 Col$_h$ 142 I	[1428]
41.7	2H	$C_{10}H_{21}$	Cr 78 Col$_h$ 133 I	[1428]
41.8	Zn	C_5H_{11}	Cr 279 Col$_h$ 292 I	[1436]
41.9	Zn	C_6H_{13}	Cr 209 Col$_{h1}$ 280 Col$_{h2}$ 285 I	[1436]
41.10	Zn	C_7H_{15}	Cr 158 Col$_{h1}$ 248 Col$_{h2}$ 272 I	[1436]
41.11	Zn	C_8H_{17}	Cr 105 Col$_{h1}$ 224 Col$_{h2}$ 258 I	[1436]
41.12	Zn	C_9H_{19}	Cr 114 Col$_{h1}$ 177 Col$_{h2}$ 242 I	[1436]
41.13	Zn	$C_{10}H_{21}$	Cr 89.6 Col$_h$ 106.6 Col$_r$ 225.1 I	[1431]
41.14	Zn	$C_{12}H_{25}$	Cr 60 Col$_h$ 88.5 Col$_r$ 205.6 I	[1431]
41.15	Zn	$C_{15}H_{31}$	Cr 65.3 Col$_h$ 88.1 Col$_{h2}$ 103.6 Col$_r$ 176.2 I	[1431]
41.16	Zn	$C_{18}H_{37}$	Cr 65.9 Col$_h$ 91 Col$_{h2}$ 108.1 Col$_r$ 160.5 I	[1431]
41.17	Cu	C_4H_9	Cr 265 I	[1428]
41.18	Cu	C_5H_{11}	Cr 261 I	[1428]
41.19	Cu	C_6H_{13}	Cr 184 Col$_h$ 235.5 Col$_{h'}$ 242 I	[1428]
41.20	Cu	C_7H_{15}	Cr 145 Col$_h$ 205 Col$_{h'}$ 236 I	[1428]
41.21	Cu	C_8H_{17}	Cr 96 Col$_h$ 156 Col$_{h'}$ 220 I	[1428]
41.22	Cu	C_9H_{19}	Cr 108 Col$_h$ 208 I	[1428]
41.23	Cu	$C_{10}H_{21}$	Cr 88 Col$_h$ 198 I	[1428]
41.24	Ni	C_5H_{11}	Cr 220 I	[1436]
41.25	Ni	C_6H_{13}	Cr 145 Col$_h$ 164 Col$_{h'}$ 169 I	[1436]
41.26	Ni	C_7H_{15}	Cr 118 Col$_h$169 I	[1436]
41.27	Ni	C_8H_{17}	Cr 66 Col$_h$153 I	[1436]
41.28	Ni	C_9H_{19}	Cr 92 Col$_h$152 I	[1436]
41.29	Ni	$C_{10}H_{21}$	Cr 64 Col$_h$136 I	[1436]
41.30	Co	C_4H_{11}	Cr 300 I	[1345]
41.31	Co	C_6H_{13}	Cr 177 Col 252 I	[1345]
41.32	Co	C_8H_{15}	Cr 100 Col 220 I	[1345]
41.33	Co	$C_{10}H_{21}$	Cr 75 Col 189 I	[1345]

Zinc complexes had the largest range of mesosphase temperature. The trend observed for different metal atoms was $Zn > Co > Cu > Ni \approx 2H$. A crystal structure of the octahexyl-Pc complex of nickel showed a striking resemblance to the crystal phase and the mesophase of the metal-free compound. In the crystalline phase, the molecules were stacked in a staggered fashion into columns, presumably to minimize the steric hindrance of the chains, with some tilt with respect to the column axis, these columns being disposed into a pseudo-hexagonal array. In contrast to the unmetallated species, the presence of the nickel has the effect of increasing the attractive interactions with the π-systems of the molecules lying above and below, and to reduce the distance between aromatic planes [1350].

The synthesis of non-peripherally substituted octaalkoxymethyl derivatives is depicted in Scheme 2.152. 2,5-Furan dimethanol **42** can be chlorinated with thionyl chloride, followed by the nucleophilic displacement of chloride with sodium alkoxide. The resulting furan **43** is reacted with fumaronitrile for about a week, and the Diels–Alder adduct **44** was aromatized by treatment

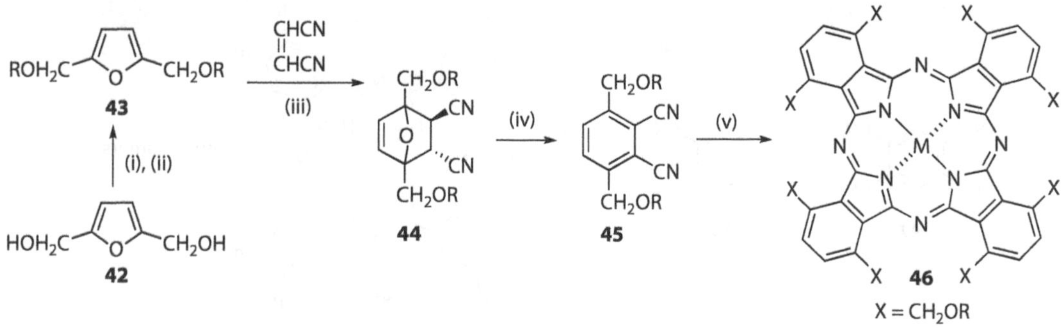

SCHEME 2.152 Synthesis of non-peripherally alkoxymethyl-substituted phthalocyanines: (i) $SOCl_2$; (ii) ROH/K_2CO_3; (iii) $LiN(SiMe_3)_2$; (iv) H_2O; (v) $LiOC_5H_{11}/C_5H_{11}OH$, metal salt.

TABLE 2.137
Phase Transition Temperatures of Non-Peripherally Alkoxymethyl Substituted Pcs (Scheme 2.152)

Structure	M	R	Transition Temperature
46.1	2H	C_4H_9	Cr 185 Col_r >300 I
46.2	2H	C_5H_{11}	Cr 123 Col_r >300 I
46.3	2H	C_6H_{13}	Cr 85 Col_r >300 I
46.4	2H	C_7H_{15}	Cr 79 Col_r >300 I
46.5	2H	C_8H_{17}	Cr 67 Col_r >300 I
46.6	2H	C_9H_{19}	Cr 75 Col_h >300 I
46.7	2H	$C_{10}H_{21}$	Cr 64 Col_h >300 I
46.8	2H	2-octyl	Col_r 106 I
46.9	2H	$CH_2CH(C_2H_5)C_4H_9$	Col_r 236 I
46.10	Zn	C_7H_{15}	Cr 70 Col >300 I
46.11	Cu	C_7H_{15}	Cr 82 Col >300 I

Source: Data from Cammidge, A.N. et al., *J. Chem. Soc., Perkin Trans.*, 3053, 1991.

with lithium bis(trimethylsilyl)amide, followed by an acidic work-up. The 3,6-bis(alkoxy-methyl) phthalonitrile **45** is cyclized under standard conditions (lithium/pentanol) to furnish the desired materials **46** [1429]. The mesophase properties of these compounds are collected in Table 2.137. Phthalocyanine compounds with alkoxymethyl groups having their central core occupied by metal-free, copper and zinc derivatives showed columnar mesophases. Two mesophases were observed for the free base compound, *viz.*, Col_h and Col_r, while the copper and zinc complex showed only one mesophase whose structure could not be identified [1429].

2.22.9 NON-SYMMETRICAL OCTA-, HEPTA-, HEXA-, AND PENTA-SUBSTITUTED PHTHALOCYANINES

A number of liquid crystalline unsymmetrical Pcs have been synthesized to visualize the effects of the number and position of peripheral chains on mesomorphism [1406,1439–1442]. The reaction of two different phthalonitriles of (nearly) the same reactivity in a 3:1 ratio will afford a mixture of products having different peripheral chains. This logic is used to synthesize Pcs with unsymmetrical substitution. The two phthalonitriles are taken in a different ratio, depending on the type of product

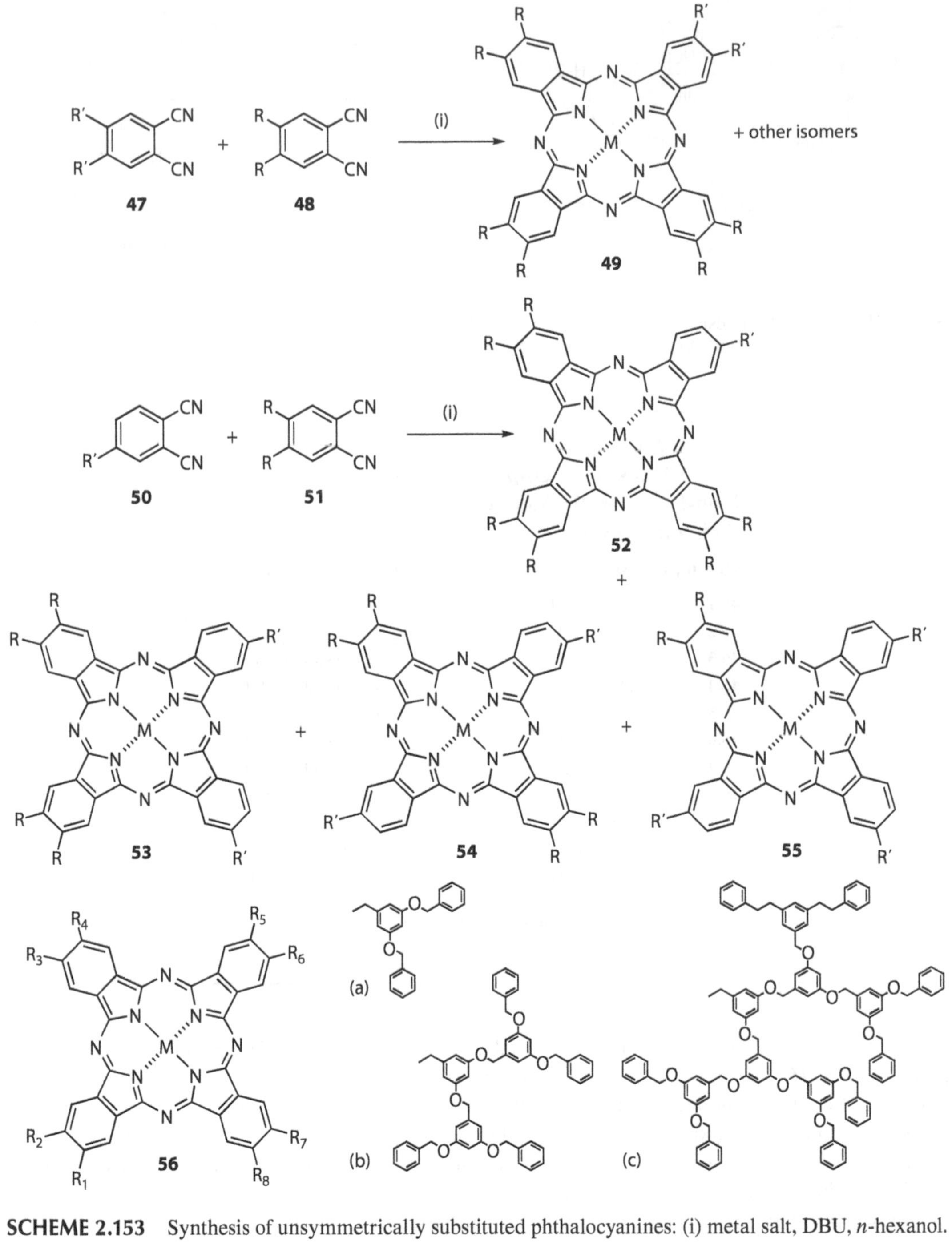

SCHEME 2.153 Synthesis of unsymmetrically substituted phthalocyanines: (i) metal salt, DBU, *n*-hexanol.

required and are refluxed in *n*-hexanol with DBU and metal salt to get the desired products [1439]. An example is shown in Scheme 2.153. Thus, the reaction of phthalonitrile **47** and **48** in 3:1 ratio yields unsymmetrical Pc **49** and other isomers [1439]. Similarly, the reaction of a mono-substituted phthalonitrile **50** with a di-substituted phthalonitrile **51** furnished a mixture of hepta- (**52**), hexa- (**53**, **54**), and penta-substituted (**55**) phthalocyanines. The phase transition temperatures of these compounds with a general structure **56** are collected in Table 2.138. Some of the derivatives exhibit only columnar

TABLE 2.138

Phase Transition Temperature of Unsymmetrically Penta-, Hexa-, Hepta-, and Octa-Substituted Pcs (Scheme 2.153)

Structure	M	R	Transition Temperature	Ref.
56.1	Ni	$R_1 = R_2 = p\text{-}NHSO_2PhCH_3$, $R_3 = R_4 = R_5 = R_6 = R_7 = R_8 = SC_6H_{13}$	Cr 165 Col$_h$	[1439]
56.2	Ni	$R_1 = R_2 = p\text{-}NHSO_2PhCH_3$, $R_3 = R_4 = R_5 = R_6 = R_7 = R_8 = SC_{12}H_{25}$	Cr 161 Col$_h$ 227 I	[1439]
56.3	Ni	$R_1 = R_2 = p\text{-}NHSO_2PhCH_3$, $R_3 = R_4 = R_5 = R_6 = R_7 = R_8 = SC_{16}H_{33}$	Cr 144 Col$_h$ 205 I	[1439]
56.4	2H	$R_3 = R_4 = R_5 = R_6 = R_7 = R_8 = C_{16}H_{33}$, $R_2 = R_1 = O(CH_2CH_2O)_3CH_3$	Cr 85 Col$_r$ 94 Col$_h$ 208 I	[1442]
56.5	2H	$R_3 = R_4 = R_7 = R_8 = C_{16}H_{33}$, $R_1 = R_2 = R_5 = R_6 = O(CH_2CH_2O)_3CH_3$	Cr 78 Col$_r$ 240 I	[1442]
56.6	2H	$R_5 = R_6 = R_7 = R_8 = C_{16}H_{33}$, $R_1 = R_2 = R_3 = R_4 = O(CH_2CH_2O)_3CH_3$	Cr 78 Col$_r$ 178 Col$_h$ 211 I	[1442]
56.7	2H	$R_3 = R_4 = R_5 = R_6 = R_7 = R_8 = O(CH_2CH_2O)_3CH_3$, $R_2 = R_1 = C_{16}H_{33}$	Cr 43 Col$_r$ 175 Col$_h$ 229 I	[1442]
56.8	2H	$R_3 = R_4 = R_7 = R_8 = C_{16}H_{33}$, $R_2 = R_6 = H$, $R_1 = R_5 = O(CH_2CH_2O)_3CH_3$	Cr 43 Col$_r$ 142 Col$_h$ 243 I	[1442]
56.9	2H	$R_5 = R_6 = R_7 = R_8 = C_{16}H_{33}$, $R_1 = R_3 = H$, $R_2 = R_4 = O(CH_2CH_2O)_3CH_3$	Cr 81 Col$_r$ 92 Col$_h$ 243 I	[1442]
56.10	2H	$R_1 = R_2 = R_5 = R_6 = C_{16}H_{33}$, $R_4 = R_8 = H$ $R_3 = R_7 = O(CH_2CH_2O)_3$trityl	Cr 41 Col$_{r1}$ 191 Col$_h$ 196 I	[1406]
56.11	2H	$R_1 = R_2 = R_5 = R_6 = C_{16}H_{33}$, $R_4 = R_8 = H$, $R_3 = R_7 = O(CH_2CH_2O)_3H$	Cr 71 Col$_{r1}$ 109 Col$_h$ 234 I	[1406]
56.12	2H	$R_1 = R_2 = R_3 = R_4 = C_{16}H_{33}$, $R_6 = R_8 = H$, $R_5 = R_7 = O(CH_2CH_2O)_3$trityl	Cr 77 Col$_{r1}$ 152 I	[1406]
56.13	2H	$R_1 = R_2 = R_3 = R_4 = C_{16}H_{33}$, $R_6 = R_8 = H$ $R_5 = R_7 = O(CH_2CH_2O)_3H$	Cr 79 Col$_L$ 132 Col$_{r2}$ 194 I	[1406]
56.14	2H	$R_1 = R_2 = C_{16}H_{33}$, $R_6 = R_4 = R_8 = H$, $R_5 = R_3 = R_7 = O(CH_2CH_2O)_3$trityl	Cr 32 Col$_r$ 76 Col$_h$ 177 I	[1406]
56.15	2H	$R_1 = R_2 = C_{16}H_{33}$, $R_6 = R_4 = R_8 = H$ $R_5 = R_3 = R_7 = O(CH_2CH_2O)_3H$	Cr 80 Col$_{r1}$ 252 Col$_h$ 273 I	[1406]
56.16	2H	$R_7 = R_8 = C_{16}H_{33}$, $R_2 = R_4 = R_6 = H$, $R_1 = R_3 = R_5 = O(CH_2CH_2O)_3CH_3$	Cr 102 Col$_h$ 302 I	[1442]
56.17	2H	$R_1\text{–}R_6 = O(CH_2CH_2O)_2Me$, $R_7 = H$, $R_8 = $ (a)	g −20 Col$_h$ >320 I	[1440]
56.18	2H	$R_1\text{–}R_6 = O(CH_2CH_2O)_2Me$, $R_7 = H$, $R_8 = $ (b)	g 115 Col$_h$ 250 I	[1440]
56.19	2H	$R_1\text{–}R_6 = O(CH_2CH_2O)_2Me$, $R_7 = H$, $R_8 = $ (c)	g 94 Col$_h$ 108 I	[1440]
56.20	2H	$R_1\text{–}R_6 = C_{16}H_{33}$, $R_7 = H$, $R_8 = O(CH_2CH_2O)_3$trityl	Cr 86 Col$_r$ 89 Col$_h$ 167 I	[1406]
56.21	2H	$R_1\text{–}R_6 = C_{16}H_{33}$, $R_7 = H$, $R_8 = O(CH_2CH_2O)_3H$	Cr 81 Col$_r$ 107 Col$_h$ 189 I	[1406]
56.22	2H	$R_1\text{–}R_6 = C_{16}H_{33}$, $R_7 = H$, $R_8 = O(CH_2CH_2O)_3CH_3$	Cr 81 Col$_r$ 86 Col$_h$ 206 I	[1442]
56.23	2H	$R_1\text{–}R_6 = OC_{12}H_{25}$, $R_7 = H$, $R_8 = O(CH_2CH_2O)_3CH_3$	Cr 72 Col$_h$ 325 I	[1441]
56.24	2H	$R_1\text{–}R_6 = OC_{12}H_{25}$, $R_7 = H$, $R_8 = O(CH_2CH_2O)_8CH_3$	Cr 71 Col$_h$ 289 I	[1441]
56.25	2H	$R_1\text{–}R_6 = OC_{12}H_{25}$, $R_7 = H$, $R_8 = O(CH_2CH_2O)_{10}CH_3$	Cr 70 Col$_h$ 234 I	[1441]
56.26	2H	$R_1\text{–}R_6 = OC_{12}H_{25}$, $R_7 = H$, $R_8 = O(CH_2CH_2O)_{50}CH_3$	Cr 74 N 195 I	[1441]

hexagonal mesophase while other derivatives exhibit polymesomorphism. Interestingly, some of the heptasubstituted derivatives freeze into glassy state of the preceding mesophase.

2.22.10   UNSYMMETRICAL NON-PERIPHERAL PHTHALOCYANINES

Similar to unsymmetrical peripherally substituted Pcs, unsymmetrical non-peripherally substituted Pcs **59** can be prepared by reacting two different 3,6-disubstituted phthalonitriles, as shown in Scheme 2.154. The mesophase behavior of these materials with a general structure **60** is summarized in Table 2.139 [1443,1444]. While some of the derivatives exhibit columnar phases with hexagonal and/or rectangular symmetry, other derivatives display mesophases whose structures have not been identified. Most of the terminally hydroxyl functionalized derivatives fail to exhibit liquid crystalline phase behavior.

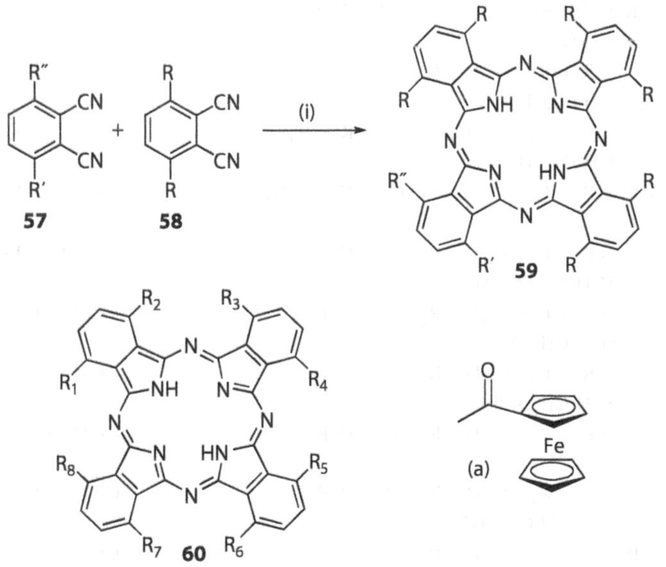

SCHEME 2.154 Synthesis of non-peripherally unsymmetrical-substituted phthalocyanines: (i) DBU, $C_5H_{11}OH$.

TABLE 2.139
Phase Transition Temperature of Non-Peripherally Unsymmetrically Substituted Phthalocyanines (Scheme 2.154)

Structure	M	R	Transition Temperature	Ref.
60.1	2H	$R_1 = (CH_2)_8OH$, $R_2 = Me$, $R_3 = R_4 = R_5 = R_6 = R_7 = R_8 = (CH_2)_7CH_3$	Cr 56 Col$_r$ 62 Col$_h$ 170 I	[1443]
60.2	2H	$R_1 = (CH_2)_8O$-(a), $R_2 = Me$, $R_3 = R_4 = R_5 = R_6 = R_7 = R_8 = (CH_2)_7CH_3$	Cr 60.1 Col$_h$ 132.7 I	[1443]
60.3	2H	$R_1 = R_2 = (CH_2)_4OH$, $R_3 = R_4 = R_5 = R_6 = R_7 = R_8 = (CH_2)_5CH_3$	Cr (158 Col) 176 I	[1444]
60.4	2H	$R_1 = R_2 = (CH_2)_4OH$, $R_3 = R_4 = R_5 = R_6 = R_7 = R_8 = (CH_2)_7CH_3$	Cr (118 Col) 133 I	[1444]
60.5	2H	$R_1 = R_2 = (CH_2)_4OH$, $R_3 = R_4 = R_5 = R_6 = R_7 = R_8 = (CH_2)_9CH_3$	Cr (107 Col) 124 I	[1444]
60.6	2H	$R_1 = (CH_2)_3OH$, $R_2 = R_3 = R_4 = R_5 = R_6 = R_7 = R_8 = (CH_2)_5CH_3$	Cr 94 Col 148 I	[1444]
60.7	2H	$R_1 = (CH_2)_3OH$, $R_2 = R_3 = R_4 = R_5 = R_6 = R_7 = R_8 = (CH_2)_6CH_3$	Cr 75 Col 116.5 Col 129 I	[1444]
60.8	2H	$R_1 = (CH_2)_3OH$, $R_2 = R_3 = R_4 = R_5 = R_6 = R_7 = R_8 = (CH_2)_7CH_3$	Cr 55 Col 73.5 Col 112 I	[1444]

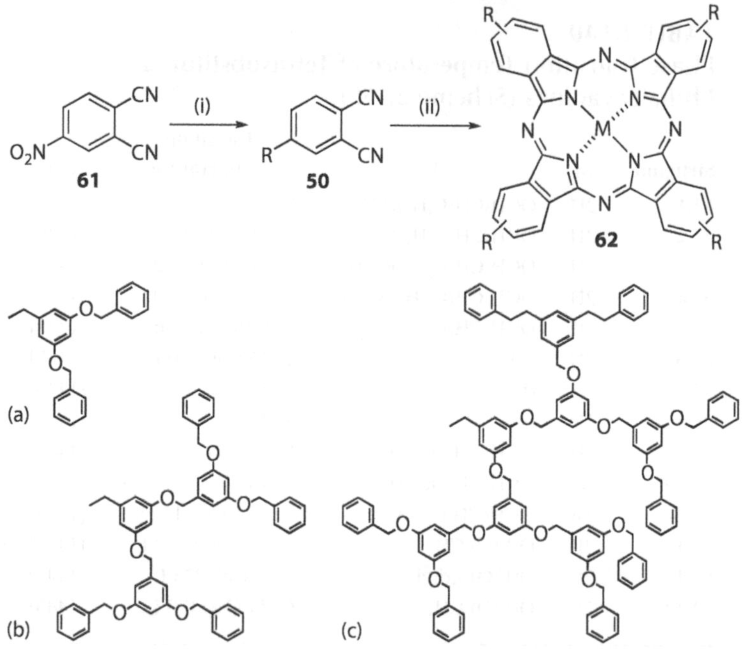

SCHEME 2.155 Synthesis of tetraalkoxy-substituted phthalocyanines: (i) ROH, DMF, K_2CO_3; (ii) $LiOC_5H_{11}$-$C_5H_{11}OH$; metal salt.

2.22.11 TETRAALKOXY-SUBSTITUTED PHTHALOCYANINES

The preparation of tetraalkoxy Pcs is outlined in Scheme 2.155. Appropriate primary alcohol is reacted with 4-nitrophthalonitrile **61** via a base-catalyzed nitro displacement reaction to yield 4-alkoxyphthalonitrile **50**. The cyclotetramerization of **50** using lithium pentaoxide in refluxing pentanol yields the metal-free Pc **62** (M = 2H), and when the phthalonitrile was refluxed in the presence of corresponding metal salt in pentanol, the metallo-derivative forms [1445]. The phase transition temperatures of these tetraderivatives are collected in Table 2.140 [756,1440,1442,1445–1448]. The branched-chain alkoxy-substituted Pcs showed mesophase behavior with low-temperature Col_r and high-temperature Col_h mesophases. Compound **62.2** deviates from the normal trend and exhibits higher melting transition than its lower homologue **62.1** [1446]. Tetra-substituted phthalocyanine with peripheral dendrimers showed Col_h mesophase only when the dendrimer was small sized, but when the dendrimer size increased, there was no mesophase and the glassy state directly cleared into the isotropic liquid [1440]. The branched derivative with $OCH(CH_2OC_{12}H_{25})_2$ substitution show Col_h mesophase. The melting behavior is not much influenced by the central metal, but clearing behavior drastically changes with metal substitution. The metal-free derivative cleared at 56°C compared to nickel and zinc complexes, which clears above 100°C [1447]. Triethyleneoxy-substituted Pcs show Col_h mesophase with very high clearing temperatures, but when it was connected to a bulky terminal group, like trityl group, the transition temperature decreases significantly. The transition temperatures also decrease on increasing the number of ethyleneoxy units [1442,1445,1448].

2.22.12 TETRATHIAALKYL- AND TETRAALKYLTHIAMETHYL-SUBSTITUTED PHTHALOCYANINES

4,5-Dichlorobenzene-1,2-dicarbonitrile when reacted with appropriate alkanethiol in the presence of mild base like potassium carbonate provides mono- and dialkylthia-substituted products, which can be separated through column chromatography. Monosubstituted product **63** ($R_1 = SR$; $R_2 = Cl$) is thus isolated and cyclotetramerized (Scheme 2.156) under usual reaction conditions to yield

TABLE 2.140
Phase Transition Temperature of Tetrasubstituted
Phthalocyanines (Scheme 2.155)

Structure	M	R	Transition Temperature	Ref.
62.1	2H	$OCH_2CH(C_4H_9)C_7H_{15}$	Col_r 79 Col_h	[1446]
62.2	2H	$OCH_2CH(C_6H_{13})C_9H_{19}$	Col_r 90.7 Col_h	[1446]
62.3	2H	$OCH_2CH(C_8H_{17})C_{11}H_{23}$	Col_r 65 Col_h 212 I	[1446]
62.4	2H	$OCH_2CH(C_{10}H_{21})C_{12}H_{25}$	Col_r 60 Col_h 180 I	[756]
62.5	2H	$OCH_2CH(C_{10}H_{21})C_{13}H_{27}$	Col_r 46 Col_h 166 I	[1446]
62.6	2H	(a)	g 115 Col_h 270 I	[1440]
62.7	2H	(b)	g 112 I	[1440]
62.8	2H	(c)	g 71 I	[1440]
62.9	2H	$OCH(CH_2OC_{12}H_{25})_2$	Cr 14.2 Col_h 56.1 I	[1447]
62.10	Ni	$OCH(CH_2OC_{12}H_{25})_2$	Cr 14.4 Col_h 116.9 I	[1447]
62.11	Zn	$OCH(CH_2OC_{12}H_{25})_2$	Cr 11.5 Col_h 122.9 I	[1447]
62.12	2H	$O(EtO)_3CH_3$	Cr 78 Col_h >330 I	[1445]
62.13	2H	$O(EtO)_4$-trityl	G 30 Col_h 225 I	[1448]
62.14	2H	$O(EtO)_8CH_3$	Cr 11 Col_h 170 I	[1445]

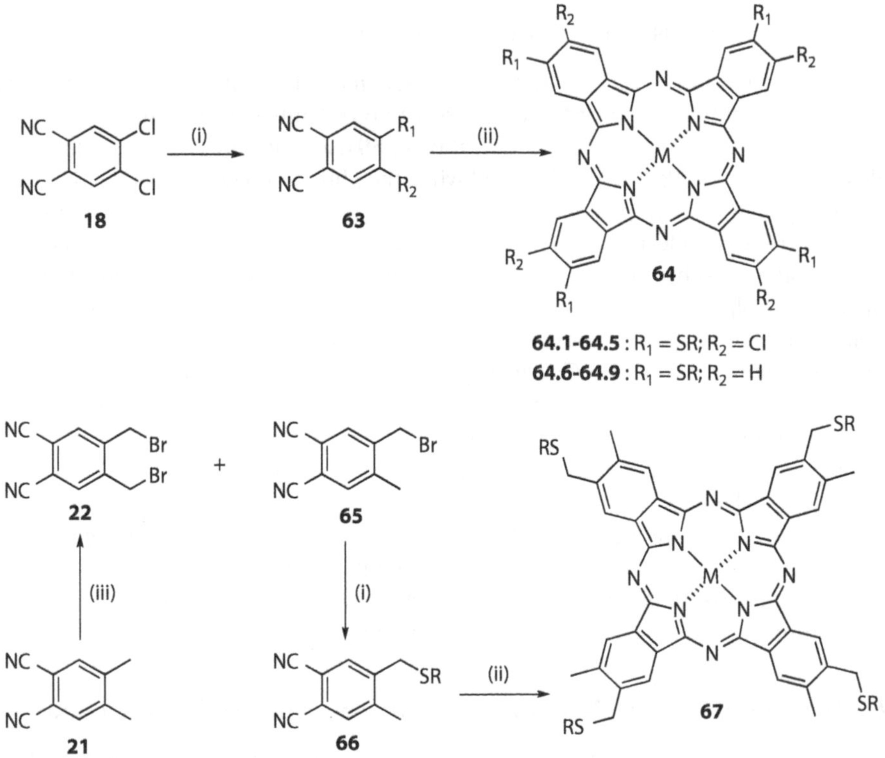

64.1-64.5 : R_1 = SR; R_2 = Cl
64.6-64.9 : R_1 = SR; R_2 = H

SCHEME 2.156 Synthesis of tetrathiaalkyl-substituted phthalocyanines: (i) RSH, DMSO, K_2CO_3; (ii) 1-pentanol, Li; metal salt; (iii) NBS.

TABLE 2.141
Phase Transition Temperature of Tetrathiaalkyl- and
Tetraalkylthiamethyl-Substituted Pcs (Scheme 2.156)

Structure	M	R	Transition Temperature	Ref.
64.1	2H	$R_1 = SC_8H_{17}$, $R_2 = Cl$	Cr −7 Col_h	[1415]
64.2	Zn	$R_1 = SC_8H_{17}$, $R_2 = Cl$	Cr 62 Col_h	[1415]
64.3	Cu	$R_1 = SC_8H_{17}$, $R_2 = Cl$	Cr −5 Col_h	[1415]
64.4	2H	$R_1 = SCH(CH_2O(EtO)_2C_2H_5)_2$, $R_2 = Cl$	Col_h 160 I	[1419]
64.5	Ni	$R_1 = SCH(CH_2O(EtO)_2C_2H_5)_2$, $R_2 = Cl$	Col_h 222.2 I	[1419]
64.6	2H	$R_1 = SCH(CH_2O(EtO)_2C_2H_5)_2$, $R_2 = H$	Col_h 198 I	[1419]
64.7	Ni	$R_1 = SCH(CH_2O(EtO)_2C_2H_5)_2$, $R_2 = H$	Col_h 230.5 I	[1419]
64.8	2H	$R_1 = SCH(CH_2OC_{12}H_{25})_2$, $R_2 = H$	Cr 13 Col_h 141 I	[1417]
64.9	Ni	$R_1 = SCH(CH_2OC_{12}H_{25})_2$, $R_2 = H$	Cr 9 Col_h 171 I	[1417]
67.1	2H	$CH(CH_2OC_{12}H_{25})_2$	Cr 8.42 Col_h 153.1 I	[1420]
67.2	Ni	$CH(CH_2O\,C_{12}H_{25})_2$	Cr 5.02 Col_h 180 I	[1420]

64 [1415]. The phase behavior of the tetrathiaalkyl-substituted Pcs is summarized in Table 2.141. The metal-free and copper complex of tetrachloro-tetrakis(octylthio)-Pc show Col_h mesophase well below room temperature, but the corresponding zinc derivative melts at significantly higher temperature [1415]. Tetrasubstituted Pcs with peripheral $SCH(CH_2O(EtO)_2C_2H_5)_2$ or bis(dodecyloxy) methanethiol chains show Col_h mesophase, and an increase in clearing temperature on complexing with nickel atom was observed [1417,1419]. When the other four peripheral positions were substituted by chlorine, the clearing temperature decreases compared to when these positions were left unsubstituted [1419].

The synthesis of liquid crystalline tetraalkylthiamethyl-substituted Pcs **67** (Scheme 2.156) essentially follow the route adopted for the synthesis of octathiaalkylmethyl-Pcs (Scheme 2.147) with the difference that here only the partial bromination of **21** was carried out. The reaction yields both monobrominated and dibrominated products from which the required monosubstituted compound **65** is separated by column chromatography and treated with alkanethiol in the presence of a mild base like potassium carbonate to give 4-methyl-5-(alkylsulfanylmethyl)phthalonitrile **66**, which is finally cyclotetramerized by lithium pentaoxide catalysis to tetraalkylthiamethyl-substituted phthalocyanine **67** [1420]. Both metal-free and Ni complex with (bis(dodecyloxy)methyl) (methyl)sulfane chains showed Col_h mesophase; the mesophase clearing temperature increases while the melting temperature decreases on complexing with nickel and thus leads to a wider mesophase range (Table 2.141) [1420].

2.22.13 TETRAESTERS OF PHTHALOCYANINE

Tetraesters of phthalocyanine with the general chemical structure **68** can be prepared by the reaction of trimellitic dianhydride or pyromellitic dianhydride with urea, CuCl, and ammonium molybdate in nitrobenzene followed by hydrolysis and esterification (Scheme 2.157) [1449, 1450]. Their phase transition temperatures are collected in Table 2.142 [1450]. Both unsubstituted tetraesters and the tetraesters containing additional tetracarboxylic acid groups display lyotropic and thermotropic mesophases. In the tetrasubstituted series, the first two members are non-liquid crystalline and mesomorphism started from the propyl ester onward. Cu complexes **68.3–68.10** exhibit a columnar phase and decompose above 300°C. The melting temperatures decrease drastically from 160°C for propyl ester to 50°C for butyl ester. The melting

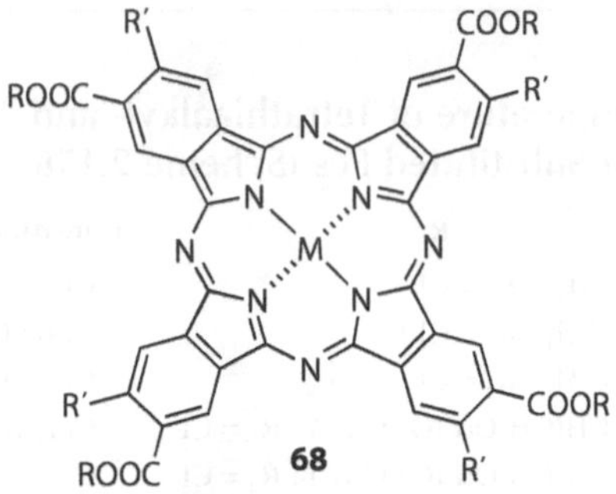

68.1–68.10: R' = H, R = C_nH_{2n+1}
68.11–68.13: R' = COOH, R = C_nH_{2n+1}

SCHEME 2.157   Molecular structure of phthalocyanine tetraesters.

TABLE 2.142
Phase Transition Temperatures of Tetraesters of Pc (Scheme 2.157)

Structure	M	R'	R	Transition Temperature
68.1	Cu	H	CH_3	Cr >300 I
68.2	Cu	H	C_2H_5	Cr >300 I
68.3	Cu	H	C_3H_7	Cr 160 Col >300 I
68.4	Cu	H	C_4H_9	Cr 50 Col >300 I
68.5	Cu	H	C_6H_{13}	Cr 37.5 Col >300 I
68.6	Cu	H	C_7H_{15}	Cr 26.6 Col >300 I
68.7	Cu	H	C_9H_{19}	Cr −4 Col >300 I
68.8	Cu	H	$C_{10}H_{21}$	Cr 21.5 Col >300 I
68.9	Cu	H	$C_{11}H_{23}$	Cr 30.5 Col >300 I
68.10	Cu	H	$C_{16}H_{33}$	Cr 80 Col >300 I
68.11	Cu	COOH	C_4H_9	Col >300 dec
68.12	Cu	COOH	C_9H_{19}	Cr 130 Col >300 I
68.13	Cu	COOH	$C_{11}H_{23}$	Cr 100 Col >300 I

Source: Data from Usol'tseva, N. et al., *Mol. Cryst. Liq. Cryst.*, 352, 45, 2000.

temperatures decrease up to C_9 derivative and then gradually increase in higher homologues. Octa-substituted derivatives **68.11–68.13** having additional free carboxylic acid groups also display columnar phase and decompose above 300°C. All these tetraesters form lyotropic mesophases in various solvents [1450].

2.22.14   CROWN-ETHER-SUBSTITUTED PHTHALOCYANINES

Crowned phthalocyanines are described as molecular cables, containing a central electron wire, four ion channels, and a surrounding insulating hydrocarbon mantle. Thus, crowned phthalocyanine discotics can be considered as systems that combine properties of electron conduction, ion

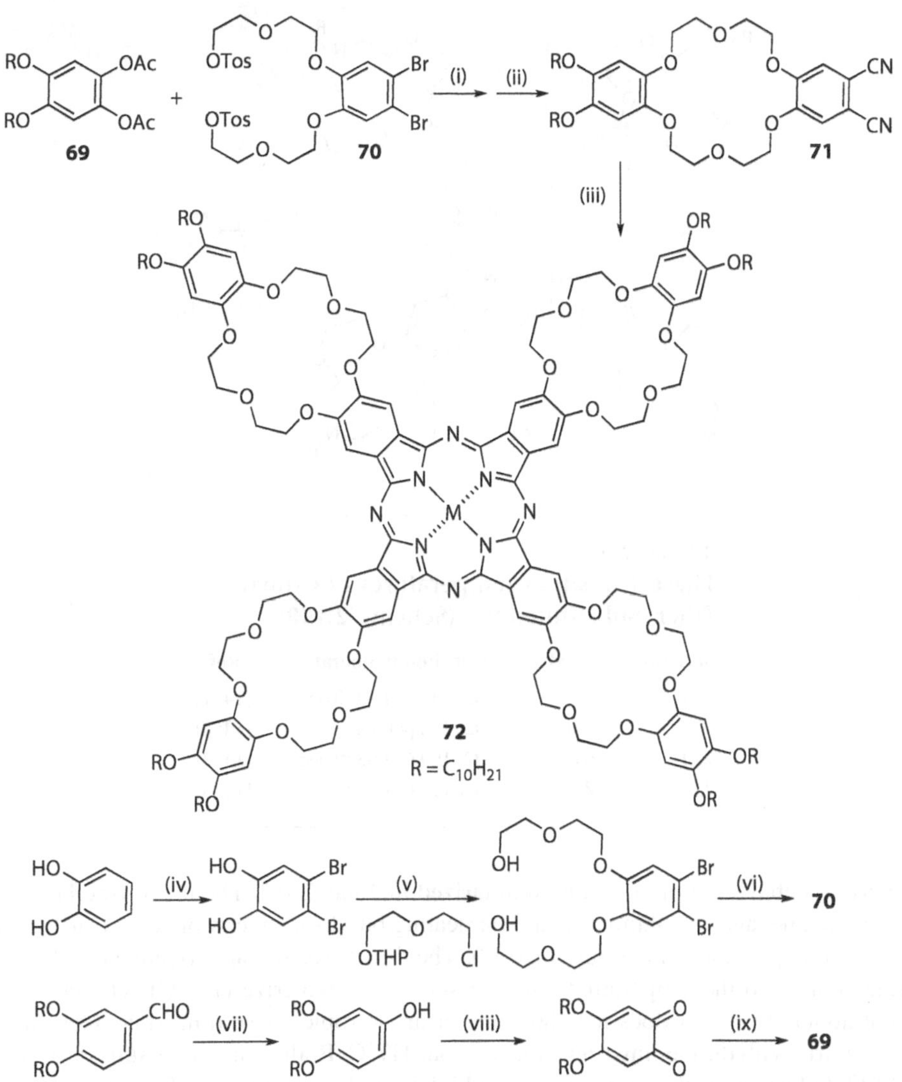

SCHEME 2.158 Synthesis of crown-substituted phthalocyanines: (i) n-BuOH, NaOH, reflux; (ii) DMF, pyridine, CuCN, NH$_4$OH; (iii) (CH$_3$)$_2$NCH$_2$CH$_2$OH, reflux; (iv) Br$_2$; (v) n-BuOH, NaOH, reflux; HCl; (vi) pyridine, pTosCl; (vii) CHCl$_3$-MeOH, H$_2$O$_2$, H$_2$SO$_4$; (viii) K(SO$_3$)$_2$NO, KH$_2$PO$_4$; (ix) AcOH/Ac$_2$O, Zn, reflux.

conduction, and liquid crystallinity [1451–1453]. These phthalocyanines form stable a monolayer at the air–water interface. These monolayers can bind alkali metal ions from the subphase and can be transferred onto solid substrates. In solution, these compounds form long fibers, and when chiral side chains are introduced into the periphery, helical long fibers are observed, which entangle to form large bundles of chiral fibers. Interestingly, when alkali metal ions are added to the solution, the helical fibers become straight and lose chirality. The synthesis of crown ether-substituted Pc **72** is shown in Scheme 2.158. Phthalocyanine **72** is prepared following the general synthetic methodology, that is, via the cyclization of crown ether **71**. Compound **71** is obtained in a simple but multi-step synthetic process (Scheme 2.158). The crowned phthalocyanine **73** can be prepared similarly from appropriate phthalonitrile.

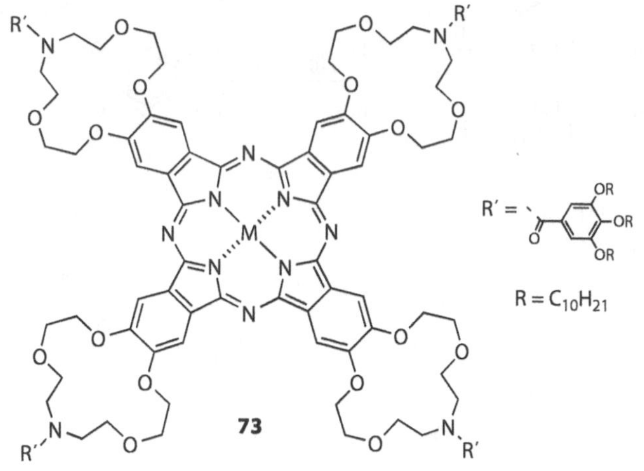

$R' = $ (structure)

$R = C_{10}H_{21}$

73

TABLE 2.143
Phase Transition Temperatures of Crown Ether–Substituted Pcs (Scheme 2.158)

Structure	M	Transiton Temperature	Ref.
72.1	2H	Cr 148 Col >320 dec	[1451]
72.2	Si(OH)$_2$	Cr 178 polymer	[1452]
73.1	Ni	Cr 109 Col >320 dec	[1453]
73.2	Zn	Cr 117 Col >320 dec	[1453]

The thermal behavior of these Pcs is summarized in Table 2.143. The metal-free Pc **72** melts at about 148°C to a hexagonal columnar phase on heating but does not exhibit any clearing transition prior to the decomposition temperature (320°C). The dihydroxysilicon derivative of **72** displays no phase transition up to the temperature where it starts to polymerize (178°C). The Ni complex of **73** melts at about 109°C but does not show any clear isotropic phase transition. The Zn complex behaves similarly with the melting transition at about 117°C. Both compounds start decomposing at about 320°C. Polarizing optical microscopy and XRD studies confirm the formation of hexagonal columnar phase in both samples.

2.22.15   CORE-EXTENDED MACRODISCOTIC PHTHALOCYANINES

The extension of Pc-core by benzofusion shifts the absorption to longer wavelength (red-shift). It also increases the temperature range over which columnar mesophase is observed. This is expected as extended planar π-system would tend to form strong co-facial aggregates. Accordingly, naphthacyanines and triphenyleno-Pcs have been designed and synthesized. As for phthalocyanines, two positions of chain substitution, peripheral and radial, are possible in naphthacyanines. The synthesis of 2,3-naphthacyanine discotics is depicted in Scheme 2.159. Peripherally substituted naphthacyanine **76** is obtained by cyclization of 6,7-disubstituted 2,3-dicyanonaphthalene **75**, under standard tetramerization conditions. For the preparation of radially substituted naphthacyanine **83**, the key precursor is **82**, which can be prepared from 1,2,4,5-tetrabromobenzene **77**. The reaction of 1,2,4,5-tetrabromobenzene with butyl lithium and 2,5-dimethylfuran yields the endoxide **78**, which, on deoxygenation, affords the dibromonaphthalene **79**. Alkyl chains are introduced in **79** via benzylic bromination followed by alkylation. The dibromide so obtained is converted to

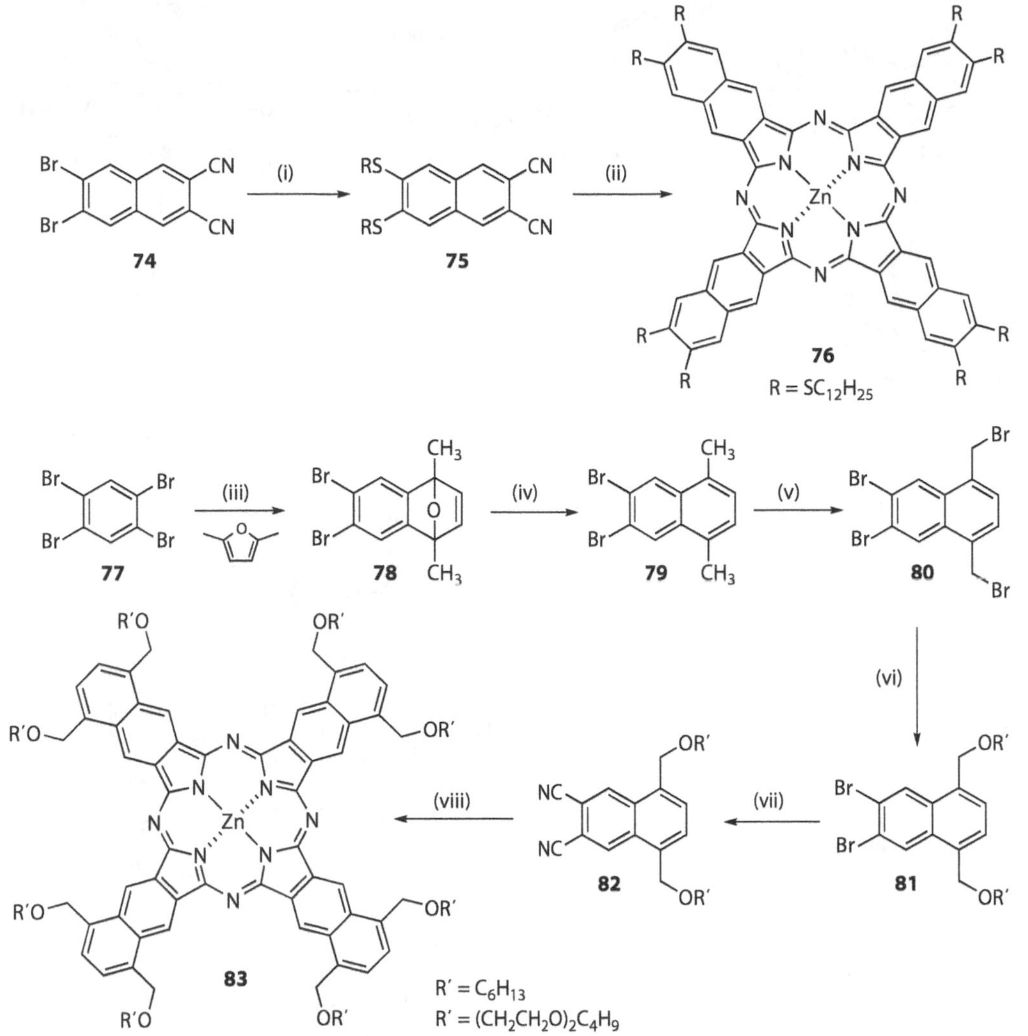

SCHEME 2.159 Synthesis of discotic naphthocyanines: (i) RSH, DMSO, K_2CO_3; (ii) $LiOC_5H_{11}$-$C_5H_{11}OH$, metal salt; (iii) BuLi; (iv) Zn, $TiCl_4$; (v) NBS; (vi) R′OH, Na; (vii) CuCN; (viii) Zn(OAc)$_2$, DBU.

dinitrile **82** by the treatment with CuCN which is cyclized to naphthacyanine **83**. Compound **76** was found to show a mesophase between 100°C and 300°C, but the full characterization of the mesophase could not be achieved. Zn complexes with radial substitution also display a mesophase above 130°C and up to the decomposition point at about 300°C. It is interesting to note that octa-alkyl- and octaalkoxy-2,3-naphthacyaninato metal complexes do not exhibit any mesomorphism [1135,1454,1455].

The triphenylenophthalocyanines **91** are prepared, as shown in Scheme 2.160 [1456]. Compounds **91** are prepared by the cyclization of 2,3-dicyanotriphenylene **90** under usual reaction conditions. The chemistry of the precursor 2,3-dicyanotriphenylene **90** is presented in Section 2.6. The thermal behavior of these triphenylenophthalocyanines is summarized in Table 2.144. Compounds **91.1–91.4** with copper metal atom exhibit multiple columnar rectangular phases before decomposition while the zinc derivatives **91.5–91.10** exhibit an unidentified columnar mesophase [1334,1456,1457].

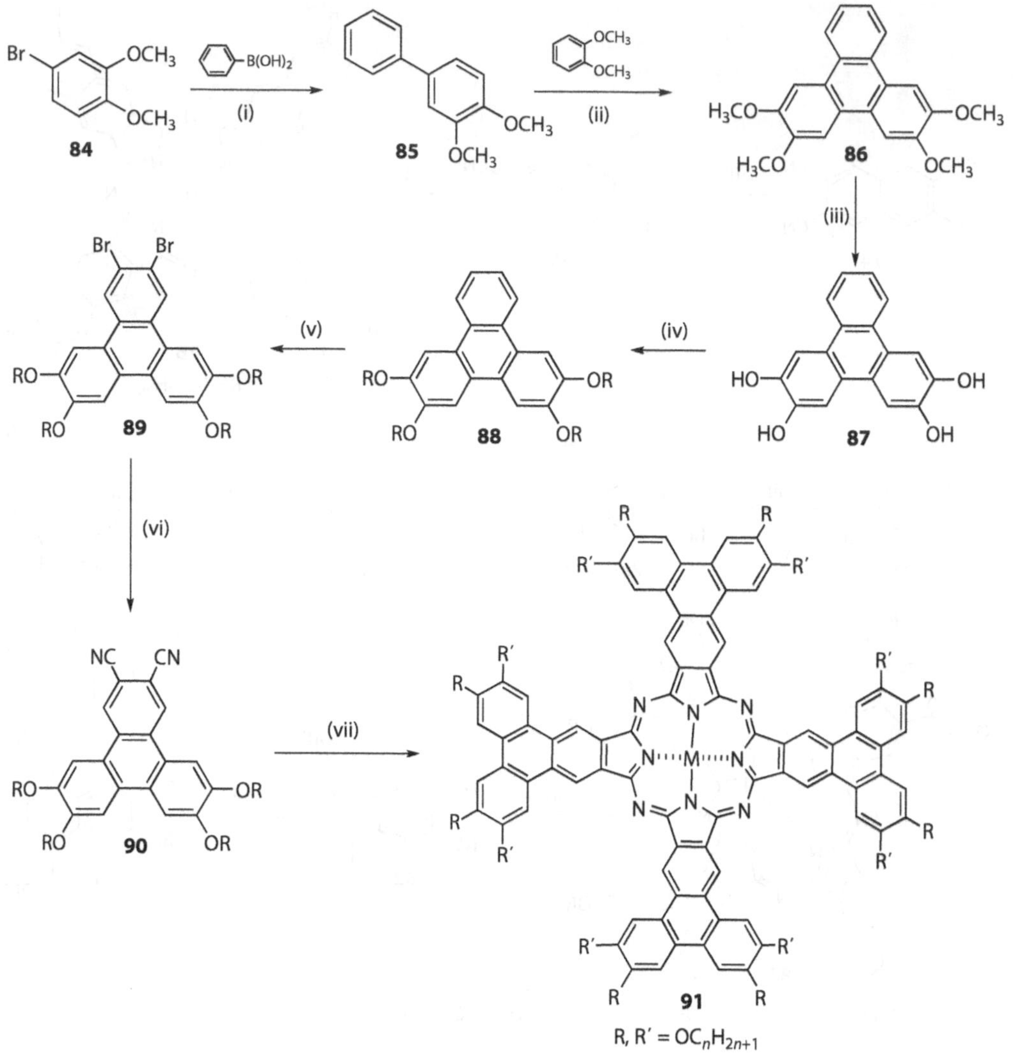

SCHEME 2.160 Synthesis of triphenylenophthalocyanines: (i) PdCl$_2$, Na$_2$CO$_3$, PPh$_3$; (ii) FeCl$_3$; (iii) HBr; (iv) C$_{12}$H$_{25}$Br, K$_2$CO$_3$; (v) Br$_2$; (vi) CuCN; (vii) DBU, 1-hexanol, metal salt.

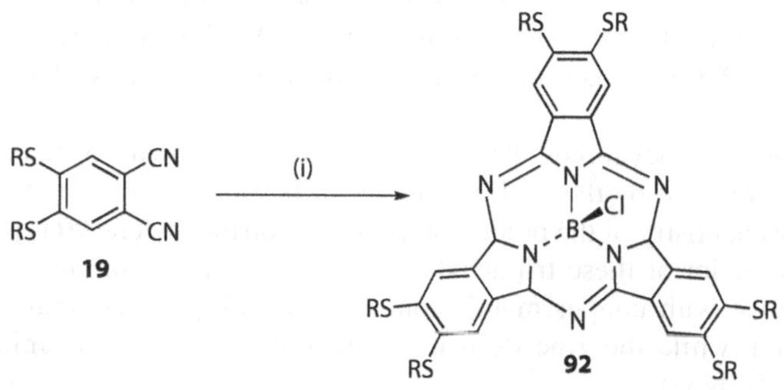

SCHEME 2.161 Synthesis of discotic subphthalocyanines: (i) BCl$_3$, 1-chloronaphthalene.

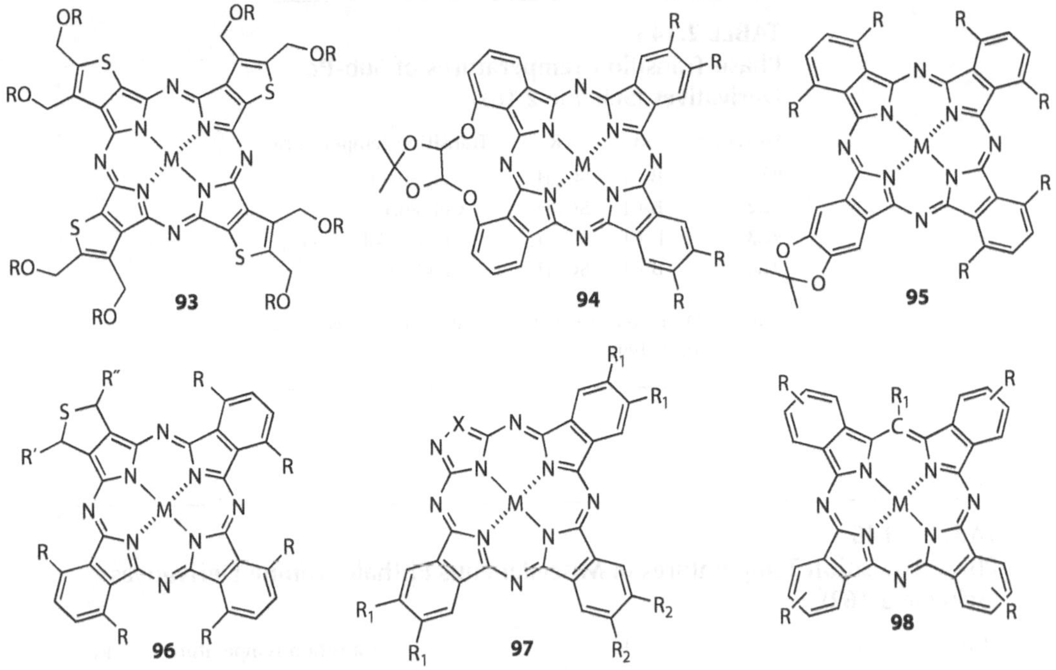

SCHEME 2.162 Molecular structures of some macrocycles structurally related to phthalocyanine.

2.22.16 SUBPHTHALOCYANINES

Subphthalocyanines are cone-shaped closed heterocycles composed of three imidazole rings coordinated to a central boron atom carrying an axial chlorine atom. Subphthalocyanines peripherally hexa-substituted with thioalkyl chains of varying chain length exhibit columnar mesophases. These compounds are synthesized, as shown in Scheme 2.161. The cyclotrimerization of 4,5-dialkylthiobenzene-1,2-dicarbonitrile in the presence of BCl_3 yields subphthalocyanines **92** in low yield [1458]. The phase-transition temperatures of these compounds are collected in Table 2.145. All

TABLE 2.144
Phase Transition Temperatures of Core-Extended Macrodiscotic Pcs (Schemes 2.159 and 2.160)

Structure	M	R	R'	Transition Temperature	Ref.
76	Zn	$SC_{12}H_{25}$	H	Cr 100 Col >300 dec	[1455]
83	Zn	H	OC_6H_{13}	Cr 133 Col >300 dec	[1455]
91.1	Cu	OC_8H_{17}	OC_8H_{17}	Col_{r1} 104.4 Col_{r2} 330 dec	[1334]
91.2	Cu	$OC_{10}H_{21}$	$OC_{10}H_{21}$	Col_{r1} 90.3 Col_{r2} 330 dec	[1334]
91.3	Cu	$OC_{12}H_{25}$	$OC_{12}H_{25}$	Col_{r1} 67.5 Col_{r2} 90.2 Col_{r3} 330 dec	[1334]
91.4	Cu	$OC_{14}H_{29}$	$OC_{14}H_{29}$	Col_{r1} 79.4 Col_{r2} 330 dec	[1334]
91.5	Zn	H	OC_6H_{13}	Col >185	[1457]
91.6	Zn	H	$OC_{12}H_{25}$	Col >60	[1457]
91.7	Zn	OC_6H_{13}	H	Col >270	[1457]
91.8	Zn	$OC_{12}H_{25}$	H	Col >100	[1457]
91.9	Zn	OC_6H_{13}	OC_6H_{13}	Col >20	[1457]
91.10	Zn	$OC_{12}H_{25}$	$OC_{12}H_{25}$	Col >20	[1457]

TABLE 2.145
Phase Transition Temperatures of Sub-Pc Derivatives (Scheme 2.161)

Structure	M	R	Transition Temperature
92.1	B-Cl	$SC_{10}H_{21}$	Col_h 90.9 I
92.2	B-Cl	$SC_{12}H_{25}$	Col_h 86.1 I
92.3	B-Cl	$SC_{16}H_{33}$	Cr 27.2 Col_h 75.3 I
92.4	B-Cl	$SC_{18}H_{37}$	Cr 43.4 Col_h 70 I

Source: Data from Kang, S.H. et al., *Chem. Commun.*, 1661, 1999.

TABLE 2.146
Phase Transition Temperatures of Miscellaneous Phthalocyanine Derivatives (Scheme 2.162)

Structure	M	R	Transition Temperature	Ref.
93.1	2H	$C_{10}H_{21}$	Col_{r1} 36 Col_{r2} 75 Col_h 278 I	[1375]
93.2	2H	$C_{14}H_{29}$	Cr 48 Col_h 223 I	[1375]
93.3	Cu	$C_{14}H_{29}$	Cr 49 Col_h 258 I	[1375]
94.1	Zn	$OC_2H_4[CH(CH_3)C_3H_7]_3CH(CH_3)_2$	Cr −68 Col_h	[1391]
94.2	Co	$OC_2H_4[CH(CH_3)C_3H_7]_3CH(CH_3)_2$	Cr −68 Col_h	[1391]
95.1	2H	$(CH_2)_5CH_3$	Cr 244 M 259 I	[1444]
95.2	2H	$(CH_2)_6CH_3$	Cr 186 M 235 M_1 237 I	[1444]
95.3	2H	$(CH_2)_7CH_3$	Cr 183 M 208 M_1 226 I	[1444]
95.4	2H	$(CH_2)_8CH_3$	Cr 155 M 192 M_1 216 I	[1444]
95.5	2H	$(CH_2)_9CH_3$	Cr 137.5 M 180 M_1 207 I	[1444]
96.1	2H	R = C_8H_{17}, R′ = R″ = H	Cr 164 Col_h 212 I	[1460]
96.2	2H	R = C_8H_{17}, R′ = R″ = CH_3	Cr 79 Col_r 170 Col_h 233 I	[1460]
96.3	2H	R = R′ = R″ = C_8H_{17}	Cr 80 Col_r 133 Col_h 212 I	[1460]
96.4	2H	R = C_8H_{17}, R′ = CH_3, R″ = $C_6H_{12}O_6$	Cr 53 Col_r 125 Col_h 208 I	[1460]
96.5	Ni	R = C_8H_{17}, R′ = CH_3, R″ = $C_6H_{12}O_6$	Cr 91 Col_h 235 I	[1460]
96.6	2H	R = C_6H_{13}, R′ = R″ = CH_3	Cr 125 Col_r 225 Col_h 248 I	[1460]
96.7	2H	R = C_6H_{13}, R′ = R″ = H	Cr 158 Col_h 231 I	[1460]
96.8	Cu	R = C_6H_{13}, R′ = R″ = H	Cr 166 Col_r 269 Col_h 307 I	[1460]
96.9	Zn	R = C_6H_{13}, R′ = R″ = H	Cr 170 Col_r 298 Col_h 337 I	[1460]
96.10	Ni	R = C_6H_{13}, R′ = R″ = H	Cr 151 Col_r 248 Col_h 267 I	[1460]
96.11	Co	R = C_6H_{13}, R′ = R″ = H	Cr 144 Col_r 278 Col_h 302 I	[1460]
97.1	Ni	X = N, R_1 = R_2 = $OCH_2CON(C_8H_{17})_2$	Non-LC	[1461]
97.2	Ni	X = N, R_1 = OC_8H_{17}, R_2 = $OCH_2CON(C_8H_{17})_2$	Col_h (glassy)	[1461]
97.3	Ni	X = $[NC_{12}H_{25}]$+Br-R_1 = R_2 = $OC_{12}H_{25}$	Cr up to 350	[1461]
97.4	Ni	X = N, R_1 = R_2 = OC_8H_{17}	Col_h 174 I	[1461]
97.5	Cu	X = N, R_1 = R_2 = OC_8H_{17}	Col_h 186 I	[1461]
97.6	Cu	X = N, R_1 = R_2 = $OC_{12}H_{25}$	Col_h 68 I	[1461]
98.1	2H	R = $C(CH_3)_3$, R_1 = $C_{15}H_{31}$	Cr 119 I	[1462]
98.2	2H	R = $OCH_2C(CH_3)_3$, R_1 = $C_{15}H_{31}$	Cr 100–125 Col >300 I	[1462]
98.3	2H	R = $OCH_2C(CH_3)_3$, R_1 = $C_{17}H_{35}$	Cr ~100 Col ~344 I	[1463]

the derivatives display a columnar mesophase. The clearing temperatures decrease with increasing chain length, while the melting temperatures increase on increasing chain length.

2.22.17 Miscellaneous Compounds Structurally Related to Phthalocyanines

Many macrocyclic compounds that are structurally closely related to Pc have been observed to exhibit mesomorphism when substituted with flexible alkyl chains. The chemical structures of these compounds are shown in Scheme 2.162, and their thermal data are collected in Table 2.146 [1331,1375,1391,1444,1459–1463]. These compounds are prepared by the cyclization of corresponding phthalonitrile or a mixture of two different phthalonitriles. Most of these exhibit columnar phases, as shown in Table 2.146.

2.23 MISCELLANEOUS DISCOTIC METALLOMESOGENS

Liquid crystalline metal complexes of organic ligands are generally known as metallomesogens and they exhibit the same type of mesophases as found in purely organic liquid crystals. Metallomesogens are becoming a very important class of mesogenic materials since new properties may be envisaged due to the introduction of metals. In additional, they offer wider possibilities for structural variations than simple organic materials. Metallomesogens combine metal-based coordination chemistry with the excellent physical properties exhibited by liquid crystals. They can exhibit thermotropic as well as lyotropic mesophases. The incorporation of transition and inner transition metal atoms brings with it fascinating features such as color and paramagnetism, redox and high polarizability properties, further extending the range of potentially useful physical properties of these intriguing mesomorphic organometallics. Moreover, metallomesogens involve the use of new methods of characterization, thereby extending the horizon of liquid crystal basic science and stimulating novel potential applications of the organometallics [1344,1345,1464–1468].

Being organometallics in nature, metallomesogens not only try to bridge the gap between the organic and inorganic liquid crystals but also truly attempt to diminish the gap between calamitic and discotic liquid crystals by exhibiting cross-over from calamitic to discotic mesophases on the appropriate substitution of peripheral chains and metal atoms. More elusive phases like cubic, biaxial smectic, and novel phases like discotic lamellar, lamella-columnar phases have been observed and even in some cases the much sought after and elusive biaxial nematic phase has been speculated [315,1469–1476]. Moreover, some metallomesogens exhibit switchable columnar phases. New molecular geometries not accessible by conventional organic liquid crystals, such as square planar, square pyramidal, trigonal bipyramidal, octahedral, etc., can be generated readily. In other words, apart from novel physical properties, novel phases and novel molecular architectures have been accessible with the help of metallomesogens.

As in the case of the organic liquid crystals, crystal structure determinations of metallomesogens have been investigated extensively to understand the relationship between the mesophase structure and the crystal organization. Most of the metallomesogens have been obtained by the coordination of two organic ligands through a metallic bridge. The ligands of the metallomesogens may be mesomorphic or nonmesomorphic in nature. Often, polydentate ligands form disk-like structures and hence exhibit columnar mesomorphism, whereas monodentate ligands form rod-like structures and hence calamitic phases.

The chemistry and mesomorphic properties of discotic metallomesogens based on porphyrins, porphyrazines, and phthalocyanines have been presented in previous sections. This section deals with metallomesogens having disk-like molecular structures and exhibiting columnar mesophases. A vast variety of such discotic metallomesogens is known in the literature. This section is not intended to systematically and exhaustively review the literature on metallomesogens but to give a flavor of the advances in columnar metallomesogens. It does not include the supramolecular, polycatenar, half-disk metallomesogens, etc., forming columnar phases. For the genesis of the field and its progress, the readers are advised to refer to the excellent texts and reviews on metallomesogens [1344,1345,1464–1468].

2.23.1 β-DIKETONATE COMPLEXES

β-Diketonate complexes are the extensively studied liquid crystals in the field of metallomesogens. The mesomorphic properties of this kind of complexes change drastically depending on the number and position of aliphatic chains on the phenyl rings. Here, only the complexes that exhibit discotic mesomorphism are considered. Owing to the square planar coordination around the metal ions, metal complexes with β-diketones possess a more disk-like shape and therefore are promising candidates for the formation of columnar phases. The preparation of the chelating ligands involves a straightforward Claisen condensation between an appropriate ketone and an ester. The versatility in the synthesis of the ligands has allowed the preparation of a wide variety of chemical structures for which the number of aliphatic chains and their positions can be controlled in a facile way to obtain the desired mesomorphism.

The first disk-like β-diketonate complexes exhibiting mesomorphism were reported by Giroud and Billard in 1981 [1477,1478]. Chemical structures of the symmetrically and unsymmetrically substituted complexes (1 and 2) are depicted in Scheme 2.163. The copper complexes containing four alkyl chains of the same or different chain lengths were found to exhibit mesomorphism. However, at this stage the structure of the mesophase was not elucidated. Following these results, Ohta and coworkers synthesized a series of tetraalkyl-substituted β-diketonate complexes 2 to understand the influence of chain length on mesomorphism and found that most of the complexes exhibit mesomorphism and even some of the members exhibit plural mesophases [1479]. Subsequently, Ohta et al. prepared a series of analogous copper complexes 3 containing four alkoxy side chains and investigated their mesophase behavior [1480,1481]. It was found that all the compounds were mesomorphic with only one mesophase. Furthermore, Ohta et al. established the structure of the mesophase by x-ray diffraction measurements that the mesophase in the complexes was a "discotic lamellar" (D_L) phase and the D_L phase has a structure in which the molecules tilt to the layers [1482]. It was for the first time that the lamellar phase in the mesophase of disk-like molecules was established and introduced. Later, many tetra-substituted copper and palladium complexes 3 containing different types of alkoxy chains were prepared by Maitlis and coworkers [1483] and found to exhibit the discotic lamellar mesophase [1484,1485]. As intermediates to the tetraalkyl and tetraalkoxy-substituted complexes, Sadashiva et al. prepared some complexes 4 containing two alkyl and two alkoxy chains, and found that they also exhibit the discotic lamellar phase [1486]. Usha et al. studied the crystal structure of tetraalkoxy and tetraalkyl-substituted

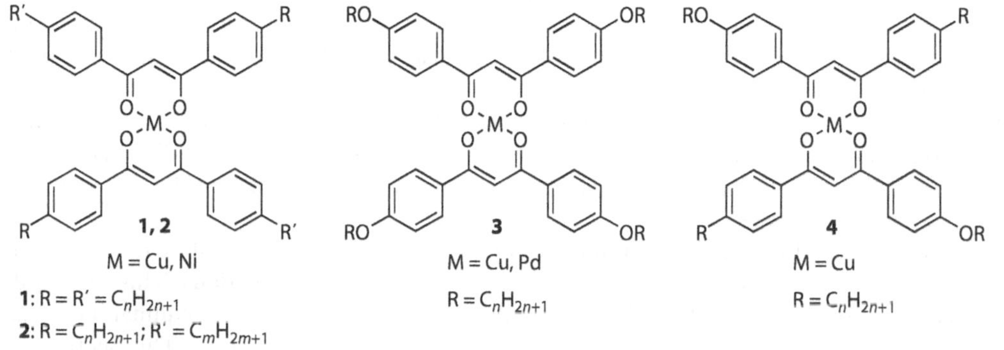

SCHEME 2.163 Molecular structures of symmetrically and unsymmetrically substituted β-diketone complexes.

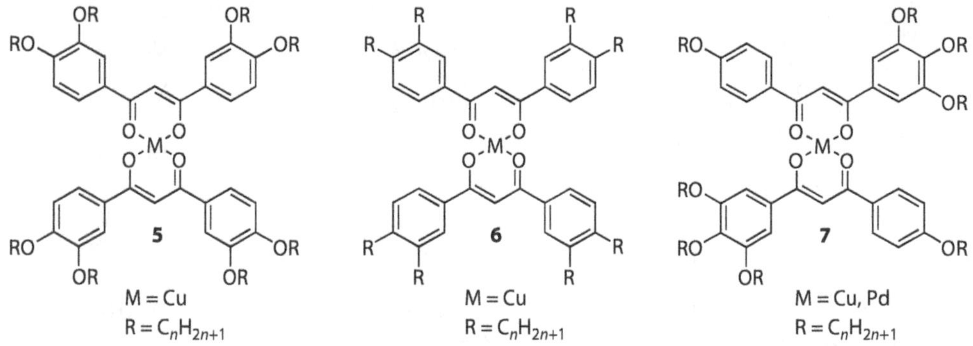

SCHEME 2.164 Chemical structures of some columnar phase forming β-diketone complexes.

complexes of copper and palladium by x-ray diffraction technique and found that the complexes **3, 4** (Scheme 2.163) possess both lamellar and columnar organization of molecules in the crystalline phase [1487,1488].

It was expected that with an increase in the number of peripheral chains, the β-diketone complexes may exhibit columnar mesophases and indeed it was found by Godquin-Giroud that octaalkoxy-substituted copper complexes **5** (Scheme 2.164) exhibit columnar phase [1489]. From miscibility and x-ray diffraction studies, the hexagonal structure of the columnar mesophase was established. Subsequently, Ohta and coworkers synthesized both octaalkoxy- **5** and octaalkyl-substituted copper complexes **6** [1490]. While the octaalkoxy derivatives were mesomorphic, the octaalkyl-substituted complex failed to exhibit mesomorphism [1490]. Recently, Lai et al. have synthesized two series of octaalkoxy-substituted copper and palladium complexes **7** with different substituents to study the effect of substituents on the columnar mesomorphism [1491]. The columnar hexagonal and columnar rectangular mesophase structures of these complexes were characterized by x-ray diffraction studies.

Swager et al. have designed and synthesized a large number of copper, palladium, and oxovanadium complexes **8** (Scheme 2.165) containing 10 side chains [1492]. All these complexes exhibit a columnar mesophase over a wide temperature range. The phase structures have been characterized by x-ray diffraction studies. It has been observed that the short chain derivatives exhibit columnar rectangular phase, while the long chain derivatives display columnar hexagonal mesophase. It is interesting to note that the oxovanadium complexes possess a square pyramidal shape but are still

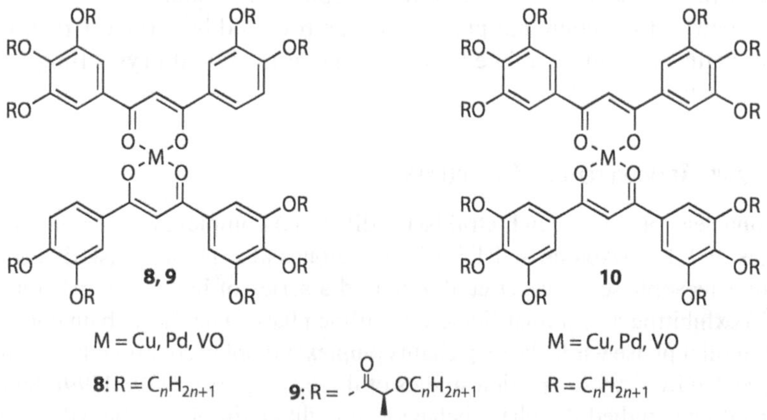

SCHEME 2.165 Chemical structures of some columnar phase forming copper, palladium, and oxovanadium β-diketone complexes with 10 or 12 peripheral chains.

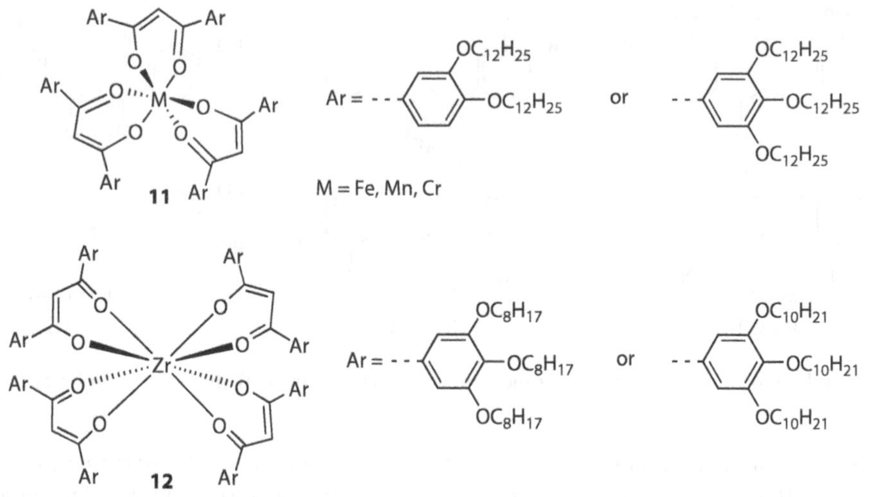

SCHEME 2.166   Chemical structures of some tris- and tetrakis-β-diketone complexes.

able to exhibit columnar mesomorphism. Later 'Serrano et al. reported a series of metallomesogens **9** containing 10 chiral side chains, which exhibit switchable columnar phase [1493]. These complexes exhibit room temperature mesomorphism with rectangular symmetry and undergo ferroelectric switching. The helical organization of the molecules within the columns has been confirmed by circular dichroism studies. Moreover, the spectral analysis of these complexes reveals that the two geometric isomers (*cis* and *trans*) exist in equimolar ratio.

Analogous to the above 10 chain complexes, 3 series of copper, palladium, and oxovanadium complexes **10** carrying 12 side chains have also been synthesized and studied by Swager and coworkers [1492]. These complexes also exhibit columnar phases with rectangular and hexagonal symmetry, depending on the length of the side chains.

Octahedral and square antiprism-shaped tris- and tetrakis-β-diketone complexes **11**, **12**, respectively (Scheme 2.166), exhibit columnar mesomorphism [1494–1497]. The octahedral-shaped liquid crystals based on diketonate complexes **11** of iron, manganese, and chromium display columnar hexagonal phases with 12 and more dodecyloxy side chains. Interestingly, cooperative chirality in the columnar liquid crystalline phase has been displayed by octahedral complexes carrying chiral side chains [1495]. Square antiprism zirconium tetrakis-β-diketonates complexes **12** with 24 alkoxy chains organize in columnar liquid crystal phases [1497]. The hexagonal symmetry of the columnar phase has been revealed by x-ray diffraction studies. The branched side chain analogue exhibited room temperature liquid crystallinity with columnar oblique phase structure.

2.23.2   TRI- AND TETRAKETONATE COMPLEXES

Like the mononuclear bis-, tris-, and tetrakis(1,3-diketone) complexes, different varieties of dinuclear complexes of 1,3,5-triketone and 1,3,5,7-tetraketones also possess disk-like shape and hence exhibit columnar mesophases. Swager et al. reported a series of bimetallic triketonates **13**, **14**, **15** (Scheme 2.167) exhibiting a columnar liquid crystalline phase. The short chain complexes exhibit a columnar rectangular phase while the long chain complexes display columnar hexagonal mesophase, as has been revealed by x-ray diffraction studies of these complexes [1498,1499]. Subsequently, Lai et al. synthesized and studied the phase behavior of eight chain bimetallic triketonates **16** [1500]. All the studied materials exhibited a columnar hexagonal mesophase over a wide range of temperatures. Tetraketonate complexes of copper **17** also exhibit columnar hexagonal mesophase and it is

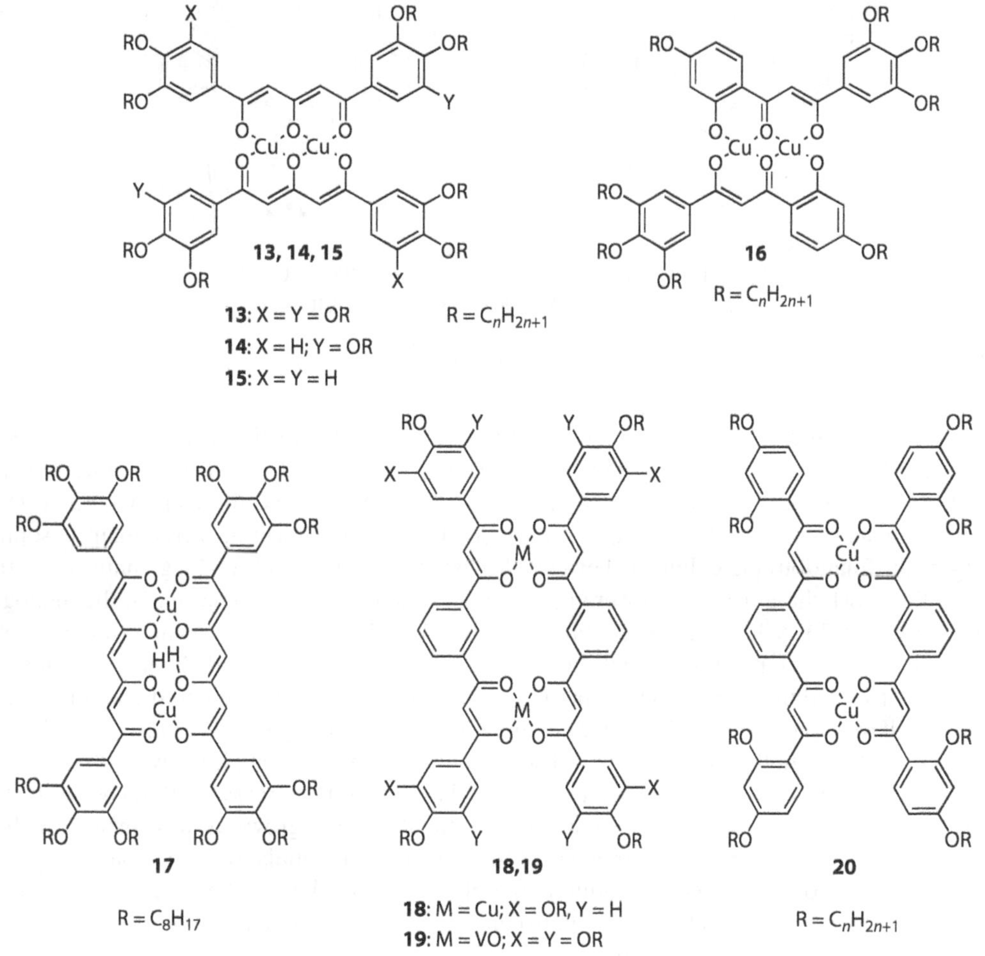

SCHEME 2.167 Molecular structures of some 1,3,5-triketone and 1,3,5,7-tetraketone complexes.

observed that with increase in the length of the side chains, the clearing temperatures of the complexes decrease [1498]. Lai et al. prepared bimetallic tetraketone complexes **18, 19** of copper and studied their mesophase behavior; unlike the above tetraketonate complexes, in these complexes the metallic centers are far apart [1501]. These complexes carrying 12 and 8 side chains also exhibit columnar hexagonal mesophase stable over a wide range of temperatures. However, another bimetallic tetraketonate complex **20** containing eight side chains has been recently reported to exhibit columnar rectangular mesophase [1502].

2.23.3 Dithiolene Complexes

The first disk-like dithiolene complexes **21** (Scheme 2.168) containing four peripheral alkoxy chains were reported independently by Veber et al. and Ohta et al. [1503,1504]. Both the groups initially reported that these nickel complexes exhibit an unidentified mesophase with high clearing enthalpies. Later, Ohta assigned this phase as discotic lamellar phase. However, subsequent x-ray diffraction study of the complexes showed that the so-called discotic mesophases of dithiolene nickel complexes were, in fact, lamellar crystalline phases with some degree of disorder [1505]. Ohta et al. have reinvestigated the mesomorphism of tetraalkoxy-substituted dithiolene nickel complexes **21** [1506]. Moreover, the dithiolene nickel complexes **22** containing four peripheral alkyl chains were

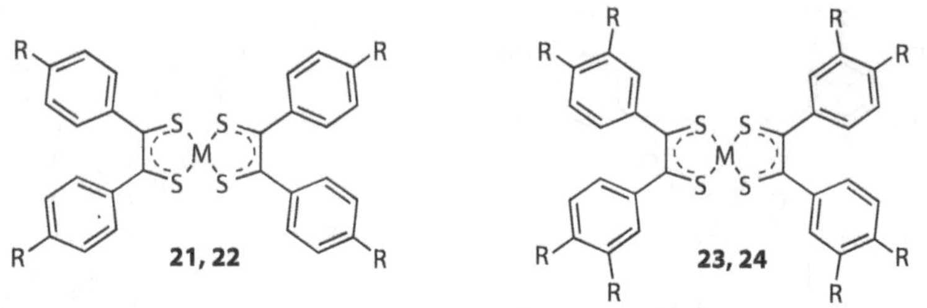

21, 22 23, 24

21: R = OC_nH_{2n+1}
22: R = C_nH_{2n+1}
M = Ni, Pd, Pt
23: R = OC_nH_{2n+1}
24: R = C_nH_{2n+1}

SCHEME 2.168 Chemical structures of some dithiolene complexes.

unambiguously found to be nonmesomorphic in nature [1506]. Interestingly, dithiolene complexes possess good π-acceptor properties, which could be applied in one-dimensional organic semiconductors. In order to realize this potential applicable property, Ohta et al. synthesized a series of dithiolene nickel complexes **23** with eight peripheral alkoxy chains forming a columnar mesophase [1507–1509]. Enantiotropic columnar hexagonal mesophase was found for long chain derivatives whereas the short chain derivatives were monotropic in nature. On the other hand, the analogous nickel complexes **24** (Scheme 2.168) containing eight alkyl chains are not mesomorphic but isotropic liquids at room temperature [1509]. In order to investigate the effect of the metal atoms, two new series of palladium and platinum complexes **23** were successfully prepared containing eight peripheral alkoxy chains [1510]. The same columnar hexagonal mesophase was present, as observed in nickel complexes, in the two new series. The mesophase of all the complexes were characterized by x-ray diffraction studies. The intracolumnar correlation length decreases with increase in chain length. The lattice parameter is independent of the metal ion and grows almost linearly with the number of methylene groups. Measurements of the reduction potentials by cyclic voltametry study have been carried out for nickel, palladium, and platinum complexes. The π-acceptor properties were revealed, and it was found that palladium complexes were the best π-acceptors while the platinum counterparts were the worst [1510].

2.23.4 Dioximato Complexes

Ohta et al. have prepared and extensively studied mesomorphism of the bis(diphenylglyoximato) metal(II)-based complexes **25**, **26** [1511,1512]. The various molecular designs are depicted in Scheme 2.169. The dioxime ligands can be prepared from the corresponding benzils by reaction

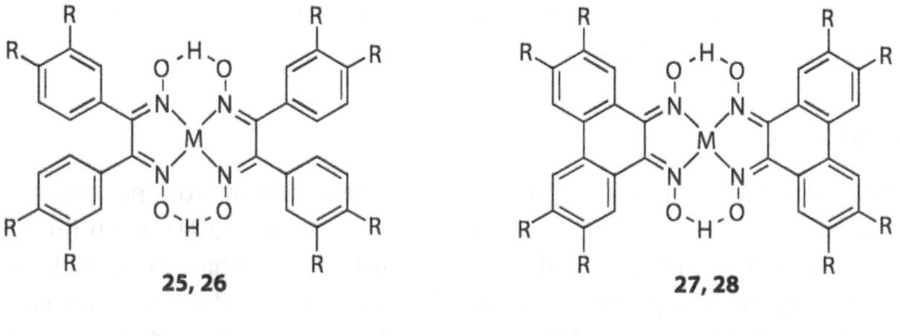

25, 26 27, 28

25: R = OC_nH_{2n+1}
26: R = C_nH_{2n+1}
M = Ni, Pd, Pt
27: R = C_nH_{2n+1}
28: R = OC_nH_{2n+1}

SCHEME 2.169 Chemical structures of some dioxime-based metal complexes.

with hydroxylamine hydrochloride. The complexes were obtained by the reaction of an alcoholic solution of metal chloride salts with the appropriate dioxime. It was found that most of the complexes display columnar hexagonal mesophase. Some of the compounds are mesomorphic at room temperature and show an extremely broad mesomorphic range. The high clearing temperatures of the complexes could be due to the possibility of intermolecular hydrogen bonding between neighboring molecules, which would promote more efficient stacking into columns. It is interesting to note that in addition to mesomorphic properties, the dioxime nickel and palladium complexes also exhibit thermochromism, solvatochromism, and gel formation [1511]. With increasing temperature, the nickel complexes turned gradually from red to yellow, the palladium complexes from orange to yellow, and the platinum complexes from green to yellow. The effect of the nature of the chains has also been investigated. The substitution of the alkoxy groups by alkyl chains has a profound effect on the mesomorphism, thermochromism, and solvatochromism [1512].

Ohta et al. have further pursued critical structural study on the nickel dioxime complexes in order to understand the effect of some molecular parameters on the mesophase structure. Particularly interesting is the columnar-to-discotic lamellar phase transformation by systematically changing the number and length of peripheral alkyl chains. Hence, two new series of complexes with non-symmetrical chain lengths were prepared to identify the molecular parameters influencing columnar-to-discotic lamellar phase transformation [1513].

In order to have more understanding of the structure–property relationships, two series of structurally related octaalkyl- and octaalkoxy-substituted dioximato metal complexes **27, 28** were prepared and studied by Mohr and coworkers (Scheme 2.169) [1514,1515]. The octaalkyl-substituted bis(phenanthrene-9,10-dionedioximato) complexes **27** of nickel, palladium, and platinum exhibit columnar mesomorphism. These complexes also possess high clearing temperatures and a broad range of temperatures. In addition, columnar rectangular phase has also been observed in these compounds; this could be due to the rigidity of the phenanthrene core. These compounds do not exhibit thermochromism. The octaalkoxy-substituted homologues **28** have been prepared and show similar behavior [1515].

2.23.5 CYCLIC PYRAZOLE–METAL COMPLEXES

Mononuclear and dinuclear metallomesogens are abundant in the metallomesogen literature; however, trinuclear metallomesogens are rare. Moreover, multinuclear metal complexes often give rise to important physical properties. Being motivated by the complex formation ability of coinage metals, which form ring-shaped trimers, Barbara et al. prepared and investigated columnar mesomorphism of Au complexes **29** with pyrazolate ligands [1516,1517]. The chemical structures of the complexes are shown in Scheme 2.170. The ligands are substituted by two aromatic rings, with a variable number of alkoxy chains. The mesophase characterization of these complexes has been carried out by optical, calorimetry, and x-ray diffraction studies. The mesophase structure of the complexes was identified by x-ray diffraction as columnar hexagonal. By increasing the number of side chains, both the melting and clearing temperatures decrease in these complexes. The unequally substituted complexes were found to exist as a mixture of two geometric isomers in 1:3 ratio. In unsymmetrical compounds, the interaction between the molecules were probably more hindered and perturbed, leading to the reduction of mesophase stability. Moreover, to understand the relationship between the molecular packing in the crystal structures and in the mesophase, x-ray diffraction study of a model compound in the crystalline state has been carried out. From the x-ray diffraction, it has been found that neighboring macrocycles are mutually rotated in such a way that the gold atoms of a molecule lie on top of the pyrazole rings of the lower neighbor.

Alkyl chain–substituted trinuclear gold complexes **30** exhibiting columnar mesomorphism was reported by Kim and coworkers [1518]. These compounds are unique in that they exhibit columnar hexagonal mesophase with only three side chains. Several compounds have been prepared but only a couple of complexes exhibit monotropic mesophase behavior. Surprisingly, neither by lengthening

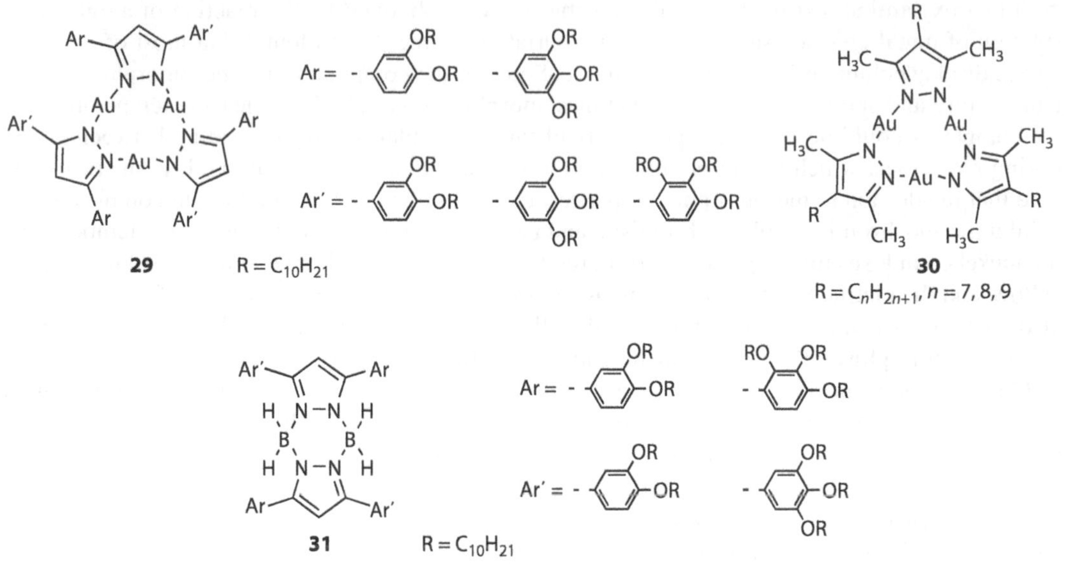

SCHEME 2.170   Molecular structures of some cyclic pyrazole–metal complexes.

nor by increasing the number of side chains in these complexes mesomorphism could be obtained, demonstrating the delicate nature of the mesophase. X-ray crystal structure reveals a weak "dimer" formation through intermolecular Au–Au interactions. This dimer behaves as a planar disk with six side chains and the subsequent stacking of dimers may lead to the hexagonal columnar structure in the mesophase.

Serrano et al. have synthesized and investigated the columnar liquid crystalline properties of pyrazaboles **31** with BH_2 bridges [1519]. They prepared 1,3,5,7-tetraphenyl pyrazaboles substituted with one, two, or three alkoxy groups on each phenyl ring by the dimerization of half-disk-shaped nonmesogenic pyrazoles. The chemical structures of some of these compounds are shown in Scheme 2.170. The pyrazaboles containing 8, 10, and 12 decyloxy chains display columnar mesomorphism. One compound displays a monotropic columnar hexagonal phase that crystallizes quickly and another compound exhibits a kinetically stable mesophase that can be kept at room temperature for several days. However, one of the mesomorphic compounds shows a thermodynamically stable mesophase at room temperature. In the last two cases, the kind of mesophase as hexagonal columnar has been confirmed by x-ray diffraction studies. From the study of several pyrazoboles with different position, length, and number of alkyl chains, it has been concluded that an increase in the number of side chains decreases the transition temperatures. Moreover, the positions of the decyloxy groups in the aromatic rings of the pyrazabole have also a crucial influence on the appearance of columnar mesophase in these derivatives, and surprisingly substitutions that reduce the molecular symmetry favor mesomorphism.

2.23.6   DIBENZOTETRAAZA[14]ANNULENE COMPLEXES

Forget and Veber synthesized and systematically investigated the mesomorphic properties of dibenzotetraaza[14]annulene nickel complexes **32** (Scheme 2.171) [1520–1523]. It was found that even some of the ligands were mesomorphic in nature [1521]. All the compounds exhibit a wide range of hexagonal columnar phases. In the series containing four alkoxy chains, with increase in alkyl chain length the clearing temperature decreases. The melting temperatures rapidly decrease for short-chain homologues while for long-chain homologues they remain almost constant. Subsequently, in order to have more insight into the mesomorphic properties, six and eight chain complexes

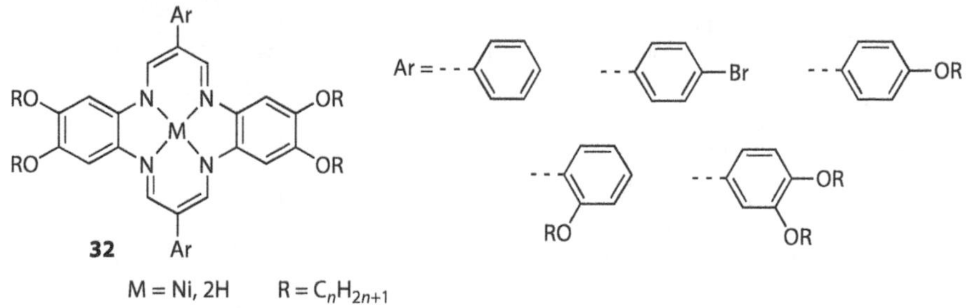

SCHEME 2.171 Chemical structures of some dibenzotetraaza[14]annulene complexes.

were synthesized and investigated. Like the four chain complexes, these nickel complexes also exhibit columnar hexagonal mesophase starting from room temperature. The influence of the number, the length, and the position of the peripheral chains on the liquid crystalline behavior have been studied and compared. As expected, as the chain length increased, the clearing temperatures decreased. Moreover, as the number of chains increased, the melting temperatures also decreased. The core–core interactions are less in six and eight chain compounds as compared to four chain complexes. Ortho substituents inhibit core–core interactions and consequently the mesomorphism was destroyed in the ortho-substituted compound [1522]. Interestingly, optical storage effect in the columnar phase of one nickel complex was observed [1523]. Exposing a uniformly aligned columnar liquid crystal to an argon ion laser beam results in a reorientation of the columnar phase, thereby changing the optical properties of the liquid crystal. The stored optical information can be erased by heating and subsequent shearing of the sample.

2.23.7 IONIC METALLOMESOGENS

Kim et al. have prepared a series of ionic liquid crystals based on TAAB (tetrabenzo[b,f,j,n][1,5,9,13] tetraazacyclohexadecine), which exhibit columnar mesomorphism. They prepared three compounds **33** with tetrafluoroborate as counterions and studied their liquid crystalline phase behavior by optical microscopy, calorimetry, and x-ray diffraction studies. All these compounds exhibit hexagonal columnar phase over a wide range of temperature and are stable down to room temperature [1524]. The chemical structures of the ionic metallomesogens are shown in Scheme 2.172. Subsequently, Lai et al. reported a series of ionic liquid crystalline metallomesogens **34** containing copper, nickel, and palladium metal ions with tetrafluoroborate as counterion [1525]. Like the metal-free derivatives, these compounds also exhibited a columnar hexagonal mesophase over a wide range of temperature. Interestingly, some derivatives with shorter chain lengths were room temperature liquid crystals. The mesomorphic behavior of these compounds was found to be dependent on the incorporated metal and chain length of the alkoxy side chains. Recently, Pucci et al. have synthesized and studied a series of bischelate ionic silver complexes **35** of 2.2′-bipyridine containing chiral side chains [1526]. Though the derivatives containing tetrafluoro borate and hexafluorophosphate counterions are non-mesomorphic, the corresponding derivatives with triflate and dodecylsulfate exhibit chiral columnar hexagonal phase starting from room temperature. The helical twisting of the mesogens about the columnar axis was confirmed by volume measurements and circular dichroism (CD) experiments in the liquid crystalline state. Moreover, the compounds exhibit unusual blue luminescence in the mesophase.

2.23.8 BIS(SALICYLALDIMINATO)METAL(II) COMPLEXES

Metal complexes of Schiff bases derived from salicylaldimine are among the best known complexes that exhibit mesomorphic properties in metallomesogenic systems. A variety of transition metals

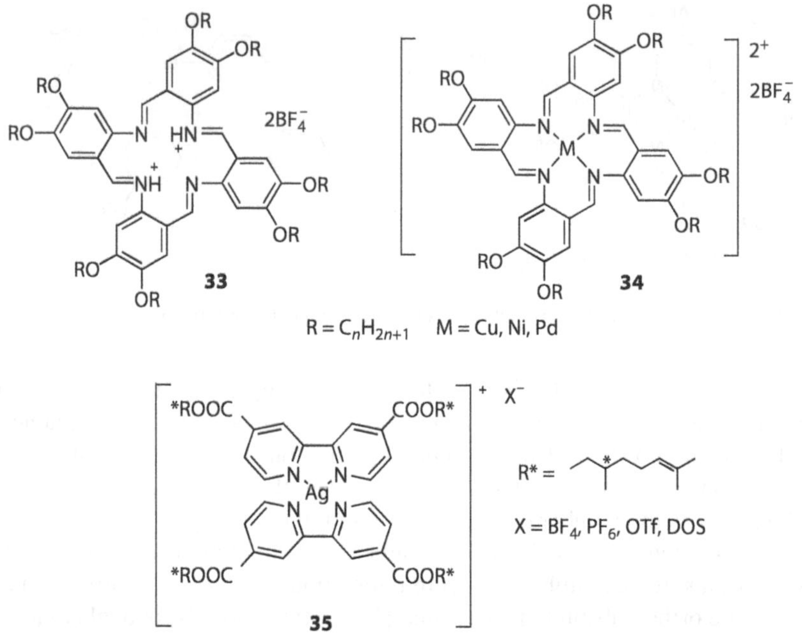

SCHEME 2.172   Molecular structures of some ionic metallomesogens.

show mesomorphism with salicylaldimine derivative Schiff bases. All of these complexes derived from salicylaldimine Schiff bases possess rod-like or brick-like shapes with two or four extended flexible side chains. These complexes thus exhibit either nematic or various smectic phases predominantly, regardless of metal center incorporated in such systems. Hence, the different mesogenic properties of these Schiff base metal complexes are attributed to the various types of geometric coordination of the metal centers. However, the first disk-like complexes of salicylaldimine Schiff bases exhibiting columnar liquid crystalline phase were reported by Lai and coworkers [1527]. They synthesized a series of copper complexes **36** (Scheme 2.173) and studied their mesomorphism by optical, calorimetric, and x-ray diffraction techniques. It was found that these complexes exhibit enantiotropic disordered columnar hexagonal mesophase over a very broad range of temperature. By comparing the mesophase of structurally similar compounds, they inferred that the mesophase structure can be changed from calamitic to discotic by changing the number of peripheral chains. Recently, Serrano et al. have reported two series of copper and palladium complexes **37** of new dendritic salicylaldimines containing eight or more aliphatic chains with the aim of inducing columnar mesophases at low temperatures [1528]. Some of the complexes indeed exhibit columnar phase with

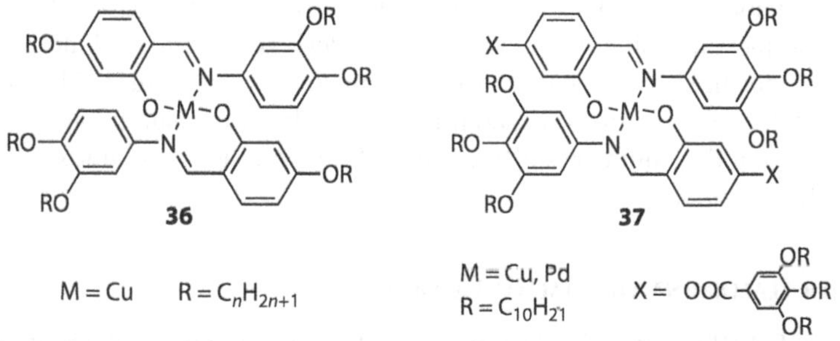

SCHEME 2.173   Chemical structures of some salicylaldimine-based metallomesogens.

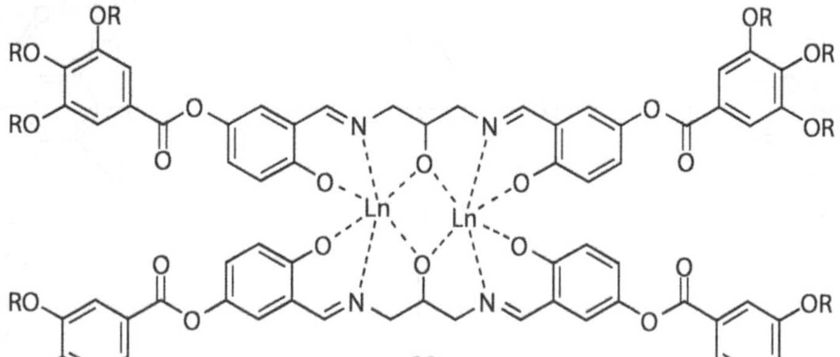

SCHEME 2.174 Molecular structures of some Schiff base lanthanide and actinide complexes.

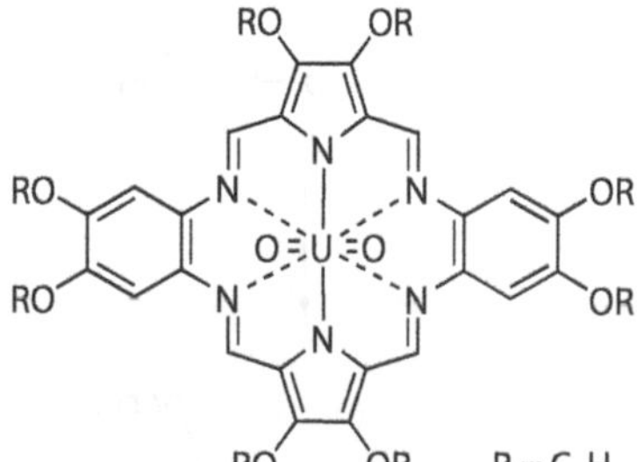

rectangular symmetry and at low temperature. The mesophase structure of the complexes has been established from x-ray diffraction studies.

2.23.9 SCHIFF BASE LANTHANIDE AND ACTINIDE COMPLEXES

The above described mesomorphic complexes are mononuclear in nature. However, recently Binnemans et al. have reported dinuclear lanthanide complexes **38** (Scheme 2.174) of a dendritic Schiff base exhibiting rectangular columnar phase over a broad temperature range [1529]. Usually, bidentate Schiff base ligands produce charged lanthanide complexes; however, they judiciously designed a tridentate ligand in such a way that it resulted in stable and neutral metallomesogens. The first uranium (an actinide) containing discotic metallomesogens have recently been realized by Bruce and coworkers [1530]. These are also believed to be the first examples of expanded porphyrin-derived liquid crystals containing a coordinated metal cation. They synthesized several alaskaphy-rin derivatives **39** (Scheme 2.174) containing peripheral aliphatic chains. These metallomesogens were found to be liquid crystalline, and from preliminary polarizing optical microscopy studies it was judged that the mesophase is columnar in nature. The identification of the mesophase on the basis of optical texture is not unequivocal, but the presence of homeotropic domains suggested that the phase is columnar hexagonal in these novel metallomesogens.

2.24 MISCELLANEOUS HETEROCYCLIC CORES

2.24.1 BENZOPYRANOBENZOPYRAN-DIONE

Ellagic acid (**1**), flavellagic acid (**2**), and coruleoellagic acid (**3**) are polyphenolic natural products that can generate a variety of discotic liquid crystals simply by attaching appropriate long alkyl chains via etherification or esterification (Scheme 2.175). While ellagic acid

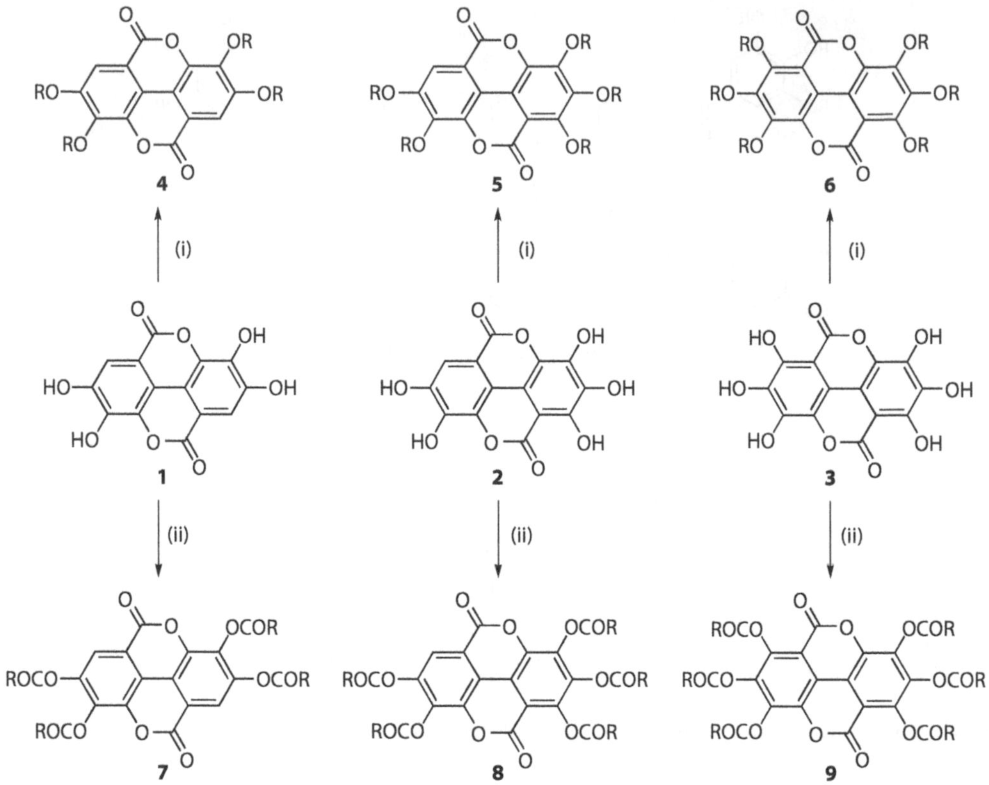

SCHEME 2.175 Synthesis of ellagic acid, flavellagic acid, and coruleoellagic acid derivatives: (i) RBr, K$_2$CO$_3$, DMA, 70°C, 4 days; (ii) RCOCl, 160°C, 20 h.

(2,3,7,8-tetrahydroxychromeno[5,4,3-*cde*]chromene-5,10-dione   or   [1]benzopyrano[5,4,3-*cde*] benzopyran-5,10-dione, 2,3,7,8-tetrahydroxy) is commercially available, flavellagic acid (1,2,3,7,8-pentahydroxychromeno[5,4,3-*cde*]chromene-5,10-dione) and coruleoellagic acid (1,2,3,6,7,8-hexahydroxychromeno[5,4,3-*cde*]chromene-5,10-dione) can be easily prepared from gallic acid [1531]. The thermal behavior of all the ethers, **4, 5, 6** and esters, **7, 8, 9** prepared from these molecules is summarized in Table 2.147. Homologous series derived from these molecules with four, five, and six chains per molecule linked via ether or ester groups are highly polymorphic with high-temperature phases being mesomorphic with the exception of ellagic acid tetraethers. Although a full characterization of mesophases has not been done, the miscibility of the high-temperature mesophase with other hexagonal columnar phases indicates its nature as Col$_h$.

2.24.2 BENZOTRISFURAN

The condensation of phloroglucinol with *p-p*-dimethoxybenzoin yields hexamethoxyphenylbenzo-trisfuran **12** (Scheme 2.176) [1532]. The demethylation of **12** with pyridinium hydrochloride followed by the esterification of the resultant hexaphenol **13** with various acid chlorides produces hexa-*n*-alkanoyloxyphenylbenzotrisfuran **14**. Six homologues varying in chain length from 5 carbon atoms to 14 carbon atoms have been synthesized this way. A hexa-*n*-alkoxybenzoyloxyphenylbenzotrisfuran derivative has also been prepared similarly. The thermal behavior of all these materials is summarized in Table 2.148. The lone hexa-*n*-alkoxybenzoyloxyphenylbenzotrisfuran derivative **14g** melts at 203°C to give a nematic phase but the material decomposes quickly. Within the hexa-*n*-alkanoyloxyphenyl-benzotrisfuran series only four derivatives are mesogenic: two exhibit enantiotropic and two display

TABLE 2.147
Transition Temperatures of Ellagic Acid, Flavellagic Acid, and Coruleoellagic Acid Derivatives

Structure	R	Phase Transition
4a	C_9H_{19}	Cr 131.2 I
4b	$C_{10}H_{21}$	Cr 128 I
4c	$C_{11}H_{23}$	Cr 128 I
4d	$C_{12}H_{25}$	Cr 126.8 I
4e	$C_{13}H_{27}$	Cr 126.5 I
4f	$C_{14}H_{29}$	Cr 125.4 I
4g	$C_{15}H_{31}$	Cr 126 I
4h	$C_{16}H_{33}$	Cr 125 I
5a	C_6H_{13}	Cr 15.8 Col$_h$ 197.6 I
5b	C_7H_{15}	Cr 36.8 Col$_h$ 186 I
5c	C_8H_{17}	Cr 23.3 Col$_h$ 175.5 I
5d	C_9H_{19}	Cr 44.6 Col$_h$ 165.6 I
5e	$C_{10}H_{21}$	Cr 42.2 Col$_h$ 145.8 I
5f	$C_{11}H_{23}$	Cr 44.7 Col$_h$ 140.6 I
5g	$C_{12}H_{25}$	Cr 45.5 Col$_h$ 133.6 I
5h	$C_{13}H_{27}$	Cr 55 Col$_h$ 125.2 I
5i	$C_{14}H_{29}$	Cr 42.4 Col$_h$ 118.2 I
5j	$C_{15}H_{31}$	Cr 44.5 Col$_h$ 116.3 I
5k	$C_{16}H_{33}$	Cr 53 Col$_h$ 108.9 I
6a	C_3H_7	Cr 142.7 Col$_h$ 159.9 I
6b	C_4H_9	Cr 72.9 Col$_h$ 170.3 I
6c	C_5H_{11}	Cr −3.7 Col$_h$ 169.7 I
6d	C_6H_{13}	Cr −85 Col$_h$ 158.9 I
6e	C_7H_{15}	Cr −77 Col$_h$ 151.8 I
6f	C_8H_{17}	Cr −34.7 Col$_h$ 138 I
6g	C_9H_{19}	Cr 3.8 Col$_h$ 123 I
6h	$C_{10}H_{21}$	Cr 6.1 Col$_h$ 117 I
7a	C_8H_{17}	Cr 198.5 I
7b	C_9H_{19}	Cr 189.4 I
7c	$C_{10}H_{21}$	Cr 183 Col$_h$ 184.5 I
7d	$C_{11}H_{23}$	Cr 173 Col$_h$ 179.2 I
7e	$C_{12}H_{25}$	Cr 166.4 Col$_h$ 176.1 I
7f	$C_{13}H_{27}$	Cr 158.1 Col$_h$ 173.1 I
7g	$C_{14}H_{29}$	Cr 155.8 Col$_h$ 170.8 I
7h	$C_{15}H_{31}$	Cr 151.5 Col$_h$ 168.2 I
7i	$C_{16}H_{33}$	Cr 151 Col$_h$ 166.2 I
8a	C_7H_{15}	Col$_h$ 180.8 I
8b	C_8H_{17}	Cr 103.9 Col$_h$ 182.5 I
8c	C_9H_{19}	Cr 79.3 Col$_h$ 175.3 I
8d	$C_{10}H_{21}$	Cr 89.8 Col$_h$ 173.9 I
8e	$C_{11}H_{23}$	Cr 71.7 Col$_h$ 170.9 I
8f	$C_{12}H_{25}$	Cr 83.1 Col$_h$ 169.7 I
8g	$C_{13}H_{27}$	Cr 88.1 Col$_h$ 167.8 I
8h	$C_{14}H_{29}$	Cr 97.3 Col$_h$ 164.8 I
8i	$C_{15}H_{31}$	Cr 93.4 Col$_h$ 163.5 I
8j	$C_{16}H_{33}$	Cr 102.3 Col$_h$ 161.5 I

(continued)

TABLE 2.147 (continued)
Transition Temperatures of
Ellagic Acid, Flavellagic Acid, and
Coruleoellagic Acid Derivatives

Structure	R	Phase Transition
9a	C_8H_{17}	Cr 159.1 Col$_h$ 217.3 I
9b	C_9H_{19}	Cr 154.3 Col$_h$ 210.8 I
9c	$C_{10}H_{21}$	Cr 146.3 Col$_h$ 208.9 I
9d	$C_{11}H_{23}$	Cr 144.5 Col$_h$ 206 I
9e	$C_{12}H_{25}$	Cr 141.1 Col$_h$ 202.2 I
9f	$C_{13}H_{27}$	Cr 140.1 Col$_h$ 198.6 I
9g	$C_{14}H_{29}$	Cr 138.1 Col$_h$ 194.6 I
9h	$C_{15}H_{31}$	Cr 136.9 Col$_h$ 190.1 I
9i	$C_{16}H_{33}$	Cr 134 Col$_h$ 186.5 I

Source: Data from Zimmermann, H. et al.,
Liq. Cryst., 12, 245, 1992.

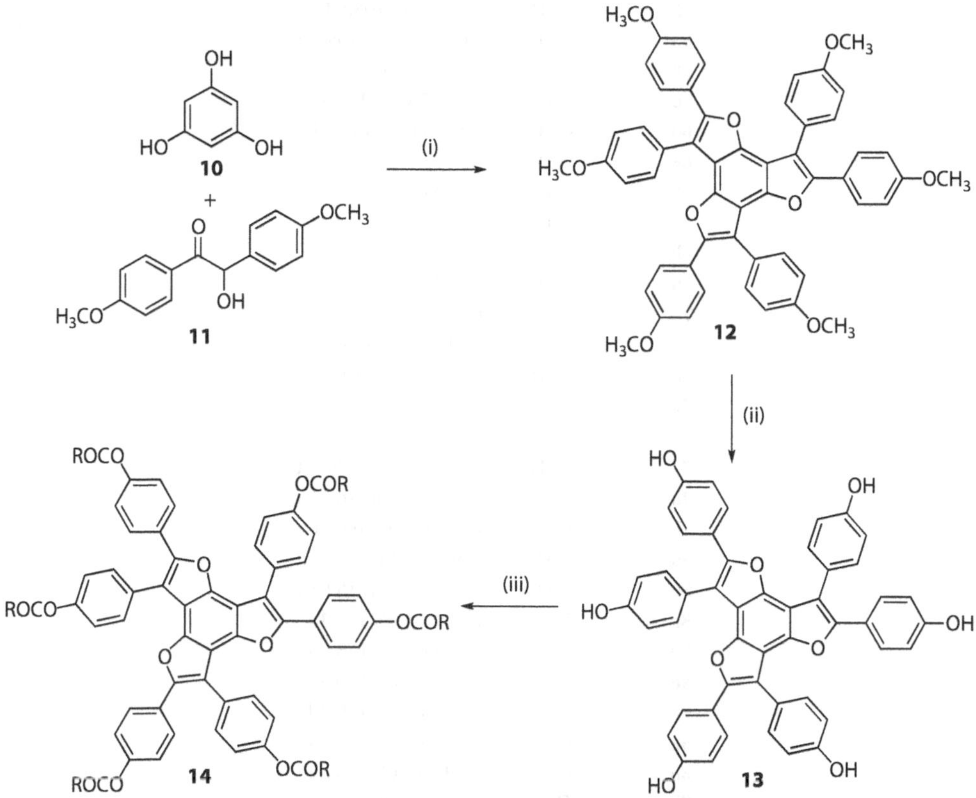

SCHEME 2.176 Synthesis of benzotrisfuran discotics: (i) H_2SO_4, 159°C, 5 min; (ii) pyridine, HCl, 220°C, 7 h; (iii) RCOCl, Mg, benzene, reflux.

TABLE 2.148
Transition Temperatures of
Benzotrisfuran Derivatives

Structure	R	Phase Transition
14a	C_5H_{11}	Cr 186 Col_h 244 I
14b	C_6H_{13}	Cr 134 Col_h 177 I
14c	C_7H_{15}	Cr (100 Col_h) 134 I
14d	C_8H_{17}	Cr (95 Col_h) 107 I
14e	C_9H_{19}	Cr 97 I
14f	$C_{14}H_{29}$	Cr 86 I
14g	$C_6H_4O\ C_8H_{17}$	Cr 203 N_D dec

Source: Data from Destrade, C. et al., *Liq. Cryst.*,
2, 229, 1987.

monotropic columnar phases. While the short chain homologues **14a** and **14b** are enantiotropic liquid crystalline, longer chain derivatives **14c** and **14d** are monotropic; **14e** and **14f** are nonmesomorphic.

2.24.3 Pyrillium and Dithiolium Salts

Pyrillium and dithiolium salts, **15** and **16**, respectively, are known to display smectic and columnar phases [1533,1534]. These architectures are essentially banana-like and, therefore, the formation of smectic and columnar phases is not surprising. The addition of a dialkoxyphenyl moiety at the 4-position of **15** leads to discotic 2,4,6-triarylpyrylium salts **21** [1535–1539]. These materials are synthesized, as shown in Scheme 2.177. The key step is the acid-catalyzed condensation between a chalcone **20** and an acetophenone derivative **18**. The thermal behavior of triaryl-2,4,6-pyrilium tetrafluoroborates **21** is presented in Table 2.149. Surprisingly, these compounds with peripheral chains as short as two or three carbon atoms display a columnar mesophase. Ionic self-assembly due to the presence of tetrafluoroborate seems to be responsible for the columnar phase formation. These salts decompose at elevated temperature, which prevents their full characterization.

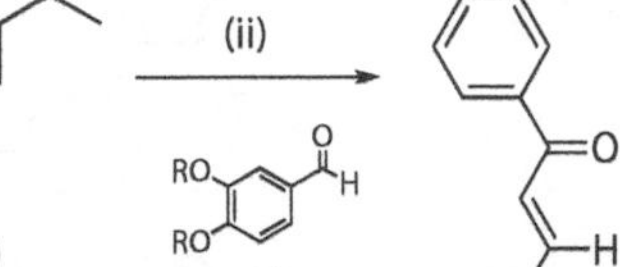

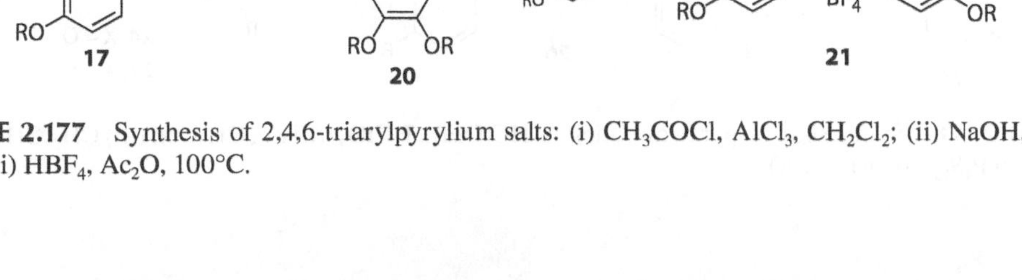

SCHEME 2.177 Synthesis of 2,4,6-triarylpyrylium salts: (i) CH_3COCl, $AlCl_3$, CH_2Cl_2; (ii) NaOH, EtOH, 60°C; (iii) HBF_4, Ac_2O, 100°C.

TABLE 2.149
Transition Temperatures of
Triarylpyrylium Salts 21

Structure	R	Phase Transition	Ref.
21a	C_2H_5	Cr 194 Col_h 213 I	[1538]
21b	C_3H_7	Cr 167 Col_h 247 I	[1538]
21c	C_4H_9	Cr 100 Col_h 277 dec	[1536]
21d	C_5H_{11}	Col_h 283 dec	[1536]
21e	C_8H_{17}	Col_h 237 dec	[1536]
21f	$C_{12}H_{25}$	Col_h 288 dec	[1536]

2.24.4 BISPYRAN AND BISTHIOPYRAN

Bispyran **24** and bisthiopyran **27** can be easily prepared from corresponding 1,6-biarylpyrillium salts by reacting with zinc, as shown in Scheme 2.178. Both oxygen and sulfur compounds display mesomorphism when substituted with appropriate peripheral chains [1540–1545]. The thermal behavior of these materials is summarized in Table 2.150. Several ionic salts and charge-transfer complexes with tetracyanoquinodimethane (TCNQ) have been prepared from these bispyran and bisthiopyran derivatives; they also exhibit columnar phases.

2.24.5 CYCLOTRIPHOSPHAZINES

A number of cyclotriphosphazines that bear polycatenar aromatic esters as promesogenic groups linked to phosphorus atoms are reported to display mesomorphism [1546,1547]. Filling the space around the core is quite important in these molecules to generate columnar phases. Mesogenic units that contain only one terminal alkyl chain give rise to calamitic mesophases [1548–1554]. On the other hand, mesogenic units possessing three long terminal chains exhibit columnar phases. Several cyclotriphosphazines bearing ester or amide groups are prepared, as shown in Scheme 2.179. Hexachlorocyclotriphosphazine was reacted with 4-benzyloxyphenol in the presence of Cs_2CO_3 to yield a protected trimer, which was converted to hydroxyl-terminated trimer **29**. Similarly, amino-terminated trimer **33** can be prepared. Treatment of **29** or **33** with an excess of the corresponding acid chloride in the presence of triethylamine in THF produces the hexaester **30** and hexaamide **34**. The partial esterification of **29** affords a liquid crystalline pentaester **31** that is further esterified

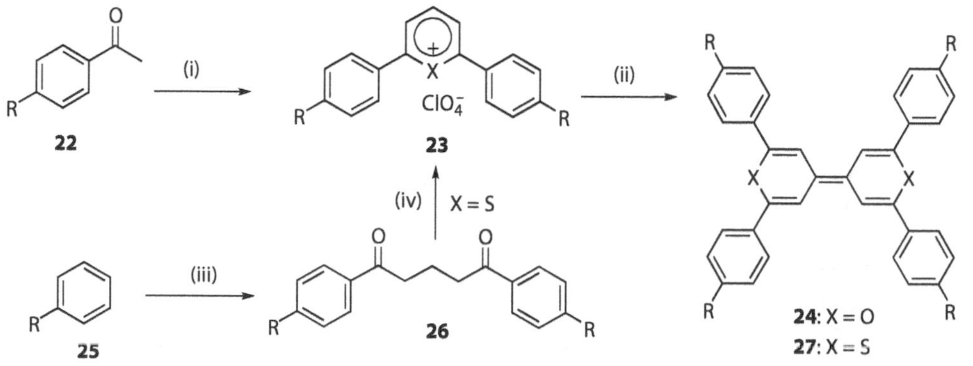

SCHEME 2.178 Synthesis of bispyran discotics: (i) HC(OEt)$_3$, HClO$_4$; (ii) Zn, CH$_3$CN; (iii) CH$_2$(CH$_2$COCl)$_2$, AlCl$_3$; (iv) P$_4$S$_{10}$, AcOH, LiClO$_4$.

TABLE 2.150
Transition Temperatures of Bispyran and Bisthiopyran Derivatives

Structure	R	Phase Transition	References
24a	C_5H_{11}	Cr 228 I	[1540]
24b	C_9H_{19}	Cr 53.5 M 171.5 I	[1540]
24c	$C_{12}H_{25}$	Cr 96 M 147 I	[1540]
27a	C_9H_{19}	Cr <40 M 167.5 I	[1542]
27b	$C_{12}H_{25}$	Cr <40 M 143 I	[1542]
27c	$OC_{12}H_{25}$	Cr 94.5 M 140 I	[1542]

with a dialkoxybenzoylchloride to produce unsymmetrical cyclotriphosphazine **32**. The thermal behavior of all these cyclotriphosphazines is summarized in Table 2.151. The mesomorphic properties strongly depend on the number of terminal alkyl chains. Cyclotriphosphazine with a single peripheral chain per phenyl unit behaves as a calamitic molecule and displays a nematic mesophase. The presence of a larger number of terminal chains causes the molecule to adopt a discotic shape and these materials display columnar phases. Columnar assembly is highly stabilized by hydrogen bonding in cyclotriphosphazine amides.

2.24.6 TETRAOXA[8]CIRCULENE

The synthesis of discotic liquid crystalline tetraoxa[8]circulene derivatives is outlined in Scheme 2.180. The key precursor **38** was obtained by the oxidation of 2,3-dialkyl-1,4-dimethoxybenzene **37**. Compound **37** can be prepared by the reduction of the triple bonds of alkylacetylene derivative **36**, which in turn was obtained from 2,3-bis(bromomethyl)-1,4-dimethoxybenzene **35** by reacting with an acetalide. Tetraoxa[8]circulene derivatives **39** are realized by the tetramerization of **38** using $AlCl_3$ in nitrobenzene or by refluxing with BF_3-Et_2O in dichloromethane [1555]. The mesomorphic nature of these materials is primarily deduced from DSC results. The thermal data are collected in Table 2.152. While the first and last members of the series were reported to be nonmesomorphic, all other derivatives exhibit liquid crystalline behavior. The exact nature of the mesophase was not revealed.

2.24.7 TETRAAZOPYRENE, BENZO[C]CINNOLINE, AND DIBENZO[C,E][1,2]THIAZINE

The synthesis of discotic derivatives derived from tetraazopyrene, benzo[c]cinnoline, and dibenzo[c,e][1,2]thiazine cores is depicted in Scheme 2.181. The parent heterocycles are substituted by four phenyl groups bearing alkoxy substituents to induce mesogenity [232,1556]. The Suzuki coupling in between tetrachloride **40** and appropriate boronic ester yielded tetraarylbiphenyl derivative **41**. The reduction of **41** with freshly prepared Raney nickel under basic conditions produces tetraazopyrene **42**. On the other hand, in case of benzo[c]cinnoline, the azo bridge was first created by the peracetic acid oxidation of diamino derivative **44** followed by the attachment of alkoxyphenyl groups via the Suzuki coupling. The N-S bridge in compound **50** was accomplished by the treatment of amino sulfide **49** with NCS. A single peripheral chain per phenyl unit is not sufficient to induce mesomorphism (Table 2.153). Compounds with 3,5-dialkoxyphenyl groups also do not display mesomorphism due to the steric interference of the alkyl chains. Tetraazopyrene and benzo[c]cinnoline having 3,4-dialkoxyphenyl groups exhibit hexagonal a columnar mesophase by striking a balance between the density of the alkyl chains and their preferred orientation. Compound **50** was liquid crystalline at ambient temperature and goes to the isotropic phase at 104°C.

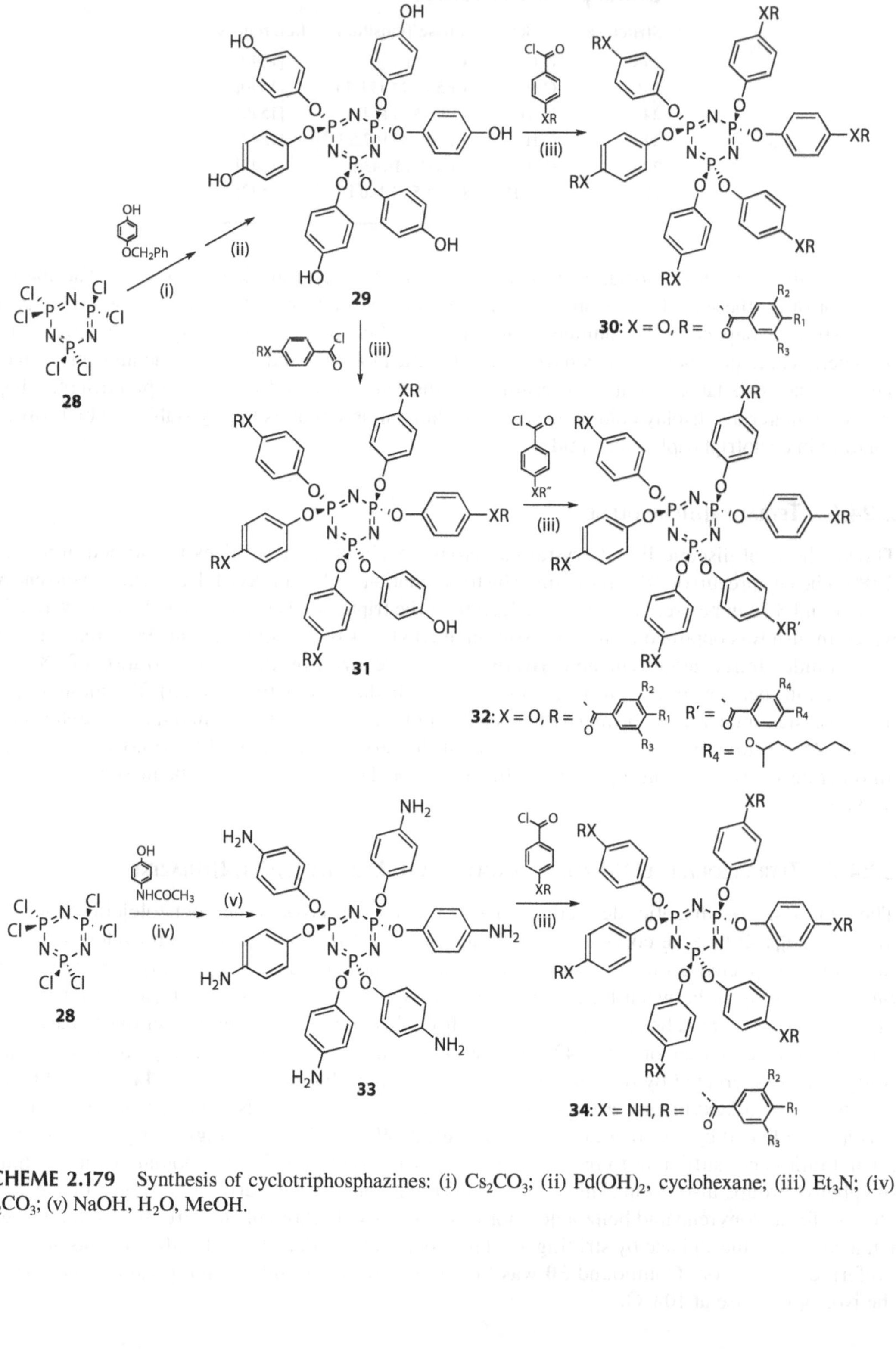

SCHEME 2.179 Synthesis of cyclotriphosphazines: (i) Cs$_2$CO$_3$; (ii) Pd(OH)$_2$, cyclohexane; (iii) Et$_3$N; (iv) K$_2$CO$_3$; (v) NaOH, H$_2$O, MeOH.

TABLE 2.151
Transition Temperatures of Cyclotriphosphazine Derivatives

Structure	R_1	R_2	R_3	Phase Transition	Ref.
30a	$OC_{10}H_{21}$	H	H	Cr 125 N 129 I	[1546]
30b	$OC_{10}H_{21}$	$OC_{10}H_{21}$	H	Cr (61 Col) 88 I	[1546]
30c	OCH_3	OCH_3	OCH_3	g 75 I	[1546]
30d	OC_6H_{13}	OC_6H_{13}	OC_6H_{13}	Col_r 67 I	[1546]
30e	$OC_{10}H_{21}$	$OC_{10}H_{21}$	$OC_{10}H_{21}$	Cr 38 Col_h 67 I	[1546]
30f	$OC_{12}H_{25}$	$OC_{12}H_{25}$	$OC_{12}H_{25}$	Cr 2 Col_h 59 I	[1546]
30g	$OC_{14}H_{29}$	$OC_{14}H_{29}$	$OC_{14}H_{29}$	Cr 32 Col_h 65 I	[1546]
31	$OC_{12}H_{25}$	$OC_{12}H_{25}$	$OC_{12}H_{25}$	Col_h 54 I	[1546]
32	$OC_{12}H_{25}$	$OC_{12}H_{25}$	$OC_{12}H_{25}$	Col_h 54 I	[1546]
34a	OC_6H_{13}	OC_6H_{13}	OC_6H_{13}	Cr 132 Col_h 177 I	[1547]
34b	$OC_{10}H_{21}$	$OC_{10}H_{21}$	$OC_{10}H_{21}$	g 61 Col_h 171 I	[1547]
34c	$OC_{12}H_{25}$	$OC_{12}H_{25}$	$OC_{12}H_{25}$	Cr 6 Col_r 87 Col_h 137 I	[1547]

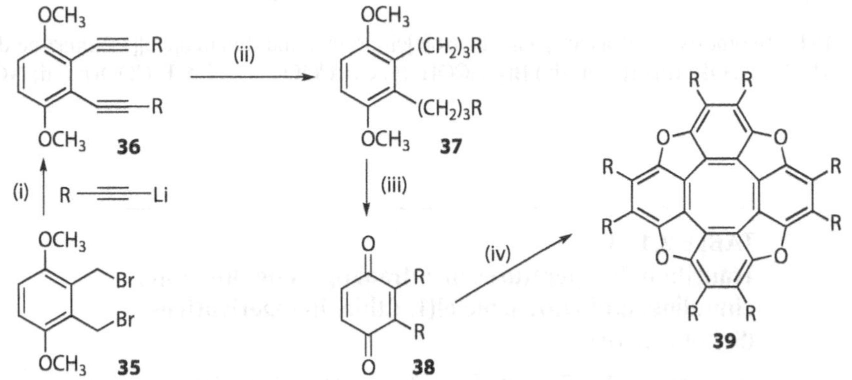

SCHEME 2.180 Synthesis of discotic liquid crystalline tetraoxa[8]circulene derivatives: (i) THF; (ii) H_2, PtO_2; (iii) $(NH_4)_2Ce(NO_3)_6$, CH_3CN, H_2O; (iv) $BF_3\text{-}Et_2O$.

TABLE 2.152
Transition Temperatures of Tetraoxa[8]circulene Derivatives (Scheme 2.180)

Structure	R	Phase Transition
39a	C_3H_7	Cr 333 I
39b	C_7H_{15}	Cr 193 M 220 I
39c	C_8H_{17}	Cr 160 M 183 I
39d	C_9H_{19}	Cr 148 M 169 I
39e	$C_{10}H_{21}$	Cr 133 M 165 I
39f	$C_{11}H_{23}$	Cr >350 I

Source: Data from Eskildsen, J. et al., *Eur. J. Org. Chem.*, 1637, 2000.

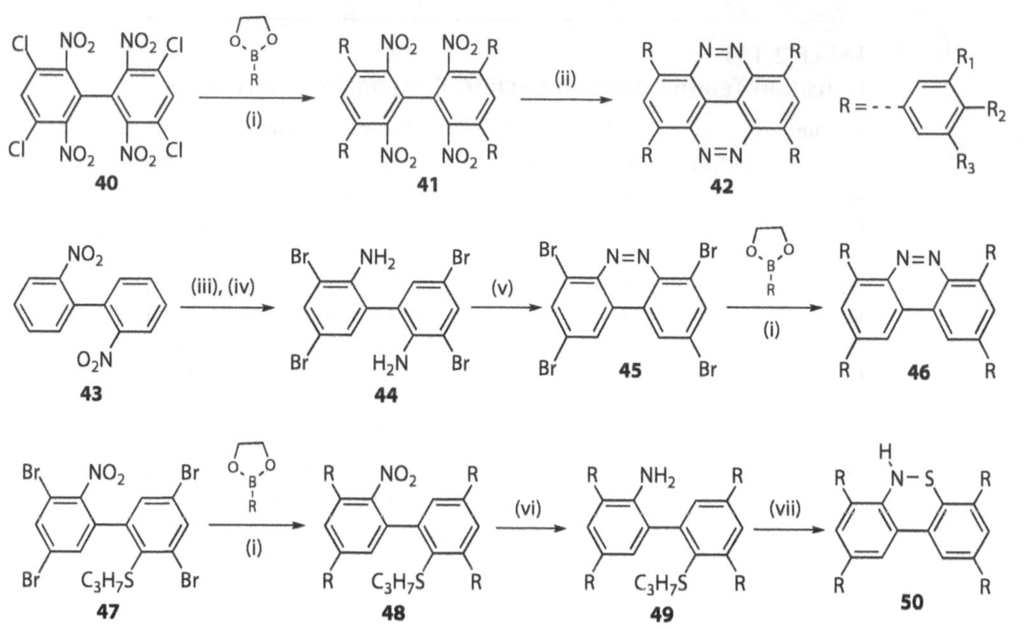

SCHEME 2.181 Synthesis of tetraazopyrene, benzo[c]cinnoline, and dibenzo[c,e][1,2]thiazine derivatives: (i) Pd(0); (ii) H$_2$, Ni, NaOH; (iii) H$_2$, Pt; (iv) Br$_2$, ACOH; (v) CH$_3$COOH; (vi) Zn, EtCOOH; (vii) NCS; NaOH; heat.

TABLE 2.153
Transition Temperatures of Tetraazopyrene, Benzo[c] cinnoline, and Dibenzo[c,e][1,2]thiazine Derivatives (Scheme 2.181)

Structure	R$_1$	R$_2$	R$_3$	Phase Transition	Ref.
42a	H	OC$_8$H$_{17}$	H	Cr 161 I	[232]
42b	OC$_8$H$_{17}$	OC$_8$H$_{17}$	H	Cr 95 Col$_h$ 199 I	[232]
42c	OC$_8$H$_{17}$	H	OC$_8$H$_{17}$	Cr 45 I	[232]
46a	H	OC$_8$H$_{17}$	H	Cr 133 I	[232]
46b	OC$_8$H$_{17}$	OC$_8$H$_{17}$	H	Cr 92 Col$_h$ 168 I	[232]
46c	OC$_8$H$_{17}$	H	OC$_8$H$_{17}$	Liquid	[232]
50	OC$_8$H$_{17}$	OC$_8$H$_{17}$	H	Col 104 I	[1556]

2.24.8 TRISTYRYLPYRIDINE

Attias et al. prepared a discotic liquid crystal based on a 3,5-dicyano-2,4,6-tristyrylpyridine core [1557,1558], as shown in Scheme 2.182. The bromination of collidine with elemental bromine afforded the 3,5-dibromo-2,4,6-trimethylpyridine **52**, which was converted into the dicyano derivative **53** by reacting with copper cyanide. The condensation of **53** with 3,4-didecyloxybenzaldehyde in the presence of toluenesulfonic acid yielded the 3′,4′-didecyloxy-2,4,6-tristyryl-3,5-dicyano-pyridine **54**. The compound exhibits a hexagonal columnar phase between 78°C and 133°C. The molecule is probably self-assembled in antiparallel alignment to fill the space for generating the columnar mesophase.

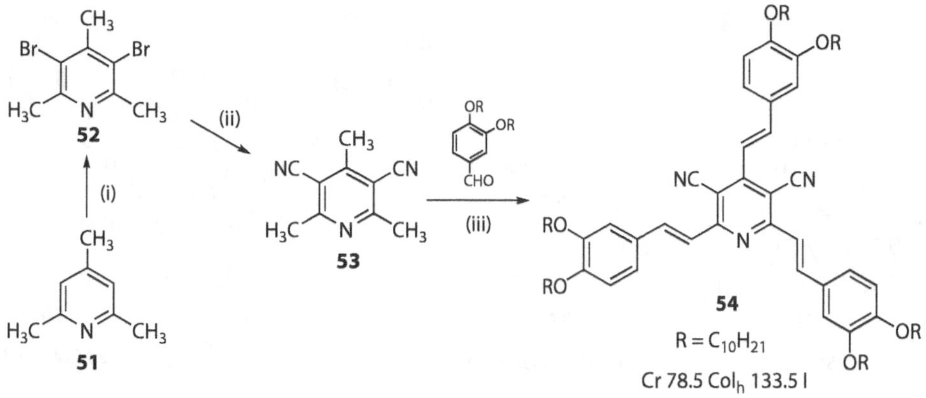

SCHEME 2.182 Synthesis of 3,5-dicyano-2,4,6-tristyrylpyridine derivatives: (i) Br$_2$, H$_2$SO$_4$; (ii) CuCN, DMF; (iii) toluenesulfonic acid.

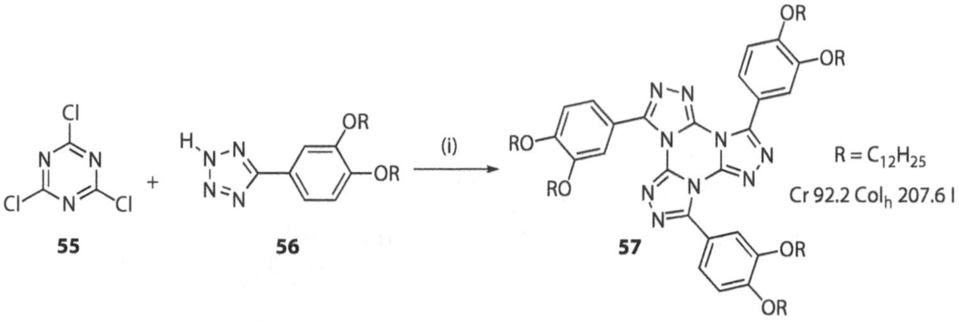

SCHEME 2.183 Synthesis of tris[1,2,4]triazolo[1,3,5]trizene derivatives: (i) K$_2$CO$_3$, butanone.

2.24.9 TRISTRIAZOLOTRIAZINE

The triphenyl-substituted tris[1,2,4]triazolo[1,3,5]triazene derivatives containing three and six peripheral aliphatic chains are synthesized, as shown in Scheme 2.183 [1559]. The reaction of aryltetrazole with cyanuric chloride in the presence of potassium carbonate afforded the trimer **57**. The required aryltetrazole can be easily obtained by reacting dialkoxybenzonitrile with sodium azide in DMF [1560]. Tristriazolotriazines bearing only one alkyl chain per phenyl unit are not liquid crystalline. Another alkyl chain is required to fill the space around the core to induce mesomorphism. Thus the compound 3,7,11-tris(3,4-bis(dodecyloxy)phenyl)tris[1,2,4]triazolo[1,3,5]triazine **57** displays a very wide hexagonal columnar phase from 92.2°C to 207.6°C.

2.24.10 BENZOTRIIMIDAZOLE

The benzotri(imidazole) core **58** bearing three alkyl chains is as such not liquid crystalline, but when treated with polymerizable alkoxybenzoic acid **59**, it forms a 1:3 supramolecular complex **60** (Scheme 2.184) through hydrogen-bonding interactions [1561]. The complex **60** exhibits a hexagonal columnar phase from 23.0°C to 74.7°C. It should be noted that the complex did not show any mesophase immediately on cooling from the isotropic phase but it appeared on standing the sample for 2–3 days at room temperature. The irradiation of the complex in the mesophase with UV light induces the polymerization of the terminal acrylate moieties to give a cross-linked polymer maintaining the same hexagonal columnar order.

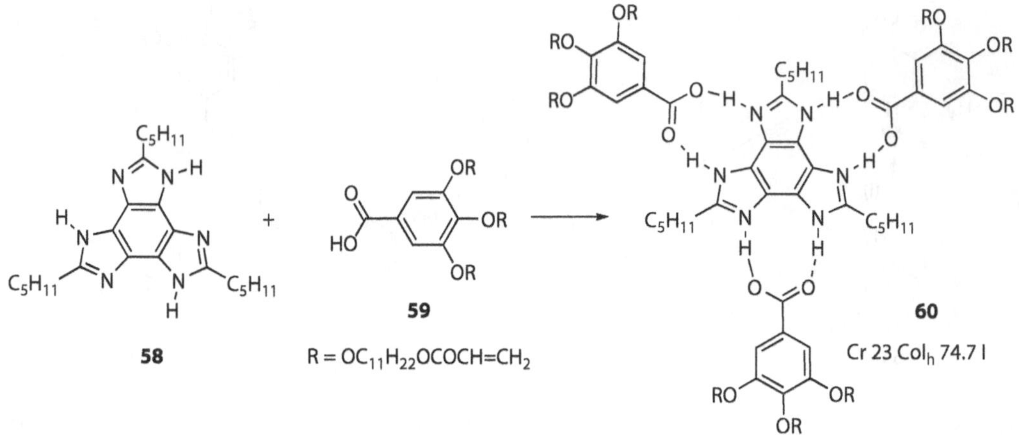

SCHEME 2.184   Formation of benzotri(imidazole)-alkoxybenzoic acid supramolecular complex.

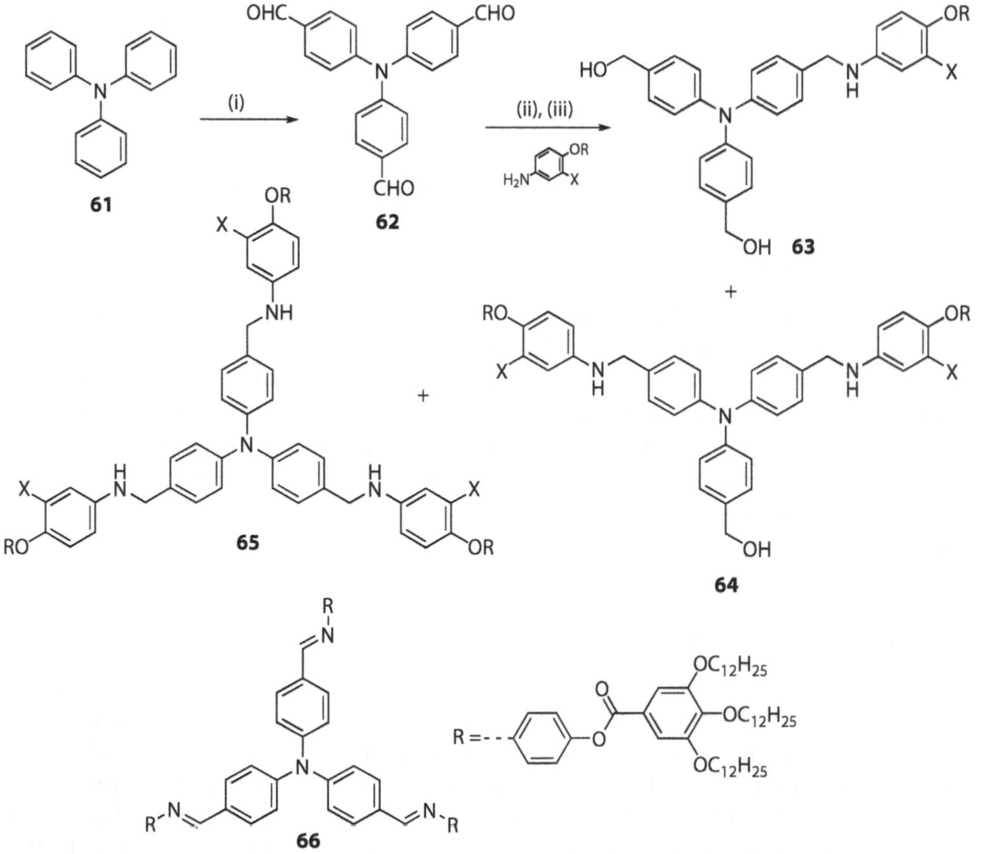

SCHEME 2.185   Synthesis of triphenylamine derivatives: (i) $POCl_3$, DMF, 100°C, 72 h; (ii) EtOH, HOAc (trace), reflux; (iii) $NaBH_4$.

TABLE 2.154
Transition Temperatures of Triphenylamine-
Derived Discotics (Scheme 2.185)

Structure	R	X	Phase Transition	Ref.
63	$C_{16}H_{33}$	—	Cr 44.7 D_L 79.2 I	[1562]
64a	$C_{14}H_{29}$	—	Cr 26.8 D_L 57.7 I	[1562]
64b	$C_{16}H_{33}$	—	Cr 39.8 D_L 65.4 I	[1562]
64c	$C_{18}H_{37}$	—	Cr 54.9 D_L 68.6 I	[1562]
64d	$C_{12}H_{25}$	$OC_{12}H_{25}$	Cr 20.5 Col_h 56.3 I	[1562]
64e	$C_{14}H_{29}$	$OC_{14}H_{29}$	Cr 36.8 Col_h 62.6 I	[1562]
64f	$C_{16}H_{33}$	$OC_{16}H_{33}$	Cr 62.2 Col_h 67.6 I	[1562]
65a	$C_{14}H_{29}$	—	Cr 30.0 D_L 52.2 I	[1562]
65b	$C_{16}H_{33}$	—	Cr 44.8 D_L 59.4 I	[1562]
65c	$C_{18}H_{37}$	—	Cr 54.1 D_L 62.1 I	[1562]
65d	$C_{10}H_{21}$	$O\ C_{10}H_{21}$	Cr 26.6 Col_h 44.2 I	[1562]
65e	$C_{12}H_{25}$	$OC_{12}H_{25}$	Cr 32.5 Col_h 52.7 I	[1562]
65f	$C_{14}H_{29}$	$OC_{14}H_{29}$	Cr 49.8 Col_h 60.0 I	[1562]
65g	$C_{16}H_{33}$	$OC_{16}H_{33}$	Cr 63.5 I	[1562]
66			Col_h 83.1 I	[1563]

2.24.11 TRIPHENYLAMINE CORE

A number of triphenylamine derivatives have been extensively investigated for their hole-transporting properties. Upon appropriate peripheral substitution, they can form columnar phases and,
therefore, are good candidates for one-dimensional charge transport. Lai and coworkers prepared
a mixture of compounds **63**, **64**, and **65** by reacting triformyltriphenylamine **62** with alkoxyphenyl amine followed by the in situ reduction of the resultant Schiff bases (Scheme 2.185) [1562].
Compounds **63**, **64**, and **65** were separated by column chromatography. The thermal behavior of all
these derivatives is summarized in Table 2.154. As expected, mesophase observed in these materials is largely dependent upon appended peripheral chains. Compounds with one, two, or three side
chains display lamellar columnar phases, while compounds with four or six side chains exhibit hexagonal columnar phases. Rao and coworkers reported a Schiff base **66** with nine peripheral chains.
Compound **66** is liquid crystalline at room temperature and goes to the isotropic phase at 83.1°C.

2.25 NON-AROMATIC CORES

2.25.1 CYCLOHEXANE

Several hexaesters and hexaethers of 1,2,3,4,5,6-hexahydroxy-cyclohexane are reported as discotic
liquid crystals [28,339,1564–1570]. These compounds can be easily prepared from naturally occurring commercially available alcohol scylloinositol. The esters are prepared by reacting scylloinositol with an excess of acid chloride and the ethers are synthesized via classical Williamson ether
synthesis using potassium hydroxide and alkyl bromide (Scheme 2.186). The thermal behavior of
these materials is summarized in Table 2.155.

It is quite interesting to note that the hexaacetate **2.1** having peripheral chains of only two carbon
atoms display columnar phase. The isotropic temperatures decrease, as expected, on increasing the
chain length. The introduction of a methyl branch into the side chain at 2-position of the octyloxy
chain leads to a destabilization of the mesophase due to the steric effect of the methyl group near
the core. On the other hand, branching in the middle of the chain reduces the melting temperatures

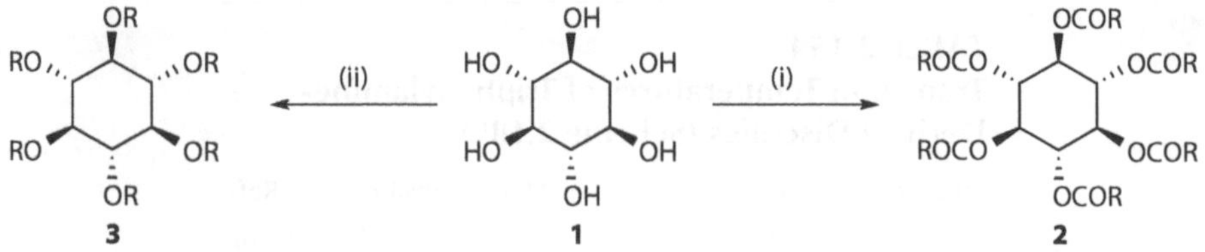

SCHEME 2.186   Synthesis of scylloinositol-based discotic liquid crystals: (i) RCOCl, TFA; (ii) KOD, RBr.

TABLE 2.155
Thermal Phase Transitions of Scylloinositol
Derivatives (Scheme 2.186)

Structure	R	Phase Transition	Ref.
2.1	CH_3	Cr 288.0 Col 292.8 I	[1566]
2.2	C_2H_5	Cr 212.0 Col 276.2 I	[1566]
2.3	C_3H_7	Cr 213.0 Col 259.4 I	[1566]
2.4	C_4H_9	Cr 184.5 Col 208.0 I	[1566]
2.5	C_5H_{11}	Cr 68.5 Col 199.5 I	[1566]
2.6	C_6H_{13}	Cr 70.0 Col 202.0 I	[1566]
2.7	C_7H_{15}	Cr 79.6 Col 199.3 I	[1566]
2.8	C_8H_{17}	Cr 81.0 Col 195.7 I	[1566]
2.9	C_9H_{19}	Cr 83.0 Col 188.7 I	[1566]
2.10	$C_{10}H_{21}$	Cr 87.7 Col 182.7 I	[1566]
2.11	$C_{11}H_{23}$	Cr 92.0 Col 176.4 I	[1566]
2.12	$CH(CH_3)C_6H_{13}$	Cr 42 Col 116 I	[28]
2.13	$CH_2CH(CH_3)C_5H_{11}$	Cr 48 Col 195 I	[28]
2.14	$(CH_2)_2CH(CH_3)C_4H_9$	Col 181 I	[28]
2.15	$(CH_2)_3CH(CH_3)C_3H_7$	Col 186 I	[28]
2.16	$(CH_2)_4CH(CH_3)C_2H_5$	Cr 54 Col 197 I	[28]
2.17	$(CH_2)_5CH(CH_3)CH_3$	Cr 87 Col 201 I	[28]
2.18	$(CH_2)_4CH(C_2H_5)_2$	Cr 38 Col 186 I	[28]
2.19	$(CH_2)_4Cp$	Cr 139 Col 199 I	[28]
2.20	$(CH_2)_3CH(CH_3)C_2H_5$	Cr 68 Col 196 I	[28]
2.21	$(CH_2)_5CH(CH_3)C_2H_5$	Cr 44 Col 192 I	[28]
2.22	$(CH_2)_2OC_4H_9$	Cr 47 Col 87 I	[1569]
2.23	$(CH_2)_5OCH_3$	Cr 44 Col 144 I	[1569]
2.24	$CH_2SC_5H_{11}$	Cr 57 Col 106 I	[1569]
2.25	$(CH_2)_2SC_4H_9$	Cr 86 Col 183 I	[1569]
2.26	$(CH_2)_3SC_3H_7$	Cr 64 Col 189 I	[1569]
2.27	$(CH_2)_5SCH_3$	Cr 56 Col 172 I	[1569]
2.28	$(CH_2)_4Cl$	Cr 205 Col 235 I	[339]
2.29	$(CH_2)_4Br$	Cr 190 Col 250 I	[339]
2.30	$(CH_2)_5Br$	Cr 93 Col 230 I	[339]
2.31	$(CH_2)_6Br$	Cr 94 Col 199 I	[339]
2.32	$(CH_2)_7Br$	Cr 92 Col 175 I	[339]
2.33	$(CH_2)_{10}Br$	Cr 90 Col 117 I	[339]

without greatly affecting the clearing temperatures. The introduction of oxygen or sulfur atom into the side chain decreases both melting and clearing temperatures. The presence of terminal halogen atom in the peripheral chains stabilizes the mesophase for lower homologues, while it destabilizes the mesophase for higher homologues.

2.25.2 GLUCOPYRANOSE

Structurally similar to scylloinositol esters are the esters of α- and β-glucopyranose [1571–1573]. Treatment of the appropriate glucopyranose with the required acid chloride in chloroform/pyridine yields penta-*O-n*-alkanoylglucopyranoses **4** and **5** (Scheme 2.187). The formation of at least one columnar mesophase in all the members of both anomeric series was reported [1571,1572]. The second heating scan data for the mesophase to isotropic transition are given in Table 2.156. In contrast, some of the compounds were reported to be non-liquid crystalline in a later publication [1573]. The attachment of 3,4-dialkoxycinnamoyl moiety to α- and β-glucopyranose produces well-defined columnar phases. Thus, 1,2,3,4,6-penta-*O*-(*trans*-3,4-dialkoxycinnamoyl)-(D)-glucopyranose derivatives bearing alkyl chains of 6–14 carbon atoms were synthesized and characterized [1574]. All these compounds exhibit enantiotropic columnar phases (Table 2.156). The irradiation of these cinnamoyl derivatives in dilute solutions leads to two intramolecular [2+2] cycloaddition reactions. While irradiation in the solid or

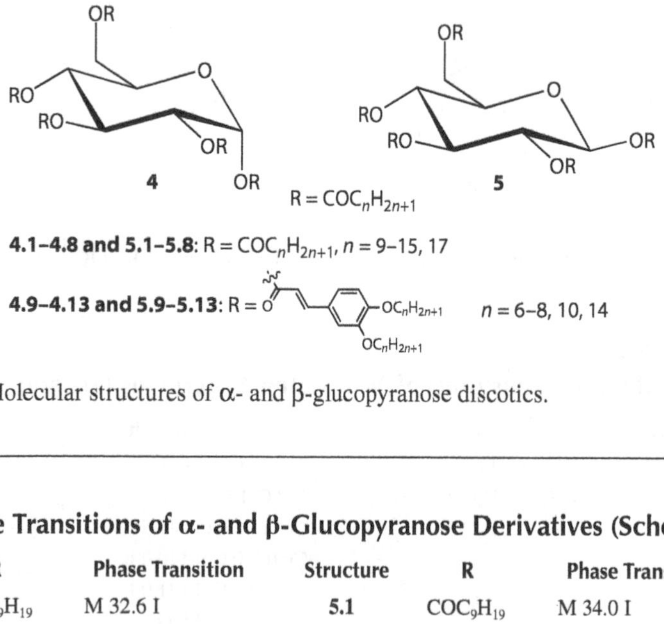

SCHEME 2.187 Molecular structures of α- and β-glucopyranose discotics.

TABLE 2.156
Thermal Phase Transitions of α- and β-Glucopyranose Derivatives (Scheme 2.187)

Structure	R	Phase Transition	Structure	R	Phase Transition	Ref.
4.1	COC_9H_{19}	M 32.6 I	**5.1**	COC_9H_{19}	M 34.0 I	[1572]
4.2	$COC_{10}H_{21}$	M 43.3 I	**5.2**	$COC_{10}H_{21}$	M 46.4 I	[1572]
4.3	$COC_{11}H_{23}$	M 46.1 I	**5.3**	$COC_{11}H_{23}$	M 52.2 I	[1572]
4.4	$COC_{12}H_{25}$	M 57.9 I	**5.4**	$COC_{12}H_{25}$	M 60.4 I	[1572]
4.5	$COC_{13}H_{27}$	M 58.3 I	**5.5**	$COC_{13}H_{27}$	M 66.1 I	[1572]
4.6	$COC_{14}H_{29}$	M 59.4 I	**5.6**	$COC_{14}H_{29}$	M 69.9 I	[1572]
4.7	$COC_{15}H_{31}$	M 67.8 I	**5.7**	$COC_{15}H_{31}$	M 74.4 I	[1572]
4.8	$COC_{17}H_{35}$	M 73.0 I	**5.8**	$COC_{17}H_{35}$	M 79.5 I	[1572]
4.9	$n=6$	Cr 58.1 Col 62.4 I	**5.9**	$n=6$	Cr 53.4 Col 84.2 I	[1574]
4.10	$n=7$	Cr 57.4 Col 81.0 I	**5.10**	$n=7$	Cr 56.1 Col 105.5 I	[1574]
4.11	$n=8$	Cr 48.7 Col 90.8 I	**5.11**	$n=8$	Cr 64.3 Col 113.6 I	[1574]
4.12	$n=10$	Cr 45.2 Col 102.9 I	**5.12**	$n=10$	Col 120.8 I	[1574]
4.13	$n=14$	Cr 54.1 Col 106.6 I	**5.13**	$n=14$	Cr 55.4 Col 119.9 I	[1574]

isotropic phases provides only short (<10 monomer units) polymers, more than 30 monomer units get linked upon irradiation in the columnar mesophase. The columnar mesophase morphology remains intact in the polymer; however, the polymer unwinds and does not retake its original form when dissolved and reprecipitated in a solvent or heated to above the clearing temperature.

2.25.3 AZAMACROCYCLES

Lehn et al. realized that appropriately substituted *N*-acylated azacrowns self-assemble to form discotic liquid crystals [1575]. Owing to the central cavity in the column, the term "tubular mesophase" was ascribed for the mesophase observed in these materials. Examples of liquid crystalline azacrown derivatives are given in Scheme 2.188 [1576–1586]. In addition, a few larger macrocycles, for example, **[24]-N_6O_2**, **[27]-N_6O_3**, **[30]-N_{10}**, are also known to exhibit monotropic mesophases.

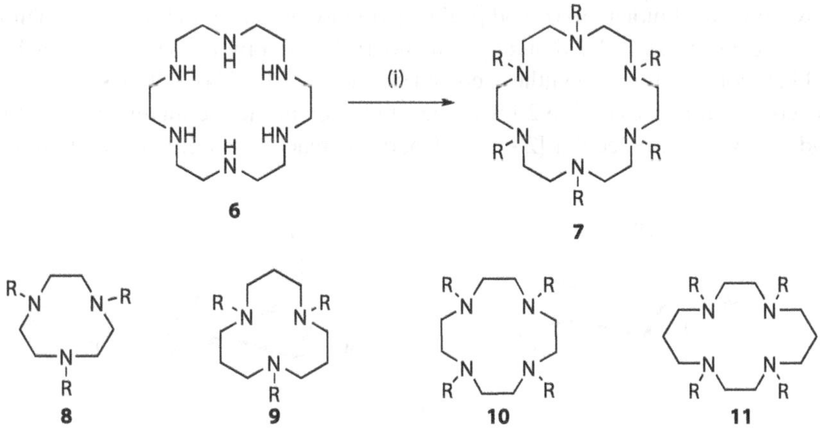

SCHEME 2.188 Synthesis of liquid crystalline azacrown derivatives: (i) RCOCl, DMA, DMAP.

TABLE 2.157
Thermal Phase Transitions of *N*-Acylated Azacrowns (Scheme 2.188)

Structure	R	Phase Transition	Ref.
7.1	$C_8H_{17}OC_6H_4CO$	Cr 120 Col 139 I	[1576]
7.2	$C_{11}H_{23}OC_6H_4CO$	Cr? Col 142 I	[1582]
7.3	$C_{12}H_{25}OC_6H_4CO$	Cr 121.5 Col 141.5 I	[1575]
		Cr 108.0 Col 140.0 I	[1578]
		Cr 102.0 Col 131.0 I	[1580]
		Cr 64 Col 140.0 I	[1582]
		Cr 110.0 Col 138.4 I	[1586]
		Cr 107.7 Col 119.4 Col 138.7 I	[1586]
		Cr 77.2 Col 138.2 I	[1586]
7.4	$C_{14}H_{29}OC_6H_4CO$	Cr 106 Col 136 I	[1578]
7.5	$C_{14}H_{29}OC_6H_4CH=CHCO$	Cr 217 Col 233 I	[1578]
7.6	$C_{10}H_{21}OC_6H_4N=NC_6H_4CO$	Cr 237 Col 245 I	[1578]
7.7	$3,4\text{-}(C_{10}H_{21}O)_2C_6H_3CO$	Cr 104 Col 140 I	[1579]
8	$3,4\text{-}(C_{10}H_{21}O)_2C_6H_3CO$	Cr 43.5 Col 66 I	[1579]
9	$3,4\text{-}(C_{10}H_{21}O)_2C_6H_3CO$	Cr 37.5 Col 59.5 I	[1579]
10	$3,4\text{-}(C_{10}H_{21}O)_2C_6H_3CO$	Cr 108 Col 154 I	[1588]
11	$3,4\text{-}(C_{10}H_{21}O)_2C_6H_3CO$	Cr 95.8 Col 132.4 I	[1577]

The synthesis of these materials is straightforward involving the condensation of the parent macrocyclic oligomer with a benzoyl chloride derivative in DMA/DMAP, as exemplified in Scheme 2.188 for the synthesis of azacrown **[18]-N6**.

The mesophase behavior of these materials is shown in Table 2.157. The mesomorphic behavior of some azacrown derivatives is still a matter of controversy. It is realized that their thermal behavior is dependent on the history/recrystallization conditions of the sample. Seven different transition temperatures are reported for the first derivative prepared (Table 2.157). The liquid crystalline behavior of these acylated azacrowns was related to their discotic shape originated from the restricted conformational mobility of the central unit by N-acylation with benzoic acid derivatives. The presence of the benzamide group appears essential for mesomorphism. A reduction of the amide groups yields non-liquid crystalline N-alkylated macrocycles due to more flexible linkage of the side chains [1587,1588]. The reinduction of a columnar phase in these cyclic amines can be achieved by complexation with transition metal ions. Interestingly, a number of N-acylated linear polyamines are also known to exhibit columnar phases [1589–1592].

REFERENCES

1. Chandrasekhar, S., Sadashiva, B. K., and Suresh, K. A. *Pramana* 9, 471–480, 1977.
2. Chandrasekhar, S., Sadashiva, B. K., and Suresh, K. A. *Mol. Cryst. Liq. Cryst.* 397, 295–305, 2003.
3. Chandrasekhar, S., Sadashiva, B. K., Suresh, K. A., Madhusudana, N. V., Kumar, S., Shashidhar, R., and Venkatesh, G. *J. Phys.* C3, 120–124, 1979.
4. Cammidge, A. N. and Bushby, R. J. In *Handbook of Liquid Crystals*, Vol. 2B, Demus, D., Goodby, J., Gray, G. W., Spiess, H. W., and Vill, V. (eds.), Wiley-VCH, Weinheim, Germany, Chap. VII, pp. 692–748, 1998.
5. Praefcke, K. In *Physical Properties of Liquid Crystals: Nematics*, Dunmur, D., Fukuda, A., and Luckhurst, G. (eds.), INSPEC, London, U.K., Chap. 1.2, pp. 17–35, 2001.
6. Kumar, S. *Chem. Soc. Rev.* 35, 83–109, 2006.
7. Kumar, S. *Pramana* 61, 199–203, 2003.
8. Bisoyi, H. K. and Kumar, S. *Chem. Soc. Rev.* 39, 264–285, 2010.
9. Laschat, S., Baro, A., Steinke, N., Giesselmann, F., Hagele, C., Scalia, G., Judele, R., Kapatsina, E., Sauer, S., Schreivogel, A., and Tosoni, M. *Angew. Chem. Int. Ed.* 46, 4832–4887, 2007.
10. Janietz, D. *J. Mater. Chem.* 8, 265–274, 1998.
11. Bushey, M. L., Nguyen, T. Q., Zhang, W., Horoszewski, D., and Nuckolls, C. *Angew. Chem. Int. Ed.* 43, 5446–5453, 2004.
12. Nguyen, T. Q., Martel, R., Bushey, M., Avouris, P., Carlsen, A., Nuckolls, C., and Brus, L. *Phys. Chem. Chem. Phys.* 9, 1515–1532, 2007.
13. Chandrasekhar, S., Prasad, S. K., Nair, G. G., Rao, D. S. S., Kumar, S., and Manickam, M. *Euro Display 99*. LateNews. 1999, 9–11.
14. Nair, G. G., Rao, D. S. S., Prasad, S. K., Chandrasekhar, S., and Kumar, S. *Mol. Cryst. Liq. Cryst.* 397, 245–252, 2003.
15. Bates, M. A. and Luckhurst, G. R. *Liq. Cryst.* 24, 229–241, 1998.
16. Sorai, M. and Saito, K. *Chem. Rec.* 3, 29–39, 2003.
17. Asahina, S. and Sorai, M. *J. Chem. Thermodyn.* 35, 649–666, 2003.
18. Wang, L. and Rey, A. D. *Liq. Cryst.* 23, 93–111, 1997.
19. Kroger, M. and Sellers, S. *Mol. Cryst. Liq. Cryst.* 300, 245–262, 1997.
20. Singh, A. P. and Rey, A. D. *Liq. Cryst.* 18, 219–230, 1995.
21. Yan, J. and Rey, A. D. *Phys. Rev. E* 65, 031713, 2002.
22. Hashim, R., Luckhurst, G. R., Prata, F., and Romano, S. *Liq. Cryst.* 15, 283–309, 1993.
23. Grecov, D. and Rey, A. D. *Rheol. Acta* 42, 590–604, 2003.
24. Tsykalo, A. L. *Mol. Cryst. Liq. Cryst.* 128, 99–110, 1985.
25. Tabushi, I., Yamamura, K., and Okada, Y. *J. Org. Chem.* 52, 2502–2505, 1987.
26. Maliniak, A., Greenbaum, S., Poupko, R., Zimmermann, H., and Luz, Z. *J. Phys. Chem.* 97, 4832–4840, 1993.
27. Collard, D. M. and Lillya, C. P. *J. Org. Chem.* 56, 6064–6066, 1991.
28. Collard, D. M. and Lillya, C. P. *J. Am. Chem. Soc.* 111, 1829–1830, 1989.

29. Goozner, R. E. and Labes, M. M. *Mol. Cryst. Liq. Cryst. Lett.* 56, 75–81, 1979.
30. Frank, F. C. and Chandrasekhar, S. *J. Phys.* 41, 1285–1288, 1980.
31. Sorai, M. and Suga, H. *Mol. Cryst. Liq. Cryst.* 73, 47–69, 1981.
32. Kardan, M., Kaito, A., Hsu, S. L., Takur, R., and Lillya, C. P. *J. Phys. Chem.* 91, 1809–1814, 1987.
33. Kardan, M., Reinhold, B. B., Hsu, S. L., Thakur, R., and Lillya, C. P. *Macromolecules* 19, 616–621, 1986.
34. Toriumi, H. and Watanabe, H. *Liq. Cryst.* 2, 553–556, 1987.
35. Rondelez, F., Baret, J. F., and Bois, A. G. *J. Phys.* 48, 1225–1234, 1987.
36. Toriumi, H., Shimmura, T., Watanabe, H., and Saito, H. *Bull. Chem. Soc. Jpn.* 61, 2569–2571, 1988.
37. Bose, M. and Sanyal, S. *Mol. Cryst. Liq. Cryst.* 185, 115–130, 1990.
38. Maliniak, A. *J. Chem. Phys.* 96, 2306–2317, 1992.
39. Fischer, H. and Karasz, F. E. *Liq. Cryst.* 15, 513–519, 1993.
40. Sandstrom, D., Zimmermann, H., and Maliniak, A. *J. Phys. Chem.* 98, 8154–8160, 1994.
41. Usoltseva, N., Praefcke, K., Smirnova, A., and Blunk, D. *Liq. Cryst.* 26, 1723–1734, 1999.
42. Lillya, C. P. and Thakur, R. *Mol. Cryst. Liq. Cryst.* 170, 179–183, 1989.
43. Akopova, O. B., Maidachenko, G. G., Tyuneva, G. A., and Shabyshev, L. S. *Zhurnal Obshchei Khimii* 54, 1861–1863, 1984 (*J. Gen. Chem. USSR* 54, 1657–1659, 1984).
44. Akopova, O. B., Tyuneva, G. A., Shabyshev, L. S., and Erykalov, Yu. G. *Zhurnal Obshchei. Khimii* 57, 650–655, 1987 (*J. Gen. Chem. USSR* 57, 570–574, 1987).
45. Lehmann, M., Gearba, R. I., Koch, M. H. J., and Ivanov, D. A. *Chem. Mater.* 16, 374–376, 2004.
46. Gearba, R. I., Bondar, A., Lehmann, M., Goderis, B., Bras, W., Koch, M. H. J., and Ivanov, D. A. *Adv. Mater.* 17, 671–676, 2005.
47. Lehmann, M., Jahr, M., and Gutmann, J. *J. Mater. Chem.* 18, 2995–3003, 2008.
48. Chang, J. Y., Baik, J. H., Lee, C. B., and Han, M. J. *J. Am. Chem. Soc.* 119, 3197–3198, 1997.
49. Chang, J. Y., Yeon, J. R., Shin, Y. S., Han, M. J., and Hong, S. K. *Chem. Mater.* 12, 1076–1082, 2000.
50. Kobayashi, Y. and Matsunaga, Y. *Bull. Chem. Soc. Jpn.* 60, 3515–3518, 1987.
51. Kawada, H. and Matsunaga, Y. *Bull. Chem. Soc. Jpn.* 61, 3083–3085, 1988.
52. Harada, Y. and Matsunaga, Y. *Bull. Chem. Soc. Jpn.* 61, 2739–2741, 1988.
53. Kawamata, J. and Matsunaga, Y. *Mol. Cryst. Liq. Cryst.* 231, 79–85, 1993.
54. Matsunaga, Y., Miyajima, N., Nakayasu, Y., Sakai, S., and Yonenaga, M. *Bull. Chem. Soc. Jpn.* 61, 207–210, 1988.
55. Palmans, A. R. A., Vekemans, J. A. J. M., Fischer, H., Hikmet, R. A., Meijer, E. W. *Chem. Eur. J.* 3, 300–307, 1997.
56. Palmans, A. R. A., Vekemans, J. A. J. M., Hikmet, R. A., Fischer, H., and Meijer, E. W. *Adv. Mater.* 10, 873–876, 1998.
57. Palmans, A. R. A., Vekemans, J. A. J. M., Havinga, E. E., and Meijer, E. W. *Angew. Chem. Int. Ed.* 36, 2648–2651, 1997.
58. Brunsveld, L., Vekemans, J. A. J. M., Janssen, H. M., and Meijer, E. W. *Mol. Cryst. Liq. Cryst.* 331, 449–456, 1999.
59. Brunsveld, L., Zhang, H., Glasbeek, M., Vekemans, J. A. J. M., and Meijer, E. W. *J. Am. Chem. Soc.* 122, 6175–6182, 2000.
60. van Gorp, J. J., Vekemans, J. A. J. M., and Meijer, E. W. *J. Am. Chem. Soc.* 124, 14759–14769, 2002.
61. van Gestel, J., Palmans, A. R. A., Titulaer, B., Vekemans, J. A. J. M., and Meijer, E. W. *J. Am. Chem. Soc.* 127, 5490–5494, 2005.
62. Toele, P., van Gorp, J. J., and Glasbeek, M. *J. Phys. Chem. A* 109, 10479–10487, 2005.
63. Brunsveld, L., Lohmeijer, B. G. G., Vekemans, J. A. J. M., and Meijer, E. W. *Chem. Commun.* 2305–2306, 2000.
64. Bushey, M. L., Hwang, A., Stephens, P. W., and Nuckolls, C. *J. Am. Chem. Soc.* 123, 8157–8158, 2001.
65. Bushey, M. L., Hwang, A., Stephens, P. W., and Nuckolls, C. *Angew. Chem. Int. Ed.* 41, 2828–2831, 2002.
66. Nguyen, T. Q., Bushey, M. L., Brus, L. E., and Nuckolls, C. *J. Am. Chem. Soc.* 124, 15051–15054, 2002.
67. Bushey, M. L., Nguyen, T. Q., and Nuckolls, C. J. *J. Am. Chem. Soc.* 125, 8264–8269, 2003.
68. Nguyen, T. Q., Martel, R., Avouris, P., Bushey, M. L., Brus, L., and Nuckolls, C. *J. Am. Chem. Soc.* 126, 5234–5242, 2004.
69. Kok, D. M., Wynberg, H., and de Jeu, W. H. *Mol. Cryst. Liq. Cryst.* 129, 53–60, 1985.
70. Poules, V. W. and Praefcke, K. *Chem. Zeit.* 107, 310–311, 1983.
71. Praefcke, K., Poules, W., Scheuble, B., Poupko, R., and Luz, Z. *Z. Naturforsch.* 39b, 950–956, 1984.
72. Sarkar, M., Spielberg, N., Praefcke, K., and Zimmermann, H. *Mol. Cryst. Liq. Cryst.* 203, 159–169, 1991.

73. Spielberg, N., Luz, Z., Poupko, R., Praefcke, K., Kohne, B., Pickardt, J., and Horn, K. *Z. Naturforsch.* 41a, 855–860, 1986.
74. Kohne, B. and Praefcke, K. *Chimia* 41, 196–198, 1987.
75. Praefcke, K., Kohne, B., and Singer, D. *Angew. Chem. Int. Ed.* 29, 177–179, 1990.
76. Ebert, M., Jungbauer, D. A., Kleppinger, R., Wendorff, J. H., Kohne, B., and Praefcke, K. *Liq. Cryst.* 4, 53–67, 1989.
77. Patel, J. S., Praefcke, K., Singer, D., and Langner, M. *Appl. Phys. B* 60, 469–472, 1995.
78. Kumar, S. and Varshney, S. K. *Angew. Chem. Int. Ed.* 39, 3140–3142, 2000.
79. Kumar, S., Varshney, S. K., and Chauhan, D. *Mol. Cryst. Liq. Cryst.* 396, 241–250, 2003.
80. Varshney, S. K., Takezoe, H., and Rao, D. S. S. *Bull. Chem. Soc. Jpn.* 81, 163–167, 2008.
81. Chien, S. C., Chen, H. H., Chen, H. C., Yang, Y. L., Hsu, H. F., Shih, T. L., and Lee, J. J. *Adv. Funct. Mater.* 17, 1896–1902, 2007.
82. Grafe, A., Janietz, D., Frese, T., and Wendorff, J. H. *Chem. Mater.* 17, 4979–4984, 2005.
83. Janietz, D., Praefcke, K., and Singer, D. *Liq. Cryst.* 13, 247–253, 1993.
84. Janietz, D., Hofmann, D., and Reiche, J. *Thin Solid Films* 244, 794–798, 1994.
85. Praefcke, K., Singer, D., Gundogan, B., Gutbier, K., and Langner, M. *Ber. Bunsenges. Phys. Chem.* 97, 1358–1361, 1993.
86. Langner, M., Praefcke, K., Kruerke, D., and Heppke, G. *J. Mater. Chem.* 5, 693–699, 1995.
87. Booth, C. J., Kruerke, D., and Heppke, G. *J. Mater. Chem.* 6, 927–934, 1996.
88. Praefcke, K., Kohne, B., Gundogan, B., Singer, D., Demus, D., Diele, S., Pelzl, G., and Bakowsky, U. *Mol. Cryst. Liq. Cryst.* 198, 393–405, 1991.
89. Sikharulidze, D., Chilaya, G., Praefcke, K., and Blunk, D. *Liq. Cryst.* 23, 439–442, 1997.
90. Praefcke, K., Kohne, B., Singer, D., Demus, D., Pelzl, G., and Diele, S. *Liq. Cryst.* 7, 589–594, 1990.
91. Janietz, D. *Chem. Commun.* 713–714, 1996.
92. Stracke, A., Wendorff, J. H., Janietz, D., and Mahlstedt, S. *Adv. Mater.* 11, 667–670, 1999.
93. Kouwer, P. H. J., Jager, W. F., Mijs, W. J., Picken, S. J., Shepperson, K. J., and Mehl, G. H. *Mol. Cryst. Liq. Cryst.* 411, 377–385, 2004.
94. Kouwer, P. H. J., Mehl, G. H., and Picken, S. J. *Mol. Cryst. Liq. Cryst.* 411, 387–396, 2004.
95. Kouwer, P. H. J., Picken, S. J., and Mehl, G. H. *J. Mater. Chem.* 17, 4196–4203, 2007.
96. Apreutesei, D. and Mehl, G. H. *Chem. Commun.* 609–611, 2006.
97. Date, R. W. and Bruce, D. W. *J. Am. Chem. Soc.* 125, 9012–9013, 2003.
98. Kouwer, P. H. J. and Mehl, G. H. *J. Am. Chem. Soc.* 125, 11172–11173, 2003.
99. Apreutesei, D. and Mehl, G. *Mol. Cryst. Liq. Cryst.* 449, 107–115, 2006.
100. Kouwer, P. H. J., Jager, W. F., Mijs, W. J., and Picken, S. J. *Macromolecules* 35, 4322–4329, 2002.
101. Kouwer, P. H. J., Jager, W. F., Mijs, W. J., and Picken, S. J. *Macromolecules* 33, 4336–4342, 2000.
102. Kouwer, P. H. J., Jager, W. F., Mijs, W. J., and Picken, S. J. *J. Mater. Chem.* 13, 458–469, 2003.
103. Fletcher, I. D. and Luckhurst, G. R. *Liq. Cryst.* 18, 175–183, 1995.
104. Kouwer, P. H. J. and Mehl, G. H. *Mol. Cryst. Liq. Cryst.* 397, 1–16, 2003.
105. Kouwer, P. H. J. and Mehl, G. H. *J. Mater. Chem.* 19, 1564–1575, 2009.
106. Kouwer, P. H. J. and Mehl, G. H. *Angew. Chem. Int. Ed.* 42, 6015–6018, 2003.
107. Kouwer, P. H. J., Welch, C. J., McRobbie, G., Dodds, B. J., Priest, L., and Mehl, G. H. *J. Mater. Chem.* 14, 1798–1803, 2004.
108. Hauser, A., Thieme, M., Saupe, A., Heppke, G., and Kruerke, D. *J. Mater. Chem.* 7, 2223–2229, 1997.
109. Heppke, G., Kitzerow, H., Oestreicher, F., Quentel, S., and Ranft, A. *Mol. Cryst. Liq. Cryst. Lett.* 6, 71–79, 1988.
110. Heppke, G., Ranft, A., and Sabaschus, B. *Mol. Cryst. Liq. Cryst. Lett.* 8, 17–25, 1991.
111. Kruerke, D., Kitzerow, H. S., and Heppke, G. *Ber. Bunsenges. Phys. Chem.* 97, 1371–1375, 1993.
112. Kruerke, D., Gough, N., Heppke, G., and Lagerwall, S. T. *Mol. Cryst. Liq. Cryst. Lett.* 351, 69–78, 2000.
113. Jutila, A., Janietz, D., Reiche, J., and Lemmetyinen, H. *Thin Solid Films* 268, 121–129, 1995.
114. Furumi, S., Janietz, D., Kidowaki, M., Nakagawa, M., Morino, S., Stumpe, J., and Ichimura, K. *Chem. Mater.* 13, 1434–1437, 2001.
115. Furumi, S., Janietz, D., Kidowaki, M., Nakagawa, M., Morino, S., Stumpe, J., and Ichimura, K. *Mol. Cryst. Liq. Cryst.* 368, 517–524, 2001.
116. Marguet, S., Markovitsi, D., Goldmann, D., Janietz, D., Praefcke, K., and Singer, D. *J. Chem. Soc. Faraday Trans.* 93, 147–155, 1997.
117. Ionov, R. and Angelova, A. *J. Phys. Chem.* 99, 17593–17605, 1995.
118. Janietz, D., Ahuja, R. C., and Mobius, D. *Langmuir* 13, 305–309, 1997.
119. Reiche, J. *Thin Solid Films* 284–285, 453–455, 1996.

120. Ionov, R. and Angelova, A. *Phys. Rev. E* 52, R21–R24, 1995.
121. Reiche, J., Dietel, R., Janietz, D., Lemmetyinen, H., and Brehmer, L. *Thin Solid Films* 226, 265–269, 1992.
122. Sabaschus, B., Singer, D., Heppke, G., and Praefcke, K. *Liq. Cryst.* 12, 863–867, 1992.
123. Praefcke, K., Singer, D., Kohne, B., Ebert, M., Liebmann, A., and Wendorff, J. H. *Liq. Cryst.* 10, 147–159, 1991.
124. Praefcke, K., Singer, D., and Eckert, A. *Liq. Cryst.* 16, 53–65, 1994.
125. Vijayaraghavan, D. and Kumar, S. *Mol. Cryst. Liq. Cryst.* 452, 11–26, 2006.
126. Yelamaggad, C. V., Achalkumar, A. S., Rao, D. S. S., and Prasad, S. K. *J. Org. Chem.* 72, 8308–8318, 2007.
127. Yelamaggad, C. V., Achalkumar, A. S., Rao, D. S. S., and Prasad, S. K. *J. Mater. Chem.* 17, 4521–4529, 2007.
128. Yelamaggad, C. V., Achalkumar, A. S., Rao, D. S. S., and Prasad, S. K. *J. Am. Chem. Soc.* 126, 6506–6507, 2004.
129. Yelamaggad, C. V. and Achalkumar, A. S., *Tetrahedron Lett.* 47, 7071–7075, 2006.
130. Meier, H. and Lehmann, M. *Angew. Chem. Int. Ed.* 37, 643–645, 1998.
131. Meier, H., Lehmann, M., and Kolb, U. *Chem. Eur. J.* 6, 2462–2469, 2000.
132. Lehmann, M., Kohn, C., Meier, H., Renker, S., and Oehlhof, A. *J. Mater. Chem.* 16, 441–451, 2006.
133. Lehmann, M., Fischbach, I., Spiess, H. W., and Meier, H. *J. Am. Chem. Soc.* 126, 772–784, 2004.
134. Meier, H., Lehmann, M., Holst, H. C., and Schwoppe, D. *Tetrahedron* 60, 6881–6888, 2004.
135. Lehmann, M., Schartel, B., Hennecke, M., and Meier, H. *Tetrahedron* 55, 13377–13394, 1999.
136. Hennrich, G., Omenat, A., Asselberghs, I., Foerier, S., Clays, K., Verbiest, T., and Serrano, J. L. *Angew. Chem. Int. Ed.* 45, 4203–4206, 2006.
137. Hennrich, G., Cavero, E., Barbera, J., Lor, B. G., Hanes, R. E., Talarico, M., Golemme, A., and Serrano, J. L. *Chem. Mater.* 19, 6068–6070, 2007.
138. Omenat, A., Barbera, J., Serrano, J. L., Houbrechts, S., and Persoons, A. *Adv. Mater.* 11, 1292–1295, 1999.
139. Yoshio, M., Mukai, T., Ohno, H., and Kato, T. *J. Am. Chem. Soc.* 126, 994–995, 2004.
140. Yoshio, M., Kagata, T., Hoshino, K., Mukai, T., Ohno, H., and Kato, T. *J. Am. Chem. Soc.* 128, 5570–5577, 2006.
141. Ito, S., Inabe, H., Morita, N., Ohta, K., Kitamura, T., and Imafuku, K. *J. Am. Chem. Soc.* 125, 1669–1680, 2003.
142. Ito, S., Ando, M., Nomura, A., Morita, N., Kabuto, C., Mukai, H., Ohta, K., Kawakami, J., Yoshizawa, A., and Tajiri, A. *J. Org. Chem.* 70, 3939–3949, 2005.
143. Lattermann, G. *Liq. Cryst.* 2, 723–728, 1987.
144. Malthete, J., Collet, A., and Levelut, A. M. *Liq. Cryst.* 5, 123–131, 1989.
145. Kim, B. G., Kim, S., and Park, S. Y. *Mol. Cryst. Liq. Cryst.* 370, 391–394, 2001.
146. Kim, B. G., Kim, S., and Park, S. Y. *Tetrahedron Lett.* 42, 2697–2699, 2001.
147. Jeong, M. J., Park, J. H., Lee, C., and Chang, J. Y. *Org. Lett.* 8, 2221–2224, 2006.
148. Geng, Y., Fechtenkotter, A., and Mullen, K. *J. Mater. Chem.* 11, 1634–1641, 2001.
149. Takenaka, S., Nishimura, K., and Kusabayashi, S. *Mol. Cryst. Liq. Cryst.* 111, 227–236, 1984.
150. Takaragi, A., Miyamoto, T., Minoda, M., and Watanabe, J. *Macromol. Chem. Phys.* 199, 2071–2077, 1998.
151. Marsh, H., Foster, J. M., Hermon, G., and Iley, M. *Fuel* 52, 234–242, 1973.
152. Jian, K., Hurt, R. H., Sheldon, B. W., and Crawford, G. P. *Appl. Phys. Lett.* 88, 163110, 2006.
153. Praefcke, K., Kohne, B., Gutbier, K., Johnen, N., and Singer, D. *Liq. Cryst.* 5, 233–249, 1989.
154. Praefcke, K., Singer, D., and Kohne, B. *Liq. Cryst.* 13, 445–454, 1993.
155. Sasaki, T., Jenkins, R. G., Eser, S., and Schobert, H. H. *Energy Fuels* 7, 1039–1046, 1993.
156. Sasaki, T., Jenkins, R. G., Eser, S., and Schobert, H. H. *Energy Fuels* 7, 1047–1053, 1993.
157. Furstner, A. and Mamane, V. *J. Org. Chem.* 67, 6264–6267, 2002.
158. Scherowsky, G. and Chen, X. H. *Liq. Cryst.* 17, 803–810, 1994.
159. Scherowsky, G. and Chen, X. H. *J. Mater. Chem.* 5, 417–421, 1995.
160. Thomson, R. H. *Naturally Occurring Quinones*, Chapman and Hall, New York, 1987.
161. Christie, R. M. *Colour Chemistry*, RSC, Cambridge, U.K., 2001.
162. Ivashchenko, A. V. *Dichroic Dyes for Liquid Crystal Displays*, CRC, Boca Raton, FL, 1994.
163. Kampmann, B., Lian, Y., Klinkel, K. L., Vecchi, P. A., Quiring, H. L., Soh, C. C., and Sykes, A. G. *J. Org. Chem.* 67, 3878–3883, 2002.
164. Terech, P. and Weiss, R. G. *Chem. Rev.* 97, 3133–3160, 1997.

165. Tietze, L. F., Gericke, K. M., and Schuberth, I. *Eur. J. Org. Chem.* 4563–4577, 2007.
166. Grimshaw, J. and Haworth, R. D. *J. Chem. Soc.* 56, 4225–4232, 1956.
167. Ignatushchenko, M. V., Winter, R. W., and Riscoe, M. *Am. J. Trop. Med. Hyg.* 62, 77–81, 2000.
168. Ziegler, J., Linck, R., and Wright, D. W. *Curr. Med. Chem.* 8, 171–189, 2001.
169. Winter, R. W., Cornell, K. A., Johnson, L. L., Isabelle, L. M., Hinrichs, D. J., and Riscoe, M. K. *Bioorg. Med. Chem. Lett.* 5, 1927–1932, 1995.
170. Martin, G. J. and Lischer, C. F. *J. Biol. Chem.* 137, 169–171, 1941.
171. Queguiner, A., Zann, A., Dubois, J. C., and Billard, J. In *Proceedings of the International Conference on Liquid Crystals*, Chandrasekhar, S. (ed.), Heyden and Son, London, U.K., p. 35, 1980.
172. Billard, J., Dubois, J. C., Vaucher, C., and Levelut, A. M. *Mol. Cryst. Liq. Cryst.* 66, 115–122, 1981.
173. Carfagna, C., Roviello, A., and Sirigu, A. *Mol. Cryst. Liq. Cryst.* 122, 151–160, 1985.
174. Carfagna, C., Iannelli, P., Roviello, A., and Sirigu, A. *Liq. Cryst.* 2, 611–616, 1987.
175. Werth, M., Leisen, J., Boeffel, C., Dong, R. Y., and Spiess, H. W. *J. Phys.* 3, 53–67, 1993.
176. Billard, J., Luz, Z., Poupko, R., and Zimmermann, H. *Liq. Cryst.* 16, 333–342, 1994.
177. Hollander, A., Hommels, J., Prins, K. O., Spiess, H. W., and Werth, M. *J. Phys. II France* 6, 1727–1741, 1996.
178. He, Z., Zhao, Y., and Caille, A. *Liq. Cryst.* 23, 317–325, 1997.
179. Raja, K. S., Ramakrishnan, S., and Raghunathan, V. A. *Chem. Mater.* 9, 1630–1637, 1997.
180. Raja, K. S., Raghunathan, V. A., and Ramakrishnan, S. *Macromol.* 31, 3807–3814, 1998.
181. Dong, R. Y. and Morcombe, C. R. *Liq. Cryst.* 27, 897–900, 2000.
182. Chenard, Y., Paiement, N., and Zhao, Y. *Liq. Cryst.* 27, 459–465, 2000.
183. Corvazier, L. and Zhao, Y. *Liq. Cryst.* 27, 137–143, 2000.
184. Krishnan, K. and Balagurusamy, V. S. K. *Liq. Cryst.* 27, 991–994, 2000.
185. Prasad, V., Krishnan, K., and Balagurusamy, V. S. K. *Liq. Cryst.* 27, 1075–1085, 2000.
186. Krishnan, K. and Balagurusamy, V. S. K. *Mol. Cryst. Liq. Cryst.* 350, 1–18, 2000.
187. Prasad, V. and Rao, D. S. S. *Mol. Cryst. Liq. Cryst.* 350, 51–65, 2000.
188. Perova, T. S., Fannin, P. C., and Perov, P. A. *Mol. Cryst. Liq. Cryst.* 352, 141–148, 2000.
189. Krishnan, K. and Balagurusamy, V. S. K. *Liq. Cryst.* 28, 321–325, 2001.
190. Prasad, V. *Liq. Cryst.* 28, 647–650, 2001.
191. Chadrasekhar, S. and Balagurusamy, V. S. K. *Proc. R. Soc. Lond. A* 458, 1783–1794, 2002.
192. Kumar, S. and Naidu, J. J. *Mol. Cryst. Liq. Cryst.* 378, 123–128, 2002.
193. Kumar, S. and Naidu, J. J. *Liq. Cryst.* 29, 1369–1371, 2002.
194. Naidu, J. J. and Kumar, S. *Mol. Cryst. Liq. Cryst.* 397, 17–24, 2003.
195. Dvinskikh, S. V., Zimmermann, H., Maliniak, A., and Sandstrom, D. *J. Magn. Reson.* 163, 46–55, 2003.
196. Dvinskikh, S. V., Zimmermann, H., Maliniak, A., and Sandstrom, D. *J. Magn. Reson.* 164, 165–170, 2003.
197. Dvinskikh, S. V., Luz, Z., Zimmermann, H., Maliniak, A., and Sandstrom, D. *J. Phys. Chem. B* 107, 1969–1976, 2003.
198. Dvinskikh, S. V., Sandstrom, D., Luz, Z., Zimmermann, H., and Maliniak, A. *J. Chem. Phys.* 119, 413–422, 2003.
199. Dong, R. Y. and Shearer, R. *Chem. Phys. Lett.* 375, 463–469, 2003.
200. Kumar, S., Naidu, J. J., and Varshney, S. K. *Liq. Cryst.* 30, 319–323, 2003.
201. Sandhya, K. L., Prasad, S. K., Nair, G. G., and Prasad, V. *Mol. Cryst. Liq. Cryst.* 396, 113–119, 2003.
202. Kumar, S., Naidu, J. J., and Varshney, S. K. *Mol. Cryst. Liq. Cryst.* 411, 355–362, 2004.
203. Maeda, Y., Yokoyama, H., and Kumar, S. *Liq. Cryst.* 32, 833–845, 2005.
204. Dhar, R., Kumar, S., Gupta, M., and Agrawal, V. K. *J. Mol. Liq.* 141, 19–24, 2008.
205. Bisoyi, H. K. and Kumar, S. *Tetrahedron Lett.* 48, 4399–4402, 2007.
206. Pal, S. K. and Kumar, S. (unpublished result).
207. Pal, S. K., Kumar, S., and Seth, J. *Liq. Cryst.* 35, 521–525, 2008.
208. Bisoyi, H. K. and Kumar, S. *Tetrahedron Lett.* 49, 3628–3631, 2008.
209. Bisoyi, H. K. and Kumar, S. *J. Mater. Chem.* 18, 3032–3039, 2008.
210. Bisoyi, H. K. and Kumar, S. *New J. Chem.* 32, 1974–1980, 2008.
211. Dhar, R., Gupta, M., Agrawal, V. K., and Kumar, S. *Phase Transitions* 81, 459–469, 2008.
212. Kumar, S. *Phase Transitions* 81, 113–128, 2008.
213. Gupta, S. K., Raghunathan, V. A., and Kumar, S. *New J. Chem.* 33, 112–118, 2009.
214. Pu, L. *Chem. Rev.* 104, 1687–1716, 2004.
215. Martinez-Manez, R. and Sancenon, F. *Chem. Rev.* 103, 4419–4476, 2003.

216. Daub, J., Engl, R., Kurzawa, J., Miller, S. E., Schneider, S., Stockmann, A., and Wasielewski, M. R. *J. Phys. Chem. A* 105, 5655–5665, 2001.

217. Baker, L. A. and Crooks, R. M. *Macromolecules* 33, 9034–9039, 2000.

218. Modrakowski, C., Flores, S. C., Beinhoff, M., and Schluter, A. D. *Synthesis* 2143–2155, 2001.

219. Chaiken, R. F. and Kearns, D. R. *J. Chem. Phys.* 49, 2846–2850, 1968.

220. Holroyd, R. A., Preses, J. M., Boettcher, E. H., and Schmidt, W. F. *J. Phys. Chem.* 88, 744–749, 1984.

221. Jones II, G. and Vullev, V. I. *Org. Lett.* 4, 4001–4004, 2002.

222. Yamana, K., Fukunaga, Y., Ohtani, Y., Sato, S., Nakamura, M., Kim, W. J., Akaike, T., and Maruyama, A. *Chem. Commun.* 2509–2511, 2005.

223. Hwang, G. T., Seo, Y. J., and Kim, B. H. *J. Am. Chem. Soc.* 126, 6528–6529, 2004.

224. Fujimoto, K., Shimizu, H., and Inouye, M. *J. Org. Chem.* 69, 3271–3275, 2004.

225. Bock, H. and Helfrich, W. *Liq. Cryst.* 18, 707–713, 1995.

226. Hirose, T., Kawakami, O., and Yasutake, M. *Mol. Cryst. Liq. Cryst.* 451, 65–74, 2006.

227. Hassheider, T., Benning, S. A., Kitzerow, H. S., Achard, M. F., and Bock, H. *Angew. Chem. Int. Ed.* 40, 2060–2063, 2001.

228. Benning, S. A., Hassheider, T., Baumann, S. K., Bock, H., Sala, F. D., Frauenheim, T., and Kitzerow, H. S. *Liq. Cryst.* 28, 1105–1113, 2001.

229. Dantras, E., Dandurand, J., Lacabanne, C., Laffont, L., Tarascon, J. M., Archambeau, S., Seguy, I., Destruel, P., Bock, H., and Fouet, S. *Phys. Chem. Chem. Phys.* 6, 4167–4173, 2004.

230. de Halleux, V., Calbert, J. P., Brocorens, P., Cornil, J., Declercq, J. P., Bredas, J. L., and Geerts, Y. *Adv. Funct. Mater.* 14, 649–659, 2004.

231. Sienkowska, M. J., Monobe, H., Kaszynski, P., and Shimizu, Y. *J. Mater. Chem.* 17, 1392–1398, 2007.

232. Sienkowska, M. J., Farrar, J. M., Zhang, F., Kusuma, S., Heiney, P. A., and Kaszynski, P. *J. Mater. Chem.* 17, 1399–1411, 2007.

233. Hayer, A., de Halleux, V., Kohler, A., El-Garoughy, A., Meijer, E. W., Barbera, J., Tant, J., Levin, J., Lehmann, M., Gierschner, J., Cornil, J., and Geerts, Y. H. *J. Phys. Chem. B* 110, 7653–7659, 2006.

234. Yasutake, M., Fujihara, T., Nagasawa, A., Moriya, K., and Hirose, T. *Eur. J. Org. Chem.* 4120–4125, 2008.

235. Sagara, Y. and Kato, T. *Angew. Chem. Int. Ed.* 47, 5175–5178, 2008.

236. Percec, V., Glodde, M., Bera, T. K., Miura, Y., Shiyanovskaya, I., Singer, K. D., Balagurusamy, V. S. K., Heiney, P. A., Schnell. I., Rapp. A., Spiess, H. W., Hudson, S. D., and Duan, H. *Nature* 419, 384–387, 2002.

237. Shiyanovskaya, I., Singer, K. D., Percec, V., Bera, T. K., Miura, Y., and Glodde, M. *Phys. Rev. B* 67, 035204, 2003.

238. Kamikawa, Y. and Kato, T. *Org. Lett.* 8, 2463–2466, 2006.

239. Kim, Y. H., Yoon, D. K., Lee, E. H., Ko, Y. K., and Jung, H. T. *J. Phys. Chem. B* 110, 20836–20842, 2006.

240. Buess, C. M. and Lawson, D. D. *Chem. Rev.* 60, 313–330, 1960.

241. Billard, J., Dubois, J. C., Tinh, N. H., and Zann, A. *J. Chim.* 2, 535–540, 1978.

242. Destrade, C., Mondon, M. C., and Malthete, J. *J. Phys. Colloque.* 40, C3–17–21, 1979.

243. Kumar, S. *Liq. Cryst.* 31, 1037–1059, 2004.

244. Kumar, S. *Liq. Cryst.* 32, 1089–1113, 2005.

245. Badjic, J. D., Balzani, V., Credi, A., Lowe, J. N., Silvi, S., and Stoddart, J. F. *Chem. Eur. J.* 10, 1926–1935, 2004.

246. Cardullo, F., Giuffrida, D., Kohnke, F. H., Raymo, F. M., Stoddart, J. F., and Williams, D. J. *Angew. Chem. Int. Ed.* 35, 339–341, 1996.

247. Fyfe, M. C. T., Lowe, J. N., Stoddart, J. F., and Williams, D. J. *Org. Lett.* 2, 1221–1224, 2000.

248. Balzani, V., Leon, M. C., Credi, A., Lowe, J. N., Badjic, J. D., Stoddart, J. F., and Williams, D. J. *Chem. Eur. J.* 9, 5348–5360, 2003.

249. Badjic, J. D., Balzani, V., Credi, A., Silvi, S., and Stoddart, J. F. *Science* 303, 1845–1849, 2004.

250. Waldvogel, S. R., Wartini, A. R., Rasmussen, P. H., and Rebek Jr., J. *Tetrahedron Lett.* 40, 3515–3518, 1999.

251. Schopohl, M. C., Faust, A., Mirk, D., Frohlich, R., Kataeva, O., and Waldvogel, S. R. *Eur. J. Org. Chem.* 2987–2999, 2005.

252. Waldvogel, S. R. and Mirk, D. *Tetrahedron Lett.* 41, 4769–4772, 2000.

253. Siering, C., Kerschbaumer, H., Nieger, M., and Waldvogel, S. R. *Org. Lett.* 8, 1471–1474, 2006.

254. Siering, C., Beermann, B., and Waldvogel, S. R. *Supramol. Chem.* 18, 23–27, 2006.

255. Lessene, G. and Feldman, K. S. In *Modern Arene Chemistry*, Astruc, D. (ed.), Wiley-VCH, Weinheim, Germany, 2002.

256. Percec, V., Okita, S., and Wang, J. H. *Macromolecules* 25, 64–74, 1992.

257. Berresheim, A. J., Muller, M., and Mullen, K. *Chem. Rev.* 99, 1747–1786, 1999.
258. Scholl, R. and Mansfeld, J. *Ber. Dtsch. Chem. Ges.* 43, 1734–1746, 1910.
259. King, B. T., Kroulik, J., Robertson, C. R., Rempala, P., Hilton, C. L., Korinek, J. D., and Gortari, L. M. *J. Org. Chem.* 72, 2279–2288, 2007.
260. Voisin, E. and Williams, V. E. *Macromolecules* 41, 2994–2997, 2008.
261. Marquardt, F. H. *J. Chem. Soc.* 1517–1518, 1965.
262. Matheson, I. M., Musgrave, O. C., and Webster, C. J. *J. Chem. Soc. Chem. Commun.* 13, 278–279, 1965.
263. Musgrave, O. C. and Webster, C. J. *J. Chem. Soc. C* 8, 1397–1401, 1971.
264. Buchan, R. and Musgrave, O. C. *J. Chem. Soc. Perkin Trans. 1*, 6, 568–572, 1975.
265. Piatelli, M., Fattorusso, E., Nicolaus, R. A., and Magno, S. *Tetrahedron* 21, 3229–3236, 1965.
266. Musgrave, O. C. *Chem. Rev.* 69, 499–531, 1969.
267. Krebs, F. C., Schiodt, N. C., Batsberg, W., and Bechgaard, K. *Synthesis* 1285–1290, 1997.
268. Kumar, S. *Mol. Cryst. Liq. Cryst.* 289, 247–253, 1996.
269. Chiang, L. Y., Safinya, C. R., Clark, N. A., Liang, K. S., and Bloch, A. N. *J. Chem. Soc. Chem. Commun.* 11, 695–696, 1985.
270. Chiang, L. Y., Stokes, J. P., Safinya, C. R., and Bloch, A. N. *Mol. Cryst. Liq. Cryst.* 125, 279–288, 1985.
271. Bechgaard, K. and Parker, V. D. *J. Am. Chem. Soc.* 94, 4749–4750, 1972.
272. Le Berre, V., Angely, L., Gueguen, N. S., and Simonet, J. *J. Chem. Soc. Chem. Commun.* 984–986, 1987.
273. Le Berre, V., Simonet, J., and Batail, P. *J. Electroanal. Chem.* 169, 325–330, 1984.
274. Chapuzet, J. M., Gueguen, N. S., Taillepied, I., and Simonet, J. *Tetrahedron Lett.* 32, 7405–7408, 1991.
275. Chapuzet, J. M. and Simonet, J. *Tetrahedron* 47, 791–798, 1991.
276. Bengs, H., Karthaus, O., Ringsdorf, H., Baehr, C., Ebert, M., and Wendorff, J. H. *Liq. Cryst.* 10, 161–168, 1991.
277. Naarmann, H., Hanack, M., and Mattmer, R. *Synthesis* 477–478, 1994.
278. Boden, N., Borner, R. C., Bushby, R. J., Cammidge, A. N., and Jesudason, M. V. *Liq. Cryst.* 15, 851–858, 1993.
279. Boden, N., Bushby, R. J., and Cammidge, A. N. *J. Chem. Soc. Chem. Commun.* 465–466, 1994.
280. Boden, N., Borner, R. C., Bushby, R. J., Cammidge, A. N., and Jesudason, M. V. *Liq. Cryst.* 33, 1443–1448, 2006.
281. Borner, R. C., Bushby, R. J., Cammidge, A. N., Boden, N., and Jesudason, M. V. *Liq. Cryst.* 33, 1439–1442, 2006.
282. Kumar, S. and Manickam, M. *Chem. Commun.* 1615–1666, 1997.
283. Kumar, S. and Varshney, S. K. *Liq. Cryst.* 26, 1841–1843, 1999.
284. Kumar, S. and Varshney, S. K. *Synthesis* 305–311, 2001.
285. Vauchier, C., Zann, A., Le Barny, P., Dubois, J. C., and Billard, J. *Mol. Cryst. Liq. Cryst.* 66, 103–114, 1981.
286. Destrade, C., Tinh, N. H., Gasparoux, H., Malthete, J., and Levelut, A. M. *Mol. Cryst. Liq. Cryst.* 71, 111–135, 1981.
287. Stackhouse, P. J. and Hird, M. *Liq. Cryst.* 35, 597–607, 2008.
288. Bushby, R. J., Boden, N., Kilner, C. A., Lozman, O. R., Lu, Z., Liu, Q., and Thornton-Pett, M. A. *J. Mater. Chem.* 13, 470–474, 2003.
289. Simmerer, J., Glüsen, B., Paulus, W., Kettner, A., Schuhmacher, P., Adam, D., Etzbach, K. H., Siemensmeyer, K., Wendorff, J. H., Ringsdorf, H., and Haarer, D. *Adv. Mater.* 8, 815–819, 1996.
290. Sergeyev, S., Pouzet, E., Debever, O., Levin, J., Gierschner, J., Cornil, J., Aspe, R. G., and Geerts, Y. H. *J. Mater. Chem.* 17, 1777–1784, 2007.
291. Pisula, W., Kastler, M., Wasserfallen, D., Mondeshki, M., Piris, J., Schnell, I., and Mullen, K. *Chem. Mater.* 18, 3634–3640, 2006.
292. Liu, C. Y., Fechtenkotter, A., Watson, M. D., Mullen, K., and Bard, A. J. *Chem. Mater.* 15, 124–130, 2003.
293. Kumar, S., Rao, D. S. S., and Prasad, S. K. *J. Mater. Chem.* 9, 2751–2754, 1999.
294. Weck, M., Dunn, A. R., Matsumoto, K., Coates, G. W., Lobkovsky, E. B., and Grubbs, R. H. *Angew. Chem. Int. Ed.* 38, 2741–2745, 1999.
295. Schultz, A., Laschat, S., Morr, M., Diele, S., Dreyer, M., and Bringmann, G. *Helv. Chim. Acta* 85, 3909–3918, 2002.
296. Dahn, U., Erdelen, C., Ringsdorf, H., Festag, R., Wendorff, J. H., Heiney, P. A., and Maliszewskyj, N. C. *Liq. Cryst.* 19, 759–764, 1995.
297. Terasawa, N., Monobe, H., Kiyohara, K., and Shimizu, Y. *Chem. Commun.* 1678–1679, 2003.
298. Terasawa, N., Monobe, H., Kiyohara, K., and Shimizu, Y. *Chem. Lett.* 32, 214–215, 2003.

299. Terasawa, N., Tanigaki, N., Monobe, H., and Kiyohara, K. *J. Fluor. Chem.* 127, 1096–1104, 2006.
300. Ikeda, M., Takeuchi, M., and Shinkai, S. *Chem. Commun.* 1354–1355, 2003.
301. Tao, K., Wu, T., Lu, D., Bai, R., Li, H., and An, L. *J. Mol. Liq.* 142, 118–123, 2008.
302. Manickam, M., Belloni, M., Kumar, S., Varshney, S. K., Rao, D. S. S., Ashton, P. R., Preece, J. A., and Spencer, N. *J. Mater. Chem.* 11, 2790–2800, 2001.
303. Cooke, G., Kaushal, N., Boden, N., Bushby, R. J., Lu, Z., and Lozman, O. *Tetrahedron Lett.* 41, 7955–7959, 2000.
304. Cooke, G., Radhi, A., Boden, N., Bushby, R. J., Lu, Z., Brown, S., and Heath, S. L. *Tetrahedron* 56, 3385–3390, 2000.
305. Zelcer, A., Donnio, B., Bourgogne, C., Cukiernik, F. D., and Guillon, D. *Chem. Mater.* 19, 1992–2006, 2007.
306. Barbera, J., Garces, A. C., Jayaraman, N., Omenat, A., Serrano, J. L., and Stoddort, J. F. *Adv. Mater.* 13, 175–180, 2001.
307. Boden, N., Bushby, R. J., Ferris, L., Hardy, C., and Sixl, F. *Liq. Cryst.* 1, 109–125, 1986.
308. Hughes, R., Smith, A., Bushby, R., Movaghar, B., and Boden, N. *Mol. Cryst. Liq. Cryst.* 332, 547–557, 1999.
309. Hughes, R. E., Hart, S. P., Smith, D. A., Movaghar, B., Bushby, R. J., and Boden, N. *J. Phys. Chem. B* 106, 6638–6645, 2002.
310. Motoyanagi, J., Fukushima, T., and Aida, T. *Chem. Commun.* 101–103, 2005.
311. Bruce, D. W. *Chem. Rec.* 4, 10–22, 2004.
312. Luckhurst, G. R. *Thin Solid Films* 393, 40–52, 2001.
313. Freiser, M. J. *Phys. Rev. Lett.* 24, 1041–1043, 1970.
314. Malthete, J., Tinh, N. H., and Levelut, A. M. *J. Chem. Soc. Chem. Commun.* 1548–1549, 1986.
315. Chandrasekhar, S., Ratna, B. R., Sadashiva, B. K., and Raja, N. V. *Mol. Cryst. Liq. Cryst.* 165, 123–130, 1988.
316. Luckhurst, G. R. *Nature* 430, 413–414, 2004.
317. Luckhurst, G. R. *Angew. Chem. Int. Ed.* 44, 2834–2836, 2005.
318. Acharya, B. R., Primak, A., and Kumar, S. *Phys. Rev. Lett.* 92, 145506, 2004.
319. Straley, J. P. *Phys. Rev. A* 10, 1881–1887, 1974.
320. Boccara, N., Mejdani, R., and de Seze, L. *J. Phys. Paris* 38, 149–151, 1977.
321. Alben, R. *J. Chem. Phys.* 59, 4299–4304, 1973.
322. Muhoray, P. P., de Bruyn, J. R., and Dunmur, D. A. *J. Chem. Phys.* 82, 5294–5295, 1985.
323. Pratibha, R. and Madhusudana, N. V. *Mol. Cryst. Liq. Cryst. Lett.* 1, 111–116, 1985.
324. Phillips, T. J., Minter, V., and Jones, J. C. *Liq. Cryst.* 21, 581–584, 1996.
325. Sharma, S. R., Muhoray, P. P., Bergersen, B., and Dunmur, D. A. *Phys. Rev. A* 32, 3752–3755, 1985.
326. Vanakaras, A. G., McGrother, S. C., Jackson, G., and Photinos, D. J. *Mol. Cryst. Liq. Cryst.* 323, 199–209, 1998.
327. Bates, M. A. and Luckhurst, G. R. *Phys. Chem. Chem. Phys.* 7, 2821–2829, 2005.
328. Hunt, J. J., Date, R. W., Timimi, B. A., Luckhurst, G. R., and Bruce, D. W. *J. Am. Chem. Soc.* 123, 10115–10116, 2001.
329. Kouwer, P. H. J., Pourzand, J., and Mehl, G. H. *Chem. Commun.* 66–67, 2004.
330. Bisoyi, H. K., Raghunathan, V. A., and Kumar, S. *Chem. Commun.* 7003–7005, 2009.
331. Silong, S., Rahman, L., Yunus, W. M. Z. W., Rahman, M. Z., Ahmad, M., and Haron, J. *Mat. Res. Soc. Proc.* 709, 105, 2002.
332. Shimizu, Y., Kurobe, A., Monobe, H., Terasawa, N., Kiyohara, K., and Uchida, K. *Chem. Commun.* 1676–1677, 2003.
333. Jeong, K. U., Jing, A. J., Mansdorf, B., Graham, M. J., Harris, F. W., and Cheng, S. Z. D. *J. Phys. Chem. B* 111, 767–777, 2007.
334. Jeong, K. U., Jing, A. J., Mansdorf, B., Graham, M. J., Yang, D. K., Harris, F. W., and Cheng, S. Z. D. *Chem. Mater.* 19, 2921–2923, 2007.
335. Imrie, C. T., Lu, Z., Picken, S. J., and Yildirim, Z. *Chem. Commun.* 1245–1247, 2007.
336. Destrade, C., Tinh, N. H., Malthete, J., and Jacques, J. *J. Phys. Lett. A* 79, 189–192, 1980.
337. Malthete, J., Jacques, J., Tinh, N. H., and Destrade, C. *Nature* 298, 46–48, 1982.
338. Tabushi, I., Yamamura, K., and Okada, Y. *Tetrahedron Lett.* 28, 2269–2272, 1987.
339. Lillya, C. P. and Collard, D. M. *Mol. Cryst. Liq. Cryst.* 182B, 201–207, 1990.
340. Bock, H. and Helfrich, W. *Liq. Cryst.* 12, 697–703, 1992.
341. Kurihara, S. and Nonaka, T. *Mol. Cryst. Liq. Cryst.* 238, 39–45, 1994.
342. Heppke, G., Kruerke, D., Muller, M., and Bock, H. *Ferroelectrics* 179, 203–209, 1996.

343. Yelamaggad, C. V., Prasad, V., Manickam, M., and Kumar, S. *Mol. Cryst. Liq. Cryst.* 325, 33–41, 1998.
344. Zniber, R., Achour, R., Cherkaoui, M. Z., Donnio, B., Gehringer, L., and Guillon, D. *J. Mater. Chem.* 12, 2208–2213, 2002.
345. Tinh, N. H., Destrade, C., and Gasparoux. H. *Phys. Lett. A* 72, 251–254, 1979.
346. Tinh, N. H., Gasparoux, H., and Destrade, C. *Mol. Cryst. Liq. Cryst.* 68, 101–111, 1981.
347. Malthete, J., Destrade, C., Tinh, N. H., and Jacques, J. *Mol. Cryst. Liq. Cryst.* 64, 233–238, 1981.
348. Destrade, C., Foucher, P., Gasparoux, H., Tinh, N. H., Levelut, A. M., and Malthete, J. *Mol. Cryst. Liq. Cryst.* 106, 121–146, 1984.
349. Beattie, D. R., Hindmarsh, P., Goodby, J. W., Haslam, S. D., and Richardson, R. M. *J. Mater. Chem.* 2, 1261–1266, 1992.
350. Hindmarsh, P., Hird, M., Styring, P., and Goodby, J. W. *J. Mater. Chem.* 3, 1117–1128, 1993.
351. Phillips, T. J., Jones, J. C., and McDonnell, D. G. *Liq. Cryst.* 15, 203–215, 1993.
352. Hindmarsh, P., Watson, M. J., Hird, M., and Goodby, J. W. *J. Mater. Chem.* 5, 2111–2123, 1995.
353. Kruerke, D., Rudquist, P., Lagerwall, S. T., Sawade, H., and Heppke, G. *Ferroelectrics* 243, 207–220, 2000.
354. Heppke, G., Kruerke, D., Lohning, C., Lotzsch, D., Moro, D., Muller, M., and Sawade, H. *J. Mater. Chem.* 10, 2657–2661, 2000.
355. Sonpatki, M. and Chien, L. C. *Mol. Cryst. Liq. Cryst.* 367, 545–553, 2001.
356. Sawade, H., Olenik, I. D., Kruerke, D., and Heppke, G. *Mol. Cryst. Liq. Cryst.* 367, 529–536, 2001.
357. Sergan, T., Sonpatki, M., Kelly, J., and Chien, L. C. *Mol. Cryst. Liq. Cryst.* 359, 259–267, 2001.
358. Tang, B. Y., Ge, J. J., Zhang, A., Calhoun, B., Chu, P., Wang, H., Shen, Z,, Hrris, F. W., and Chang, S. Z. D. *Chem. Mater.* 13, 78–86, 2001.
359. Kohmoto, S., Mori, E., and Kishikawa, K. *J. Am. Chem. Soc.* 129, 13364–13365, 2007.
360. Sasada, Y., Monabe, H., Ueda, Y., and Shimizu, Y. *Chem. Lett.* 36, 584–585, 2007.
361. Sasada, Y., Monobe, H., Ueda, Y., and Shimizu, Y. *Chem. Commun.* 1452–1454, 2008.
362. Wu, L. H., Janarthanan, N., and Hsu, C. S. *Liq. Cryst.* 28, 17–24, 2001.
363. Tinh, N. H., Bernaud, M. C., Sigaud, G., and Destrade, C. *Mol. Cryst. Liq. Cryst.* 65, 307–316, 1981.
364. Kreuder, W. and Ringsdorf, H. *Makromol. Chem. Rapid Commun.* 4, 807–815, 1983.
365. Werth, M., Vallerien, S. U., and Spiess, H. W. *Liq. Cryst.* 10, 759–770, 1991.
366. Wenz, G. *Makromol. Chem. Rapid Commun.* 6, 577, 1985; Ringsdorf, H., Schlarb, B., Venzmer, J. *Angew. Chem. Int. Ed.* 27, 113–158, 1988.
367. Boden, N., Bushby, R. J., Cammidge, A. N., and Headdock, G. *Synthesis* 31–32, 1995.
368. Boden, N., Bushby, R. J., and Cammidge, A. N. *J. Am. Chem. Soc.* 117, 924–927, 1995.
369. Boden, N., Bushby, R. J., and Lu, Z. B. *Liq. Cryst.* 25, 47–58, 1998.
370. Boden, N., Bushby, R. J., Lu, Z. B., and Cammidge, A. N. *Liq. Cryst.* 26, 495–499, 1999.
371. Goodby, J. W., Hird, M., Toyne, K. J., and Watson, T. *J. Chem Soc. Chem. Commun.* 1701–1702, 1994.
372. Cross, S. J., Goodby, J. W., Hall, A. W., Hird, M., Kelly, S. M., Toyne, K. J., and Wu, C. *Liq. Cryst.* 25, 1–11, 1998.
373. Closs, F., Hausseling, L., Henderson, P., Ringsdorf, H., and Schuhmacher, P. *J. Chem. Soc. Perkin Trans. 1*, 829–837, 1995.
374. Henderson, P., Ringsdorf, H., and Schuhmacher, P. *Liq. Cryst.* 18, 191–195, 1995.
375. Stewart, D., McHattie, G. S., and Imrie, C. T. *J. Mater. Chem.* 8, 47–51, 1998.
376. Borner, R. C. and Jackson, R. F. W. *J. Chem. Soc. Chem. Commun.* 845–846, 1994.
377. Perez, D. and Guitian, E. *Chem. Soc. Rev.* 33, 274–283, 2004.
378. Kumar, S. and Naidu, J. J. *Liq. Cryst.* 29, 899–906, 2002.
379. Rose, B. and Meier, H. *Z. Naturforsch.* 53B, 1031–1034, 1998.
380. Meier, H. and Rose, B. *J. Prakt. Chem.* 340, 536–543, 1998.
381. Brenna, E., Fuganti, C., and Serra, S. *J. Chem. Soc. Perkin Trans. 1*, 901–904, 1998.
382. Cammidge, A. N. and Gopee, H. *J. Mater. Chem.* 11, 2773–2783, 2001.
383. Cammidge, A. N. and Gopee, H. *Mol. Cryst. Liq. Cryst.* 397, 117–128, 2003.
384. Cammidge, A. N. *Phil. Trans. Soc. A* 364, 2697–2708, 2006.
385. Boden, N., Bushby, R. J., Cammidge, A. N., El-Mansoury, A., Martin, P. S., and Lu, Z. *J. Mater. Chem.* 9, 1391–1402, 1999.
386. Rose, A., Lugmair, C. G., and Swager, T. M. *J. Am. Chem. Soc.* 123, 11298–11299, 2001.
387. Bushby, R. J. and Lu, Z. *Synthesis* 763–767, 2001.
388. Kumar, S. and Lakshmi, B. *Tetrahedron Lett.* 46, 2603–2605, 2005.
389. Kumar, S. and Manickam, M. *Synthesis* 1119–1122, 1998.

390. Pal, S. K., Bisoyi, H. K., and Kumar, S. *Tetrahedron* 63, 6874–6878, 2007.

391. Moller, M., Tsukruk, V., Wendorff, J. H., Bengs, H., and Ringsdorf, H. *Liq. Cryst.* 12, 17–36, 1992.

392. Moller, M., Wendorff, J. H., Werth, M., and Spiess, H. W. *J. Non-Cryst. Solids* 170, 295–299, 1994.

393. Henderson, P., Kumar, S., Rego, J. A., Ringsdorf, H., and Schuhmacher. P. *J. Chem. Soc. Chem. Commun.* 1059–1060, 1995.

394. Boden, N., Bushby, R. J., Cammidge, A. N., and Martin, P. S. *J. Mater. Chem.* 5, 1857–1860, 1995.

395. Maliszewskyj, N. C., Heiney, P. A., Josefowicz, J. Y., Plesnivy, T., Ringsdorf, H., and Schuhmacher, P. *Langmuir* 11, 1666–1674, 1995.

396. Kumar, S., Schuhmacher, P., Henderson, P., Rego, J. A., and Ringsdorf, H. *Mol. Cryst. Liq. Cryst.* 288, 211–222, 1996.

397. Rego, J. A., Kumar, S., Dmochowski, I. J., and Ringsdorf, H. *Chem. Commun.* 1031–1032, 1996.

398. Rego, J. A., Kumar, S., and Ringsdorf, H. *Chem. Mater.* 8, 1402–1409, 1996.

399. Glusen, B., Heitz, W., Kettner, A., and Wendorff, J. H. *Liq. Cryst.* 20, 627–633, 1996.

400. Tsukruk, V. V., Bengs, H., and Ringsdorf, H. *Langmuir* 12, 754–757, 1996.

401. Henderson, P., Beyer, D., Jonas, U., Karthaus, O., Ringsdorf, H., Heiney, P. A., Maliszewskyj, N. C., Ghosh, S. S., Mindyuk, O. Y., and Josefowicz, J. Y. *J. Am. Chem. Soc.* 119, 4740–4748, 1997.

402. Gidalevitz, D., Mindyuk, O. Y., Heiney, P. A., Ocko, B. M., Henderson, P., Ringsdorf, H., Boden, N., Bushby, R. J., Martin, P. S., Strzalka, J., McCauley Jr., J. P., and Smith III, A. B. *J. Phys. Chem. B* 101, 10870–10875, 1997.

403. Allinson, H., Boden, N., Bushby, R. J., Evans, S. D., and Martin, P. S. *Mol. Cryst. Liq. Cryst.* 303, 273–278, 1997.

404. Glusen, B., Kettner, A., and Wendoff, J. H. *Mol. Cryst. Liq. Cryst.* 303, 115–120, 1997.

405. Wright, P. L., Gillies, I., and Kilburn, J. D. *Synthesis* 1007–1009, 1997.

406. Schulte, J. L., Laschat, S., Vill, V., Nishikawa, E., Finkelmann, H., and Nimtz, M. *Eur. J. Org. Chem.* 2499–2506, 1998.

407. Kumar, S. and Manickam, M. *Chem. Commun.* 1427–1428, 1998.

408. Glusen, B., Kettner, A., Kopitzke, J., and Wendorff, J. H. *Non-Cryst. Solids* 241, 113–120, 1998.

409. Boden, N., Bushby, R. J., and Martin, P. S. *Langmuir* 15, 3790–3797, 1999.

410. Kumar, S. and Manickam, M. *Liq. Cryst.* 26, 1097–1099, 1999.

411. Manickam, M. and Kumar, S. *Mol. Cryst. Liq. Cryst.* 326, 165–176, 1999.

412. Kumar, S. and Manickam, M. *Liq. Cryst.* 26, 939–941, 1999.

413. Kettner, A. and Wendorff, J. H. *Liq. Cryst.* 26, 483–487, 1999.

414. Manickam, M., Kumar, S., Preece, J. A., and Spencer, N. *Liq. Cryst.* 27, 703–706, 2000.

415. Mahlstedt, S., Janietz, D., Stracke, A., and Wendroff, J. H. *Chem. Commun.* 15–16, 2000.

416. Kumar, S. and Manickam, M. *Mol. Cryst. Liq. Cryst.* 338, 175–179, 2000.

417. Kumar, S., Manickam, M., Varshney, S. K., Rao, D. S. S., and Prasad, S. K. *J. Mater. Chem.* 10, 2483–2489, 2000.

418. Kopitzke, J., Wendorff, J. H., and Glusen, B. *Liq. Cryst.* 27, 643–648, 2000.

419. Schultz, A., Laschat, S., Abbott, A. P., Langner, M., and Reeve, T. B. *J. Chem. Soc. Perkin Trans. 1,* 3356–3361, 2000.

420. Allen, M. T., Harris, K. D. M., Kariuki, B. M., Kumari, N., Preece, J. A., Diele, S., Lose, D., Hegmann, T., and Tschierske, C. *Liq. Cryst.* 27, 689–692, 2000.

421. Allen, M. T., Diele, S., Harris, K. D. M., Hegmann, T., Kariuki, B. M., Lose, D., Peece, J. A., and Tschierske, C. *J. Mater. Chem.* 11, 302–311, 2001.

422. Schönherr, H., Kremer, F. J. B., Kumar, S., Rego, J. A., Wolf, H., Ringsdorf, H., Jaschke, M., Butt, H. J., and Bamberg, E. *J. Am. Chem. Soc.* 118, 13051–13057, 1996.

423. Kumar, S. and Sanjay, K. V. *Liq. Cryst.* 28, 161–163, 2001.

424. Bayer, A., Kopitzke, J., Noll, F., Seifert, A., and Wendorff, J. H. *Macromolecules* 34, 3600–3606, 2001.

425. Boden, N., Bushby, R. J., Lu, Z., and Lozman, O. R. *Liq. Cryst.* 28, 657–661, 2001.

426. Zimmermann, S., Wendorff, J. H., and Weder, C. *Chem. Mater.* 14, 2218–2223, 2002.

427. Schonherr, H., Manickam, M., and Kumar, S. *Langmuir* 18, 7082–7085, 2002.

428. Abeysekera, R., Bushby, R. J., Caillet, C., Hamley, I. W., Lozman, O. R., Lu, Z., and Robards, A. W. *Macromolecules* 36, 1526–1533, 2003.

429. Manickam, M., Cooke, G., Kumar, S., Ashton, P. R., Preece, J. A., and Spencer, N. *Mol. Cryst. Liq. Cryst.* 397, 99–116, 2003.

430. Setoguchi, Y., Monobe, H., Wan, W., Terasawa, N., Kiyohara, K., Nakamura, N., and Shimizu, Y. *Mol. Cryst. Liq. Cryst.* 412, 9–18, 2004.

431. Bushby, R. J., Liu, Q., Lozman, O. R., Lu, Z., and McLaren, S. R. *Mol. Cryst. Liq. Cryst.* 411, 293–304, 2004.
432. McKenna, M. D., Barbera, J., Marcos, M., and Serrano, J. L. *J. Am. Chem. Soc.* 127, 619–625, 2005.
433. Yildirim, Z., Wubbenhorst, M., Mendes, E., Picken, S. J., Paraschiv, I., Marcelis, A. T. M., Zuilhof, H., and Sudholter, E. J. R. *J. Non-Cryst. Solids* 351, 2622–2628, 2005.
434. Yildirim, Z., Mendes, E., Picken, S. J., Paraschiv, I., Zuilhof, H., de Vos, W. M., and Sudholter, E. J. R. *Mol. Cryst. Liq. Cryst.* 439, 237–243, 2005.
435. Kumar, S. and Pal, S. K. *Tetrahedron Lett.* 46, 2607–2610, 2005.
436. Kumar, S. and Pal, S. K. *Tetrahedron Lett.* 46, 4127–4130, 2005.
437. Zhao, K. Q., Wang, B. Q., Hu, P., Li, Q., and Zhang, L. F. *Chin. J. Chem.* 23, 767–774, 2005.
438. Zhao, K. Q., Gao, C. Y., Hu, P., Wang, B. Q., and Li, Q. *Acta Chim. Sin.* 64, 1051–1062, 2006.
439. Paraschiv, I., Giesbers, M., Van Lagen, B., Grozema, F. C., Abellon, R. D., Siebbeles, L. D. A., Marcelis, A. T. M., Zuilhof, H., and Sudholter, E. J. R. *Chem. Mater.* 18, 968–974, 2006.
440. Yamada, M., Shen, Z., and Miyake, M. *Chem. Commun.* 2569–2571, 2006.
441. Cui, L. and Zhu, L. *Liq. Cryst.* 33, 811–818, 2006.
442. Ma, X. J., Yang, Y. L., Deng, K., Zeng, Q. D., Wang, C., Zhao, K. Q., Hu, P., and Wang, B. Q. *ChemPhysChem.* 8, 2615–2620, 2007.
443. Ichihara, M., Suzuki, H., Mohr, B., and Ohta, K. *Liq. Cryst.* 34, 401–410, 2007.
444. Bisoyi, H. K. and Kumar, S. *J. Phys. Org. Chem.* 21, 47–52, 2008.
445. Shen, Z., Yamada, M., and Miyake, M. *J. Am. Chem. Soc.* 129, 14271–14280, 2007.
446. Yu, W. H., Zhao, K. Q., Wang, B. Q., and Hu, P. *Acta Chim. Sin.* 65, 1140–1148, 2007.
447. Chen, H. M., Zhao, K. Q., Hu, P., and Wang, B. Q. *Acta Chim. Sin.* 65, 1368–1376, 2007.
448. Wang, B. Q., Zhao, K. Q., Hu, P., Yu, W. H., Gao, C. Y., and Shimizu, Y. *Mol. Cryst. Liq. Cryst.* 479, 135–150, 2007.
449. Wang, B. Q., Lei, B. L., Yang, G. F., Zhao, K. Q., Yu, W. H., Hu, P., and Ding, F. J. *Chin. J. Org. Chem.* 27, 1552–1557, 2007.
450. Zhang, C., He, Z., Wang, J., Wang, Y., and Ye, S. *J. Mol. Liq.* 138, 93–99, 2008.
451. Song, Z. Q., Zhao, K. Q., Hu, P., and Wang, B. Q. *Acta Chim. Sin.* 66, 1344–1352, 2008.
452. Jian, Z. B., Zhao, K. Q., Hu, P., and Wang, B. Q. *Acta Chim. Sin.* 66, 1353–1360, 2008.
453. Yelamaggad, C. V., Achalkumar, A. S., Rao, D. S. S., Nobusawa, M., Akutsu, H., Yamada, J. I., and Nakatsuji, S. *J. Mater. Chem.* 18, 3433–3437, 2008.
454. Paraschiv, I., de Lange, K., Giesbers, M., Van Lagen, B., Grozema, F. C., Abellon, R. D., Siebbeles, L. D. A., Sudholter, E. J. R., Zuilhof, H., and Marcelis, A. T. M. *J. Mater. Chem.* 18, 5475–5481, 2008.
455. Kumar, S. and Lakshminarayanan, V. *Chem Commun.* 1600–1601, 2004.
456. Kumar, S., Pal, S. K., and Lakshminarayanan, V. *Mol. Cryst. Liq. Cryst.* 434, 251–258, 2005.
457. Kumar, S., Pal, S. K., Kumar, P. S., and Lakshminarayanan, V. *Soft Matter* 2, 896–900, 2007.
458. Kumar, S. *Syn. React. Inorg. Metal-Org. Nano-Metal Chem.* 37, 327–331, 2007.
459. Holt, L. A., Bushby, R. J., Evans, S. D., Burgess, A., and Seeley, G. *J. Appl. Phys.* 103, 063712, 2008.
460. Kumar, S. and Bisoyi, H. K. *Angew. Chem. Int. Ed.* 46, 1501–1503, 2007.
461. Breslow, R., Jaun, B., Kluttz, R. Q., and Xia, C. Z. *Tetrahedron* 38, 863–867, 1982.
462. Boden, N., Bushby, R. J., Cammidge, A. N., and Headdock, G. *J. Mater. Chem.* 5, 2275–2281, 1995.
463. Boden, N., Bushby, R. J., and Cammidge, A. N. *Liq. Cryst.* 18, 673–676, 1995.
464. Boden, N., Bushby, R. J., and Cammidge, A. N. *Mol. Cryst. Liq. Cryst.* 260, 307–313, 1995.
465. Kumar, S. and Manickam, M. *Mol. Cryst. Liq. Cryst.* 309, 291–295, 1998.
466. Kumar, S., Manickam, M., Balagurusamy, V. S. K., and Schonherr, H. *Liq. Cryst.* 26, 1455–1466, 1999.
467. Akopova, O. B., Bronnikova, A. A., Kruvchinskii, A., Kotovich, L. N., Shabyshev, L. S., and Valkova, L. A. *J. Struct. Chem.* 39, 376–382, 1998.
468. Zemtsova, O., Akopova, O., Usoltseva, N., and Erdelen, C. *Mol. Cryst. Liq. Cryst.* 364, 625–634, 2001.
469. Zemtsova, O. V., Syromyatnikova, O. K., Kotovich, L. N., and Akopova, O. B., *J. Struct. Chem.* 42, 38–42, 2001.
470. Akopova, O. B. *J. Struct. Chem.* 47, 120–129, 2006.
471. Akopova, O. B., Kurbatova, E. V., and Gruzdev, M. S. *Rus. J. Gen. Chem.* 78, 1902–1909, 2008.
472. Boden, N., Bushby, R. J., Cammidge, A. N., and Headdock, G. *Tetrahedron Lett.* 36, 8685–8686, 1995.
473. Boden, N., Bushby, R. J., Cammidge, A. N., Duckworth, S., and Headdock, G. *J. Mater. Chem.* 7, 601–605, 1997.
474. Praefcke, K., Eckert, A., and Blunk, D. *Liq. Cryst.* 22, 113–119, 1997.

475. Muccioli, L., Berardi, R., Orlandi, S., Ricci, M., and Zannoni, C. *Theor. Chem. Acc.* 117, 1085–1092, 2007.
476. Schulte, J. L., Laschat, S., Ladbeck, R. S., Von Arnim, V., Schneider, A., and Finkelmann, H. *J. Organomet. Chem.* 552, 171–176, 1998.
477. Paraschiv, I., Delforterie, P., Giesbers, M., Posthumus, M. A., Marcelis, A. T. M., Zuilhof, H., and Sudholter, E. J. R. *Liq. Cryst.* 32, 977–983, 2005.
478. Boden, N., Bushby, R. J., Lu, Z. B., and Eichhorn, H. *Mol. Cryst. Liq. Cryst.* 332, 281–291, 1999.
479. Adam, D., Schuhmacher, P., Simmerer, J., Häussling, L., Paulus, W., Siemensmeyer, K., Etzbach, K. H., Ringsdorf, H., and Haarer, D. *Adv. Mater.* 7, 276–280, 1995.
480. Bleyl, I., Erdelen, C., Etzbach, K. H., Paulus, W., Schmidt, H. W., Siemensmeyer, K., and Haarer, D. *Mol. Cryst. Liq. Cryst.* 299, 149–155, 1997.
481. Kreuder, W., Ringsdorf, H., Schönherr, O. H., and Wendorff, J. H. *Angew. Chem. Int. Ed.* 26, 1249–1252, 1987.
482. Kimura, M., Moriyama, M., Kishimoto, K., Yoshio, M., and Kato, T. *Liq. Cryst.* 34, 107–112, 2007.
483. Kumar, S. and Varshney, S. K. *Org. Lett.* 4, 157–159, 2002.
484. Paraschiv, I., Tomkinson, A., Giesbers, M., Sudholter, E. J. R., Zuilhof, H., and Marcelis, A. T. M. *Liq. Cryst.* 34, 1029–1038, 2007.
485. Kranig, W., Hüser, B., Spiess, H. W., Kreuder, W., Ringsdorf, H., and Zimmermann, H. *Adv. Mater.* 2, 36–40, 1990.
486. Boden, N., Bushby, R. J., Liu, Q., and Lozman, O. R. *J. Mater. Chem.* 11, 1612–1617, 2001.
487. Wendorff, J. H., Christ, T., Glüsen, B., Greiner, A., Kettner, A., Sander, R., Stümpflen, V., and Tsukruk, V. V. *Adv. Mater.* 9, 48–52, 1997.
488. Wan, W., Monobe, H., Sugino, T., Tanaka, Y., and Shimizu, Y. *Mol. Cryst. Liq. Cryst.* 364, 597–603, 2001.
489. Setoguchi, Y., Monobe, H., Wan, W., Terasawa, N., Kiyohara, K., Nakamura, N., and Shimizu, Y. *Thin Solid Films* 438–439, 407–413, 2003.
490. Mao, H., He, Z., Wang, J., Zhang, C., Xie, P., and Zhang, R. *J. Lumin.* 122–123, 942–945, 2007.
491. Disch, S., Finkelmann, H., Ringsdorf, H., and Schuhmacher, P. *Macromolecules* 28, 2424–2428, 1995.
492. Boden, N., Bushby, R. J., Cooke, G., Lozman, O. R., and Lu, Z. *J. Am. Chem. Soc.* 123, 7915–7916, 2001.
493. Kohne, V. B., Poules, W., and Praefcke, K. *Chem. Z.* 108, 113, 1984.
494. Gramsbergen, E. F., Hoving, H. J., de Jeu, W. H., Praefcke, K., and Kohne, B. *Liq. Cryst.* 1, 397–400, 1986.
495. Marguet, S., Markovitsi, D., Millie, P., Sigal, H., and Kumar, S. *J. Phys. Chem. B* 102, 4697–4710, 1998.
496. Khone, R. B., Praefcke, K. Derz, T., Frischmuth, W., and Gansau, C. *Chem. Z.* 108, 408, 1984.
497. Fontes, E., Heiney, P. A., and de Jeu, W. H. *Phys. Rev. Lett.* 61, 1202–1205, 1988.
498. Lee, W. K., Heiney, P. A., Mccauley Jr., J. P., and Smith III, A. B. *Mol. Cryst. Liq. Cryst.* 198, 273–284, 1991.
499. Idziak, S. H. J., Heiney, P. A., Mccauley Jr., J. P., Carroll, P., and Smith III, A. B. *Mol. Cryst. Liq. Cryst.* 237, 271–275, 1993.
500. Kumar, S. Unpublished result.
501. Arikainen, E. O., Boden, N., Bushby, R. J., Lozman, O. R., Vinter, J. G., and Wood, A. *Angew. Chem. Int. Ed.* 39, 2333–2336, 2000.
502. Yatabe, T., Harbison, M. A., Brand, J. D., Wagner, M., Mullen, K., Samori, P., and Rabe, J. P. *J. Mater. Chem.* 10, 1519–1525, 2000.
503. Boden, N., Bushby, R. J., Headdock, G., Lozman, O. R., and Wood, A. *Liq. Cryst.* 28, 139–144, 2001.
504. Bushby, R. J., Evans, S. D., Lozman, O. R., McNeill, A., and Movaghar, B. *J. Mater. Chem.* 11, 1982–1984, 2001.
505. Bushby, R. J., Fisher, J., Lozman, O. R., Lange, S., Lydon, J. E., and Mclaren, S. R. *Liq. Cryst.* 33, 653–664, 2006.
506. Vijayaraghavan, D. and Kumar, S. *Mol. Cryst. Liq. Cryst.* 508, 101–114, 2009.
507. Jeong, K. U., Jing, A. J., Mansdorf, B., Graham, M. J., Tu, Y. F., Harris, F. W., and Cheng, S. Z. D. *Chin. J. Polym. Sci.* 25, 57–71, 2007.
508. Brandl, B. and Wendorff, J. H. *Liq. Cryst.* 32, 553–563, 2005.
509. Tang, B. Y., Jing, A. J., Li, C. Y., Shen, Z., Wang, H., Harris, F. W., and Cheng, S. Z. D. *Cryst. Growth Des.* 3, 375–382, 2003.
510. Prasad, S. K., Rao, D. S. S., Chandrasekhar, S., and Kumar, S. *Mol. Cryst. Liq. Cryst.* 396, 121–139, 2003.
511. Andresen, T. L., Krebs, F. C., Thorup, N., and Bechgaard, K. *Chem. Mater.* 12, 2428–2433, 2000.

512. Li, G., Luo, J., Wang, T., Zhou, E., Huang, J., Bengs, H., and Ringsdorf, H. *Mol. Cryst. Liq. Cryst.* 309, 73–91, 1998.
513. Wang, T., Yan, D., Luo, J., Zhou, E., Karthaus, O., and Ringsdorf, H. *Liq. Cryst.* 23, 869–878, 1997.
514. Voigt-Martin, I. G., Garbella, R. W., and Schumacher, M. *Liq. Cryst.* 17, 775–801, 1994.
515. Voigt-Martin, I. G., Schumacher, M., Honig, M., Simon, P., and Garbella, R. W. *Mol. Cryst. Liq. Cryst.* 254, 299–320, 1994.
516. Oswald, P. *J. Phys. France* 49, 1083–1089, 1988.
517. Sheldrick, B. *Mol. Cryst. Liq. Cryst.* 6, 9–13, 1988.
518. Boden, N., Bushby, R. J., Hardy, C., and Sixl, F. *Chem. Phys. Lett.* 123, 359–364, 1986.
519. Safinya, C. R., Clark, N. A., Liang, K. S., Varady, W. A., and Chiang, L. Y. *Mol. Cryst. Liq. Cryst.* 123, 205–216, 1985.
520. Safinya, C. R., Liang, K. S., Varady, W. A., Clark, N. A., and Andersson, G. *Phys. Rev. Lett.* 53, 1172–1175, 1984.
521. Takabatake, M. and Iwayanagi, S. *Jpn. J. Appl. Phys.* 21, L685–L686, 1982.
522. Levelut, A. M., Hardouin, F., Gasparoux, H., Destrade, C., and Tinh, N. H. *J. Phys.* 42, 147–152, 1981.
523. Beguin, A., Billard, J., Dubois, J. C., Tinh, N. H., and Zann, A. *J. de Phys.* 40, C3–15, 1979.
524. Davidson, P., Clerc, M., Ghosh, S. S., Maliszewskyj, N. C., Heiney, P. A., Hynes Jr., J., and Smith, III, A. B. *J. Phys. II France* 5, 249–262, 1995.
525. Heiney, P. A., Fontes, E., De Jeu, W. H., Riera, A., Carroll, P., and Smith III, A. B. *J. Phys. France* 50, 461–483, 1989.
526. Cotrait, M., Marsau, P., Destrade, C., and Malthete, J. *J. Phys. Lett.* 19, L519–L522, 1979.
527. Dvinskikh, S. V., Thaning, J., Stevensson, B., Jansson, K., Kumar, S., Zimmermann, H., and Maliniak, A. *Phys. Rev. E* 74, 021703, 2006.
528. Zhang, J. and Dong, R. Y. *Phys. Rev. E* 73, 061704, 2006.
529. Kearley, G. J., Mulder, F. M., Picken, S. J., Kouwer, P. H. J., and Stride, J. *Chem. Phys.* 292, 185–190, 2003.
530. Mulder, F. M., Stride, J., Picken, S. J., Kouwer, P. H. J., de Haas, M. P., Siebbeles, L. D. A., and Kearley, G. J. *J. Am. Chem. Soc.* 125, 3860–3866, 2003.
531. Ribeiro, A. C., Sebastiao, P. J., and Cruz, C. *Pramana* 61, 205–218, 2003.
532. Dvinskikh, S. V., Furo, I., Zimmermann, H., and Maliniak, A. *Phys. Rev. E* 65, 050702(R), 2002.
533. Buchanan, G. W., Rastegar, M. F., and Yap, G. P. A. *Can. J. Chem.* 79, 195–200, 2001.
534. Dong, R. Y., Boden, N., Bushby, R. J., and Martin, P. S. *Mol. Phys.* 97, 1165–1171, 1999.
535. Cruz, C., Ribeiro, A. C., and Biforked. *Mol. Cryst. Liq. Cryst.* 331, 75–80, 1999.
536. Shen, X., Dong, R. Y., Boden, N., Bushby, R. J., Martin, P. S., and Wood, A. *J. Chem. Phys.* 108, 4324–4332, 1998.
537. Zamir, S., Spielberg, N., Zimmermann, H., Poupko, R., and Luz, Z. *Liq. Cryst.* 18, 781–786, 1995.
538. Hagemeyer, A., Tarroni, R., and Zannoni, C. *J. Chem. Soc. Faraday Trans.* 90, 3433–3442, 1994.
539. Moller, M., Wendorff, J. H., Werth, M., Spiess, H. W., Bengs, H., Karthaus, O., and Ringsdorf, H. *Liq. Cryst.* 17, 381–395, 1994.
540. Zamir, S., Poupko, R., Luz, Z., Huser, B., Boeffel, C., and Zimmermann, H. *J. Am. Chem. Soc.* 116, 1973–1980, 1994.
541. Photinos, D. J., Luz, Z., Zimmermann, H., and Samulski, E. T. *J. Am. Chem. Soc.* 115, 10895–10900, 1993.
542. Leisen, J., Werth, M., Boeffel, C., and Spiess, H. W. *J. Chem. Phys.* 97, 3749–3759, 1992.
543. Hirschinger, J., Kranig, W., and Spiess, H. W. *Colloid Polym. Sci.* 269, 993–1002, 1991.
544. Kranig, W., Boeffel, C., Spiess, H. W., Karthaus, O., Ringsdorf, H., and Wustefeld, R. *Liq. Cryst.* 8, 375–388, 1990.
545. Kranig, W., Boeffel, C., and Spiess, H. W. *Macromolecules* 23, 4061–4067, 1990.
546. Vallerien, S. U., Werth, M., Kremer, F., and Spiess, H. W. *Liq. Cryst.* 8, 889–893, 1990.
547. Goldfarb, D., Dong, R. Y., Luz, Z., and Zimmermann, H. *Mol. Phys.* 54, 1185–1202, 1985.
548. Goldfarb, D., Lifshitz, E., Zimmermann, H., and Luz, Z. *J. Chem. Phys.* 82, 5155–5163, 1985.
549. Dong, R. Y., Goldfarb, D., Moseley, M. E., Luz, Z., and Zimmermann, H. *J. Phys. Chem.* 88, 3148–3152, 1984.
550. Goldfarb, D., Poupko, R., Luz, Z., and Zimmermann, H. *J. Chem. Phys.* 79, 4035–4047, 1983.
551. Goldfarb, D., Luz, Z., and Zimmermann, H. *J. Chem. Phys.* 78, 7065–7072, 1983.
552. Goldfarb, D., Luz, Z., and Zimmermann, H. *J. Phys.* 43, 421–430, 1982.
553. Goldfarb, D., Luz, Z., and Zimmermann, H. *J. Phys.* 43, 1255–1258, 1982.
554. Rutar, V., Blinc, R., Vilfan, M., Zann, A., and Dubois, J. C. *J. Phys.* 43, 761–765, 1982.

555. Vilfan, M., Lahajnar, G., Rutar, V., Blinc, R., Topic, B., Zann, A., and Dubois, J. C. *J. Chem. Phys.* 75, 5250–5255, 1981.
556. Goldfarb, D., Luz, Z., and Zimmermann, H. *J. Phys.* 42, 1303–1311, 1981.
557. Dong, R. Y. *Thin Solid Films* 517, 1367–1379, 2008.
558. Ngai, K. L. *J. Non-Cryst. Solids* 197, 1–7, 1996.
559. Joghems, E. A., Biesheuvel, C. A., and Bulthuis, J. *Liq. Cryst.* 13, 427–443, 1993.
560. Richards, G. M. and Dong, R. Y. *Liq. Cryst.* 5, 1011–1018, 1989.
561. Monobe, H., Awazu, K., and Shimizu, Y. *Thin Solid Films* 516, 2677–2681, 2008.
562. Terasawa, N. and Monobe, H. *Liq. Cryst.* 34, 447–455, 2007.
563. Monobe, H., Hori, H., Shimizu, Y., and Awazu, K. *Mol. Cryst. Liq. Cryst.* 475, 13–22, 2007.
564. Monobe, H., Awazu, K., and Shimizu, Y. *Adv. Mater.* 18, 607–610, 2006.
565. Le, K. V., Amemiya, K., Takanishi, Y., Ishikawa, K., and Takezoe, H. *Jpn. J. Appl. Phys.* 45, 5149–5150, 2006.
566. Monobe, H., Hori, H., Heya, M., Awazu, K., and Shimizu, Y. *Thin Solid Films* 499, 259–262, 2006.
567. Furumi, S., Kidowaki, M., Ogawa, M., Nishiura, Y., and Ichimura, K. *J. Phys. Chem. B* 109, 9245–9254, 2005.
568. Shimizu, Y., Monobe, H., Heya, M., and Awazu, K. *Mol. Cryst. Liq. Cryst.* 443, 49–58, 2005.
569. Monobe, H., Shimizu, Y., Heya, M., and Awazu, K. *Mol. Cryst. Liq. Cryst.* 443, 211–217, 2005.
570. Monobe, H., Terasawa, N., Shimizu, Y., Kiyohara, K., Heya, M., and Awazu, K. *Mol. Cryst. Liq. Cryst.* 437, 81–88, 2005.
571. Shimizu, Y., Monobe, H., Heya, M., and Awazu, K. *Mol. Cryst. Liq. Cryst.* 441, 287–295, 2005.
572. Monobe, H., Terasawa, N., Kiyohara, K., Shimizu, Y., Azehara, H., Nakasa, A., and Fujihira, M. *Mol. Cryst. Liq. Cryst.* 412, 229–236, 2004.
573. Monobe, H., Kiyohara, K., Shimizu, Y., Heya, M., and Awazu, K. *Mol. Cryst. Liq. Cryst.* 410, 29–38, 2004.
574. Monobe, H., Kiyohara, K., Terasawa, N., Heya, M., Awazu, K., and Shimizu, Y. *Adv. Funct. Mater.* 13, 919–924, 2003.
575. Monobe, H., Kiyohara, K., Terawasa, N., Heya, M., Awazu, K., and Shimizu, Y. *Chem. Lett.* 32, 870–871, 2003.
576. Perova, T. S., Astrova, E. V., Tsvetkov, S. E., Tkachenko, A. G., Vij, J. K., and Kumar, S. *Phys. Solid State* 44, 1196–1202, 2002.
577. Sergan, T., Sonpatki, M., Kelly, J., and Chien, L. C. *Mol. Cryst. Liq. Cryst.* 359, 245–257, 2001.
578. Monobe, H., Azehara, H., Shimizu, Y., and Fujihira, M. *Chem. Lett.* 1268–1269, 2001.
579. Monobe, H., Mima, S., Sugino, T., Shimizu, Y., and Ukon, M. *Liq. Cryst.* 28, 1253–1258, 2001.
580. Monobe, H., Mima, S., Ukon, M., Sugino, T., and Shimizu, Y. *Mol. Cryst. Liq. Cryst.* 370, 241–244, 2001.
581. Monobe, H., Awazu, K., and Shimizu, Y. *Mol. Cryst. Liq. Cryst.* 364, 453–460, 2001.
582. Monobe, H., Awazu, K., and Shimizu, Y. *Adv. Mater.* 12, 1495–1499, 2000.
583. Averyanov, E. M., Gunyakov, V. A., Korets, A. Y., and Akopova, O. B. *Opt. Spectrosc.* 88, 891–897, 2000.
584. Ichimura, K., Furumi, S., Morino, S., Kidowaki, M., Nakagawa, M., Ogawa, M., and Nishiura, Y. *Adv. Mater.* 12, 950–953, 2000.
585. Perova, T., Tsvetkov, S., Vij, J., and Kumar, S. *Mol. Cryst. Liq. Cryst.* 351, 95–102, 2000.
586. Orgasinska, B., Kocot, A., Merkel, K., Wrzalik, R., Ziolo, J., Perova, T., and Vij, J. K. *J. Mol. Struct.* 511–512, 271–276, 1999.
587. Orgasinska, B., Perova, T. S., Merkel, K., Kocot, A., and Vij, J. K. *Mater. Sci. Eng. C* 8–9, 283–289, 1999.
588. Ikeda, S., Takanishi, Y., Ishikawa, K., and Takezoe, H. *Mol. Cryst. Liq. Cryst.* 329, 589–595, 1999.
589. Perova, T. S. and Vij, J. K. *Adv. Mater.* 7, 919–922, 1995.
590. Perova, T., Kocot, A., and Vij, J. K. *Mol. Cryst. Liq. Cryst.* 301, 111–121, 1997.
591. Burhanudin, Z. and Etchegoin, P. *Chem. Phys. Lett.* 336, 7–12, 2001.
592. Perova, T. S., Kocot, A., and Vij, J. K. *Supramol. Sci.* 4, 529–534, 1997.
593. Vij, J. K., Kocot, A., and Perova, T. S. *Mol. Cryst. Liq. Cryst.* 397, 231–244, 2003.
594. Monobe, H., Kiyohara, K., Heya, M., Awazu, K., and Shimizu, Y. *Mol. Cryst. Liq. Cryst.* 397, 59–65, 2003.
595. Wu, L. H., Lee, W. C., Hsu, C. S., and Wu, S. T. *Liq. Cryst.* 28, 317–320, 2001.
596. Wu, L. H., Lee, W. C., Hsu, C. S., and Wu, S. T. *Jpn. J. Appl. Phys.* 39, 5899–5903, 2000.
597. Tsvetkov, S. E., Perova, T. S., Vij, J. K., Simpson, D., Kumar, S., and Vladimirov, F. *Mol. Mater.* 11, 267–276, 1999.
598. Schaper, A. K., Yoshioka, T., Ogawa, T., and Tsuji, M. *J. Microsc.* 223, 88–95, 2006.
599. Palma, M., Pace, G., Roussel, O., Geerts, Y., and Samori, P. *Aus. J. Chem.* 59, 376–380, 2006.
600. Kubo, K., Takahashi, H., and Takechi, H. *J. Oleo Sci.* 55, 545–549, 2006.

601. Perronet, K. and Charra, F. *Surf. Sci.* 551, 213–218, 2004.
602. Katsonis, N., Marchenko, A., and Fichou, D. *Synth. Met.* 147, 73–77, 2004.
603. Yu, X., Fu, J., Han, Y., and Pan, C. *Macromol. Rapid Commun.* 24, 742–747, 2003.
604. Xu, S., Zeng, Q., Lu, J., Wang, C., Wan, L., and Bai, C. L. *Surf. Sci.* 538, L451–L459, 2003.
605. Acharya, S., Parichha, T. Kr., and Talapatra, G. B. *J. Lumin.* 96, 177–184, 2002.
606. Katsonis, N., Marchenko, A., and Fichou, D. *J. Am. Chem. Soc.* 125, 13682–13683, 2003.
607. Robinson, A. P. G., Palmer, R. E., Tada, T., Kanayama, T., Allen, M. T., Preece, J. A., and Harris, K. D. M. *J. Vac. Sci. Technol.* 18, 2730–2736, 2000.
608. Clements, J., Boden, N., Gibson, T. D., Chandler, R. C., Hulbert, J. N., and Keene, E. A. R. *Sens. Actuat. B* 47, 37–42, 1998.
609. Charra, F. and Cousty, J. *Phys. Rev. Lett.* 80, 1682–1685, 1998.
610. Vaes, A., Van der Auweraer, M., De Schryver, F. C., Laguitton, B., Jonas, A., Henderson, P., and Ringsdorf, H. *Langmuir* 14, 5250–5254, 1998.
611. Gidalevitz, D., Mindyuk, O. Y., Heiney, P. A., Ocko, B. M., Kurnaz, M. L., and Schwartz, D. K. *Langmuir* 14, 2910–2915, 1998.
612. Charra, F. and Cousty, J. *Opt. Mater.* 9, 386–389, 1998.
613. Gallivan, J. P. and Schuster, G. B. *J. Org. Chem.* 60, 2423–2429, 1995.
614. Maliszewskyj, N. C., Heiney, P. A., Josefowicz, J. Y., McCauley, J. P., and Smith III, A. B. *Science* 264, 77–79, 1994.
615. Tsukruk, V. V., Reneker, D. H., Bengs, H., and Ringsdorf, H. *Langmuir* 9, 2141–2144, 1993.
616. Josefowicz, J. Y., Maliszewskyj, N. C., Idziak, S. H. J., Heiney, P. A., McCauley Jr., J. P., and Smith III, A. B. *Science* 260, 323–326, 1993.
617. Tsukruk, V. V., Wendorff, J. H., Karthaus, O., and Ringsdorf, H. *Langmuir* 9, 614–618, 1993.
618. Vandevyver, M., Albouy, P. A., Mingotaud, C., Perez, J., Barraud, A., Karathaus, O., and Ringsdorf, H. *Langmuir* 9, 1561–1567, 1993.
619. Askadskaya, L., Boeffel, C., and Rabe, J. P. *Bunsenges. Phys. Chem.* 97, 517–521, 1993.
620. Karthaus, O., Ringsdorf, H., Tsukruk, V. V., and Wendorff, J. H. *Langmuir* 8, 2279–2283, 1992.
621. Voigt-Martin, I. G., Garbella, R. W., and Schumacher, M. *Macromolecules* 25, 961–971, 1992.
622. Auweraer, M. V. D., Catry, C., Chi, L. F., Karathaus, O., Knoll, W., Ringsdorf, H., Sawodny, M., and Urban, C. *Thin Solid Films* 210–211, 39–41, 1992.
623. Karthaus, O., Ringsdorf, H., and Urban, C. *Makromol. Chem. Macromol. Symp.* 46, 347–352, 1991.
624. Sheu, E. Y., Liang, K. S., and Chiang, L. Y. *J. Phys. France* 50, 1279–1295, 1989.
625. Albrecht, O., Cumming, W., Kreuder, W., Laschewsky, A., and Ringsdorf, H. *Colloid Polym. Sci.* 264, 659–667, 1986.
626. Gabriel, J. C., Larsen, N. B., Larsen, M., Harrit, N., Pedersen, J. S., Schaumburg, K., and Bechgaard, K. *Langmuir* 12, 1690–1692, 1996.
627. Schaper, A. K., Kurata, H., Yoshioka, T., and Tsuji, M. *Microsc. Microanal.* 13, 336–341, 2007.
628. Li, J., Fruchey, K., and Fayer, M. D. *J. Chem. Phys.* 125, 194901, 2006.
629. Averyanov, E. M. *Phys. Solid State* 46, 1554–1559, 2006.
630. Steinhart, M., Murano, S., Schaper, A. K., Ogawa, T., Tsuji, M., Gosele, U., Weder, C., and Wendorff, J. H. *Adv. Funct. Mater.* 15, 1656–1664, 2005.
631. Karthaus, O., Honma, Y., Taguchi, D., and Fujiwara, Y. *J. Surf. Sci. Nanotechnol.* 3, 156–158, 2005.
632. Sugita, A., Suzuki, K., and Tasaka, S. *Phys. Rev. B* 69, 212201, 2004.
633. Moriyama, M., Mizoshita, N., and Kato, T. *Polym. J.* 36, 661–664, 2004.
634. Gunyakov, V. A. and Shibli, S. M. *Mol. Cryst. Liq. Cryst.* 409, 409–420, 2004.
635. Gunyakov, V. A. and Shibli, S. M. *Mol. Cryst. Liq. Cryst.* 397, 273–283, 2003.
636. Gunyakov, V. A. and Shibli, S. M. *Liq. Cryst.* 30, 59–64, 2003.
637. Gunyakov, V. A., Shestakov, N. P., and Shibli, S. M. *Liq. Cryst.* 30, 871–875, 2003.
638. Olenik, I. D., Spindler, L., Copic, M., Sawade, H., Kruerke, D., and Heppke, G. *Phys. Rev. E* 65, 011705–011709, 2001.
639. Derbel, N., Rais, K., Tounsi, N., Othman, T., Gharbia, M., Gharbi, A., Tinh, N. H., and Malthete, J. *Polym. Int.* 50, 778–783, 2001.
640. Bayer, A., Hbner, J., Kopitzke, J., Oestreich, M., Rhle, W., and Wendorff, J. H. *J. Phys. Chem. B* 105, 4596–4602, 2001.
641. Nollmann, M. and Etchegoin, P. *Phys. Rev. E* 61, 5345–5348, 2000.
642. Schartel, B., Kettner, A., Kunze, R., Wendorff, J. H., and Hennecke, M. *Adv. Mater. Opt. Electron.* 9, 55–64, 1999.
643. Averyanov, E. M., Gunyakov, V. A., Korets, A. Y., and Akopova, O. B. *JETP Lett.* 70, 29–35, 1999.

644. Kevenhorster, B., Kopitzke, J., Seifert, A. M., Tsukruk, V., and Wendorff, J. H. *Adv. Mater.* 11, 246–250, 1999.
645. Nair, G. G., Prasad, S. K., Kumar, S., and Manickam, M. *Mol. Cryst. Liq. Cryst.* 319, 89–99, 1998.
646. Phillips, T. J. and Minter, V. *Liq. Cryst.* 20, 243–245, 1996.
647. Daoud, M., Louis, G., Quelin, X., Gharbia, M., Gharbi, A., and Peretti, P. *Liq. Cryst.* 19, 833–837, 1995.
648. Groothues, H., Kremer, F., Collard, D. M., and Lillya, C. P. *Liq. Cryst.* 18, 117–121, 1995.
649. Phillips, T. J. and Jones, J. C. *Liq. Cryst.* 16, 805–812, 1994.
650. Daoud, M., Gharbia, M., and Gharbi, A. *J. Phys. II France* 4, 989–1000, 1994.
651. Kruk, G., Kocot, A., Wrzalik, R., Vij, J. K., Karthaus, O., and Ringsdorf, H. *Liq. Cryst.* 14, 807–819, 1993.
652. Lafon, M. R. and Hemida, A. T. *Mol. Cryst. Liq. Cryst.* 178, 33–51, 1990.
653. Sorai, M., Asahina, S., Destrade, C., and Tinh, N. H. *Liq. Cryst.* 7, 163–180, 1990.
654. Vallerien, S. U., Kremer, F., Huser, B., and Spiess, H. W. *Colloid Polym. Sci.* 267, 583–586, 1989.
655. Warmerdam, T. W., Frenkel, D., and Zijlstra, R. J. J. *Liq. Cryst.* 3, 1105–1114, 1988.
656. Braitbart, O., Sasson, R., and Weinreb, A. *Mol. Cryst. Liq. Cryst.* 159, 233–242, 1988.
657. Warmerdam, T. W., Frenkel, D., and Zijlstra, R. J. J. *Liq. Cryst.* 3, 369–380, 1988.
658. Sasson, R., Braitbart, O., and Weinreb, A. *J. Lumin.* 40/41, 207–208, 1988.
659. Hecke, G. R. V., Kaji, K., and Sorai, M. *Mol. Cryst. Liq. Cryst.* 136, 197–219, 1986.
660. Lafon, M. R., Destrade, C., and Hemida, A. T. *Mol. Cryst. Liq. Cryst.* 137, 381–390, 1986.
661. Meirovitch, E., Luz, Z., and Zimmermann, H. *J. Phys. Chem.* 88, 2870–2874, 1984.
662. Gonen, O. and Levanon, H. *J. Phys. Chem.* 78, 2214–2218, 1983.
663. Destrade, C., Tinh, N. H., Malthete, J., and Levelut, A. M. *J. Phys.* 44, 597–602, 1983.
664. Mourey, B., Perbet, J. N., Hareng, M., and Le Berre, S. *Mol. Cryst. Liq. Cryst.* 84, 193–199, 1982.
665. Vandewal, K., Goris, L., Haenen, K., Geerts, Y., and Manca, J. V. *Eur. Phys. J. Appl. Phys.* 36, 281–283, 2007.
666. Gharbia, M., Gharbi, A., Cagnon, M., and Durand, G. *J. Phys. France* 51, 1355–1365, 1990.
667. Markovitsi, D., Marguet, S., Bondkowski, J., and Kumar, S. *J. Phys. Chem. B* 105, 1299–1306, 2001.
668. Markovitsi, D., Marguet, S., Gallos, L. K., Sigal, H., Millie, P., Argyrakis, P., Ringsdorf, H., and Kumar, S. *Chem. Phys. Lett.* 306, 163–167, 1999.
669. Sigal, H., Markovitsi, D., Gallos, L. K., and Argyrakis, P. *J. Phys. Chem.* 100, 10999–11004, 1996.
670. Markovitsi, D., Germain, A., Millie, P., Lecuyer, P., Gallos, L. K., Argyrakis, P., Bengs, H., and Ringsdorf, H. *J. Phys. Chem.* 99, 1005–1017, 1995.
671. Markovitsi, D., Lecuyer, I., Lianos, P., and Malthete, J. *J. Chem. Soc. Faraday Trans.* 87, 1785–1790, 1991.
672. Markovitsi, D., Rigaut, F., Mouallem, M., and Malthete, J. *Chem. Phys. Lett.* 135, 236–242, 1987.
673. Kruglova, O., Mendes, E., Yildirim, Z., Wubbenhorst, M., Mulder, F. M., Stride, J. A., Picken, S. J., and Kearley, G. J. *ChemPhysChem* 8, 1338–1344, 2007.
674. Boden, N., Bushby, R. J., and Lozman, O. R. *Mol. Cryst. Liq. Cryst.* 411, 345–354, 2004.
675. Park, L. Y., Hamilton, D. G., McGehee, E. A., and McMenimen, K. A. *J. Am. Chem. Soc.* 125, 10586–10590, 2003.
676. Boden, N., Bushby, R. J., and Lozman, O. R. *Mol. Cryst. Liq. Cryst.* 400, 105–113, 2003.
677. Okabe, A., Fukushima, T., Ariga, K., and Aida, T. *Angew. Chem. Int. Ed.* 41, 3414–3417, 2002.
678. Calucci, L., Zimmermann, H., Wachtel, E. J., Poupko, R., and Luz, Z. *Liq. Cryst.* 22, 621–630, 1997.
679. Boden, N., Bushby, R. J., and Hubbard, J. F. *Mol. Cryst. Liq. Cryst.* 304, 195–200, 1997.
680. Fimmen, W., Glusen, B., Kettner, A., Wittenberg, M., and Wendorff, J. H. *Liq. Cryst.* 23, 569–573, 1997.
681. Praefcke, K. and Singer, D. *Mol. Mater.* 3, 265–270, 1994.
682. Bengs, H., Renkel, R., Ringsdorf, H., Baehr, C., Ebert, M., and Wendorff, J. H. *Makromol. Chem. Rapid Commun.* 12, 439–446, 1991.
683. Brandl, B. and Wendorff, J. H. *Liq. Cryst.* 32, 425–430, 2005.
684. Rao, D. S. S., Prasad, S. K., Bamezai, R. K., and Kumar, S. *Mol. Cryst. Liq. Cryst.* 397, 143–159, 2003.
685. Rao, D. S. S., Gupta, V. K., Prasad, S. K., Manickam, M., and Kumar, S. *Mol. Cryst. Liq. Cryst.* 31, 193–206, 1998.
686. Raja, V. N., Shashidhar, R., Chandrasekhar, S., Boehm, R. E., and Martire, D. E. *Pramana J. Phys.* 25, L119–L122, 1985.
687. Maeda, Y., Rao, D. S. S., Prasad, S. K., Chandrasekhar, S., and Kumar, S. *Liq. Cryst.* 28, 1679–1690, 2001.
688. Maeda, Y., Rao, D. S. S., Prasad, S. K., Chandrasekhar, S., and Kumar, S. *Mol. Cryst. Liq. Cryst.* 397, 129–142, 2003.

689. Merekalova, N. D., Merekalov, A. S., Otmakhova, O. A., and Talroze, R. V. *Polym. Sci. Ser. A* 50, 84–90, 2008.
690. Sun, D. G., Ding, F. J., and Zhao, K. Q. *Acta Chim. Sin.* 66, 738–744, 2008.
691. Wang, B. Q., Ding, F. J., and Zhao, K. Q. *Acta Chim. Sin.* 66, 627–632, 2008.
692. Ding, F. J. and Zhao, K. Q. *Acta Chim. Sin.* 65, 1454–1458, 2007.
693. Hu, J. D., Li, Q., and Zhao, K. Q. *Acta Chim. Sin.* 65, 1784–1788, 2007.
694. Miglioli, I., Muccioli, L., Orlandi, S., Ricci, M., Berardi, R., and Zannoni, C. *Theo. Chim. Acta* 118, 203–210, 2007.
695. Orlandi, S., Muccioli, L., Ricci, M., Berardi, R., and Zannoni, C. *Chem. Cent. J.* 1–15, 2007.
696. Kruglova, O., Mulder, F. M., Kotlewski, A., Picken, S. J., Parker, S., Johnson, M. R., and Kearley, G. J. *Chem. Phys.* 330, 360–364, 2006.
697. Ding, F. J. and Zhao, K. Q. *Acta Chim. Sin.* 64, 117–120, 2006.
698. Bayer, A., Stillings, C., and Wendorff, J. H. *Liq. Cryst.* 33, 1103–1111, 2006.
699. Cinacchi, G. *J. Phys. Chem. B* 109, 8125–8131, 2005.
700. Cinacchi, G., Colle, R., and Tani, A. *J. Phys. Chem. B* 108, 7969–7977, 2004.
701. Pecchia, A., Kelsall, R. W., Movaghar, B., Bourlange, A., Evans, S. D., Hickey, B. J., and Boden, N. *J. Comput. Electr.* 1, 399–403, 2002.
702. Akopov, D. A. and Akopova, O. B. *J. Struct. Chem.* 43, 1050–1052, 2002.
703. Keszthelyi, T., Balakrishnan, G., Wilbrandt, R., Yee, W. A., and Negri, F. *J. Phys. Chem. A* 104, 9121–9129, 2000.
704. Hebert, M. *Phys. Rev. E* 55, 7063–7067, 1997.
705. Hebert, M. and Caille, A. *Phys. Rev. E* 53, 1714–1724, 1996.
706. Bast, T. and Hentschke, R. *J. Phys. Chem.* 100, 12162–12171, 1996.
707. Rey, A. D. *Liq. Cryst.* 19, 325–331, 1995.
708. Ghose, D., Bose, T. R., Saha, J., Mukherjee, C. D., Roy, M. K., and Saha, M. *Mol. Cryst. Liq. Cryst.* 264, 165–179, 1995.
709. Phillips, T. J. *J. Phys. I France* 5, 1667–1680, 1995.
710. Edwards, R. G., Henderson, J. R., and Pinning, R. L. *Mol. Phys.* 86, 567–598, 1995.
711. Horiguchi, T. and Fukui, Y. *Phys. Rev. B* 50, 7140–7143, 1994.
712. Zarragoicoechea, G. J., Levesque, D., and Weis, J. J. *Mol. Phys.* 78, 1475–1492, 1993.
713. Oswald, P. *J. Phys.* 50, C3-127–C3-132, 1989.
714. Oswald, P. *J. Phys. France* 49, 2119–2124, 1988.
715. Chandrasekhar, S., Savithramma, K. L., and Madhusudana, N. V. *Liq. Cryst. Order. Fluid.* 4, 299–309, 1984.
716. Cagnon, M., Gharbia, M., and Durand, G. *Phys. Rev. Lett.* 53, 938–940, 1984.
717. Samulski, E. T. and Toriumi, H. *J. Chem. Phys.* 79, 5194–5199, 1983.
718. Cotrait, M., Marsau, P., Pesquer, M., and Volpilhac, V. *J. Phys.* 43, 355–359, 1982.
719. Oswald, P. *J. Phys. Lett.* 42, L171–L173, 1981.
720. Oswald, P. and Kleman, M. *J. Phys.* 42, 1461–1472, 1981.
721. Pesquer, M., Cotrait, M., Marsau, P., and Volpilhac, V. *J. Phys.* 41, 1039–1043, 1980.
722. Bouligand, Y. *J. Phys.* 41, 1307–1315, 1980.
723. Ghose, D., Bose, T. R., Mukherjee, C. D., Roy, M. K., and Saha, M. *Mol. Cryst. Liq. Cryst.* 173, 17–29, 1989.
724. Lamoureux, G., Caille, A., and Senechal, D. *Phys. Rev. E* 58, 5898–5908, 1998.
725. Tang, C. W. *Appl. Phys. Lett.* 48, 183–185, 1986.
726. Shi, M. M., Chen, H. Z., Sun, J. Z., Ye, J., and Wang, M. *Chem. Phys. Lett.* 381, 666–671, 2003.
727. Horowitz, G., Kouki, F., Spearman, P., Fichou, D., Nogues, C., Pan, X., and Garnier, F. *Adv. Mater.* 8, 242–245, 1996.
728. Malenfant, P. R. L., Dimitrakopoulos, C. D., Gelorme, J. D., Kosbar, L. L., Graham, T. O., Curioni, A., and Andreoni, W. *Appl. Phys. Lett.* 80, 2517–2519, 2002.
729. Haas, U., Thalacker, C., Adams, J., Fuhrmann, J., Riethmuller, S., Beginn, U., Ziener, U., Moller, M., Dobrawa, R., and Wurthner, F. *J. Mater. Chem.* 13, 767–772, 2003.
730. Ranke, P., Bleyl, I., Simmerer, J., Haarer, D., Bacher, A., and Schmidt, H. W. *Appl. Phys. Lett.* 71, 1332–1334, 1997.
731. Mitchell, R. H., Chaudhary, M., Williams, R. V., Fyles, R., Gibson, J., Smith, M. J. A., and Fry, A. J. *Can. J. Chem.* 70, 1015–1021, 1992.
732. Stolarski, R. and Fiksinski, K. *J. Dyes Pigm.* 24, 295–303, 1994.
733. Benning, S., Kitzerow, H. S., Bock, H., and Achard, M. F. *Liq. Cryst.* 27, 901–906, 2000.

734. Archambeau, S., Seguy, I., Jolinat, P., Farenc, J., Destruel, P., Nguyen, T. P., Bock, H., and Grelet, E. *Appl. Surf. Sci.* 253, 2078–2086, 2006.
735. Fouet, S. A., Seguy, I., Bobo, J. F., Destruel, P., and Bock, H. *Chem. Eur. J.* 13, 1746–1753, 2007.
736. Takahashi, M., Suzuki, Y., Ichihashi, Y., Yamashita, M., and Kawai, H. *Tetrahedron Lett.* 48, 357–359, 2007.
737. Mo, X., Chen, H. Z., Shi, M. M., and Wang, M. *Chem. Phys. Lett.* 417, 457–460, 2006.
738. Mo, X., Shi, M. M., Huang, J. C., Wang, M., and Chen, H. Z. *Dyes Pigm.* 76, 236–242, 2008.
739. Wurthner, F. *Chem. Commun.* 1564–1579, 2004.
740. Struijk, C. W., Sieval, A. B., Dakhorst, J. E. J., van Dijk, M., Kimkes, P., Koehorst, R. B. M., Donker, H., Schaafsma, T. J., Picken, S. J., van de Craats, A. M., Warman, J. M., Zuilhof, H., and Sudholter, E. J. R. *J. Am. Chem. Soc.* 122, 11057–11066, 2000.
741. Langhals, H., Karolin, J., and Johansson, L. B. A. *J. Chem. Soc. Faraday Trans.* 94, 2919–2922, 1998.
742. Mende, L. S., Fechtenkotter, A., Mullen, K., Moons, E., Friend, R. H., and MacKenzie, J. D. *Science* 293, 1119–1122, 2001.
743. Breeze, A. J., Salomon, A., Ginley, D. S., Gregg, B. A., Tillmann, H., and Horhold, H. H. *Appl. Phys. Lett.* 81, 3085–3087, 2002.
744. Kraft, A., Grimsdale, A. C., and Holmes, A. B. *Angew. Chem. Int. Ed.* 37, 402–428, 1998.
745. Pan, J., Zhu, W., Li, S., Zeng, W., Cao, Y., and Tian, H. *Polymer* 46, 7658–7669, 2005.
746. Karapire, C., Zafer, C., and Icli, S. *Synth. Met.* 145, 51–60, 2004.
747. Wurthner, F. and Schmidt, R. *ChemPhysChem.* 7, 793–797, 2006.
748. Dimitrakopoulos, C. D. and Malenfant, P. R. L. *Adv. Mater.* 14, 99–117, 2002.
749. Jones, B. A., Ahrens, M. J., Yoon, M. H., Facchetti, A., Marks, T. J., and Wasielewski, M. R. *Angew. Chem. Int. Ed.* 43, 6363–6366, 2004.
750. Cormier, R. A. and Gregg, B. A. *J. Phys. Chem. B* 101, 11004–11006, 1997.
751. Cormier, R. A. and Gregg, B. A. *Chem. Mater.* 10, 1309–1319, 1998.
752. Wurthner, F., Thalacker, C., Diele, S., and Tschierske, C. *Chem. Eur. J.* 7, 2245–2253, 2001.
753. van Herrikhuyzen, J., Syamakumari, A., Schenning, A. P. H. J., and Meijer, E. W. *J. Am. Chem. Soc.* 126, 10021–10027, 2004.
754. Debije, M. G., Chen, Z., Piris, J., Neder, R. B., Watson, M. M., Mullen, K., and Wurthner, F. *J. Mater. Chem.* 15, 1270–1276, 2005.
755. An, Z., Yu, J., Jones, S. C., Barlow, S., Yoo, S., Domercq, B., Prins, P., Siebbeles, L. D. A., Kippelen, B., and Marder, S. R. *Adv. Mater.* 17, 2580–2583, 2005.
756. Zucchi, G., Donnio, B., and Geerts, Y. H. *Chem. Mater.* 17, 4273–4277, 2005.
757. Nolde, F., Pisula, W., Muller, S., Kohl, C., and Mullen, K. *Chem. Mater.* 18, 3715–3725, 2006.
758. Li, X. Q., Stepanenko, V., Chen, Z., Prins, P., Siebbeles, L. D. A., and Würthner, F. *Chem. Commun.* 3871–3873, 2006.
759. Wurthner, F., Chen, Z., Dehm, V., and Stepanenko, V. *Chem. Commun.* 1188–1190, 2006.
760. Percec, V., Aqad, E., Peterca, M., Imam, M. R., Glodde, M., Bera, T. K., Miura, Y., Balagurusamy, V. S. K., Ewbank, P. C., Wurthner, F., and Heiney, P. A. *Chem. Eur. J.* 13, 3330–3345, 2007.
761. Chen, Z., Baumeister, U., Tschierske, C., and Wurthner, F. *Chem. Eur. J.* 13, 450–465, 2007.
762. Chen, Z., Stepanenko, V., Dehm, V., Prins, P., Siebbeles, L. D. A., Seibt, J., Marquetand, P., Engel, V., and Wurthner, F. *Chem. Eur. J.* 13, 436–449, 2007.
763. Dehm, V., Chen, Z., Baumeister, U., Prins, P., Siebbeles, L. D. A., and Wurthner, F. *Org. Lett.* 9, 1085–1088, 2007.
764. Iden, R. and Seybold, G. *Gen. Pat. Appl.* 103, 3869, 1985.
765. Seybold, G. and Wagenblast, G. *Dyes Pigm.* 11, 303–317, 1989.
766. Göltner, C., Pressner, D., Müllen, K., and Spiess, H. W. *Angew. Chem. Int. Ed.* 32, 1660–1662, 1993.
767. Pressner, D., Göltner, C., Spiess, H. W., and Müllen, K. *Ber. Buns. Phys. Chem.* 97, 1362–1365, 1993.
768. Biasutti, M. A., de Feyter, S., de Backer, S., Dutt, G. B., de Schryver, F. C., Ameloot, M., Schlichting, P., and Müllen, K. *Chem. Phys. Lett.* 248, 13–19, 1996.
769. Müller, G. R. J., Meiners, C., Enkelmann, V., Geerts, Y., and Müllen, K. *J. Mater. Chem.* 8, 61–64, 1998.
770. Fouet, S. A., Dardel, S., Bock, H., Oukachmih, M., Archambeau, S., Seguy, I., Jolinat, P., and Destruel, P. *ChemPhysChem* 4, 983–985, 2003.
771. Charlet, E., Grelet, E., Brettes, P., Bock, H., Saadaoui, H., Cisse, L., Destruel, P., Gherardi, N., and Seguy, I. *Appl. Phys. Lett.* 92, 024107, 2008.
772. Yamaguchi, S. and Swager, T. M. *J. Am. Chem. Soc.* 123, 12087–12088, 2001.
773. Clar, E., Vuilleme, J. F. G., and Stephen, J. F. *Tetrahedron* 20, 2107–2117, 1964.
774. Talapatra, S. K., Chakrabarti, S., Mallik, A. K., and Talapatra, B. *Tetrahedron* 46, 6047–6052, 1990.

775. Alder, R. W. and Harvey, J. N. *J. Am. Chem. Soc.* 126, 2490–2494, 2004.

776. Thulstrup, E. W., Larsen, J. S., and Waluk, J. *J. Chem. Soc. Perkin Trans.* 2 712–713, 1975.

777. Klumpp, D. A., Baek, D. N., Prakash, G. K. S., and Olah, G. A. *J. Org. Chem.* 62, 6666–6671, 1997.

778. Tokito, S., Noda, K., Fujikawa, H., Taga, Y., Kimura, M., Shimada, K., and Sawaki, Y. *Appl. Phys. Lett.* 77, 160–162, 2000.

779. Li, C. W., Wang, C. I., Liao, H. Y., Chaudhuri, R., and Liu, R. S. *J. Org. Chem.* 72, 9203–9207, 2007.

780. Kumar, S. and Varshney, S. K. *Mol. Cryst. Liq. Cryst.* 378, 59–64, 2002.

781. Herbstein, F. H. *Acta Cryst.* B35, 1661–1670, 1979.

782. Larock, R. C., Doty, M. J., Tian, Q., and Zenner, J. M. *J. Org. Chem.* 62, 7536–7537, 1997.

783. Schultz, A., Diele, S., Laschat, S., and Nimtz, M. *Adv. Funct. Mater.* 11, 441–446, 2001.

784. Chaudhuri, R., Hsu, M. Y., Li, C. W., Wang, C. I., Chen, C. J., Lai, C. K., Chen, L. Y., Liu, S. H., Wu, C. C., and Liu, R. S. *Org. Lett.* 10, 3053–3056, 2008.

785. Schultz, A., Laschat, S., Diele, S., and Nimtz, M. *Eur. J. Org. Chem.* 2829–2839, 2003.

786. Bock, H. and Helfrich, W. *Liq. Cryst.* 18, 387–399, 1995.

787. Meyer, R. B., Liebert, L., Strzelecki, L., and Keller, P. *J. Phys. Lett.* 36, 69–71, 1975.

788. Prost, J. In *Symmetries and Broken Symmetries*, Boccara, N. (ed.), IDSET, Paris, France, p. 159, 1981.

789. Musgrave, O. C. and Webster, C. J. *Chem. Commun.* 712–713, 1969.

790. Musgrave, O. C. and Webster, C. J. *J. Chem. Soc. C* 8, 1393–1397, 1971.

791. Ormsby, J. L., Black, T. D., Hilton, C. L., Bharat., and King, B. T. *Tetrahedron* 64, 11370–11378, 2008.

792. Zamir, S., Singer, D., Spielberg, N., Wachtel, E. J., Zimmermann, H., Poupko, R., and Luz, Z. *Liq. Cryst.* 21, 39–50, 1996.

793. Jakli, A., Muller, M., Kruerke, D., and Heppke, G. *Liq. Cryst.* 24, 467–472, 1998.

794. Kitzerow, H. S. and Bock, H. *Mol. Cryst. Liq. Cryst.* 299, 117–128, 1997.

795. Perova, T. S., Vij, J. K., and Bock, H. *Mol. Cryst. Liq. Cryst.* 263, 293–303, 1995.

796. Shilov, S. V., Muller, M., Kruerke, D., Heppke, G., Skupin, H., and Kremer, F. *Phys. Rev. E* 65, 021707, 2002.

797. Uznanski, P., Marguet, S., Markovitsi, D., Schuhmacher, P., and Ringsdorf, H. *Mol. Cryst. Liq. Cryst.* 293, 123–133, 1997.

798. Uznanski, P. *Acta Phys. Pol. A* 94, 825–834, 1998.

799. Kumar, S. and Naidu, J. J. *Liq. Cryst.* 28, 1435–1437, 2001.

800. Kumar, S., Naidu, J. J., and Rao, D. S. S. *J. Mater. Chem.* 12, 1335–1341, 2002.

801. Gabriel, S. and Michael, A. *Chem. Ber.* 10, 1551, 1881.

802. Husmann, J. *Chem. Ber.* 22, 2019, 1889.

803. Lor, B. G., Cantalapiedra, E. G., Ruiz, M., de Frutos, O., Cardenas, D. J., Santos, A., and Echavarren, A. M. *Chem. Eur. J.* 10, 2601–2608, 2004.

804. Boorum, M. M. and Scott, L. T. In *Modern Arene Chemistry*, Astruc. D. (ed.), WILEY-VCH, Weinheim, Germany, Chap. 1, pp. 20–31, 2002.

805. Omer, K. M., Kanibolotsky, A. L., Skabara, P. J., Perepichka, I. F., and Bard, A. J. *J. Phys. Chem. B* 111, 6612–6619, 2007.

806. Yuan, M. S., Liu, Z. Q., and Fang, Q. *J. Org. Chem.* 72, 7915–7922, 2007.

807. Wang, J. L., Duan, X. F., Jiang, B., Gan, L. B., Pei, J., He, C., Li, Y. F. *J. Org. Chem.* 71, 4400–4410, 2006.

808. Wang, J. L., Luo, J., Liu, L. H., Zhou, Q. F., Ma, Y., and Pei, J. *Org. Lett.* 8, 2281–2284, 2006.

809. Kimura, M., Kuwano, S., Sawaki, Y., Fujikawa, H., Noda, K., Taga, Y., and Takagi, K. *J. Mater. Chem.* 15, 2393–2398, 2005.

810. Sun, Y., Xiao, K., Liu, Y., Wang, J., Pei, J., Yu, G., and Zhu, D. *Adv. Funct. Mater.* 15, 818–822, 2005.

811. Cao, X. Y., Liu, X. H., Zhou, X. H., Zhang, Y., Jiang, Y., Cao, Y., Cui, Y. X., and Pei, J. *J. Org. Chem.* 69, 6050–6058, 2004.

812. Kanibolotsky, A. L., Berridge, R., Skabara, P. J., Perepichka, I. F., Bradley, D. D. C., and Koeberg, M. *J. Am. Chem. Soc.* 126, 13695–13702, 2004.

813. Cao, X. Y., Zhou, X. H., Zi, H., and Pei, J. *Macromolecules* 37, 8874–8882, 2004.

814. Cao, X. Y., Zhang, W., Zi, H., and Pei, J. *Org. Lett.* 26, 4845–4848, 2004.

815. Cao, X. Y., Zhang, W. B., Wang, J. L., Zhou, X. H., Lu, H., and Pei, J. *J. Am. Chem. Soc.* 125, 12430–12431, 2003.

816. Pei, J., Wang, J. L., Cao, X. Y., Zhou, X. H., and Zhang, W. B. *J. Am. Chem. Soc.* 125, 9944–9945, 2003.

817. Lor, B. G., de Frutos, O., Ceballos, P. A., Granier, T., and Echavarren, A. M. *Eur. J. Org. Chem.* 2107–2114, 2001.

818. Yuan, M. S., Fang, Q., Liu, Z. Q., Guo, J. P., Chen, H. Y., Yu, W. T., Xue, G., and Liu, D. S. *J. Org. Chem.* 71, 7858–7861, 2006.
819. Destrade, C., Malthete, J., Tinh, N. H., and Gasparoux, H. *Phys. Lett.* 78A, 82–84, 1980.
820. Destrade, C., Gasparoux, H., Babeau, A., Tinh, N. H., and Malthete, J. *Mol. Cryst. Liq. Cryst.* 67, 37–48, 1981.
821. Tinh, N. H., Malthete, J., and Destrade, C. *Mol. Cryst. Liq. Cryst.* 64, 291–298, 1981.
822. Tinh, N. H., Malthete, J., and Destrade, C. *J. Phys. Lett.* 42, L417–L419, 1981.
823. Destrade, C., Foucher, P., Malthete, J., and Tinh, N. H. *Phys. Lett.* 88A, 187–190, 1982.
824. Goldfarb, D., Belsky, I., and Luz, Z. *J. Chem. Phys.* 79, 6203, 1983.
825. Lejay, J. and Pesquer, M. *Mol. Cryst. Liq. Cryst.* 95, 31–43, 1983.
826. Baumann, C., Marcerou, J. P., Rouillon, J. C., and Prost, J. *J. Phys.* 45, 451–458, 1984.
827. Lejay, J. and Pesquer, M. *Mol. Cryst. Liq. Cryst.* 111, 293–310, 1984.
828. Foucher, P., Destrade, C., Tinh, N. H., Malthete, J., and Levelut, A. M. *Mol. Cryst. Liq. Cryst.* 108, 219–229, 1984.
829. Tinh, N. H., Foucher, P., Destrade, C., Levelut, A. M., and Malthete, J. *Mol. Cryst. Liq. Cryst.* 111, 277–292, 1984.
830. Raghunathan, V. A., Madhusudana, N. V., Chandrasekhar, S., and Destrade, C. *Mol. Cryst. Liq. Cryst.* 148, 77–83, 1987.
831. Buisine, J. M., Cayuela, R., Destrade, C., and Tinh, N. H. *Mol. Cryst. Liq. Cryst.* 144, 137–160, 1987.
832. Warmerdam, T. W., Nolte, R. J. M., Drenth, W., van Miltenburg, J. C., Frenkel, D., and Zijlstra, R. J. J. *Liq. Cryst.* 3, 1087–1104, 1988.
833. Warmerdam, T., Frenkel, D., and Zijlstra, R. J. J. *Liq. Cryst.* 3, 149–152, 1988.
834. Fontes, E., Heiney, P. A., Ohba, M., Haseltine, J. N., and Smith III, A. B. *Phys. Rev. A* 37, 1329–1334, 1988.
835. Lee, W. K., Winter, B. A., Fontes, E., Heiney, P. A., Ohba, M., Haseltine, J. N., and Smith III, A. B. *Liq. Cryst.* 4, 87–102, 1989.
836. Lee, W. K., Heiney, P. A., Ohba, M., Haseltine, J. N., and Smith III, A. B. *Liq. Cryst.* 8, 839–850, 1990.
837. Maliszewskyj, N. C., Heiney, P. A., Blasie, J. K., McCauley Jr., J. P., and Smith III, A. B. *J. Phys. II France* 2, 75–85, 1992.
838. Sandstroem, D., Nygren, M., Zimmermann, H., and Maliniak, A. *J. Phys. Chem.* 99, 6661–6669, 1995.
839. Othman, T., Gharbia, M., Gharbi, A., Destrade, C., and Durand, G. *Liq. Cryst.* 18, 839–842, 1995.
840. Averyanov, E. M. *JETP* 83, 1000–1011, 1996.
841. Averyanov, E. M. *JETP Lett.* 63, 33–37, 1996.
842. Averyanov, E. M. *J. Struct. Chem.* 38, 71–77, 1997.
843. Amick, A. W., Griswold, K. S., and Scott, L. T. *Can. J. Chem.* 84, 1268–1272, 2006.
844. Iglesias, B., Pena, D., Perez, D., Guitian, E., and Castedo, L. *Synlett.* 486–488, 2002.
845. Zimmermann, K. and Haenel, M. W. *Synlett.* 609–611, 1997.
846. Dziewonski, K. *Chem. Ber.* 36, 962–971, 1903.
847. Ruland, W. *Carbon* 2, 365–378, 1965.
848. Fitzer, E., Muller, K., and Schaeffer, W. *Chemistry and Physics of Carbon*, Marcel Dekker, New York, p. 259, 1971.
849. Imamura, T. and Nakamizo, M. *Carbon* 17, 507–508, 1979.
850. Kokaji, K., Oya, A., Maruyama, K., Yamada, Y., and Shiraishi, M. *Carbon* 35, 253–258, 1997.
851. Kimura, T., Koizumi, H., Kinoshita, H., and Ichikawa, T. *Jpn. J. Appl. Phys.* 46, 703–707, 2007.
852. Kimura, T., Koizumi, H., and Ichikawa, T. *Surf. Coating. Technol.* 201, 8506–8510, 2007.
853. Taylor, G. H. *Fuel* 40, 465–472, 1961.
854. Brooks, J. D. and Taylor, G. H. *Nature* 206, 697–699, 1965.
855. Brooks, J. D. and Taylor, G. H. *Carbon* 3, 185–193, 1965.
856. Brooks, J. D. and Taylor, G. H. *Carbon* 4, 243–286, 1968.
857. Zimmer, J. E. and White, J. L. *Adv. Liq. Cryst.* 5, 157–213, 1982.
858. Gasparoux, H. *Mol. Cryst. Liq. Cryst.* 63, 231–248, 1981.
859. Otani, S. *Mol. Cryst. Liq. Cryst.* 63, 249–263, 1981.
860. Levelut, A. M. *J. Chim. Phys.* 80, 149–161, 1983.
861. Malthete, J. and Collet, A. *J. Am. Chem. Soc.* 109, 7544–7545, 1987.
862. Keinan, E., Kumar, S., Moshenberg, R., Ghirlando, R., and Wachtel, E. J. *Adv. Mater.* 3, 251–254, 1991.
863. Saji, T. and Aoyanui, S. *J. Electroanal. Chem. Interfacial Electrochem.* 102, 139–141, 1979.
864. Heinze, J. *Angew. Chem.* 96, 823–840, 1984.

865. Munakata, M., Wu, L. P., Ning, G. L., Sowa, T. K., Maekawa, M., Suenaga, Y., and Maeno, N. *J. Am. Chem. Soc.* 121, 4968–4976, 1999.

866. Schneider, J. J., Czap, N., Spickermann, D., Lehmann, C. W., Fontani, M., Laschi, F., and Zanello, P. *J. Organomet. Chem.* 590, 7–14, 1999.

867. Schneider, J. J., Spickermann, D., Labahn, T., Magull, J., Fontani, M., Laschi, F., and Zanello, P. *Chem. Eur. J.* 6, 3686–3691, 2000.

868. Schneider, J. J., Spickermann, D., Lehmann, C. W., Magull, J., Krüger, H. J., Ensling, J., and Gütlich, P. *Chem. Eur. J.* 12, 1427–1435, 2006.

869. Sano, T., Fujii, H., Nishio, Y., Hamada, Y., Takahashi, H., and Shibata, K. *Synth. Met.* 91, 27–30, 1997.

870. Acharya, S., Parichha, T. K., and Talapatra, G. B. *Mol. Cryst. Liq. Cryst. Mol. Mater.* 12, 91–100, 2000.

871. Das, S., Chowdhury, A., and Pal, A. J. *Phys. Stat. Sol.* 185, 383–389, 2001.

872. Li, C., Zeng, Q., Wu, P., Xu, S., Wang, C., Qiao, Y., Wan, L., and Bai, C. *J. Phys. Chem. B* 106, 13262–13267, 2002.

873. Hirota, K., Tajima, K., and Hashimoto, K. *Synth. Met.* 157, 290–296, 2007.

874. Wu, J., Pisula, W., and Mullen, K. *Chem. Rev.* 107, 718–747, 2007.

875. Simpson, C. D., Wu, J., Watson, M. D., and Mullen, K. *J. Mater. Chem.* 14, 494–504, 2004.

876. Watson, M. D., Fechtenkotter, A., Mullen, K. *Chem. Rev.* 101, 1267–1300, 2001.

877. Grimsdale, A. C. Wu, J., and Mullen, K. *Chem. Commun.* 2197–2204, 2005.

878. Mullen, K. and Rabe, J. P. *Acc. Chem. Res.* 41, 511–520, 2008.

879. Sergeyev, S., Pisula, W., and Geerts, Y. H. *Chem. Soc. Rev.* 36, 1902–1929, 2007.

880. Sarhan, A. A. O. and Bolm, C. *Chem. Soc. Rev.* 38, 2730–2744, 2009.

881. Li, J., Kastler, M., Pisula, W., Robertson, J. W. F., Wasserfallen, D., Grimsdale, A. C., Wu, J., and Mullen, K. *Adv. Funct. Mater.* 17, 2528–2533, 2007.

882. Pisual, W., Menon, A., Stepputat, M., Lieberwirth, I., Kolb, U., Tracz, A., Sirringhaus, H., Pakula, T., and Mullen, K. *Adv. Mater.* 17, 684–689, 2005.

883. Shklyarevskiy, I. O., Jonkheijm, P., Stutzman, N., Wasserberg, D., Wondergem, H. J., Christianen, P. C. M., Schenning, A. P. H. J., de Leeuw, D. M., Tomovic, Z., Wu, J., Mullen, K., and Maan, J. C. *J. Am. Chem. Soc.* 127, 16233–16237, 2005.

884. Van de Craats, A. M., Stutzmann, N., Bunk, O., Nielsen, M. M., Watson, M., Mullen, K., Chanzy, H. D., Sirringhaus, H., and Friend, R. H. *Adv. Mater.* 15, 495–499, 2003.

885. Fischbach, I., Ebert, F., Spiess, H. W., and Schnell, I. *ChemPhysChem* 5, 895–908, 2004.

886. Brown, S. P., Schnell, I., Brand, J. D., Mullen, K., and Spiess, H. W. *J. Mol. Str.* 521, 179–195, 2000.

887. Warman, J. M., Piris, J., Pisula, W., Kastler, M., Wasserfallen, D., and Mullen, K. *J. Am. Chem. Soc.* 127, 14257–14262, 2005.

888. Jackel, F., Wang, Z., Watson, M. D., Mullen, K., and Rabe, J. P. *Chem. Phys. Lett.* 387, 372–376, 2004.

889. Piris, J., Debije, M. G., Stutzmann, N., Laursen, B. W., Pisula, W., Watson, M. D., Bjornholm, T., Mullen, K., and Warman, J. M. *Adv. Funct. Mater.* 14, 1053–1061, 2004.

890. Piris, J., Debije, M. G., Watson, M. D., Mullen, K., and Warman, J. M. *Adv. Funct. Mater.* 14, 1047–1052, 2004.

891. Gherghel, L., Kubel, C., Lieser, G., Rader, H. J., and Mullen, K. *J. Am. Chem. Soc.* 124, 13130–13138, 2002.

892. Piot, L., Marchenko, A., Wu, J., Mullen, K., and Fichou, D. *J. Am. Chem. Soc.* 127, 16245–16250, 2005.

893. Friedlein, R., Crispin, X., Osikowicz, W., Braun, S., de Jong, M. P., Simpson, C. D., Watson, M. D., von Kieseritzky, F., Samori, P., Jonsson, S. K. M., Fahlman, M., Jackel, F., Rabe, J. P., Hellberg, J., Mullen, K., and Salaneck, W. R. *Synth. Met.* 147, 79–83, 2004.

894. Piris, J., Pisula, W., and Warman, J. M. *Synth. Met.* 147, 85–89, 2004.

895. Jung, J., Rybak, A., Slazak, A., Bialecki, S., Miskiewicz, P., Glowacki, I., Ulanski, J., Rosselli, S., Yasuda, A., Nelles, G., Tomovic, Z., Watson, M. D., and Mullen, K. *Synth. Met.* 155, 150–156, 2005.

896. Pisula, W., Kastler, M., Wasserfallen, D., Robertson, J. W. F., Nolde, F., Kohl, C., and Mullen, K. *Angew. Chem. Int. Ed.* 45, 819–823, 2006.

897. Pisula, W., Kastler, M., Hamaoui, B. E., Gutierrez, M. C. G., Davies, R. J., Riekel, C., and Mullen, K. *ChemPhysChem* 8, 1025–1028, 2007.

898. Breiby, D. W., Hansteen, F., Pisula, W., Bunk, O., Kolb, U., Andreasen, J. W., Mullen, K., and Nielsen, M. M. *J. Phys. Chem. B* 109, 22319–22325, 2005.

899. Feng, X., Pisula, W., Zhi, L., Takase, M., and Mullen, K. *Angew. Chem. Int. Ed.* 47, 1703–1706, 2008.

900. Kastler, M., Pisula, W., Wasserfallen, D., Pakula, T., and Mullen, K. *J. Am. Chem. Soc.* 127, 4286–4296, 2005.

901. Pisula, W., Kastler, M., Wasserfallen, D., Pakula, T., and Mullen, K. *J. Am. Chem. Soc.* 126, 8074–8075, 2004.
902. Samori, P., Fechtenkotter, A., Reuther, E., Watson, M. D., Severin, N., Mullen, K., and Rabe, J. P. *Adv. Mater.* 18, 1317–1321, 2006.
903. Feng, X., Pisula, W., Ai, M., Groper, S., Rabe, J. P., and Mullen, K. *Chem. Mater.* 20, 1191–1193, 2008.
904. Chebny, V. J., Gwengo, C., Gardinier, J. R., and Rathore, R. *Tetrahedron Lett.* 49, 4869–4872, 2008.
905. van de Craats, A. M. and Waraman, J. M. *Synth. Met.* 121, 1287–1288, 2001.
906. Wang, Z., Dotz, F., Enkelmann, V., and Mullen, K. *Angew. Chem. Int. Ed.* 44, 1247–1250, 2005.
907. Keil, M., Samori, P., dos Santos, D. A., Kugler, T., Stafstrom, S., Brand, J. D., Mullen, K., Bredas, J. L., Rabe, J. P., and Salaneck, W. R. *J. Phys. Chem. B* 104, 3967–3975, 2000.
908. Ruffieux, P., Groning, O., Bielmann, M., Simpson, C., Mullen, K., Schlapbach, L., and Groning, P. *Phys. Rev. B* 66, 073409, 2002.
909. Hosoya, H., Tsukano, Y., Nakada, K., Iwata, S., and Nagashima, U. *Croat. Chim. Acta* 77, 89–95, 2004.
910. El Hamaoui, B., Laquai, F., Baluschev, S., Wu, J., and Mullen, K. *Synth. Met.* 156, 1182–1186, 2006.
911. Glowatzki, H., Gavrila, G. N., Seifert, S., Johnson, R. L., Rader, J., Mullen, K., Zahn, D. R. T., Rabe, J. P., and Koch, N. *J. Phys. Chem. C* 112, 1570–1574, 2008.
912. Treier, M., Ruffieux, P., Schillinger, R., Greber, T., Mullen, K., and Fasel, R. *Surf. Sci.* 602, L84–L88, 2008.
913. Wu, J., Fechtenkotter, A., Gauss, J., Watson, M. D., Kastler, M., Fechtenkotter, C., Wagner, M., and Mullen, K. *J. Am. Chem. Soc.* 126, 11311–11321, 2004.
914. Reitzel, N., Hassenkam, T., Balashev, K., Jensen, T. R., Howes, P. B., Kjaer, K., Fechtenkotter, A., Tchebotareva, N., Ito, S., Mullen, K., and Bjornholm, T. *Chem. Eur. J.* 7, 4894–4901, 2001.
915. Pisula, W., Tomovic, Z., Stepputat, M., Kolb, U., Pakula, T., and Mullen, K. *Chem. Mater.* 17, 2641–2647, 2005.
916. Pisula, W., Kastler, M., Wasserfallen, D., Davies, R. J., Gutierrez, M. C. G., and Mullen, K. *J. Am. Chem. Soc.* 128, 14424–14425, 2006.
917. Friedlein, R., Crispin, X., Simpson, C. D., Watson, M. D., Jackel, F., Osikowicz, W., Marciniak, S., de Jong, M. P., Samori, P., Jonsson, S. K. M., Fahlman, M., Mullen, K., Rabe, J. P., and Salaneck, W. R. *Phys. Rev. B* 68, 195414, 2003.
918. Duati, M., Grave, C., Tcbeborateva, N., Wu, J., Mullen, K., Shaporenko, A., Zharnikov, M., Kriebel, J. K., Whitesides, G. M., and Rampi, M. A. *Adv. Mater.* 18, 329–333, 2006.
919. Tracz, A., Jeszka, J. K., Watson, M. D., Pisula, W., Mullen, K., and Pakula, T. *J. Am. Chem. Soc.* 125, 1682–1683, 2003.
920. Cristadoro, A., Lieser, G., Rader, H. J., and Mullen, K. *ChemPhysChem* 8, 586–591, 2007.
921. Kubowicz, S., Pietsch, U., Watson, M. D., Tchebotareva, N., Mullen, K., and Thunemann, A. F. *Langmuir* 19, 5036–5041, 2003.
922. Ai, M., Groeper, S., Zhuang, W., Dou, X., Feng, X., Mullen, K., and Rabe, J. P. *Appl. Phys. A* 93, 277–283, 2008.
923. Yang, X., Dou, X., and Mullen, K. *Chem. Asian J.* 3, 759–766, 2008.
924. Zheng, Q., Ohulchanskyy, T. Y., Sahoo, Y., and Prasad, P. N. *J. Phys. Chem. C* 111, 16846–16851, 2007.
925. Kirkpatrick, J., Marcon, V., Kremer, K., Nelson, J., and Andrienko, D. *J. Chem. Phys.* 129, 094506, 2008.
926. Marcon, V., Vehoff, T., Kirkpatrick, J., Jenog, C., Yoon, D. Y., Kremer, K., and Andrienko, D. *J. Chem. Phys.* 129, 094505, 2008.
927. Kirkpatrick, J., Marcon, V., Kremer, K., Nelson, J., and Andrienko, D. *Phys. Stat. Sol. B* 245, 835–838, 2008.
928. Andrienko, D., Marcon, V., and Kremer, K. *J. Chem. Phys.* 125, 124902, 2006.
929. Clar, E. and Ironside, C. T. *Proc. Chem. Soc.* 150–151, 1958.
930. Clar, E., Ironside, C. T., and Zander, M. *J. Chem. Soc.* 142–147, 1959.
931. Clar, E. and Stephen, J. F. *Tetrahedron* 21, 467–470, 1965.
932. Kokkin, D. L., Troy, T. P., Nakajima, M., Nauta, K., Varberg, T. D., Metha, G. F., Lucas, N. T., and Schmidt, T. W. *Astrophys. J.* 681, L49–L51, 2008.
933. Hendel, W., Khan, Z. H., and Schmidt, W. *Tetrahedron* 42, 1127–1134, 1986.
934. Halleux, A., Martin, R. H., and King, G. S. D. *Helv. Chim. Acta* 129, 1177–1183, 1958.

935. Herwig, P., Kayser, C. W., Mullen, K., and Spiess, H. W. *Adv. Mater.* 8, 510–513, 1996.
936. Weiss, K., Beernink, G., Dotz, F., Birkner, A., Mullen, K., and Woll, C. H. *Angew Chem. Int. Ed.* 38, 3748–3752, 1999.
937. Fechtenkotter, A., Saalwachter, K., Harbison, M. A., Mullen, K., and Spiess, H. W. *Angew. Chem. Int. Ed.* 38, 3039–3042, 1999.
938. Stabel, A., Herwig, P., Mullen, K., and Rabe, J. P. *Angew. Chem. Int. Ed.* 34, 1609–1611, 1995.
939. van de Craats, A. M., Warman, J. M., Mullen, K., Geerts, Y., and Brand, J. D. *Adv. Mater.* 10, 36–38, 1998.
940. van de Craats, A. M., Warman, J. M., Fechtenkotter, A., Brand, J. D., Harbison, M. A., and Mullen, K. *Adv. Mater.* 11, 1469–1472, 1999.
941. Brown, S. P., Schnell, I., Brand, J. D., Mullen, K., and Spiess, H. W. *J. Am. Chem. Soc.* 121, 6712–6718, 1999.
942. Rakotondradany, F., Fenniri, H., Rahimi, P., Gawrys, K. L., Kilpatrick, P. K., and Gray, M. R. *Energy Fuels* 20, 2439–2447, 2006.
943. Pisula, W., Tomovic, Z., Simpson, C., Kastler, M., Pakula, T., and Mullen, K. *Chem. Mater.* 17, 4296–4303, 2005.
944. Wadumethrige, S. H. and Rathore, R. *Org. Lett.* 10, 5139–5142, 2008.
945. Fechtenkotter, A., Tchebotareva, N., Watson, M., and Mullen, K. *Tetrahedron* 57, 3769–3783, 2001.
946. Samori, P., Fechtenkotter, A., Jackel, F., Bohme, T., Mullen, K., and Rabe, J. P. *J. Am. Chem. Soc.* 123, 11462–11467, 2001.
947. Kastler, M., Pisula, W., Laquai, F., Kumar, A., Davies, R. J., Baluschev, S., Gutierrez, M. C. G., Wasserfallen, D., Butt, H. J., Riekel, C., Wegner, G., and Mullen, K. *Adv. Mater.* 18, 2255–2259, 2006.
948. Pisula, W., Tomovic, Z., El Hamaoui, B., Watson, M. D., Pakula, T., and Mullen. K. *Adv. Funct. Mater.* 15, 893–904, 2005.
949. Pisula, W., Tomovic, Z., Watson, M. D., Mullen, K., Kussmann, J., Ochsenfeld, C., Metzroth, T., and Gauss, J. *J. Phys. Chem. B* 111, 7481–7487, 2007.
950. Kikuzawa, Y., Mori, T., and Takeuchi, H. *Org. Lett.* 9, 4817–4820, 2007.
951. Zhang, Q., Prins, P., Jones, S. C., Barlow, S., Kondo, T., An, Z., Siebbeles, L. D. A., and Marder, S. R. *Org. Lett.* 7, 5019, 5022, 2005.
952. Alameddine, B., Aebischer, O. F., Amrein, W., Donnio, B., Deschenaux, R., Guillon, D., Savary, C., Scanu, D., Scheidegger, O., and Jenny, T. A. *Chem. Mater.* 17, 4798–4807, 2005.
953. Aebischer, O. F., Tondo, P., Alameddine, B., and Jenny, T. A. *Synthesis* 17, 2891–2896, 2006.
954. Aebischer, O. F., Aebischer, A., Tondo, P., Alameddine, B., Dadras, M., Gudel, H. U., and Jenny, T. A. *Chem. Commun.* 4221–4223, 2006.
955. Aebischer, O. F., Aebischer, A., Donnio, B., Alameddine, B., Dadras, M., Gudel, H. U., Guillon, D., and Jenny, T. A. *J. Mater. Chem.* 17, 1262–1267, 2007.
956. Grimsdale, A. C., Bauer, R., Weil, T., Tchebotareva, N., Wu, J., Watson, M., and Mullen, K. *Synthesis* 1229–1238, 2002.
957. Elmahdy, M. M., Dou, X., Mondeshki, M., Floudas, G., Butt, H. J., Spiess, H. W., and Mullen, K. *J. Am. Chem. Soc.* 130, 5311–5319, 2008.
958. Tchebotareva, N., Yin, X., Watson, M. D., Samori, P., Rabe, J. P., and Mullen, K. *J. Am. Chem. Soc.* 125, 9734–9739, 2003.
959. Samori, P., Yin, X., Tchebotareva, N., Wang, Z., Pakula, T., Jackel, F., Watson, M. D., Venturini, A., Mullen, K., and Rabe, J. P. *J. Am. Chem. Soc.* 126, 3567–3575, 2004.
960. Dou, X., Pisula, W., Wu, J., Bodwell, G. J., and Mullen, K. *Chem. Eur. J.* 14, 240–249, 2008.
961. Brand, J. D., Kubel, C., Ito, S., and Mullen, K. *Chem. Mater.* 12, 1638–1647, 2000.
962. Ito, S., Wehmeier, M., Brand, J. D., Kubel, C., Epsch, R., Rabe, J. P., and Mullen, K. *Chem. Eur. J.* 6, 4327–4342, 2000.
963. Kastler, M., Pisula, W., Davies, R. J., Gorelik, T., Kolb, U., and Mullen, K. *Small* 3, 1438–1444, 2007.
964. Wu, J., Watson, M. D., Zhang, L., Wang, Z., and Mullen, K. *J. Am. Chem. Soc.* 126, 177–186, 2004.
965. Wu, J., Watson, M. D., and Mullen, K. *Angew. Chem. Int. Ed.* 42, 5329–5333, 2003.
966. Wu, J., Baumgarten, M., Debije, M. G., Warman, J. M., and Mullen, K. *Angew. Chem. Int. Ed.* 43, 5331–5335, 2004.
967. Lee, M., Kim, J. W., Peleshanko, S., Larson, K., Yoo, Y. S., Vaknin, D., Markutsya, S., and Tsukruk, V. V. *J. Am. Chem. Soc.* 124, 9121–9128, 2002.
968. Wu, J., Li, J., Kolb, U., and Mullen, K. *Chem. Commun.* 48–50, 2006.

969. Watson, M. D., Debije, M. G., Warman, J. M., and Mullen, K. *J. Am. Chem. Soc.* 126, 766–771, 2004.

970. Wang, Z., Watson, M. D., Wu, J., and Mullen, K. *Chem. Commun.* 336–337, 2004.

971. Feng, X., Pisula, W., Takase, M., Dou, X., Enkelmann, V., Wagner, M., Ding, N., and Mullen, K. *Chem. Mater.* 20, 2872–2874, 2008.

972. El Hamaoui, B., Zhi, L., Pisula, W., Kolb, U., Wu, J., and Mullen, K. *Chem. Commun.* 2384–2386, 2007.

973. Jin, W., Yamamoto, Y., Fukushima, T., Ishii, N., Kim, J., Kato, K., Takata, M., and Aida, T. *J. Am. Chem. Soc.* 130, 9434–9440, 2008.

974. Hill, J. P., Jin, W., Kosaka, A., Fukushima, T., Ichihara, H., Shimomura, T., Ito, K., Hashizume, T., Ishii, N., and Aida, T. *Science* 304, 1481–1483, 2004.

975. Jin, W., Fukushima, T., Niki, M., Kosaka, A., Ishii, N., and Aida, T. *Proc. Natl. Acad. Sci. USA* 102, 10801–10806, 2005.

976. Motoyanagi, J., Fukushima, T., Ishii, N., and Aida, T. *J. Am. Chem. Soc.* 128, 4220–4221, 2006.

977. Yamamoto, Y., Fukushima, T., Suna, Y., Ishii, N., Saeki, A., Seki, S., Tagawa, S., Taniguchi, M., Kawai, T., and Aida, T. *Science* 314, 1761–1764, 2006.

978. Yamamoto, Y., Fukushima, T., Saeki, A., Seki, S., Tagawa, S., Ishii, N., and Aida, T. *J. Am. Chem. Soc.* 129, 9276–9277, 2007.

979. Xiao, S., Myers, M., Miao, Q., Sanaur, S., Pang, K., Steigerwald, M. L., and Nuckolls, C. *Angew. Chem. Int. Ed.* 44, 7390–7394, 2005.

980. Plunkett, K. N., Godula, K., Nuckolls, C., Tremblay, N., Whalley, A. C., and Xiao, S. *Org. Lett.* 11, 2225–2228, 2009.

981. Harris, K. D., Xiao, S., Lee, C. Y., Strano, M. S., Nuckolls, C., and Blanchet, G. B. *J. Phys. Chem. C* 111, 17947–17951, 2007.

982. Cohen, Y. S., Xiao, S., Steigerwald, M. L., Nuckolls, C., and Kagan, C. R. *Nano Lett.* 6, 2838–2841, 2006.

983. Xiao, S., Tang, J., Beetz, T., Guo, X., Tremblay, N., Siegrist, T., Zhu, Y., Steigerwald, M., and Nuckolls, C. *J. Am. Chem. Soc.* 128, 10700–10701, 2006.

984. Guo, X., Myers, M., Xiao, S., Lefenfeld, M., Steiner, R., Tulevski, G. S., Tang, J., Baumert, J., Leibfarth, F., Yardley, J. T., Steigerwald, M. L., Kim, P., and Nuckolls, C. *Proc. Natl. Acad. Sci. USA* 103, 11452–11456, 2006.

985. Feng, X., Pisula, W., and Mullen, K. *J. Am. Chem. Soc.* 129, 14116–14117, 2007.

986. Kastler, M., Schmidt, J., Pisula, W., Sebastiani, D., and Mullen, K. *J. Am. Chem. Soc.* 128, 9526–9534, 2006.

987. Tomovic, Z., Watson, M. D., and Mullen, K. *Angew. Chem. Int. Ed.* 43, 755–758, 2004.

988. Feng, X., Liu, M., Pisula, W., Takase, M., Li, J., and Mullen, K. *Adv. Mater.* 20, 2684–2689, 2008.

989. Wasserfallen, D., Kastler, M., Pisula, W., Hofer, W. A., Fogel, Y., Wang, Z., and Mullen, K. *J. Am. Chem. Soc.* 128, 1334–1339, 2006.

990. Iyer, V. S., Yoshimura, K., Enkelmann, V., Epsch, R., Rabe, J. P., and Mullen, K. *Angew. Chem. Int. Ed.* 37, 2696–2699, 1998.

991. Debije, M. G., Piris, J., de Hass, M. P., Warman, J. M., Tomovic, Z., Simpson, C. D., Watson, M. D., and Mullen, K. *J. Am. Chem. Soc.* 126, 4641–4645, 2004.

992. Zimmermann, H., Tolstoy, P., Limbach, H. H., Poupko, R., and Luz, Z. *J. Phys. Chem. B* 108, 18772–18778, 2004.

993. Lutz Jr., M. R., French, D. C., Rehage, P., and Becker, D. P. *Tetrahedron Lett.* 48, 6368–6371, 2007.

994. Luz, Z., Poupko, R., Wachtel, E. J., Zheng, H., Friedman, N., Cao, X., Freedman, T. B., Nafie, L. A., and Zimmermann, H. *J. Phys. Chem. A* 111, 10507–10516, 2007.

995. Collet, A. In *Comprehensive Supramolecular Chemistry*, Atwood, J. L., Davies, J. E. D., MacNicol, D. D., Vogtle, F., and Lehn, J. M. (eds.), Pergamon, Oxford, U.K., pp. 281–303, 1996.

996. Burlinson, N. E. and Ripmeester, J. A. *J. Inclusion Phenom.* 1, 403–409, 1984.

997. Steed, J. W., Zhang, H., and Atwood, J. L. *Supramol. Chem.* 7, 37–45, 1996.

998. Ahmad, R. and Hardie, M. J. *Supramol. Chem.* 18, 29–38, 2006.

999. Caira, M. R., Jacobs, A., and Nassimbeni, L. R. *Supramol. Chem.* 16, 337–342, 2004.

1000. Huber, J. G., Dubois, L., Desvaux, H., Dutasta, J. P., Brotin, T., and Berthault, P. *J. Phys. Chem. A* 108, 9608–9615, 2004.

1001. Ahmad, R., Dix, I., and Hardie, M. J. *Inorg. Chem.* 42, 2182–2184, 2003.

1002. Holman, K. T., Orr, G. W., Atwood, J. L., and Steed, J. W. *Chem. Commun.* 2109–2110, 1998.

1003. Steed, J. W., Junk, P. C., Atwood, J. L., Barnes, M. J., Raston, C. L., and Burkhalter, R. S. *J. Am. Chem. Soc.* 116, 10346–10347, 1994.

1004. Konarev, D. V., Khasanov, S. S., Vorontsov, I. I., Saito, G., Antipin, M. Y., Otsuka, A., and Lyubovskaya, R. N. *Chem. Commun.* 2548–2549, 2002.

1005. Zhang, S. and Echegoyen, L. *J. Am. Chem. Soc.* 127, 2006–2011, 2005.

1006. Zhang, S., Palkar, A., Fragoso, A., Prados, P., de Mendoza, J., and Echegoyen, L. *Chem. Mater.* 17, 2063–2068, 2005.

1007. Zhang, S. and Echegoyen, L. *Comptes Rendus Chim.* 9, 1031–1037, 2006.

1008. Collet, A., Dutasta, J. P., Lozach, B., and Canceill, J. *Top. Curr. Chem.* 165, 103–129, 1993.

1009. Ahmad, R. and Hardie, M. J. *Cryst. Eng. Comm.* 4, 227–231, 2002.

1010. Zimmermann, H., Poupko, R., Luz, Z., and Billard, J. *Z. Naturforsch* 40a, 149–160, 1985.

1011. Malthete, J. and Collet, A. *Nouv. J. Chem.* 9, 151–153, 1985.

1012. Robinson, G. M. *J. Chem. Soc.* 107, 267–276, 1915.

1013. Lindsey, A. S. *Chem. Ind. (Lond.)* 823–824, 1963.

1014. Lindsey, A. S. *J. Chem. Soc.* 1685–1692, 1965.

1015. White, J. D. and Gesner, B. D. *Tetrahedron* 30, 2273–2277, 1974.

1016. Scott, J. L., MacFarlane, D. R., Raston, C. L., and Teoh, C. M. *Green Chem.* 2, 123–126, 2000.

1017. Zimmermann, H., Poupko, R., Luz, Z., and Billard, J. *Z. Naturforsch* 41a, 1137–1140, 1986.

1018. Poupko, R., Luz, Z., Spielberg, N., and Zimmermann, H. *J. Am. Chem. Soc.* 111, 6094–6105, 1989.

1019. Zamir, S., Luz, Z., Poupko, R., Alexander, S., and Zimmermann, H. *J. Chem. Phys.* 94, 5927–5938, 1991.

1020. Budig, H., Lunkwitz, R., Paschke, R., Tschierske, C., Nutz, U., Diele, S., and Pelzl, G. *J. Mater. Chem.* 6, 1283–1289, 1996.

1021. Jakli, A., Muller, M., and Heppke, G. *Liq. Cryst.* 26, 945–952, 1999.

1022. Zimmermann, H., Bader, V., Poupko, R., Wachtel, E. J., and Luz, Z. *J. Am. Chem. Soc.* 124, 15286–15301, 2002.

1023. Luz, Z., Poupko, R., Wachtel, E. J., Zimmermann, H., and Bader, V. *Mol. Cryst. Liq. Cryst.* 397, 67–77, 2003.

1024. Percec, V., Cho, C. G., and Pugh, C. *Macromolecules* 24, 3227–3234, 1991.

1025. Zimmermann, H., Poupko, R., Luz, Z., and Billard, J. *Liq. Cryst.* 3, 759–770, 1988.

1026. Zimmermann, H., Poupko, R., Luz, Z., and Billard, J. *Liq. Cryst.* 6, 151–166, 1989.

1027. Kranig, W., Spiess, H. W., and Zimmermann, H. *Liq. Cryst.* 7, 123–129, 1990.

1028. Percec, V., Cho, C. G., and Pugh, C. *J. Mater. Chem.* 1, 217–222, 1991.

1029. Spielberg, N., Sarkar, M., Luz, Z., Poupko, R., Billard, J., and Zimmermann, H. *Liq. Cryst.* 15, 311–330, 1993.

1030. Percec, V., Cho, C. G., Pugh, C., and Tomazos, D. *Macromolecules* 25, 1164–1176, 1992.

1031. Swager, T. M. and Xu, B. J. *Inclusion Phenom. Mol. Recognit. Chem.* 19, 389–398, 1994.

1032. Kuebler, S. C., Boeffel, C., and Spiess, H. W. *Liq. Cryst.* 18, 309–318, 1995.

1033. Budig, H., Diele, S., Göring, P., Paschke, R., Sauer, C., and Tschierske, C. *J. Chem. Soc. Chem. Comm.* 2359–2360, 1994.

1034. Budig, H., Diele, S., Göring, P., Paschke, R., Sauer, C., and Tschierske, C. *J. Chem. Soc. Perkin Trans 2* 767–775, 1995.

1035. Lunkwitz, R., Tschierske, C., and Diele, S. *J. Mater. Chem.* 7, 2001–2011, 1997.

1036. Cometti, G., Dalcanale, E., Du Vosel, A., and Levelut, A. M. *J. Chem. Soc. Chem. Commun.* 163–165, 1990.

1037. Bonsignore, S., Cometti, G., Dalcanale, E., and Du Vosel, A. *Liq. Cryst.* 8, 639–649, 1990.

1038. Dalcanale, E., Du Vosel, A., Levelut, A. M., and Malthete, J. *Liq. Cryst.* 10, 185–198, 1991.

1039. Abis, L., Arrighi, V., Cometti, G., Dalcanale, E., and Du Vosel, A. *Liq. Cryst.* 9, 277–284, 1991.

1040. Cometti, G., Dalcanale, E., Du Vosel, A., and Levelut, A. M. *Liq. Cryst.* 11, 93–100, 1992.

1041. Bonsignore, S., Du Vosel, A., Guglielmetti, G., Dalcanale, E., and Ugozzli, F. *Liq. Cryst.* 13, 471–482, 1993.

1042. Xu, B. and Swager, M. *J. Am. Chem. Soc.* 115, 1159–1160, 1993.

1043. Akopova, O. B. and Bronnikova, A. A. *J. Struct. Chem.* 39, 384–387, 1998.

1044. Zhang, J. and Moore, J. S. *J. Am. Chem. Soc.* 116, 2655–2656, 1994.

1045. Mindyuk, O. Y., Stetzer, M. R., Heiney, P. A., Nelson, J. C., and Moore, J. S. *Adv. Mater.* 10, 1363–1369, 1998.

1046. Hoger, S., Meckenstock, A. D., and Muller, S. *Chem. Eur. J.* 4, 2423–2434, 1998.

1047. Zhang, D., Tessier, C. A., and Youngs, W. J. *Chem. Mater.* 11, 3050–3057, 1999.

1048. Hoger, S., Enkelmann, V., Bonrad, K., and Tschierske, C. *Angew. Chem. Int. Ed.* 39, 2268–2270, 2000.

1049. Hoger, S., Morrison, D. L., Bonrad, K., Enkelmann, V., and Tschierske, C. *Mol. Cryst. Liq. Cryst.* 389, 73–78, 2002.

1050. Hoger, S. *Chem. Eur. J.* 10, 1320–1329, 2004.
1051. Hoger, S., Cheng, X. H., Ramminger, A. D., Enkelmann, V., Rapp, A., Mondeshki, M., and Schnell, I. *Angew. Chem. Int. Ed.* 44, 2801–2805, 2005.
1052. Ziegler, A., Mamdouh, W., ver Heyen, A., Surin, M., Uji-i, H., Mottaleb, M. M. S. A., de Schryver, F. C., de Feyter, S., Lazzaroni, R., and Hoger, S. *Chem. Mater.* 17, 5670–5683, 2005.
1053. Seo, S. H., Jones, T. V., Seyler, H., Peters, J. O., Kim, T. H., Chang, J. Y., and Tew, G. N. *J. Am. Chem. Soc.* 128, 9264–9265, 2006.
1054. Pisula, W., Kastler, M., Yang, C., Enkelmann, V., and Müllen, K. *Chem. Asian J.* 2, 51–56, 2007.
1055. Hoger, S., Weber, J., Leppert, A., and Enkelmann, V. *Beil. J. Org. Chem.* 4, 1–8, 2008.
1056. Staab, H. A. and Neunhoeffer, K. *Synthesis* 6, 424, 1974.
1057. Barbera, J., Rakitin, O. A., Ros, M. B., and Torroba, T. *Angew. Chem. Int. Ed.* 37, 296–299, 1998.
1058. Basurto, S., Garcia, S., Neo, A. G., Torroba, T., Marcos, C. F., Miguel, D., Barbera, J., Ros, M. B., and de la Fuente, M. R. *Chem. Eur. J.* 11, 5362–5376, 2005.
1059. Lau, K., Foster, J., and Williams, V. *Chem. Commun.* 2172–2173, 2003.
1060. Wuckert, E., Dix, M., Laschat, S., Baro, A., Schulte, J. L., Hagele, C., and Giesselmann, F. *Liq. Cryst.* 31, 1305–1309, 2004.
1061. Wuckert, E., Laschat, S., Baro, A., Hagele, C., Giesselmann, F., and Luftmann, H. *Liq. Cryst.* 33, 103–107, 2006.
1062. Artal, M. C., Toyne, K. J., Goodby, J. W., Barbera, J., and Photinos, D. J. *J. Mater. Chem.* 11, 2801–2807, 2001.
1063. Nuckolls, C., Katz, T. J., and Castellanos, L. *J. Am. Chem. Soc.* 118, 3767–3768, 1996.
1064. Lovinger, A. J., Nuckolls, C., and Katz, T. J. *J. Am. Chem. Soc.* 120, 264–268, 1998.
1065. Nuckolls, C. and Katz, T. J. *J. Am. Chem. Soc.* 120, 9541–9544, 1998.
1066. Nuckolls, C., Katz, T. J., Verbiest, T., van Elshocht, S., Kuball, H. G., Kiesewalter, S., Lovinger, A. J., and Persoons, A. *J. Am. Chem. Soc.* 120, 8656–8660, 1998.
1067. Nuckolls, C., Katz, T. J., Katz, G., Collings, P. J., and Castellanos, L. *J. Am. Chem. Soc.* 121, 79–88, 1999.
1068. Verbiest, T., Sioncke, S., Persoons, A., Vyklicky, L., and Katz, T. J. *Angew. Chem. Int. Ed.* 41, 3882–3884, 2002.
1069. Phillips, K. E. S., Katz, T. J., Jockusch, S., Lovinger, A. J., and Turro, N. J. *J. Am. Chem. Soc.* 123, 11899–11907, 2001.
1070. Nuckolls, C., Shao, R., Jang, W. G., Clark, N. A., Walba, D. M., and Katz, T. J. *Chem. Mater.* 14, 773–776, 2002.
1071. Vyklicky, L., Eichhorn, S. H., and Katz, T. J. *Chem. Mater.* 15, 3594–3601, 2003.
1072. Pegenau, A., Hegmann, T., Tschierske, C., and Diele, S. *Chem. Eur. J.* 5, 1643–1660, 1999.
1073. Pegenau, A., Goring, P., Diele, S., and Tschierske, C. In *Liquid Crystals: Chemistry and Structure*, Tykarska, M., Dabroski, R., and Zielinski, J. (eds.), *Proc. SPIE* 3319, 70, 1998.
1074. Dudley, J. R., Thurston, J. T., Schaefer, F. C., Hansen, D. H., Hull, C. J., and Adams, P. *J. Am. Chem. Soc.* 73, 2986–2989, 1951.
1075. Gamez, P. and Reedijk, J. *Eur. J. Inorg. Chem.* 29–42, 2006.
1076. Blotny, G. *Tetrahedron* 62, 9507–9522, 2006.
1077. Giacomelli, G., Porcheddu, A., and de Luca, L. *Curr. Org. Chem.* 8, 1497–1519, 2004.
1078. Lehn, J. M. In *Supramolecular Chemistry: Concepts and Perspectives*, VCH, Weinheim, Germany, 1995.
1079. Lawrence, D. S., Jiang, T., and Levett, M. *Chem. Rev.* 95, 2229–2260, 1995.
1080. Vora, R. A. and Patel, N. C. *Mol. Cryst. Liq. Cryst.* 129, 159–168, 1985.
1081. Janietz, D., Sundholm, F., Leppanen, J., Karhinen, H., and Bauer, M. *Liq. Cryst.* 13, 499–505, 1993.
1082. Mormann, W. and Zimmermann, J. G. *Liq. Cryst.* 19, 481–488, 1995.
1083. Lattermann, G. and Hocker, H. *Mol. Cryst. Liq. Cryst.* 133, 245–257, 1986.
1084. Paleos, C. M., Tsiourvas, D., Fillipakis, S., and Fillipaki, L. *Mol. Cryst. Liq. Cryst.* 242, 9–15, 1994.
1085. Barclay, G. G., Ober, C. K., Papathomas, K. I., and Wang, D. W. *Macromolecules* 25, 2947–2954, 1992.
1086. Wang, Y. H., Hong, Y. L., Yen, F. S., and Hong, J. L. *Mol. Cryst. Liq. Cryst.* 287, 109–113, 1996.
1087. Lo, W. J., Hong, Y. L., Lin, R. H., and Hong, J. L. *Mol. Cryst. Liq. Cryst.* 308, 133–146, 1997.
1088. Goldmann, D., Janietz, D., Schmidt, C., and Wendorff, J. H. *Liq. Cryst.* 25, 711–719, 1998.
1089. Gibson, H. W., Dotson, D. L., Marand, H., and Swager, T. M. *Mol. Cryst. Liq. Cryst.* 326, 113–138, 1999.
1090. Huang, S. J., Feldman, J. A., and Cercena, J. L. *Polym. Prepr.* 30, 348–349, 1989.
1091. Goldmann, D., Janietz, D., Festag, R., Schmidt, C., and Wendorff, J. H. *Liq. Cryst.* 21, 619–623, 1996.
1092. Goldmann, D., Dietel, R., Janietz, D., Schmidt, C., and Wendorff, J. H. *Liq. Cryst.* 24, 407–411, 1998.
1093. Janietz, D., Goldmann, D., Schmidt, C., and Wendorff, J. H. *Mol. Cryst. Liq. Cryst.* 332, 141–148, 1999.

1094. Goldmann, D., Janietz, D., Schmidt, C., and Wendorff, J. H. *Angew. Chem. Int. Ed.* 39, 1851–1854, 2000.

1095. Goldmann, D., Janietz, D., Schmidt, C., and Wendorff, J. H. *J. Mater. Chem.* 14, 1521–1525, 2004.

1096. Barbera, J., Puig, L., Serrano, J. L., and Sierra, T. *Chem. Mater.* 16, 3308–3317, 2004.

1097. Barbera, J., Puig, L., Romero, P., Serrano, J. L., and Sierra, T. *J. Am. Chem. Soc.* 127, 458–464, 2005.

1098. Kohlmeier, A. and Janietz, D. *Liq. Cryst.* 34, 289–294, 2007.

1099. Alvarez, L., Barbera, J., Puig, L., Romero, P., Serrano, J. L., and Sierra, T. *J. Mater. Chem.* 16, 3768–3773, 2006.

1100. Barbera, J., Puig, L., Romero, P., Serrano, J. L., and Sierra, T. *J. Am. Chem. Soc.* 128, 4487–4492, 2006.

1101. Xu, M., Chen, L., Zhou, Y., Yi, T., Li, F., and Huang, C. *J. Colloid Interface Sci.* 326, 496–502, 2008.

1102. Yagai, S., Nakajima, T., Kishikawa, K., Kohmoto, S., Karatsu, T., and Kitamura, A. *J. Am. Chem. Soc.* 127, 11134–11139, 2005.

1103. Akopova, O. B., Frolova, T. V., and Usol'tseva, N. V. *Rus. J. Gen. Chem.* 77, 103–110, 2007.

1104. Bai, R., Li, S., Zou, Y., Pan, C., Xie, P., Kong, B., Zhou, X., and Zhang, R. *Liq. Cryst.* 28, 1873–1876, 2001.

1105. Lee, C. J., Lee, S. J., and Chang, J. Y. *Tetrahedron Lett.* 43, 3863–3866, 2002.

1106. Lee, S. J. and Chang, J. Y. *Tetrahedron. Lett.* 44, 7493–7497, 2003.

1107. Mahlstedt, S. and Bauer, M. Abstracts of papers of the *Am. Chem. Soc.* 208, 801–802, 1994.

1108. Schaefer, F. C., Thurston, J. T., and Dudley, J. R. *J. Am. Chem. Soc.* 73, 2990–2992, 1951.

1109. Lee, C. H. and Yamamoto, T. *Tetrahedron Lett.* 42, 3993–3996, 2001.

1110. Lee, C. H. and Yamamoto, T. *Mol. Cryst. Liq. Cryst.* 378, 13–21, 2002.

1111. Ishi-i, T., Kuwahara, R., Takata, A., Jeong, Y., Sakurai, K., and Mataka, S. *Chem. Eur. J.* 12, 763–776, 2006.

1112. Kotha, S., Kashinath, D., and Kumar, S. *Tetrahedron Lett.* 49, 5419–5423, 2008.

1113. Kohlmeier, A. and Janietz, D. *Chem. Mater.* 18, 59–68, 2006.

1114. Kohlmeier, A., Janietz, D., and Diele, S. *Chem. Mater.* 18, 1483–1489, 2006.

1115. Maly, K. E., Dauphin, C., and Wuest, J. D. *J. Mater. Chem.* 16, 4695–4700, 2006.

1116. Bubulak, T. V., Buchs, J., Kohlmeier, A., Bruma, M., and Janietz, D. *Chem. Mater.* 19, 4460–4466, 2007.

1117. Lee, H., Kim, D., Lee, H. K., Qiu, W., Oh, N. K., Zin, W. C., and Kim, K. *Tetrahedron Lett.* 45, 1019–1022, 2004.

1118. Shu, W. and Valiyaveettil, S. *Chem. Commun.* 1350–1351, 2002.

1119. Pieterse, K., Lauritsen, A., Schenning, A. P. H. J., Vekemans, J. A. J. M., and Meijer, E. W. *Chem. Eur. J.* 9, 5597–5604, 2003.

1120. Meier, H., Holst, H. C., and Oehlhof, A. *Eur. J. Org. Chem.* 4173–4180, 2003.

1121. Holst, H. C., Pakula, T., and Meier, H. *Tetrahedron* 60, 6765–6775, 2004.

1122. Meier, H., Karpuk, E., and Holst, H. C. *Eur. J. Org. Chem.* 2609–2617, 2006.

1123. Irle, C. and Mormann, W. *Liq. Cryst.* 21, 295–305, 1996.

1124. Porter, A. E. A. In *Comprehensive Heterocyclic Chemistry*, Vol. 3, Katritzky, A. R., Rees, C. W., Boulton, J. J., and Mckillop, A. (eds.), Pergamon Press, New York, 1984.

1125. Hu, J., Zhang, D., Jin, S., Cheng, S. Z. D., and Harris, F. W. *Chem. Mater.* 16, 4912–4915, 2004.

1126. Lee, D. C., Jang, K., McGrath, K. K., Uy, R., Robins, K. A., and Hatchett, D. W. *Chem. Mater.* 20, 3688–3695, 2008.

1127. Kaafarani, B. R., Lucas, L. A., Wex, B., and Jabbour, G. E. *Tetrahedron. Lett.* 48, 5995–5998, 2007.

1128. Gao, B., Wang, M., Cheng, Y., Wang, L., Jing, X., and Wang, F. *J. Am. Chem. Soc.* 130, 8297–8306, 2008.

1129. Jradi, F. M., Al-Sayah, M. H., and Kaafarani, B. R. *Tetrahedron. Lett.* 49, 238–242, 2008.

1130. Foster, E. J., Lavigueur, C., Ke, Y. C., and Williams, V. E. *J. Mater. Chem.* 15, 4062–4068, 2005.

1131. Foster, E. J., Jones, R. B., Lavigueur, C., and Williams, V. E. *J. Am. Chem. Soc.* 128, 8569–8574, 2006.

1132. Lavigueur, C., Foster, E. J., and Williams, V. E. *Liq. Cryst.* 34, 833–840, 2007.

1133. Lavigueur, C., Foster, E. J., and Williams, V. E. *J. Am. Chem. Soc.* 130, 11791–11800, 2008.

1134. Ong, C. W., Hwang, J. Y., Tzeng, M. C., Liao, S. C., Hsu, H. F., and Chang, T. H. *J. Mater. Chem.* 17, 1785–1790, 2007.

1135. Mohr, B., Wegner, G., and Ohta, K. *J. Chem. Soc. Chem. Commun.* 995–996, 1995.

1136. Babuin, J., Foster, J., and Williams, V. E. *Tetrahedron. Lett.* 44, 7003–7005, 2003.

1137. Foster, E. J., Babuin, J., Nguyen, N., and Williams, V. E. *Chem. Commun.* 2052–2053, 2004.

1138. Kohne, B. and Praefcke, K. *Leibigs. Ann. Chem.* 522–528, 1985.

1139. Rogers, D. Z. *J. Org. Chem.* 51, 3904–3905, 1986.

1140. Kanakarajan, K. and Czarnik, A. W. *J. Org. Chem.* 51, 5241–5243, 1986.

1141. Rademacher, J. T., Kanakarajan, K., and Czarnik, A. W. *Synthesis* 378–380, 1993.

1142. Beeson, J. C., Lawrence, J. F., Gallucci, J. C., Gerkin, R. E., Rademacher, J. T., and Czarnik, A. W. *J. Am. Chem. Soc.* 116, 4621–4622, 1994.
1143. Aumiller, W. D., Dalton, C. R., and Czarnik, A. W. *J. Org. Chem.* 60, 728–729, 1995.
1144. Ishi-i, T., Murakami, K. I., Imai, Y., and Mataka, S. *J. Org. Chem.* 71, 5752–5760, 2006.
1145. Ishi-i, T., Murakami, K. I., Imai, Y., and Mataka, S. *Org. Lett.* 7, 3175–3178, 2005.
1146. Ishi-i, T., Yaguma, K., Kuwahara, R., Taguri, Y., and Mataka, S. *Org. Lett.* 8, 585–588, 2006.
1147. Kim, Y. K., Kim, J. W., and Park, Y. *Appl. Phys. Lett.* 94, 063305, 2009.
1148. Secondo, P. and Fages, F. *Org. Lett.* 8, 1311–1314, 2006.
1149. Gao, B., Liu, Y., Geng, Y., Cheng, Y., Wang, L., Jing, X., and Wang, F. *Tetrahedron Lett.* 50, 1649–1652, 2009.
1150. Salomon, E., Zhang, Q., Barlow, S., Marder, S. R., and Kahn, A. *J. Phys. Chem. C* 112, 9803–9807, 2008.
1151. Zhao, W., Salomon, E., Zhang, Q., Barlow, S., Marder, S. R., and Kahn, A. *Phys. Rev B.* 77, 165336, 2008.
1152. Cho, B. R., Lee, S. K., Kim, K. A., Son, K. N., Kang, T. I., and Jeon, S. J. *Tetrahedron. Lett.* 39, 9205–9208, 1998.
1153. Furukawa, S., Okubo, T., Masaoka, S., Tanaka, D., Chang, H. C., and Kitagawa, S. *Angew. Chem. Int. Ed.* 44, 2700–2704, 2005.
1154. Mascaros, J. R. G. and Dunbar, K. R. *Chem. Commun.* 217–218, 2001.
1155. D'Alessandro, D. M., Davies, M. S., and Keene, F. R. *Inorg. Chem.* 45, 1656–1666, 2006.
1156. Herrera, J. M., Ward, M. D., Adams, H., Pope, S. J. A., and Faulkner, S. *Chem. Commun.* 1851–1853, 2006.
1157. Bu, X. H., Biradha, K., Yamaguchi, T., Nishimura, M., Ito, T., Tanaka, K., and Shionoya, M. *Chem. Commun.* 1953–1954, 2000.
1158. Rademacher, J. T. and Czarnik, A. W. *J. Am. Chem. Soc.* 115, 3018–3019, 1993.
1159. Lozman, O. R., Bushby, R. J., and Vinter, J. G. *J. Chem. Soc. Perkin Trans. 2* 2, 1446–1452, 2001.
1160. Pieterse, K., van Hal, P. A., Kleppinger, R., Vekemans, J. A. J. M., Janssen, R. A. J., and Meijer, E. W. *Chem. Mater.* 13, 2675–2679, 2001.
1161. Chang, T. H., Wu, B. R., Chiang, M. Y., Liao, S. C., Ong, C. W., Hsu, H. F., and Lin, S. Y. *Org. Lett.* 7, 4075–4078, 2005.
1162. Roussel, O., Kestemont, G., Tant, J., de Halleux, V., Aspe, R. G., Levin, J., Remacle, A., Gearba, I. R., Ivanov, D., Lehmann, M., and Geerts, Y. *Mol. Cryst. Liq. Cryst.* 396, 35–39, 2003.
1163. Gearba, R. I., Lehmann, M., Levin, J., Ivanov, D. A., Koch, M. H. J., Barbera, J., Debije, M. G., Piris, J., and Geerts, Y. H. *Adv. Mater.* 15, 1614–1618, 2003.
1164. Kanakarajan, K. and Czarnik, A. W. *J. Heterocycl. Chem.* 25, 1869–1872, 1988.
1165. Palma, M., Levin, J., Debever, O., Geerts, Y., Lehmann, M., and Samori, P. *Soft Matter* 4, 303–310, 2008.
1166. Palma, M., Levin, J., Lemaur, V., Liscio, A., Palermo, V., Cornil, J., Geerts, Y., Lehmann, M., and Samori, P. *Adv. Mater.* 18, 3313–3317, 2006.
1167. Ishi-i, T., Hirayama, T., Murakami, K. I., Tashiro, H., Thiemann, T., Kubo, K., Mori, A., Yamasaki, S., Akao, T., Tsuboyama, A., Mukaide, T., Ueno, K., and Mataka, S. *Langmuir* 21, 1261–1268, 2005.
1168. Barlow, S., Zhang, Q., Kaafarani, B. R., Risko, C., Amy, F., Chan, C. K., Domercq, B., Starikova, Z. A., Antipin, M. Y., Timofeeva, T. V., Kippelen, B., Bredas, J. L., Kahn, A., and Marder, S. R. *Chem. Eur. J.* 13, 3537–3547, 2007.
1169. Kaafarani, B. R., Kondo, T., Yu, J., Zhang, Q., Dattilo, D., Risko, C., Jones, S. C., Barlow, S., Domercq, B., Amy, F., Kahn, A., Bredas, J. L., Kippelen, B., and Marder, S. R. *J. Am. Chem. Soc.* 127, 16358–16359, 2005.
1170. Xu, Q. M., Ma, H., Tucker, N., Bardecker, J. A., and Jen, A. K. Y. *Nanotechnology* 18, 335302, 2007.
1171. Ha, S. D., Kaafarani, B. R., Barlow, S., Marder, S. R., and Kahn, A. *J. Phys. Chem. C* 111, 10493–10497, 2007.
1172. Lind, S. J., Walsh, T. J., Blackman, A. G., Polson, M. I. J., Irwin, G. I. S., and Gordon, K. C. *J. Phys. Chem. A* 113, 3566–3575, 2009.
1173. Piglosiewicz, I. M., Beckhaus, R., Wittstock, G., Saak, W., and Hasse, D. *Inorg. Chem.* 46, 7610–7620, 2007.
1174. Piglosiewicz, I. M., Beckhaus, R., Saak, W., and Hasse, D. *J. Am. Chem. Soc.* 127, 14190–14191, 2005.
1175. Yip, H. L., Zou, J., Ma, H., Tian, Y., Tucker, N. M., and Jen, A. K. Y. *J. Am. Chem. Soc.* 128, 13042–13043, 2006.
1176. Budd, P. M., Ghanem, B., Msayib, K., McKeown, N. B., and Tattershall, C. *J. Mater. Chem.* 13, 2721–2726, 2003.
1177. Skujins, S. and Webb, G. A. *Tetrahedron* 25, 3935–3945, 1969.

1178. Kestemont, G., de Halleux, V., Lehmann, M., Ivanov, D. A., Watson, M., and Geerts, Y. H. *Chem. Commun.* 2074–2075, 2001.

1179. Lehmann, M., Kestemont, G., Aspe, R. G., Herman, C. B., Koch, M. H. J., Debije, M. G., Piris, J., de Haas, M. P., Warman, J. M., Watson, M. D., Lemaur, V., Cornil, J., Geerts, Y. H., Gearba, R., and Ivanov, D. A. *Chem. Eur. J.* 11, 3349–3362, 2005.

1180. Domercq, B., Yu, J., Kaafarani, B. R., Kondo, T., Yoo, S., Haddock, J. N., Barlow, S., Marder, S. R., and Kippelen, B. *Mol. Cryst. Liq. Cryst.* 481, 80–93, 2008.

1181. Clarke, J., Archer, R., Redding, T., Foden, C., Tant, J., Geerts, Y., Friend, R. H., and Silva, C. *J. Appl. Phys.* 103, 124510, 2008.

1182. Crispin, X., Cornil, J., Friedlein, R., Okudaira, K. K., Lemaur, V., Crispin, A., Kestemont, G., Lehmann, M., Fahlman, M., Lazzaroni, R., Geerts, Y., Wendin, G., Ueno, N., Bredas, J. L., and Salaneck, W. R. *J. Am. Chem. Soc.* 126, 11889–11899, 2004.

1183. Ong, C. W., Liao, S. C., Chang, T. H., and Hsu, H. F. *Tetrahedron. Lett.* 44, 1477–1480, 2003.

1184. Ong, C. W., Liao, S. C., Chang, T. H., and Hsu, H. F. *J. Org. Chem.* 69, 3181–3185, 2004.

1185. Bock, H., Bebeau, A., Seguy, I., Jolinat, P., and Destruel, P. *ChemPhysChem.* 532–535, 2002.

1186. Mamlok, L., Malthete, J., Tinh, N. H., Destrade, C., and Levelut, A. M. *J. Phys. Lett.* 43, 641–647, 1982.

1187. Destrade, C., Tinh, N. H., Mamlok, L., and Malthete, J. *Mol. Cryst. Liq. Cryst.* 114, 139–150, 1984.

1188. Tinh, N. H., Cayuela, R., Destrade, C., and Malthete, J. *Mol. Cryst. Liq. Cryst.* 122, 141–149, 1985.

1189. Cayuela, R., Tinh, N. H., Destrade, C., and Levelut, A. M. *Mol. Cryst. Liq. Cryst.* 177, 81–91, 1989.

1190. Bergman, J. and Eklund, N. *Tetrahedron* 36, 1445–1450, 1980.

1191. Bocchi, V. and Palla, G. *Tetrahedron* 42, 5019–5024, 1986.

1192. Robertson, N., Parsons, S., MacLean, E. J., Coxallb, R. A., and Mount, A. R. *J. Mater. Chem.* 10, 2043–2047, 2000.

1193. Lor, B. G., Alonso, B., Omenat, A., and Serrano, J. L. *Chem. Commun.* 5012–5014, 2006.

1194. Talarico, M., Termine, R., Frutos, E. M. G., Omenat, A., Serrano, J. L., Lor, B. G., and Golemme, A. *Chem. Mater.* 20, 6589–6591, 2008.

1195. Frutos, E. M. G. and Lor, B. G. *J. Am. Chem. Soc.* 130, 9173–9177, 2008.

1196. Baldwin, R. W., Butler, K., Cooper, F. C., Partridge, M. W., and Cunningham, G. J. *Nature* 181, 838–839, 1958.

1197. Baldwin, R. W., Cunningham, G. J., and Partridge, M. W. *Br. J. Cancer.* 13, 94–98, 1959.

1198. Baldwin, R. W., Cunningham, G. J., Partridge, M. W., and Vipond, H. J. *Br. J. Cancer.* 16, 275–282, 1962.

1199. Brunswick, D. J., Partridge, M. W., and Vipond, H. J. *J. Chem. Soc. C* 2641–2647, 1970.

1200. Nagata, C., Kodama, M., Imamura, A., and Tagashira, Y. *GANN* 57, 75–84, 1966.

1201. Butler, K. and Partridge, M. W. *J. Chem. Soc.* 2396–2400, 1959.

1202. Iball, J. and Motherwell, W. D. S. *Acta Cryst.* B25, 882–888, 1969.

1203. Baldwin, R. W., Palmer, H. C., and Partridge, M. W. *Br. J. Cancer* 16, 740–748, 1962.

1204. Baldwin, R. W., Cunningham, G. J., Davey, A. T., Partridge, M. W., and Vipond, H. J. *Br. J. Cancer* 17, 266–271, 1963.

1205. Pakrashi, S. C. *J. Org. Chem.* 36, 642–645, 1971.

1206. Mklow, K. *Chimia* 40, 389, 1986.

1207. Yamaguchi, H., Akimoto, Y., Ikeda, T., and Yoneda, F. *J. Chem. Soc. Faraday Trans. II* 75, 1506–1514, 1979.

1208. Cundall, R. B., Grant, D. J. W., and Shulman, N. H. *J. Chem. Soc. Faraday Trans. II* 78, 27–37, 1982.

1209. Cundall, R. B., Grant, D. J. W., and Shulman, N. H. *J. Chem. Soc. Faraday Trans. II* 78, 737–750, 1982.

1210. Leupin, W., Magde, D., Persy, G., and Wirz, J. *J. Am. Chem. Soc.* 108, 17–22, 1986.

1211. Cooper, F. C. and Partridge, M. W. *J. Chem. Soc.* 3429–3435, 1954.

1212. Ponomarev, I. I. and Sinichkin, M. K. *Polym. Sci. Ser. A* 38, 951–953, 1996.

1213. Yoneda, F. and Mera, K. *Chem. Pharm. Bull.* 21, 1610–1611, 1973.

1214. Keinan, E., Kumar, S., Singh, S. P., Ghirlando, R., and Wachtel, E. J. *Liq. Cryst.* 11, 157–173, 1992.

1215. Boden, N., Borner, R. C., Bushby, R. J., and Clements, J. *J. Am. Chem. Soc.* 116, 10807–10808, 1994.

1216. Uznanski, P. and Kryszewski, M. *Proc. SPIE Int. Soc. Optic. Eng.* 3318, 398–401, 1997.

1217. Hiesgen, R., Schonherr, H., Kumar, S., Ringsdorf, H., and Meissner, D. *Thin Solid Films* 358, 241–249, 2000.

1218. Kumar, S., Wachtel, E. J., and Keinan, E. *J. Org. Chem.* 58, 3821–3827, 1993.

1219. Boden, N., Bushby, R. J., Donovan, K., Liu, Q., Lu, Z., Kreouzis, T., and Wood, A. *Liq. Cryst.* 28, 1739–1748, 2001.

1220. Bushby, R. J., Lozman, O. R., Mason, L. A., Taylor, N., and Kumar, S. *Mol. Cryst. Liq. Cryst.* 410, 171–181, 2004.
1221. Kadam, J., Faul, C. F. J., and Scherf, U. *Chem. Mater.* 16, 3867–3871, 2004.
1222. Shimizu, Y., Miyake, Y., and Kumar, S. *Poster 3-MA53, ILCC.* 2008.
1223. Battersby, A. R., Fookes, C. J. R., Matcham, G. W. J., and McDonald, E. *Nature* 285, 17–21, 1980.
1224. Falk, J. E. In *Porphyrins and Metalloporphyrins*, Elsevier, Amsterdam, the Netherlands, 1964.
1225. Wrobel, D. and Dudkowiak, A. *Mol. Cryst. Liq. Cryst.* 448, 15–38, 2006.
1226. Drain, C. M., Varotto, A., and Radivojevic, I. *Chem. Rev.* 109, 1630–1658, 2009.
1227. Macdonald, I. J. and Dougherty, T. J. *J. Porphyr. Phthalocyanins* 5, 105–129, 2001.
1228. Donnio, B. *Curr. Opin. Colloid Interface Sci.* 7, 371–394, 2002.
1229. Usoltseva, N. V. and Bykova, V. V. *Mol. Cryst. Liq. Cryst.* 215, 89–100, 1992.
1230. Bykova, V., Usol'tseva, N., Ananjeva, G., Semeiken, A., and Karmanova, T. *Mol. Cryst. Liq. Cryst.* 265, 651–657, 1995.
1231. Kashitsin, A., Usol'tseva, N., Bykova, V., Ananjeva, G., and Zhukova, L. *Mol. Cryst. Liq. Cryst.* 260, 595–603, 1995.
1232. Usol'tseva, N. *Mol. Cryst. Liq. Cryst.* 288, 201–210, 1996.
1233. Goodby, J. W., Robinson, P. S., Teo, B. K., and Clad, P. E. *Mol. Cryst. Liq. Cryst.* 56, 303–309, 1980.
1234. Gregg, B. A., Fox, M. A., and Bard, A. J. *J. Chem. Soc. Chem. Commun.* 1134–1135, 1987.
1235. Gregg, B. A., Fox, M. A., and Bard, A. J. *J. Am. Chem. Soc.* 111, 3024–3029, 1989.
1236. Knorr, L. *Chem. Ber.* 17, 1635–1642, 1884.
1237. Paganuzzi, V., Guatteri, P., Riccardi, P., Sacchelli, T., Barbera, J., Costa, M., and Dalcanale, E. *Eur. J. Org. Chem.* 1527–1539, 1999.
1238. Lee, K. S., Shin, J., Shim, Y. S., Shim, Y. K., Isoda, S., and Jin, J. *Bull. Korean Chem. Soc.* 15, 817–819, 1994.
1239. Caughey, W. S., Eberspaecher, H., Fuchsman, W. H., McCoy, S., and Alben, J. O. *Ann. N. Y. Acad. Sci.* 153, 722–737, 1969.
1240. Abraham, R. J., Eivazi, F., Pearson, H., and Smith, K. M. *J. Chem. Soc. Chem. Commun.* 699–701, 1976.
1241. Shearman, G. C., Yahioglu, G., Kirstein, J., Milgrom, L. R., and Seddon, J. M. *J. Mater. Chem.* 19, 598–604, 2009.
1242. Sartori, E., Fontana, M. P., Costa, M., Dalcanale, E., and Paganuzzi, V. *Thin Solid Films* 284–285, 204–207, 1996.
1243. Facci, P., Fontana, M. P., Dalcanale, E., Costa, M., and Sacchelli, T. *Langmuir* 16, 7726–7730, 2000.
1244. Liu, C. Y., Pan, H. L., Tang, H., Fox, M. A., and Bard, A. J. *J. Phys. Chem.* 99, 7632–7636, 1995.
1245. Schouten, P. G., Warman, J. M., de Haas, M. P., Fox, M. A., and Pan, H. L. *Nature* 353, 736–737, 1991.
1246. Liu, C. Y., Pan, H. L., Fox, M. A., and Bard, A. J. *Science* 261, 897–899, 1993.
1247. Liu, C. Y., Pan, H. L., Fox, M. A., and Bard, A. J. *Chem. Mater.* 9, 1422–1429, 1997.
1248. Gregg, B. A., Fox, M. A., and Bard, A. J. *J. Phys. Chem.* 93, 4227–4234, 1989.
1249. Gregg, B. A. *Mol. Cryst. Liq. Cryst.* 257, 219–227, 1994.
1250. Gregg, B. A., Fox, M. A., and Bard, A. J. *J. Phys. Chem.* 94, 1586–1598, 1990.
1251. Murashima, T., Wakamori, N., Uchihara, Y., Ogawa, T., Uno, H., and Ono, N. *Mol. Cryst. Liq. Cryst.* 278, 165–171, 1996.
1252. Castella, M., Lopez-Calahorra, F., Velasco, D., and Finkelmann, H. *Liq. Cryst.* 29, 559–565, 2002.
1253. Castella, M., Lopez-Calahorra, F., Velasco, D., and Finkelmann, H. *Chem. Commun.* 2348–2349, 2002.
1254. Segade, A., Lopez-Calahorra, F., and Velasco, D. *Mol. Cryst. Liq. Cryst.* 439, 201–208, 2005.
1255. Segade, A., Castella, M., Lopez-Calahorra, F., and Velasco, D. *Chem. Mater.* 17, 5366–5374, 2005.
1256. Dorphin, D. (ed.). *The Porphyrins*, Vol. 1, Academic Press, New York, pp.1–7, 1978.
1257. Rothemund, P. and Menotti, A. R. *J. Am. Chem. Soc.* 63, 267–270, 1941.
1258. Adler, A. D., Longo, F. R., Finarelli, J. D., Goldmacher, J., Assour, J., and Korsakoff, L. *J. Org. Chem.* 32, 476, 1967.
1259. Lindsey, J. S., Schreiman, I. C., Hsu, H. C., Kearney, P. C., and Marguerettaz, A. M. *J. Org Chem.* 52, 827–836, 1987.
1260. Arsenault, G. P., Bullock, E., and Macdonald, S. F. *J. Am. Chem. Soc.* 82, 4384–8389, 1960.
1261. Burrows, H. D., Gonsalves, A. M. R., Leitao, M. L. P., da G. Miguel, M., and Pereira, M. M. *Supramol. Sci.* 4, 241–246, 1997.
1262. Fox, M. A., Grant, J. V., Melamed, D., Torimoto, T., Liu, C. Y., and Bard, A. J. *Chem. Mater.* 10, 1771–1176, 1998.
1263. Shimizu, Y., Miya, M., Nagata, A., Ohta, K., Yamamoto, I., and Kusabayashi, S. *Liq. Cryst.* 14, 795–805, 1993.

1264. Shimizu, Y., Ishikawa, A., Kusabayashi, S., Miya, M., and Nagata, A. *J. Chem. Soc. Chem. Commun.* 656–658, 1993.

1265. Shimizu, Y., Matsuno, J. Y., Miya, M., and Nagata, A. *J. Chem. Soc. Chem. Commun.* 2411–2412, 1994.

1266. Nagata, A., Shimizu, Y., Nagamoto, H., and Miya, M. *Inorg. Chim. Acta* 238, 169–171, 1995.

1267. Shimizu, Y., Matsuno, J. Y., Nakao, K., Ohta, K., Miya, M., and Nagata, A. *Mol. Cryst. Liq. Cryst.* 260, 491–497, 1995.

1268. Kroon, J. M., Koehorst, R. B. M., van Dijk, M., Sanders, G. M., and Sudholter, E. J. R. *J. Mater. Chem.* 7, 615–624, 1997.

1269. Shimizu, Y., Higashiyama, T., and Fuchita, T. *Thin Solid Films* 331, 279–284, 1998.

1270. Sugino, T., Santiago, J., Shimizu, Y, Heinrich, B., and Guillon, D. *Mol. Cryst. Liq. Cryst.* 330, 15–22, 1999.

1271. Hill, J., Sugino, T., and Shimizu, Y. *Mol. Cryst. Liq. Cryst.* 332, 119–125, 1999.

1272. Shimizu, Y., Fuchita, T., Higashiyama, T., and Sugino, T. *Mol. Cryst. Liq. Cryst.* 331, 575–582, 1999.

1273. Monobe, H., Mima, S., and Shimizu, Y. *Chem. Lett.* 29, 1004–1005, 2000.

1274. Shimizu, Y., Santiago, J., Sugino, T., and Monobe, H. *Mol. Cryst. Liq. Cryst.* 364, 235–242, 2001.

1275. Monobe, H., Mima, S., Sugino, T., and Shimizu, Y. *J. Mater. Chem.* 11, 1383–1392, 2001.

1276. Monobe, H., Miyagawa, Y., Mima, S., Sugino, T., Uchida, K., and Shimizu, Y. *Thin Solid Films* 393, 217–224, 2001.

1277. Shimizu, Y. *Mol. Cryst. Liq. Cryst.* 370, 83–91, 2001.

1278. Sugino, T., Santiago, J., Shimizu, Y., Heinrich, B., and Guillon, D. *Liq. Cryst.* 31, 101–108, 2004.

1279. Ohta, K., Ando, M., and Yamamoto. I. *J. Porphyr. Phthalocyanins* 3, 249–258, 1999.

1280. Kugimiya, S. I. and Takemura, M. *Tetrahedron Lett.* 31, 3157–3160, 1990.

1281. Griesar, K., Athanassopoulou, M. A., Bustamante, E. S., Haase, W., Tomkowicz, Z., and Zaleski, A. J. *Adv. Mater.* 9, 45–48, 1997.

1282. Shimizu, Y., Miya, M., Nagata, A., Ohta, K., Matsumura, A., Yamamoto, I., and Kusabayashi, S. *Chem. Lett.* 20, 25–28, 1991.

1283. Ohta, K., Ando, N., and Yamamoto, I. *Liq. Cryst.* 26, 663–668, 1999.

1284. Zhao, Z. and Liu, G. *Liq. Cryst.* 29, 1335–1337, 2002.

1285. Binnemans, K. and Walrand, C. G. *Chem. Rev.* 102, 2303–2346, 2002.

1286. Yu, M., Liu, G., Cheng, Y., and Xu, W. *Liq. Cryst.* 32, 771–780, 2005.

1287. Ramasseul, R., Maldivi, P., and Marchon, J. C. *Liq. Cryst.* 13, 729–733, 1993.

1288. Liu, W., Shi, Y., Shi, T., Liu, G., Liu, Y., Wang, C., and Zhang, W. *Liq. Cryst.* 30, 1255–1257, 2003.

1289. Liu, W. and Shi, T. *Sci. China B* 50, 488–493, 2007.

1290. Patel, B. R. and Suslick, K. S. *J. Am. Chem. Soc.* 120, 11802–11803, 1998.

1291. Li, J., Tang, T., Li, F., and Li, M. *Dyes Pigm.* 77, 395–401, 2008.

1292. Li, L., Kang, S. W., Harden, J., Sun, Q., Zhou, X., Dai, L., Jakli, A., Kumar, S., and Li, Q. *Liq. Cryst.* 35, 233–239, 2008.

1293. Kang, S. W., Li, Q., Chapman, B. D., Pindak, R., Cross, J. O., Li, L., Nakata, M., and Kumar, S. *Chem. Mater.* 19, 5657–5663, 2007.

1294. Zhou, X., Kang, S. W., Kumar, S., Kulkarni, R. R., Cheng, S. Z. D., and Li, Q. *Chem. Mater.* 20, 3551–3553, 2008.

1295. Zhou, X., Kang, S. W., Kumar, S., and Li, Q. *Liq. Cryst.* 36, 269–274, 2009.

1296. Li, J., Xin, H., and Li, M. *Liq. Cryst.* 33, 913–919, 2006.

1297. Kruper Jr., W. J., Chamberlin, T. A., and Kochanny, M. *J. Org. Chem.* 54, 2753–2756, 1989.

1298. Sun, E., Sun, Z., Yuan, M., Wang, D., and Shi, T. *Dyes Pigm.* 81, 124–130, 2009.

1299. Sun, E., Shi, Y., Zhang, P., Zhou, M., Zhang, Y., Tang, X., and Shi, T. *J. Mol. Struct.* 889, 28–34, 2008.

1300. Sun, E., Cheng, X., Wang, D., Tang, X., Yu, S., and Shi, T. *Solid State Sci.* 9, 1061–1068, 2007.

1301. Milgrom, L. R., Yahioglu, G., Bruce, D. W., Morrone, S., Henari, F. Z., and Blau, W. *J. Adv. Mater.* 9, 313–316, 1997.

1302. Ohta, K., Yamaguchi, N., and Yamamoto, I. *J. Mater. Chem.* 8, 2637–2650, 1998.

1303. Wenz, G. *Makromol. Chem. Rapid Commun.* 6, 577–584, 1985.

1304. Kimura, M., Saito, Y., Ohta, K., Hanabusa, K., Shirai, H., and Kobyashi, N. *J. Am. Chem. Soc.* 124, 5274–5275, 2002.

1305. Camerel, F., Ulrich, G., Barbera, J., and Ziessel, R. *Chem. Eur. J.* 13, 2189–2200, 2007.

1306. Yu, M., Chen, G. J., and Liu, G. F., *J. Phys. Chem. Solids* 68, 541–548, 2007.

1307. Yu, M., Zhang, W. Y., Fan, Y., Jian, W. P., and Liu, G. F. *J. Phys. Org. Chem.* 20, 229–235, 2007.

1308. Cui, X. L., Liu, G. F., and Yu, M. *J. Coord. Chem.* 59, 1361–1370, 2006.

1309. Bruce, D. W., Dunmur, D. A., Santa, L. S., and Wali, M. A. *J. Mater. Chem.* 2, 363–364, 1992.

1310. Bruce, D. W., Wali, M. A., and Wang, Q. M. *Chem. Commun.* 2089–2090, 1994.

1311. Wang, Q. M. and Bruce, D. W. *Chem. Commun.* 2505–2506, 1996.
1312. Wang, Q. M. and Bruce, D. W. *Tetrahedron Lett.* 37, 7641–7644, 1996.
1313. Wang, Q. M. and Bruce, D. W. *Angew. Chem. Int. Ed. Engl.* 36, 150–152, 1997.
1314. Bhyrappa, P., Arunkumar, C., Varghese, B., Rao, D. S. S., and Prasad, S. K. *J. Porphyr. Phthalocyanines* 12, 54–64, 2008.
1315. Sessler, J. L., Callaway, W., Dudek, S. P., Date, R. W., Lynch, V., and Bruce, D. W. *Chem. Commun.* 2422–2423, 2003.
1316. Vogel, E., Kocher, M., Schmickler, H., and Lex, J. *Angew. Chem. Int. Ed.* 25, 257–259, 1986.
1317. Stepien, M., Donnio, B., and Sessler, J. L. *Chem. Eur. J.* 13, 6853–6863, 2007.
1318. Qi, M. H. and Liu, G. F. *J. Phys. Chem. B* 107, 7640–7646, 2003.
1319. Qi, M. H. and Liu, G. F. *J. Mater. Chem.* 13, 2479–2484, 2003.
1320. Doppelt, P. and Huille, S. *New J. Chem.* 14, 607–609, 1990.
1321. Morelli, G., Ricciardi, G., and Roviello, A. *Chem. Phys. Lett.* 185, 468–472, 1991.
1322. Schramm, C. J. and Hoffman, B. M. *Inorg. Chem.* 19, 383–385, 1980.
1323. Lelj, F., Morelli, G., Ricciardi, G., Roviello, A., and Sirigu, A. *Liq. Cryst.* 12, 941–960, 1992.
1324. Bonosi, F., Ricciardi, G., Lelj, F., and Martini, G. *Thin Solid Films* 243, 335–338, 1994.
1325. Dye, D. F., Raghavachari, K., and Zaleski, J. M. *Inorg. Chim. Acta* 361, 1177–1186, 2008.
1326. Eichhorn, S. H., Bruce, D. W., Guillon, D., Gallani, J. L., Fischer, T., Stumpe, J., and Geue, T. *J. Mater. Chem.* 11, 1576–1584, 2001.
1327. Belviso, S., Ricciardi, G., and Lelj, F. *J. Mater. Chem.* 10, 297–304, 2000.
1328. Belviso, S., Ricciardi, G., Lelj, F., Scolaro, L. M., Bencini, A., and Carbonera, C. *J. Chem. Soc. Dalton Trans.* 1143–1150, 2001.
1329. Belviso, S., Amati, M., de Bonis, M., and Lelj, F. *Mol. Cryst. Liq. Cryst.* 481, 56–72, 2008.
1330. Belviso, S., Giugliano, A., Amati, M., Ricciardi, G., Lelj, F., and Scolaro, L. M. *Dalton Trans.* 305–312, 2004.
1331. Ohta, K., Watanabe, T., Hasebe, H., Morizumi, Y., Fujimoto, T., Yamamoto, I., Lelievre, D., and Simon, J. *Mol. Cryst. Liq. Cryst.* 196, 13–26, 1991.
1332. Dini, D., Hanack, M., and Meneghetti, M. *J. Phys. Chem. B* 109, 12691–12696, 2005.
1333. Ohta, K., Azumane, S., Kawahara, W., Kobayashi, N., and Yamamoto, I. *J. Mater. Chem.* 9, 2313–2320, 1999.
1334. Ichihara, M., Miida, M., Mohr, B., and Ohta, K. *J. Porphyr. Phthalocyanines* 10, 1145–1155, 2006.
1335. Fernandez, O., de la Torre, G., Lazaro, F. F., Barbera, J., and Torres, T. *Chem. Mater.* 9, 3017–3022, 1997.
1336. Ohta, K., Hatsusaka, K., Sugibayashi, M., Ariyoshi, M., Ban, K., Maeda, F., Naito, R., and Nishizawa, K. *Mol. Cryst. Liq. Cryst.* 397, 25–45, 2003.
1337. Eichhorn, H. *J. Porphyr. Phthalocyanines* 4, 88–102, 2000.
1338. McKeown, N. B. *Phthalocyanine Materials: Synthesis, Structure and Function*, Cambridge University Press, Cambridge, U.K., 1998.
1339. van Nostrum, C. F. and Nolte, R. J. M. *Chem. Commun.* 2385–2392, 1996.
1340. Elemans, J. A. A. W., van Hameren, R., Nolte, R. J. M., and Rowan, A. E. *Adv. Mater.* 18, 1251–1266, 2006.
1341. Claessens, C. G., Hahn, U., and Torres, T. *Chem. Rec.* 8, 75–97, 2008.
1342. de la Torre, G., Vazquez, P., Lopez, F. A., and Torres, T. *J. Mater. Chem.* 8, 1671–1683, 1998.
1343. de la Torre, G., Vazquez, P., Lopez, F. A., and Torres, T. *Chem. Rev.* 104, 3723–3750, 2004.
1344. Serrano, J. L. (ed.). *Metallomesogens, Synthesis, Properties, and Applications*, VCH, Weinhem, Germany, 1996.
1345. Donnio, G., Guillon, D., Deschenaux, R., and Bruce, D. W. *Comprehensive Co-ordination Chemistry II*, Vol. 7, McCleverty, J. A. and Meyer, T. J. (eds.), Elsevier, Oxford, U.K., Chap. 7.9, pp. 357–627, 2003.
1346. Piechocki, C., Simon, J., Skoulios, A., Guillon, D., and Weber, P. *J. Am. Chem. Soc.* 104, 5245–5247, 1982.
1347. Gaspard, S. A., Hochapfel, M., and Viovy, C. R. *Hebd. Sciences Acad. Sci. C* 289, 387–390, 1979.
1348. Schouten, P. G., Warman, J. M., de Haas, M. P., van Nostrum, C. F., Gelinck, G. H., Nolte, R. J. M., Copyn, M. J., Zwikker, J. W., Engel, M. K., Hanack, M., Chang, Y. H., and Ford, W. T. *J. Am. Chem. Soc.* 116, 6880–6894, 1994.
1349. Weber, P., Guillon, D., and Skoulios, A. *Liq. Cryst.* 9, 369–382, 1991.
1350. Chambrier, I., Cook, M. J., Helliwell, M., and Powell, A. K. *J. Chem. Soc. Chem. Commun.* 444–445, 1992.

1351. Helliwell, M., Deacon, A., Moon, K. J., Powell, A. K., and Cook, M. J. *Acta Cryst.* B53, 231–240, 1997.
1352. Weber, P., Guillon, D., and Skoulios, A. *J. Phys. Chem.* 91, 2242–2243, 1987.
1353. Andre, J. J., Bernard, M., Piechocki, C., and Simon, J. *J. Phys. Chem.* 90, 1327–1330, 1986.
1354. Markovitsi, D., Thi, T. H. T., and Briois, V. *J. Am. Chem. Soc.* 110, 2001–2002, 1988.
1355. Markovitsi, D. and Lecuyer, I. *Chem. Phys. Lett.* 149, 330–333, 1988.
1356. Markovitsi, D., Lecuyer, I., and Simon, J. *J. Phys. Chem.* 95, 3620–3626, 1991.
1357. Blasse, G., Dirksen, G. J., Meijerink, A., van der Pol, J. F., Neeleman, E., and Drenth, W. *Chem. Phys. Lett.* 154, 420–424, 1989.
1358. Belarbi, Z., Sirlin, C., Simon, J., and Andre, J. J. *J. Phys. Chem.* 93, 8105–8110, 1989.
1359. Schutte, W. J., Rehbach, M. S., and Sluyters, J. H. *J. Phys. Chem.* 97, 6069–6073, 1993.
1360. Osburn, E. J., Schmidt, A., Chau, L. K., Chen, S. Y., Smolenyak, P., Armstrong, N. R., and O'Brien, D. F. *Adv. Mater.* 8, 926–928, 1996.
1361. Duzhko, V. and Singer, K. D. *J. Phys. Chem. C* 111, 27–31, 2007.
1362. Nalwa, H. S., Engel, M. K., Hanack, M., and Schultz, H. *Appl. Organomet. Chem.* 10, 661–664, 1996.
1363. Sato, M., Takeuchi, A., Yamada, T., Hoshi, H., Ishikawa, K., Mori, T., and Takezoe, H. *Phys. Rev. E* 56, R6264–R6266, 1997.
1364. Cook, M. J., McKeown, N. B., Thomson, A. J., Harrison, K. J., Richardson, R. M., Davies, A. N., and Roser, S. J. *Chem. Mater.* 1, 287–289, 1989.
1365. Poynter, R. H., Cook, M. J., Chesters, M. A., Slater, D. A., McMurdo, J., and Welford, K. *Thin Solid Films* 243, 346–350, 1994.
1366. Kumaran, N., Donley, C. L., Mendes, S. B., and Armstrong, N. R. *J. Phys. Chem. C* 112, 4971–4977, 2008.
1367. Qiu, X., Wang, C., Zeng, Q., Xu, B., Yin, S., Wang, H., Xu, S., and Bai, C. *J. Am. Chem. Soc.* 122, 5550–5556, 2000.
1368. Mouthuy, P. O., Melinte, S., Geerts, Y. H., Nysten, B., and Jonas, A. M. *Small* 4, 728–732, 2008.
1369. Tracz, A., Makowski, T., Masirek, S., Pisula, W., and Geerts, Y. H. *Nanotechnology* 18, 485303, 2007.
1370. Belushkin, A. V., Cook, M. J., Frezzato, D., Haslam, S. D., Ferrarini, A., Martin, D., McMurdo, J., Nordio, P. L., Richardson, R. M., and Stafford, A. *Mol. Phys.* 93, 593–607, 1998.
1371. Orti, E., Bredas, J. L., and Clarisse, C. *J. Chem. Phys.* 92, 1228–1235, 1990.
1372. Leznoff, C. C. and Lever, A. B. P. (eds.). *Phthalocyanines, Properties and Applications*, Vol. 2, VCH, New York, p. 226, 1993.
1373. Ohta, K., Jacquemin, L., Sirlin, C., Bosio, L., and Simon, J. *New J. Chem.* 12, 751–754, 1988.
1374. Piechocki, C. and Simon, J. *J. Chem. Soc. Chem. Commun.* 259–260, 1985.
1375. Knawby, D. M. and Swager, T. M. *Chem. Mater.* 9, 535–538, 1997.
1376. He, N., Chen, Y., Doyle, J., Liu, Y., and Blau, W. J. *Dyes Pigm.* 76, 569–573, 2008.
1377. Piechocki, C., Boulou, J. C., and Simon, J. *Mol. Cryst. Liq. Cryst.* 149, 115–120, 1987.
1378. Guillon, D., Weber, P., Skoulios, A., Piechocki, C., and Simon, J. *Mol. Cryst. Liq. Cryst.* 130, 223–229, 1985.
1379. Hanack, M., Beck, A., and Lehmann, H. *Synthesis* 703–705, 1987.
1380. Sirlin, C., Bosio, L., and Simon, J. *Mol. Cryst. Liq. Cryst.* 155, 231–238, 1988.
1381. Sirlin, C., Bosio, L., and Simon, J. *J. Chem. Soc. Chem. Commun.* 379–380, 1987.
1382. Cho, I. and Lim, Y. *Mol. Cryst. Liq. Cryst.* 154, 9–26, 1988.
1383. Sleven, J. Mesomorphism and optical properties of peripherally substituted phthalocyanines; Influence of chain length, linking group and central metal ion, PhD thesis, The Katholieke Universiteit, Leuven, Belgium, 2002.
1384. Cho, I. and Lim, Y. *Bull. Korean Chem. Soc.* 9, 98–101, 1988.
1385. van der Pol, J. F., Neeleman, E., Zwikker, J. W., Nolte, R. J. M, Drenth, W., Aerts, J., Visser, R., and Picken, S. J. *Liq. Cryst.* 33, 1378–1387, 2006.
1386. Sleven, J., Cardinaels, T., Binnemans, K., Nelis, D., Mullens, J., Hübner, D. H., and Meyer, G. *Liq. Cryst.* 30, 143–148, 2003.
1387. Haisch, P., Knecht, S., Schlick, U., Subramanian, L. R., and Hanack, M. *Mol. Cryst. Liq. Cryst.* 270, 7–16, 1995.
1388. Vacus, J., Doppelt, P., Simon, J., and Memetzidis, G. *J. Mater. Chem.* 2, 1065–1068, 1992.
1389. Sauer, T. and Wegner, G. *Mol. Cryst. Liq. Cryst.* 162B, 97–118, 1988.
1390. Severs, L. M., Underhill, A. E., Edwards, D., Wight, P., and Thetford, D. *Mol. Cryst. Liq. Cryst.* 234, 235–240, 1993.
1391. Kimura, M., Ueki, H., Ohta, K., Shirai, H., and Kobayashi, N. *Langmuir* 22, 5051–5056, 2006.
1392. Nolte, R. J. M, van der Pol, J. F., Neeleman, E., Zwikker, J. W., Nolte, R. J. M., Drenth, W., Aerts, J., Visser, R., and Picken, S. J. *Liq. Cryst.* 33, 1373–1377, 2006.

1393. Schouten, P. G., van der Pol, J. F., Zwikker, J. W., Drenth, W., and Picken, S. J. *Mol. Cryst. Liq. Cryst.* 195, 291–305, 1991.
1394. Komatsu, T., Ohta, K., Watanabe, T., Ikemoto, H., Fujimoto, T., and Yamamoto, I. *J. Mater. Chem.* 4, 537–540, 1994.
1395. van der Pol, J. F., Neeleman, E., van Miltenburg, J. C., Zwikker, J. W., Nolte, R. J. M., and Drenth, W. *Macromolecules* 23, 155–162, 1990.
1396. Kimura, M., Wada, K., Ohta, K., Hanabusa, K., Shirai, H., and Kobayashi, N. *J. Am. Chem. Soc.* 123, 2438–2439, 2001.
1397. Kobayashi, N., Higashi, R., Ishii, K., Hatsusaka, K., and Ohta, K. *Bull. Chem. Soc. Jpn.* 72, 1263–1271, 1999.
1398. van Nostrum, C. F. *Mol. Cryst. Liq. Cryst.* 302, 303–308, 1997.
1399. Minch, B. A., Xia, W., Donley, C. L., Hernandez, R. M., Carter, C., Carducci, M. D., Dawson, A., O'Brien, D. F., and Armstrong, N. R. *Chem. Mater.* 17, 1618–1627, 2005.
1400. Duro, J. A., de la Torre, G., Barberá, J., Serrano, J. L., and Torres, T. *Chem. Mater.* 8, 1061–1066, 1996.
1401. Ford, W. T., Sumner, L., Zhu, W., Chang, Y. H., Um, P. J., Choi, K. H., Heiney, P. A., and Maliszewskyj, N. C. *New J. Chem.* 18, 495–505, 1994.
1402. Sleven, J., Cardinaels, T., Binnemans, K., Guillon, D., and Donnio, B. *Liq. Cryst.* 29, 1425–1433, 2002.
1403. Lelievre, D., Petit, M. A., and Simon, J. *Liq. Cryst.* 4, 707–710, 1989.
1404. Cuellar, E. A. and Marks, T. J. *Inorg. Chem.* 20, 3766–3770, 1981.
1405. Engel, M. K., Bassoul, P., Bosio, L., Lehmann, H., Hanack, M., and Simon, J. *Liq. Cryst.* 15, 709–722, 1993.
1406. Adib, Z. A, Clarkson, G. J., McKeown, N. B., Treacher, K. E., Gleeson, H. F., and Stennett, A. S. *J. Mater. Chem.* 8, 2371–2378, 1998.
1407. Kimura, M., Narikawa, H., Ohta, K., Hanabusa, K., Shirai, H., and Kobayashi, N. *Chem. Mater.* 14, 2711–2717, 2002.
1408. Haisch, P., Winter, G., Hanack, M., Luer, L., Egelhaaf, H, J., and Oelkrug, D. *Adv. Mater.* 9, 316–321, 1997.
1409. Henari, F., Davey, A., Blau, W., Haisch, P., and Hanack, M. *J. Porphyr. Phthalocyanines* 3, 331–338, 1999.
1410. Ban, K., Nishizawa, K., Ohta, K., and Shirai, H. *J. Mater. Chem.* 10, 1083–1090, 2000.
1411. Lux, A., Rozenberg, G. G., Petritsch, K., Moratti, S. C, Holmes, A. B, and Friend, R. H. *Synth. Met.* 102, 1527–1528, 1999.
1412. Basova, T. V., Gurek, A. G., and Ahsen, V. *Mater. Sci. Eng. C* 22, 99–104, 2002.
1413. Tantrawong, S., Sugino, T., Shimizu, Y., Takeuchi, A., Kimura, S., Mori, T., and Takezoe, H. *Liq. Cryst.* 24, 783–785, 1998.
1414. de la Escosura, A., Díaz, M. V. M., Barbera, J., and Torres, T. *J. Org. Chem.* 73, 1475–1480, 2008.
1415. Eichhorn, H., Wohrle, D., and Pressner, D. *Liq. Cryst.* 22, 643–653, 1997.
1416. Santiago, J., Sugino, T., and Shimizu, Y. *Mol. Cryst. Liq. Cryst.* 332, 497–504, 1999.
1417. Gürek, A. G., Ahsen, V., Heinemann, F., and Zugenmaier, P. *Mol. Cryst. Liq. Cryst.* 338, 75–97, 2000.
1418. Basova, T. V., Gürek, A. G., Atilla, D., Hassan, A. K., and Ahsen, V. *Polyhedron* 26, 5045–5052, 2007.
1419. Gurek, A. G., Durmus, M., and Ahsen, V. *New J. Chem.* 28, 693–699, 2004.
1420. Atilla, D., Aslibay, G., Gürek, A. G., Can, H., and Ahsen, V. *Polyhedron* 26, 1061–1069, 2007.
1421. Ohta, K., Watanabe, T., Tanaka, S., Fujimoto, T., Yamamoto, I., Bassoul, P., Kucharczyk, N., and Simon, J. *Liq. Cryst.* 10, 357–368, 1991.
1422. Ohta, K., Azumane, S., Watanabe, T., Tsukada, S., and Yamamoto, I. *Appl. Organomet. Chem.* 10, 623–635, 1996.
1423. Ichihara, M., Suzuki, A., Hatsusaka, K., and Ohta, K. *Liq. Cryst.* 34, 555–567, 2007.
1424. Hatsusaka, K., Ohta, K., Yamamoto, I., and Shirai, H. *J. Mater. Chem.* 11, 423–433, 2001.
1425. Dulog, L. and Gittinger, A. *Mol. Cryst. Liq. Cryst.* 213, 31–42, 1992.
1426. Dulog, L. and Gittinger, A. *Mol. Cryst. Liq. Cryst.* 237, 235–242, 1993.
1427. Tylleman, B., Aspe, R. G., Gbabode, G., Geerts, Y. H., and Sergeyev, S. *Tetrahedron* 64, 4155–4161, 2008.
1428. Cook, M. J., Daniel, M. F., Harrison, K. J., McKeown, N. B., and Thomson, A. J. *J. Chem. Soc. Chem. Commun.* 1086–1088, 1987.
1429. Cammidge, A. N., Cook, M. J., Harrison, K. J., and McKeown, N. B. *J. Chem. Soc. Perkin Trans.* 3053–3058, 1991.
1430. Cherodian, A. S., Davies, A. N., Richardson, R. M., Cook, M. J., McKeown, N. B., Thomson, A. J., Feijoo, J., Ungar, G., and Harrison, K. J. *Mol. Cryst. Liq. Cryst.* 196, 103–114, 1991.
1431. Swarts, J. C., Langner, E. H. G., Hove, N. K., and Cook, M. J. *J. Mater. Chem.* 11, 434–443, 2001.
1432. Joraid, A. A., Alamri, S. N., Al-Raqa, S. Y., and Mohamed, A. A. *Liq. Cryst.* 35, 351–356, 2008.

1433. Iino, H., Hanna, J. I., Bushby, R. J., Movaghar, B., Whitaker, B. J., and Cook, M. J. *Appl. Phys. Lett.* 87, 132102, 2005.
1434. Cammidge, A. N., Cook, M. J., Haslam, S. D., Richardson, R. M., and Harrison, K. J. *Liq. Cryst.* 14, 1847–1862, 1993.
1435. Cook, M. J. *J. Mater. Sci. Mater. Electron.* 3, 117–128, 1994.
1436. Cook, M. J., Cracknell, S. J., and Harrison, K. J. *J. Mater. Chem.* 1, 703–704, 1991.
1437. Critchley, S. M., Willis, M. R., Maruyama, Y., Bandow, S., Cook, M. J., and McMurdo, J. *Mol. Cryst. Liq. Cryst.* 229, 47–51, 1993.
1438. Laschewsky, A. *Angew. Chem. Int. Ed.* 28, 1574–1577, 1989.
1439. Yuksel, F., Atilla, D., and Ahsen, V. *Polyhedron* 26, 4551–4556, 2007.
1440. Brewis, M., Clarkson, G. J., Holder, A. M., and McKeown, N. B. *Chem. Commun.* 969–970, 1998.
1441. Clarkson, G. J., Cook, A., McKeown, N. B., Treacher, K. E., and Adib, Z. A. *Macromolecules* 29, 913–917, 1996.
1442. Treacher, K. E., Clarkson, G. J., and McKeown, N. B. *Mol. Cryst. Liq. Cryst.* 260, 255–260, 1995.
1443. Cook, M. J., Cooke, G., and Fini, A. J. *J. Chem. Soc. Chem. Commun.* 1715–1718, 1995.
1444. Chambrier, I., Cook, M. J., Cracknell, S. J., and McMurdo, J. *J. Mater. Chem.* 3, 841–849, 1993.
1445. Mckeown, N. B. and Painter, J. *J. Mater. Chem.* 4, 1153–1156, 1994.
1446. Gearba, R. I., Bondar, A. I., Goderis, B., Bras, W., and Ivanov, D. A. *Chem. Mater.* 17, 2825–2832, 2005.
1447. Durmuş, M., Lebrun, C., and Ahsen, V. J. *Porphyr. Phthalocyanines* 8, 1175–1186, 2004.
1448. Treacher, K. E., Clarkson, G. J., and McKeown, N. B. *Liq. Cryst.* 19, 887–889, 1995.
1449. Ogawa, K., Kinoshita, S. I., Yonehara, H., Nakahara, H., and Fukuda, K. *J. Chem. Soc. Chem. Commun.* 477–479, 1989.
1450. Usol'tseva, N., Bykova, V., Ananjeva, G., Smirnova, A., Shaposhnikov, G., Maizlish, V., Kudrik, E., and Shirokov, A. *Mol. Cryst. Liq. Cryst.* 352, 45–57, 2000.
1451. van Nostrum, C. F., Picken, S. J., and Nolte, R. J. M. *Angew. Chem. Int. Ed. Engl.* 33, 2173–2175, 1994.
1452. van Nostrum, C. F., Picken, S. J., Schouten, A. J., and Nolte, R. J. M. *J. Am. Chem. Soc.* 117, 9957–9965, 1995.
1453. Yılmaz, F., Atilla, D., and Ahsen, V. *Polyhedron* 23, 1931–1937, 2004.
1454. Cammidge, A. N., Chambrier, I., Cook, M. J., Garland, A. D., Heeney, M. J., and Welford, K. *J. Porphyr. Phthalocyanines* 1, 77–86, 1997.
1455. Ng, D. K. P., Yeung, Y. O., Chan, W. K., and Yu, S. C. *Tetrahedron Lett.* 38, 6701–6704, 1997.
1456. Cammidge, A. N. and Gopee, H. *Chem. Commun.* 966–967, 2002.
1457. Cammidge, A. N. and Gopee, H. *J. Porphyr. Phthalocyanines* 13, 235–246, 2009.
1458. Kang, S. H., Kang, Y. S., Zin, W. C., Olbrechts, G., Wostyn, K., Clays, K., Persoons, A., and Kim, K. *Chem. Commun.* 1661–1662, 1999.
1459. Cook, M. J. and Fini, A. J. *J. Mater. Chem.* 7, 5–7, 1997.
1460. Cook, M. J. and Fini, A. J. *Tetrahedron* 56, 4085–4094, 2000.
1461. Cabezón, B., Nicolau, M., Barberá, J., and Torres, T. *Chem. Mater.* 12, 776–781, 2000.
1462. McKeown, N. B., Leznoff, C. C., Richardson, R. M., and Cherodian, A. S. *Mol. Cryst. Liq. Cryst.* 213, 91–98, 1992.
1463. Corsellis, E. A., Coles, H. J., McKeown, N. B., Weber, P., Guillon, D., and Skoulios, A. *Liq. Cryst.* 23, 475–479, 1997.
1464. Donnio, B. and Bruce, D. W. In *Structure and Bonding*, Vol. 95, Mingos, D. M. P. (ed.), Springer-Verlag, New York, pp. 193–247, 1999.
1465. Hudson, S. A. and Maitlis, P. M. *Chem. Rev.* 93, 861–885, 1993.
1466. Godquin, A. M. G. and Maitlis, P. M. *Angew. Chem. Int. Ed.* 30, 375–402, 1991.
1467. Serrano, J. L. and Sierra, T. *Chem. Eur. J.* 6, 759–766, 2000.
1468. Godquin, A. M. G. In *Handbook of Liquid Crystals*, Demus, D., Goodby, J., Gray, G. W., Spiess, H. W., and Vill, V. (eds.), Wiley-VCH, Weinheim, Germany, Chap. XIV, 1998.
1469. Hegmann, T., Kain, J., Diele, S., Pelzl, G., and Tschierske, C. *Angew. Chem. Int. Ed.* 40, 887–890, 2001.
1470. Hegmann, T., Peidis, F., Diele, S., and Tschierske, C. *Liq. Cryst.* 27, 1261–1265, 2000.
1471. Eran, B. B., Tschierske, C., Diele, S., and Baumeister, U. *J. Mater. Chem.* 16, 1136–1144, 2006.
1472. Eran, B. B., Tschierske, C., Diele, S., and Baumeister, U. *J. Mater. Chem.* 16, 1145–1153, 2006.
1473. Omnes, L., Circu, V., Hutchins, P. T., Coles, S. J., Horton, P. N., Hursthouse, M. B., and Bruce, D. W. *Liq. Cryst.* 32, 1437–1447, 2005.

1474. Glebowska, A., Przybylski, P., Winek, M., Krzyczkowska, P., Krowczynski, A., Szydlowska, J., Pociecha, D., and Gorecka, E. *J. Mater. Chem.* 19, 1395–1398, 2009.
1475. Steinke, N., Frey, W., Baro, A., Laschat, S., Drees, C., Nimtz, M., Hagele, C., and Giesselmann, F. *Chem. Eur. J.* 12, 1026–1035, 2006.
1476. Praefcke, K., Diele, S., Pickardt, J., Gundogan, B., Nutz, U., and Singer, D. *Liq. Cryst.* 18, 857–865, 1995.
1477. Giroud, A. M. G. and Billard, J. *Mol.* Cryst. *Liq. Cryst.* 66, 147–150, 1981.
1478. Giroud, A. M. G. and Billard, J. *Mol. Cryst. Liq. Cryst.* 97, 287–295, 1983.
1479. Ohta, K., Muroki, H., Takagi, A., Yamamoto, I., and Matszaki, K. *Mol. Cryst. Liq. Cryst.* 135, 247–264, 1986.
1480. Ohta, K., Ishii, A., Yamamoto, I., and Matsuzaki, K. *J. Chem. Soc. Chem. Commun.* 1099–1101, 1984.
1481. Ohta, K., Muroki, H., Takagi, A., Hatada, K. I., Ema, H., Yamamoto, I., and Matsuzaki, K. *Mol. Cryst. Liq. Cryst.* 140, 131–152, 1986.
1482. Sakashita, H., Nishitani, A., Sumiya, Y., Terauchi, H., Ohta, K., and Yamamoto, I. *Mol. Cryst. Liq. Cryst.* 163, 211–219, 1988.
1483. Poelsma, S. N., Servante, A. H., Fanizzi, F. P., and Maitlis, P. M. *Liq. Cryst.* 16, 675–685, 1994.
1484. Yang, X., Lu, Q., Dong, S., Liu, D., Zhu, S., Wu, F., and Zhang, R. *J. Phys. Chem.* 97, 6726–6730, 1993.
1485. Zhaohui, Z., Youzhi, W., Fuzhou, W., Ping, X., and Rongben, Z. *Polym. Adv. Technol.* 7, 662–666, 1996.
1486. Sadashiva, B. K. and Ramesha, S. *Mol. Cryst. Liq. Cryst.* 141, 19–24, 1986.
1487. Usha, K., Vijayan, K., and Sadashiva, B. K. *Mol. Cryst. Liq. Cryst. Lett.* 5, 67–71, 1987.
1488. Usha, K., Sadashiva, B. K., and Vijayan, K. *Mol. Cryst. Liq. Cryst.* 241, 91–102, 1994.
1489. Giroud, A. M. G., Sigaud, G., Achard, M. F., and Hardouin, F. *J. Phys. Lett.* 45, L387–L392, 1984.
1490. Ohta, K., Ema, H., Muroki, H., Yamamoto, I., and Matsuzaki, K. *Mol. Cryst. Liq. Cryst.* 147, 61–78, 1987.
1491. Chien, C. W., Liu, K. T., and Lai, C. K. *J. Mater. Chem.* 13, 1588–1595, 2003.
1492. Zheng, H., Lai, C. K., and Swager, T. M. *Chem. Mater.* 7, 2067–2077, 1995.
1493. Barbera, J., Iglesias, R., Serrano, J. L., Sierra, T., de la Fuente, M. R., Palacios, B., Jubindo, M. A. P., and Vazquez, J. T. *J. Am. Chem. Soc.* 120, 2908–2918, 1998.
1494. Swager, T. M. and Zheng, H. *Mol. Cryst. Liq. Cryst.* 260, 301–306, 1995.
1495. Trzaska, S. T., Hsu, H. F., and Swager, T. M. *J. Am. Chem. Soc.* 121, 4518–4519, 1999.
1496. Kakegawa, N., Hoshino, N., Matsuoka, Y., Wakabayashi, N., Nishimura, S. I., and Yamagishi, A. *Chem. Commun.* 2375–2377, 2005.
1497. Trzaska, S. T., Zheng, H., and Swager, T. M. *Chem. Mater.* 11, 130–134, 1999.
1498. Serrette, A. G., Lai, C. K., and Swager, T. M. *Chem. Mater.* 6, 2252–2268, 1994.
1499. Lai, C. K., Serrette, A. G., and Swager, T. M. *J. Am. Chem. Soc.* 114, 7948–7949, 1992.
1500. Lai, C. K., Chen, F. G., Ku, Y. J., Tsai, C. H., and Lin, R. *J. Chem. Soc. Dalton Trans.* 4683–4687, 1997.
1501. Lai, C. K. and Lin, F. J. *J. Chem. Soc. Dalton Trans.* 17–19, 1996.
1502. Jiang, J., Shen, Z., Lu, J., Fu, P., Lin, Y., Tang, H., Gu, H., Sun, J., Xie, P., and Zhang, R. *Adv. Mater.* 16, 1534–1538, 2004.
1503. Veber, M., Fugnitto, R., and Strzelecka, H. *Mol. Cryst. Liq. Cryst.* 96, 221–227, 1983.
1504. Ohta, K., Takagi, A., Muroki, H., Yamamoto, I., Matsuzaki, K., Inabe, T., and Maruyama, Y. *J. Chem. Soc. Chem. Commun.* 883–885, 1986.
1505. Veber, M., Davidson, P., Jallabert, C., Levelut, A. M., and Strzelecka, H. *Mol. Cryst. Liq. Cryst. Lett.* 5, 1–7, 1987.
1506. Horie, H., Takagi, A., Hasebe, H., Ozawa, T., and Ohta, K. *J. Mater. Chem.* 11, 1063–1071, 2001.
1507. Ohta, K., Hasebe, H., Ema, H., Fujimoto, T., and Yamamoto, I. *J. Chem. Soc. Chem. Commun.* 1610–1611, 1989.
1508. Ohta, K., Hasebe, H., Moriya, M., Fujimoto, T., and Yamamoto, I. *Mol. Cryst. Liq. Cryst.* 208, 33–41, 1991.
1509. Ohta, K., Hasebe, H., Ema, H., Moriya, M., Fujimoto, T., and Yamamoto, I. *Mol. Cryst. Liq. Cryst.* 208, 21–32, 1991.
1510. Ohta, K., Oka, Y. I., Hasebe, H., and Yamamoto, I. *Polyhedron* 19, 267–274, 2000.
1511. Ohta, K., Hasebe, H., Moriya, M., Fujimoto, T., and Yamamoto, I. *J. Mater. Chem.* 1, 831–834, 1991.
1512. Ohta, K., Ikejima, M., Moriya, M., Hasebe, H., and Yamamoto, I. *J. Mater. Chem.* 8, 1971–1977, 1998.
1513. Ohta, K., Higashi, R., Ikejima, M., Yamamoto, I., and Kobayashi, N. *J. Mater. Chem.* 8, 1979–1991, 1998.
1514. Mohr, B., Enkelmann, V., and Wegner, G. *Mol. Cryst. Liq. Cryst.* 281, 215–228, 1996.
1515. Mohr, B., Enkelmann, V., and Wegner, G. *J. Org. Chem.* 59, 635–638, 1994.

1516. Barbera, J., Elduque, A., Gimenez, R., Oro, L. A., and Serrano, J. L. *Angew. Chem. Int. Ed.* 35, 2832–2834, 1996.

1517. Barbera, J., Elduque, A., Gimenez, R., Lahoz, F. J., Lopez, J. A., Oro, L. A., and Serrano, J. L. *Inorg. Chem.* 37, 2960–2967, 1998.

1518. Kim, S. J., Kang, S. H., Park, K. M., Kim, H., Zin, W. C., Choi, M. G., and Kim, K. *Chem. Mater.* 10, 1889–1893, 1998.

1519. Barbera, J., Gimenez, R., and Serrano, J. L. *Chem. Mater.* 12, 481–489, 2000.

1520. Forget, S., Veber, M., and Strzelecka, H. *Mol. Cryst. Liq. Cryst.* 258, 263–275, 1995.

1521. Forget, S. and Veber, M. *Mol. Cryst. Liq. Cryst.* 300, 229–243, 1997.

1522. Forget, S. and Veber, M. *Mol. Cryst. Liq. Cryst.* 308, 27–42, 1997.

1523. Forget, S. and Kitzerow, H. S. *Liq. Cryst.* 23, 919–922, 1997.

1524. Kang, S. H., Kim, M., Lee, H. K., Kang, Y. S., Zin, W. C., and Kim, K. *Chem. Commun.* 93–94, 1999.

1525. Wu, H. C., Sung, J. H., Yang, C. D., and Lai, C. K. *Liq. Cryst.* 28, 411–415, 2001.

1526. Pucci, D., Barberio, G., Bellusci, A., Crispini, A., Donnio, B., Giorgini, L., Ghedini, M., la Deda, M., and Szerb, E. I. *Chem. Eur. J.* 12, 6738–6747, 2006.

1527. Lai, C. K., Chang, C. H., and Tsai, C. H. *J. Mater. Chem.* 8, 599–602, 1998.

1528. Barbera, J., Gimenez, R., Gimeno, N., Marcos, M., del Carmen Pina, M., and Serrano, J. L. *Liq. Cryst.* 30, 651–661, 2003.

1529. Binnemans, K., Lodewyckx, K., Cardinaels, T., Vogt, T. N. P., Bourgogne, C., Guillon, D., and Donnio, B. *Eur. J. Inorg. Chem.* 150–157, 2006.

1530. Sessler, J. L., Callaway, W. B., Dudek, S. P., Date, R, W., and Bruce, D. W. *Inorg. Chem.* 43, 6650–6653, 2004.

1531. Zimmermann, H., Billard, J., Gutman, H., Wachtel, E. J., Poupko, R., and Luz, Z. *Liq. Cryst.* 12, 245–262, 1992.

1532. Destrade, C., Tinh, N. H., Gasparoux, H., and Mamlok, L. *Liq. Cryst.* 2, 229–233, 1987.

1533. Strzelecka, H., Jallabert, C., and Veber, M. *Mol. Cryst. Liq. Cryst.* 156, 355–359, 1988.

1534. Veber, M., Jallabert, C., Strzelecka, H., Jullien, O., and Davidson, P. *Liq. Cryst.* 8, 775–785, 1990.

1535. Davidson, P., Jallabert, C., Levelut, A. M., Strzelecka, H., and Veber, M. *Liq. Cryst.* 3, 133–137, 1988.

1536. Strzelecka, H., Jallabert, C., Veber, M., Davidson, P., and Levelut, A. M. *Mol. Cryst. Liq. Cryst.* 161, 395–401, 1988.

1537. Strzelecka, H., Jallabert, C., Veber, M., Davidson, P., Levelut, A. M., Malthete, J., Sigaud, G., Skoulios, A., and Weber, P. *Mol. Cryst. Liq. Cryst.* 161, 403–411, 1988.

1538. Veber, M., Sotta, P., Davidson, P., Levelut, A. M., Jallabert, C., and Strzelecka, H. *J. Phys. France* 51, 1283–1301, 1990.

1539. Albouy, P. A., Vandevyver, M., Perez, X., Ecoffet, C., Markovitsi, D., Veber, M., Jallabert, C., and Strzelecka, H. *Langmuir* 8, 2262–2268, 1992.

1540. Fugnitto, R., Strzelecka, H., Zann, A., Dubois, J. C., and Billard, J. *J. Chem. Soc. Chem. Commun.* 271–272, 1980.

1541. Saeva, F. D., Reynolds, G. A., and Kaszczuk, L. *J. Am. Chem. Soc.* 104, 3524–3525, 1982.

1542. Strzelecka, H., Gionis, V., Rivory, J., and Flandrois, S. *J. Phys.* 44, C3-1201–C3-1206, 1983.

1543. Davidson, P., Levelut, A. M., Strzelecka, H., and Gionis, V. *J. Phys. Lett.* 44, L-823–L-828, 1983.

1544. Saeva, F. D. and Reynolds, G. A. *Mol. Cryst. Liq. Cryst.* 132, 29–34, 1986.

1545. Kormann, R., Zuppiroli, L., Gionis, V., and Strzelecka, H. *Mol. Cryst. Liq. Cryst.* 133, 283–290, 1986.

1546. Barbera, J., Bardaji, M., Jimenez, J., Laguna, A., Martinez, M. P., Oriol, L., Serrano, J. L., and Zaragozano, I. *J. Am. Chem. Soc.* 127, 8994–9002, 2005.

1547. Barbera, J., Jimenez, J., Laguna, A., Oriol, L., Perez, S., and Serrano, J. L. *Chem. Mater.* 18, 5437–5445, 2006.

1548. Moriya, K., Suzuki, T., Kawanishi, Y., Masuda, T., Mizusaki, H., Nakagawa, S., Ikematsu, H., Mizuno, K., Yano, S., and Kajiwara, M. *Appl. Organomet. Chem.* 12, 771–779, 1998.

1549. Moriya, K., Suzuki, T., Yano, S., and Kajiwara, M. *Liq. Cryst.* 19, 711–713, 1995.

1550. Moriya, K., Suzuki, T., Yano, S., and Kajiwara, M. *Trans. Mater. Res. Soc. Jpn.* 24, 481–484, 1999.

1551. Moriya, K., Kawanishi, Y., Yano, S., and Kajiwara, M. *Chem. Commun.* 1111–1112, 2000.

1552. Kim, C. and Allcock, H. R. *Macromolecules* 20, 1726–1727, 1987.

1553. Singler, R. E., Willingham, R. A., Lenz, R. W., Furukawa, A., and Finkelmann, H. *Macromolecules* 20, 1727–1728, 1987.

1554. Jaglowski, A. J., Singler, R. E., and Atkins, E. D. T. *Macromolecules* 28, 1668–1672, 1995.

1555. Eskildsen, J., Reenberg, T., and Christensen, J. B. *Eur. J. Org. Chem.* 1637–1640, 2000.

1556. Sienkowska, M. J., Farrar, J. M., and Kaszynski, P. *Liq. Cryst.* 34, 19–24, 2007.

1557. Attias, A. J., Cavalli, C., Donnio, B., Guillon, D., Hapiot, P., and Malthete, J. *Chem. Mater.* 14, 375–384, 2002.

1558. Attias, A. J., Cavalli, C., Donnio, B., Guillon, D., Hapiot, P., and Malthete, J. *Mol. Cryst. Liq. Cryst.* 415, 169–177, 2004.

1559. Cristiano, R., Gallardo, H., Bortoluzzi, A. J., Bechtold, I. H., Campos, C. E. M., and Longo, R. L. *Chem. Commun.* 5134–5136, 2008.

1560. Gallardo, H., Magnago, R., and Bortoluzzi, A. J. *Liq. Cryst.* 28, 1343–1352, 2001.

1561. Lee, H. K., Lee, H., Ko, Y. H., Chang, Y. J., Oh, N. K., Zin, W, C., and Kim, K. *Angew. Chem.* 113, 2741–2743, 2001.

1562. Wang, Y. J., Sheu, H. S., and Lai, C. K. *Tetrahedron* 63, 1695–1705, 2007.

1563. Majumdar, K. C., Pal, N., Debnath, P., and Rao, N. V. S. *Tetrahedron Lett.* 48, 6330–6333, 2007.

1564. Kohne, B. and Praefcke, K. *Angew. Chem. Int. Ed.* 23, 82–83, 1984.

1565. Kohne, B., Praefcke, K., Stephan, W., and Nurnberg, P. *Z. Naturforsch.* 40b, 981–986, 1985.

1566. Kohne, B., Praefcke, K., and Billard, J. *Z. Naturforsch.* 41b, 1036–1044, 1986.

1567. Neuling, H. W., Stegemeyer, H., Praefcke, K., and Kohne, B. *Z. Naturforsch.* 42a, 631–635, 1987.

1568. Praefcke, K., Kohne, B., Stephan, W., and Marquardt, P. *Chimica* 43, 380–382, 1989.

1569. Collard, D. M. and Lillya, C. P. *J. Am. Chem. Soc.* 56, 6064–6066, 1991.

1570. Praefcke, K., Psaras, P., and Kohne, B. *Chem. Ber.* 124, 2523–2529, 1991.

1571. Zimmermann, R. G., Jameson, G. B., Weiss, R. G., and Demailly, G. *Mol. Cryst. Liq. Cryst. Lett.* 1, 183–189, 1985.

1572. Morris, N. L., Zimmermann, R. G., Jameson, G. B., Dalziel, A. W., Reuss, P. M., and Weiss, R. G. *J. Am. Chem. Soc.* 110, 2177–2185, 1988.

1573. Vill, V. and Thiem, J. *Liq. Cryst.* 9, 451–455, 1991.

1574. Mukkamala, R., Burns Jr., C. L., Catchings III, R. M., and Weiss, R. G. *J. Am. Chem. Soc.* 118, 9498–9508, 1996.

1575. Lehn, J. M., Malthete, J., and Levelut, A. M. *J. Chem. Soc. Chem. Commun.* 1794–1796, 1985.

1576. Malthete, J., Poupinet, D., Vilanove, R., and Lehn, J. M. *J. Chem. Soc. Chem. Commun.* 1016–1019, 1989.

1577. Lattermann, G. *Liq. Cryst.* 6, 619–625, 1989.

1578. Mertesdorf, C. and Ringsdorf, H. *Liq. Cryst.* 5, 1757–1772, 1989.

1579. Lattermann, G. *Mol. Cryst. Liq. Cryst.* 182B, 299–311, 1990.

1580. Tatarsky, D., Banerjee, K., and Ford, W. T. *Chem. Mater.* 2, 138–141, 1990.

1581. Mertesdorf, C., Ringsdorf, H, and Stumpe, J. *Liq. Cryst.* 9, 337–357, 1991.

1582. Idziak, S. H. J., Maliszewskyj, N. C., Heiney, P. A., McCauley Jr., J. P., Sprengeler, P. A., and Smith III, A. B. *J. Am. Chem. Soc.* 113, 7666–7672, 1991.

1583. Idziak, S. H. J., Maliszewskyj, N. C., Vaughan, G. B. M., Heiney, P. A., Mertesdorf, C., Ringsdorf, H., McCauley Jr., J. P., and Smith III, A. B. *J. Chem. Soc. Chem. Commun.* 98–99, 1992.

1584. Malthete, J., Levelut, A. M., and Lehn, J. M. *J. Chem. Soc. Chem. Commun.* 1434–1436, 1992.

1585. Bauer, S., Plesnivy, T., Ringsdorf, H., and Schuhmacher, P. *Makromol. Chem. Macromol. Symp.* 64, 19–32, 1992.

1586. Zhao, M., Ford, W. T., Idziak, S. H. J., Maliszewskyj, N. C., Heiney, P. A. *Liq. Cryst.* 1994, 16, 583–599.

1587. Liebmann, A., Mertesdorf, C., Plesnivy, T., Ringsdorf, H, and Wendorff, J. H. *Angew. Chem. Int. Ed. Engl.* 30, 1375–1377, 1991.

1588. Lattermann, G., Schmidt, S., Kleppinger, R., and Wendorff, J. H. *Adv. Mater.* 4, 30–33, 1992.

1589. Fischer, H., Ghosh, S. S., Heiney, P. A., Maliszewskyj, N. C., Plesnivy, T., Ringsdorf, H., and Seitz, M. *Angew. Chem. Int. Ed.* 34, 795–798, 1995.

1590. Seitz, M., Plesnivy, T., Schimossek, K., Edelmann, M., Ringsdorf, H., Fischer, H., Uyama, H., and Kobayashi, S. *Macromolecules* 29, 6560–6574, 1996.

1591. Stebani, U., Lattermann, G., Wittenberg, M., and Wendorff, J. H. *J. Mater. Chem.* 7, 607–614, 1997.

1592. Fischer, H., Plesnivy, T., Ringsdorf, H., and Seitz, M. *J. Mater. Chem.* 8, 343–351, 1998.

3 Discotic Dimers

A liquid crystal dimer is composed of molecules containing two mesogenic groups linked, often, via a flexible spacer or, rarely, through a rigid spacer. The interest in these materials stems not only from their ability to act as model compounds for semiflexible main chain liquid crystalline polymers, but also because they differ quite substantially in their properties from the conventional low molar mass mesogens. Liquid crystalline dimers show interesting mesomorphic behavior depending on the length of the spacer and on the structure of the linking group [1,2]. Although the first examples of liquid crystalline dimers were reported by Vorlander in 1927 [3], they attracted particular attention during the 1980s. Griffin and Britt showed that a liquid crystalline dimer can be prepared by coupling two mesogenic units with an aliphatic spacer [4]. Subsequently, several classes of dimeric liquid crystals have been prepared and studied extensively [2].

Compared to the large number of liquid crystal dimers consisting of two rod-like mesogens, the number of discotic dimers is very low. The various structural possibilities for discotic dimers are shown schematically in Figure 3.1. Dimers in which two identical discotic mesogens are connected via a flexible spacer (Figure 3.1a) are the most widely synthesized and studied discotic dimers. A few examples of symmetric discotic dimers with a rigid spacer instead of a flexible spacer are also known. It is generally difficult to link two different discotic mesogens and, therefore, only a few examples of unsymmetrical dimers (Figure 3.1b) are realized. Dimers in which two discogens are connected laterally or linearly to a calamitic molecule are depicted in Figure 3.1c and d, respectively. Cyclic dimers (Figure 3.1e) in which two mesogenic units are connected to each other via more than one spacer are also known. A number of double-decker dimers (Figure 3.1f) in which two discotic mesogens are linked via a metal or any other atom have been documented. In addition to these possibilities, several other combinations, such as rod-disk dimers, in which a rod-like mesogen is connected to a discotic mesogen, phasmids, or polycatenar mesogens that fill the gap between rod-like and disk-like liquid crystals exhibiting nematic, smectic, and columnar phases, are also known. However, this chapter covers only dimeric liquid crystals having two discotic mesogens. Other architectures, though they form columnar phases, have not been treated here. Discotic cores that have been explored to create dimers are benzene, scylloinositol, pyranose sugars, anthraquinone, triphenylene, cyclotetraveratrylene, phthalocyanine, porphyrin, and hexabenzocoronene (HBC). Discotic dimers derived from these cores are discussed in the following sections.

3.1 BENZENE-BASED DISCOTIC DIMERS

The first discotic twin based on benzene core was introduced by Lillya and Murthy [5]. Two penta-n-heptanoyloxy benzene rings were connected via a methylene spacer to produce dimer 3 (Scheme 3.1). An excess amount of hexahydroxy benzene 1 was treated with appropriate diacid chloride, and the resulting mixture containing 2 was reacted with excess of heptanoyl chloride to afford 3. Zamir et al. followed another route to prepare these dimers (Scheme 3.1) [6]. Tetrahydroxy-1,4-benzoquinone 4 was heated at 90°C with n-heptanoyl chloride to give 2,5-diheptanoyloxy-1,4-benzoquinone 5. Tetraheptanoyloxy-1,4-benzoquinone 6 was obtained by heating 5 with n-heptanoyl chloride and a few grains of magnesium at 180°C. Reduction of the quinone with $SnCl_2$ in HCl furnished tetraheptanoyloxy-1,4-dihydroxybenzene 7. The transesterification reaction of 7 with hexaheptanoyloxybenzene in the presence of dimethylaminopyridine (DMAP) provided pentaheptanoyloxyphenol 8. The desired dimer 3 was achieved by reacting 8 with an appropriate diacid chloride in dichloromethane and dry pyridine.

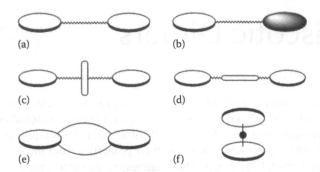

FIGURE 3.1 Sketches of some possible molecular architectures for discotic liquid crystal dimers: (a) symmetric discotic dimer, (b) nonsymmetric discotic dimer, (c) two discogens laterally linked to a calamitic molecule, (d) two discogens terminally linked to a calamitic molecule, (e) cyclic dimers in which two mesogenic units are connected to each other via more than one spacer, and (f) double-decker dimers in which two discotic mesogens are tethered via a metal or other atom.

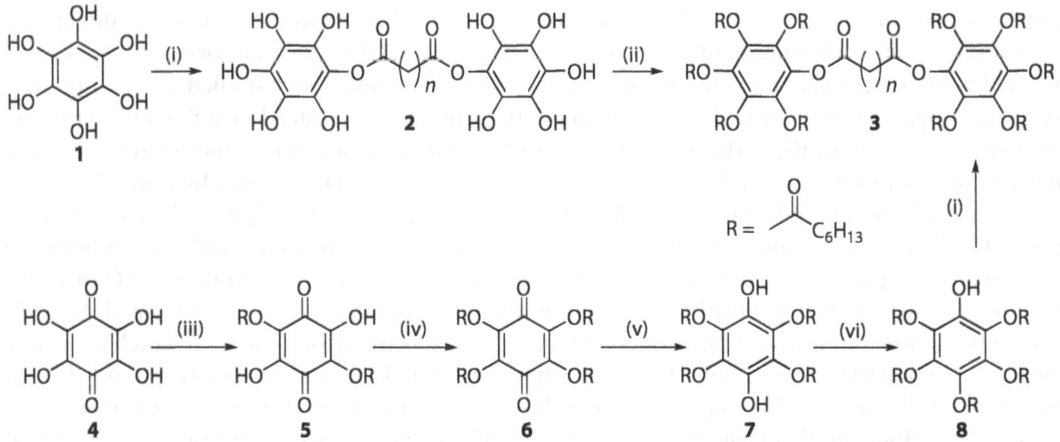

SCHEME 3.1 Synthetic routes to benzene-based discotic dimers: (i) ClOC(CH$_2$)$_n$COCl, pyridine, CH$_2$Cl$_2$; (ii) n-C$_6$H$_{13}$COCl; (iii) n-C$_6$H$_{13}$COCl, 90°C; (iv) n-C$_6$H$_{13}$COCl, Mg, 180°C; (v) SnCl$_2$, HCl; (vi) hexaheptanoyloxybenzene, DMAP.

All prepared dimers have a heptanoyloxy periphery while the spacer length is varied from 4 carbons to 18 carbons. The thermal behavior of these dimers is presented in Table 3.1. It can be seen from this table that on increasing the spacer length, the tendency of symmetrical dimers to display liquid crystallinity increases. Higher homologues of the dimer series **3d–f** having 10 or more carbon atoms in the spacer exhibited mesomorphic behavior. Derivatives with smaller spacer **3a–c** failed to show liquid crystalline behavior. Whereas compounds **3d** and **3e** with a decamethylene and dodecamethylene spacer showed a columnar rectangular mesophase, the longest spacer derivative **3f** displayed a hexagonal stacking of molecules. The thermal stability of the mesophase of these dimers was found to be greater than their monomeric counterparts. Thus, it was concluded that if the spacer is sufficiently long, discotic dimers can arrange themselves in columns. If the spacer length is about twice the length of peripheral chains or longer, the mesophases obtained are similar to that of the monomers.

3.1.1 Alkynylbenzene-Based Discotic Dimers

Hexaalkynylbenzenes or multiynes are well-known discotic nematic liquid crystals. Connecting two bulkier pentaynes via a proper spacer may restrict the free rotation of monomesogenic units

TABLE 3.1
Thermal Behavior of Benzene-Based Dimers (Scheme 3.1)

Structure	n	Phase Transition	Ref.
3a	4	Cr 115.6 I	[6]
3b	6	Cr 94 I	[6]
3c	8	Cr 128 I	[5]
3d	10	Cr 75 Col$_r$ 96.8 I	[6]
3e	12	Cr 77 Col$_r$ 91.3 I	[6]
3f	18	Cr 112 Col$_h$ 120 I	[5]

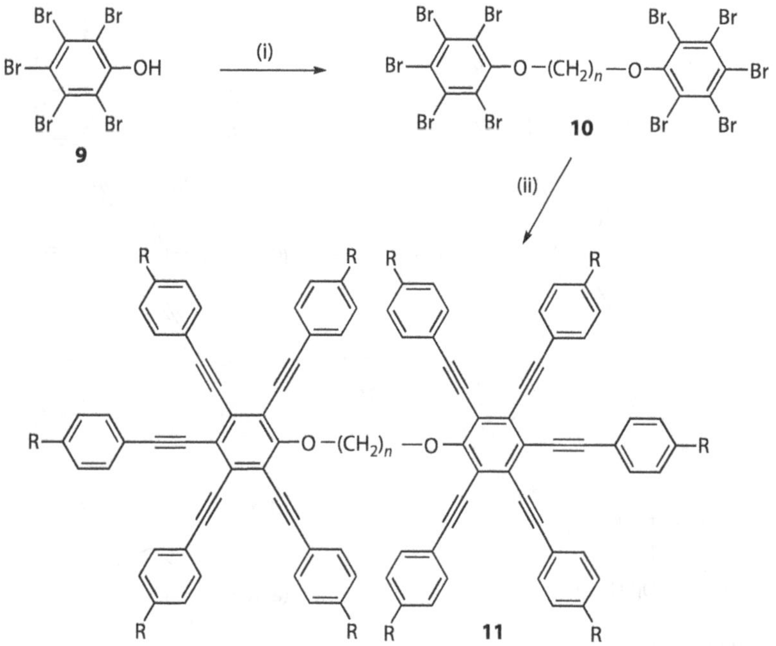

SCHEME 3.2 Synthesis of multiyne-based twin ethers: (i) Br(CH$_2$)$_n$Br, NaH; (ii) 4-alkylphenylacetylene, PdCl$_2$(PPh$_3$)$_2$, PPh$_3$, CuI, NEt$_3$.

around the long axis of the dimer molecule and, therefore, may induce a biaxial nematic mesophase in such dimeric systems. This prompted Praefcke and coworkers to design and synthesize a dimer of radial pentayne **11a** (Scheme 3.2) [7]. Subsequently, many other related dimers were prepared and characterized [8–14]. Twin ethers **11** were prepared in a simple two-step synthesis (Scheme 3.2). Commercially available pentabromophenol **9** was etherified with appropriate α,ω-dibromoalkane in the presence of sodium hydride in dimethylformamide to yield α,ω-bis(pentabromophenyloxy) alkane **10**. Tenfold palladium-catalyzed alkynylation of the twin ether **10** with suitable 4-alkyl-phenylacetylene furnished the desired pentayne ether dimers **11a–g** in moderate yields. Kouwer et al. used hydroxyl- or carboxyl-functionalized monomers to prepare a number of multiyne-based discotic dimers (Scheme 3.3) [13,14]. These monomers were prepared from a straightforward reaction from **9** with the commercially available alcohol- or carboxylate-terminated spacers. These terminally functionalized monomers were connected together via various functional groups using

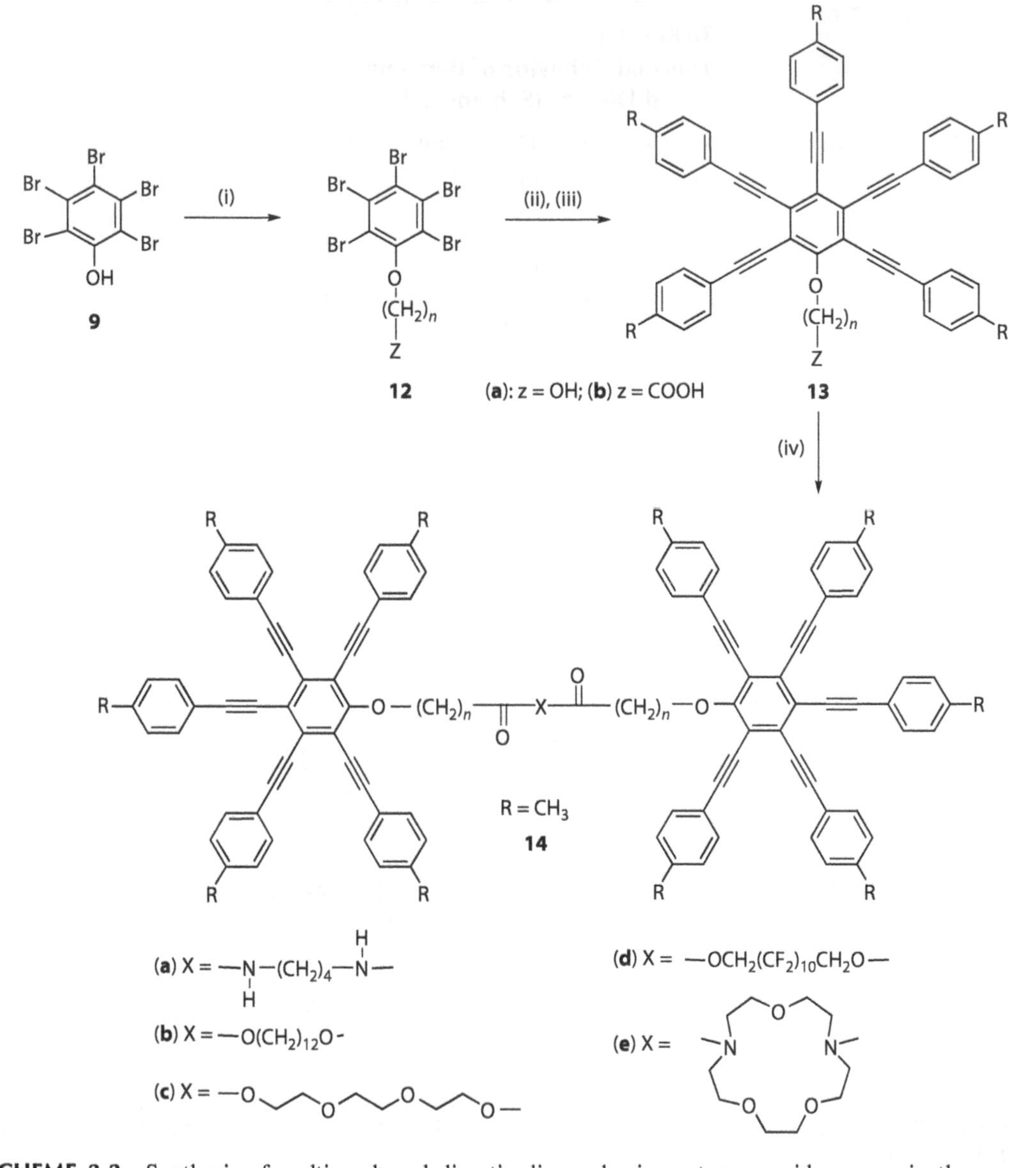

SCHEME 3.3 Synthesis of multiyne-based discotic dimers having ester or amide groups in the spacer: (i) $Br(CH_2)_{10}COOEt$, K_2CO_3, KI, DMF; (ii) $HC≡CPhCH_3$, $PdCl_2(PPh_3)_2$, PPh_3, CuI, NEt_3, piperidine; (iii) aq KOH, EtOH; (iv) HOXOH or H_2NXNH_2, DCC, DMAP, p-TSA, CH_2Cl_2.

classical coupling reactions. Thus, esterification of **13b** with α,ω-diols furnished dimers **14b–d**, while its reaction with diamino compounds yielded amides **14a** and **14e**. Compounds **15** and **16** (Figure 3.2) were prepared from alcohol precursor **13a** via an esterification reaction with an appropriate diacid.

The thermal behavior of pentayne dimers is presented in Table 3.2. All dimers displayed a discotic nematic phase. As can be seen from this table, the ether dyads with short spacers, **11a** and **11b**, showed very narrow or unstable mesophase. The mesophase range was comparatively broader for dimers **11c–e** with longer spacers. The clearing temperatures of twins **11a–e** show a significant odd–even effect. For a constant spacer length (compare **11e, f, g**), the length of a peripheral chain had a reciprocal effect on crystalline-to-nematic point, clearing temperature, as well as the

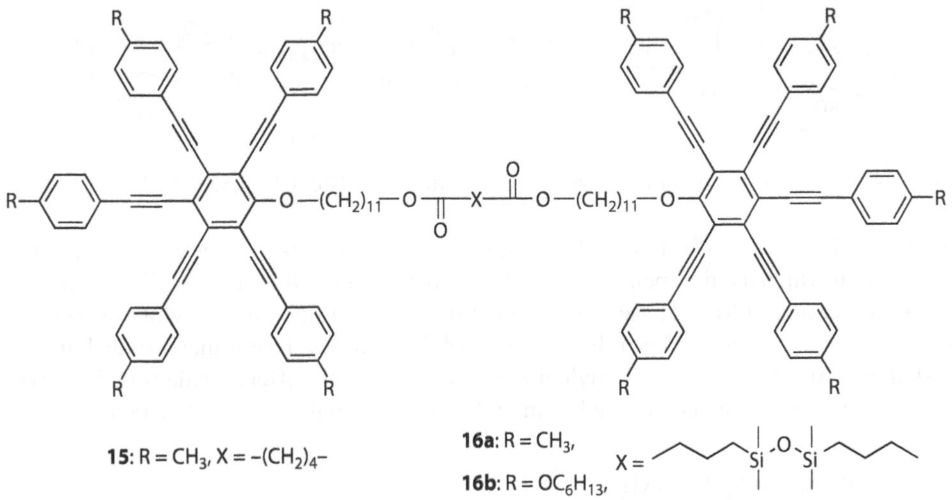

15: R = CH₃, X = –(CH₂)₄–

16a: R = CH₃,
16b: R = OC₆H₁₃,

X =

FIGURE 3.2 Multiyne-based dimers having ester or siloxane groups in the spacer.

TABLE 3.2
Thermal Behavior of Pentayne Dimers
(Schemes 3.2 and 3.3 and Figure 3.2)

Structure	R	n	Phase Transition	Ref.
11a	C_5H_{11}	8	Cr 127.0 N_D 127.8 I	[8]
11b	C_5H_{11}	9	Cr (112.5 N_D) 131.0 I	[8]
11c	C_5H_{11}	10	Cr 129.1 N_D 153.5 I	[8]
11d	C_5H_{11}	11	Cr 118.2 N_D 140.8 I	[8]
11e	C_5H_{11}	12	Cr 121.4 N_D 156.2 I	[7]
11f	C_6H_{13}	12	Cr 92 N_D 107 I	[9]
11g	C_7H_{15}	12	Cr 76 N_D 85 I	[12]
14a			Cr 202 N_D 230 I	[14]
14b			Cr 159 N_D 180 I	[14]
14c			Cr 167 N_D 197 I	[14]
14d			Cr 166 N_D 196 I	[14]
14e			Cr 158 N_D 172 I	[14]
15			Cr 169 N_D 209 I	[13]
16a			Cr 142 I	[13]
16b			Cr 38 N_D 55 I	[13]

mesophase range. Some early studies described the biaxial behavior of nematic phase in **11e**, but it was later concluded that none of these dimers exhibit a biaxial nematic phase.

The introduction of an ester or amide linkage in spacer tends to increase the melting and clearing temperatures (Table 3.2). Dimer **14a** exhibits the highest value of melting as well as isotropic temperature. This is most likely because of the hydrogen bonding caused by amide linkage. The insertion of a siloxane unit in the middle of a spacer causes a significant decrease in transition temperatures. The peripheral methyl group substituted material **16a** (Figure 3.2) was not liquid crystalline; however, the replacement of the methyl group with the hexyloxy group (**16b**) allowed the occurrence of a low-temperature discotic nematic phase.

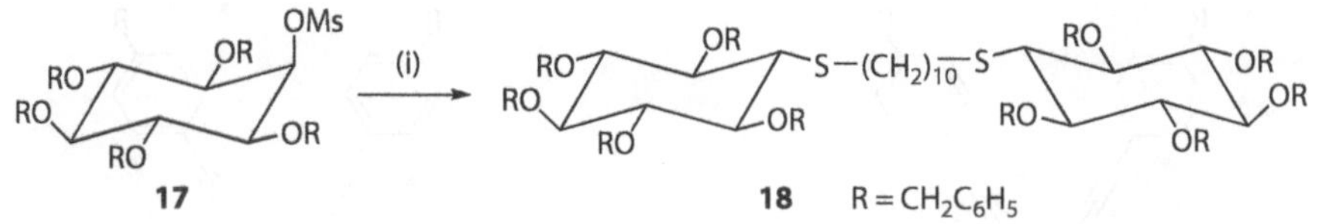

SCHEME 3.4 Synthesis of scylloinositol-based discotic dimer: (i) $NaS(CH_2)_{10}SNa$, DMEU, 100°C.

There are a few reports discussing the synthesis and characterization of unsymmetrical charge transfer twin, in which a radial pentayne moiety is linked to a nonliquid crystalline electron acceptor trinitrofluorenone (TNF). These compounds cannot be considered as discotic dimers, as only one unit is discotic mesogenic. The other fragment of these dimers is non-mesogenic. Further, some disk-rod dimers, in which a pentaalkynylbenzene is tethered with different calamitic liquid crystals, are also known. They have been treated as monofunctionalized discotics in Chapter 2.

3.2 SCYLLOINOSITOL DIMER

Scylloinositol is a cyclohexanehexol stereoisomer and is known to form liquid crystals on appropriate peripheral substitutions. Though the core is really not disk-like, its few derivatives have been classified as DLCs. A dimer, namely, S, S'-decamethylene-bis(penta-O-benzyl-monothioscyllitol) **18**, was prepared from a reaction between 2-O-mesyl-1,3,4,5,6-penta-O-benzyl-myo-inositol **17** and disodium salt of decandithiol in the presence of N,N'-dimethylethyleneurea (DMEU) at 100°C (Scheme 3.4). It was reported to exhibit a monotropic mesomorphism between 121°C and 149°C [15].

3.3 DISCOTIC DIMERS DERIVED FROM PYRANOSE SUGARS

Cellobiose is a readymade dimer that can form liquid crystals upon appropriate substitutions. It is a disaccharide derived from the condensation of two glucose molecules. Though the molecule does not possess a disk shape, its derivatives have been reported as discotic liquid crystals. Vill and Thiem prepared the perlauroyl cellobioside **19a** (Figure 3.3), which gives a columnar mesophase [16]. The acylation of dodecyl-β-D-cellobioside with an excess of acylating agent produces a mixture of hepta-(**19b**) and hexa-(p-dodecyloxy)benzoate **20** that can be separated by chromatography. Both compounds **19b** and **20** form columnar phases similar to the laurate **19a** (Table 3.3). A series of cellobiose octaalkanoates **21** were prepared by Watanabe and coworkers [17,18]. The short chain compound with $n = 6$ forms a rectangular columnar phase, while its homologue with $n = 7$ exhibits

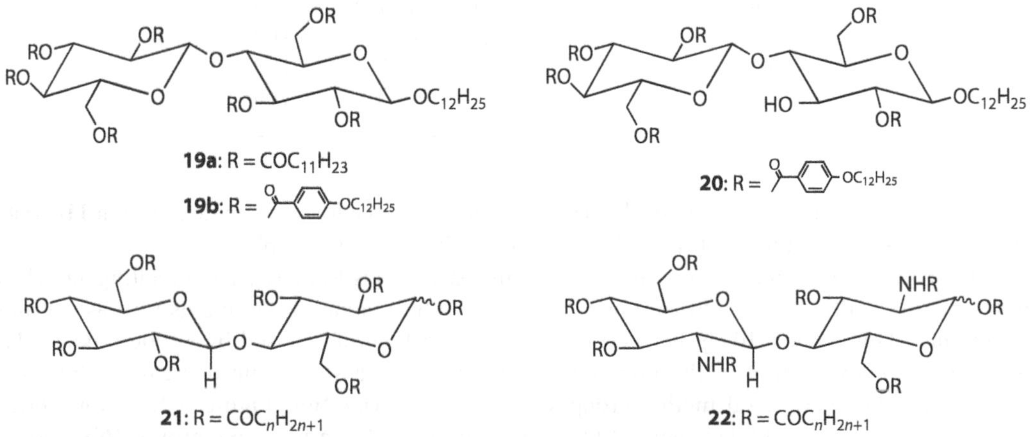

FIGURE 3.3 Discotic dimers derived from pyranose sugars.

TABLE 3.3
Thermal Behavior of Pyranose Sugar–
Based Dimers (Figure 3.3)

Structure	n	Phase Transition	Ref.
19a		Cr 74 Col 92.7 I	[16]
19b		Col 120 I	[16]
20		Col 145 I	[16]
21a	6	Cr 4 Col$_r$ 91 I	[18]
21b	7	Cr 25 Col$_r$ 62 Col$_h$ 87 I	[18]
21c	8	Cr 37 Col$_h$ 86 I	[18]
21d	9	Cr 43 Col$_h$ 83 I	[18]
		Cr 66 Col$_h$ 90 I	[17]
21e	11	Cr 49 Col$_h$ 81 I	[18]
21f	13	Cr 56 Col$_h$ 73 I	[18]
22a	9	Cr 65 Col 207 I	[19]
22b	13	Cr 70 Col 196 I	[19]
22c	17	Cr 73 Col 185 I	[19]

a hexagonal columnar phase at a higher temperature and a rectangular columnar phase at a lower temperature. Homologues with $n = 8$–13 display only a hexagonal columnar phase. Similarly, esters derived from chitobiose 22 also exhibit an enantiotropic mesophase at temperatures ranging from *ca.* 50°C to *ca.* 200°C [19]. As expected, the amido groups in the chito-derivatives stabilize the mesophase significantly. The exact nature of the columnar phase of these dimers has not been revealed.

3.4 RUFIGALLOL-BASED DISCOTIC DIMERS

An anthraquinone-based symmetrical dimer 25 (Scheme 3.5), in which two pentasubstituted anthraquinone units were connected through a methylene spacer, was reported by Krishnan and Balagurusamy [20]. The dimer was derived from 5-hydroxy-1-(4-nitrobenzyloxy)-2,3,6,7-tetrapentyloxy-9,10-anthraquinone 24, in which the four non-hydrogen-bonded hydroxyl groups of rufigallol were substituted with normal alkyl chains and one hydrogen-bonded hydroxyl group was coupled with a 4-nitrobenzyl group. The remaining hydrogen-bonded hydroxyl group was used to prepare dimer 25. The primary reason for attaching with 4-nitrobenzyl group instead of an alkyl group could be its polar nature. It is easier to purify the 5-hydroxy-1-(4-nitrobenzyloxy)-2,3,6,7-tetrapentyloxy-9,10-anthraquinone than 1-hydroxy-2,3,5,6,7-pentaalkoxyanthra-9,10-quinone from the reaction mixture. Dimer 25 was reported to exhibit two transitions—one at 147°C and the other

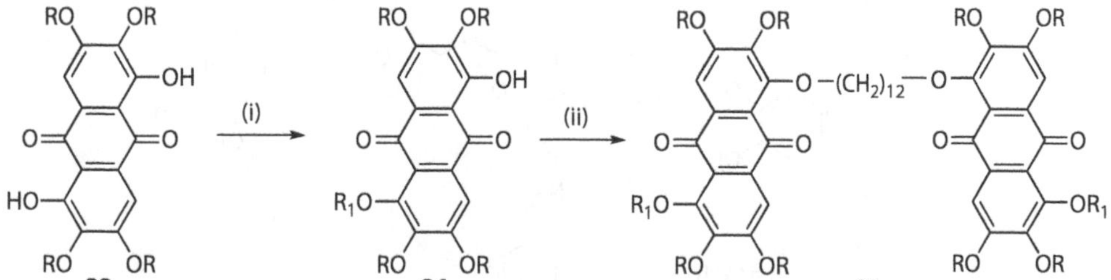

SCHEME 3.5 Synthesis of rufigallol-based dimer: (i) DEAD, TPP, 4-nitrobenzyl alcohol; (ii) K$_2$CO$_3$, dibromododecane.

at 180.4°C during the first heating. However, during the cooling cycle, only one transition at 176.6°C corresponding to isotropic to columnar phase appears. In the second heating run, the first transition shifts to 114.7°C. The actual reason for the first transition is uncertain. X-ray diffraction of dimer **25** recorded at 125°C showed features of a hexagonal columnar mesophase.

3.5 TRIPHENYLENE-BASED DISCOTIC DIMERS

3.5.1 SYMMETRICAL TRIPHENYLENE DISCOTIC DIMERS

The pioneering work of Ringsdorf's group on the synthesis of triphenylene-based polymeric side-chain and main-chain discotic liquid crystals [21,22] has also opened routes to triphenylene-based discotic dimers and oligomers. They prepared mono- and di-functionalized triphenylenes using a statistical approach and converted them into side-chain and main-chain polymers. The potential utility of discotic dimers, oligomers, and polymers has not yet been fully explored, primarily due to the difficulties in preparing functionalized discotic precursors. However, in recent years, there have been many advances in the synthesis of mono-functionalized triphenylenes and, consequently, a number of triphenylene-based discotic dimers, oligomers, and polymers have been realized [23].

Given a readily available supply of monohydroxy-pentaalkoxytriphenylenes, it is straightforward to prepare various triphenylene dimers. Although the simplest way to prepare a dimer is attaching the two monomers via a flexible spacer, the chemistry of triphenylene dimers actually began with a rather complicated molecule in which the two triphenylene units were connected through a calamitic molecule. Thus, to prepare the chemical equivalent of the "Wheel of Mainz," **26**, Ringsdorf and coworkers linked two triphenylene units laterally to a calamitic mesogen [24]. Compound **26** was found to be a spherolithic crystalline compound melting at 167°C. Attaching the two triphenylene molecules linearly to an azobenzene moiety resulted in only an amorphous material **27**, but replacing the azobenzene by a mesogenic azobiphenyl moiety gave a liquid crystalline material **28**. It melts at 72°C to a highly viscous mesophase and clears at 120°C. The optical texture of this mesophase resembles the S_B mesophase. X-ray studies on this compound indicate a layer structure typical for smectic phases, but in which the molecular disks are regularly stacked as observed for columnar phases [24]. As two discotic and one calamitic mesogens are involved in these molecules, they have also been described as "trimesogens" in some literature.

26: Cr 167 I

R = C₅H₁₁ **27**: Amorphous

R = C₅H₁₁ **28**: Cr 72 LC 120 I

The most commonly prepared triphenylene-based dimers contain two identical triphenylene moieties connected via a spacer. Often the spacer is a polymethylene chain [25–31], but in some cases an ester or an amide linkage in the middle of the spacer or at the terminal positions have also been used [31–34]. The thermal behavior of symmetrical triphenylene dimers with a polymethylene spacer is summarized in Table 3.4. These symmetrical dimers can be synthesized either in a single step, by reacting the monohydroxytriphenylene with 0.5 equivalent of the appropriate α,ω-dibromoalkane under classical etherification reaction conditions, or in two steps. In the two-step procedure, the monohydroxy-pentaalkoxytriphenylene is first reacted with an excess of the appropriate α,ω-dibromoalkane to obtain the ω-brominated product **31** (Scheme 3.6), which may be reacted further with monohydroxytriphenylene to obtain the desired dimer. This two-step process is particularly important for preparing nonsymmetrical dimers. The intermediate ω-brominated product **31** can also be prepared using the biphenyl–phenyl oxidative coupling route [35–38].

The length of the spacer has a dramatic influence on the thermal behavior of dimers **32**. Symmetrical triphenylene dimers with a short spacer are not liquid crystalline. Modeling studies

TABLE 3.4
Thermal Behavior of Symmetrical
Triphenylene Twins (Scheme 3.6)

Structure	R	x	Phase Transition	Ref.
32a	C₄H₉	6	Col$_p$ 154 I	[31]
32b	C₄H₉	8	Col$_p$ 148.5 I	[29]
32c	C₅H₁₁	8	Cr 67.0 Col$_h$ 135.6 I	[25]
32d	C₆H₁₃	1	Cr 81 I	[31]
32e	C₆H₁₃	3	Cr 98 I	[31]
32f	C₆H₁₃	5	Cr 69 I	[31]
32g	C₆H₁₃	6	Cr 58 Col$_h$ 91 I	[31]
32h	C₆H₁₃	7	Cr 72 Col$_h$ 92 I	[31]
32i	C₆H₁₃	8	Cr 50 Col$_h$ 104 I	[31]
32j	C₆H₁₃	10	Cr 68 Col$_h$ 107 I	[31]
32k	C₆H₁₃	14	Cr 41 Col$_h$ 84 I	[31]

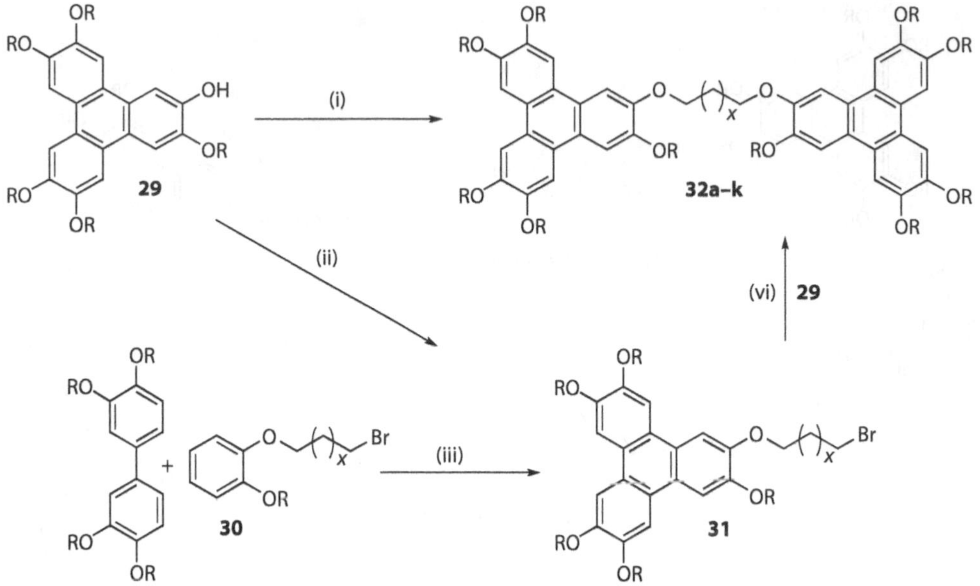

SCHEME 3.6 Synthetic routes to symmetrical triphenylene twins: (i) $Br(CH_2)_nBr$ (0.5 equiv.), K_2CO_3; (ii) $Br(CH_2)_nBr$ (excess), K_2CO_3; (iii) $FeCl_3$, CH_2Cl_2; (iv) K_2CO_3.

suggest that the two triphenylene units are not coplanar due to steric crowding. Only those dimers in which the spacer is more than seven methylene units form a columnar mesophase. In most of these dimers, the columnar phase freezes into a glassy state on cooling. The stability of the glassy state depends on the spacer length as well as on the symmetry of the molecule. The nonsymmetrical dimers usually give longer-lived glasses. As the length of the linking chain increases, in general, the lifetime of the glassy state decreases [31].

Kimura et al. have designed and synthesized five novel triphenylene twins. In these dimers, the two triphenylene cores were connected with the help of a flexible ethylene oxide spacer [39]. The liquid crystallinity of these triphenylene-oligo(ethylene oxide)-triphenylene dimers **33a–e** was notably affected by the length of the ethylene oxide spacer. Compound **33a**, with the smallest spacer, displayed a monotropic hexagonal columnar phase. On heating, this compound melted to the isotropic phase at 77°C, while on cooling, the mesophase appeared at 54°C (Table 3.5). Compounds **33b** and **33c** with a medium spacer length showed an enantiotropic hexagonal columnar mesophase. These materials exhibited a mesophase from 50°C to 93°C. In contrast, higher homologues **33d** and **e** failed to exhibit any liquid crystalline property.

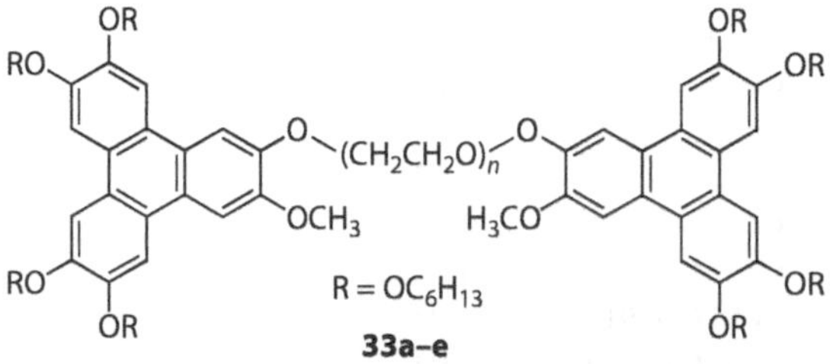

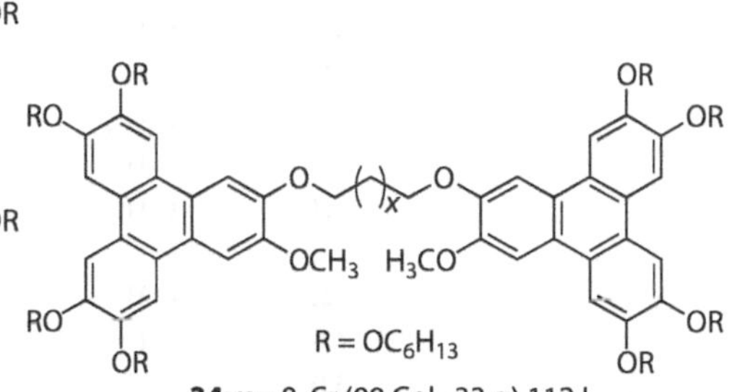

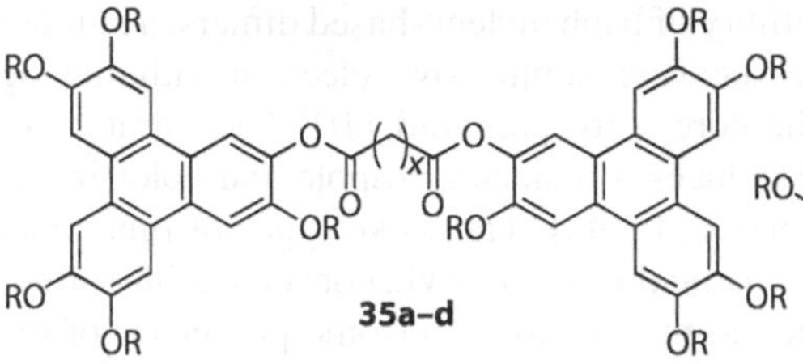

35a–d

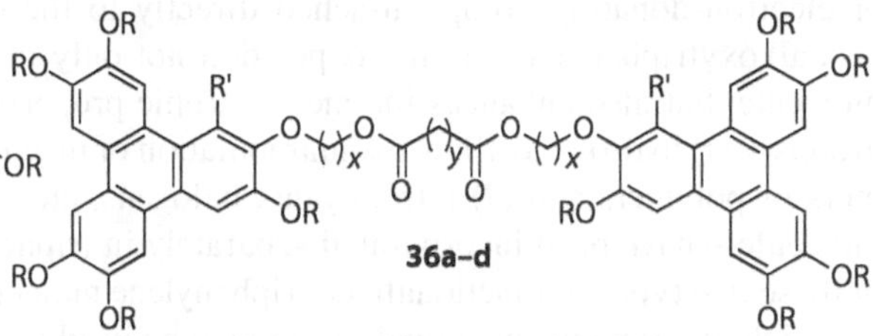

36a–d

TABLE 3.5
**Thermal Behavior of
Triphenylene-Ethylene
Oxide-Based Twins**

Structure	n	Phase Transition
33a	3	Cr (54 Col$_h$) 77 I
33b	4	Cr 50 Col$_h$ 93 I
33c	5	Cr 50 Col$_h$ 83 I
33d	6	Cr 43 I
33e	8	Cr 38 I

Source: Data from Kimura, M. et al., *Liq. Cryst.*, 34, 107, 2007.

Boden and coworkers reported another symmetrical dimer **34** that differs from the above twins in the sense that in **34** instead of five long alkoxy chains in the monomer, one peripheral chain is only a short methoxy group. The synthesis of such dimers is relatively easy, as the required monofunctionalized precursor can be conveniently prepared via phenyl–biphenyl coupling in between a tetraalkoxybiphenyl and 2-methoxyphenol. On heating, this compound shows a crystal-to-isotropic transition at 112°C, but on cooling, exhibits a monotropic columnar phase at 98°C that persists down to a glass transition at 33°C [27–31].

Four symmetrical triphenylene dimers **35a–d** in which two pentaalkoxytriphenylene units are connected via an ester linkage instead of an ether linkage were reported by Kranig et al. in 1990 [34]. These four compounds were reported to be liquid crystalline at room temperature (Table 3.6). These dimers exhibit a much broader mesophase range compared with the dimers linked via ether bridges. This behavior is in accord with that observed for the nonsymmetrical monomeric compounds. Thus, monoalkanoyloxy-pentaalkoxytriphenylenes (having one ester-bonded and five ether-bonded peripheral chains) show a broader mesophase range than the parent hexaalkoxy- or hexaalkanoyloxytriphenylenes due to the steric hindrance of the ester group [40].

While the hydroxy-functional group has often been utilized to prepare dimers, in order to tune the electronic properties of the molecule, core functionalization with electron-withdrawing or electron-donating groups is also necessary. Most triphenylene discotic liquid crystals are colorless, low fluorescence materials, and thus their use in many applications is limited. In an effort to

TABLE 3.6
**Thermal Behavior of Symmetrical
Triphenylene Dimers Linked via an
Ester Chain**

Structure	R	x	Phase Transition
35a	C$_5$H$_{11}$	10	Col$_h$ 180 I
35b	C$_5$H$_{11}$	12	Col$_h$ 168 I
35c	C$_5$H$_{11}$	14	Col$_h$ 147 I
35d	C$_7$H$_{15}$	14	Col$_h$ 147 I

Source: Data from Kranig, W. et al., *Adv. Mater.*, 2, 36, 1990.

enhance the fluorescence as well as the liquid crystallinity of triphenylene-based dimers, a number of monofunctionalized triphenylene-based discotics bearing conjugative electron-withdrawing or electron-donating groups attached directly to the core were prepared [41]. The nitration of hexaalkoxytriphenylenes at the α-position not only induces a molecular dipole and color in the molecule, but also enhances the mesomorphic properties [42–44]. Thus, two types of functional groups—(1) hydroxy to allow the manipulation of mechanical properties via conversion into oligomers or polymers and (2) nitro, cyano, halogens, etc., to modify the electronic properties of the molecule—have been incorporated separately in triphenylene-based molecules. The combination of these two types of functionalized triphenylene molecules leads to novel bifunctional derivatives, in which one type of functional group may be used to tune the electronic nature of the molecule, while at the same time the other may be utilized to convert them into a processable oligomer or polymer. Three such nitro-functionalized triphenylene dimers **36a–c** have been prepared starting from 2-hydroxy-3,6,7,10,11-pentaalkoxytriphenylene [32]. Unfortunately, the synthesis of the precursor mononitro-monohydroxy-pentaalkoxytriphenylene is difficult, and often the nitration of monohydroxy-pentaalkoxytriphenylene results in only the oxidized products 3,6,7,10,11-pentaalkoxytriphenylene-1,2-dione. Under highly controlled and careful reaction conditions, the 3,6,7,10,11-pentaalkoxy-1-nitro-2-triphenylenol can be isolated, and this has to be alkylated immediately with a ω-hydroxybromoalkane. The resulting alcohol can be coupled with various diacid chlorides or diacids to obtain the dimers. Thus, these dimers have an ester group in the middle of the spacer. A similar dimer **36d**, but without the nitro group in the triphenylene unit, was reported by Manickam et al. [33]. The thermal behavior of these dimers is summarized in Table 3.7. As may be seen from this table, the spacer length of the diacid plays a crucial role in determining the phase behavior of these dimers. The presence of two or more methylene units in the spacer gives a liquid crystalline dimer with a very broad mesophase range; while spacers with less than two methylene units result in a monotropic mesophase (compound **36a**) or a crystalline phase (compound **36d**). This may be attributed to the steric hindrance of the two carbonyl groups, which is less pronounced if they are separated by two or more methylene units. The presence of the nitro group makes these materials yellow-colored. Although, these nitro-substituted dimers have very interesting properties, such as stability of the mesophase at ambient temperature, very broad mesophase range, yellow color, a highly polar nitro group in the core, etc., because of synthetic problems they could not be prepared in large amounts for various physical studies.

The formation of highly ordered self-assembled monolayers (SAMs) by mesogenic thiols, disulfides, and thioethers on gold-covered substrates is well known [45]. Schonherr et al. studied the SAM formation of the disulfide-bridged triphenylene dimer **37**. The dimer can be easily prepared by coupling bis-(11,11'-undecanol)disulfide with 2-hydroxy-3,6,7,10,11-pentapentyloxytriphenylene using DEAD and triphenylphosphine. This waxy solid dimer shows an edge-on orientation on the gold surface [46]. A similar dimer with a short spacer and with a disiloxane unit at the center of the spacer was prepared by Zelcer et al. [47]. This dimer exhibits hexagonal columnar mesophase between 45°C and 88°C.

TABLE 3.7

Thermal Behavior of Symmetrical Triphenylene Dimers Linked via an Ether Spacer Having an Ester Group in the Middle

Structure	R	R'	x	y	Phase Transition	Ref.
36a	C_4H_9	NO_2	2	0	Cr (169.1 Col) 198.8 I	[32]
36b	C_4H_9	NO_2	2	2	Col_h 163.0 I	[32]
36c	C_4H_9	NO_2	2	4	Col_h 144.9 I	[32]
36d	C_5H_{11}	H	7	1	Cr 33.5 I	[33]

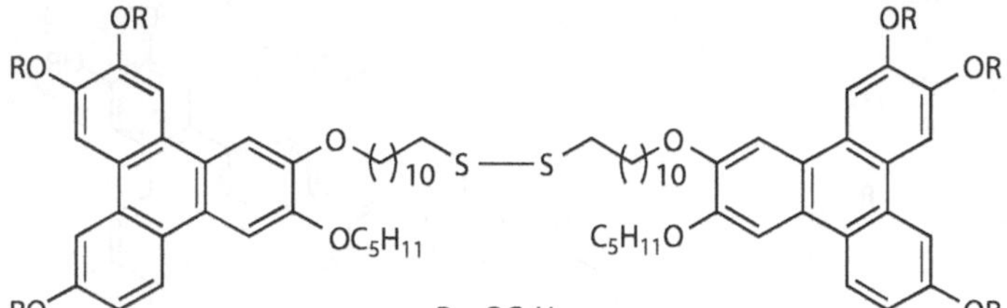

37: Waxy solid 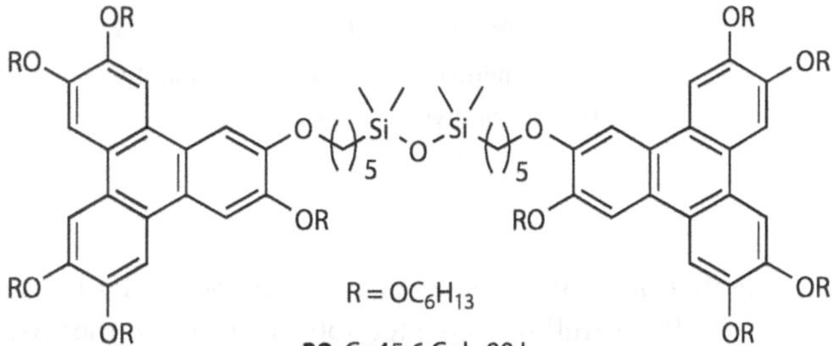
R = OC5H11

38: Cr 45.6 Col_h 88 I
R = OC6H13

Fullerene (C_{60}) and its derivatives have received considerable interest since the discovery of C_{60} in 1985 [48]. A number of chemical and physical properties of this carbon allotrope have been explored by many researchers throughout the world. Several studies have been devoted to the incorporation of calamitic mesogens into the C_{60} framework [49–55]. To explore the possibility that the insertion of C_{60} into the columnar mesophase may result in novel materials useful in device applications, Preece and coworkers synthesized the C_{60} Bingel cyclopropanation-bistriphenylene adduct **39** [33], but, as expected, the fullerene-bridged triphenylene dimer does not show any liquid crystalline properties.

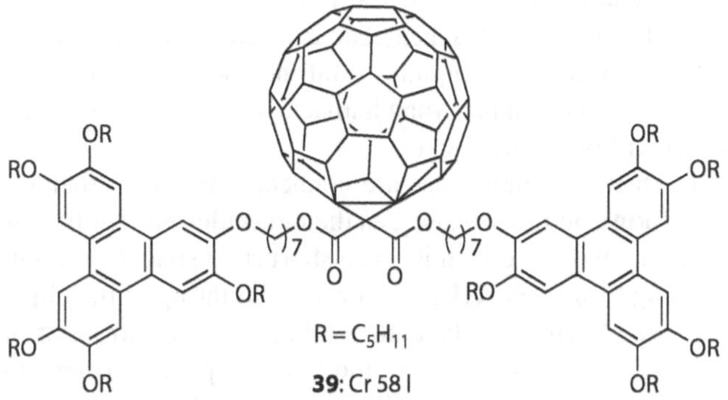

39: Cr 58 I 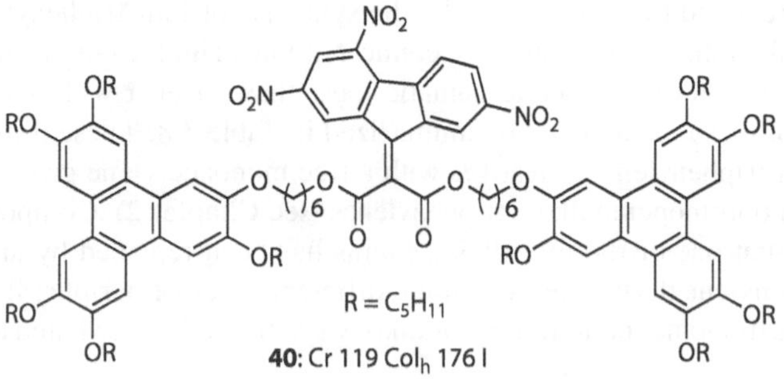
R = C5H11

40: Cr 119 Col_h 176 I
R = C5H11

41a–d: Cr > 330 I

41a: R = methyl **41c:** R = isopentyl
41b: R = isobutyl **41d:** R = hexyl

The stabilization and induction of mesophases due to charge-transfer interactions between donor and acceptor molecules have been well documented [56]. In most of the cases, the TNF molecule has been used as the electron-acceptor. In order to understand the molecular stacking in liquid crystalline donor–acceptor dimers, the TNF molecule has also been covalently connected to the mesogenic moiety [56–58]. In a novel triphenylene-based discotic donor–acceptor–donor type dimer, two triphenylene fragments were chemically connected to a TNF-based unit. Dimer **40**, a condensation product of the corresponding acceptor diethyl ester and the free alcohol derivative of the triphenylene-based monomer, forms a columnar mesophase in the temperature range of 119°C–176°C. The Langmuir–Blodgett film formed by this compound shows an edge-on orientation of the molecules (the columns lie parallel to a solid substrate) [58].

Hanack and coworkers have designed and synthesized four novel conjugated-bridged triphenylene-based dimers **41a–d** for the construction of organic light-emitting diodes (OLEDs) [59]. In these compounds, an electron-withdrawing cyano-substituted *p*-phenylenevinylene unit was linked to two triphenylene units. These dimers were prepared by a double Knoevenagel reaction of the mono-functionalized cyanomethyltriphenylene derivatives and a terphthalaldehyde derivative. The four dimers were found not to be liquid crystalline and melted above 330°C. They exhibited orange to red photoluminescence and a strong bathachromic shift of more than 250 nm when compared with a monomeric triphenylene system.

Discotic dimers in which two molecules are connected to each other via a long flexible alkyl chain spacer generally form columnar phases, as the molecules have sufficient flexibility to stack in adjacent columns. Linking two discotic units via a short rigid spacer would be expected to give rise to steric hindrance arising from the overlapping or interdigitating of the aliphatic side chains; while the weak distortion in the planarity of the core would reduce the strong π–π interactions between the disks. Thus, the rigid molecules may adopt more or less parallel orientation but lose their long-range translational order and, therefore, are likely to form discotic nematic phase. This concept has recently been realized by Kumar et al. by the synthesis of four triphenylene dimers **42a–d**, in which two identical triphenylene units were connected via a rigid π-conjugated diacetylene spacer [60–62]. These dimers form a discotic nematic mesophase over a wide temperature range. The thermal behavior of these compounds is summarized in Table 3.8. These dimers were prepared by the dimerization of triphenylene derivatives with a free monoacetylene group; these in turn can be prepared from monobromopentaalkoxytriphenylenes (see Chapter 2). Compound **42d** differs from the other three in that one of the five alkoxy chains has been replaced by an alkylsulfanyl chain. This unusual triphenylene derivative, with three different types of peripheral chains, was prepared from a triphenylene monomer containing four alkoxy groups, a thioalkyl, and a free acetylene group

TABLE 3.8
Thermal Behavior of Diacetylene-Bridged
Triphenylene Dimers (Scheme 3.7)

Structure	R	R′	Phase Transition	Ref.
42a	C_4H_9	OC_4H_9	Cr 188.6 N_D 243.5 I	[60]
42b	C_5H_{11}	OC_5H_{11}	Cr 161.0 N_D 215.9 I	[60]
42c	C_6H_{13}	OC_6H_{13}	Cr 135.3 N_D 172.8 I	[60]
42d	C_5H_{11}	SC_5H_{11}	Cr 182.5 N_D 196.3 I	[62]

via a multistep synthesis [62]. As can be seen from the thermal data, this change destabilizes the nematic phase.

Another π-conjugated discotic dimeric system comprises the metal-bridged triphenylene dimers. Although mercury-bridged linear calamitic dimers were described by Vorlander in 1923 [63], similar discotic structures appeared only in 2001 [64]. Two organometallic discotic liquid crystals in which a mercury atom is located at the center of two substituted triphenylene molecules were prepared by Kumar et al. [64]. The crystalline compound **43a** transforms into a highly viscous but shearable fluid phase at about 150°C. On further heating, this mesophase changes to an isotropic phase at about 210°C but with decomposition. In order to reduce the clearing temperature, one of the normal alkyl chains was replaced by a branched chain (compound **43b**). On heating, this material shows a solid-to-solid transition at about 110°C, and this soft solid transforms into isotropic phase at 186°C. On cooling the isotropic phase, a metastable nematic discotic phase appears at 185°C. This monotropic mesophase has a strong tendency to crystallize, which begins in the mesophase and is completed at 180°C. On subsequent heating, the first transition at 110°C was not observed and the crystal melts to isotropic liquid at 186°C.

During the preparation of main-chain triphenylene polymers, Boden and coworkers isolated a low-molar mass compound that was identified as a cyclic dimer having structure **44a** or **44b**. The similarity of the ¹H NMR spectrum of the dimer with that of polymer supports structure **44a** over **44b**. The dimer shows a mesophase range of 130°C–140°C [35].

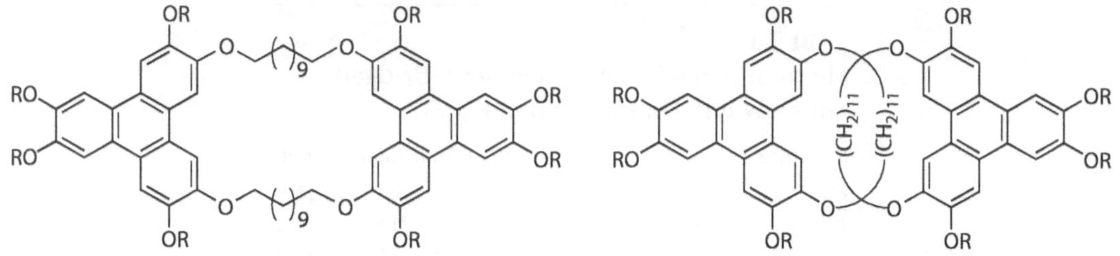

44a   R = C₆H₁₁: Cr 130 Col 140 I   **44b**

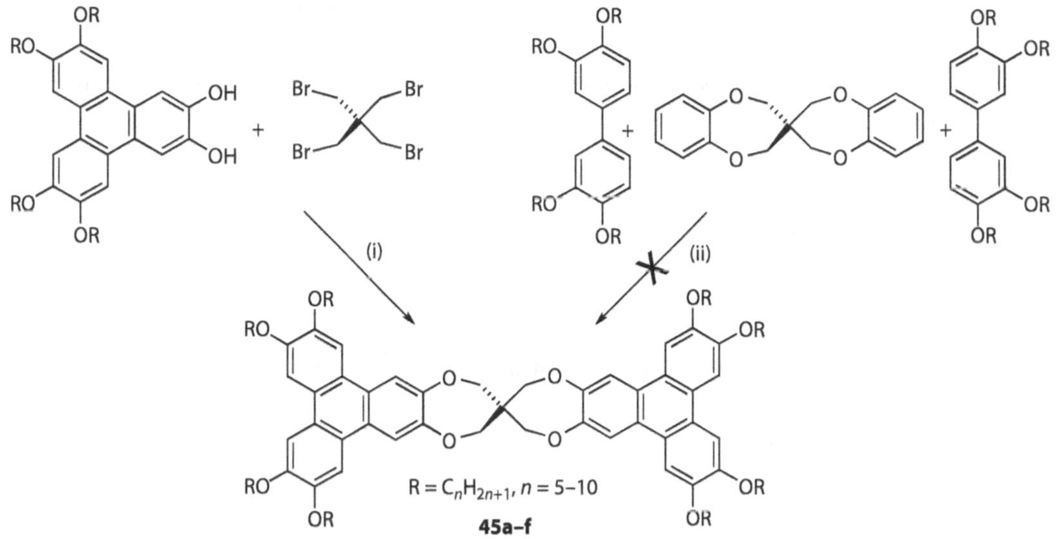

45a–f   R = CₙH₂ₙ₊₁, n = 5–10

SCHEME 3.7   Synthesis of triphenylene spiro-twins: (i) K₂CO₃, DMF, 120°C; (ii) FeCl₃.

Schulte et al. prepared six novel triphenylene-based spiro-twins **45a–f** by condensing 1,2-dihydroxytriphenylenes with a tetrabromide (Scheme 3.7) [65]. Although the phenyl–biphenyl coupling route to prepare a variety of symmetrical and unsymmetrical triphenylenes is well established [35–38], efforts to prepare these spiro-twins by this technique failed [65]. Whereas compounds **45a** and **45b** containing pentyloxy and hexyloxy chains, respectively, showed only crystals to isotropic phase transition, the higher homologues **45c–f** displayed columnar mesophase. Their thermal behavior is summarized in Table 3.9.

Mao et al. connected two triphenylene fragments covalently to a benzene ring via a carbamate linkage to produce dimer **46** [66]. The key step of synthesis involves a reaction between one molecule of 1,4-phenylene diisocyanate and two molecules of hydroxyl-terminated triphenylene derivative. The high value of melting and clearing temperatures can be attributed to the presence of hydrogen bonding, which makes the molecular packing more efficient. In addition to the above phenomenon, hydrogen bonding supported this dimer to exhibit an excellent film-forming property [66].

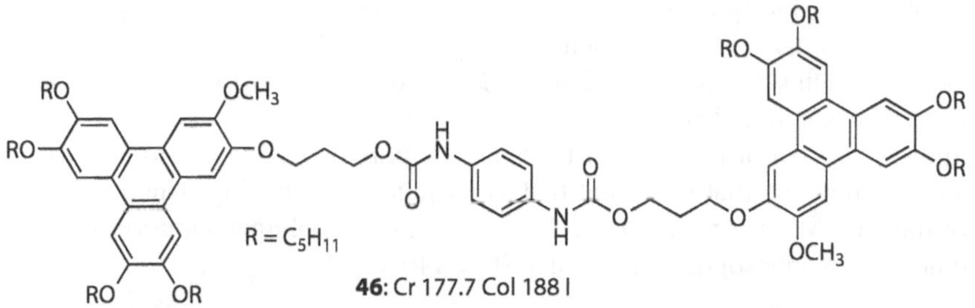

R = C₅H₁₁   **46**: Cr 177.7 Col 188 I

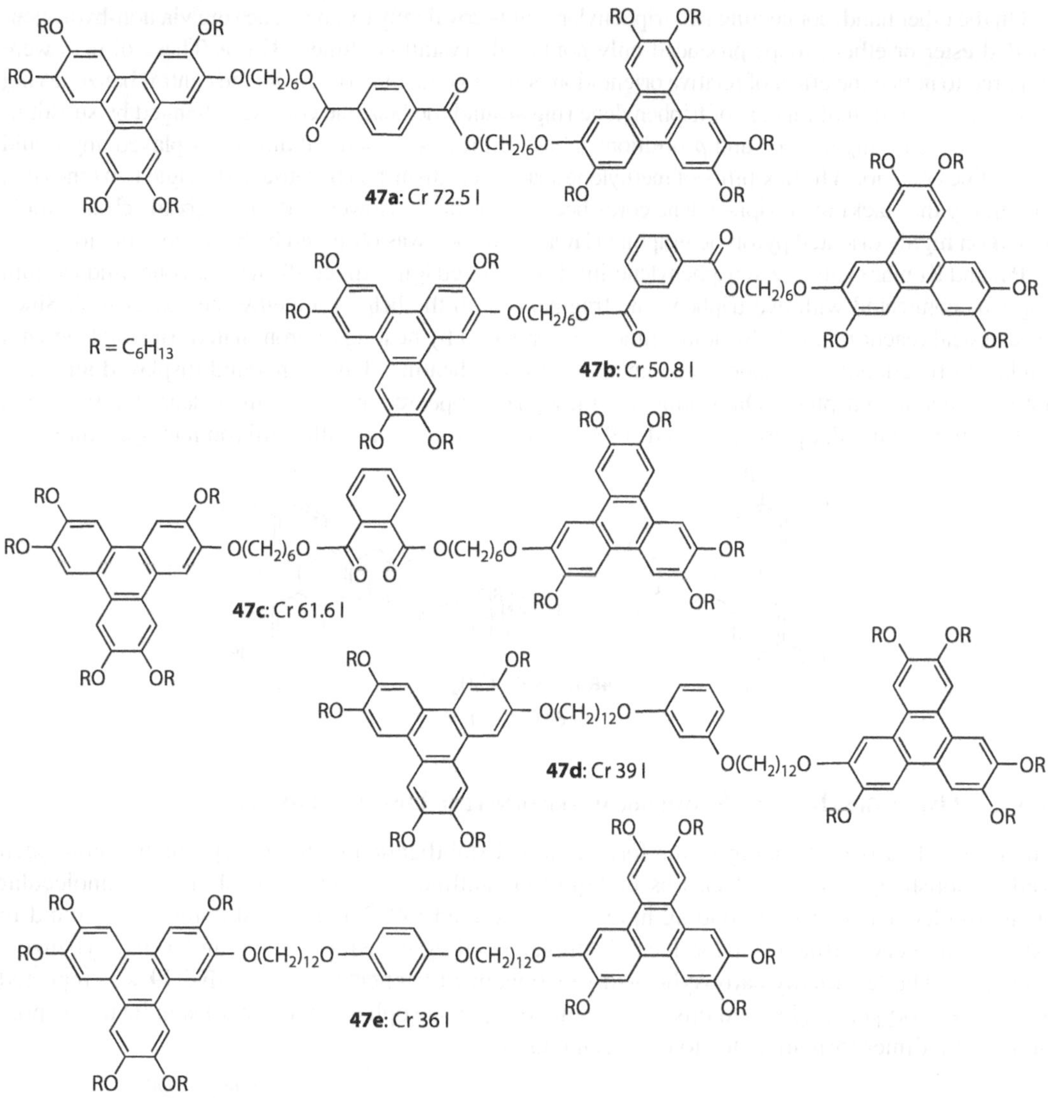

47a: Cr 72.5 I

47b: Cr 50.8 I

R = C₆H₁₃

47c: Cr 61.6 I

47d: Cr 39 I

47e: Cr 36 I

TABLE 3.9
Thermal Behavior of Triphenylene-Based Spiro-Twins

Structure	R	Phase Transition
45a	C₅H₁₁	Cr 222 I
45b	C₆H₁₃	Cr 155 I
45c	C₇H₁₅	Cr 70 Col 103 I
45d	C₈H₁₇	Cr 65 Col 109 I
45e	C₉H₁₉	Cr 60 Col 121 I
45f	C₁₀H₂₁	Cr 56 Col 106 I

Source: Data from Schulte, J.L. et al., *Eur. J. Org. Chem.*, 2499, 1998.

On the other hand, connecting two triphenylene units covalently to a benzene ring via non-hydrogen-bonded ester or ether groups produced only nonliquid crystalline dimers **47a–e**. These dimers were prepared to notice the effect of relative orientation of mesogenic moiety around the central benzene ring [67]. The orientation of linkage of triphenylene ring around the benzene core was changed by substituting the benzene ring at *o*, *m*, and *p*-positions. None of these symmetrical dimers displayed any liquid crystalline behavior. The flexibility of methylene spacer gets disturbed because of the rigid benzene ring, and finally the stacking of triphenylene cores becomes difficult. However, the hexagonal self-assembly of **47d** on highly oriented pyrolytic graphite (HOPG) surface was observed by STM imaging [67].

Pal and Kumar synthesized triphenylene-imidazole-based ionic dimer **48**, in which one imidazolium ring was connected with two triphenylene fragments with the help of a methylene spacer [68]. Since the classical reaction failed, this ionic dimer was prepared by heating ω-brominated triphenylene with imidazole functionalized triphenylene under microwave heating. This compound displayed an enantiotropic rectangular phase. On heating, the mesophase appeared at 84°C, which cleared at 95°C. On cooling, the rectangular phase appeared at 92°C, which does not crystallize till room temperature [68].

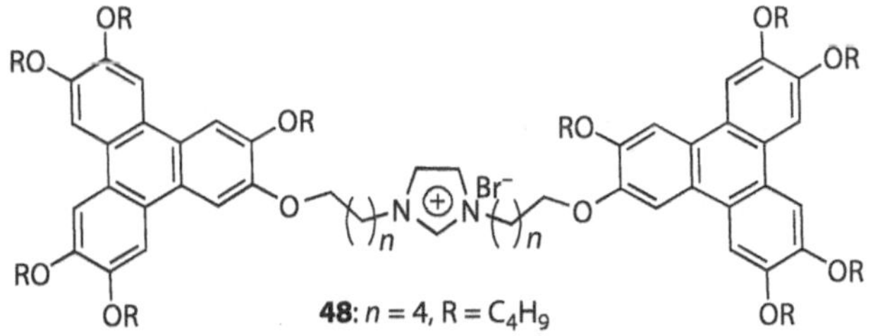

48: *n* = 4, R = C$_4$H$_9$
Cr 84 Col$_r$ 95 I

3.5.2 Hydrogen-Bonded Symmetrical Triphenylene Discotic Dimers

The role of hydrogen bonding in the formation and stabilization of liquid crystals has now been well established, and several classes of liquid crystalline compounds involving intramolecular or intermolecular hydrogen bonding have been described [69–73]. The first compounds found to exhibit liquid crystalline behavior due to hydrogen bonding were the 4-alkyl- and 4-alkoxybenzoic acids [74]. The terminally carboxylic acid-functionalized triphenylene derivative **49** was reported by Maliszewskyj et al. [75], and this shows a monotropic mesophase. X-ray studies indicate the possibility of a dimer formation due to hydrogen bonding.

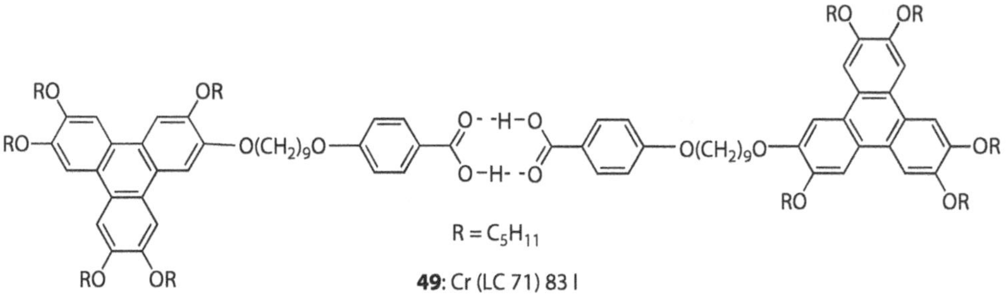

R = C$_5$H$_{11}$
49: Cr (LC 71) 83 I

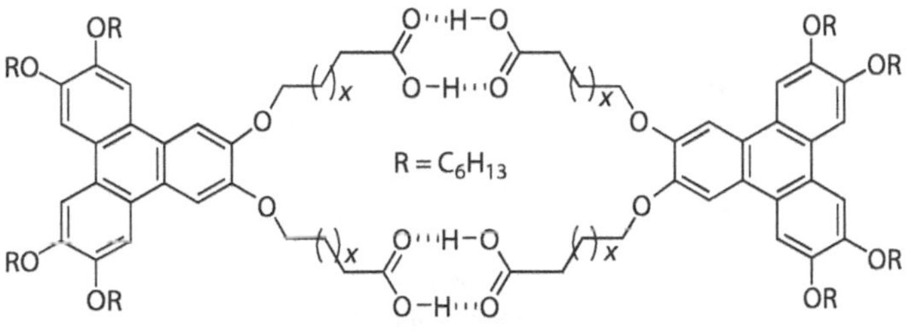

R = C$_6$H$_{13}$

50a–f: *x* = 3–7, 10

TABLE 3.10
Thermal Behavior of H-Bonded
Triphenylene Dimers

Structure	x	Phase Transition
50a	3	Cr (117 Col$_h$) 130 I
50b	4	Cr 94 Col$_h$ 105.3 I
50c	5	Cr (78 Col$_h$) 103.8 I
50d	6	Cr (68 Col$_h$) 106.1 I
50e	7	Cr 95.4 I
50f	10	Cr 69.1 I

Source: Data from Wan, W. et al., *Mol.*
Cryst. Liq. Cryst., 364, 597, 2001.

Six homologues of the triphenylene-based series, **50a–f**, containing two terminal carboxylic acids were prepared by Shimizu and coworkers [76,77]. Using temperature-dependent IR spectroscopy, it was observed that the band intensity of hydrogen-bonded carbonyl in the carboxylic acid dimer increases at the isotropic to columnar phase transition on cooling. The mesomorphic behavior of these materials depends significantly on the spacer length connecting the triphenylene core and the carboxylic acid group. Transition temperatures of these compounds are given in Table 3.10.

3.5.3 NONSYMMETRICAL TRIPHENYLENE DISCOTIC DIMERS

The concept of reducing the molecular symmetry in order to reduce the melting point in monomeric DLCs has been adopted by several workers [78]. It is also known that the crystallization of glassy columnar phase shown by monomeric DLCs is retarded by reducing the symmetry of the disk [31]. Nonsymmetrical discotic dimers can be prepared by linking two different triphenylene units using a two-step synthesis procedure as shown in Scheme 3.6. Two such nonsymmetric triphenylene-based dimers, **51a** and **51b**, have been reported. In dimer **51a**, a pentahexyloxytriphenylene unit was connected to a pentabutyloxytriphenylene unit via a polymethylene spacer. This compound shows a Col$_h$ phase to isotropic phase transition at 98°C. On cooling, the columnar phase freezes into a glassy state at 30°C [31].

51a: R = C$_6$H$_{13}$, R′ = C$_4$H$_9$: Col$_h$ (g 30) 98 I

51b: R = (structure) R′ = C$_{10}$H$_{21}$: Cr 34 Col$_h$ 45 I

R = C$_6$H$_{13}$

52: Cr (g 30) 89 Col$_h$ 99 I

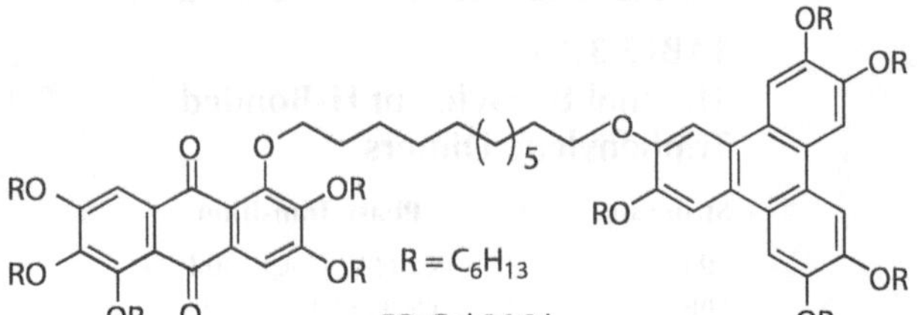

53: Col 96.9 I

In compound **51b**, a pentadecyloxytriphenylene subunit was tethered to a chiral pentakis(3,7-dimethyloctyloxy)triphenylene via a decyl spacer. This compound shows a crystal-to-columnar phase transition at 34°C and clears at 45°C in the first heating run of the DSC. On cooling, the columnar phase appears at 44°C and is stable to 0°C. Despite the presence of a chiral triphenylene unit, no chiral mesophase was observed in this dimer [79].

The synthesis of dimer **52**, in which there is an amide group in the linking chain, was reported by Boden et al. [31]. The compound was obtained by condensing an amine-terminated triphenylene-based derivative with an acid chloride-functionalized triphenylene unit. It exhibits an enantiotropic columnar phase and forms a glass on cooling at about 30°C. The glassy columnar state is stable for several months at room temperature [31].

A few examples of unsymmetrical charge-transfer twins, in which a triphenylene unit is connected to an electron-deficient nonliquid crystalline TNF or anthraquinone molecule, have been reported. Truly speaking, these are not discotic dimers as only one discotic fragment is involved. The other part of these dimers is either a nonliquid crystalline molecule or a calamitic molecule. These compounds can be best described as monofunctional triphenylenes [78], and their physical properties have been presented in Chapter 2.

A triphenylene-anthraquinone-based dimer **53**, in which both electron-rich and electron-deficient units are surrounded by peripheral alkyl chains, was prepared by Kumar et al. [80]. On first heating, this material showed a glass to mesophase transition at 40°C, and this mesophase transformed into isotropic phase at 96.9°C. On cooling the liquid phase, the columnar phase appears at 94.3°C. On subsequent heating, the first transition at 40°C was not observed and the liquid crystal melted to isotropic liquid.

3.6 CYCLOTETRAVERATRYLENE DIMER

Percec and coworkers have reported the synthesis and mesomorphic behavior of a cyclotetraveratrylene (tetrabenzocyclododecatetraene)-based symmetric dimer **56** [81]. The two cyclotetraveratrylene rings were linked through a hexadecane spacer in the twin. The condensation of alcohol **54** and dimeric alcohol **55** in the presence of a very strong acid (CF₃COOH) produces dimer **56** (Scheme 3.8) and branched oligomers of cyclotetraveratrylene mesogens. The dimer exhibits liquid crystalline behavior at ambient

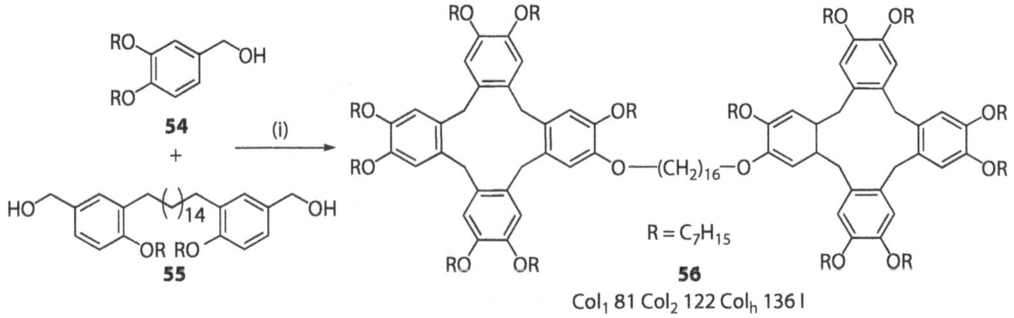

Col₁ 81 Col₂ 122 Col_h 136 I

SCHEME 3.8 Synthesis of CTTV-based discotic dimer **56**: (i) CF₃COOH.

temperature. It showed three enantiotropic columnar phases. The high-temperature mesophase was characterized as hexagonal columnar, but the nature of the other two phases could not be revealed. The clearing point of this twin was lower than that of the corresponding monomeric compound.

3.7 PHTHALOCYANINE-BASED DISCOTIC DIMERS

3.7.1 PHTHALOCYANINE DIMERS CONNECTED THROUGH FLEXIBLE SPACERS

A number of dinuclear phthalocyanine complexes **59** were reported by Bryant et al. [82,83]. In these phthalocyanine dimers, two octasubstituted phthalocyanine cores were connected via a flexible spacer. The key monomeric phthalocyanine precursor was prepared following the general method for nonsymmetric phthalocyanine derivatives (Chapter 2) [84]. The esterification of the mono-hydroxyalkyl phthalocyanines **57** with excess of oxalyl chloride yields a phthalocyanine derivative with a terminal carbonyl chloride group **58**. This was followed by further esterification of **58** with one equivalent of **57** to produce the desired dinuclear complex **59** (Scheme 3.9).

Data related to the mesomorphic behavior of dinuclear complexes **59** are listed in Table 3.11. As can be seen from this table, dimer **59a** with the smallest linking spacer is not liquid crystalline. Higher homologues, **59b–d**, **59f** exhibit a rectangular and hexagonal arrangement in their lower and higher temperature mesophases, respectively. On increasing the spacer length (**59b–d**), a decrease in Col_r–Col_h transition temperature can be seen. Compound **59e** with the longest spacer displayed only a single columnar mesophase. On heating, this compound melted to the isotropic phase at 175°C. It is worth mentioning that the insertion of copper metal in the phthalocyanine cavity (compare **59f** vs. **59c**) caused a considerable enhancement in the melting point, the isotropic temperature, as well as the mesophase range of the dinuclear complex. The dimers showed a broader mesophase range than their monomeric precursors.

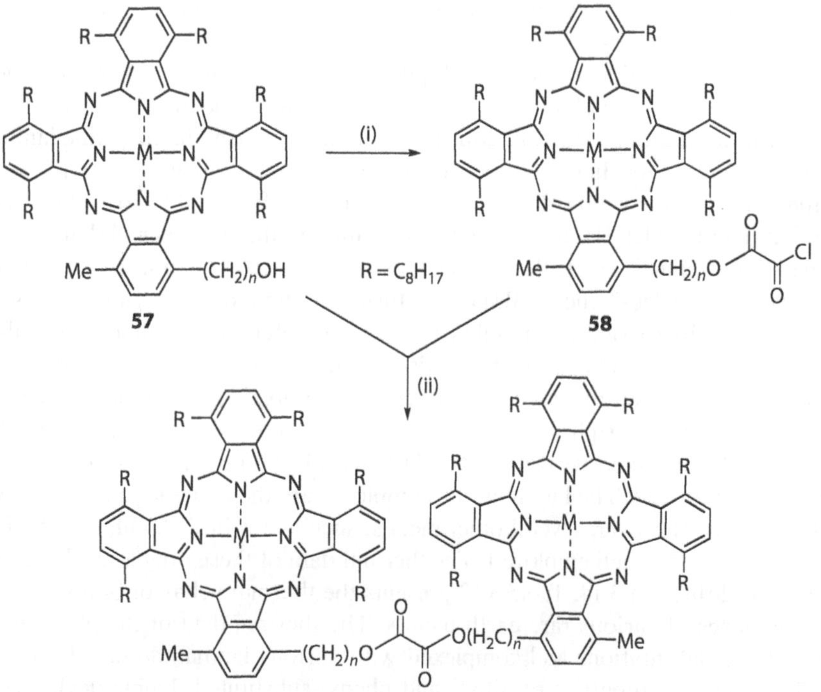

SCHEME 3.9 Synthesis of dinuclear phthalocyanine complexes **59**: (i) ClCOCOCl, CH_2Cl_2, K_2CO_3; (ii) 1, 2-dichloroethane, K_2CO_3.

TABLE 3.11
Thermal Behavior of Dinuclear
Complexes 59 (Scheme 3.9)

Structure	n	M	Phase Transition
59a	3	2H	Cr 160 I
59b	4	2H	Col$_r$ 144 Col$_h$ 163 I
59c	5	2H	Col$_r$ 135 Col$_h$ 180 I
59d	6	2H	Col$_r$ 123 Col$_h$ 188 I
59e	8	2H	Col 175 I
59f	5	Cu	Col$_r$ 161 Col$_h$ 254 I

Source: Data from Bryant, G.C. et al., *J. Mater. Chem.*, 4, 209, 1994; Bryant, G.C. et al., *Tetrahedron*, 52, 809, 1996.

3.7.2 PHTHALOCYANINE DOUBLE-DECKERS

In phthalocyanine double-decker complexes, two phthalocyanine macrocycles are connected to each other via a rare earth metal. These bis[2,3,9,10,16,17,23,24-octasubstituted-phthalocyaninato] lanthanide(III) complexes are face-to-face stacking dimers. The two phthalocyanine rings are staggered with respect to each other with an angle of 45°. Because of the trivalent nature of the lanthanide atom, one of the phthalocyanine rings is dianionic, whereas the other one is radical. This results in more distortion of the radical phthalocyanine macrocycle as compared to the dianionic one. These "intrinsic organic semiconductors" have recently gained much interest due to their very high charge carrier mobilities (see Chapter 6). Phthalocyanine-based double-deckers have also been employed as resistive and voltametric sensors.

The first example of phthalocyanine-based liquid crystalline "double-decker" or "sandwich complex" was reported by Simon et al. in 1985 [85], a few years after the discovery of its single-decker counterpart by the same group [86]. Subsequently, a number of bisphthalocyanine lanthanum complexes have been reported for their mesomorphism and electrochromism [87–102].

The synthesis of phthalocyanine double-decker complexes, bis[2,3,9,10,16,17,23,24-octa(alkoxy) phthalocyaninato]lanthanide(III), is essentially the same as that of other phthalocyanine derivatives. The reaction of phthalonitrile derivatives **60** with rare earth metal acetate in the presence of 1,8-diazabicyclo[5,4,0]undec-7-ene (DBU) easily furnish double-decker complexes (Scheme 3.10). The reaction is generally carried out in refluxing hexanol. Alternatively, first the metal-free phthalocyanines can be prepared and these compounds can then be coordinated with lanthanide.

Based on peripheral substitutions, the phthalocyanine double-decker complexes can be grouped into three classes: (1) with peripheral alkoxy and phenoxy-substituted derivatives **61–92**, (2) with peripheral thioalkyl-substituted derivatives **93–119**, and (3) with peripheral alkyl and phenyl-substituted derivatives **120–124**. The lutetium(III) complexes are the first and most widely synthesized and studied materials. However, several other metals, such as Er, Pr, Nd, Gd, Tb, Dy, Ho, Tm, Yb, Eu, Ce, and Sm, have also been employed. The thermal data of these double-decker complexes are listed in Tables 3.12 through 3.14. Table 3.12 presents the thermal behavior of alkoxy- or phenoxy-substituted complexes of various rare earth metals. The thermal data of the phthalocyanine with thioalkyl peripheral substitutions and complexed with various lanthanide metals are compiled in Table 3.13. The thermal properties of alkyl- and phenyl-substituted double-decker complexes of various rare earth metals are presented in Table 3.14.

The lowest homologue of bis(octaalkoxyphthalocyaninato)lutetium(III) series, **65**, did not display any liquid crystalline behavior. Other derivatives, **68**, **70**, **81**, **84**, **86**, were reported to exhibit

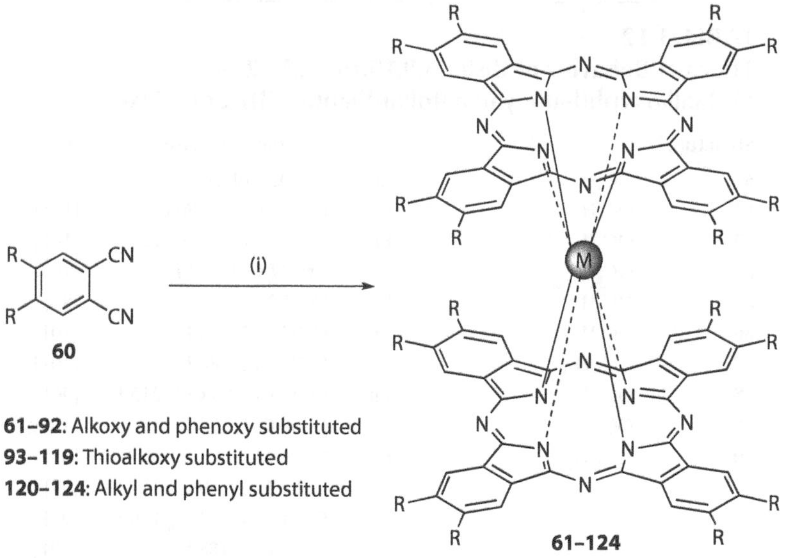

SCHEME 3.10 Synthesis of phthalocyanine double-decker complexes: (i) M(OAc)$_3$, DBU, hexanol, reflux.

rectangular columnar and hexagonal columnar arrangements in their lower and higher temperature mesophases, respectively. Longer chain derivatives **70, 81, 84, 86** displayed crystal-to-crystal transformation at lower temperatures as well. The temperature of isotropization decreases on increasing the peripheral chain length, and the peripheral chain length shows reciprocal effect on the mesophase range. In compounds **81, 84, 86**, a rare superheated transition from a crystalline phase to a discotic liquid crystalline phase was observed. This was attributed to the high viscosity of the mesophase. Hydrophilic poly(oxyethylene)-based lutetium double-deckers **87–90** were found to be soluble in water. Compound **89** displays a tetragonal phase with a lattice constant of 26.7 Å; all other derivatives of this series were not liquid crystalline.

The bis(octaalkoxyphthalocyaninato)erbium(III) complexes **61–64, 66, 67, 69, 80, 82, 83, 85** exhibit a viscous mesophase over a broad temperature range. As usual, the melting and clearing temperatures decrease on increasing peripheral chain length. In order to examine the effect of various lanthanide metal ions on the mesomorphic behavior of phthalocyanine double-decker complexes, Binnemans et al. prepared a series of double-decker compounds **69–79**, in which the peripheral alkyl chain was kept constant [101]. As can be seen from Table 3.12, the variation of rare earth metal ions has a very small effect on the transition temperatures. Only a trivial contraction of the mesophase range was observed on going from a lighter metal complex **71** to a heavier metal complex **79, 70**. This behavior can be explained on the basis of the fact that the rare earth metals are buried and packed in between the two phthalocyanine disks. Therefore, the lanthanide atom is isolated from the nearest double-decker molecule, and hence is not able to play a significant role in the intramolecular interaction.

The thermal behavior of octathioalkyl derivatives of various rare earth metals, Eu (**98, 102, 106, 110, 114**), Tb (**99, 103, 107, 111, 115, 118**), Lu (**93, 97, 101, 105, 109, 113, 117**), Ce (**100, 104, 108, 112, 116, 119**), Gd (**94**), Dy (**95**), and Sm(**96**) is summarized in Table 3.13. Within a lanthanide complex series, the clearing temperature decreases as the thioalkyl peripheral chain length increases. An additional mesophase was observed for decylthio (**101–104**), dodecylthio (**105–108**), and tetradecylthio (**111, 112**) substituted derivatives. A detailed x-ray diffraction study of a lower temperature mesophase of **106** demonstrated properties of a rectangular as well as a hexagonal columnar mesophase. The core-to-core distance was found to be 3.3 Å. On the basis of co-facial distance of 3.3 Å and the ratio of the first and second peak [1:1/3$^{1/2}$], this phase was assigned as a novel pseudo-hexagonal mesophase. This unique behavior was explained on the basis of the fact that

TABLE 3.12
Thermal Behavior of Bis[(2,3,9,10,16,17,23,24)-Octaalkoxyphthalocyaninato]lanthanum(III) Complexes

Structure	R	M	Phase Transition	Ref.
61	OC_4H_9	Er	Cr 202 Col$_h$ >280 dec	[101]
62	OC_5H_{11}	Er	Cr 174 Col$_h$ >280 dec	[101]
63	OC_6H_{13}	Er	Cr 147 Col$_h$ >280 dec	[101]
64	OC_8H_{17}	Er	Cr 137 Col$_h$ 263 I	[101]
65	OC_8H_{17}	Lu	Cr 25 I	[85]
66	OC_9H_{19}	Er	Cr 93 Col$_h$ 239 I	[101]
67	$OC_{10}H_{21}$	Er	Cr 71 Col$_h$ 180 I	[101]
68	$OC_{10}H_{21}$	Lu	Cr 43 Col$_r$ 96 Col$_h$ 215 I	[96]
69	$OC_{12}H_{25}$	Er	Cr 68 Col$_h$ 174 I	[101]
70	$OC_{12}H_{25}$	Lu	Cr 24 M 30 I	[85]
			Cr 86 Col$_h$ 189 I	[87]
			Cr 61 Col$_r$ 90 Col$_h$ 196 I	[96]
			Cr 92 Col$_h$ 188 I	[101]
71	$OC_{12}H_{25}$	Pr	Cr 74 Col$_h$ 208 I	[101]
72	$OC_{12}H_{25}$	Nd	Cr 66 Col$_h$ 206 I	[101]
73	$OC_{12}H_{25}$	Eu	Cr 65 Col$_h$ 203 I	[101]
74	$OC_{12}H_{25}$	Gd	Cr 68 Col$_h$ 218 I	[101]
75	$OC_{12}H_{25}$	Tb	Cr 72 Col$_h$ 205 I	[101]
76	$OC_{12}H_{25}$	Dy	Cr 83 Col$_h$ 205 I	[101]
77	$OC_{12}H_{25}$	Ho	Cr 85 Col$_h$ 190 I	[101]
78	$OC_{12}H_{25}$	Tm	Cr 68 Col$_h$ 182 I	[101]
79	$OC_{12}H_{25}$	Yb	Cr 84 Col$_h$ 192 I	[101]
80	$OC_{14}H_{29}$	Er	Cr 58 Col$_h$ 180 I	[101]
81	$OC_{14}H_{29}$	Lu	Cr 51 Col$_r$ 72 Col$_h$ 171 I	[96]
82	$OC_{15}H_{31}$	Er	Cr 44 Col$_h$ 170 I	[101]
83	$OC_{16}H_{33}$	Er	Cr 43 Col$_h$ 163 I	[101]
84	$OC_{16}H_{33}$	Lu	Cr 51 Col$_r$ 61 Col$_h$ 149 I	[96]
85	$OC_{18}H_{37}$	Er	Cr 65 Col$_h$ 151 I	[101]
86	$OC_{18}H_{37}$	Lu	Cr 51 M 56 I	[85]
			Cr 54 Col$_r$ 64 Col$_h$ 131 I	[96]
87	$OCH_2CH_2OCH_3$	Lu	Cr 246 I	[90]
88	$O(CH_2CH_2O)_2CH_3$	Lu	Cr 138 I	[90]
89	$O(CH_2CH_2O)_3CH_3$	Lu	Cr 53 Col$_t$ 57.6 I	[90]
90	$O(CH_2CH_2O)_4CH_3$	Lu	Liquid	[90]
91	$OC_6H_3[m, p\text{-}(OC_{12}H_{25})_2]$	Lu	Cr 35 Col$_h$ 139 Cub$_1$ 164 Cub$_2$ 216 Col$_t$ 239 I	[102]
92	$OC_6H_3[m, p\text{-}(OC_{13}H_{27})_2]$	Lu	Cr 31 Col$_h$ 122 Cub$_1$ 143 Cub$_2$ 166 Col$_t$ 222 I	[102]

non-tilted disks were arranged in a hexagonal pattern, whereas a molecule as a whole had an overall rectangular arrangement. The stacking distance of around 7 Å and around 3.5 Å observed in XRD corresponds to double-decker and single-decker, respectively, for some of the smaller chain derivatives in their low-temperature mesophase. Because of distortion in planarity caused by the radical nature, one of the phthalocyanine rings acquires a dome shape. At higher temperature, faster trampoline movement of the dome gives time-averaged, single-decker packing distance. The replacement of alkoxy substituents with thioalkyl substituents resulted in a decreased value of melting points.

TABLE 3.13
Thermal Behavior of Bis[(2,3,9,10,16,17,23,24)-Octathioalkylphthalocyaninato]lanthanum(III) Complexes

Structure	R	M	Phase Transition	Ref.
93	SC_6H_{13}	Lu	Cr 120 Col$_h$ 242 I	[97]
94	SC_6H_{13}	Gd	Cr 52 Col$_h$ 258 I	[99]
95	SC_6H_{13}	Dy	Cr 55 Col$_h$ 250 I	[99]
96	SC_6H_{13}	Sm	Cr 60 Col$_h$ 240 I	[99]
97	SC_8H_{17}	Lu	Cr 89 Col$_h$ 219 I	[94]
98	SC_8H_{17}	Eu	Cr 98 Col$_h$ 252 I	[94]
99	SC_8H_{17}	Tb	Cr 102 Col$_h$ 242 I	[94]
100	SC_8H_{17}	Ce	Cr 95 Col$_h$ 257 I	[100]
101	$SC_{10}H_{21}$	Lu	Cr 6 Col 40 Col$_h$ 192 I	[94]
102	$SC_{10}H_{21}$	Eu	Cr -1 Col 68 Col$_h$ 209 I	[94]
103	$SC_{10}H_{21}$	Tb	Cr 3 Col 56 Col$_h$ 206 I	[94]
104	$SC_{10}H_{21}$	Ce	Cr 4 Col$_h$ 56 Col$_h$ 217 I	[100]
105	$SC_{12}H_{25}$	Lu	Cr 27 Col 38 Col$_h$ 167 I	[94]
106	$SC_{12}H_{25}$	Eu	Cr 22 Col 64 Col$_h$ 181 I	[94]
107	$SC_{12}H_{25}$	Tb	Cr 20 Col 53 Col$_h$ 172 I	[94]
108	$SC_{12}H_{25}$	Ce	Cr -3 Col$_h$ 47 Col$_h$ 180 I	[100]
109	$SC_{14}H_{29}$	Lu	Cr 38 Col$_h$ 146 I	[94]
110	$SC_{14}H_{29}$	Eu	Cr 13 Col$_h$ 152 I	[94]
111	$SC_{14}H_{29}$	Tb	Cr 22 M 47 Col$_h$ 144 I	[94]
112	$SC_{14}H_{29}$	Ce	Cr 31 Col$_h$ 47 Col$_h$ 156 I	[100]
113	$SC_{16}H_{33}$	Lu	Cr 46 Col$_h$ 126 I	[94]
114	$SC_{16}H_{33}$	Eu	Cr 50 Col$_h$ 134 I	[94]
115	$SC_{16}H_{33}$	Tb	Cr 46 Col$_h$ 132 I	[94]
116	$SC_{16}H_{33}$	Ce	Cr 49 Col$_h$ 139 I	[100]
117	$SC_{18}H_{37}$	Lu	Cr 53 Col$_h$ 110 I	[94]
118	$SC_{18}H_{37}$	Tb	Cr 33 Col$_h$ 116 I	[94]
119	$SC_{18}H_{37}$	Ce	Cr 53 Col$_h$ 119 I	[98]

TABLE 3.14
Thermal Behavior of Alkyl/Phenyl-Substituted Lutetium Phthalocyanine Double-Deckers

Structure	R	Phase Transition	Ref.
120	C_8H_{17}	Cr 79 Col$_h$ 82 I	[88]
121	$C_{12}H_{25}$	Col$_{ob}$ 32 I	[88]
122	$C_{18}H_{37}$	Col$_L$ 30 I	[88]
123	$CH_2OC_{18}H_{37}$	Cr 51 Col$_h$ 56 I	[87]
124	$C_6H_4(p\text{-}OC_{18}H_{37})$	Col$_t$ 47 Col$_h$ 242 I	[89]

Contrary to other lanthanide metal double-decker complexes, cerium-based sandwiched compounds exhibit different behavior. Both phthalocyanine rings are dianionic, and hence cerium atom is in the tetravalent oxidation state. Therefore, the extent of distortion from planarity is the same for both phthalocyanine rings. This causes equivalence in both macrocyclic rings. Cerium derivatives **91, 104, 108, 112, 116, 119** were reported to show a wider mesophase range as compared to those of other rare earth metal sandwiched compounds. This can be explained by the similar concave nature of both phthalocyanine ligands. The dialkoxyphenoxy-substituted phthalocyanine derivatives, **91** and **92**, exhibit a remarkably broad mesophase range and rich polymorphism. Compounds **91** and **92** have been accounted as the first instance of phthalocyanine derivatives manifesting two optically isotropic cubic mesophases. The low-temperature hexagonal phase gave a stacking distance of 10.1 Å. This larger distance can be linked to steric hindrance of the eight peripheral phenoxy groups. These compounds (**91** and **92**) were also the first mesomorphic phthalocyanine-based double-decker complexes to exhibit an unprompted uniform homeotropic alignment [104].

The phase transition data of octaalkyl- and alkoxyphenyl-substituted phthalocyanine lutetium complexes, **120–124**, are represented in Table 3.14. These derivatives displayed a variety of mesophases. It is worth mentioning that the mesophase range of derivatives **120–123** in which alkyl chains are directly linked to the phthalocyanine ring was very narrow. However, the introduction of a phenyl ring resulted in broadening the liquid crystalline phase range of **124**. The complex **121** was the first phthalocyanine compound in which a disordered oblique columnar phase was observed at room temperature. On heating, this oblique phase goes to an isotropic phase at 32°C, which relaxes into a columnar phase on leaving it as such. Further heating of the sample resulted in isotropization at 44°C. A similar behavior was demonstrated by **122**. Compound **124** possesses a disordered tetragonal columnar arrangement of molecules in its pure state at room temperature. On heating, the stacking of molecules in the column becomes loose due to vigorous thermal mobility of peripheral alkyl chains. This resulted in the rotation of phthalocyanine disks and hence the transformation of the lattice into a more stable hexagonal one.

3.8 PORPHYRIN DOUBLE-DECKERS

Ohta and coworkers studied mesomorphism in cerium metal-based porphyrin double-decker complexes **125a–g** [103]. These materials were synthesized by reacting metal-free porphyrin derivatives with cerium acetate. Table 3.15 summarizes the thermal behavior of porphyrin-based double-deckers **125a–g**. The number of alkoxy chains around the porphyrin ring played a significant role in mesophase formation. The eight alkoxy chain substituted derivative **125a** showed columnar lamellar mesophase. On the other hand, 16 alkoxy chain substituted double-deckers **125b, c, f**, and **g** displayed a columnar rectangular phase. Compound **125a** was crystalline in its virgin state and had the highest value of isotropization temperature, while the 16 alkoxy chain substituted derivatives were either in liquid crystalline (**125b, c, f, g**) or in isotropic liquid state (**125d, e**) in their pure form. Because of very high viscosity of the isotropic phase of **125b, c, f, g**, it was not possible to

TABLE 3.15
Thermal Behavior of Porphyrin Dinuclear Complexes

Structure	R	Phase Transition
125a	$C_{14}H_{29}$	Cr 93.3 Col_L 105.8 I
125b	C_8H_{17}	Col_r 71.0 I
125c	$C_{10}H_{21}$	Col_r 70.3 I
125d	$C_{12}H_{25}$	Liquid
125e	$C_{14}H_{29}$	Liquid
125f	$C_{16}H_{33}$	Col_r 43.5 I
125g	$C_{18}H_{37}$	Col_r 51.3 I

Source: Data from Nakai, T. et al., *J. Mater. Chem.*, 12, 844, 2002.

get back the rectangular phase while cooling. The material remained in supercooled isotropic liquid state. The stacking distances corresponding to double-decker (8.6 Å) and single-decker (3.7 Å) were observed in x-ray for the lamellar phase of **125a**.

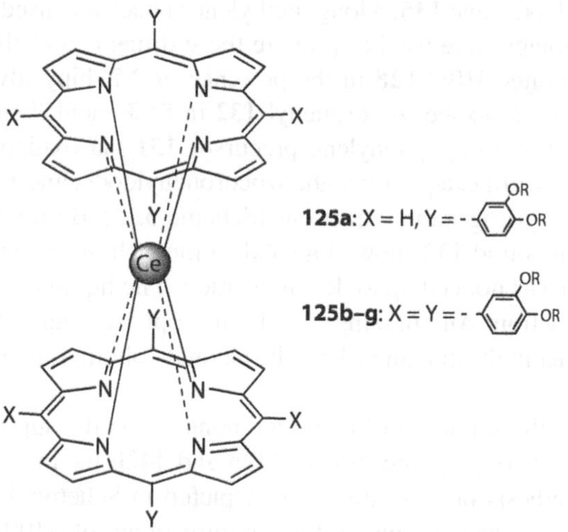

125a: X = H, Y = $-\langle\bigcirc\rangle$OR

125b–g: X = Y = $-\langle\bigcirc\rangle$OR

Shimizu et al. obtained the μ-oxo-dimer **127** by heating the tetraphenyl porphyrin **126** above its isotropic temperature (Scheme 3.11) [104]. This conversion was proved by the absence of any peak corresponding to the hydroxyl group in IR and NMR. The monomeric compound **126** is crystalline at room temperature, but exhibits disordered hexagonal phase after heating to 84°C. However, dimer **127** was liquid crystalline at room temperature. It exhibited two columnar lamellar phases in the temperature range of 10°C–148°C.

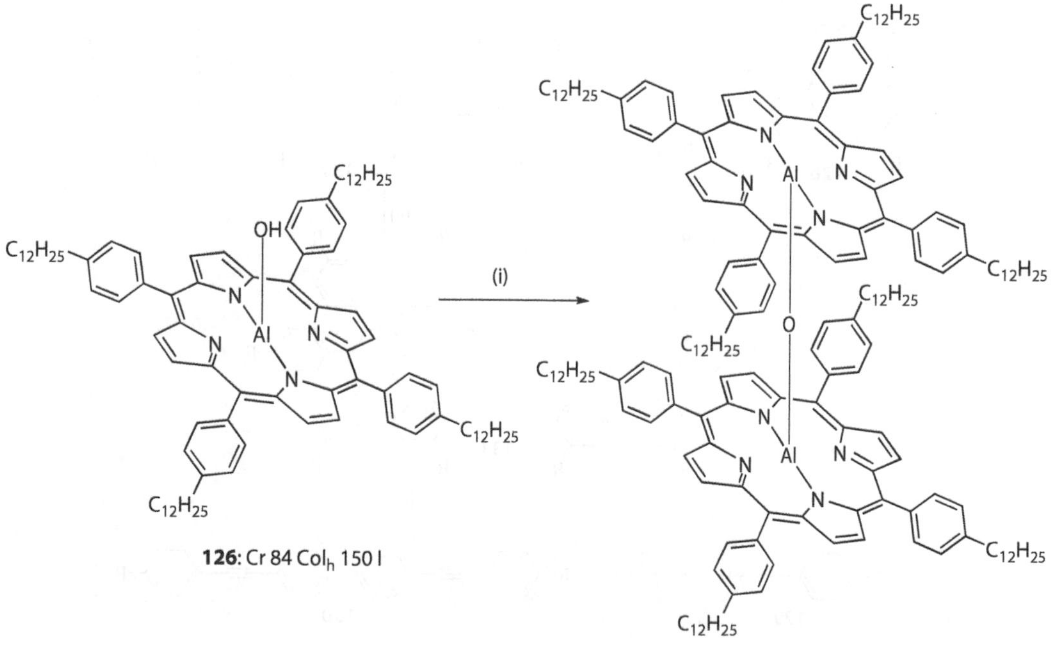

126: Cr 84 Col$_h$ 150 I

127: Cr −3 M 10 Col$_L$ 69 Col$_L$′ 148 I

SCHEME 3.11 Synthesis of porphyrin μ-oxo-dimer **127**: (i) heat.

3.9   HEXABENZOCORONENE-BASED DIMERS

Mullen and coworkers reported the synthesis of HBC dimers **132** and **136** [105]. In the bishexa-*peri*-hexabenzocoronenyl **132**, two HBC rings were directly connected to each other, while in the dihexa-*peri*-hexabenzocoronenyldodecane **136**, a long methylene spacer was used to tether two HBC disks. Two different synthetic routes were used to prepare these dimers. Nickel-catalyzed reductive coupling of the bromo-substituted HBC **128** in the presence of 2,2′-bipyridyl and 1,5-cyclooctadiene (COD) afforded bishexa-*peri*-hexabenzocoronenyl **132** in 85% yield (Scheme 3.12). Alternatively, this can be obtained from the oligophenylene precursor **131** via oxidative cyclodehydrogenation (Scheme 3.12). Similarly, the dihexa-*peri*-hexabenzocoronenyldodecane **136** was prepared using the cycloaddition and cyclodehydrogenation reactions (Scheme 3.13). Both HBC dimers were found to be liquid crystalline. Compound **132** showed a solid-to-mesophase transition at 124°C on heating, but no clearing transition was noticed up to decomposition. The liquid crystalline nature of **132** was confirmed by x-ray diffraction. An amazingly wider mesophase range of 317°C was observed for compound **136**. This dimer melts into an ordered hexagonal columnar phase at 53°C and goes to the isotropic phase at 370°C.

In order to understand the influence of hydrogen bonding on the supramolecular order of HBC dimers, Mullen and coworkers prepared dimers **142a** and **142b** having amide and ester groups in the spacer [106]. The synthesis of these dimers is depicted in Scheme 3.14. Dimer **142a** stayed in the pseudocrystalline phase over the whole temperature range of −100°C to 300°C. Compound **142b** was found to be liquid crystalline at room temperature with ordered hexagonal columnar arrangement. It did not show any signs of isotropization up to 300°C on heating. Solid state NMR investigation of dimer **142b** revealed a very unstable pseudocrystalline phase below 14°C. The formation of disordered 3D network due to the flexibility of the ester linkage was assumed to be the reason for the absence of any long-range order in the pseudocrystalline phase of **142b** below 14°C.

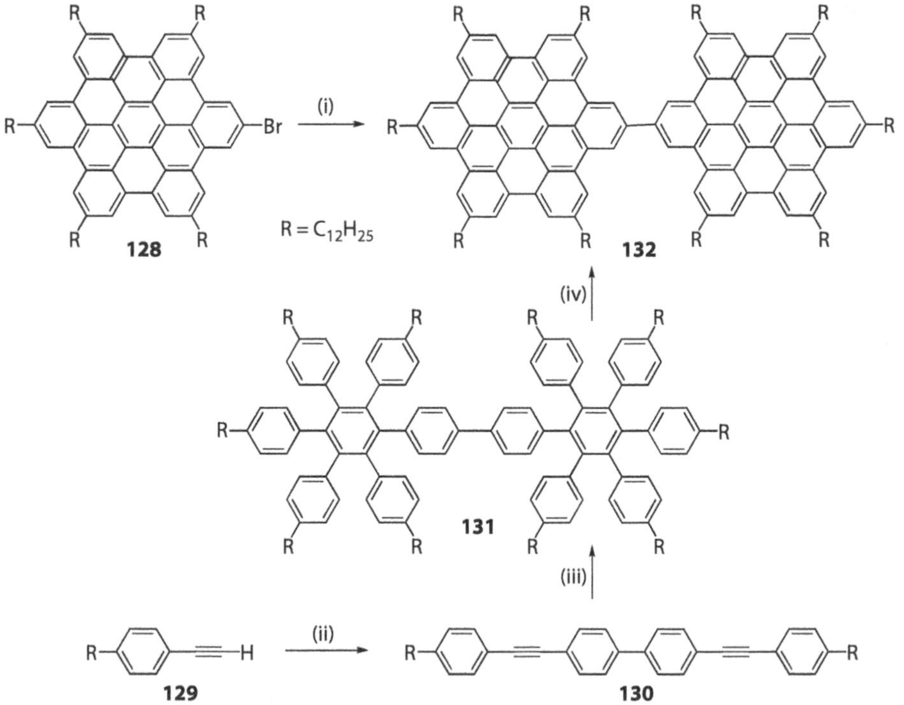

SCHEME 3.12   Synthesis of HBC dimer **132**: (i) [Ni(COD)$_2$], 2,2′-bipyridyl, toluene, 60°C; (ii) 4,4-dibromobiphenyl, [Pd(PPh$_3$)$_4$], CuI, piperidine, 80°C; (iii) tetra(4-tert-butylphenyl)cyclopentadienone, diphenyl ether, reflux; (iv) FeCl$_3$, CH$_2$Cl$_2$.

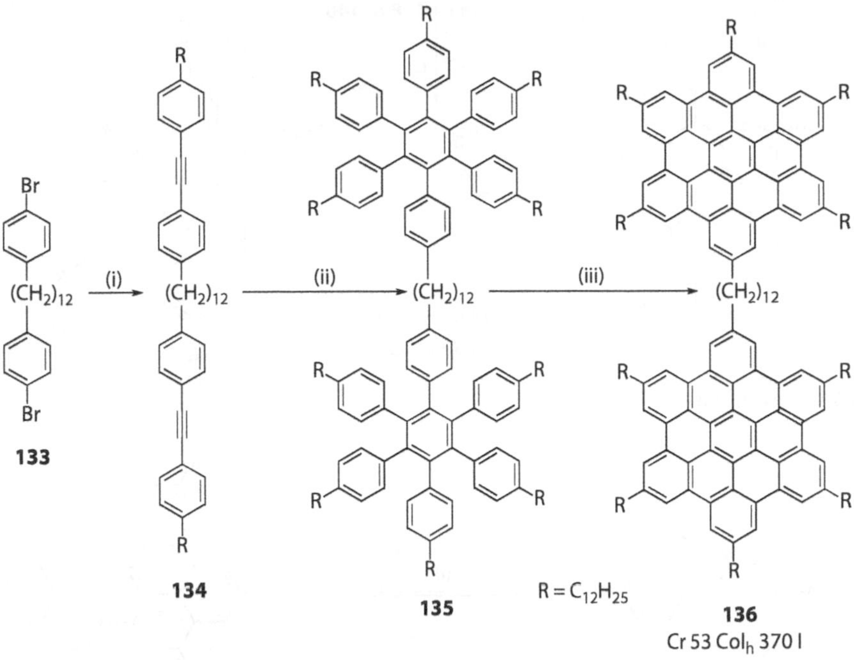

SCHEME 3.13 Synthesis of HBC dimer **136**: (i) 4-*n*-dodecylphenylacetylene, [Pd(PPh₃)₄], CuI, piperidine, 80°C; (ii) tetra(4-*n*-dodecylphenyl)cyclopentadienone (**22**), diphenyl ether, reflux; (iii) FeCl₃, CH₂Cl₂.

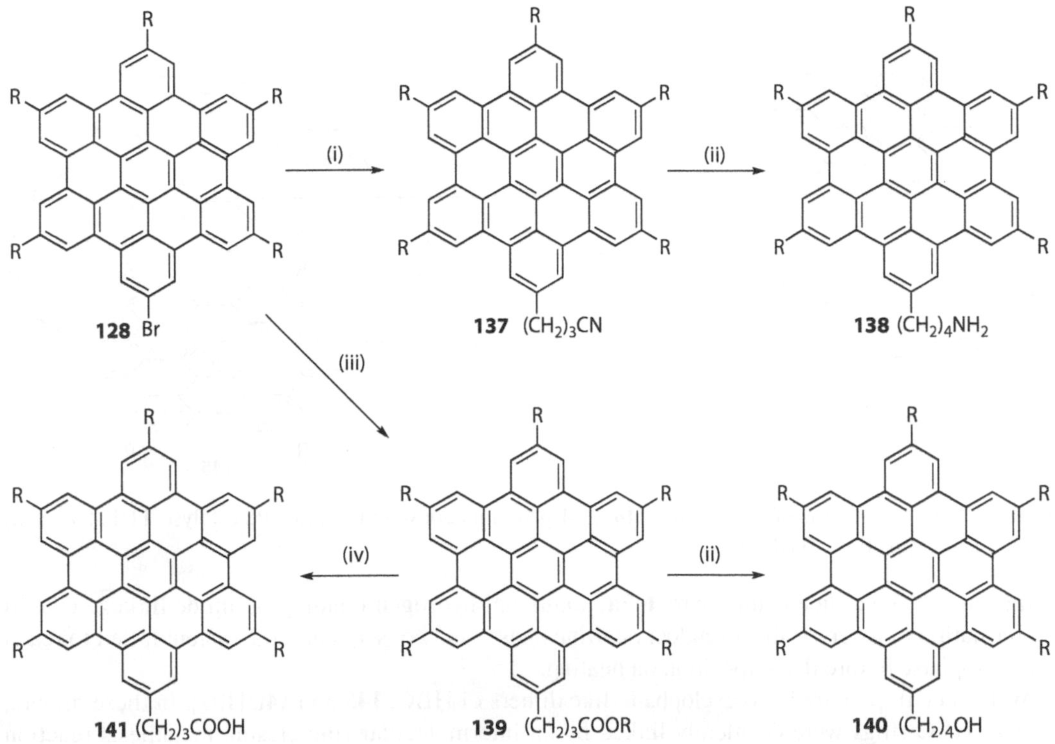

SCHEME 3.14 Synthesis of HBC dimer **142**: (i) BrZn(CH₂)₃CN, Pd catalyst; (ii) LiAlH₄, THF; (iii) BrZn(CH₂)₃COOR, Pd catalyst; (iv) KOH, MeOH-H₂O; (v) EDC, CH₂Cl₂.

(continued)

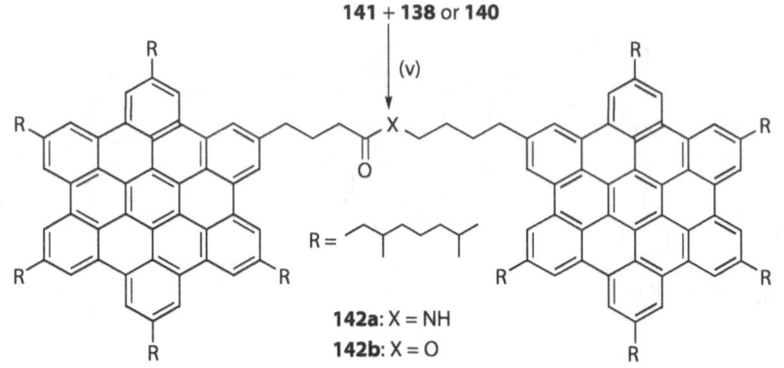

141 + 138 or 140

(v)

R =

142a: X = NH
142b: X = O

SCHEME 3.14 (continued)

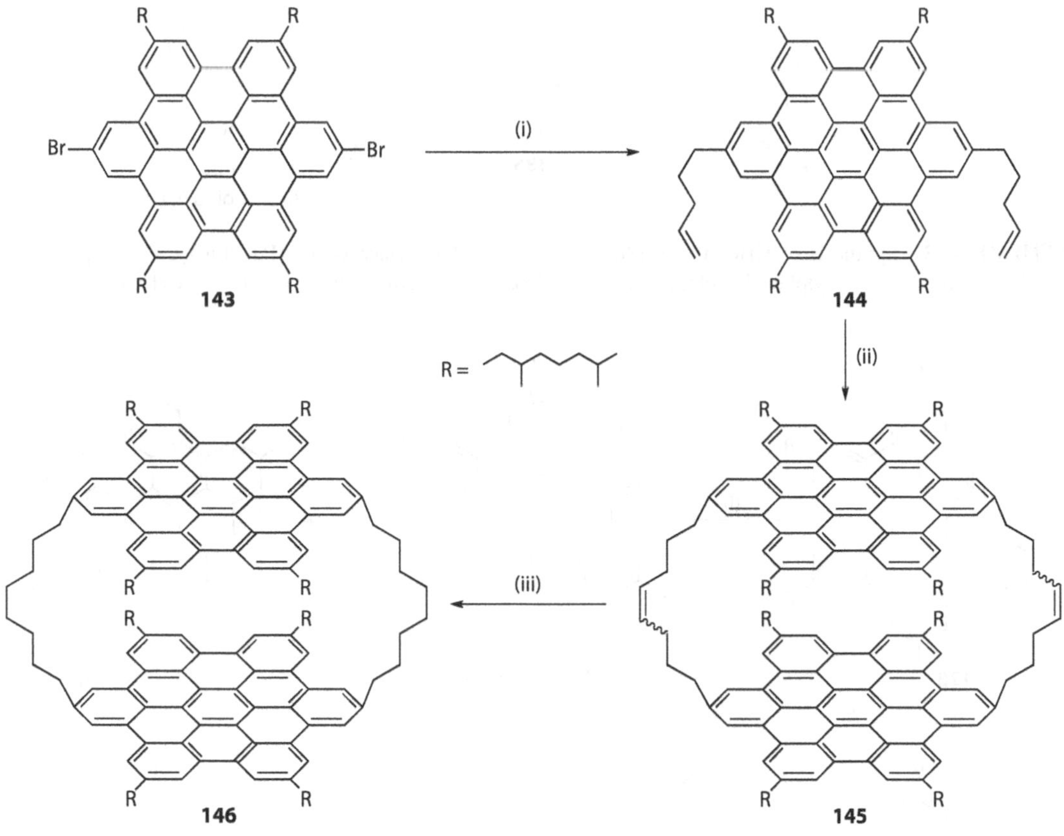

R =

143   **144**   **145**   **146**

SCHEME 3.15 Synthesis of HBC dimer **146**: (i) 4-pentyl magnesium bromide, Pd catalyst, THF, 55°C; (ii) RuL$_n$, toluene; (iii) Pd/C, THF, H$_2$.

In the case of the amide counterpart **142a**, additional hydrogen bonding of amide linkage caused the formation of a very stable pseudocrystalline phase, and consequently the system failed to reach any mesophase before decomposition on heating.

Watson et al. prepared two cyclophane-like dimers of HBC, **145** and **146** [107]. In these dimers, the two HBC rings were covalently linked by an intermolecular ring-closing metathesis reaction of diene derivative **144** to yield an isomeric mixture (*cis* and *trans*) of **145**. The linear oligomers formed in the reaction can be easily separated by passing through a short silica gel column. The palladium-catalyzed hydrogenation of **145** furnished **146** (Scheme 3.15).

Both cyclophane HBC dimers **145** and **146** assembled in 2D hexagonal columnar mode. The DSC thermogram of **145** displayed a first-order transition at 40°C on heating. However, a very weak second-order glass transition at −10°C was observed for compound **146.** The mesomorphic behavior of these HBC dimers persists up to 400°C on heating, as monitored by POM. Intracolumnar face-to-face stacking distance was found to be 3.6 Å. Dimer **146**, which does not contain any double bond in the spacer, did not show any phase transition other than glass transition at −10°C. On the other hand, cooling of sample **145** below 40°C resulted in the tilting of HBC disks with respect to the columnar axis. It is notable that the conversion of a double bond (in **145**) to a saturated single bond (in **146**) resulted in inhibition in the bulk crystallization tendency of rigid HBC cores.

REFERENCES

1. Imrie, C. T. and Luckhurst, G. R. In *Handbook of Liquid Crystal*, Vol. 2B, Demus, D., Goodby, J., Gray, G. W., Spiess, H. W., and Vill, V. (eds.), Wiley-VCH, Weinheim, Germany, pp. 801–833, 1998.
2. Imrie, C. T. and Henderson, P. A. *Chem. Soc. Rev.* 36, 2096–2124, 2007.
3. Vorlander, D. *Z. Phys. Chem.* 126, 449–472, 1927.
4. Griffin, A. C. and Britt, T. R. *J. Am. Chem. Soc.* 103, 4957–4959, 1981.
5. Lillya, C. P. and Murthy, Y. L. N. *Mol. Cryst. Liq. Cryst. Lett.* 2, 121–122, 1985.
6. Zamir, S., Wachtel, E. J., Zimmermann, H., Dai, S., Spielberg, N., Poupko, R., and Luz, Z. *Liq. Cryst.* 23, 689–698, 1997.
7. Praefcke, K., Kohne, B., Singer, D., Demus, D., Pelzl, G., and Diele, S. *Liq. Cryst.* 7, 589–594, 1990.
8. Praefcke, K., Kohne, B., Gundogan, B., Singer, D., Demus, D., Diele, S., Pelzl, G., and Bakowsky, U. *Mol. Cryst. Liq. Cryst.* 198, 393–405, 1991.
9. Langner, M., Praefcke, K., Kruerke, D., and Heppke, G. *J. Mater. Chem.* 5, 693–699, 1995.
10. Booth, C. J., Kruerke, D., and Heppke, G. *J. Mater. Chem.* 6, 927–934, 1996.
11. Contzen, J., Heppke, G., Kitzerow, H. S., Kruerke, D., and Schmid, H. *Appl. Phys. B* 63, 605–608, 1996.
12. Patel, J. S., Praefcke, K., Singer, D., and Langner, M. *Appl. Phys. B* 60, 469–472, 1995.
13. Kouwer, P. H. J., Jager, W. F., Mijs, W. J., Picken, S. J., Shepperson, K. J., and Mehl, G. H. *Mol. Cryst. Liq. Cryst.* 411, 377–385, 2004.
14. Kouwer, P. H. J., Mehl, G. H., and Picken, S. J. *Mol. Cryst. Liq. Cryst.* 411, 387–396, 2004.
15. Kohne, B., Marquardt, P., Praefcke, K., Psaras, P., Stephan, W., and Turgay, K. *Chimia* 40, 360–362, 1986.
16. Vill, V. and Thiem, J. *Liq. Cryst.* 9, 451–455, 1991.
17. Itoh, T., Takada, A., Fukuda, T., Miyamoto, T., Yakoh, Y., and Watanabe, J. *Liq. Cryst.* 9, 221–228, 1991.
18. Takada, A., Fukuda, T., Miyamoto, T., Yakoh, Y., and Watanabe, J. *Liq. Cryst.* 12, 337–345, 1992.
19. Sugiura, M., Minoda, M., Watanabe, J., Fukuda, T., and Miyamoto, T. *Bull. Chem. Soc. Jpn.* 65, 1939–1943, 1992.
20. Krishnan, K. and Balagurusamy, V. S. K. *Liq. Cryst.* 28, 321–325, 2001.
21. Kreuder, W. and Ringsdorf, H. *Makromol. Chem. Rapid Commun.* 4, 807–815, 1983.
22. Kreuder, W., Ringsdorf, H., and Tschirner, P. *Makromol. Chem. Rapid Commun.* 6, 367–373, 1985.
23. Kumar, S. *Liq. Cryst.* 32, 1089–1113, 2005.
24. Kreuder, W., Ringsdorf, H., Schonherr, O. H., and Wendorff, J. H. *Angew. Chem. Int. Ed. Engl.* 26, 1249–1252, 1987.
25. Zamir, S., Poupko, R., Luz, Z., Hueser, B., Boeffel, C., and Zimmermann, H. *J. Am. Chem. Soc.* 116, 1973–1980, 1994.
26. Adam, D., Schuhmacher, P., Simmerer, J., Haussling, L., Paulus, W., Siemensmeyer, K., Etzbach, K. H., Ringsdorf, H., and Haarer, D. *Adv. Mater.* 7, 276–280, 1995.
27. Boden, N., Bushby, R. J., Cammidge, A. N., and Martin, P. S. *J. Mater. Chem.* 5, 1857–1860, 1995.
28. Kumar, S., Schuhmacher, P., Henderson, P., Rego, J., and Ringsdorf, H. *Mol. Cryst. Liq. Cryst.* 288, 211–222, 1996.
29. Bacher, A., Bleyl, I., Erdelen, C. H., Haarer, D., Paulus, W., and Schmidt, H. W. *Adv. Mater.* 9, 1031–1035, 1997.
30. van de Craats, A. M., Siebbeles, L. D. A., Bleyl, I., Haarer, D., Berlin, Y. A., Zharikov, A. A., and Warman, J. M. *J. Phys. Chem. B* 102, 9625–9634, 1998.

31. Boden, N., Bushby, R. J., Cammidge, A. N., El-Mansoury, A., Martin, P. S., and Lu, Z. *J. Mater. Chem.* 9, 1391–1402, 1999.
32. Kumar, S., Manickam, M., and Schonherr, H. *Liq. Cryst.* 26, 1567–1571, 1999.
33. Manickam, M., Smith, A., Belloni, M., Shelley, E. J., Ashton, P. R., Spencer, N., and Preece, J. A. *Liq. Cryst.* 29, 497–504, 2002.
34. Kranig, W., Huser, B., Spiess, H. W., Kreuder, W., Ringsdorf, H., and Zimmermann, H. *Adv. Mater.* 2, 36–40, 1990.
35. Boden, N., Bushby, R. J., and Cammidge, A. N. *J. Am. Chem. Soc.* 117, 924–927, 1995.
36. Henderson, P., Ringsdorf, H., and Schuhmacher, P. *Liq. Cryst.* 18, 191–195, 1995.
37. Kumar, S. and Manickam, M. *Chem. Commun.* 1615–1616, 1997.
38. Kumar, S. and Varshney, S. K. *Synthesis* 305–311, 2001.
39. Kimura, M., Moriyama, M., Kishimoto, K., Yoshio, M., and Kato, T. *Liq. Cryst.* 34, 107–112, 2007.
40. Werth, M., Vallerien, S. U., and Spiess, H. W. *Liq. Cryst.* 10, 759–770, 1991.
41. Rego, J. A., Kumar, S., and Ringsdorf, H. *Chem. Mater.* 8, 1402–1409, 1996.
42. Boden, N., Bushby, R. J., Cammidge, A. N., Duckworth, S., and Headdock, G. *J. Mater. Chem.* 7, 601–605, 1997.
43. Praefcke, K., Eckert, A., and Blunk, D. *Liq. Cryst.* 22, 113–119, 1997.
44. Kumar, S., Manickam, M., Balagurusamy, V. S. K., and Schonherr, H. *Liq. Cryst.* 26, 1455–1466, 1999.
45. Ulman, A. (ed.). *An Introduction to Ultrathin Films: From Langmuir–Blodgett Films to Self Assembly*, Academic Press, Boston, MA, 1991.
46. Schonherr, H., Kremer, F. J. B., Kumar, S., Rego, J. A., Wolf, H., Ringsdorf, H., Jaschke, M., Butt, H. J., and Bamberg, E. *J. Am. Chem. Soc.* 1996, 118, 13051–13057.
47. Zelcer, A., Donnio, B., Bourgogne, C., Cukiernik, F. D., and Guillon, D. *Chem. Mater.* 19, 1992–2006, 2007.
48. Kroto, H. W., Heath, J. R., O'Brien, S. C., Curl, R. F., and Smalley, R. E. *Nature* 318, 162–163, 1985.
49. Chuard, T. and Deschenaux, R. *Helv. Chim. Acta* 79, 736–741, 1996.
50. Deschenaux, R., Even, M., and Guillon, D. *Chem. Commun.* 537–538, 1998.
51. Chuard, T. and Deschenaux, R. *Chimia* 52, 547–550, 1998.
52. Tirelli, N., Cardullo, F., Habicher, T., Suter, U. W., and Diederich, F. *J. Chem. Soc. Perkin Trans. 2* 2, 193–198, 2000.
53. Felder, D., Heinrich, B., Guillon, D., Nicoud, J. F., and Nierengarten, J. F. *Chem. Eur. J.* 6, 3501–3507, 2000.
54. Campidelli, S. and Deschenaux, R. *Helv. Chim. Acta* 84, 589–593, 2001.
55. Campidelli, S., Eng, C., Saez, I. M., Goodby, J. W., and Deschenaux, R. *Chem. Commun.* 1520–1521, 2003.
56. Praefcke, K. and Singer, D. In *Handbook of Liquid Crystal*, Vol. 2B, Demus, D., Goodby, J., Gray, G. W., Spiess, H. W., and Vill, V. (eds.), Wiley-VCH, Weinheim, Germany, 1998, Chap. XVI.
57. Mahlstedt, S., Janietz, D., Stracke, A., and Wendorff, J. H. *Chem. Commun.* 15–16, 2000.
58. Tsukruk, V. V., Bengs, H., and Ringsdorf, H. *Langmuir* 12, 754–757, 1996.
59. Freudenmann, R., Behnisch, B., and Hanack, M. *J. Mater. Chem.* 11, 1618–1624, 2001.
60. Kumar, S. and Varshney, S. K. *Org. Lett.* 4, 157–159, 2002.
61. Kumar, S. *Pramana* 61, 199–203, 2003.
62. Kumar, S. and Naidu, J. J. *Liq. Cryst.* 29, 899–906, 2002.
63. Vorlander, D. *Z. Phys. Chem.* 105, 211–254, 1923.
64. Kumar, S. and Varshney, S. K. *Liq. Cryst.* 28, 161–163, 2001.
65. Schulte, J. L., Laschat, S., Vill, V., Nishikawa, E., Finkelmann, H., and Nimtz, M. *Eur. J. Org. Chem.* 2499–2506, 1998.
66. Mao, H., He, Z., Wang, J., Zhang, C., Xie, P., and Zhang, R. *J. Lumin.* 122–123, 942–945, 2007.
67. Gupta, S. K., Raghunathan, V. A., Lakshminarayanan, V., and Kumar, S. *J. Phys. Chem. B* 113, 12887–12895, 2009.
68. Pal, S. K. and Kumar, S. *Tetrahedron Lett.* 47, 8993–8997, 2006.
69. Kato, T. In *Handbook of Liquid Crystal*, Vol. 2B, Demus, D., Goodby, J., Gray, G. W., Spiess, H. W., and Vill. V. (eds.), Wiley-VCH, Weinheim, Germany, 1998, Chap. XVII.
70. Paleos, C. M. and Tsiourvas, D. *Angew. Chem. Int. Ed.* 34, 1696–1711, 1995.
71. Stewart, D., Paterson, B. J., and Imrie, C. T. *Eur. Poly. J.* 33, 285–290, 1997.
72. Percec, V., Ahn, C. H., Bera, T. K., Ungar, G., and Yeardley, D. J. P. *Chem. Eur. J.* 5, 1070–1083, 1999.
73. Barbera, J., Puig, L., Serrano, J. L., and Sierra, T. *Chem. Mater.* 16, 3308–3317, 2004.

74. Gray, G. W. (ed.). *Molecular Structure and the Properties of Liquid Crystals*, Academic Press, London, U.K., 1962, pp. 139–196.
75. Maliszewskyj, N. C., Heiney, P. A., Josefowicz, J. Y., Plesnivy, T., Ringsdorf, H., and Schuhmacher, P. *Langmuir* 11, 1666–1674, 1995.
76. Wan, W., Monobe, H., Sugino, T., Tanaka, Y., and Shimizu, Y. *Mol. Cryst. Liq. Cryst.* 364, 597–603, 2001.
77. Setoguchi, Y., Monobe, H., Wan, W., Terasawa, N., Kiyohara, K., Nakamura, N., and Shimizu, Y. *Thin Solid Films* 438–439, 407–413, 2003.
78. Kumar, S. *Liq. Cryst.* 31, 1037–1059, 2004.
79. Hirst, D., Diele, S., Laschat, S., and Nimtz, M. *Helv. Chim. Acta* 84, 1190–1196, 2001.
80. Kumar, S., Naidu, J. J., and Varshney, S. K. *Mol. Cryst. Liq. Cryst.* 411, 355–362, 2004.
81. Percec, V., Cho, C. G., Pugh, C., and Tomazos, D. *Macromolecules* 25, 1164–1176, 1992.
82. Bryant, G. C., Cook, M. J., Haslam, S. D., Richardson, R. M., Ryan, T. G., and Thorne, A. J. *J. Mater. Chem.* 4, 209–216, 1994.
83. Bryant, G. C., Cook, M. J., Ryan, T. G., and Thorne, A. J. *Tetrahedron* 52, 809–824, 1996.
84. McKeown, N. B., Chambrier, I., and Cook, M. J. *J. Chem. Soc. Perkin Trans. 1* 1169–1177, 1990.
85. Piechocki, C., Simon, J., Andre, J. J., Guillon, D., Petit, P., Skoulios, A., and Weber, P. *Chem. Phys. Lett.* 122, 124–128, 1985.
86. Piechocki, C., Simon, J., Skoulios, A., Guillon, D., and Weber, P. *J. Am. Chem. Soc.* 104, 5245–5247, 1982.
87. Belarbi, Z., Sirlin, C., Simon, J., and Andre, J. J. *J. Phys. Chem.* 93, 8105–8110, 1989.
88. Komatsu, T., Ohta, K., Fujimoto, T., and Yamamoto, I. *J. Mater. Chem.* 4, 533–536, 1994.
89. Komatsu, T., Ohta, K., Watanabe, T., Ikemoto, H., Fujimoto, T., and Yamamoto, I. *J. Mater. Chem.* 4, 537–540, 1994.
90. Toupance, T., Bassoul, P., Mineau, L., and Simon, J. *J. Phys. Chem.* 100, 11704–11710, 1996.
91. van de Craats, A. M., Warman, J. M., Hasebe, H., Naito, R., and Ohta, K. *J. Phys. Chem. B* 101, 9224–9232, 1997.
92. Yoshino, K., Sonoda, T., Lee, S., Hidayat, R., Nakayama, H., Tong, L., Fujii, A., Ozaki, M., Ban, K., Nishizawa, K., and Ohta, K. *Proceedings of 13th International Conference on Dielectric Liquids*, 1999, Nara, Japan, 598–601.
93. Naito, R., Ohta, K., and Shirai, H. *J. Porphyr. Phthalocyanines* 5, 44–50, 2001.
94. Ban, K., Nishizawa, K., Ohta, K., van de Craats, A. M., Warman, J. M., Yamamoto, I., and Shirai, H. *J. Mater. Chem.* 11, 321–331, 2001.
95. Sleven, J., Walrand, C. G., and Binnemans, K. *J. Mater. Sci. Eng. C* 18, 229–238, 2001.
96. Maeda, F., Hatsusaka, K., Ohta, K., and Kimura, M. *J. Mater. Chem.* 13, 243–251, 2003.
97. Basova, T., Kol'tsov, E., Hassan, A. K., Nabok, A., Ray, A. K., Gurek, A. G., and Ahsen, V. *J. Mater. Sci. Mater. Electron.* 2004, 15, 623–628.
98. Nekelson, F., Monobe, H., and Shimizu, Y. *Chem. Commun.* 3874–3876, 2006.
99. Gurek, A. G., Basova, T., Luneau, D., Lebrun, C., Kol'tsov, E., Hassan, A. K., and Ahsen, V. *Inorg. Chem.* 45, 1667–1676, 2006.
100. Nekelson, F., Monobe, H., and Shimizu, Y. *Mol. Cryst. Liq. Cryst.* 479, 205–211, 2007.
101. Binnemans, K., Sleven, J., de Feyter, S., de Schryver, F. C., Donnio, B., and Guillon, D. *Chem. Mater.* 15, 3930–3938, 2003.
102. Hatsusaka, K., Kimura, M., and Ohta, K. *Bull. Chem. Soc. Jpn.* 76, 781–787, 2003.
103. Nakai, T., Ban, K., Ohta, K., and Kimura, M. *J. Mater. Chem.* 12, 844–850, 2002.
104. Shimizu, Y., Matsuno, J. Y., Nakao, K., Ohta, K., Miya, M., and Nagata, A. *Mol. Cryst. Liq. Cryst.* 260, 491–497, 1995.
105. Ito, S., Herwig, P. T., Bohme, T., Rabe, J. P., Rettig, W., and Mullen, K. *J. Am. Chem. Soc.* 122, 7698–7706, 2000.
106. Wasserfallen, D., Fischbach, I., Chebotareva, N., Kastler, M., Pisula, W., Jackel, F., Watson, M. D., Schnell, I., Rabe, J. P., Spiess, H. W., and Mullen, K. *Adv. Funct. Mater.* 15, 1585–1594, 2005.
107. Watson, M. D., Jackel, F., Severin, N., Rabe, J. P., and Mullen, K. *J. Am. Chem. Soc.* 126, 1402–1407, 2004.

4 Discotic Oligomers

Liquid crystalline oligomers are composed of more than two similar or different mesogenic moieties connected together via flexible spacers. The overall shape of an oligomesogen may vary from linear to star shaped, depending on the molecular topology, that is, the manner in which the mesogenic fragments are connected to each other or one another. In star-shaped oligomers, mesogens are grafted to a central unit (mesogenic or non-mesogenic) through spacers, while in linear oligomers, discotic mesogens are connected to each other or one another via spacers. In contrast to a variety of calamitic oligomers [1–5], only a few discotic oligomers are known. This chapter presents the chemistry and physical properties of discotic trimers, tetramers, pentamers, hexamers, heptamers, and other higher discotic oligomers.

4.1 DISCOTIC TRIMERS

Three types of discotic trimers have been realized: linear, branched (star-shaped), and metal-bridged triple decker (Figure 4.1). In a linear trimer, three discotic units are connected with two flexible spacers (Figure 4.1a); in a star-shaped trimer, three discotic fragments are linked to a central core (Figure 4.1b); while the triple deckers consist of three disks connected to each other via two central atoms (Figure 4.1c). Triphenylene, pyranose sugars, alkynylbenzene, phthalocyanine, porphyrin, and hexabenzocoronene (HBC)-based discotic mesogens have been exploited to prepare discotic trimers. The chemistry and physical properties of these trimers are presented in the following sections.

4.1.1 TRIPHENYLENE-BASED DISCOTIC TRIMERS

Ringsdorf and coworkers prepared a linear triphenylene trimer **1** by reacting a monofunctionalized triphenylene (3,6,7,10,11-pentakis(heptyloxy)-2-triphenylenyl acetate) with an excess of the hexadecanoic diacid, and then condensing the resultant acid-functionalized triphenylene derivative with a difunctionalized triphenylene molecule [6]. The trimer was liquid crystalline at room temperature and showed a mesophase to isotropic phase transition at 157°C. It was possible that it vitrified at lower temperatures, but the glass transition could not be detected by differential scanning calorimetry (DSC).

R = C$_7$H$_{15}$ **1**   Col$_h$ 157 I

A linear trimer **2** in which three triphenylene units were connected via ethereal linkage was reported by Boden and coworkers [7]. The compound was prepared by reacting the ω-brominated triphenylene with 1,2-dihydroxy-6,7,10,11-tetraalkoxytriphenylene. On first heating, it exhibited a crystal to mesophase transition at 60°C, a mesophase-to-mesophase transition at 92°C, and then cleared at 109°C. On cooling, the columnar mesophase appeared at 105°C, which transformed into a glassy state at 32°C. The glass crystallized slowly over a period of months.

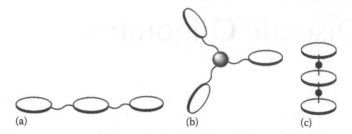

FIGURE 4.1 Sketches of (a) a linear discotic trimer, (b) a star-shaped discotic trimer, and (c) a metal-bridged discotic triple decker.

The synthesis of star-shaped trimers is relatively straightforward as it involves the preparation of just monofunctionalized triphenylenes, which may be attached to a trifunctional central nucleus to obtain the trimer. Two such trimers have been prepared by Kumar and Manickam [8]. These molecular architectures consist of trisubstituted benzene as the central core and three triphenylene-based or nitro-functionalized triphenylene-based units as peripheral mesogenic cores. These outer cores are linked to the central core via ester linkages through an alkyl chain spacer. While the nitro-functionalized trimer **3** exhibited a monotropic columnar mesophase, the unfunctionalized trimer **4** was found to be non-mesomorphic. Trimer **3** showed two peaks on the first DSC heating run: one weak transition at 161.1°C and the other broad peak at 181.3°C. On cooling, the columnar phase appeared at 158.7°C and remained stable down to room temperature. In comparison, trimer **4** melted at 185°C and crystallized at 166°C on cooling.

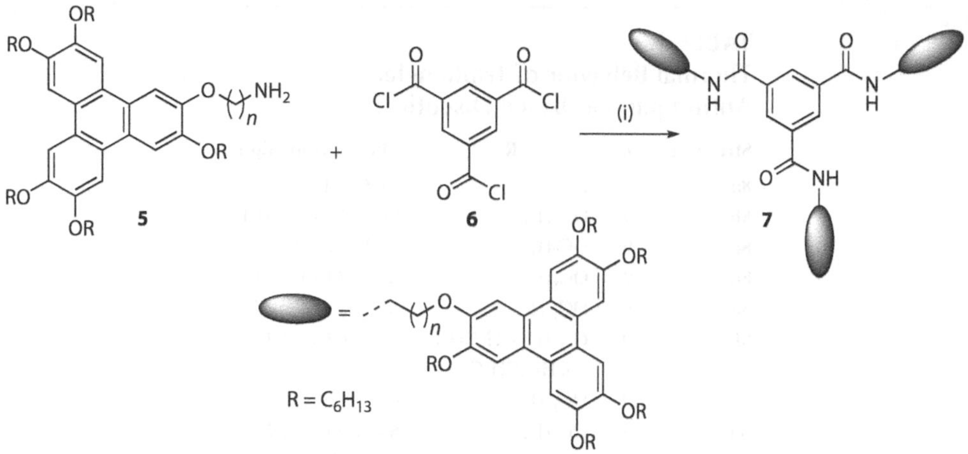

SCHEME 4.1 Synthesis of star-shaped triphenylene trimers: (i) CH$_2$Cl$_2$, Et$_3$N.

Paraschiv et al. reported a series of 1,3,5-benzenetrisamide derivatives **7a–f** having three hexaalkoxytriphenylene pendent groups [9,10]. These trimers were prepared by reacting an amino-terminated hexaalkoxytriphenylene **5** with trimesoyl chloride **6** (Scheme 4.1). The amino-functionalized triphenylene derivative **5** can be prepared from monofunctionalized triphenylene. The alkylation of monohydroxy-pentaalkoxytriphenylene with α,ω-dibromoalkane gives the bromo-terminated triphenylene, which can be converted to an azide by treatment with sodium azide in ethanol. The reduction of the azide with lithium aluminum hydride furnished the required amino-terminated triphenylene **5**, which can be coupled with 1,3,5-benzenetricarbonyl trichloride **6** in the presence of triethylamine. In these materials, the columnar phase is stabilized via intermolecular hydrogen bonding. This way, a high directionality of interaction can be achieved. These materials form ordered columnar phases, in which the charge carrier mobility is several times higher compared to monomeric triphenylene discotics. The thermal behavior of these trimers is summarized in Table 4.1. All compounds exhibit a liquid crystalline behavior with high isotropization temperatures due to H-bonding. Interestingly, these materials show a variety of columnar phases (Col$_h$, Col$_p$, Col$_{ob}$, Col$_r$, and unidentified Col phases).

Two series of triphenylene-anthraquinone-based symmetric discotic liquid crystalline trimers **8a–m** have recently been prepared by Gupta et al. [11]. These triads were prepared using microwave

TABLE 4.1
Thermal Behavior of Benzenetrisamide
Derivatives 7

Structure	n	R	Phase Transition
7a	3	C$_6$H$_{13}$	Col$_h$ 172 I
7b	4	C$_4$H$_9$	Col$_p$ 199 I
7c	4	CH$_2$CH(CH$_3$)CH$_2$CH$_3$	Col 135 Col$_r$ 182 I
7d	4	C$_6$H$_{13}$	Col$_p$ 208 I
7e	5	C$_6$H$_{13}$	Col 162 Col$_{ob}$ 190 I
7f	6	C$_6$H$_{13}$	Col 183 I

Source: Data from Paraschiv, I. et al., *J. Mater. Chem.*, 18, 5475, 2008.

TABLE 4.2
Thermal Behavior of Triphenylene-
Anthraquinone-Based Discotic Trimers 8

Structure	n	R	Phase Transition
8a	9	H	Cr 39.7 I
8b	9	OC_6H_{13}	Ss 59.1 Col_h 104.1 I
8c	9	OC_7H_{15}	Ss 37 Col_h 89.6 I
8d	9	OC_8H_{17}	Ss 47.3 Col_h 83 I
8e	9	$OC_{10}H_{21}$	Ss 51.2 Col_h 59.1 I
8f	9	$O(CH_2)_2CH(CH_3)$ $(CH_2)_3CH(CH_3)_2$	Ss 45.6 Col_h 69 I
8g	9	$OC_{14}H_{29}$	Cr 47 I
8h	7	OC_6H_{13}	Ss 41.5 Col_h 69.3 I
8i	7	OC_7H_{15}	Ss 45 Col_h 65.8 I
8j	7	$OC_{10}H_{21}$	Cr 63 I
8k	7	$O(CH_2)_2CH(CH_3)$ $(CH_2)_3CH(CH_3)_2$	Cr 70.7 I
8l	7	$OC_{12}H_{25}$	Cr 44.4 I
8m	7	$OC_{14}H_{29}$	Cr 60.1 I

Source: Data from Gupta, S.K. et al., *New J. Chem.*, 33, 112, 2009.

dielectric heating. It may be noted that the etherification of H-bonded hydroxyl groups of tetraalkoxy rufigallol with bulky ω-bromo-substituted triphenylenes failed to produce the desired triads under classical reaction conditions. These were the first donor–acceptor–donor triads in which all three components represent discotic mesogenic moieties. They exhibit a columnar mesophase over a wide range of temperatures. The columnar hexagonal mesophase structure of these discotic oligomers was elucidated with the help of x-ray diffraction studies. The thermal behavior of these trimers is presented in Table 4.2. Mesomorphism in these trimers is critically dependent on the peripheral substitution around the central core as well as on the spacer length connecting discotic cores. Trimer **8a**, in which the central anthraquinone core is not surrounded by peripheral alkyl chains, failed to exhibit any mesomorphic behavior. Further, trimers do not show mesomorphism if the peripheral alkyl chains around the central anthraquinone core are large but the spacer is short. The thermal data indicate that longer spacer length, smaller peripheral alkyl chain length, and branching in the peripheral alkyl chain of anthraquinone are in favor of the liquid crystalline property in these symmetrical trimers.

8   R' = C₆H₁₃

TABLE 4.3
Thermal Behavior of Trisaccharide-Based Discotic Liquid Crystals

Structure	n	Phase Transition	Ref.
9		Cr 38 Col$_h$ 93 I	[12]
10b	8	Cr 60 Col 195 I	[13]
10c	12	Cr 65 Col 175 I	[13]
10d	16	Cr 75 Col 170 I	[13]

4.1.2 TRISACCHARIDE DISCOTICS

Although it is difficult to visualize a discotic shape in these oligomers, peripherally substituted trisaccharides have been described as discotic liquid crystals (DLCs) [12,13]. The liquid crystalline undecaalkanoates of cellotriose [12] and chitotriose [13] can be easily prepared via esterification of natural oligosaccharides. The hendecadecanoate of cellotriose 9 melts at 38°C to form a hexagonal columnar phase that goes to the isotropic phase at 93°C. The hexagonal columnar nature of the mesophase was confirmed by x-ray diffraction studies. Four chitotriose undecaalkanoates 10 have been prepared via esterification of chitotriose with hexanoic, decanoic, tetradecanoic, and octadecanoic acid chlorides. A number of partially esterified side-products form in the reaction, but the desired product can be isolated by means of preparative size-exclusion chromatography. The thermal behavior of these pyranose sugar trimers is summarized in Table 4.3. The trimer with hexyl chains does not show mesomorphism, but all other higher homologues display columnar phases. Based on polarizing optical microscopy (POM) texture resemblance with that of cellotriose discotic, the nature of the columnar phase formed by chitotriose esters was described as hexagonal columnar.

4.1.3 MULTIYNE-BASED DISCOTIC TRIMERS

Janietz and coworkers prepared a series of star-shaped pentayne-based trimers 11a–d by coupling a hydroxyl-terminated radial pentaalkynylbenzene derivative with 1,3,5-benzenetricarbonyltrichloride in the presence of pyridine and dimethylaminopyridine (DMAP) [14–16]. The mesomorphic behavior of these star-like trimers is given in Table 4.4. All the derivatives exhibit a columnar nematic phase. The methyl-substituted trimer 11a displayed a monotropic columnar nematic phase. It transforms from the crystalline phase into the isotropic phase at 173.6°C while heating, but on cooling, the mesophase appears at 172.9°C. As can be seen from Table 4.4, the peripheral alkyl chain length has a reciprocal effect on melting as well as on clearing temperatures. When cooled from the liquid crystalline phase, the lower homologues 11a and 11b form crystals. On the other

TABLE 4.4
Thermal Behavior of Pentayne-Based Trimers 11

Structure	R	Phase Transition	Ref.
11a	CH$_3$	Cr 173.6 (N$_{Col}$ 172.9) I	[15]
11b	C$_2$H$_5$	Cr 110.8 N$_{Col}$ 157.9 I	[16]
11c	C$_3$H$_7$	g 24 Cr 66.1 N$_{Col}$ 141.9 I	[15]
11d	C$_5$H$_{11}$	g -12.43 N$_{Col}$ 23.6 I	[15]

hand, the mesophase of higher homologues, **11c** and **11d**, is quenched into a glassy state on cooling. The columnar nematic arrangement is stabilized by the combined effect of the reduced mobility of pentayne disks within the column due to the star-like covalent linkage, and steric repulsion between the long spacer, the periphery of the mesogenic core, and the bulky benzene connecting group, which prevents the regular arrangement of the columns into a highly ordered stack. To avoid the steric repulsion, out-of-plane tilting of the pentayne core occurs, which further hinders the parallel alignment of columns. On extending the peripheral chain length, steric repulsion increases, which results in reduced mesophase stability for the pentyl-substituted oligomer **11d**. The composites of these oligomers with trinitrofluorenone also displayed a columnar nematic mesophase [15].

11

4.1.4 Phthalocyanine and Porphyrin Discotic Trimers

Bryant et al. prepared three phthalocyanine-based discotic trimers **24a–c** [17,18], as shown in Scheme 4.2. The condensation of two different types of phthalonitriles yields the monofunctionalized

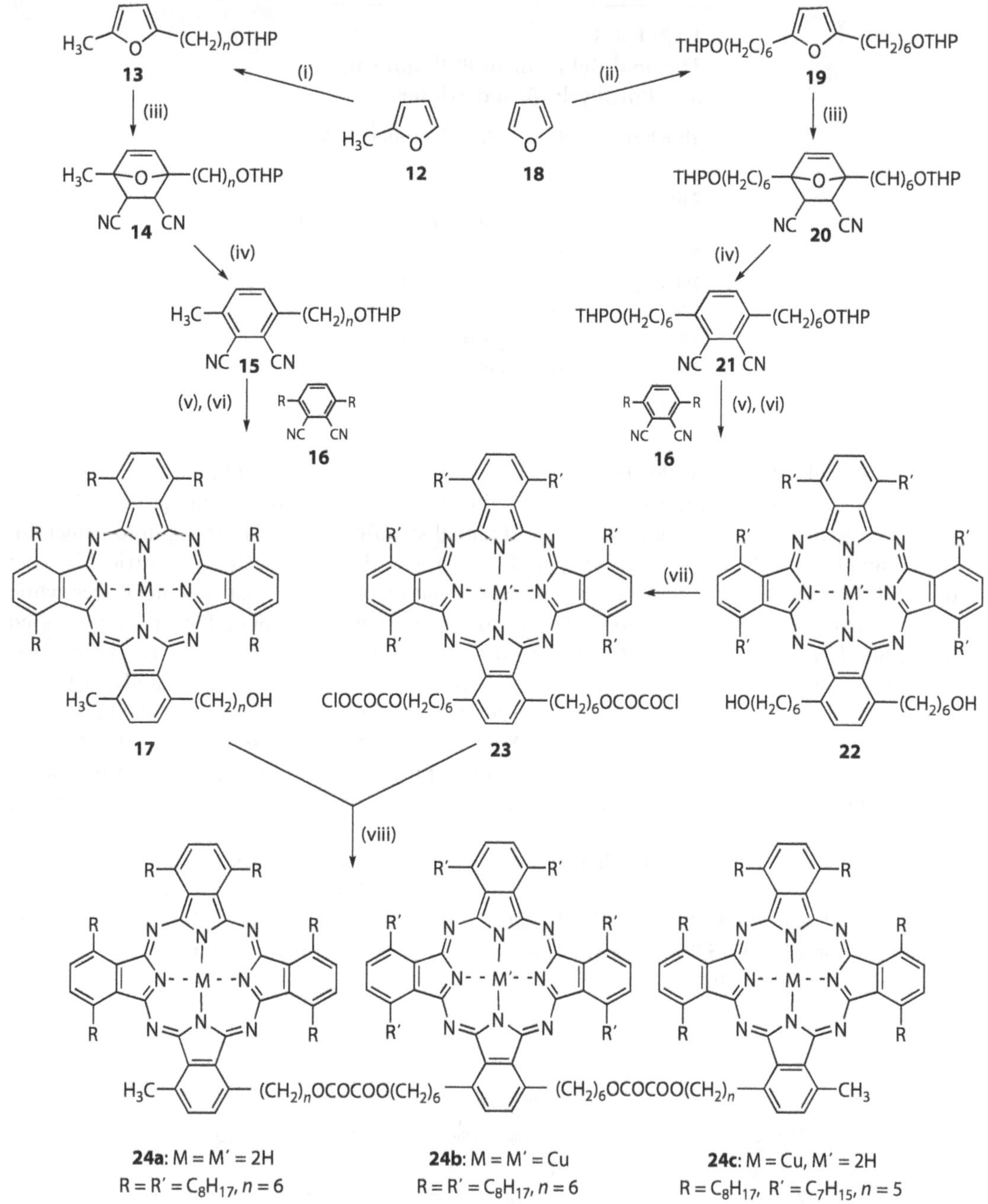

SCHEME 4.2 Synthesis of Pc trimers: (i) BuLi, THPO(CH$_2$)$_n$I; (ii) 2×BuLi, THPO(CH$_2$)$_n$I; (iii) NCCH=CHCN; (iv) LiN(SiMe$_3$)$_2$; (v) C$_5$H$_{11}$OLi, C$_5$H$_{11}$OH, reflux; (vi) Cu(OAc)$_2$, C$_5$H$_{11}$OH, reflux; (vii) ClCOCOCl, CH$_2$Cl$_2$, K$_2$CO$_3$; (viii) CH$_2$Cl$_2$, K$_2$CO$_3$.

phthalocyanine **17** and the difunctionalized phthalocyanine **22**. The 3,6-bis(6-hydroxyalkyl) phthalonitrile **21** and 3-hydroxyalkyl-6-methylphthalonitrile **15** were prepared from furan **18** and 2-methylfuran **12**, respectively (Scheme 4.2). Similarly, the 3,6-dialkylphthalonitrile **16** can be prepared from furan or thiophene [19,20]. In the condensation reaction of two phthalonitriles to prepare the functionalized phthalocyanines, the dialkylphthalonitrile **16** is used in excess. The reaction produces a mixture of products from which the desired difunctionalized phthalocyanine

TABLE 4.5
**Thermal Behavior of Phthalocyanine-
and Porphyrin-Based Trimers**

Structure	R	Phase Transition	Ref.
24a		Col 177 I	[18]
24b		Col 252 I	[18]
24c		Col 222 I	[18]
25a	$C_{10}H_{21}$	D_L 104 Col_r 148 I	[21]
25b	$C_{12}H_{25}$	D_L 100 Col_r 143 I	[21]
25c	$C_{14}H_{29}$	D_L 106 Col_r 136 I	[21]
25d	$C_{12}H_{25}$	g 104 Col_h 158 I	[21]
25e	$C_{16}H_{33}$	Col_h 49 Col_h 127 I	[21]

can be isolated, albeit in low yield. The bis-hydroxylalkyl phthalocyanine **22** on esterification with oxalyl chloride affords the phthalocyanine **23** with two terminal acid chloride groups. This on treatment with an excess of monofunctionalized phthalocyanine **17** yields the required trimer **24**. Three compounds in this series were prepared in this way. Two contained symmetrical linking chains and the same alkyl groups on all three rings; out of these two, one was metal free while the other metallated with copper ions. The third copper-complexed phthalocyanine possessed different alkyl chains on the unmetallated central ring from those on the outer rings (only the outer phthalocyanines are metallated; the central phthalocyanine is non-metallated). All three trimers are liquid crystalline at room temperature and display a disordered hexagonal columnar phase when cooled from the isotropic phase. While the compounds **24b** and **24c** exhibit only a hexagonal columnar phase throughout the mesophase range (Table 4.5), the nature of this phase changed from hexagonal columnar to rectangular columnar at room temperature on cooling in the case of **24a**.

Ohta and coworkers prepared porphyrin-based triple deckers by complexing the metal-free porphyrin with cerium ions [21,22]. Interestingly, the strip-like diarylporphyrins, **25**, exhibited lamellar and columnar phases (Table 4.5) [21]. On the other hand, the discotic triple decker **26**, prepared by the complexation of 5,10,15,20-tetrakis[3,4-bis(octyloxy)phenyl]porphyrin with two cerium metal ions, was not liquid crystalline [22].

4.1.5 Hexabenzocoronene Discotic Trimers

A linear hexa-*peri*-benzocoronene (HBC) trimer **34** was prepared by Mullen and coworkers, as shown in Scheme 4.3 [23]. The synthesis involved a twofold Suzuki coupling between dibromo-substituted hexaphenylbenzene **27** and 4-(trimethylsilyl)benzene boronic acid **28** to generate the oligophenylene **29**. The trimethylsilyl groups were then replaced by iodo groups by reacting with iodine monochloride in chloroform. A Sonogashira coupling between the diiodo compound **30** and alkyl-substituted phenylacetylene furnished the extended hexaphenylbenzene **31**. A Diels–Alder cycloaddition of **31** with the cyclopentadione derivative **32** yielded the oligophenylene **33**, which was cyclized via oxidative cyclodehydrogenation using FeCl₃ to obtain the desired linear trimer **34**.

In order to realize the cyclic trimer **38**, the ortho-connected HBC trimers **37** were also prepared following a similar synthetic strategy (Scheme 4.4). However, the cyclization of **37** to the planar trimer **38** could not be realized due to the twisted nature of the former. Because of a strong steric hindrance, intramolecular coupling did not take place. As expected, the ortho-connected HBC trimers **37** did not self-assemble into a columnar phase due to the twisted conformation, but the linear trimer **34** exhibited a columnar phase. The trimer did not show any transition up to 300°C, but its mesomorphic nature was confirmed by x-ray diffraction studies.

4.2 DISCOTIC TETRAMERS

While linear calamitic tetramers have been realized in 1999 [24], probably because of synthetic difficulties, no linear discotic tetramer has so far been prepared. However, a few star-like tetramers have been reported [25–34]. In all cases, four monofunctionalized discotic units were connected to a central tetrafunctionalized unit. Ringsdorf and coworkers prepared triphenylene tetramers **39** and **40** by attaching the monofunctionalized triphenylene to an azacrown or siloxane core, and studied their various physical properties [25–29]. The tetramer **39** shows a glass transition at 77°C and a mesophase to isotropic transition at 98°C [26]. On the other hand, the tetramer **40** exhibits a very broad columnar phase between the glass transition at −48°C and the isotropic temperature of 141°C (Table 4.6) [25].

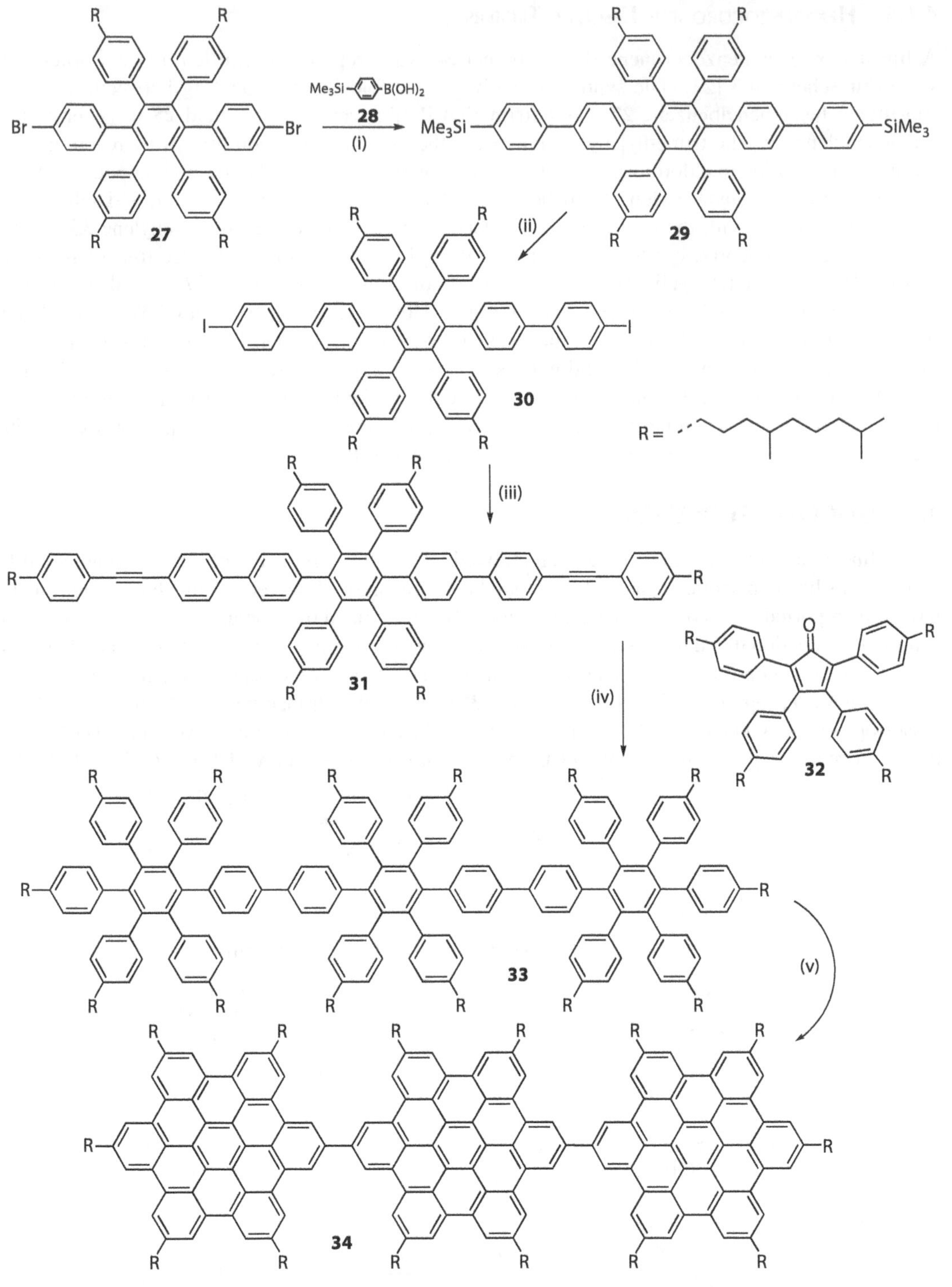

SCHEME 4.3 Synthesis of linear HBC-based trimers: (i) Pd(PPh$_3$)$_4$, K$_2$CO$_3$, toluene, 95°C; (ii) ICl, CHCl$_3$; (iii) 4-(3,7-dimethyloctanoyl)phenylacetylene, Pd(PPh$_3$)$_4$, CuI, piperidine; (iv) C$_6$H$_5$OC$_6$H$_5$, reflux; (v) FeCl$_3$, CH$_3$NO$_2$, CH$_2$Cl$_2$.

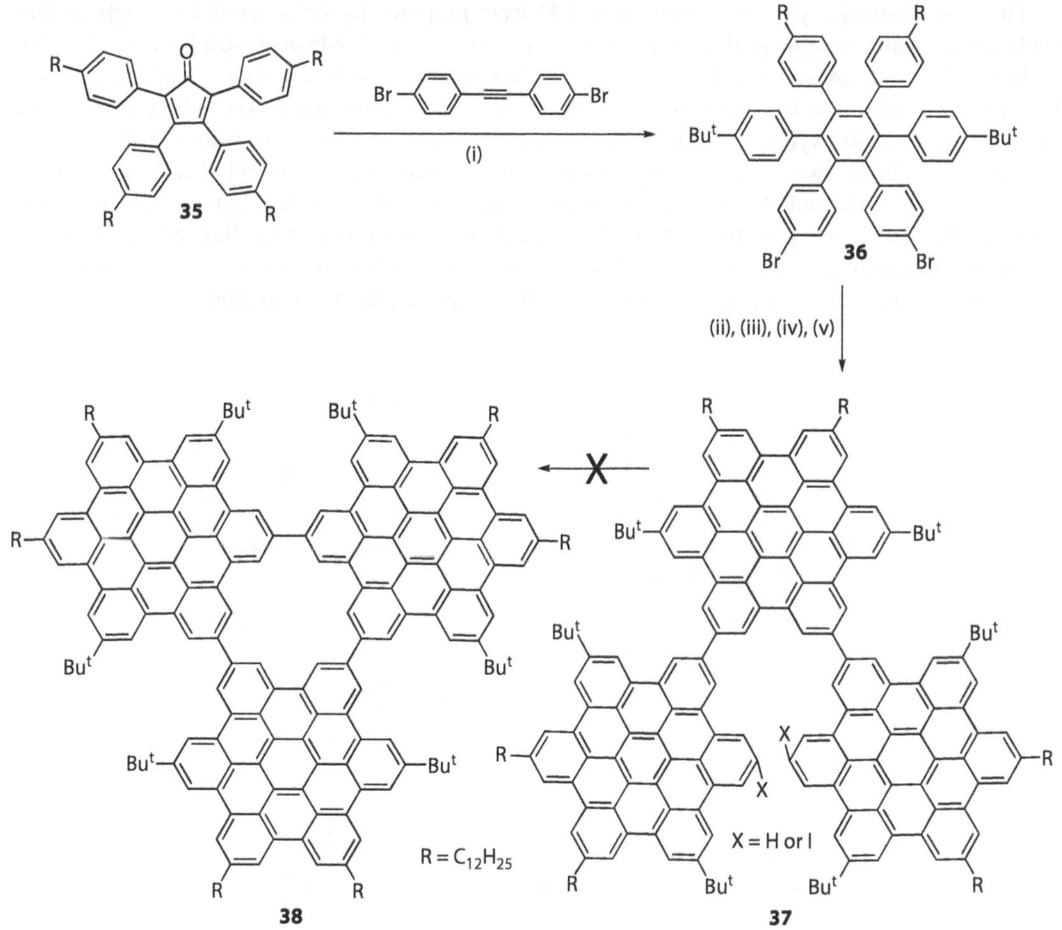

SCHEME 4.4 Synthesis of cyclic HBC trimers: (i) $C_6H_5OC_6H_5$, reflux; (ii) Pd(PPh$_3$)$_4$, K$_2$CO$_3$, toluene; (iii) ICl, CHCl$_3$; (iv) phenylacetylene or 4-trimethylsilylphenylacetylene, Pd(PPh$_3$)$_4$, CuI, piperidine; (v) **35**, Ph$_2$O, reflux.

TABLE 4.6
Thermal Behavior of Discotic Tetramers

Structure	R	Phase Transition	Ref.
39		g 77 Col 98 I	[26]
40		g −48 Col 141 I	[25]
46		g 54.1 Col$_h$ 106.1 I	[30]
47		Cr 30.5 Col$_h$ 108 I	[30]
51		Cr 34 N 45 I	[31]
52a	C_5H_{11}	Cr 142 I	[33]
52b	C_6H_{13}	Cr 116 I	[33]
52c	C_7H_{15}	g −40 Col$_h$ 80 I	[33]
52d	C_8H_{17}	g −59 Col$_h$ 79 I	[33]
52e	C_9H_{19}	g −69 Col$_h$ 59 I	[33]
52f	$C_{10}H_{21}$	g −61 Col$_h$ 58 I	[33]
53		Cr 61 Col$_r$ 79 I	[34]

The star-like triphenylene tetramers **46** and **47** were prepared by Zelcer et al. by attaching three triphenylene units to a central symmetrical or nonsymmetrical trihydroxy-trialkoxytriphenylene (Scheme 4.5) [30]. A statistical cleavage of the hexyl chains of hexahexyloxytriphenylene with B-bromocatecholborane (see Section 2.6) easily furnishes the symmetrical (**41**) and nonsymmetrical (**42**) trihydroxy-trialkoxytriphenylenes. The alkylation of these trihydroxyl derivatives with 5-bromo-pent-1-ene yields the two corresponding trifunctional triphenylenes **43** and **44**. The hydrosilylation of these alkenes with monofunctionalized triphenylene-silane **45** furnishes the tetramers **46** and **47**, respectively. Both tetramers show a broad hexagonal columnar phase. The siloxane-containing C_3 symmetric tetramer **46** is partially crystalline at room temperature and forms a columnar phase at about 30°C on heating. When cooled from the isotropic phase (108°C), it exhibits a columnar phase

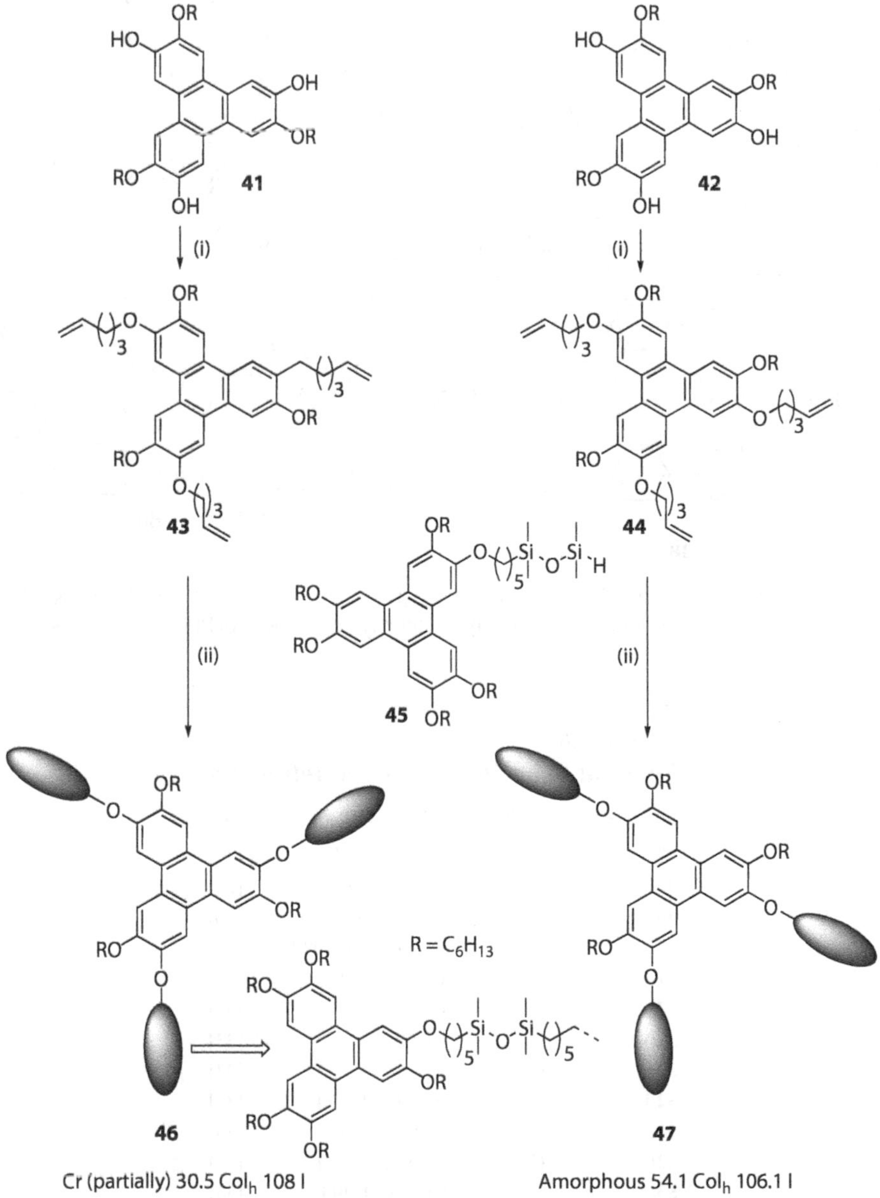
41
42
(i)
(i)
43
44
(ii)
(ii)
45
R = C6H13
46
47
Cr (partially) 30.5 Colh 108 I
Amorphous 54.1 Colh 106.1 I

SCHEME 4.5 Synthesis of symmetrical and nonsymmetrical triphenylene tetramers: (i) C_5H_9Br, K_2CO_3, MEK, reflux; (ii) Pt, toluene.

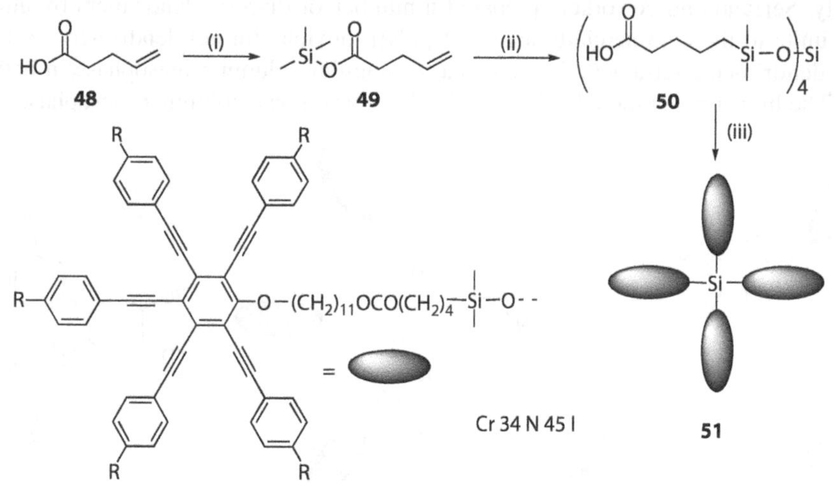

SCHEME 4.6 Synthesis of the multiyne-based discotic tetramer **51**: (i) hexamethylsilazane, THF; (ii) tetrakis(dimethylsiloxy)silane; (iii) Al, diisopropylcarbodiimide, dimethylaminopyridine, CH_2Cl_2.

that vitrifies at about −45°C. The nonsymmetric tetramer also behaves similarly. The virgin material is amorphous at room temperature and forms a columnar phase at about 54°C, and then goes to the isotropic phase at 106°C (Table 4.6). On cooling, the material vitrifies at −43°C.

Kouwer et al. prepared a siloxane-based tetramer **51** by attaching four hydroxyl-functionalized multiyne-based discotic monomers to tetrakis(dimethylsiloxy)silane, as shown in Scheme 4.6 [31]. Pentenoic acid **48** was protected with hexamethyldisilazane, and this protected acid **49** was coupled with tetrakis(dimethylsiloxy)silane via a hydrosilylation reaction. After deprotection of the acid, the resultant tetraacid **50** was esterified with hydroxyl-functionalized multiyne to yield the siloxane tetramer **51**. The tetramer shows a nematic phase at 34°C and goes to the isotropic phase at 45°C.

Contrary to the general belief that a flat or nearly flat core is essential to form columnar meso-phases, it has been shown that molecules with a tetrahedral core unit can also organize into columns exhibiting mesogenity [32]. The tetrafunctional alcohol pentaerythritol is a popular core in dendrimer chemistry; it has been modified into triphenylene tetramers by Schulte et al. [33]. The tetramers **52a** and **52b** with pentyloxy and hexyloxy chains are not liquid crystalline, but higher homologues **52c–f** display a columnar mesophase. Their thermal behavior is summarized in Table 4.6.

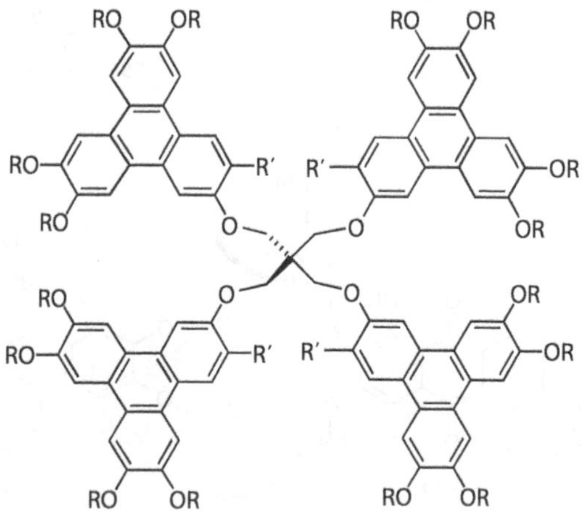

52a–f: R′ = CH₃, R = C_nH_{2n+1}, n = 5–10

Similarly, Serrano and coworkers prepared a number of discotic dendrimers by attaching triphenylene units to the commercially available poly(propylene imine) dendrimers [34]. The first-generation dendrimer, a tetramer **53**, exhibits a rectangular columnar mesophase in between 61°C and 79°C. The higher-generation dendrimers display a hexagonal columnar mesophase.

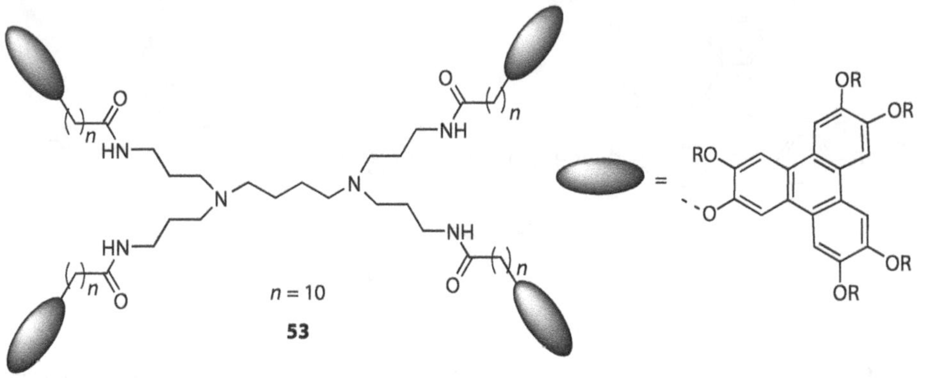

4.3 DISCOTIC PENTAMERS

The synthesis and mesomorphism of the star-shaped discotic pentamer **56** and its diacetate are reported by Bisoyi and Kumar [35]. The molecular design is such that it contains the well-studied electron-rich triphenylene moiety as the periphery and the electron-deficient anthraquinone as the central core (Scheme 4.7). These molecular double-cables may stack one on top of the other in the columns to give columnar double-cables, which could eventually provide side-by-side percolation pathways for electrons and holes in solar cells. Since these star-shaped oligomers are difunctional in nature, mixed-chain discotic polymers and novel star-shaped heptamers can be realized containing two different kinds of discotic cores possessing opposite electronic properties. The synthesis of these discotic pentamers is shown in Scheme 4.7. Rufigallol **54** was alkylated under mild etherification

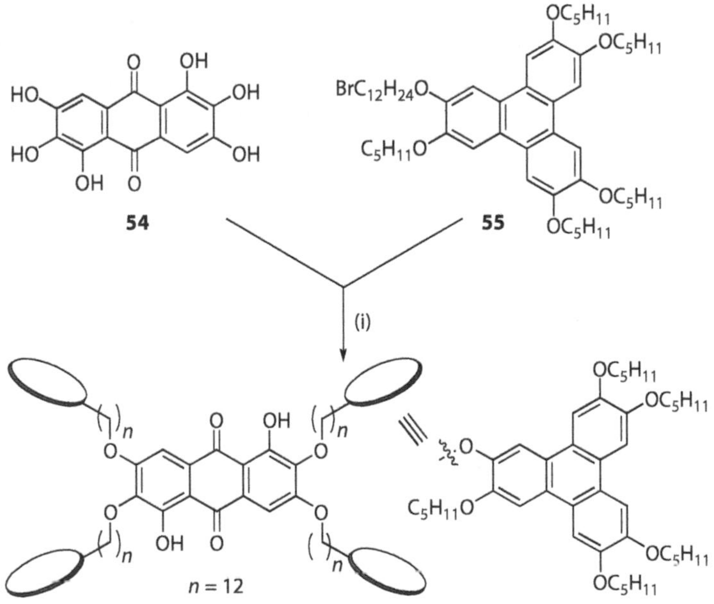

SCHEME 4.7   Synthesis of the discotic pentamer **56**: (i) DMSO, NaOH, 90°C.

conditions with ω-bromo-substituted triphenylene **55** to furnish the pentamer **56**, leaving the less reactive intramolecular hydrogen-bonded hydroxyl groups at the 1- and 5-positions unreacted. The dihydroxyl-functionalized pentamer **56** can be acetylated to its corresponding diacetate under classical conditions. Pentamer **56** showed the transition from columnar to isotropic at 143.6°C, while on cooling it exhibited an isotropic to columnar transition at 139°C. There was no other detectable transition down to −40°C. The diacetate of **56** shows a columnar phase to isotropic phase transition at 112.5°C on heating. The isotropic phase to columnar phase transition occurs at 106°C on cooling. The nature of the mesophase as a hexagonal columnar was established by x-ray diffraction studies.

Two triphenylene side-chain oligomers, **57a** and **57b**, in which pentyloxytriphenylene side groups were attached to amino-substituted 1,3,5-triazine moieties in the main chain, were reported by Janietz et al. [36]. The gel permeation chromatography (GPC) analysis indicates coexistence of up to pentameric structures in these triazine oligomers. Oligomer **57a** displays a glass transition at 41°C and the mesophase to isotropic phase transition at 121°C. These transitions occur at 71°C and 136°C, respectively, in the case of oligomer **57b**. A lamellar S_A-like arrangement of the molecules was reported in these oligomers. Doping of **57a** and **57b** with trinitrofluorinone (TNF) results in an increase of the clearing temperature of the binary system, and the mesogenic layer structure of the oligomeric triphenylene changes to a rectangular columnar arrangement by charge transfer interaction [36].

57a: $n = 6$, g 41 Lm 121 I
57b: $n = 11$, g 71 Lm 136 I

4.4 DISCOTIC HEXAMERS

The linear hexamer **58** was obtained as one of the products from the condensation reaction of the difunctional triphenylene derivative with a diacid. It shows a glass transition at 35°C and a columnar mesophase to isotropic phase transition at 195°C [6].

4.5 STAR-SHAPED DISCOTIC HEPTAMERS

Ringsdorf and coworkers prepared several star-shaped oligomers (Figure 4.2) by attaching periph-eral triphenylene units to a central benzene, azacrown, or triphenylene unit [26,37]. The benzene-centered compound **59** shows a columnar mesophase to isotropic phase transition at 109°C, while the azacrown-centered compound **60** exhibits this transition at 171°C [26]. Compounds **61** contain seven triphenylene units, one of which acts as the central core and the other six as peripheral units. These compounds exhibit a columnar hexagonal phase by a statistical arrangement of triphenylenes without any super-column or superlattice formation in the mesophase. Interestingly, these com-pounds undergo conformational changes at the air–water interface such that the peripheral triphe-nylene substituents sit perpendicular (edge-on) and the central core sits parallel (face-on) to the

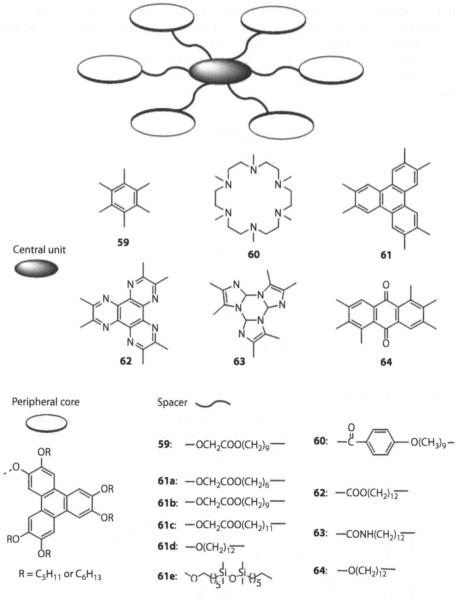

FIGURE 4.2 Structural model of a star-shaped discotic heptamer. The central core is surrounded by six peripheral discotic mesogens. Peripheral units may or may not be the same as the central unit.

TABLE 4.7
Thermal Behavior of Discotic Heptamers

Structure	R	Phase Transition	Ref.
59	C_5H_{11}	Col 109 I	[26]
60	C_5H_{11}	Col 171 I	[26]
61a	C_5H_{11}	Col_h 137 I	[37]
61b	C_5H_{11}	Col_h 152 I	[37]
61c	C_5H_{11}	Col_h 132 I	[37]
61d	C_5H_{11}	Col_h 136.8 I	[38]
61e	C_6H_{13}	Cr 38 Col_h 111 I	[30]
62	C_5H_{11}	g 40 Col_h 155 I	[39]
64	C_5H_{11}	Col_h 123.8 I	[38]
65	$C_{12}H_{25}$	Cr 35 Col_h >400	[40]

interface [26]. This conformation of the molecules is preserved upon transfer to a solid substrate. Transition temperatures of these heptamers are listed in Table 4.7. The three heptamers **61** exhibit a hexagonal columnar phase. Their clearing temperatures were slightly higher than the clearing temperature of the corresponding monomeric triphenylene discotic. No sign of crystallization in any heptamer was observed on cooling down to room temperature. Probably, these materials vitrify at low temperatures, but no actual glass transition was detected in the temperature range between −30°C and the clearing points. Bisoyi et al. prepared a similar heptamer **61d** by attaching six peripheral triphenylene units to the central triphenylene core via ethereal linkage instead of ester linkage [38]. It displayed a columnar phase to isotropic phase transition at about 137°C.

Another similar triphenylene-based heptamer **61e** was reported by Zelcer et al. [30]. In this heptamer, six peripheral triphenylene discotics were connected to the central triphenylene core via an alkyl spacer having a disiloxane unit. The heptamer was prepared following the synthetic route similar to that adopted for the synthesis of tetramers **46** and **47**, involving a hydrosilylation reaction in between a hexaalkoxytriphenylene having terminal olefinic groups and another monofunctionalized triphenylene with terminal siloxane units. The virgin material was partially crystalline at ambient temperature and formed a hexagonal columnar phase at 38°C, which cleared at 111°C. Interestingly, this compound exhibited a super-lattice formation, where each heptamer shares one peripheral triphenylene with six adjacent neighbors so that the distance between planes containing the cores is twice that of the regular hexagonal lattice.

Eichhorn et al. reported a discotic liquid crystalline heptamer **62** based on electron-rich and electron-deficient triphenylene units [39]. The central part of the molecule contains electron-deficient hexaazatriphenylene surrounded by six normal triphenylene units. This compound shows a tendency to form a spontaneous homeotropic alignment between two glass plates, which is a requirement for such compounds to be employed in photovoltaic applications. The heptamer **62** exhibits a glass transition at 40°C and goes from the hexagonal columnar phase to the isotropic phase at 155°C. The hexaamide **63** based on the trisimidazole central core was also prepared by these researchers, but the material was not characterized fully [38]. On the other hand, Kumar and coworkers prepared the heptamer **64** containing an electron-deficient anthraquinone core and an electron-rich triphenylene periphery from rufigallol and ω-bromo-substituted triphenylene in a single step by irradiating the reaction mixture in a microwave oven [38]. Classical reaction conditions to obtain the heptamer **64** resulted in an inseparable mixture of less substituted by-products. The hetero-heptamer **64** exhibits a columnar phase to isotropic phase transition at about 124°C.

Similar to triphenylene heptamers, Mullen and coworkers prepared an HBC heptamer **65**, in which seven HBC units are connected by flexible aliphatic chains [40]. The material was prepared following the synthetic route developed by these authors for the synthesis of HBC and other large-core discotics, as described in Chapter 2. This compound showed a strong tendency to aggregate in solution and in bulk state. A physical gel formation was observed due to the presence of covalent intercolumnar interactions. The heptamer **65** melts at about 35°C to a columnar phase, but it does not clear up to 400°C. The columnar nature of the mesophase was established by x-ray diffraction studies.

$R = C_{12}H_{25}$

65: ∿∿ = - - - $(CH_2)_{11}$ - -

REFERENCES

1. Imrie, C. T. and Henderson, P. A. *Curr. Opin. Colloid Interface Sci.* 7, 298–311, 2002.
2. Imrie, C. T. In *Physical Properties of Liquid Crystals: Nematics*, Dunmur, D. A., Fukuda, A., and Luckhurst, G. R. (eds.), IET, London, U.K., pp. 36–54, 2001.
3. Imrie, C. T., Henderson, P. A., and Yeap, G. Y. *Liq. Cryst.* 36, 755–777, 2009.
4. Imrie, C. T. and Luckhurst, G. R. In *Handbook of Liquid Crystal*, Vol. 2B, Demus, D., Goodby, J., Gray, G. W., Spiess, H. W., and Vill, V. (eds.), Wiley-VCH, Weinheim, Germany, 1998, Chap. X.
5. Imrie, C. T., Henderson, P. A., and Seddon, J. M. *J. Mater. Chem.* 14, 2486–2488, 2004.
6. Kranig, W., Huser, B., Spiess, H. W., Kreuder, W., Ringsdorf, H., and Zimmermann, H. *Adv. Mater.* 2, 36–40, 1990.
7. Boden, N., Bushby, R. J., Cammidge, A. N., and Martin, P. S. *J. Mater. Chem.* 5, 1857–1860, 1995.
8. Kumar, S. and Manickam, M. *Liq. Cryst.* 26, 939–941, 1999.
9. Paraschiv, I., Giesbers, M., van Lagen, B., Grozema, F. C., Abellon, R. D., Siebbeles, L. D. A., Marcelis, A. T. M., Zuilhof, H., and Sudholter, E. J. R. *Chem. Mater.* 18, 968–974, 2006.

10. Paraschiv, I., de Lange, K., Giesbers, M., van Lagen, B., Grozema, F. C., Abellon, R. D., Siebbeles, L. D. A., Sudholter, E. J. R., Zuilhof, H., and Marcelis, A. T. M. *J. Mater. Chem.* 18, 5475–5481, 2008.
11. Gupta, S. K., Raghunathan, V. A., and Kumar, S. *New J. Chem.* 33, 112–118, 2009.
12. Itoh, T., Takada, A., Fukuda, T., Miyamoto, T., Yakoh, Y., and Watanabe, J. *Liq. Cryst.* 9, 221–228, 1991.
13. Sugiura, M., Minoda, M., Watanabe, J., Fukuda, T., and Miyamoto, T. *Bull. Chem. Soc. Jpn.* 65, 1939–1943, 1992.
14. Janietz, D. *Mol. Cryst. Liq. Cryst.* 396, 251–264, 2003.
15. Grafe, A. and Janietz, D. *Mol. Cryst. Liq. Cryst.* 411, 477–482, 2004.
16. Grafe, A., Janietz, D., Frese, T., and Wendorff, J. H. *Chem. Mater.* 17, 4979–4984, 2005.
17. Bryant, G. C., Cook, M. J., Ryan, T. G., and Thorne, A. J. *Tetrahedron* 52, 809–824, 1996.
18. Bryant, G. C., Cook, M. J., Haslam, S. D., Richardson, R. M., Ryan, T. G., and Thorne, A. J. *J. Mater. Chem.* 4, 209–216, 1994.
19. Chambrier, I., Cook, M. J., Cracknell, S. J., and McMurdo, J. *J. Mater. Chem.* 3, 841–849, 1993.
20. McKeown, N. B., Chambrier, I., and Cook, M. J. *J. Chem. Soc. Perkin Trans. 1* 1169–1177, 1990.
21. Miwa, H., Kobayashi, N., Ban, K., and Ohta, K. *Bull. Chem. Soc. Jpn.* 72, 2719–2728, 1999.
22. Nakai, T., Ban, K., Ohta, K., and Kimura, M. *J. Mater. Chem.* 12, 844–850, 2002.
23. Wu, J., Watson, M. D., Tchebotareva, N., Wang, Z., and Mullen, K. *J. Org. Chem.* 69, 8194–8204, 2004.
24. Imrie, C. T., Stewart, D., Remy, C., Christie, D. W., Hamley, I. W., and Harding, R. *J. Mater. Chem.* 9, 2321–2325, 1999.
25. Kumar, S., Schuhmacher, P., Henderson, P., Rego, J., and Ringsdorf, H. *Mol. Cryst. Liq. Cryst.* 288, 211–222, 1996.
26. Maliszewskyj, N. C., Heiney, P. A., Josefowicz, J. Y., Plesnivy, T., Ringsdorf, H., and Schuhmacher, P. *Langmuir* 11, 1666–1674, 1995.
27. Marguet, S., Markovitsi, D., Millie, P., Sigal, H., and Kumar, S. *J. Phys. Chem. B* 102, 4697–4710, 1998.
28. Markovitsi, D., Marguet, S., Gallos, L. K., Sigal, H., Millie, P., Argyrakis, P., Ringsdorf, H., and Kumar, S. *Chem. Phys. Lett.* 306, 163–167, 1999.
29. Haarer, D., Simmerer, J., Adam, D., Schuhmacher, P., Paulus, W., Etzbach, K. H., Siemensmeyer, K., and Ringsdorf, H. *Mol. Cryst. Liq. Cryst.* 283, 63–68, 1996.
30. Zelcer, A., Donnio, B., Bourgogne, C., Cukiernik, F. D., and Guillon, D. *Chem. Mater.* 19, 1992–2006, 2007.
31. Kouwer, P. H. J., Jager, W. F., Mijs, W. J., Picken, S. J., Shepperson, K. J., and Mehl, G. H. *Mol. Cryst. Liq. Cryst.* 411, 377–385, 2004.
32. Tschierske, C. *J. Mater. Chem.* 8, 1485–1508, 1998.
33. Schulte, J. L., Laschat, S., Vill, V., Nishikawa, E., Finkelmann, H., and Nimtz, M. *Eur. J. Org. Chem.* 2499–2506, 1998.
34. McKenna, M. D., Barbera, J., Marcos, M., and Serrano, J. L. *J. Am. Chem. Soc.* 127, 619–625, 2005.
35. Bisoyi, H. K. and Kumar, S. *Tetrahedron Lett.* 49, 3628–3631, 2008.
36. Janietz, D., Festag, R., Schmidt, C., and Wendorff, J. H. *Liq. Cryst.* 20, 459–467, 1996.
37. Plesnivy, T., Ringsdorf, H., Schuhmacher, P., Nutz, U., and Diele, S. *Liq. Cryst.* 18, 185–190, 1995.
38. Bisoyi, H. K. and Kumar, S. *Chem. Soc. Rev.* 39, 264–285, 2010.
39. Eichhorn, S. H., Fox, N., and Bornais, B. *Mater. Res. Soc. Symp. Proc.* 836, L4.6.1–L4.6.3, 2004.
40. Zhi, L., Wu, J., and Mullen, K. *Org. Lett.* 7, 5761–5764, 2005.

5 Discotic Polymers

Conventional polymeric materials are well known for their processability, and mechanical and thermal stabilities. However, these are generally highly disordered materials and therefore devoid of interesting properties associated with the ordered systems. The incorporation of self-assembling properties of liquid crystalline systems in polymers may generate materials useful for many device applications. Accordingly, liquid crystalline polymers containing rod-like moieties have been extensively investigated to explore their technological potential [1–11]. A range of synthetic methods, including polymer homologous reactions, free radical polymerizations, optical polymerizations, ring-opening metathesis polymerizations, and condensation polymerizations, have been used to prepare such polymers. By contrast, polymers derived from discotic mesogens have been relatively less explored, largely because only a few functionalized discogens are available from which polymers could be prepared. Polymeric discotic liquid crystals were first realized by the Ringsdorf group in 1983 [12], and subsequently many other discotic polymers have been prepared and studied for various physical properties [12–119]. As with polymers containing rod-like units, discotic polymers can also be subclassified into (a) main-chain polymers, (b) side-chain polymers, (c) networks, and (d) elastomers (Figure 5.1). Benzene, rufigallol, triphenylene, cyclotetraveratrylene, phthalocyanine, and hexabenzocoronene cores have been exploited to create discotic polymers. The synthesis and physical properties of these polymers are discussed in the following sections.

5.1 BENZENE-BASED DISCOTIC POLYMERS

The first examples of benzene-based liquid crystalline main-chain polymers were reported by Ringsdorf and coworkers in 1985 [13], shortly after the discovery of their triphenylene counterpart by the same group [12]. They prepared tetraalkanoyloxybenzene **1** by a method described in Chapter 3. The final polymer was achieved by the solution polycondensation of **1** with alkane-α,ω-dioic acid dichloride [1:1 molar ratio] in pyridine and toluene (Scheme 5.1).

The polymers **2a** and **2b** with a short methylene spacer were found to be nonliquid crystalline (Table 5.1). The isotropization temperatures as well as the mesophase range of the liquid crystalline derivatives **2c–e** are higher as compared to the monomeric benzene hexaalkanoate. The elongation of the spacer to 20 carbon methylene units resulted in the disappearance of mesomorphic behavior in **2f**. The exact nature of the mesophase was not revealed in these polymers.

Similarly, benzoate-based polyesters can be derived by condensing **1** with various substituted terephthaloyl chloride (Scheme 5.1). The solution polycondensation of tetrakis(alkanoyloxy)-1,4-hydroquinone **1** and terphthaloyl dichloride **3** in a mixture of pyridine and toluene as solvent leads to the formation of polymer of the general structure **4**. The synthesis of corresponding tetraalkoxybenzene-derived polybenzoates is shown in Scheme 5.2. Tetramethoxy-1,4-benzoquinone **5** was heated with 1-alkanol at 140°C in presence of sodium hydroxide to yield tetraalkoxy-1,4-benzoquinone **6**. The reduction of **6** with sodium thiosulfate in a mixture of diethyl ether and water furnished tetraalkoxy-1,4-hydroquinone **7**. The solution polycondensation of tetraalkoxy-1,4-hydroquinone **7** and terephthaloyl dichloride **3** in a mixture of pyridine and toluene as solvent yields polymers of the general structure **8** [14,15].

To prepare benzene-based discotic polyamides, the diamino-tetraalkanoyloxybenzene **15** was prepared, as shown in Scheme 5.3. The stepwise oxidation and nitration of durene **9** gives 3,6-dinitro-1,2,4,5-benzenetetracarboxylic acid **12**. The silver salt of this acid **13** was now treated with appropriate alkyl iodide to furnish dinitrotetraester derivative **14**. The reduction of **14** with zinc and

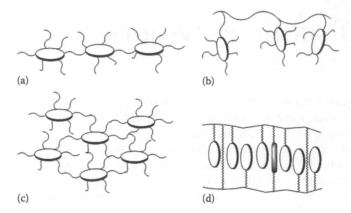

FIGURE 5.1   Sketches of some possible molecular architecture for discotic polymers: (a) main-chain discotic polymer, (b) side-chain discotic polymer, (c) discotic network, and (d) discotic elastomer.

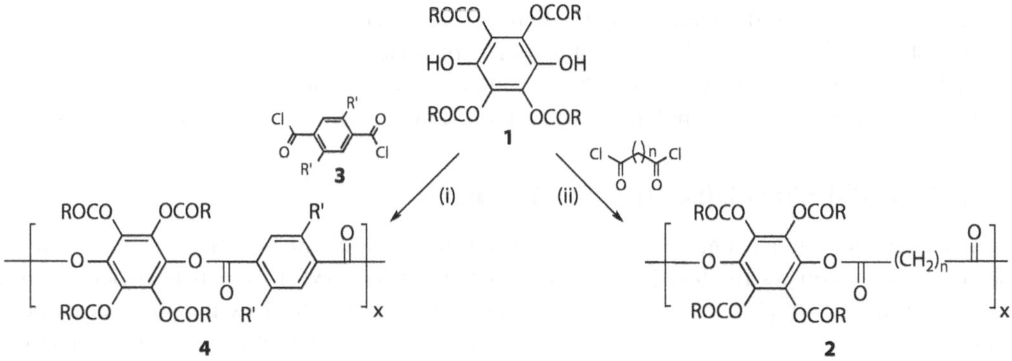

SCHEME 5.1   Synthesis of tetraalkanoyloxy benzene polyesters: (i) pyridine, toluene, ClOOC(CH$_2$)$_n$COOCl, −20°C, 2h, r.t., 5 days; (ii) pyridine, toluene, r.t., 7 days.

TABLE 5.1
Thermal Behavior of Benzene-Methylene-Based Polymers

Structure	n	M_n	Phase Transition
2a	7	—	g 54 I
2b	8	10,400	g 55 I
2c	10	9,800	g 55 M 95 I
2d	12	12,400	g 54 M 97 I
2e	14	12,100	Cr 64 M 92 I
2f	20	6,600	g 30 I

Source: Data from Kreuder, W. et al., *Makromol. Chem. Rapid Commun.*, 6, 367, 1985.

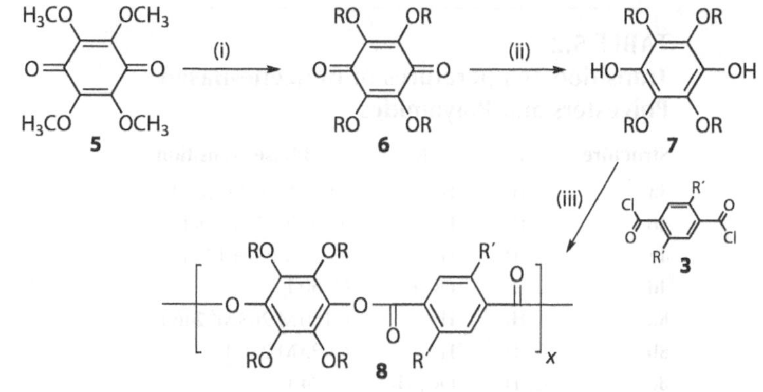

SCHEME 5.2 Synthesis of tetraalkoxy benzene polyesters: (i) ROH, NaOH, 140°C; (ii) $Na_2S_2O_4$, diethyl ether, H_2O, 2 h; (iii) pyridine, toluene, 60°C, 24 h.

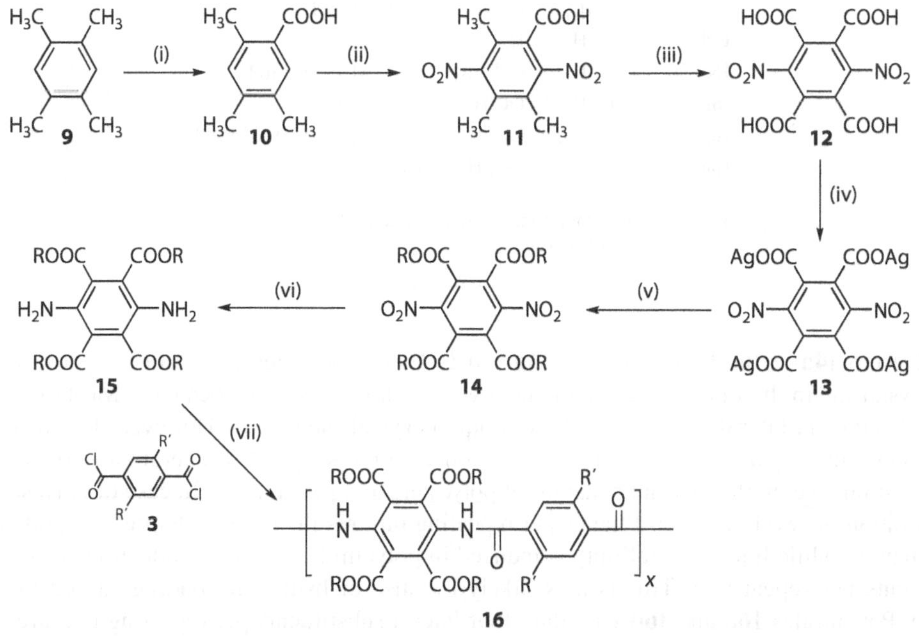

SCHEME 5.3 Synthesis of benzene polyamides: (i) HNO_3; (ii) H_2SO_4, KNO_3; (iii) $KMnO_4$; (iv) $AgNO_3$; (v) RI, diethyl ether, r.t., 3 days, reflux, 5 h; (vi) AcOH, H_2O, Zn dust, 100°C; (vii) N-methyl-2-pyrrolidone, r.t., 2–4 h, 60°C–70°C, 24 h.

acetic acid produces tetrasubstituted 1,4-phenylenediamine **15**. The polyamide **16** was created by the solution polycondensation of **15** with **3** in N-methyl-2-pyrrolidone as solvent.

The thermal behavior of these polymers is presented in Table 5.2. The polyesters **4a–c** in which the repeat unit was surrounded by four lateral alkanoyloxy groups were reported to exhibit liquid crystalline behavior with unidentified texture. The temperature of isotropization decreases on increasing lateral chain length. The polyester **4d** having hexasubstituted repeat unit did not display any mesomorphism. This polymer was crystalline at room temperature and cleared at 50°C. Similarly, polyesters **8c** and **8d** with six lateral substituents per repeat unit failed to demonstrate mesomorphic behavior. On the other hand, polymers **8a** and **8b** with four lateral substituents per repeat unit were reported to show liquid crystalline behavior. The mesophase range was broader for dodecyl-substituted polymer **8b** as compared to the polyester **8a** with shorter substituents. Unlike

TABLE 5.2
Transition Temperatures of Benzene-Based
Polyesters and Polyamides

Structure	R	R′	Phase Transition
4a	C_7H_{15}	H	g 97 M 146 M′ 260 I
4b	C_9H_{19}	H	g 50 Cr 95 M 138 I
4c	$C_{11}H_{23}$	H	g 55 Cr 63 M 123 I
4d	C_7H_{15}	OC_8H_{17}	Cr 50 I
8a	C_8H_{17}	H	g 150 M 208 M′ 246 I
8b	$C_{12}H_{25}$	H	Cr 25 M 210 I
8c	$C_{12}H_{25}$	$OC_{12}H_{25}$	Cr 74 I
8d	$C_{12}H_{25}$	OC_8H_{17}	Cr 67 I
16a	C_8H_{17}	H	Cr 273 I
16b	C_8H_{17}	OC_8H_{17}	Cr 39 M 117 I
16c	C_8H_{17}	$OC_{12}H_{25}$	Cr 51 M 69 I
16d	$C_{12}H_{25}$	H	Cr 270 I
16e	$C_{12}H_{25}$	$OC_{12}H_{25}$	Cr 17 M 70 M′ 130 I
16f	$C_{12}H_{25}$	OC_8H_{17}	Cr 61 M 122 I
16g	$C_{14}H_{29}$	$OC_{14}H_{29}$	Cr 38 M 74 M′ 88 M′ 106 I
16h	$C_{16}H_{33}$	$OC_{16}H_{33}$	Cr 86 I

Source: Data from Ringsdorf, H. et al., *Makromol. Chem.*, 188, 1431, 1987.

the polyesters (**4a–c**, **8a**, **b**), polyamides **16a** and **16d** with four lateral substituents per repeat unit were crystalline in their pure state and did not exhibit liquid crystalline behavior. Most of the polyamides having six lateral substituents showed liquid crystalline phases. However, the exact nature of the mesophase could not be confirmed by optical microscopy. The lateral substituents had a different influence on the thermal behavior of polyesters and polyamides. In case of polyesters, the incorporation of two lateral substituents per repeating unit results in the collapse of liquid crystalline behavior, while liquid crystallinity is induced in polyamides on the introduction of two lateral substituents per repeat unit. This is most likely because of hydrogen bonding caused by amide linkage. Polyamides **16a** and **16d** with only four lateral substituents per repeating unit are able to form strong intermolecular hydrogen bonding and thus have a tendency to crystallize. The attachment of two more lateral substituents per repeating unit causes weaker intermolecular interactions and hence favors the formation of mesomorphic phase. The exact nature of liquid crystalline phase could not be deduced in these polyamides.

5.2 ALKYNYLBENZENE-BASED DISCOTIC POLYMERS

Side-chain liquid crystalline polymers bearing nematogenic radial multiyne have been widely investigated [16–28]. In order to mingle the suppressed crystallization tendency of a polymer backbone with propensity of radial alkynylbenzene to form nematic mesophase, Kouwer et al. synthesized a series of side-chain polyacrylates based on radial pentayne. The combination of the above two properties was expected to result in a low-temperature liquid crystalline macromolecular superstructure with a broader mesophase range.

AIBN-catalyzed free radical polymerization of acryloyl chloride **17** furnished the polyacryloyl chloride **18**. The final step involves a straightforward esterification of **18** with hydroxyl-terminated

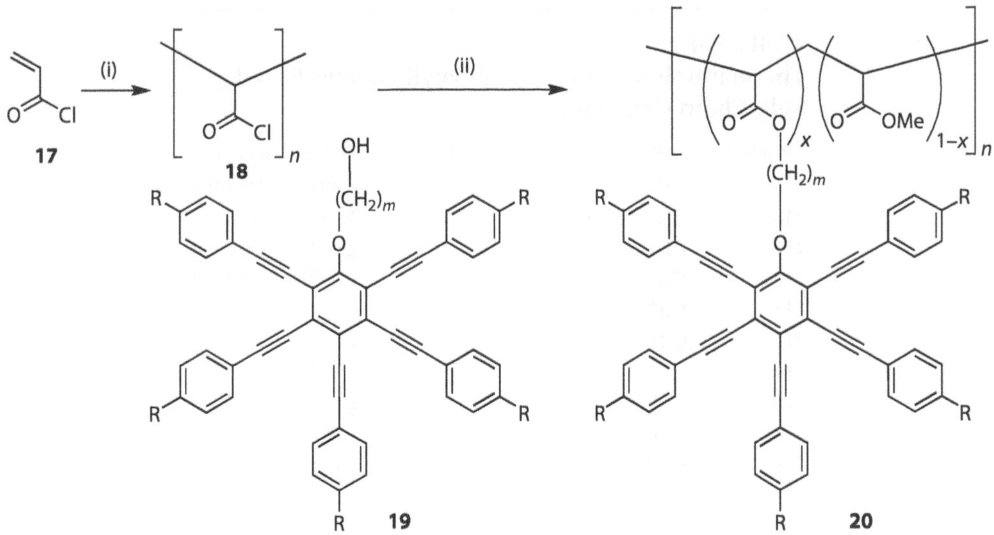

SCHEME 5.4 Synthesis of alkynylbenzene-based side-chain polyacrylates: (i) AIBN, dioxane; (ii) **19**, C_5H_5N, DMAP, CH_2Cl_2.

radial pentaalkynylbenzene derivative **19** in the presence of pyridine and catalytic amount of *N,N*-dimethylaminopyridine in dry dichloromethane. The reaction was quenched by esterifying the remaining acyl groups by adding an excess of dry methanol to achieve the desired polymer **20** (Scheme 5.4).

The thermal behavior of multiyne-based side-chain discotic polyacrylates is summarized in Table 5.3. As can be seen from this table, the polymers **20a** and **20g** are non-mesomorphic. Changing the connecting group from methylene to oxygen (compare **20a** vs. **20b/c**) resulted in the formation of a nematic phase. These polymers are in a glassy state near room temperature and exhibit a very broad nematic phase. Polymers with short methoxy peripheral groups decompose at higher temperature before reaching the isotropic phase. The replacement of the methoxy groups with long alkoxy (hexyloxy) chains reduces the isotropic temperature significantly. The polymer **20b** melts at 25°C to form a nematic phase that goes to the isotropic phase at 107°C.

Similarly, a series of side-chain pentaalkynylbenzene co-polyacrylates **21a–j** have been prepared by Kouver and coworkers [21]. The synthesis of these copolymers starts in a similar way to that of **20**, but the quenching of esterification reaction involves two steps. First, an excess of alcohol (other

TABLE 5.3
Thermal Behavior of Pentaalkynylbenzene-Based Side-Chain Polymers (Scheme 5.4)

Structure	R	x	Phase Transition	Ref.
20a	C_7H_{15}	0.50	g −15 I	[17]
20b	OC_6H_{13}	0.60	g 25 N 107 I	[17]
20c	OC_6H_{13}	0.63	g 24 N 113 I	[17]
20d	OCH_3	0.63	g 41 N >200 dec	[18]
20e	OCH_3	0.70	g 50 N >200 dec	[18]
20f	OCH_3	0.83	g 45 N >200 dec	[18]
20g	H	0.72	g 53 I	[19]

TABLE 5.4
Thermal Behavior of Pentaalkynylbenzene-Based Side-Chain Copolymers

Structure	R	x	y	$1-x-y$	Phase Transition
21a	CH_3	0.90	0	0.10	g 43 N_c 164 N 235 I
21b	CH_3	0.64	0	0.36	N_c 165 N 233 I
21c	CH_3	0.48	0	0.52	g 92 N_c 158 N 204 I
21d	CH_3	0.31	0	0.69	g 62 N_c 149 N 148 I
21e	C_5H_{11}	0.69	0.31	0	g 77 N_c 157 N 209 I
21f	C_5H_{11}	0.41	0.43	0.16	g 70 N_c 169 I
21g	C_5H_{11}	0.42	0.50	0.08	g 51 N_c 157 I
21h	$C_{12}H_{25}$	0.61	0.11	0.28	g 86 N_c 155 N 210 I
21i	$C_{12}H_{25}$	0.45	0.45	0.10	g 50 N_c 154 I
21j	$C_{12}H_{25}$	0.33	0.42	0.25	g 49 N_c 136 I

Source: Data from Kouwer, P.H.J. et al., *Mol. Cryst. Liq. Cryst.*, 364, 225, 2001.

than methanol) was added to the reaction mixture. This was followed by the addition of methanol to afford the desired co-polyester.

The thermal behavior of these side-chain polyacrylates is presented in Table 5.4. All copolymers displayed nematic columnar phase at lower temperatures. The thermal behavior of these polymers is strongly dependent on the weight fraction of mesogen in the copolymer. It is notable that increasing the mesogen fraction tends to an increase in glass transition as well as clearing temperature. Higher percentage of mesogen (compare **21e** vs. **21f/g** and **21h** vs. **21i/j**) in the copolymer was in favor of the formation of nematic phase at higher temperature.

21

22
M_n = 22,000, g 35 N 140 I

Suzuki et al. synthesized pentayne-based side-chain polyacrylates **22** containing five biphenyl units [27]. This polymer with number average molecular weight (M_n) 22,000 g mol^{-1} displays a nematic phase in the range of 35°C–140°C. Similarly, the monomeric hexayne **23** having six terminal double-bond containing biphenyl units can be photopolymerized to yield the polymer **24** (Scheme 5.5). The glass transition temperature of this polymer is 51°C. The liquid crystalline phase

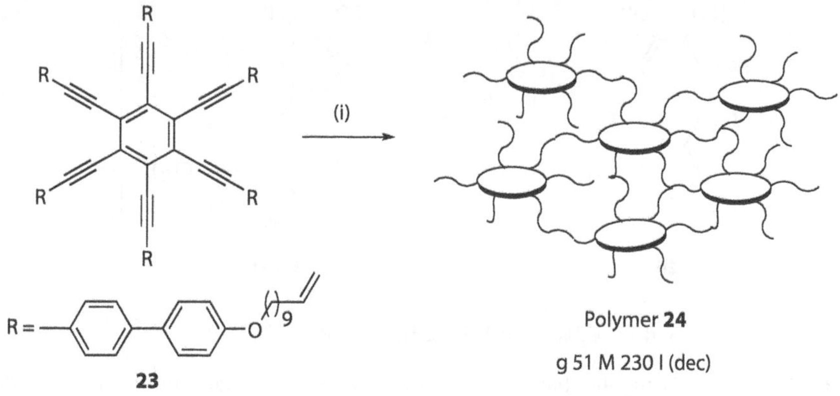

SCHEME 5.5 Photopolymerization of hexayne with six terminal double bonds: (i) UV.

persists above 230°C up to decomposition on heating. This could be due to the highly cross-linked network formation. The exact nature of the mesophase was not identified [27].

Efforts have been made to prepare side-chain liquid crystalline polymers **25** having pentaalkyne side groups connected to amino-substituted triazine units in the main chain [28]. However, only non-mesomorphic oligomers (up to pentamers) could be realized. Being amphiphilic in nature, these oligomers form stable edge-on orientation at the air–water interface.

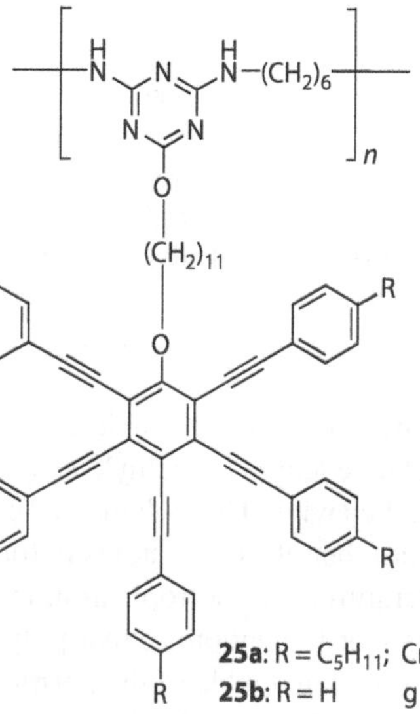

25a: R = C$_5$H$_{11}$; Cr 158 I
25b: R = H   g 112 I

5.3 RUFIGALLOL-BASED MAIN-CHAIN DISCOTIC POLYMERS

Anthraquinone-based main-chain polyethers **27** with different peripheral chains and methylene spacers in the main chain have been prepared by Raja et al. [29] and Bisoyi et al. [30]. These rufigallol-based main-chain polyethers were prepared from 1,5-dihydroxy-2,3,6,7-tetraalkoxy-9,10-anthraquinone (Scheme 5.6). Potassium carbonate (aq) mediated polyetherification reaction of dihydroxy-tetra-alkoxyanthraquinone **26** with equimolar amount of α,ω-dibromoalkane under nitrogen atmosphere in o-dichlorobenzene and in the presence of a phase-transfer catalyst at 90°C for 2 weeks afforded the respective polymers **27a–h** [29]. Such polyethers, for example, **27i**, can also be prepared by heating an equimolar mixture of dihydroxy-tetraalkoxyanthraquinone **26** and 1,12-dibromododecane in the presence of cesium carbonate in o-dichlorobenzene at 90°C for 10 days [30].

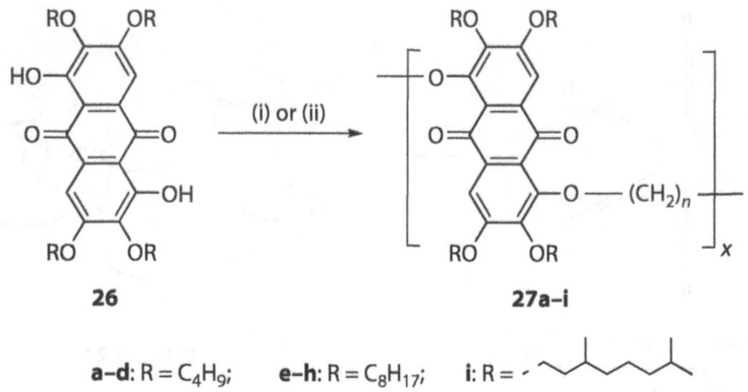

a–d: R = C_4H_9;   **e–h**: R = C_8H_{17};   **i**: R =

SCHEME 5.6 Synthesis of rufigallol-based main-chain polyethers: (i) α,ω-dibromoalkane, o-dichloroben-zene, aq K_2CO_3, tetrabutylammonium bromide (TBAB), 90°C, 14 days; (ii) α,ω-dibromoalkane, o-dichloro-benzene, Cs_2CO_3, 90°C, 10 days.

TABLE 5.5
**Thermal Behavior of Rufigallol-Based Main-Chain Polyethers
(Scheme 5.6)**

Structure	R	n	M_w	Phase Transition	Ref.
27a	C_4H_9	6	6,800	g 72 Col_r 164 $Col_{r'}$ 175 I	[29]
27b	C_4H_9	8	17,000	g 72 Col_h 164 $Col_{h'}$ 176 I	[29]
27c	C_4H_9	10	6,400	Col_h 133 $Col_{h'}$ 144 I	[29]
27d	C_4H_9	12	5,400	Col_h 138 I	[29]
27e	C_8H_{17}	6	7,400	Col_r 132 I	[29]
27f	C_8H_{17}	8	14,500	g 54 Col_r 96 I	[29]
27g	C_8H_{17}	10	8,800	Col_r 108 I	[29]
27h	C_8H_{17}	12	7,600	Col_r 127 I	[29]
27i	$CH_2CH_2CH(CH_3)$ $(CH_2)_3CH(CH_3)_2$	12	13,691	Col_r 56 I	[30]

The thermal data of these polymers are listed in Table 5.5. These polyethers are liquid crystal-line at ambient temperature. The relative length of methylene spacer and peripheral alkyl chain play a significant role in the mesophase behavior. The polymer **27b** with $n/R = 2$, where n = number of carbon atoms in the spacer and R = number of carbon atoms in the side chain, was reported to exhibit the highest value of clearing temperature. The isotropization temperature was lower in the case of other polymers having $n/R \neq 2$. It is worth mentioning that polymers **27b**, **27c**, and **27d** with alkyl spacer length double ($n/R = 2$) or more than double of the peripheral alkyl chain length exhibit hex-agonal columnar phase, while polymers **27a** and **27e–i** with spacer length smaller than double the alkyl side-chain length ($n/R < 2$) show rectangular columnar phase. The isotropization temperature was minimum for the polymer **27i** having branched peripheral chain. This polymer has been used to disperse nanoparticles in its supramolecular structure [30]. The insertion of carbon nanotubes into the columnar lattice of **27i** did not affect the mesophase structure; however, the mesophase-to-isotropic transition temperature was influenced in a reciprocal way on increasing the carbon nanotube proportion.

5.4 TRIPHENYLENE-BASED POLYMERS

Like monomeric triphenylenes, polymeric triphenylene derivatives have also been extensively investigated for their mesomorphic and physical properties. A number of side-chain polymers, main-chain polymers, elastomers, hyperbranched polymers, and networks have been derived from functionalized triphenylene discotics. The chemistry and mesomorphism of these polymers are presented in Sections 5.4.1 through 5.4.5 [13,31–97].

5.4.1 TRIPHENYLENE-BASED SIDE-CHAIN POLYMERS

A mono-functionalized triphenylene (TP) derivative is primarily required to prepare side-chain polymers. Several methods have been developed to prepare monohydroxy-pentaalkoxy-TPs (see Section 2.6). Ringsdorf and coworkers prepared the side-chain triphenylene polymers **29** and **30** by alkylating the monofunctionalized-TP with 10-undecenyl bromide followed by the hydrosilylation of the resultant 3,6,7,10,11-pentakis(pentyloxy)-2-(10-undecenyoxy)-TP **28** with Si–H containing polymers (Scheme 5.7). The homopolysiloxane **29** have an average degree of polymerization (DP) of 35 while the copolysiloxane **30** have a DP,70 and a 1:1 ratio of the two repeating units. Both polymers show a broad glass transition and a clearing transition. Polymer **29a** exhibits the phase sequence, g −24 Col 27 Col 35 Col 42 I, but on annealing the sample at 20°C for 2 days, it shows the phase behavior, g −19 Col 39 I. In a later report, the thermal behavior g −41 Col 75 I was reported for the same polymer. The clearing temperature of these polymers increases on decreasing the spacer length [31]. The thermal behavior of polymer **30** was reported to be g −31 Col 26 Col 32 Col 41 I and on annealing, g −29 Col 36 I. Other triphenylene-based polysiloxanes are shown in Figure 5.2 and their thermal behavior is summarized in Table 5.6. The deuterated triphenylene-containing polysiloxane **31**, with an average molecular weight of 45,000 g mol^{-1}, was prepared for use in NMR studies [32]. It shows a glass transition at 35°C and a columnar phase to isotropic phase transition at 75°C; a columnar to columnar phase transition at 45°C was also observed [32]. The corresponding undeuterated polymer **29a** (DP ~ 50, M_w ~ 45,000 g mol^{-1}) shows the same phase behavior [33], but the columnar phase to columnar phase transition at 45°C is seen only in the first heating run. In polymers **29–31** and in the ladder-like polysiloxane **33**, the spacer is linked to the disks via ether groups, while in polymer **32** it is joined via ester linkages [31]. As expected, the polymers with ester linkages exhibit higher clearing temperatures than those with ether links. The ladder-like polymer **33** (M_n ~ 16,000 g mol^{-1}) shows a glass transition at 20.6°C and a columnar to isotropic phase transition at 62.1°C. The formation of a board-like columnar structure in this polymer was revealed using XRD and AFM [34].

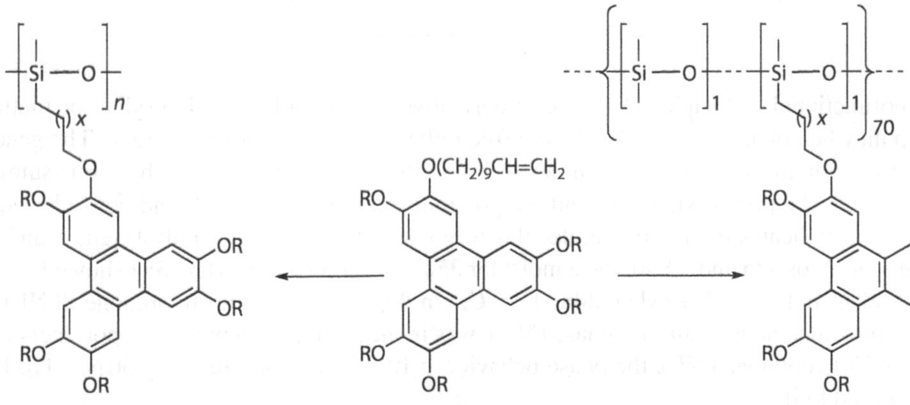

SCHEME 5.7 Synthesis of triphenylene-based side-chain polysiloxanes: (i) poly(methylsiloxane), toluene, H_2PtCl_6; (ii) poly(dimethylsiloxane-*co*-methylsiloxane), toluene, H_2PtCl_6.

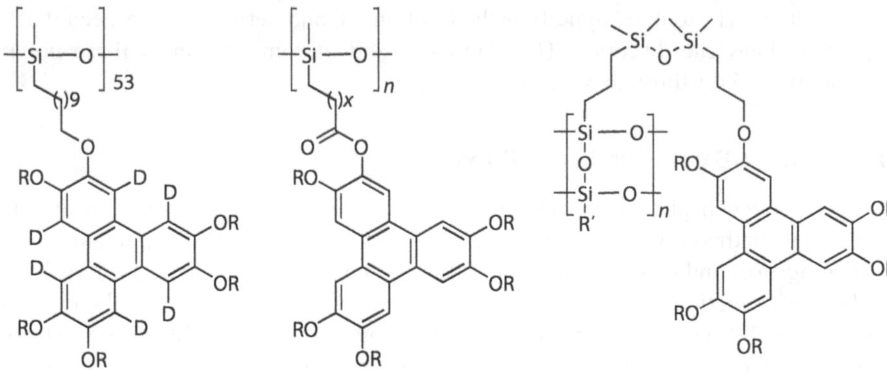

FIGURE 5.2 Chemical structures of triphenylene-containing polysiloxanes.

TABLE 5.6
Thermal Behavior of Triphenylene-Containing
Polysiloxanes (Scheme 5.7, Figure 5.2)

Structure	R	x	Phase Transition	Ref.
29a	C_5H_{11}	9	g −19 Col 39 I	[12]
			g −41 Col_h 75 I	[31]
29a′	C_5H_{11}	9	g −35 Col 75 I	[33]
29b	C_5H_{11}	3	g −53 Col_h 141 I	[31]
29c	C_5H_{11}	6	g −52 Col_h 131 I	[31]
30	C_5H_{11}	9	g −29 Col 36 I	[12]
31	C_5H_{11}	9	g −35 Col 75 I	[32]
32a	C_5H_{11}	3	g −54 Col_h 171 I	[31]
32b	C_5H_{11}	6	g −51 Col_h 186 I	[31]
32c	C_5H_{11}	9	g −37 Col_h 112 I	[31]
32d	C_7H_{15}	9	g −62 Col_h 112 I	[31]
32e	$CH_2CH_2CH(CH_3)_2$	9	Col_h 81 I	[31]
33	C_5H_{11}		g 20.6 Col 62.1 I	[34]

The monofunctionalized triphenylene-based derivatives having a terminal acrylate or methacrylate group may be polymerized readily by the free radical polymerization technique. The general structures of the polymers obtained are shown in Figure 5.3 and their thermal behavior is summarized in Table 5.7. The polyacrylate **34a** and the polymethacrylate **35a** were found not to be liquid crystalline; GPC indicates the relative molecular weights, 380,000 g mol^{-1} (polystyrene standard, THF, RI detection) for **34a** and 3,800,000 g mol^{-1} for **35a**. The polymethacrylate **35a** showed a glass transition at 30°C and the polyacrylate **34a** at −6°C. On doping with trinitrofluorenone (TNF), an electron acceptor, a nematic columnar phase (N_{col}), was induced in these amorphous polymers [35]. At a doping of 3:1 (polymer:TNF), the phase behavior, g $10 N_{col}$ 88 I for **34a** and g $50 N_{col}$ 170 I for **35a** was observed [35].

Among the polyacrylates and polymethacrylates having both identical triphenylene-based units and spacers (**34b** and **35b**), only the polyacrylate shows a mesophase [31]. The non-mesomorphic nature of the polymethacrylate was attributed to the rigidity of the methacrylate-based polymer backbone. In contrast to this observation, however, the Imrie and Boden groups have shown that appropriately substituted triphenylene containing polymethacrylates can form columnar phases

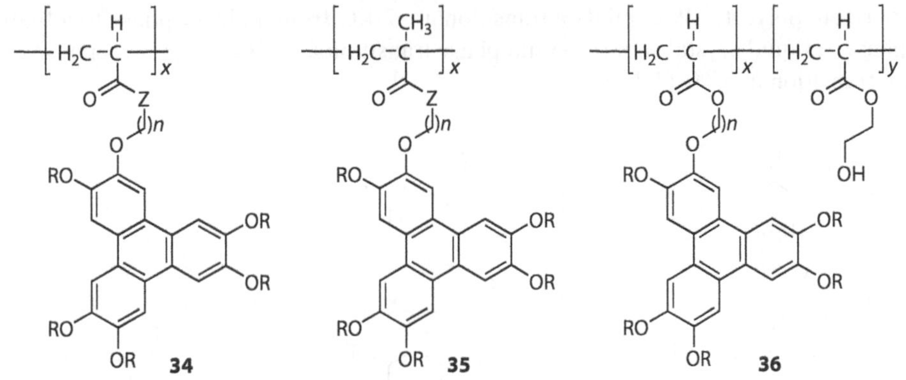

FIGURE 5.3 Chemical structures of triphenylene-based polyacrylates and polymethacrylates.

TABLE 5.7
Thermal Behavior of Triphenylene-Containing Polyacrylates and Methacrylates (Figure 5.3)

Structure	R	n	Z	M_w	Phase Transition	Ref.
34a	$(CH_2)_2O(CH_2)_2OCH_3$	0	$(CH_2CH_2O)_3$	380,000	g –6 I	[35]
34b	C_5H_{11}	11	O		g –8 Col$_h$ 39 I	[31]
34c	C_6H_{13}	6	O	7,350	g 5 Col 84 I	[36]
34d	C_6H_{13}	0	$(CH_2CH_2O)_2$	8,960	Col 117 I	[36]
35a	C_5H_{11}	6	O	3,800,000	g 30 I	[35]
35b	C_5H_{11}	11	O		g –16	[31]
35c	C_6H_{13}	11	O	61,000	g 2 Col 45 I	[37]
35d	C_6H_{13}	6	O	148,000	No transition	[36]
35e	C_6H_{13}	0	$(CH_2CH_2O)_2$	8,890	Col 110 I	[36]

[36,37]. Polymer **35c** having a weight average molar mass of 61,000 g mol^{-1} but a number average molar mass of only 9,000 g mol^{-1}, due to a high polydispersity, exhibits a glass transition at 2°C and a mesophase to isotropic phase transition at 45°C [37]. Similarly, polymethacrylate **35e** displays a columnar phase at room temperature and clears at 110°C [36]. Bleyl et al. prepared a number of polyacrylates by the photopolymerization of acrylate-functionalized triphenylene-based monomers to study their charge transport properties, but only partial polymerization was achieved [38]. The thermal behavior of these polymers was not described. The polymerization of a nitro-functionalized triphenylene-based acrylate was reported to give polyacrylates with molar masses up to 8000 g mol^{-1}, but the thermal behavior of those polyacrylates was not reported [39].

The free radical copolymerization of an acrylate-terminated triphenylene-based monomer with 2-hydroxyethyl acrylate leads to the formation of copolymers of the general structure **36**. These copolymers exhibit mesophase to isotropic transitions between 104°C and 165°C depending upon their composition. The average molecular weight of these polymers ranges from 51,000 to 1,000,000 g mol^{-1}. The formation of networks having columnar order in these polymers was reported [40]; the copolymerization of a triphenylene-based acrylate with methyl acrylate is reported to give a copolymer that does not show liquid crystallinity [36].

The condensation of the malonate-terminated triphenylene-based monomer **37** with a, α,ω-diols resulted in the synthesis of triphenylene-based side-chain polyesters **38** and **39**. Polymer **38** showed a glass transition at −10°C and a crystal to isotropic phase transition at 20°C [35].

By comparison, polyester **39** exhibits a transition at −20°C from a glassy phase to a liquid crystalline phase, a liquid crystal–liquid crystal phase transition at 20°C, and a mesophase to isotropic phase transition at 35°C [41].

37   **38**: g −10 Cr 20 I   **39**: M_n = 15,000, g −20 LC 35 I

The polycondensation of malonate **37** with dihydroxy-functionalized calamitic mesogens furnished polymers **40** and **41** having a triphenylene derivative as a side group discotic mesogen and azobenzene or biphenyl derivatives as rod-like mesogens in the main chain; neither polymer exhibits liquid crystalline behavior. The homopolymer **40** with an average molecular weight of 64,800 g mol^{-1} showed a crystal to crystal transition at 50°C and crystal to isotropic transition at 74°C. The copolymer **41** having an average molecular weight of 52,000 g mol^{-1} was found to be amorphous with a glass transition temperature of 20°C [42].

40: M_w = 64,000, Cr 74 I

41: M_w = 52,000, g 20 I

Imrie and coworkers prepared a triphenylene-based side-chain polymer by melt polycondensation of the triphenylene-substituted isophthalate with polyethylene glycol (PEG 300) [43]. The polymer **42** exhibits a weak glass transition at −17°C and a mesophase to isotropic transition at 47°C. An enhancement in the mesophase range was observed upon complexation of the polymer with lithium perchlorate. The 14:1 (**42**:LiClO₄) complex displays a glass transition at −11°C and a clearing transition at 52°C [43].

Efforts have been made to realize side-chain liquid crystalline polymers **43** having triphenylene pendant groups connected to amino-substituted triazine units in the main chain [44,45]. However, only oligomers (up to pentamers) could be realized (see Chapter 4). Being amphiphilic in nature, these oligomers form stable edge-on orientation at the air–water interface [44].

The ring-opening metathesis polymerization of discotic liquid crystalline monomers has been reported, leading to polynorbornenes **46a**, **46b** and polybutadienes **48a**, **48b** bearing pentyloxy- or decyloxy-substituted triphenylene-based units in the side chain [46] (Scheme 5.8). While decyloxytriphenylene-based polymers exhibit columnar mesophases, pentyloxytriphenylene derivatives were found not to be liquid crystalline. The reduction of polybutadiene backbone gives triphenylene-substituted polybutane derivatives **49a**, **49b**. Again, only the decyloxytriphenylene-substituted polybutane **49b** shows mesomorphism. The thermal behavior of these side-chain polymers is summarized in Table 5.8.

The side-chain diblock polymers **50a–c** containing a side group triphenylene-based block and a *p*-methoxystyrene block were prepared by Boden and coworkers by alkylating the diblock copolymer, poly[(*p*-methoxystyrene)-*b*-(*p*-hydroxystyrene)] with ώ-bromo terminated alkoxytriphenylenes [47]. *p*-Methoxystyrene and *p*-*tert*-butoxystyrene were used in 18:1 and 18:3 ratios, respectively, to prepare the diblock copolymers. No indication of columnar phase formation was observed in these polymers. The three polymers (**50a**: R = C₆H₁₃, x:y = 18:1, *n* = 6, M_n ~ 24,000 g mol⁻¹, **50b**: R = C₆H₁₃, x:y = 18:3, *n* = 6; and **50c**: R = C₄H₉, x:y = 18:3, *n* = 10, M_n ~ 21,000 g mol⁻¹) show birefringent streaks in the range 85°C–95°C and a very broad clearing range above 140°C. Although microphase separation was not apparent in these copolymers, block copolymers prepared from a triphenylene-based macroinitiator **51** and tertbutylacrylate, using atom transfer radical polymerization, show clear microphase segregation. The five copolymers **52a–e** with poly-*tert*-butylacrylate blocks of lengths ranging from DP ~ 6 to DP ~ 101 were prepared using the triphenylene macroinitiator **51,** which was prepared from 2,6,7,10,11-pentapentyloxy-3-(3-acryloylpropyloxy)triphenylene using the atom transfer radical polymerization technique [48]. Thermal transitions (g 36 Col 163 I) consistent with separate phases of poly-*tert*-butylacrylate (Tg 40°C–45°C) and polytriphenylene acrylate (g 83 Col 163 I) were observed in the DSC traces of these block copolymers. The formation of a columnar phase in the polytriphenylene acrylate block was confirmed by x-ray studies [48].

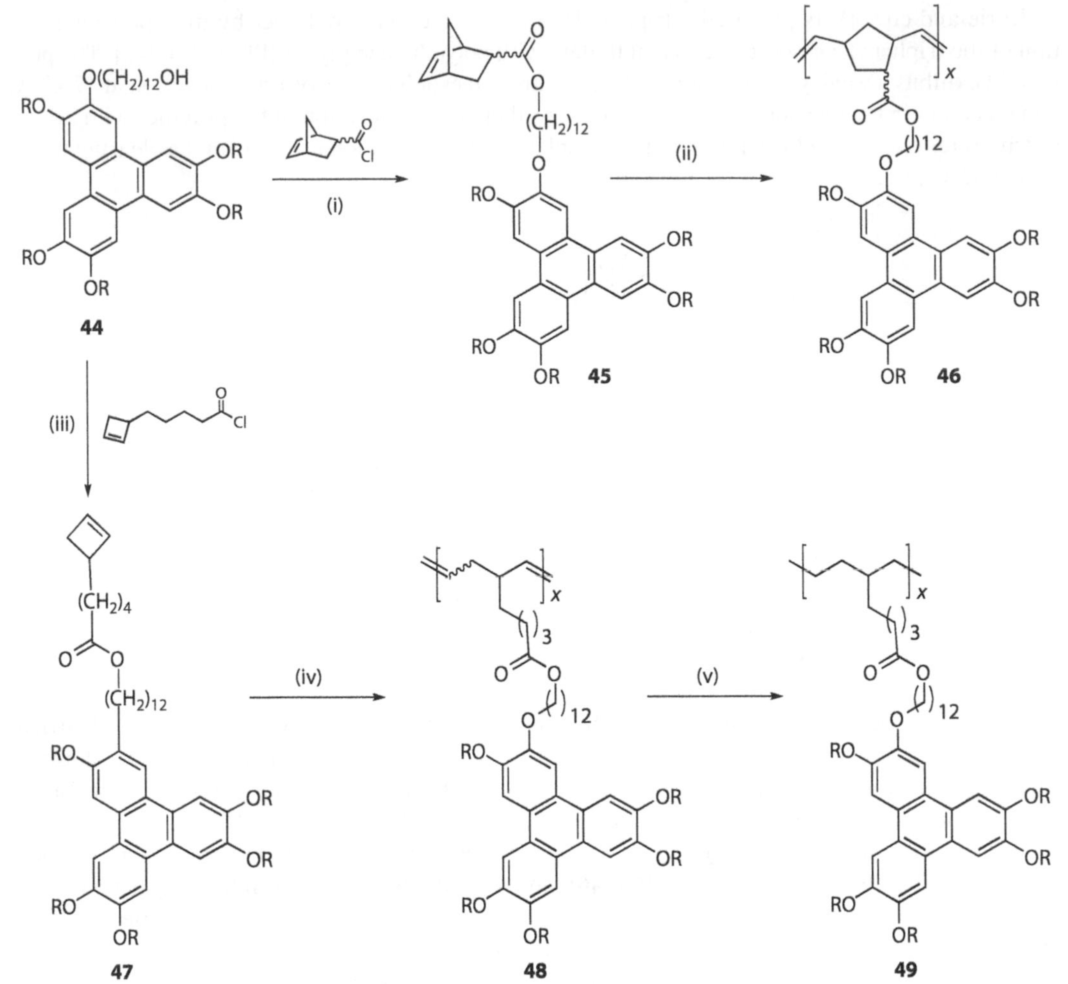

SCHEME 5.8 Synthesis of triphenylene-based side-chain polynorbornene and polybutadiene: (i) THF, triethylamine, reflux, 16 h; (ii) $Cl_2(PCy_3)_2Ru=CPhH$, CH_2Cl_2; (iii) $SOCl_2$, THF, triethylamine, r.t., 16 h; (iv) $Cl_2(PCy_3)_2Ru=CPhH$, CH_2Cl_2; (v) $[Ir(COD)(Cy_3)(py)]PF_6$, CH_2Cl_2, 55°C, 16 h.

TABLE 5.8
Thermal Behavior of Polynorbornene-, Polybutadiene-, and Polybutane-Bearing Alkoxytriphenylenes (Scheme 5.8)

Structure	R	M_n	Phase Transition
46a	C_5H_{11}	48,500	g −4 I
46b	$C_{10}H_{21}$	46,500	g −3 Col$_h$ 36 Col$_h$ 42 I
48a	C_5H_{11}	157,000	g −12 I
48b	$C_{10}H_{21}$	33,000	g −17 Col$_h$ 37 Col$_h$ 45 I
49a	C_5H_{11}	125,000	g −17 I
49b	$C_{10}H_{21}$	50,000	g −18 Col$_h$ 34 Col$_h$ 43 I

Source: Data from Weck, M. et al., *Macromolecules*, 30, 6430, 1997.

50

a: R = C$_6$H$_{13}$, x:y = 18:1,
  n = 6, M$_n$ = 24,000
b: R = C$_6$H$_{13}$, x:y = 18:3,
  n = 6
c: R = C$_4$H$_9$, x:y = 18:3,
  n = 10, M$_n$ = 21,000

51

R = C$_5$H$_{11}$, M$_n$ = 6,560
g 83 Col 163 I

52

R = C$_5$H$_{11}$, M$_n$ = 9,990
g 36 Col 163 I

Recently, Zhao and coworkers reported triphenylene-containing polyalkynes **54a–f** [49]. In these polymers, the triphenylene units with variable peripheral alkyl chain length were incorporated into the polyacetylene main chain (Scheme 5.9). The alkyne-terminated triphenylene derivative **53** can be prepared by reacting the monohydroxy-pentaalkoxy-TP with acetylene-substituted undecynoyl chloride. Though polymerization of **53** catalyzed by molybdenum and tungsten-based catalysts failed to produce polymers **54** in sufficient amount, stirring a mixture of **53** and catalytic amount of rhodium-based catalyst, [Rh(norbornadiene)Cl]$_2$, in THF/Et$_3$N yielded the desired polymer **54a–f** in good yield. The thermal stability of these triphenylene-bearing polyacetylenes was found to be much better than unsubstituted poly(1-alkyne)s.

Table 5.9 shows the phase transition temperatures of polyacetylene polymers. All polymers displayed hexagonal columnar mesophase. As can be seen from Table 5.9, isotropization temperature increases on going from butyl (**54a**) to hexyl (**54c**), but starts decreasing on moving from hexyl to nonyl (**54f**) peripheral chain length. This phenomenon can be attributed to the competition between packing effect of triphenylene disks and destructive plasticization effect of side chains. Additionally, the hexagonal structure of mesophase was also affected by alkyl side chains. The mesophase of polymers with small and long peripheral chains has homogenous hexagonal columnar lattice. On the other hand, polymers with intermediate alkyl length displayed hexagonal lattice with mixed structures.

With the intention of combining the ionic conductivity and the other interesting physical properties of ionic liquids with the self-organization tendency of discogens, Pal et al. prepared triphenylene-imidazole-based side-chain ionic liquid crystalline polymers [50]. Triphenylene-imidazolium-based ionic polymer **58** was prepared according to Scheme 5.10. The ω-brominated triphenylene **55** was reacted with 1-vinylimidazole to provide the triphenylene-bearing imidazolium salt **57**. The photopolymerization of **57** was performed using 2,2-dimethoxy-2-phenylacetophenone as photoinitiator.

Polymer **58** is a glassy material at room temperature, which goes to rectangular columnar phase at 224°C on heating. The mesophase persists up to 244°C. Cooling of the polymer from isotropic phase did not give any sign of crystallization down to room temperature.

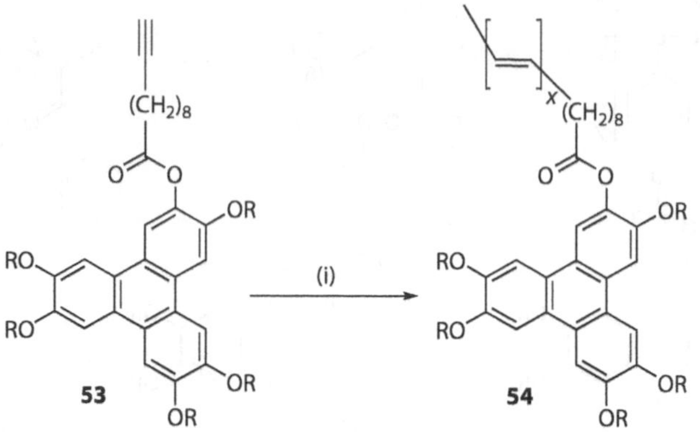

SCHEME 5.9 Synthesis of triphenylene-based side-chain polyalkynes: (i) [Rh(norbornadiene)Cl]$_2$, THF: Et$_3$N (3:1 v/v), r.t., 24 h.

TABLE 5.9
Thermal Behavior of Triphenylene-Based Poly(1-Alkyne)s (Scheme 5.9)

Structure	R	M_w	Phase Transition
54a	C$_4$H$_9$	~20,000	Col$_h$ 132.9 I
54b	C$_5$H$_{11}$	~20,000	g 113.8 Col$_h$ 156 I
54c	C$_6$H$_{13}$	~20,000	g 115.1 Col$_h$ 160.3 I
54d	C$_7$H$_{15}$	~20,000	Col$_h$ 154.3 I
54e	C$_8$H$_{17}$	16,000	Col$_h$ 144.6 I
54f	C$_9$H$_{19}$	~20,000	g 115.7 Col$_h$ 139.1 I

Source: Data from Xing, C. et al., *J. Polym. Sci. A Polym. Chem.*, 46, 2960, 2008.

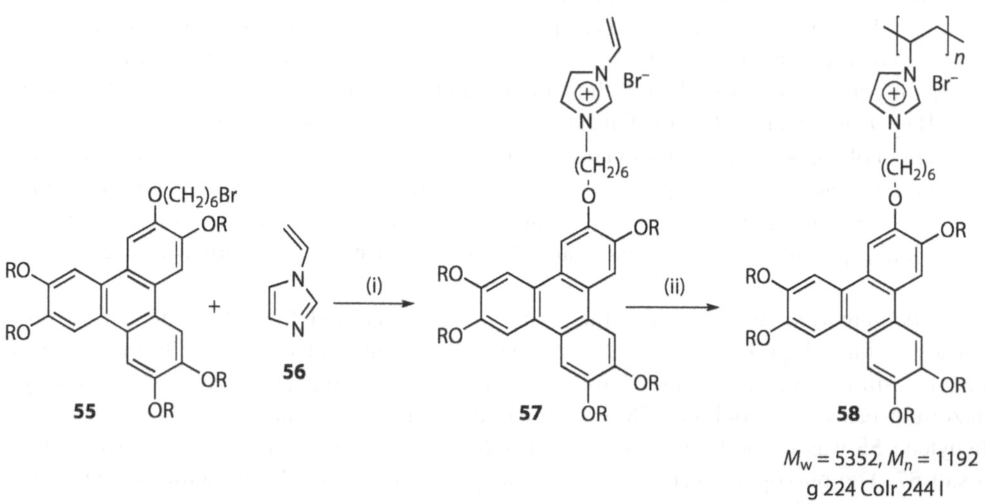

$M_w = 5352, M_n = 1192$
g 224 Colr 244 I

SCHEME 5.10 Synthesis of triphenylene-imidazole-based side-chain ionic polymers: (i) toluene, reflux, 8 h; (ii) 2,2-dimethoxy-2-phenylacetophenone, 100°C, r.t., *hv.*

5.4.2 TRIPHENYLENE-BASED MAIN-CHAIN POLYMERS

Difunctional triphenylenes are the precursor molecules for the preparation of main-chain polymers. The functional groups are often hydroxyl moieties that can easily be converted to ethers or esters. There are four different isomers of a dihydroxytetraalkoxy-TP in which the two hydroxyl groups could be present at 2,3-; 2,6-; 2,7-; or 2,11-positions (see Section 2.6). In addition to these dihydroxy-functionalized triphenylenes, a 2,3-dicarboxylic acid-functionalized triphenylene has also been used to prepare discotic polymers.

Following their synthesis of triphenylene-based side-chain polymers, the Ringsdorf group reported the synthesis of triphenylene-based main-chain discotic liquid crystal polymers in 1985 [13]. The melt polycondensation of bifunctionalized triphenylene derivatives (a mixture of 2,6- and 2,7-functionalized isomers) with various α,ω-diacids yielded polymers 59a–n. Subsequently, polymers derived from this 2,6- and 2,7-isomeric mixture have been structurally shown to be derivatives of only the 2,6-isomer 60 [32–35,51–56].

The thermal behavior of these polymers is presented in Table 5.10. The polymers having 10 or 14 methylene units in the spacer, except for 59m with chiral peripheral chains and a tetra-decyl spacer, show mesomorphism. The mesophase range of these alkoxytriphenylene containing polyesters was higher than that of the monomeric triphenylene hexaethers or esters. A similar mesophase stabilization was observed in triphenylene-based mixed ether–ester monomers and side-chain polymers. Polymers with very long spacers containing 20 methylene units (59g and 59n) were found not to be liquid crystalline. The deuteriation of the side-chain or aromatic groups does not significantly change the thermal behavior of these polymers. Polymer 59l, having chiral peripheral chains and a short spacer, shows mesomorphism while those with longer spacers, 59m and 59n, were found to be amorphous in nature [60]. Compound 59n on doping with TNF exhibits a nematic columnar phase. The formation of a helical superstructure in these chiral discotic polymers has been reported [60].

The main-chain discotic-calamitic polymer 61 derived from a mixture of 2,6- and 2,7-functionalized triphenylene isomers and an azobiphenyl-based calamitic mesogen was found to be amorphous in nature, exhibiting only a glass transition at 130°C [57].

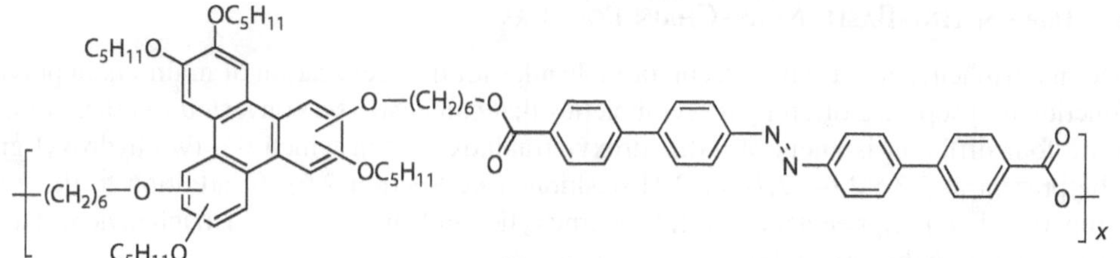

61: g 130 l

Wenz's rational synthesis of a 2,3-difunctionalized triphenylene derivative followed by its poly-condensation with various α,ω-diols having 8, 10, 12, 14, and 16 methylene units furnished poly-esters **62a–e**. These polymers melt at about 93°C and clear between 100°C and 200°C [52]. The polycondensation of malonic acid diethyl ester with the 2,3-difunctionalized triphenylene derivative 2,3-bis(6-hydroxyhexyl)oxy-6,7,10,11-tetrakis(pentyloxy)triphenylene afforded polymer **63** [58]. In contrast to the majority of triphenylene-based polymers, it shows a discotic nematic phase between the glass transition at −10°C and isotropic transition at 31°C (Table 5.11). The nematic phase trans-forms into a columnar phase on doping with the electron-acceptor TNF. The complex exhibits the thermal behavior, Col$_h$ 130 I [59].

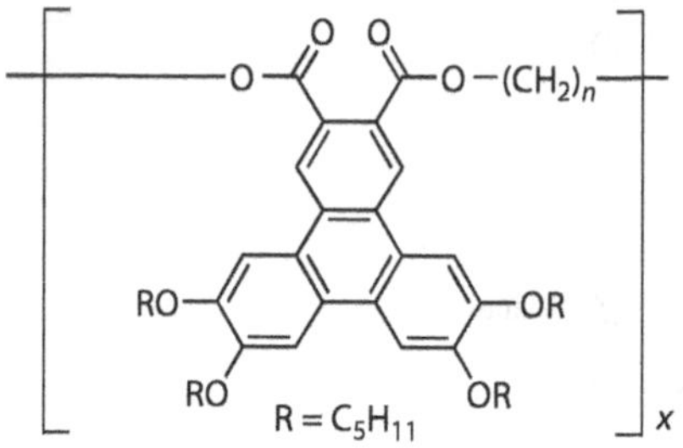

62

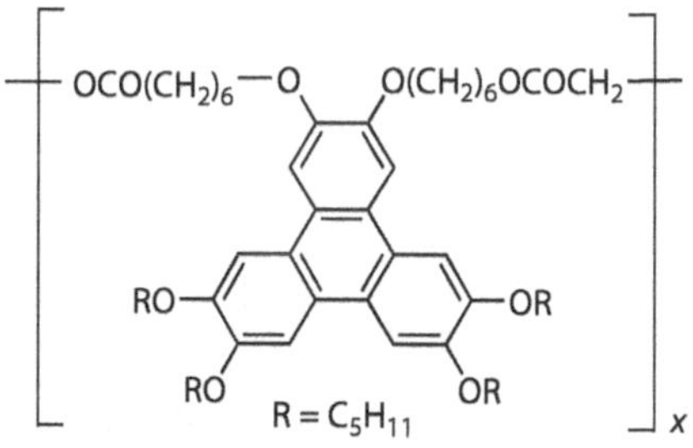

63: M_w = 9,300, g −10 N 31 l

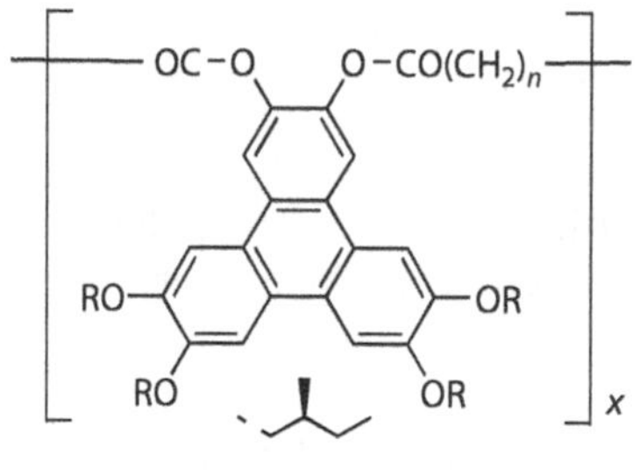

64

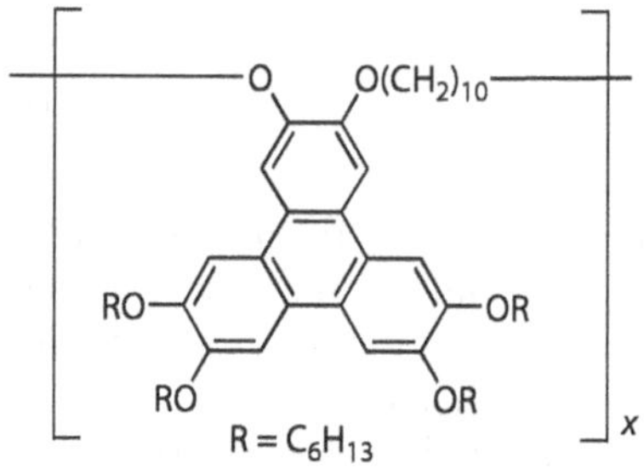

65: M_w = 7,000, Cr 100 Col 115 l

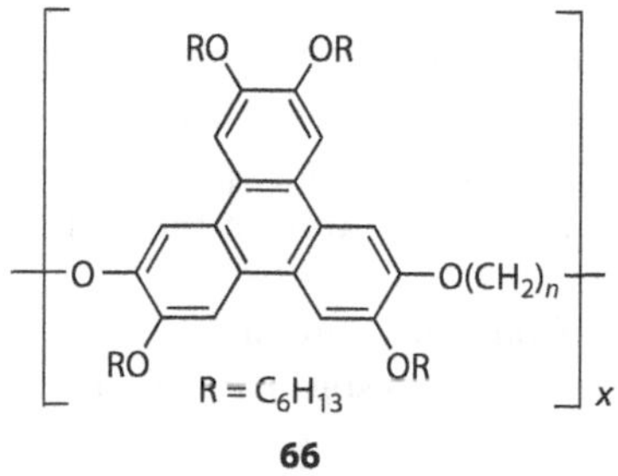

66

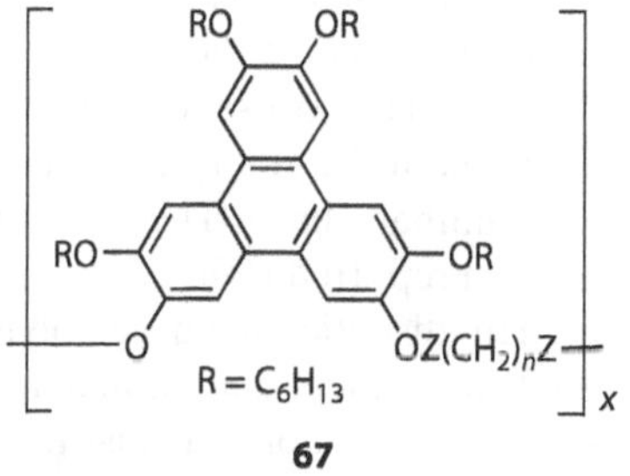

67

TABLE 5.10
Thermal Behavior of Mixtures of 2,6-Positions and 2,7-Positions Linked Triphenylene Main-Chain Polymers

Structure	R	Spacer	M_w	Phase Transition	Ref.
59a	C_5H_{11}	$-(CH_2)_{10}-$		g 35 Col 195 I	[13]
59b	C_5H_{11}	$-CD_2(CH_2)_8CD_2-$		g 50 Col 220 I	[13]
59c	C_5H_{11}	$-(CH_2)_{12}-$	20,000	g 50 Col 220 I	[51]
59d	C_5H_{11}	$-(CH_2)_{14}-$		g 60 Col 150 I	[13]
59e	C_5H_{11}	$-(CH_2)_{14}-$	19,000	g 57 Col_h 143 I	[51]
59f	C_5H_{11}	$-(CH_2)_{14}-$	51,000	g 57 Col_h 143 I	[32]
59g	C_5H_{11}	$-(CH_2)_{20}-$	13,000	g 35 I	[35]
59h	$CD_2C_4H_9$	$-(CH_2)_{14}-$	28,000	g 58 Col_h 140 I	[32]
59i	C_7H_{15}	$-(CH_2)_{14}-$		g 47 Col_h 182 I	[55]
59j	C_7H_{15}	$-(CH_2)_{14}-$	17,000	g 50 Col_h 182 I	[51]
59k	$(CH_2)_2CD_2C_4H_9$	$-(CH_2)_{14}-$		g 50 Col_h 180 I	[56]
59l	$CH_2C^*H(CH_3)C_2H_5$	$-(CH_2)_{10}-$	19,000	g 140 Col_h 192 I	[60]
59m	$CH_2C^*H(CH_3)C_2H_5$	$-(CH_2)_{14}-$	16,000	g 108 I	[60]
59n	$CH_2C^*H(CH_3)C_2H_5$	$-(CH_2)_{20}-$	18,000	g 79 I	[60]

TABLE 5.11
Thermal Behavior of Triphenylene Main-Chain Polymers 61–66

Structure	n	M_w	Phase Transition	Ref.
61			g 130 I	[42]
62	12	39,722	g 93 Col 163 I	[52]
63		9,300	g −10 N 31 I	[59]
64a	14	17,000	g 105 I	[60]
64b	20	15,000	g 75 I	[60]
65		7,000	Cr 100 Col 115 I	[61]
66	11	24,000	Cr 98 Col 118 I	[61]

In polymers **64a** ($n = 14$) and **64b** ($n = 20$), triphenylene-based units with chiral peripheral chains ($CH_2C^*H(CH_3)C_2H_5$) were linked together via ester linkages. Both the polymers, having relative molecular weights 17,000 and 15,000 g mol^{-1}, respectively, were amorphous in nature. Polymer **64a** shows a glass transition temperature at 105°C and polymer **64b** at 75°C [60]. The polyetherification of 2,3-dihydroxy-6,7,10,11-tetrahexyloxytriphenylene with 1,10-dibromodecane yielded the polymer **65** with an average molecular weight of 7000 g mol^{-1}. This polymer showed a columnar phase to isotropic transition at 115°C [61].

The polyetherification of 2,7-dihydroxy-3,6,10,11-tetrahexyloxy-triphenylene and 2,11-dihydroxy-3,6,7,10-tetrahexyloxy-triphenylene (also referred to as 3,6-dihydroxy-2,7,10,11-tetrahexyloxy-triphenylene) with α,ω-dibromoalkanes furnished polymers **66** and **67a** [61]. The polyesterification of 2,11-dihydroxytetrahexyloxytriphenylene with α,ω-diacids produced polymers **67b–d**. These polymers form an ordered columnar mesophase between the glass transition and clearing temperatures [62]. The thermal data of these polymers are listed in Table 5.12.

Discotic main-chain diblock and triblock copolymers containing one triphenylene block and either one or two polyethyleneoxy blocks, respectively, were prepared by Bushby and coworkers [63].

TABLE 5.12
Thermal Behavior of Triphenylene Main-Chain Polymers 67

Structure	Z	n	M_w	Phase Transition	Ref.
67a	CH$_2$	9	9,000	Cr 93 Col 120 I	[61]
67b	CO	8	39,797	g 56 Col$_h$ 211 I	[62]
67c	CO	10	9,659	g 40 Col$_h$ 198 I	[62]
67d	CO	12	8,818	g 25 Col$_h$ 162 I	[62]

TABLE 5.13
Thermal Behavior of Triphenylene Main-Chain Block Copolymers

Structure	n	M	Z	M_w	Phase Transition	Ref.
68	16	10–12		~10,000	<LC 86.5 I	[63]
69a	16	10–12	CH$_2$	~10,000	<LC 66.4 I	[63]
69b	~12	~4–13	CO		Cr 43.9 Col$_h$ 72.1 I	[64]
69c	~33	~4–13	CO		Cr 38–48 Col$_h$ 85.7 I	[64]
69d	~45	~12	CO	18,700	Cr 39.2 Col$_h$ 81.4 I	[65]
					Cr 48.6 Col$_h$ 85 I	[94]
69e	~45	~12	CO	18,500	Cr 38.4 Col$_h$ 85.7 I	[65]

The polyetherification of 2,7-dihydroxy-3,6,10,11-tetrahexyloxy-triphenylene was carried out using 0.8 equivalent of 1,12-dibromododecane to ensure the termination of the polymerization with hydroxyfunctionalized triphenylene moieties. The product, having an average of 11 triphenylene units in the discotic main-chain block, was reacted in situ with 2-methoxypoly(ethyleneoxy)ethyl tosylate (PEG750). The chromatographic separation of the products furnished the diblock copolymer **68** and the triblock copolymer **69a**. Both polymers were found to be liquid crystalline at room temperature. The columnar phase of copolymer **68** clears at 86°C, and that of **69a** at 66°C. The triblock copolymers **69b–d**, having polyethyleneoxide (PEO2000) blocks at both ends of the triphenylene block and linked via ester linkages, were also prepared by Bushby and coworkers [64,65]. The thermal behavior of these main-chain block polymers is summarized in Table 5.13.

5.4.3 TRIPHENYLENE-BASED DISCOTIC ELASTOMERS

Elastomers are loosely bound polymers. Liquid crystalline elastomers exhibit many entirely new properties that are not associated with monomeric or polymeric liquid crystals and, therefore, there are many potential applications of these materials to be explored [66–68]. While the field of calamitic elastomers has been well studied, only Ringsdorf and coworkers have reported discotic elastomers [69,70]. Elastomers are usually prepared in one or two steps from monomers having one terminal reactive functional group and a small amount of a difunctional reactive cross-linker. The first discotic elastomer was prepared using an olefin-terminated monofunctionalized triphenylene derivative and two cross-linkers with different reactive groups [69]. One of the cross-linkers contained two olefinic groups and the other possessed an olefinic group and a methacryloyl group. The critical feature of this reaction is the difference in the reactivity of the functional double bonds. During the first step of this reaction, the methacryloyl group remains essentially unreacted while the addition of the vinyl groups is almost complete. After this first cross-linking step, the weakly cross-linked network is mechanically deformed to obtain a macroscopic alignment of the liquid crystalline phase. In the second slow cross-linking step, additional bonds are introduced chemically to lock-in this anisotropic network. The formation of these elastomers is shown schematically in Figure 5.4 [97]. Elastomer **70a** was prepared using this methodology, while elastomer **70b** was prepared without applying the mechanical field to understand the effect of mechanical strain prior to the second cross-linking process. Both elastomers show very low clearing temperatures (g −50 Col 28 I) compared with the uncross-linked triphenylene polymer (g −51Col$_r$ 131 I). This is attributed to the disturbed columnar packing due to the bifunctional calamitic cross-linker [69]. To compensate for this, the bifunctional calamitic cross-linker was replaced by a bifunctional triphenylene-based cross-linker. Five triphenylene-based elastomers, **71a–e**, with different concentrations of discotic cross-linker and load applied during the synthesis were prepared [70]. The thermal behavior of these elastomers is summarized in Table 5.14.

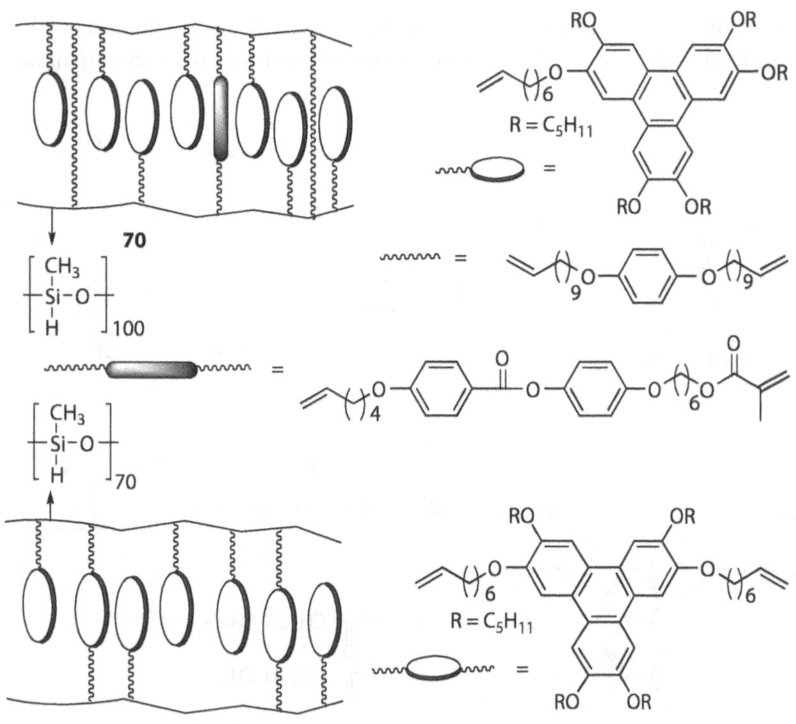

FIGURE 5.4 Schematic representation of triphenylene-based discotic elastomers.

TABLE 5.14
Thermal Behavior of Triphenylene-Based Discotic Elastomers

Structure	Phase Transition	Ref.
70a	g −50 Col 28 I	[69]
70b	g −50 Col 28 I	[69]
71a	g −50 Col 106 I	[70]
71b	g −49 Col 94 I	[70]
71c	g −50 Col 94 I	[70]
71d	g −45 Col 78 I	[70]
71e	g −44 Col 76 I	[70]

5.4.4 TRIPHENYLENE-BASED HYPERBRANCHED POLYMER

Zhang et al. obtained triphenylene-based hyperbranched polymer **73** by reacting 2,6,10-trihydroxy-3,7,11-tripentyloxytriphenylene **72** with adipoyl chloride in pyridine and THF. This was followed by quenching the terminal phenolic hydroxyl groups by adding butyryl chloride to the reaction mixture (Scheme 5.11) [71].

This hyperbranched polymer was reported to be liquid crystalline in the temperature range of 165°C–180°C. Constraint in the motion of triphenylene disks caused by branching interconnections resulted in the high value of clearing point as compared to its monomeric analog.

5.4.5 TRIPHENYLENE-BASED DISCOTIC COMPENSATION FILMS

The negative birefringence film formed by polymerized nematic discotic liquid crystals is their first successful commercial application. These films are used to improve the viewing angle of commonly

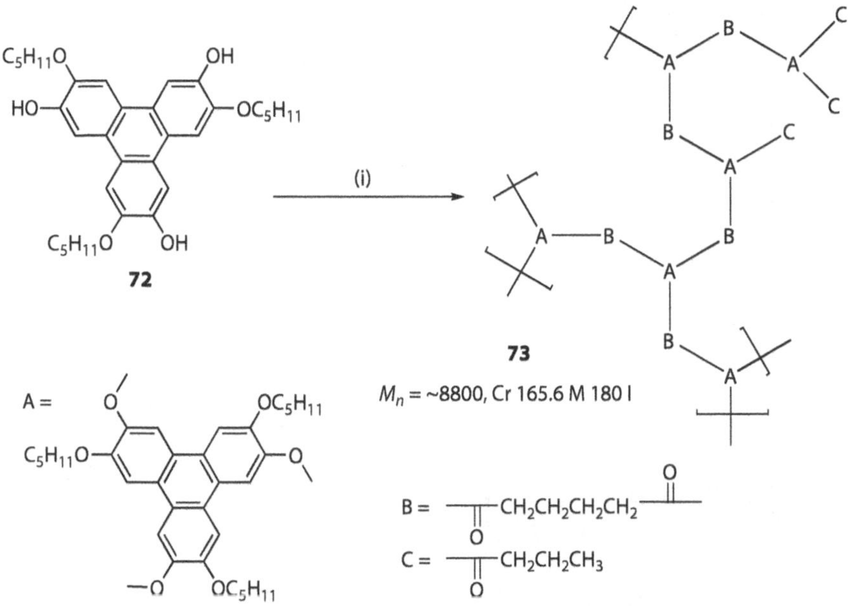

SCHEME 5.11 Synthesis of triphenylene-based hyperbranched polymer: (i) ClOC(CH$_2$)$_4$COCl, pyridine, THF, r.t., 4 h, 40°C, 2 h; CH$_3$CH$_2$CH$_2$COCl, 2 h.

used LCDs (see Chapter 6). These phase compensation films are usually prepared by aligning (hybrid alignment) the reactive monomer in the nematic phase followed by photopolymerization. Triphenylene benzoate derivatives with one to six acrylate end groups or epoxide groups have been used to prepare anisotropic networks. These cross-linked polymers showed no thermal transition up to 200°C [36,72–77].

5.5 CYCLOTETRAVERATRYLENE-BASED POLYMERS

Percec et al. have reported the synthesis and liquid crystalline behavior of octasubstituted cyclotetraveratrylene (tetrabenzocyclododecatetraene)-based branched polymers **76** [98]. Different proportions of 3,4-bis-(n-alkoxy)benzyl alcohol **74** were co-cyclotetramerized with α,ω-bis{[2-(n-alkoxy)-5-(hydroxymethyl)phenyl]oxy}alkane **75** to supply branched polymers **76a–m** (Scheme 5.12).

Table 5.15 shows the transition temperatures corresponding to branched polymers. As can be seen from this table, the molar ratio of **74** to **75** plays a significant role in determining the molecular weight of the branched polymers. These polymers display hexagonal columnar mesophase. For a constant spacer and peripheral alkyl chain length, the clearing temperature decreases on decreasing the ratio of **74** to **75**, that is, the isotropization temperature is reciprocally affected by the proportion of the reactant **75** in the reaction mixture. In other words, we can say that it is the amount of spacer (integral part of **75**) that is affecting the clearing point.

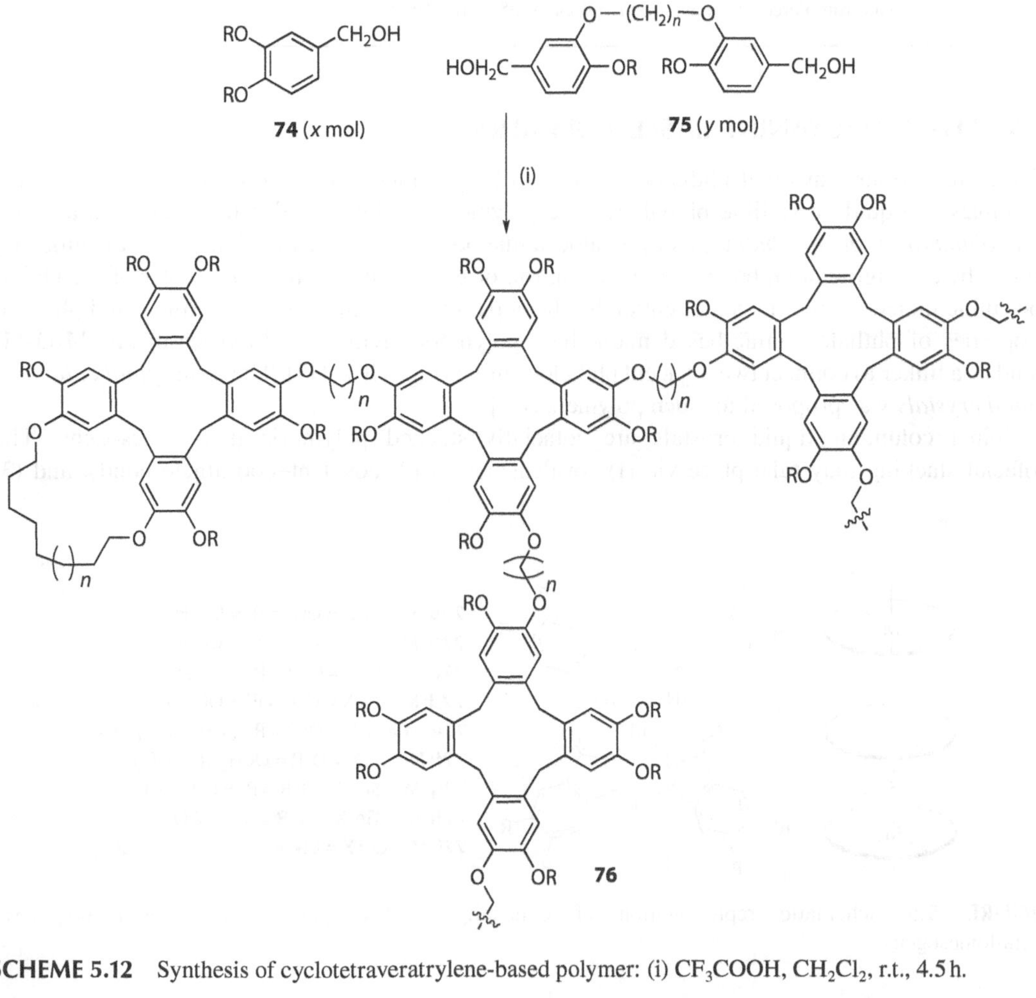

SCHEME 5.12 Synthesis of cyclotetraveratrylene-based polymer: (i) CF$_3$COOH, CH$_2$Cl$_2$, r.t., 4.5 h.

TABLE 5.15
Thermal Behavior of Cyclotetraveratrylene-Based Branched Polymers

Structure	R	n	Ratio between 74 and 75 [x:y]	M_n	Phase Transition
76a	C_6H_{13}	12	6:1	2800	g 15 Col 102 Col 120 Col$_h$ 145 I
76b	C_6H_{13}	12	4:1	2800	g 18 Col 99 Col 120 Col$_h$ 139 I
76c	C_6H_{13}	12	3:1	2600	g 20 Col 98 Col 120 Col$_h$ 136 I
76d	C_6H_{13}	12	2:1	3000	g 20 Col 95 Col$_h$ 118 I
76e	C_7H_{15}	12	6:1	1800	Col 90 Col 103 Col 116 Col$_h$ 141 I
76f	C_7H_{15}	12	4:1	2100	Col 103 Col 114 Col$_h$ 139 I
76g	C_7H_{15}	12	3:1	2200	Col 64 Col 92 Col$_h$ 110 I
76h	C_7H_{15}	12	2:1	2700	Col 89 Col$_h$ 103 I
76i	C_7H_{15}	16	6:1	3700	g 14 Col 90 Col 105 Col 120 Col$_h$ 127 Col$_h$ 142 I
76j	C_7H_{15}	16	4:1	4300	g 18 Col 87 Col 103 Col 117 Col$_h$ 134 I
76k	C_7H_{15}	16	3:1	4800	g 18 Col 86 Col 102 Col 112 Col$_h$ 129 I
76l	C_7H_{15}	16	2:1	5800	g 19 Col 83 Col 99 Col 107 Col$_h$ 120 I
				7300	g 20 Col 83 Col 109 Col$_h$ 122 I
76m	C_7H_{15}	16	3:2	3400	—
				7800	g 20 Col$_h$ 102 I

Source: Data from Percec, V. et al., *Macromolecules*, 25, 1164, 1992.

5.6 PHTHALOCYANINE-BASED POLYMERS

There are various ways that phthalocyanines can be polymerized [99,100]; however, only a few examples of liquid crystalline phthalocyanine polymers are known. The first examples are *spinal columnar* or *shish kebab* type of phthalocyanine polymers where the phthalocyanine units are linked by a bridging atom bonded to a silicon, tin, or germanium atom at the center of the phthalocyanine molecule. Significant attention has been paid to examine various chemical and physical properties of phthalocyanine-based main-chain polymers having metal-oxygen-metal (M-O-M) bond as a linker to connect two adjacent phthalocyanine rings [101–113]. The term *spinal columnar liquid crystals* was proposed for such polymers [111].

Spinal columnar liquid crystals are cofacially stacked polymeric metallomesogens. The cofacial stacking may take place via (1) covalent bonds, (2) covalent–coordinate bonds, and (3)

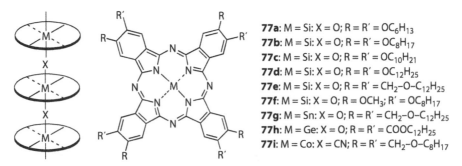

77a: M = Si; X = O; R = R′ = OC$_6$H$_{13}$
77b: M = Si; X = O; R = R′ = OC$_8$H$_{17}$
77c: M = Si; X = O; R = R′ = OC$_{10}$H$_{21}$
77d: M = Si; X = O; R = R′ = OC$_{12}$H$_{25}$
77e: M = Si; X = O; R = R′ = CH$_2$–O–C$_{12}$H$_{25}$
77f: M = Si; X = O; R = OCH$_3$; R′ = OC$_8$H$_{17}$
77g: M = Sn; X = O; R = R′ = CH$_2$–O–C$_{12}$H$_{25}$
77h: M = Ge; X = O; R = R′ = COOC$_{12}$H$_{25}$
77i: M = Co; X = CN; R = R′ = CH$_2$–O–C$_8$H$_{17}$

FIGURE 5.5 Schematic representation of cofacially stacked phthalocyanine-based polymeric metallomesogens.

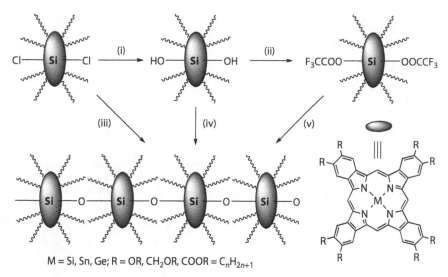

M = Si, Sn, Ge; R = OR, CH$_2$OR, COOR = C$_n$H$_{2n+1}$

SCHEME 5.13 Synthesis of phthalocyanine-based polymer having metal–oxygen–metal bond as a linker: (i) ion exchange; (ii) CF$_3$COOCOCF$_3$, pyridine; (iii) AgSO$_3$CF$_3$, TlSO$_3$CF$_3$, [Cu(CH$_3$CN)$_4$]SO$_3$CF$_3$; (iv) (a) 180°C, 7 h or (b) catalyst (FeCl$_3$, ZnCl$_2$, AlCl$_3$, CdCl$_2$, CaCl$_2$, CsCl); (v) 200°C.

coordinate–coordinate bonds. However, in the field of discotic liquid crystals, mainly covalent **77a–g** and covalent–coordinate **77h** bonds derived polymers have been realized (Figure 5.5). Particular attention has been paid to polysiloxane derivatives. The close cofacial packing of the phthalocyanine along the polysiloxane chain creates strong interactions between monomeric units, which results in a Si–O–Si bond angle of nearly 180°.

Synthetic routes adopted to prepare these polymers are shown in Scheme 5.13. The simplest method for the synthesis of oxo-bridged phthalocyanine polymers is the thermal bulk condensation of dihydroxy precursors. The dihydroxy-functionalized discotic metallomesogens are heated in the mesophase at about 200°C for 1–7 h. Generally, a mixture of oligomers and polymers is formed. The degree of polymerization is largely dependent upon peripheral substitution, temperature, and time of polymerization. A number of catalysts, such as FeCl$_3$, ZnCl$_2$, AlCl$_3$, CdCl$_2$, CaCl$_2$, etc., have been used to facilitate the reaction. The reaction can also be accomplished without using a catalyst but replacing the hydroxyl groups by better leaving groups such as trifluoroacetate. The polymerization can also be carried out at lower temperature (about 100°C) by taking a dichloro monomer and using halogenophilic condensation agents (Scheme 5.13).

The thermal behavior of these polymers is not very straightforward as most of these do not display any clear isotropic transition. Short side-chain derivatives do not exhibit any phase transition up to the decomposition temperature. Medium side-chain polymers show melting transitions below 100°C but no further phase transition is observed up to 300°C. The long side-chain derivatives under polarizing microscope exhibit a non-birefringent isotropic state at higher temperature without decomposing, but this transition is not observed in DSC. The formation of columnar mesophases in these polymers was established by x-ray diffraction studies.

An interesting example of this class of materials is the phthalocyaninato polysiloxanes substituted with crown ether moieties **78** (Figure 5.6). The synthesis of these polymers was realized by Nolte and coworkers [111–113]. Crown ether phthalocyanine can be prepared from 4,5-dicyanobenzo-crown ether under standard phthalocyanine synthesis reaction conditions. These phthalocyanines can be easily converted into dihydroxysilicon derivatives, which, on heating in the presence of a catalyst, yields polysiloxanes with crown ether substituents. These discotic stacks are expected to transport electrons, via central phthalocyanine nucleus, and ions, via

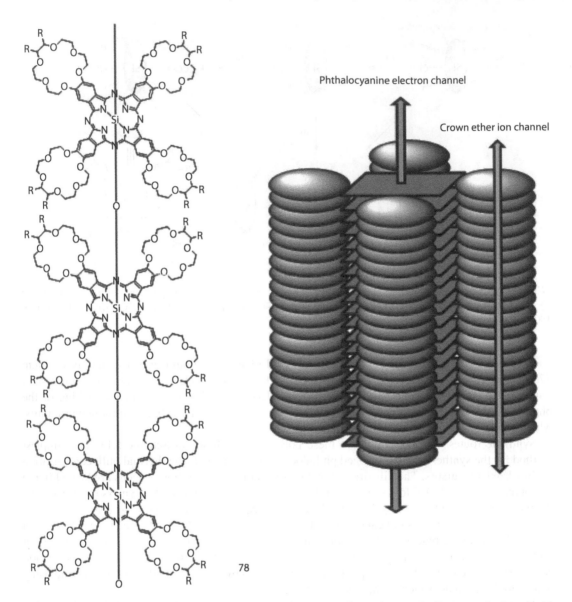

FIGURE 5.6 Schematic representation of spinal columnar phthalocyaninato polysiloxanes substituted with crown ether moieties. These discotic stacks are expected to transport electrons, via central phthalocyanine nucleus, and ions, via crown ether periphery, in the direction of the columnar axis.

crown ether periphery, in the direction of columnar axis (Figure 5.6). Connecting long hydrocarbon chains to the crown ether periphery generates liquid crystalline phthalocyanine, which displays a hexagonal columnar phase [112]. However, its silicon derivative with two axial hydroxyl groups is non-mesomorphic because of the steric hindrance of axial hydroxyl groups, which prevent the molecules to aggregate. This dihydroxysilicon tetrakis[4'5'-bis(decyloxy)benzo-18-crown-6]phthalocyanine can be polymerized to form a polysiloxane but the exact nature of the polymer is not revealed [112].

Cook et al. incorporated phthalocyanine disks as part of main-chain polymer [114,115]. These polymers were prepared by condensing a dihydroxy-functionalized phthalocyanine **79** with oxalyl chloride (Scheme 5.14). The phthalocyanine monomer **79** was first reacted with excess oxalyl

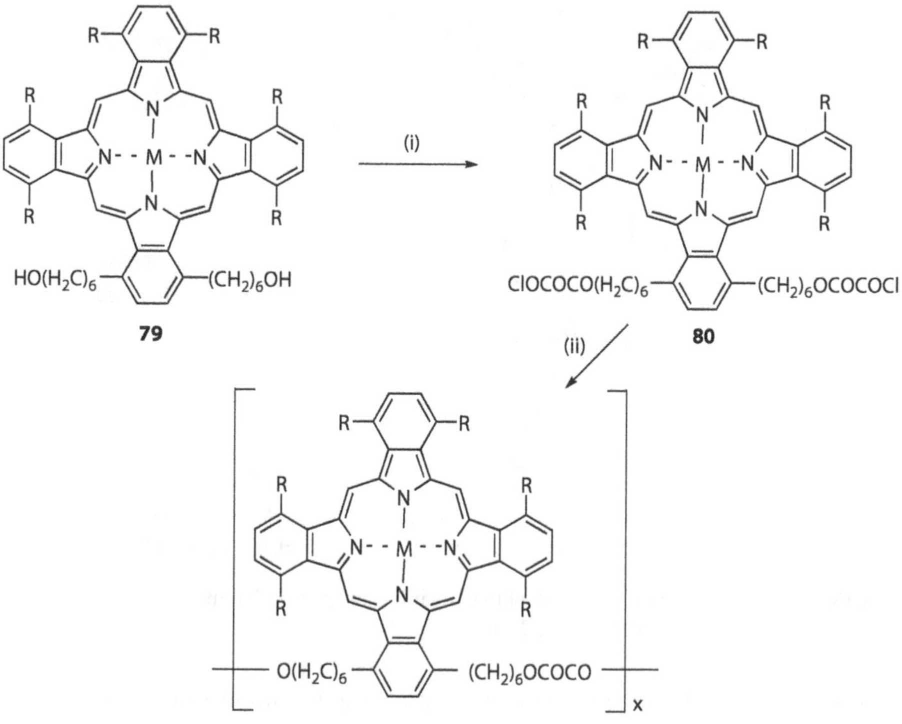

81a: $M_w = 7,180$; Col_h 158 I
81b: $M_w = 157,000$; Col_h 161 I

SCHEME 5.14 Synthesis of phthalocyanine-based main-chain polymer: (i) ClOCCOCl; (ii) **79**, 48 h, 90°C (a) precipitated by acetone to get **81a**, (b) used pyridine and 4-dimethylaminopyridine to yield **81b**.

chloride to form the bis-acid chloride **80**, which was coupled with starting diol at 90°C in dichloromethane to yield the polymer **81**. Two polymers were obtained under different reaction conditions. Polymer **81a** was precipitated by adding acetone while polymer **81b** was prepared by adding pyridine at the first stage and DMAP at the second stage. Both polymers are liquid crystalline at ambient temperature and exhibit hexagonal columnar mesophase over a broad mesophase range. Polymer **81a** goes to the isotropic phase at 158°C on heating while this transition occurs at 161°C for **81b**.

Phthalocyanines substituted with eight terminally acryloyl or methacryloyl functionalized alkoxy chains were prepared by van der Pol et al. [116]. While photopolymerization of the acrylate failed to produce any polymer, the thermal polymerization of the acrylate as well as methacrylate in the columnar mesophase leads to liquid crystalline polymers **83** (Scheme 5.15). These polymers do not show any phase transition in DSC probably due to the formation of highly cross-linked network. The columnar nature of the mesophase formed by these polymers was revealed by x-ray diffraction studies.

5.7 HEXABENZOCORONENE-BASED POLYMERS

HBC-based side-chain ionic polymers were obtained by complexing carboxylic acid functionalized HBC derivatives with various polymeric backbones such as polyethyleneimine [117], polysiloxane [118], and poly(ethylene oxide)-block-poly(L-lysine) [119]. Mullen and coworkers prepared three types of HBC-based side-chain ionic polymers **85**, **86**, and **87** by complexing

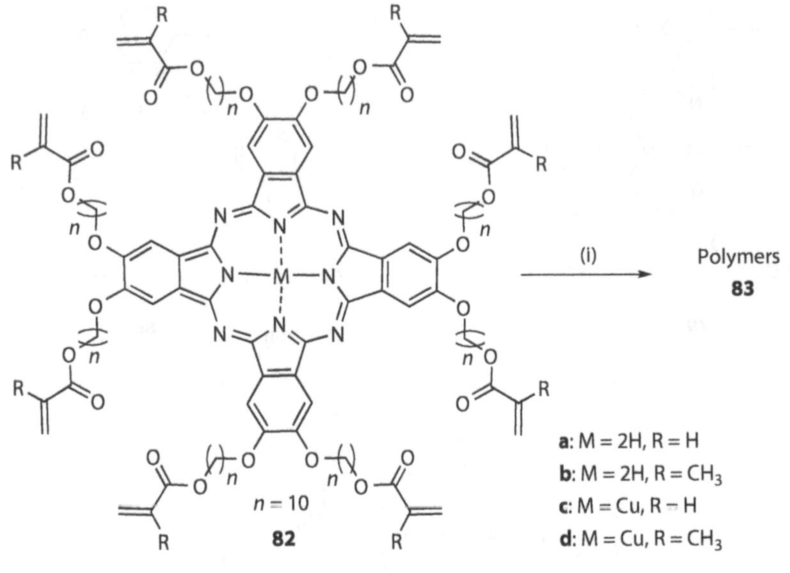

a: M = 2H, R = H
b: M = 2H, R = CH₃
c: M = Cu, R = H
d: M = Cu, R = CH₃

SCHEME 5.15 Thermal polymerization of phthalocyanine derivative having eight terminal acryloyl or methacryloyl groups: (i) AIBN, CHCl₃, 93°C, 2.5 h.

the carboxylic acid functionalized HBC **84** with various amino-functionalized molecules (Scheme 5.16).

In the first polymer **85**, carboxylic acid functionalized HBC **84** was complexed with hydrophobically modified polyethyleneimine. The mesomorphic behavior of **85** was investigated by x-ray diffraction. At room temperature, the morphology of mesophase was rectangular. At higher temperature (110°C), the phase changes to ordered hexagonal columnar. Intracolumnar long range order in the polyethyleneimine complex of HBC **85** was reported to be higher compared to the noncomplexed carboxylic acid functionalized HBC derivative **84**. In contrast, the intercolumnar order was more in case of **84** than in the complex **85** [117].

The second polymer **86** is the outcome of complexation of carboxylate terminated hexabenzocoronene **84** with an amino-functionalized polysiloxane backbone [118]. This polymer complex displays two highly ordered columnar mesophases. The columnar-to-columnar (Col₁-to-Col₂) transition was noticed at 90°C on heating. On cooling, the Col₂-to-Col₁ transition occurs at 57°C. The low-temperature Col₁ phase possesses short-range intracolumnar order with tilted arrangement of HBC disks with respect to the column axis. Contrary to this, HBC disks were oriented perpendicular to the column axis in the high-temperature Col₂ phase with a high range of intracolumnar order. The intracolumnar ordering in the macromolecular complex **86** was again stronger than the noncomplex HBC derivative **84**.

In the complex **87**, the amphiphilic HBC derivative **84** was connected ionically with poly(ethylene oxide)-block-poly(L-lysine). This is an example of the amalgamation of columnar phase–forming feature of HBC core with the potential of poly-L-lysine to form well-defined secondary structure [119]. The incorporation of polyethylene oxide into poly-L-lysine stabilizes the α-helical conformation of the later. The Col₍h1₎-to-Col₍h2₎ transition was observed at 54°C while heating. The reversible transition Col₍h2₎-to-Col₍h1₎ took place at −30°C. Both the hexagonal columnar structures are defined by cylinders containing α-helical structure of poly(L-lysine) blocks, which are hexagonally surrounded by hexabenzocoronene columns. The unexpected higher intra-columnar order in the higher temperature phase as compared to the lower temperature mesophase in the above three cases can be attributed to packing constraints caused by polymer chains [117–119].

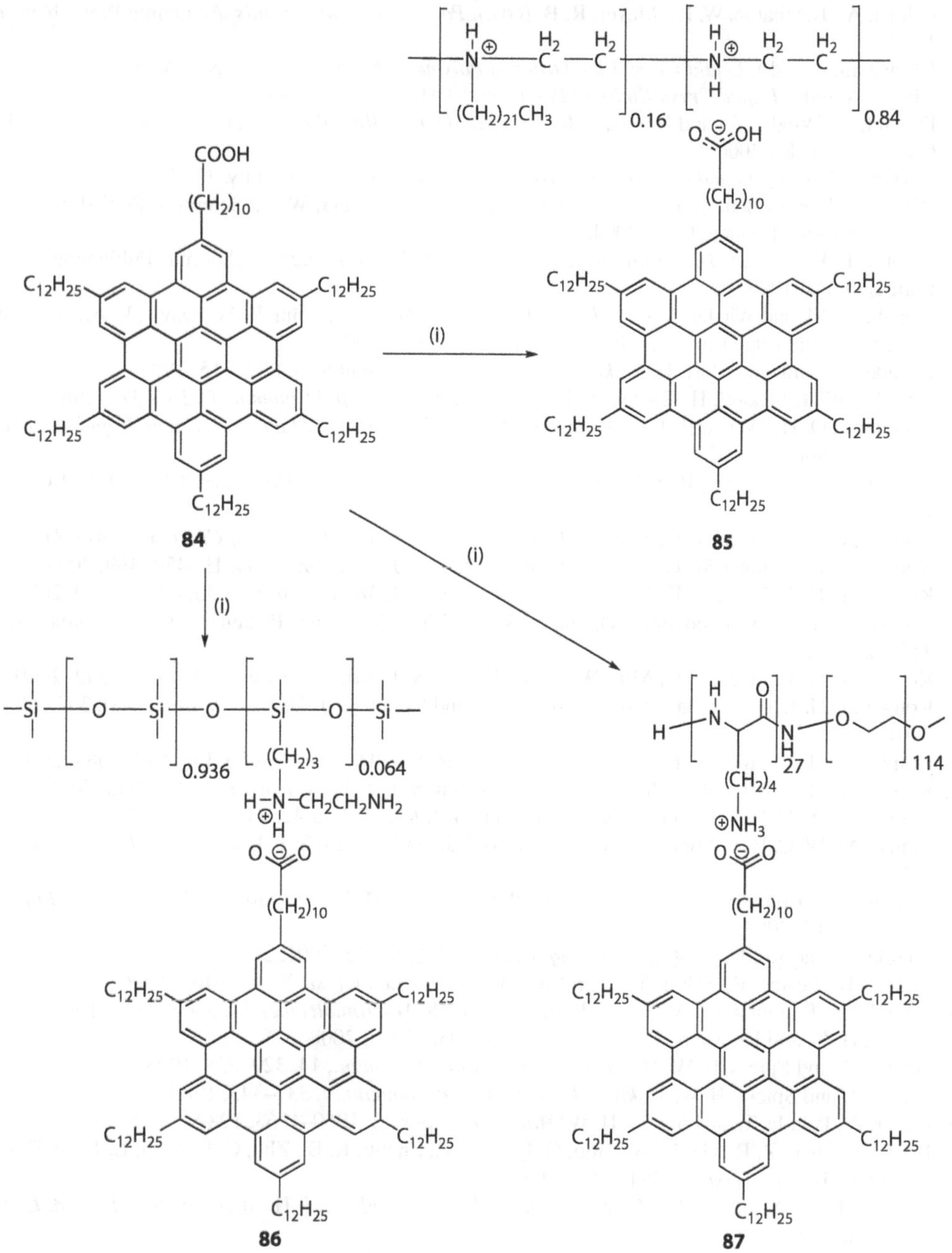

SCHEME 5.16 Synthesis of hexabenzocoronene-based side-chain ionic polymers: (i) complexation in between amino and carboxylic acid groups.

REFERENCES

1. Demus, D., Goodby, J., Gray, G.W., Spiess, H. W., and Vill, V. (eds.). *Handbook of Liquid Crystals*, Vol. 3, Wiley-VCH, Weinheim, Germany, 1998.
2. Shibaev, V. P. and Lam, L. (eds.). *Liquid Crystalline and Mesomorphic Polymers*, Springer Verlag, New York, 1994.
3. McArdle, C. B. (ed.). *Side-Chain Liquid Crystal Polymers*, Blackie and Sons, Glasgow, U.K., 1989.

4. Ciferri, A., Krigbaum, W. R., Meyer, R. B. (eds.). *Polymer Liquid Crystals*, Academic Press, New York, 1982.
5. Blumstein, A. (ed.). *Liquid Crystalline Order in Polymers*, Academic Press, New York, 1978.
6. Ciferri, A. (ed.). *Liquid Crystallinity in Polymers*, VCH, New York, 1991.
7. Donald, A., Windle, A., and Hanna, S. (eds.). *Liquid Crystalline Polymers*, Cambridge University Press, Cambridge, U.K., 2006.
8. Carfagna, C. (ed.). *Liquid Crystalline Polymers*, Elsevier Science Ltd., Italy, 1994.
9. Wang, X. J. and Zhou, Q. F. (eds.). *Liquid Crystalline Polymers*, World Scientific Publishing Co. Pte. Ltd., Singapore, London, U.K., 2004.
10. Mantia, F. P. L. (ed.). *Thermotropic Liquid Crystal Polymer Blends*, Technomic Publishing Co. Inc., Lancaster, PA, 1993.
11. Donald, A. M. and Windle, A. H. In *Liquid Crystalline Polymers*, Cahn, R. W., Davis, E. A., and Ward, I. M. (eds.), Cambridge University Press, Cambridge, U.K., 1992.
12. Kreuder, W. and Ringsdorf, H. *Makromol. Chem., Rapid Commun.* 4, 807–815, 1983.
13. Kreuder, W., Ringsdorf, H., Tschirner, P. *Makromol. Chem. Rapid Commun.* 6, 367–373, 1985.
14. Schonherr, O. H., Wendorff, J. H., Ringsdorf, H., and Tschirner, P. *Makromol. Chem. Rapid Commun.* 7, 791–796, 1986.
15. Ringsdorf, H., Tschirner, P., Schonherr, O. H., and Wendorff, J. H. *Makromol. Chem.* 188, 1431–1445, 1987.
16. Kouwer, P. H. J., Jager, W. F., Mijs, W. J., and Picken, S. J. *Mol. Cryst. Liq. Cryst.* 411, 305–312, 2004.
17. Kouwer, P. H. J., Jager, W. F., Mijs, W. J., and Picken, S. J. *J. Mater. Chem.* 13, 458–469, 2003.
18. Kouwer, P. H. J., Jager, W. F., Mijs, W. J., and Picken, S. J. *Macromolecules* 35, 4322–4329, 2002.
19. Kouwer, P. H. J., van den Berg, O., Jager, W. F., Mijs, W. J., and Picken, S. J. *Macromolecules* 35, 2576–2582, 2002.
20. Kouwer, P. H. J., Jager, W. F., Mijs, W. J., and Picken, S. J. *Macromolecules* 33, 4336–4342, 2000.
21. Kouwer, P. H. J., Gast, J., Jager, W. F., Mijs, W. J., and Picken, S. J. *Mol. Cryst. Liq. Cryst.* 364, 225–234, 2001.
22. Kouwer, P. H. J., Jager, W. F., Mijs, W. J., and Picken, S. J. *Macromolecules* 34, 7582–7584, 2001.
23. Kouwer, P. H. J., Mijs, W. J., Jager, W. F., and Picken, S. J. *J. Am. Chem. Soc.* 123, 4645–4646, 2001.
24. Franse, M. W. C. P., te Nijenhuis, K., and Picken, S. J. *Rheol. Acta* 42, 443–453, 2003.
25. Franse, M. W. C. P., te Nijenhuis, K., Groenewold, J., and Picken, S. J. *Macromolecules* 37, 7839–7845, 2004.
26. Picken, S. J., Kouwer, P. H. J., Jager, W. F., Wubbenhorst, M. R., and Mijs, W. J. *Mol. Cryst. Liq. Cryst.* 411, 503–513, 2004.
27. Suzuki, D. and Koide, N. *Mol. Cryst. Liq. Cryst.* 364, 635–645, 2001.
28. Janietz, D., Festag, R., Schmidt, C., and Wendorff, J. H. *Liq. Cryst.* 20, 459–467, 1996.
29. Raja, K. S., Raghunathan, V. A., and Ramakrishnan, S. *Macromolecules* 31, 3807–3814, 1998.
30. Bisoyi, H. K. and Kumar, S. *J. Mater. Chem.* 18, 3032–3039, 2008.
31. Werth, M. and Spiess, H. W. *Makromol. Chem. Rapid Commun.*, 14, 329–338, 1993.
32. Huser, B. and Spiess, H. W. *Makromol. Chem. Rapid Commun.* 9, 337–343, 1988.
33. Huser, B., Pakula, T., and Spiess, H. W. *Macromolecules* 22, 1960–1963, 1989.
34. Ba, C. Y., Shen, Z. R., Gu, H. W., Guo, G. Q., Xie, P., Zhang, R. B., Zhu, C. F., Wan, L. J., Li, F. Y., and Huang, C. H. *Liq. Cryst.* 30, 391–397, 2003.
35. Ringsdorf, H., Wustefeld, R., Zerta, E., Ebert, M., and Wendorff, J. H. *Angew. Chem. Int. Ed. Engl.* 28, 914–918, 1989.
36. Boden, N., Bushby, R. J., and Lu, Z. B. *Liq. Cryst.* 25, 47–58, 1998.
37. Stewart, D., McHattie, G. S., and Imrie, C. T. *J. Mater. Chem.* 8, 47–51, 1998.
38. Bleyl, I., Erdelen, C., Etzbach, K. H., Paulus, W., Schmidt, H. W., Siemensmeyer, K., and Haarer, D. *Mol. Cryst. Liq. Cryst.* 299, 149–155, 1997.
39. Schonherr, H., Manickam, M., and Kumar, S. *Langmuir* 18, 7082–7085, 2002.
40. Talroze, R. V., Otmakhova, O. A., Koval, M. A., Kuptsov, S. A., Plate, N. A., and Finkelmann, H. *Macromol. Chem. Phys.* 201, 877–881, 2000.
41. Catry, C., van der Auweraer, M., de Schryver, F. C., Bengs, H., Haussling, L., Karthaus, O., and Ringsdorf, H. *Makromol. Chem.* 194, 2985–2999, 1993.
42. Karthaus, O., Ringsdorf, H., Ebert, M., and Wendorff, J. H. *Makromol. Chem.* 193, 507–513, 1992.
43. Imrie, C. T., Inkster, R. T., Lu, Z., and Ingram, M. D. *Mol. Cryst. Liq. Cryst.* 408, 33–43, 2004.
44. Janietz, D., Festag, R., Schmidt, C., Wendorff, J. H., and Tsukruk, V. V. *Thin Solid Films* 284–285, 289–292, 1996.

45. Tsukruk, V. V. and Janietz, D. *Langmuir* 12, 2825–2829, 1996.

46. Weck, M., Mohr, B., Maughon, B. R., and Grubbs, R. H. *Macromolecules* 30, 6430–6437, 1997.

47. Boden, N., Bushby, R. J., Lu, Z. B., and Eichhorn, H. *Mol. Cryst. Liq. Cryst.* 332, 281–291, 1999.

48. Otmakhova, O. A., Kuptsov, S. A., Talroze, R. V., and Patten, T. E. *Macromolecules* 36, 3432–3435, 2003.

49. Xing, C., Lam, J. W. Y., Zhao, K., and Tang, B. Z. *J. Polym. Sci. A Polym. Chem.* 46, 2960–2974, 2008.

50. Pal, S. K. and Kumar, S. *Liq. Cryst.* 35, 381–384, 2008.

51. Kranig, W., Huser, B., Spiess, H. W., Kreuder, W., Ringsdorf, H., and Zimmermann, H. *Adv. Mater.* 2, 36–40, 1990.

52. Wenz, G. *Makromol. Chem. Rapid Commun.* 6, 577–584, 1985.

53. Engel, M., Hisgen, B., Keller, R., Kreuder, W., Reck, B., Ringsdorf, H., Schmidt, H. W., and Tschirner, P. *Pure Appl. Chem.* 57, 1009–1014, 1985.

54. Schonherr, O. H., Wendorff, J. H., Kreuder, W., and Ringsdorf, H. *Makromol. Chem. Rapid Commun.* 7, 97–101, 1986.

55. Bauer, S., Plesnivy, T., Ringsdorf, H., and Schuhmacher, P. *Makromol. Chem. Macromol. Symp.* 64, 19–32, 1992.

56. Voigt-Martin, I. G., Schumacher, M., Honig, M., Simon, P., and Garbella, R. W. *Mol. Cryst. Liq. Cryst.* 254, 299–320, 1994.

57. Ebert, M., Frick, G., Baehr, CH., Wendorff, J. H., Wüstefeld, R., and Ringsdorf, H. *Liq. Cryst.* 11, 293–309, 1992.

58. Voigt-Martin, I. G., Garbella, R. W., and Schumacher, M. *Macromolecules* 25, 961–971, 1992.

59. Vandevyver, M., Albouy, P. A., Mingotaud, C., Perez, J., Barraud, A., Karthaus, O., and Ringsdorf, H. *Langmuir* 9, 1561–1567, 1993.

60. Green, M. M., Ringsdorf, H., Wagner, J., and Wustefeld, R. *Angew Chem. Int. Ed. Engl.* 29, 1478–1481, 1990.

61. Boden, N., Bushby, R. J., and Cammidge, A. N. *J. Am. Chem. Soc.* 117, 924–927, 1995.

62. Wan, W., Monobe, H., Tanaka, Y., and Shimizu, Y. *Liq. Cryst.* 30, 571–578, 2003.

63. Boden, N., Bushby, R. J., Eichhorn, H., Lu, Z. B., Abeysekera, R., and Robardes, A. W. *Mol. Cryst. Liq. Cryst.* 332, 293–302, 1999.

64. Boden, N., Bushby, R. J., Cooke, G., Lozman, O. R., and Lu, Z. *J. Am. Chem. Soc.* 123, 7915–7916, 2001.

65. Abeysekera, R., Bushby, R. J., Caillet, C., Hamley, I. W., Lozman, O. R., Lu, Z., and Robards, A. W. *Macromolecules* 36, 1526–1533, 2003.

66. Warner, M. and Terentjev, E. M. (eds.). *Liquid Crystal Elastomers*, Clarendon Press, Oxford, NY, 2003.

67. Zentel, R. *Angew. Chem. Int. Ed. Engl.* 28, 1407–1415, 1989.

68. Finkelmann, H. and Rehage, G. *Adv. Polym. Sci.* 60–61, 99–172, 1984.

69. Bengs, H., Finkelmann, H., Küpfer, J., Ringsdorf, H., and Schuhmacher, P. *Makromol. Chem. Rapid Commun.* 14, 445–450, 1993.

70. Disch, S., Finkelmann, H., Ringsdorf, H., and Schuhmacher, P. *Macromolecules* 28, 2424–2428, 1995.

71. Zhang, C., He, Z., Mao, H., Wang, J., Wang, D., Wang, Y., Li, Z., and Pu, J. *J. Lumin.* 122–123, 931–935, 2007.

72. Favre-Nicolin, C. D. and Lub, J. *Macromolecules* 29, 6143–6149, 1996.

73. Favre-Nicolin, C. D., Lub, J., and van der Sluis, P. *Mol. Cryst. Liq. Cryst.* 299, 157–162, 1997.

74. Braun, C. D. and Lub, J. *Liq. Cryst.* 26, 1501–1509, 1999.

75. Wu, L. H., Lee, W. C., Hsu, C. S., and Wu, S. T. *Liq. Cryst.* 28, 317–320, 2001.

76. Inoue, M., Ukon, M., Monobe, H., Sugino, T., and Shimizu, Y. *Mol. Cryst. Liq. Cryst.* 365, 439–446, 2001.

77. Sergan, T., Sonpatki, M., Kelly, J., and Chien, L. C. *Mol. Cryst. Liq. Cryst.* 359, 245–257, 2001.

78. Kumar, S., Schuhmacher, P., Henderson, P., Rego, J., and Ringsdorf, H. *Mol. Cryst. Liq. Cryst.* 288, 211–222, 1996.

79. Bacher, A., Bleyl, I., Erdelen, C. H., Haarer, D., Paulus, W., and Schmidt, H. W. *Adv. Mater.* 9, 1031–1035, 1997.

80. Hsu, T. C., Hüser, B., Pakula, T., Spiess, H. W., and Stamm, M. *Makromol. Chem.* 191, 1597–1609, 1990.

81. Kranig, W., Boeffel, C., and Spiess, H. W. *Macromolecules* 23, 4061–4067, 1990.

82. Kranig, W., Boeffel, C., Spiess, H. W., Karthaus, O., Ringsdorf, H., and Wustefeld, R. *Liq. Cryst.* 8, 375–388, 1990.

83. Ringsdorf, H., Wustefeld, R., and Hodge, P. *Phil. Trans. R. Soc. Lond. A* 330, 95–108, 1990.

84. Karthaus, O., Ringsdorf, H., and Urban, C. *Makromol. Chem. Macromol. Symp.* 46, 347–352, 1991.

85. Karthaus, O., Ringsdorf, H., Tsukruk, V. V., and Wendorff, H. *Langmuir* 8, 2279–2283, 1992.
86. Auweraer, M. V. D., Catry, C., Chi, L. F., Karthaus, O., Knoll, W., Ringsdorf, H., Sawodny, M., and Urban, C. *Thin Solid Films* 210–211, 39–41, 1992.
87. Tsukruk, V. V., Wendorff, J. H., Karthaus, O., and Ringsdorf, H. *Langmuir* 9, 614–618, 1993.
88. Spiess, H. W. *Ber. Bunsenges. Phys. Chem.* 97, 1294–1305, 1993.
89. Wang, T., Yan, D., Zhou, E., Karthaus, O., and Ringsdorf, H. *Polymer* 39, 4509–4513, 1998.
90. Boden, N., Bushby, R. J., and Lozman, O. R. *Mol. Cryst. Liq. Cryst.* 400, 105–113, 2003.
91. Kawata, K. *Chem. Rec.* 2, 59–80, 2002.
92. Mori, H., Itoh, Y., Nishiura, Y., Nakamura, T., and Shinagawa, Y. *Jpn. J. Appl. Phys.* 36, 143–147, 1997.
93. Bunning, J. C., Donovan, K. J., Bushby, R. J., Lozman, O. R., and Lu, Z. *Chem. Phys.* 312, 145–150, 2005.
94. Voigt-Martin, I. G., Durst, H., Brzezinski, V., Krug, H., Kreuder, W., and Ringsdorf, H. *Angew. Chem. Int. Ed.* 28, 323–325, 1989.
95. Kruk, G., Vij, J. K., Karthaus, O., and Ringsdorf, H. *Supramol. Sci.* 2, 51–58, 1995.
96. Voigt-Martin, I. G., Krug, H., and Van Dyck, D. *J. Phys. France* 51, 2347–2371, 1990.
97. Kumar, S. *Liq. Cryst.* 32, 1089–1113, 2005.
98. Percec, V., Cho, C. G., Pugh, C., and Tomazos, D. *Macromolecules* 25, 1164–1176, 1992.
99. Oriol, L. and Serrano, J. L. *Adv. Mater.* 7, 348–369, 1995.
100. Leznoff, C. C. and Lever, A. B. P. (eds.). *Phthalocyanine. Properties and Application*, VCH, New York, 1989.
101. Sirlin, C., Bosio, L., and Simon, J. *J. Chem. Soc. Chem. Commun.* 379–380, 1987.
102. Orthmann, E. and Wegner, G. *Angew. Chem. Int. Ed. Engl.* 25, 1105–1107, 1986.
103. Sirlin, C., Bosio, L., and Simon, J. *J. Chem. Soc. Chem. Commun.* 236–237, 1988.
104. Sauer, T. *Macromolecules* 26, 2057–2063, 1993.
105. Sirlin, C., Bosio, L., and Simon, J. *Mol. Cryst. Liq. Cryst.* 155, 231–238, 1988.
106. Hanack, M., Beck, A., and Lehmann, H. *Synthesis* 703–705, 1987.
107. Sauer, T. and Wegner, G. *Macromolecules* 24, 2240–2252, 1991.
108. Caseri, W., Sauer, T., and Wegner, G. *Makromol. Chem. Rapid Commun.* 9, 651–657, 1988.
109. Schouten, P. G., Warman, J. M., de Hass, M. P., van der Pol, J. F., and Zwikker, J. W. *J. Am. Chem. Soc.* 114, 9028–9034, 1992.
110. Dulog, L., Gittinger, A., Roth, S., and Wagner, T. *Makromol. Chem.* 194, 493–500, 1993.
111. Sielcken, O. E., van de Kuil, L. A., Drenth, W., Schoonman, J., and Nolte, R. J. M. *J. Am. Chem. Soc.* 112, 3086–3093, 1990.
112. van Nostrum, C. F., Picken, S. J., Schouten, A. J., and Nolte, R. J. M. *J. Am. Chem. Soc.* 117, 9957–9965, 1995.
113. van Nostrum, C. F. and Nolte, R. J. M. *Chem. Commun.* 2385–2392, 1996.
114. Cook, M. J. *Adv. Mater.* 7, 877–880, 1995.
115. Bryant, G. C., Cook, M. J., Ryan, T. G., and Thorne, A. J. *J. Chem. Soc. Chem. Commun.* 467–468, 1995.
116. van der pol, J. F., Neeleman, E., van Miltenburg, J. C., Zwikker, J. W., Nolte, R. J. M., and Drenth, W. *Macromolecules* 23, 155–162, 1990.
117. Thunemann, A. F., Ruppelt, D., Ito, S., and Mullen, K. *J. Mater. Chem.* 9, 1055–1057, 1999.
118. Thunemann, A. F., Ruppelt, D., Burger, C., and Mullen, K. *J. Mater. Chem.* 10, 1325–1329, 2000.
119. Thunemann, A. F., Kubowicz, S., Burger, C., Watson, M. D., Tchebotareva, N., and Mullen, K. *J. Am. Chem. Soc.* 125, 352–356, 2003.

6 Perspectives

The field of discotic liquid crystals (DLCs) is at a relatively nascent stage. Nevertheless, many significant advances have been achieved in a short span of time. Products derived from DLCs have reached the market. A typical example is the optical compensation films for wide viewing liquid crystal displays (LCDs). The field, in recent times, is experiencing an exponential growth. More than 3000 DLCs, from monomers to polymers, have been synthesized and extensively studied for various physical properties. Several potential applications of DLCs, particularly as one-dimensional (1D) organic semiconductors, are being explored with intense expectations. Some of the applications of DLCs are discussed in the following sections.

6.1 DISCOTICS FOR WIDE VIEWING DISPLAYS

6.1.1 OPTICAL COMPENSATION FILMS FOR LIQUID CRYSTAL DISPLAYS

The nematic phase of DLCs has made very significant progress over the last three decades since their discovery. It has made its way from a mere scientific curiosity to application in commodities and has emerged as a potential candidate for many technological applications. Although liquid crystals (LCs) have diverse applications, such as temperature sensing, solvents in chemical reactions, in chromatography, in spectroscopy, in holography, etc., they are primarily recognized by their ubiquitous presence in electro-optical display devices such as watches, calculators, telephones, personal organizers, laptops, flat panel televisions, etc. [1]. The twisted nematic (TN) and supertwisted nematic (STN) display devices have been dominating commercial displays since their invention [2]. The LC layer in these devices is exclusively composed of a mixture of calamitic LCs [3,4]. Simple TN displays were directly addressed followed by multiplex-addressing for complex information display. However, the LC molecules constituting the display cannot respond fast enough to such addressing, which results in poor contrast. STN displays followed in the mid-1980s to address this problem but suffered from the same limitations along with the inherent generation of interference colors. However, these problems were solved by the development of active matrix (AM) addressing of the twisted nematic device through the use of thin film transistors (TFTs). Such technology allows the advantages of multiplex-addressing with no loss of contrast but is complex and expensive. Hence, cheaper and easily constructed multiplex-addressed TN and STN devices were dominating in most of the applications in the early 1990s. TN active matrix TFT technology developed steadily, eventually became much cheaper and much more reliable and consistent, and is invaluable in satisfying the needs of small portable devices, such as personal organizers, cameras, mobile telephones, laptops, desktop monitors, and some small televisions. However, TN displays suffered from two major problems: the narrow-viewing-angle characteristics and the slow optical response speed. These are severe limitations for large-area television and fast-moving graphics displays. To improve the viewing angle characteristics and response speed, other LCD formats like in-plane switching (IPS) [5,6], multi-domain vertical alignment (VA) LCDs [7–9], fringe-field switching [10–12], etc., were introduced into the market. But these are not very cost effective for smaller displays. When there was intense cost competition among the various LCD modes, a negative birefringence optical compensation film was introduced by Fuji photo film laboratory to widen the viewing angle characteristics and to increase the contrast ratio (CR) of TN TFT-LCDs owing to the advantages of high light transmittance, good process margin, and cost effectiveness of TN modes [13]. The optical compensation film is nothing but a film made from a hybrid alignment of discotic nematic LCs by photopolymerization.

To realize this alignment, in the compensation film, the tilt angle of DLC layers changes linearly from horizontal to vertical alignment without having a twist angle (Figure 6.1). The poor viewing angle performance of LCDs is caused by various factors such as the optical anisotropy of LCs, the off-axis light leakage from crossed polarizers and light scattering on the surface of polarizers and color filters, etc. The discotic optical compensation film widens the viewing angle by compensating the positive optical anisotropy of the calamitic LCs by the negative optical anisotropy of DLCs since for a positive uniaxial medium, a negative uniaxial medium compensates the birefringence (Figure 6.2).

FIGURE 6.1 Schematic representation of hybrid alignment of discotic nematic LC. The tilt angle of DLC layers changes linearly from horizontal to vertical alignment without having a twist angle.

In the TN-LCD, the rod-shaped LCs orient horizontally on the orientation film of the electrode substrate when the voltage is not applied and tend to orient vertically to the substrate when voltage is applied. In principle, therefore, during switching, it should have a bright OFF state and black ON state. However, the LCs in touch with the surface of the orientation film interact strongly with it and stay almost horizontal, and gradually incline and change to vertical orientation in thickness direction when voltage is applied. As a result, a hybrid alignment region exists in close vicinity of the alignment layer. When linearly polarized light transmits through this so-called hybrid region of the rod-like LCs in the on-state of TN-LCD, it gets elliptically polarized. Because of the fact that an elliptically polarized light cannot be completely extinguished by a linear polarizer regardless of the nature of the polarizer, there is a light leakage leading to poor CR. Moreover, owing to the positive birefringence of the rod-like LCs, light leakages occurs at oblique viewing directions. The narrowness of the viewing angle is considered to be an unacceptable aspect of TN-LCD performance. From the optical anisotropy point of view, rod-like and disk-like LCs are in antisymmetric relationship, that is, they have opposite optical anisotropy. So it was logically envisaged that the optical anisotropy of hybrid-oriented rod-like molecules can be compensated by the optical anisotropy of disk-like molecules similarly oriented, that is, their combination will be optically equivalent to the isotropic space (Figure 6.2). Then linearly polarized light passing through the combination will not suffer from any distortion and hence there will not be any viewing angle dependence. This compensation configuration minimizes the light leakage by the birefringence of the TN-LC layer, leading to high CR in all viewing angle and to wide-viewing angle characteristics. A typical cell structure for the compensated TN-LC cells with a polymerized discotic hybrid aligned layer is shown in Figure 6.3. Such discotic compensated TN-LC cells have succeeded in giving a wider viewing angle (Figure 6.4). The recent technological development of the optical compensation film for the optically compensated bend mode promises the development of a fast optical response speed and wide viewing angle LCD TV.

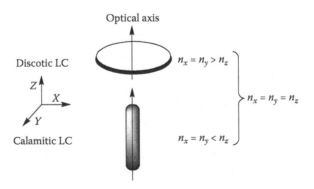

FIGURE 6.2 Principle of optical compensation using rod-like and disk-like LCs.

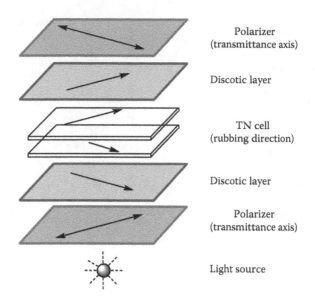

FIGURE 6.3 A typical cell structure of a TN-LCD with the discotic optical compensator.

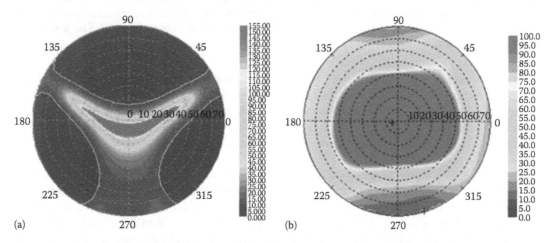

FIGURE 6.4 (See color insert following page 240.) Measured iso-CR plots for TN-LCDs without (a) and with (b) the discotic optical compensation films. Clearly, there is a remarkable widening of the viewing angle characteristics of the TN-LCD with the negative optical compensation film. (Reproduced from Mori, H., *J. Display. Tech.*, 1, 179, 2005. With permission. Copyright @ 2005 IEEE.)

Having been motivated by the excellent performance of the optical compensating discotic film, many other possibilities were also speculated on the film to improve device performance. Some of them were theoretically simulated and experimentally demonstrated [14–30]. Compensation of both normally white and normally black modes was optimized for wide viewing cone, gray scale stability, and an achromatic dark state [17]. To reduce the cost of LCDs, the use of a single discotic compensation film with twist alignment has been proposed for low-voltage, high-CR, and wide viewing cones [20,21].

6.1.2 Discotic Nematic Materials as Active Component in LCDs

Chandrasekhar and coworkers prepared a novel LCD device employing discotic nematic materials instead of calamitic nematic materials [31,32]. The LCD prepared using hexalkynylbenzene-based

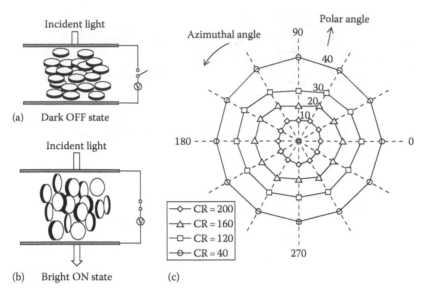

FIGURE 6.5 LCD device configuration in the (a) OFF state and (b) ON state and the measured iso-CR plot (c) using discotic nematic LCs. Note that the LCD exhibits wide and symmetrical viewing angle characteristics. (Reproduced from Chandrasekhar, S. et al., *EuroDisplay'99, The 19th International Display Research Conference Late-News Papers*, Berlin, Germany, 1999, pp. 9–11. With permission from The Society for Information Display.)

discotic nematic LC exhibits many improvements over a conventional TN display device using a calamitic nematic material. The device is simple to fabricate, has excellent viewing angle characteristics showing a wide and symmetrical viewing angle profile (Figure 6.5), and has much less difference in the pixel capacitance between the ON and OFF states, resulting in a reduced cross talk problem. Initially, a high-temperature discotic nematic material was used to fabricate the device but later it was also demonstrated with a novel room temperature discotic nematic as the active switching material [32]. However, the response time of the device was slower than the conventional TN devices due to the high viscosity of the N_D material. To overcome this problem, the discotic nematic material was doped, in small concentrations, with a long-chain alkane compound. Systematic studies on the switching response of the mixture show that both the ON and OFF response times are decreased significantly, leading to a faster switching action. It should be mentioned that wide and symmetrical viewing angle characteristics remain unaffected by the addition of the dopant material. However, both the switch ON and OFF response times are still an order of magnitude slower compared to that of conventional TN devices. Considering the fact that these response times are not very different from those for the STN displays, the achievement of symmetric and wide-viewing angle characteristics with a simple fabrication process makes this device quite interesting. To make them attractive enough to be considered for commercial applications, new discotic nematic materials with faster response times have to be developed or alternatively novel dopants should be sought. In this regard, single-walled carbon nanotubes (CNTs) seem to be promising candidates as they can be added as dopants to existing room temperature discotic nematic LCs to improve the device performance since they have been successfully demonstrated in conventional calamitic display devices to dramatically reduce the rotational viscosity and increase the response times [33].

6.1.3 Thin Film E-Polarizer from Discotic Nematic Lyo-Mesophases

Polarizers are key components in LCD devices. They are made of anisotropic media, which converts an unpolarized beam of electromagnetic wave into a single polarization beam by transmitting one

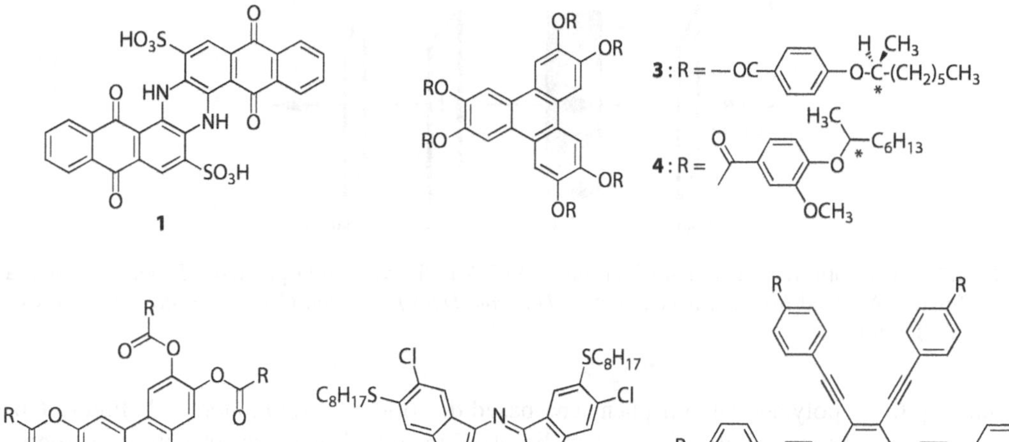

FIGURE 6.6 Chemical structure of some discoids forming discotic lyotropic nematic phases.

component of the electric field vector and absorbing or redirecting the other component. Polarizing prisms, typically made of anisotropic crystalline material, are well known to produce plane polarized light by double refraction which, however, are not suitable for LCDs. The polarizers for LCDs are generally made of dichroic materials, which transmit one polarization component and absorb the other component [34]. Yeh and Paukshto developed thin crystal film polarizer (E-polarizer) by processing from the lyotropic nematic phase of indathrone disulfonate. Indanthrone disulfonate **1** (Figure 6.6), a non-mesomorphic discotic compound, forms columnar nematic LC phase in water. It has been demonstrated that the thin crystal film E-polarizer can be used to increase the CR and viewing angle of vertically aligned LCD devices (VA-LCD) in combination with the O-polarizer and negative birefringent plate (Figure 6.7) [35].

It is worth mentioning here that lyotropic nematic LCs are important from the technological point of view since they constitute high strength fibers like Kevlar and natural silk. Thermotropic mesomorphic and non-mesomorphic discotic compounds can also self-assemble in solutions to exhibit lyotropic nematic mesomorphism, that is, anisotropic solutions can be obtained by dissolving discotic mesogens in suitable organic solvents. However, discotic nematic lyo-mesophase-forming compounds are limited in number as compared to the thermotropic discotic nematic

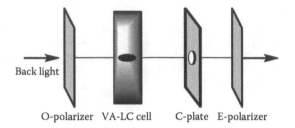

FIGURE 6.7 Schematic representation of an improved VA-LCD with an O-polarizer, E-polarizer, and a negative C-plate. (Adapted from Ivashchenko, A.V., *Dichroic Dyes for Liquid Crystal Displays*, CRC Press, Boca Raton, FL, 1994.)

compounds [36]. A polymerizable triphenylene-based discotic nematic monomer **2** (Figure 6.6) shows room temperature nematic phase when dissolved in suitable organic solvents at an appropriate concentration (xylene, alkyl benzene and 1,3-dichlorobenzene, conc. >60%) [37]. In the case of discotic cholesteric monomer **3**, the cholesteric phase is observed at room temperature in xylene at a concentration of 55 wt%. The lyotropic cholesteric phase exhibits selective reflection like the thermotropic cholesteric phase and is also temperature dependent. The mixture of both monomers prepared in one solvent also exhibits cholesteric lyo-mesophase. The lyotropic phase of the mixture can be aligned and photopolymerized in the mesophase to obtain thin films with patterned colors. The selective reflection of the thin films so obtained is temperature independent owing to the formation of a cross-linked network. Interestingly, another chiral triphenylene-based compound **4** has been found to form a cholesteric columnar nematic mesophase in dodecane solution at about 25% concentration [38]. The helical pitch of the phase depends on temperature and concentration, which have been studied by the selective reflection technique. A few octasubstituted phthalocyanines **5** have been observed to exhibit columnar nematic mesophase in hexadecane solution [39]. A hexakis(phenylethynyl)benzene derivative **6** with chiral alanine pendate groups forms lyotropic helical nematic columnar phase because of the intermolecular hydrogen bonding in hexane but not in H-bond-forming solvents [40]. Certain halogen and thiocyanato-bridged tetranuclear palladium complexes **7** that exhibit thermotropic columnar phases also forms lyotropic nematic phases in chloroform or pentadecane, rendering these metallomesogens amphotropic [41]. Other discoid amphiphiles that exhibit lyotropic nematic phase behavior include ethyleneoxy side-chain-substituted triphenylenes (**8** and **9**) [42] and chromonic LCs of drugs, dyes, and nucleic acids [43]. Several ethyleneoxy side-chain-substituted triphenylenes have been designed and synthesized, but only **8** and **9** are found to exhibit columnar nematic phase in aqueous solution over a wide concentration range by stacking of amphiphilic triphenylenes [41]. However, by both increasing and decreasing the ethyleneoxy groups in the side chains, no mesophase formation was observed in these amphiphilic triphenylenes. If the hydrophilic ethyleneoxy chains are too short, the compounds are not water soluble. If the chains are made too long, the compounds tend to dissolve in water and no mesophase is observed. So it is essential to maintain a balance between the hydrophobic and hydrophilic properties of such molecules to exhibit lyo-mesomorphism.

6.2 DISCOTICS FOR HIGH-QUALITY CARBON PRODUCTS

6.2.1 CARBONACEOUS MESOPHASE

In the mid-1960s, it was observed that there is a mesophase transformation at high temperatures during the carbonization of certain graphitizable organic materials such as petroleum and coal tar pitches [44–46]. This mesophase with characteristic nematic texture is called carbonaceous mesophase. The carbonaceous mesophase, which usually forms at temperatures of 400°C–500°C, is a key intermediate in the process of coke and graphite manufacture via carbonization of organic precursors [47,48]. The structural features of this chemically unstable lamellar LC are of great industrial interest because they critically influence the structure and quality of the final carbon products, for example, carbon fibers [49]. The mesophase has been modeled as a flat, polynuclear aromatic structure of the order of 25 Å in diameter with a ring of hydrogen atoms and methyl groups at the edge [44–50]. Carbonaceous mesophase is a complex multicomponent system composed of large disk-like polyaromatic molecules having a wide range of molecular weights rather than well-defined organic molecules. It has a transient existence, its lifetime being limited by its hardening into semicoke. A number of theoretical studies have been done on discotic carbonaceous phase [51–71] due to their potential industrial applications. However, carbonaceous mesophase is not well understood as many other LC systems. Its complex composition and chemical instability have discouraged the systematic investigation of its key fundamental properties. Although carbonaceous mesophase was discovered in 1965, the synthesis of the first discotic pure compound exhibiting LC behavior did not occur until 1977, with the first pure nematic discotics following in 1979. The optical texture of carbonaceous mesophase identifies it as a discotic nematic phase, but an analogous phase has rarely been demonstrated with any pure compound truly representative of the constituents of pitch. In practice, the carbonaceous mesophase typically forms upon heating, in contrast to most LC systems that form ordered configurations upon cooling the isotropic liquid. Carbonaceous mesophases are opaque, a practical difference that precludes many optical device applications involving transmitted light. However, carbonaceous mesophases are used in the industrial manufacturing of high-performance carbon fibers, carbon foams, carbon fiber–carbon mesophase composites, and carbon nanotube–carbon mesophase nanocomposites [72,73]. This relatively new carbon fiber is more competitive than the conventional one made from the acrylic precursors in several application areas. Their extremely high thermal conductivity combined with their relative low density make the mesophase-pitch-based fibers attractive. There are a variety of potential applications for these new fibers in chemical sensing, adsorption, thermal protection material, and electronic devices, such as battery electrodes, because they should have a high density of surface-active sites, easy access to interlayer spacers, etc.

6.2.2 Carbon Nanostructures from Discotics

The facile formation of ordered columnar superstructures from the discotic molecules in the bulk state and their high stability in the mesophase qualify them as precursors toward novel carbon nanostructures. Therefore, pyrolysis under controlled conditions may maintain the order existing in the mesophase during the formation of carbonaceous materials. The pyrolysis of well-defined discotic molecules in the bulk state produces novel carbon nano- and microstructures. CNTs without metal catalysts and with the desired graphene layer orientations have been prepared by the carbonization of the discotic columnar phase. The temperatures are much lower than the normally used graphitization temperature (2000°C–3000°C). Recently, a template method has been used to fabricate uniform CNTs by the pyrolysis of graphitic molecule hexabenzocoronene (HBC) in porous alumina membranes. Upon carbonization under a controlled heating process, the preorganized ordered columnar superstructures can be converted into nanotubes. Mullen et al. have produced nanotubes from thermotropic DLCs [74–76]. A unique approach to self-assembled graphitic nanotubes from an amphiphilic HBC is documented [77,78]. The nanotubes consist of graphitic walls formed from numerous molecular graphene sheets stacked parallel to the longer axis of the tube. The proposed structure of the nanotube consists of helically rolled up bilayer tapes composed of π-stacked HBC units, where the inner and outer HBC layers are connected by the interdigitation of the hydrophobic alkyl chains, while the hydrophilic ethylene oxide chains are located on both sides of the tubular wall. The π-stacked HBC units provide a charge-carrier transport pathway. Suitable chemical modifications of the HBC amphiphile result in the formation of nanotubes with various interesting properties [77,78]. Another interesting thing about these nanotubes is the formation of discotic columnar mesophase upon heating.

CNTs and nanoribbons have also been obtained from the columnar nematic lyo-mesophase of indanthrone disulfonate by template-directed LC assembly and subsequent covalent capture [79]. A superphenylene-based DLC has recently been used to prepare transparent conducting electrodes for organic solar cells [80].

6.3   DISCOTIC LIQUID CRYSTALS AS MATERIALS FOR A NEW GENERATION OF ORGANIC ELECTRONICS

Organic electronics, or, the use of organic molecules as active components in electronic devices, is a field of immense scientific research because of the prospect of the creation of new industries dealing with electronic devices such as light-emitting diodes (LEDs), field effect transistors (FET), sensors, photovoltaic (PV) solar cells, etc. The interest in devices is not only because of their potential market but also to understand the basic structure–property relationships that are governed by the physics and chemistry of organic materials. New organic materials with innovative design and semiconducting behavior that deviate from conventional conjugated materials are required to develop these devices. Two of the leading contenders for application in organic (opto)electronic devices are π-conjugated polymers and π–π-stacked discotic materials. Molecules with hierarchical self-assembly into supramolecular systems, like DLCs, which bring order as well as fluidity, are currently viewed to have potential as a new generation of organic semiconductors in electronics. Electronic conduction in DLCs has recently been covered in several review articles [81–110]. The columnar order provides properties similar to the organic single crystal, while the mobility is vital for the processability and self-healing of structural defects.

A majority of DLCs form columnar mesophases probably due to intense π–π interactions of polycyclic aromatic cores. The core–core (intracolumnar) separation in a columnar mesophase is usually of the order of 3.6 Å so that there is considerable overlap of π-orbitals. As flexible long aliphatic chains surround the core, the intercolumnar distance is usually 20–40 Å, depending on the number of lateral chains and their lengths. The number of aliphatic chains around the discotic core generally varies from 3 to 8 to produce columnar mesophases. Therefore, interactions between neighboring molecules within the same column (intracolumnar) are much stronger than interactions between neighboring columns (intercolumnar). DLCs, like any other organic molecule, are insulators in pure form due to the large energy gap and low intrinsic charge concentration. However, they can be made conducting by generating charges via chemical or photochemical doping. The columnar phase provides a facile path for the movement of generated charges. As the molecules are packed closely, charges can migrate easily via hopping from one molecule to another. Due to the 1D stacking of molecules in the columnar phase, charge migration in these materials is expected to be quasi-1D. Conductivity along the columns in columnar mesophases has been reported to be several orders of magnitude greater than in the perpendicular direction. Thus, the columns may be described as molecular wires or, more appropriately, molecular cables since the conducting aromatic cores are surrounded by insulating aliphatic chains in the columnar phase as shown in Figure 6.8. The electronic properties of organic materials are correlated with the highly ordered single crystalline structure, and single crystals are the best organic material from the charge transport point of view. However, single crystals are difficult and expensive to prepare, particularly as thin film on an electrode surface, which is desirable in electronic cell configuration. The supramolecular order of discotics on surfaces, (substrates) can be controlled macroscopically by various techniques (see Chapter 1) and thus, devices like PV solar cells, LEDs (homeotropic alignment), and TFTs (homogeneous or planar alignment) can be realized on large surface area. The self-assembling/self-healing property of LCs is analogous to highly efficient biological systems. Therefore, one can visualize highly efficient electronic devices based on self-ordering supramolecularly ordered discotic materials.

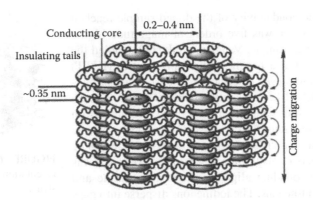

FIGURE 6.8 One-dimensional charge transport in a columnar phase.

6.3.1 ONE-DIMENSIONAL ELECTRICAL AND PHOTOCONDUCTIVITY IN DLCs

As already described, columns in DLCs would form molecular wires with conductive channels surrounded by insulating peripheral chains. Such an arrangement in the columnar LC can facilitate 1D charge migration. A number of theoretical and experimental studies have been carried out on charge migration properties of DLCs [111–265]. Though a large number of DLCs have been realized, only a few were experimentally studied for their electrical and photoconducting properties.

Columnar phases of metallomesogens exhibit electrical properties of molecular semiconductors. Simon and coworkers envisaged 1D conducting systems based on liquid crystalline metallophthalocyanine [111]. Lutetium bisphthalocyanine and lithium phthalocyanine exhibit intrinsic molecular semiconducting properties [266–268]. Belarbi et al. described the AC conductivity of discotic liquid crystalline bis(octa-octadecyloxymethylphthalocyanato)lutetium over the frequency range $10^{-3}–10^{-5}$ Hz as a function of temperature [115]. The conductivity value of 1.8×10^{-9} Ω^{-1} cm^{-1} at 10^4 Hz was observed. Several other liquid crystalline phthalocyanine derivatives displayed conductivity in the range of $10^{-10}–10^{-14}$ Ω^{-1} cm^{-1} [117]. Significantly high conductivity of the order of 10^{-5} Ω^{-1} m^{-1} for heat-treated films of dysprosium-phthalocyanine-based DLC was observed by Basova et al. [250]. The electrical properties of octaalkoxyphthalocyanines were studied by van der Pol et al. [119]. The conductivity is of the order of 5×10^{-8} Sm^{-1} at 175°C and increases with increasing temperature. Upon doping the sample with iodine, the conductivity increases by more than four orders of magnitude.

Triphenylene derivatives as model systems have been extensively studied for charge migration. As triphenylene derivatives are insulators in their pure form, charges were created by chemical doping or through photogeneration to investigate the charge transport properties along the columns in these DLCs. Among several dopants, bromine, iodine, $AlCl_3$, $NOBF_4$, TNF, ferrocenium ions, and gold nanoparticles are the most studied systems in the columnar liquid crystalline matrix. Chiang et al. showed that hexakis(pentyloxy)triphenylene doped with bromine has semiconducting properties in aligned single crystals [112]. Freely suspended fibers of hexaalkoxytriphenylenes (HATs) were drawn using the strand technique [269,270], consisting of a mechanically operated pin and a reservoir cup. Fibers were exposed with bromine vapors. The conductivity of these fibers was found to increase by four orders of magnitude over the undoped sample to a semiconducting value of $10^{-4}–10^{-5}$ Ω^{-1} cm^{-1}. The conductivity was two to three orders of magnitude higher than that realized for powder samples of HATs doped with bromine.

It is well known that the doping of various polycyclic aromatic hydrocarbons with iodine increases the electrical conductivity by several orders of magnitude [271]. Iodine acts as an acceptor to form charge-transfer complexes and arranges itself into $(I_3)^-$ stacks [272]. Drenth and coworkers doped HATs with approximately 70 mol% of iodine by keeping the discotic-filled cell in an iodine atmosphere [114]. The conductivity of the doped sample reached 10^{-3}–10^{-4} Ω^{-1} cm^{-1}, which was five orders of magnitude higher than that of the undoped sample. Vaughan et al. [123] studied the DC conductivities of doped hexahexylthiotriphenylene (HHTT) with iodine to saturation level. Thermogravimetric analysis indicated an approximately 1:1.55 mol ratio of HHTT:I$_2$ in the composite. The phase transition temperatures decrease significantly on doping. The composite retains the crystal, columnar, and isotropic phases but the helical phase is lost. The conductivity increases by about four to five orders of magnitude as a result of doping. It was proposed that the surrounding aliphatic chains move aside and give room to adjust iodine ions. The iodine ions disperse into gaps between columns of HHTT molecules (Figure 6.9).

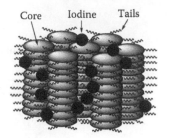

FIGURE 6.9 Schematic representation of the proposed structure of the iodine-doped columnar phase.

Boden et al. measured the AC conductivity of hexakis(hexyloxy)triphenylene doped with a small amount of Lewis acid AlCl$_3$, as a function of frequency (10^{-3}–10^{-7} Hz) and temperature in its crystalline, columnar, and isotropic phases [134,138,146]. At low frequencies, the conductivity was independent of frequency while it was a power law dependent on frequency at higher frequencies. This was attributed to a typical hopping behavior of charge carriers. The conductivity measured parallel to the column axis ($\sigma_\parallel$) was found to be about 10^3 times greater than that in the perpendicular direction ($\sigma_\perp$). Doping of HATs with NOBF$_4$ gives similar results. In a homologous series, the value of $\sigma_\perp$ decreases on increasing peripheral chain length, which can be rationalized as due to an increase in the distance between the columns. However, the value of $\sigma_\parallel$ also decreases with increasing peripheral chain length, which is attributed to the increasing degree of disorder in the packing of molecules within the columns [146].

Chandrasekhar and coworkers studied the 1D electrical conductivity of well-aligned samples of HHTT in the pure as well as doped states [167,195]. Trinitrofluorenone (TNF) was used as an electron acceptor. Unlike the saturated doping of HHTT with iodine, the addition of TNF in small amounts (0.62 mol%) does not change the phase behavior of the material drastically. Figure 6.10 presents the AC conductivity parallel to the columns of pure and TNF-doped HHTT at 1 kHz. The conductivity parallel to the columnar axis ($\sigma_\parallel$) increased by a factor of 10^7 or more on doping with TNF. The conductivity reaches a value of 10^{-2} S m^{-1}, which was the upper measurable limit of the experimental setup. The DC conductivity of doped samples exhibits an enormous anisotropy, $\sigma_\parallel/\sigma_\perp \geq 10^{10}$.

The dispersion of electron-deficient ferrocenium ions in HHTT and HATs was studied by Lakshminarayanan and coworkers [261]. The idea was to form a donor–acceptor pair, since the electron-rich triphenylene can act as a donor while the ferrocenium cation is a good electron acceptor. The composite dispersion retains the columnar liquid crystalline order even after doping up to about 10% of ferrocenium. The UV-vis absorption spectra of the composites confirm the formation of a charge-transfer complex. Figure 6.11 presents the conductivity of HAT and HHTT as a function of temperature after doping. The DC conductivity values for the neat HAT and HHTT were very low (10^{-12} S cm^{-1}). The conductivity increases by several orders of magnitude after doping for both unaligned and aligned samples with the latter showing significantly higher conductivity. This large enhancement in the DC conductivity is attributed to the field-induced electron hopping from the donor triphenylene to the acceptor ferrocenium cation. Interestingly, the ionic conductivity measured under low AC field and no DC bias was found to be about two orders of magnitude lower than the corresponding DC conductivity values at all temperatures. This shows that the mechanism operating in these systems is mainly electron transport by hopping rather than ion transport.

In a novel approach, Kumar and coworkers doped triphenylene discotics with monolayer-protected gold nanoparticles and studied their conductivities [251]. A dramatic increase in DC conductivity by more than 10^6 times was observed upon doping 1% gold nanoparticles in hexaheptyloxytriphenylene (Figure 6.12). It is worth noting that the measurements were carried out under ambient conditions

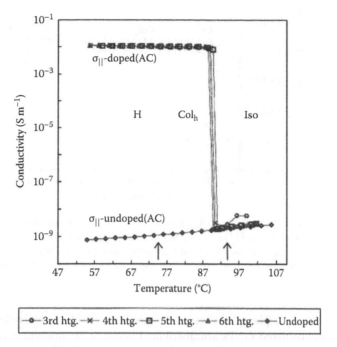

FIGURE 6.10 The measured AC electrical conductivity at 1 kHz parallel to the columns, $\sigma_{\parallel}$, in doped HHTT in the cooling mode over several thermal cycles. The transition temperatures (DSC peak values) are indicated by the arrows. (Reproduced from Balagurusamy, V.S.K. et al., *Pramana*, 53, 3, 1999. With permission from *Journal of Physics*, Indian Academy of Sciences.)

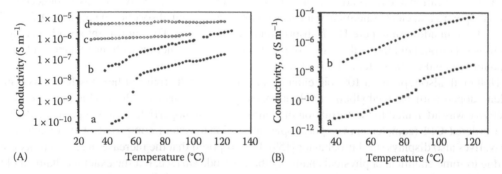

FIGURE 6.11 DC conductivity plots of different composites as a function of temperature for (A) (a) 1% Fc+/HAT6, (b) 10% Fc+/HAT6 unaligned samples, (c) 1% Fc+/HAT6, and (d) 10% Fc+/HAT6 aligned samples; and (B) (a) 1% Fc+/HHTT and (b) 10% Fc+/HHTT unaligned samples. (Reproduced from Kumar, P.S. et al., *J. Phys. Chem. B*, 112, 4865, 2008. With permission from ACS.)

without taking any precaution to eliminate atmospheric moisture and oxygen, which are known to inhibit the conductivity of donor–acceptor systems. This makes the system very interesting, as in most other discotic systems, the conductivity experiments were carried out under rigorous anhydrous conditions in the presence of either nitrogen or argon atmosphere or under applied magnetic field, which is used for the alignment of the phase.

The enhancement of conductivity in this system was attributed to the electron hopping process by multistep tunneling from gold core to gold core along the nanoparticle array on the application of the electric field. The highly delocalized π electrons of triphenylene bonded to the gold nanoparticles provide a facile pathway for the electron hopping and the gold nanoparticles as relay

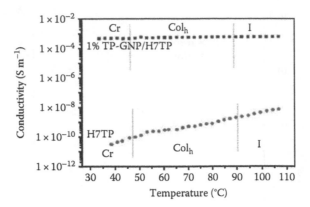

FIGURE 6.12 The variation of measured DC conductivity values as a function of temperature for 1% TP-GNPs mixed with H7TP and neat H7TP. The conductivities shown are obtained while cooling from the isotropic phase. The vertical lines denote the phase transition temperatures obtained from the DSC studies. (Reproduced from Kumar, S. et al., *Soft Matter*, 3, 896, 2007. With permission from The Royal Society of Chemistry.)

centers. In this process, the triphenylene acts as a donor while gold nanoparticles act as acceptors. The triphenylene-functionalized GNPs are distributed between the domain gaps of the columnar phase in a random disordered manner (Figure 6.13). Interestingly, HAT and gold tetrachloride complex (AuCl$_4^-$) form nanocomposites exhibiting higher electronic and ionic conductivities of several orders of magnitude under ambient conditions vis-à-vis the respective pure compounds and broad light absorption [265]. The doped HAT6 retains the columnar order in the liquid crystalline phase. The mixture of triphenylene donor and gold tetrachloride acceptor results in the formation of HAT$^+$ radical cation and reduced complex of monovalent gold complex, AuCl$_2^-$, as evidenced by the UV-vis spectra and electrochemical data. The composites exhibit absorption peaks at 430, 580, 630, and 830 nm in addition to pure HAT absorption peak at 280 nm. The significant absorbance by the composites throughout the visible range of the spectrum in conjunction with the enhanced electrical conductivity makes the system highly attractive for PV applications.

Holt et al. also reported a 10^6-fold enhancement in the conductivity of hexakis(hexyloxy)triphenylene doped with 1% methylbenzene thiol-coated gold nanoparticles [262]. This increase in conductivity was attributed to the formation of chains of gold nanoparticles upon applying a DC field. The ordered CPI (complementary polytropic interactions) system composed of two triphenylene derivatives also displays similar behavior [159]. It is proposed that the increase in conductivity could be due to some chemical or physical change at the electrode surface, but the exact mechanism of the origin of the enhanced conductivity in the field-annealed system is not yet clear.

Similar to above-mentioned electron rich systems, which can be doped with electron-deficient molecules, electron-deficient discotics such as tricycloquinazoline (TCQ), anthraquinone, hexaazotriphenylene, etc., can be doped with electron-rich materials to generate charge. Boden et al.

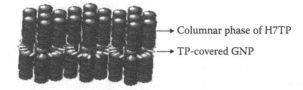

FIGURE 6.13 Schematic diagram illustrating arrangement of TP-GNPs in the inter-domain spacing formed between the discotic columns. (Reproduced from Kumar, S. et al., *Soft Matter*, 3, 896, 2007. With permission from The Royal Society of Chemistry.)

prepared the first *n*-doped quasi-1D electronically conducting discotic liquid crystalline system by doping 6 mol% of potassium metal in hexahexylthio-TCQ [139]. The AC conductivity of this composite at 191°C in the frequency range 5 Hz–13 MHz was measured. The doped system displayed conductivity 2.9×10^{-5} S m^{-1} in the columnar phase with an anisotropy $(\sigma_{\parallel}/\sigma_{\perp})$ of 518. In order to enhance the level of potassium doping, TCQ discotic with polar peripheral chains was prepared, which displayed substantially higher conductivity upon 10 mol% potassium doping [189].

The electron-deficient 1,2,3,5,6,7-hexakis(pentyloxy)-9,10-anthraquinone was doped with electron-rich anthracene molecule and the electrical conductivity $\sigma_{\parallel}$ was measured by Chandrasekhar and Balagurusamy [195]. The composite exhibits an increase of about seven orders of magnitude in conductivity in the columnar phase, similar to that observed by these authors in the HHTT/TNF system.

6.3.2 STEADY-STATE PHOTOCONDUCTIVITY

Photoconducting organic materials have received great attention in the past few years due to their commercial application in xerography and laser printing. A majority of technically used xerographic photoreceptors are organic photoconductors [273]. Steady-state photoconductivity measurement is a convenient technique to evaluate the photoconducting behavior of organic molecules. Steady-state photoconductivity experiments have been carried out on a few DLCs. The method is quite straightforward but suffers from a lack of knowledge of the exact value of the carrier density and, therefore, the charge-carrier mobility cannot be determined accurately. Generally, a sandwich-type cell consisting of two ITO-coated glass plates filled with a discotic sample is used to measure the dark- and photocurrents. A typical setup is shown in Figure 6.14.

Porphyrins, being natural photoconductive materials, have been extensively studied for photoconducting properties. Steady-state photoconductivity measurement on a metal-free discotic liquid crystalline porphyrin derivative was first reported by Shimizu et al. in 1986 [113]. The phase-transition dependence of photocurrent on the positive electrode illumination clearly showed a marked increase of photocurrent at the crystalline-to-mesophase transition. Figure 6.15 shows the dark- and photocurrents as a function of temperature for a 3.3 μm thick cell filled with a discotic liquid crystalline tetraphenylporphyrin oxovanadium(IV) complex [186]. The dependence of the dark current obeyed the Arrhenius equation in isotropic phase as well as in mesophases. The dark current continuously increases with temperature. Thermal excitation of charge carriers from traps

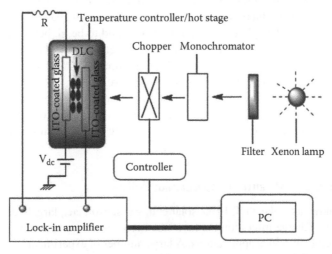

FIGURE 6.14 Schematic diagram of the experimental setup commonly used for photoconductivity measurements.

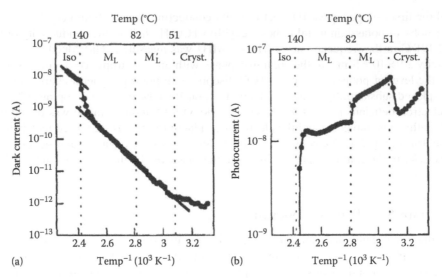

FIGURE 6.15 Temperature dependence of (a) dark- and (b) photocurrents in the positive electrode illumination under 8000 V cm^{-1}. (Reproduced from Monobe, H. et al., *Mol. Cryst. Liq. Cryst.*, 368, 311, 2001. With permission from Taylor & Francis.)

created by structural disorders can increase the population of charge carriers for electric conduction. An increase in the mobility of ionic impurities is expected at elevated temperature and this can enhance the conductivity. The behavior of photocurrent deviates drastically from that of dark current. At the crystal-to-mesophase transition, the photocurrent increases stepwise and then gradually decreases in the mesophase. At the mesophase-to-isotropic transition, the photocurrent drastically decreases to an undetectable level. The tetraphenylporphyrin oxovanadium(IV) complex displayed much higher photoconductivity than metal-free porphyrins [86]. This could be due to the higher efficiency of the carrier generation of oxovanadium(IV) complex than that of metal-free porphyrin owing to the difference of ionization potential. The *I–V* characteristics of photocurrent exhibits rectification behavior for all phases (Figure 6.16). A positive electrode illumination causes a larger photocurrent, which increases with bias.

Steady-state photoconductivity measurements have also been carried out on triphenylene-based DLCs [128]. Several HAT discotics have been investigated for their photoconducting properties. All samples display photoconductivity within the mesophase, whereas in the isotropic phase the photocurrent drops to zero. The photoconductivity was found to be higher during the cooling from isotropic melt. This can be corroborated due to highly ordered hexagonal columnar structure formation during the slow cooling from the isotropic phase. Figure 6.17 shows the photocurrent behavior of hexapentyloxytriphenyle during heating and cooling cycles. It is interesting to note that these materials exhibit photoconductivity only in the mesophase. The photocurrent increases rapidly at the crystal-to-columnar phase transition, followed by gradual increase in the mesophase upon increasing the temperature to its maximum before collapsing at the columnar phase–isotropic phase transition.

6.3.3 CHARGE-CARRIER MOBILITY IN COLUMNAR PHASES

The performance of devices like TFT, LED, solar cell, gas sensor, etc., largely depends on the mobility of charge carriers. Charge migration is one of the most important factors in visualizing DLCs as organic semiconductors in device applications. A large number of experiments has been carried out to study charge-carrier mobility in columnar phases. The charge mobility (μ) can be calculated from conductivity (σ) experiments using the formula, $\sigma = ne\mu$, where n is the charge density and e is the value

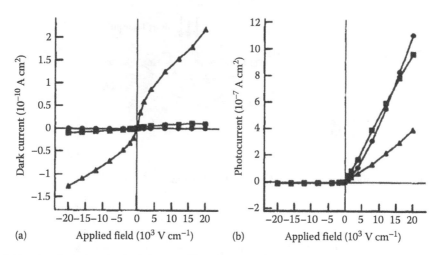

(a)   Applied field (10^3 V cm^{-1})   (b)   Applied field (10^3 V cm^{-1})

FIGURE 6.16 Applied electric field dependence of (a) dark- and (b) photocurrents in an ITO/porphyrin LC/ITO cell for the crystalline (25°C; ●), LC phase 1 (60°C; ■), and LC phase 2 (100°C; ▲) of tetraphenyl-porphyrin oxovanadium(IV) complex under 550 nm with 0.34 mW cm^{-2} light illumination. (Reproduced from Monobe, H. et al., *Mol. Cryst. Liq. Cryst.*, 368, 311, 2001. With permission.)

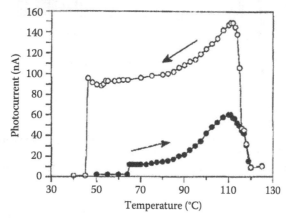

FIGURE 6.17 Photocurrent of hexapentoxytriphenylene and its dependence upon temperature. (Reproduced from Bengs, H. et al., *Liq. Cryst.*, 15, 565, 1993. With permission.)

of unit electrostatic charge. However, the exact charge-carrier concentration is usually not known and, therefore, mobility cannot be calculated accurately using this method. Therefore, the methods that have been applied to measure charge mobility in discotics are the time-of-flight (TOF) technique, the pulse-radiolysis time-resolved microwave conductivity (PR-TRMC) method, the field-effect transistor (FET) fabrication, and the space-charge limited current (SCLC) technique. Out of these, the TOF and PR-TRMC techniques have been extensively used to study charge mobility in DLCs.

6.3.3.1 Time-of-Flight Technique

The TOF technique provides direct, absolute measurement of charge-carrier mobilities in columnar phases. A typical setup of the TOF experiments is depicted in Figure 6.18. In the TOF method, charges are generated using a pulse of laser light. A common sandwich cell configuration is used and DLC is filled between transparent conducting electrodes. It is essential to align the sample homeotropically (columns aligned perpendicular to the electrodes). Any defect in the path (reminiscent of grain boundaries in crystalline materials), severely reduces the mobility. Therefore, discotics

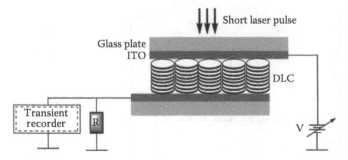

FIGURE 6.18 Schematic representation of a typical TOF setup.

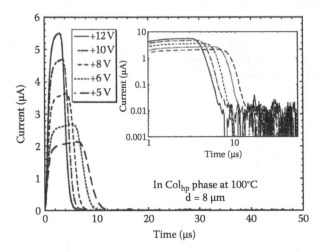

FIGURE 6.19 Linear plot of typical transient photocurrents for positive charge carrier as a function of time in Col$_{hp}$ phase of H4T at 100°C. The inset shows the logarithm plots of the same transient photocurrents. (Reproduced from Iino, H. et al., *Mol. Cryst. Liq. Cryst.*, 436, 217, 2005. With permission.)

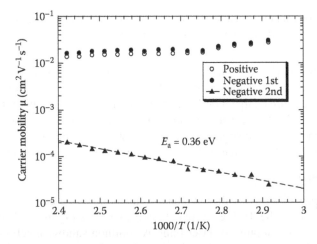

FIGURE 6.20 Arrhenius plots of the charge-carrier mobility for positive charge carrier and for the fast and slow negative charge carriers corresponding to the first and second transits in Col$_{hp}$ phase of H4T. (Reproduced from Iino, H. et al., *Mol. Cryst. Liq. Cryst.*, 436, 217, 2005. With permission.)

that cannot be aligned properly are not suitable for TOF measurements. A short light pulse of definite wavelength is passed through the sample that generates charges at the interface layer. When an electric field is applied across the electrodes, there is a flow of charges. Depending on the polarity of the applied field, electron or holes (radical cations) move through the columnar stacks of molecules. This induces a transient current, which is recorded using an oscilloscope. The mobility of the charge carriers is estimated on the basis of the time these charges take to travel between the electrodes according to the equation $\mu = v/E = d^2/Vt_t$, where d is the film thickness, E is the applied electric field, and v is the drift velocity.

Figure 6.19 presents a typical transient photocurrent decay curve for positive charge carriers as a function of time in the columnar phase of hexabutyloxytriphenylene obtained by the TOF method. A similar curve is observed for negative charge carriers. Arrhenius plots of the charge-carrier mobility for positive and negative charge carriers are shown in Figure 6.20 [232]. Charge-carrier mobility data of various discotics are provided in Tables 6.1 through 6.5 and are discussed after the PR-TRMC measurement technique.

6.3.3.2 Pulse-Radiolysis Time-Resolved Microwave Conductivity Technique

When it is difficult to align the sample properly, the PR-TRMC technique is used. This technique measures the sum of positive (hole) and negative (electron) mobilities. An oversimplified PR-TRMC setup is shown in Figure 6.21. The material is filled in a copper cell and ionized by a nanosecond pulse of high-energy electrons from a Van de Graaff accelerator. Generally, about 200 mg of the sample is required to fill the cell, but a smaller cell cavity may be obtained by filling the cell with polymer sample, which does not contribute to the conductivity signal. Irradiation of the sample creates charge carriers with a uniform and known concentration of the order of micromolar. Mobile charge carriers lead to an increase in the conductivity of the sample, which is measured as a transient decrease in the power of microwaves that propagate through the irradiated sample [97]. The microwave frequency band used is 28–39 GHz. The mobility is calculated using the equation, $\Delta\sigma = eN\Sigma\mu$, where $\Sigma\mu$ is the sum of positive and negative charge-carrier mobilities [μ(hole) + μ(electron)], N is the concentration of the charge-carrier pairs present at the end of the pulse, and e is the electrostatic charge. The change in conductivity ($\Delta\sigma$) on irradiation of the sample is measured experimentally with nanosecond time resolution as the increase in microwave power absorbed by the sample. The

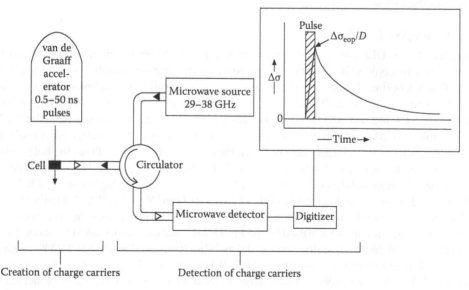

FIGURE 6.21 A much simplified schematic of the PR-TRMC equipment. (Reproduced from Warman, J.M. and van de Craats, A.M., *Mol. Cryst. Liq. Cryst.*, 396, 41, 2003. With permission.)

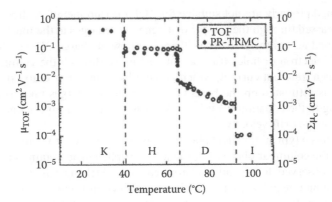

FIGURE 6.22 The mobility of holes in HHTT measured by TOF (open circles): 1D intracolumnar mobility sum estimated by PR-TRMC (filled circles). Both sets of data apply to a cooling trajectory beginning at ca. 100°C with the isotropic liquid. (Reproduced van de Craats, A.M. et al., *Adv. Mater.*, 8, 823, 1996. With permission. Copyright Wiley-VCH Verlag GmbH & Co.)

major advantage of the PR-TRMC technique is that it can be applied to any material irrespective of its morphology or optical properties and, therefore, a variety of materials can be investigated. On the other hand, the major disadvantage of the technique is that the individual contributions of the positive and negative charge carriers cannot be separately determined.

The PR-TRMC technique measures the charge-carrier mobility on a nanometer-length scale, so that traps due to structural disorder and impurities do not interfere with the mobility measurement. On the other hand, TOF measures the mobility on a micrometer scale and, therefore, defects in the film cause divesting effects. Therefore, the mobility measured by the PR-TRMC method indicates an upper limit that could only be reached in a device having an ideal ordered film. However, from a device application point of view, the direct TOF measurements are more reliable. Though it is expected that the PR-TRMC method should show higher charge mobility, comparative studies using TOF and PR-TRMC techniques on the same discotic sample indicate that both methods give more or less similar results (Figure 6.22) [149]. Charge transport properties of some important DLCs are discussed in the following sections.

6.3.3.2.1 Triphenylene Discotics

Triphenylene-based DLCs are easy to align homeotropically to obtain monodomain of a columnar phase on a macroscopic scale. Most triphenylene discotics have reasonably low isotropic temperature (<200°C). Cooling slowly from the isotropic phase generally gives a homeotropically aligned sample between two substrates. Therefore, triphenylene discotics (Figure 6.23) have been extensively studied by the TOF method, and the mobility data were compared with the data from the PR-TRMC method (Table 6.1). Most of the discotics exhibit field- and temperature-independent mobility but, as expected, the mobility is largely dependent on the order of mesophase. Thus, the hole mobility in the Col_h phase of triphenylene-based discotics is of the order of 10^{-3} cm^2 V^{-1} s^{-1}. More ordered Col_p phase exhibits the hole mobility of the order of 10^{-2} cm^2 V^{-1} s^{-1}, and the highly ordered helical phase of HHTT (**10**) displays the hole mobility of the order of 10^{-1} cm^2 V^{-1} s^{-1} [136]. Shortening the peripheral chain length, generally increases the order of the columnar phase, and thus the charge-carrier mobility. The charge mobility measured by the PR-TRMC method was found to be about 20% lower in the H phase [149]. In the crystalline phase, the mobility (measured by the PR-TRMC method) further increased three to four times and reached 0.3–0.4 cm^2 V^{-1} s^{-1} [154]. Based on these triphenylene discotics mobility data, an order-mobility diagram as shown in Figure 6.24 can be depicted. Iino and Hanna measured the hole mobility of HHTT in the polycrystalline phase by the TOF method. The mobility (0.3 cm^2 V^{-1} s^{-1}) was comparable with that measured by the PR-TRMC method and,

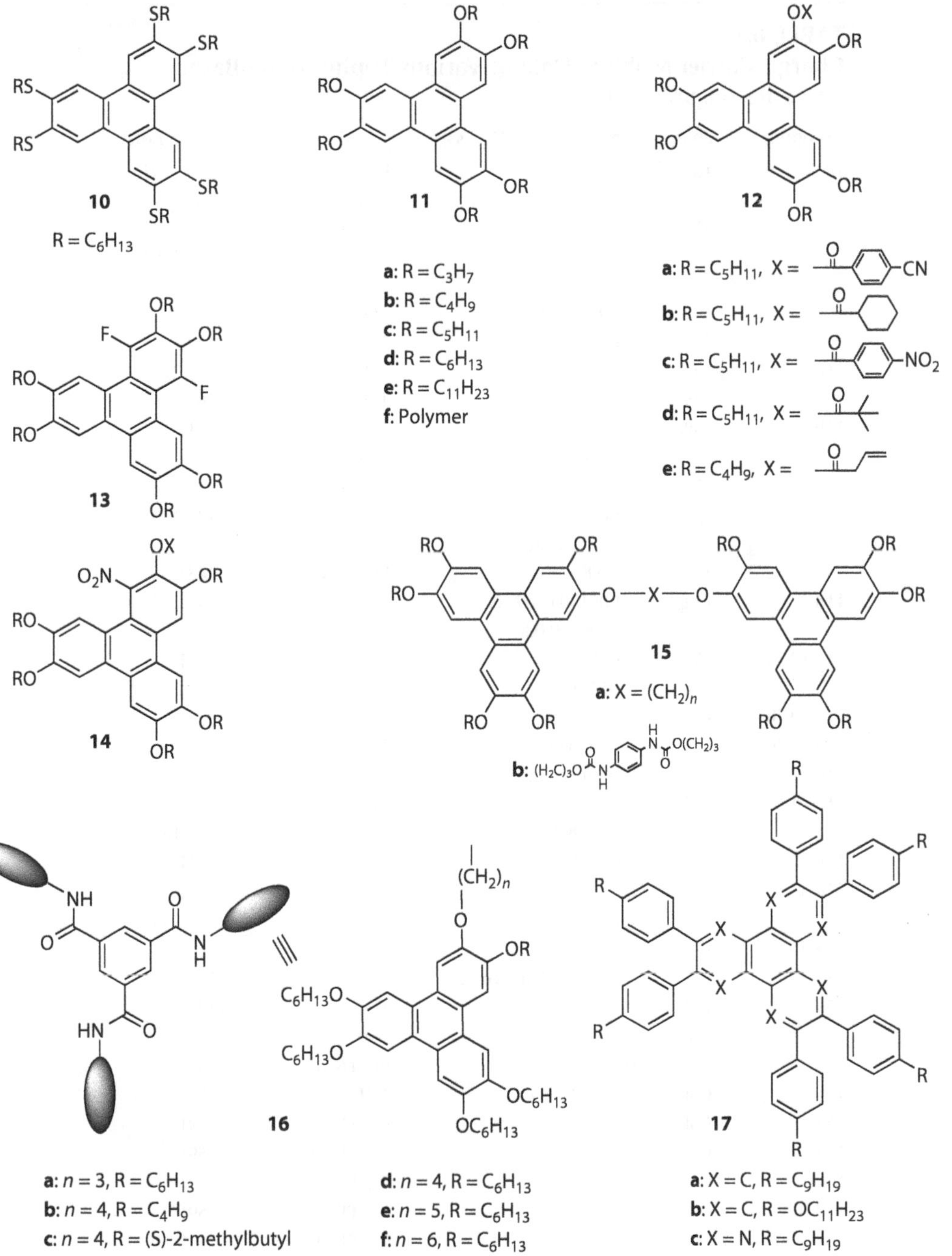

FIGURE 6.23 Chemical structure of triphenylene-based DLCs studied for charge-carrier mobility by TOF or PR-TRMC methods.

therefore, they concluded that the grain boundaries do not have much divesting effect on the charge-carrier mobility [241]. However, more experiments are required to reach this conclusion.

The effects of C60 and C70 doping in HATs, **11**, have been studied [155,201]. However, no significant effect of these nanoparticles doping on the hole mobility was realized. This is not surprising as these carbon nanoparticles are known scavengers of electrons but not holes. Surprisingly, the negative carrier mobility decreased upon C60 doping but the activation energy of the negative carriers

TABLE 6.1
Charge-Carrier Mobility Data of Various Triphenylene-Based Discotics (Figure 6.23)

Compound	Phase	Mobility (cm² V⁻¹ s⁻¹)	Method	Ref.
10	Cr	0.4	PR-TRMC	[149,154]
		0.3 (hole)	TOF	[241]
	H	0.08	PR-TRMC	[149]
		0.1 (hole)	TOF	[136,149]
		0.08 (electron)	TOF	[230]
	Col$_h$	0.001–0.006	TOF, PR-TRMC	[149]
	I	0.0001 (hole)	TOF	[149]
11a	Cr	0.013 (hole)	TOF	[202,219]
	Col$_p$	0.012 (hole)	TOF	[202,219]
11b	Col$_p$	0.02 (hole)	TOF	[151,232,233,242]
		0.02–0.025 (electron)	TOF	
11c	Cr	0.005	PR-TRMC	[143]
	Col$_h$	0.001–0.002 (hole)	TOF	[132,133,137,233]
		0.002 (electron)		
	I	<0.0002	PR-TRMC	[143]
11d	Col$_h$	0.0001–0.0007 (hole),	TOF	[145,233,242]
		0.0002–0.0004 (electron)		
11d+17a	Col	0.023 (hole)	TOF	[188]
	glass	0.014 (hole)		[188]
11d+17b	Col	0.02 (hole)	TOF	[202]
11d+17c	Col	0.016 (hole)	TOF	[202]
11e	Col$_h$	0.0001 (hole)	TOF	[174,187,219]
11e+17a	Col$_h$	0.029	TOF	[174,187,219]
	glass	0.016 (hole)		[174,187,219]
11f	N$_D$	0.001 (hole)	TOF	[231]
12a–e	Col$_h$	0.03–0.005 (hole)	TOF	[169]
13	Col$_h$	0.0016 (hole)	TOF	[202,234]
13+17a	Col$_h$	0.015 (hole)	TOF	[202]
14	Col$_h$	0.0002–0.00002 (hole)	TOF	[244]
15a	Col$_h$	0.001 (hole)	TOF	[142]
	glass	0.000001 (hole)	TOF	[159]
		0.015	PR-TRMC	[159]
15b	Col$_h$	0.00002 (hole)	TOF	[254]
16a	Col$_h$	0.017	PR-TRMC	[259]
16b	Col$_p$	0.06	PR-TRMC	[259]
16c	Col$_p$	0.09	PR-TRMC	[259]
16d	Col$_r$	0.23	PR-TRMC	[259]
16e	Col$_{ob}$	0.016	PR-TRMC	[259]
16f	Col$_{ob}$	0.018	PR-TRMC	[259]

did not change much. It was concluded that the migrating negative carriers in these carbon nanoparticle doped systems are not electrons but anions. On the other hand, doping of HATs with a small amount (3 wt %) of an amino acid derived from L-valine generates liquid crystalline gels [192,259]. The hole mobility in these gels was much higher (up to about 0.003 cm² V⁻¹ s⁻¹) than in pure triphenylene discotics. This could be because of the low molecular fluctuation of discotic molecules due to the dispersion of fibrous network of gelator in the supramolecular order of the columnar phase.

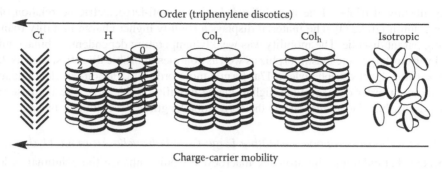

FIGURE 6.24 Mobility-order diagram for triphenylene discotics. In general, the mobility increases as the order of mesophase increases.

The charge mobility in the high-temperature Col_h phase of triphenylene dimer **15a** was comparable to that of a monomer (on the order of 10^{-3} cm^2 V^{-1} s^{-1}), but it is drastically decreased in the glassy phase (on the order of 10^{-6} cm^2 V^{-1} s^{-1}) [140,150,159]. This was attributed to the presence of structural disorder in the columnar phase. At higher temperature due to a dynamical healing of defects, the mobility increases by about three orders of magnitude. In comparison to the above-mentioned triphenylene discotic dimer, the triphenylene discotic dimer **15b** prepared by coupling two triphenylene units via phenylene carbonate linkage [254] displays a very low charge-carrier mobility (2×10^{-5} cm^2 V^{-1} s^{-1}) measured by the TOF technique. This could be primarily due to the unaligned nature of the sample.

Triphenylene-based discotic polymer prepared by the photopolymerization of acrylate-terminated triphenylene monomer was reported to exhibit a carrier mobility of 10^{-3} cm^2 V^{-1} s^{-1} in the nematic phase [231]. The value is rather too high for a discotic polymer having nematic phase. As a clear transient decay curve was not observed, the mobility data need verification by other techniques.

Early studies on TOF measurements looked only at the hole mobility in these electron-rich triphenylene discotics, but later electron mobility was also measured. Iino et al. investigated the electron transport in HAT derivatives [242]. Transient photocurrents exhibited two transits of negative carriers that appeared in different time ranges. The first transit was for electron mobility, while the second transit was attributed to ionic mobility. The electron mobility in HATs was almost the same as hole mobility for each triphenylene derivative (Table 6.1). The electron mobility in the highly ordered helical phase of HHTT displays similar behavior [230].

Paraschiv et al. reported an interesting class of intramolecular H-bond-stabilized columnar liquid crystalline materials by synthesizing a series of 1,3,5-benzenetrisamide derivatives **16** with three HAT pendent groups [243,260]. This way, a high directionality of interaction could be achieved. These materials form ordered columnar phases in which the charge-carrier mobility was several times higher compared to monomeric triphenylene discotics and close to the mobility observed in large-core HBC derivatives.

Unlike most triphenylene-based DLCs, the charge mobility in a polar nitro-functionalized triphenylene discotic **14** was found to be dependent on both the electric field and temperature [244]. The hole mobility in the Col_h phase of mononitro-hexahexyloxytriphenylene **14** at room temperature was about 2×10^{-5} cm^2 V^{-1} s^{-1} at a field of 1×10^5 V cm^{-1}, which increases to 2×10^{-4} cm^2 V^{-1} s^{-1} at a temperature of 135°C. Monoester-functionalized triphenylenes **12** also behave similarly [169]. The mobility in these ester-substituted triphenylenes is not independent of temperature but follows an $\ln \mu \propto 1/T^2$ law, and the field dependence of their mobility obeys an $\ln \propto E^{1/2}$ law. Thus, these molecules exhibit a disorder-dominated hopping process similar to dipole-substituted amorphous molecular solids. These materials exhibit much lower charge mobility than the corresponding parent discotics. On the other hand, the introduction of the smaller fluoro substituent at the α position of triphenylene **13** does not disturb the columnar ordering. The columnar structure of fluoro-substituted triphenylene is stabilized due to an anti-ferroelectric correlation of dipoles along the column [202,234]. The material displays a slightly higher charge mobility than the corresponding parent discotic. The mobility was weakly temperature dependent. A binary mixture of HAT and various large-core mesomorphic as well as non-mesomorphic discotics, such as 2,3,6,7,10, 11-hexakis(4-nonylphenyl)triphenylene **17a**, form a more ordered CPI non-covalent adduct, which exhibits about an order of magnitude higher charge mobility (Table 6.1). The increase in carrier mobility was attributed to an increase in the correlation length of the columnar phase.

6.3.3.2.2 Hexabenzocoronene and Other Large-Core Polycyclic Aromatic Hydrocarbons

It is expected that enlarging the aromatic macrocycle would enhance the columnar order due to intense π–π interactions and, thus, charge mobility. Discotic dibenzopyrene, tricycloquinazoline, etc., were designed to visualize this concept [274,275]. However, it was the Mullen group that exploited this concept by extensively studying the charge transport properties of HBC and other large size polyaromatic hydrocarbons (Figure 6.25). The charge-carrier mobility data of these discotics are presented in Table 6.2.

In the crystalline phase of hexaalkyl-HBC discotics, **19a–c**, the mobility increases slightly on increasing the temperature. At the crystal-to-mesophase transition, an abrupt decrease in the mobility occurs, but it increases again gradually on further heating. During the cooling process, the

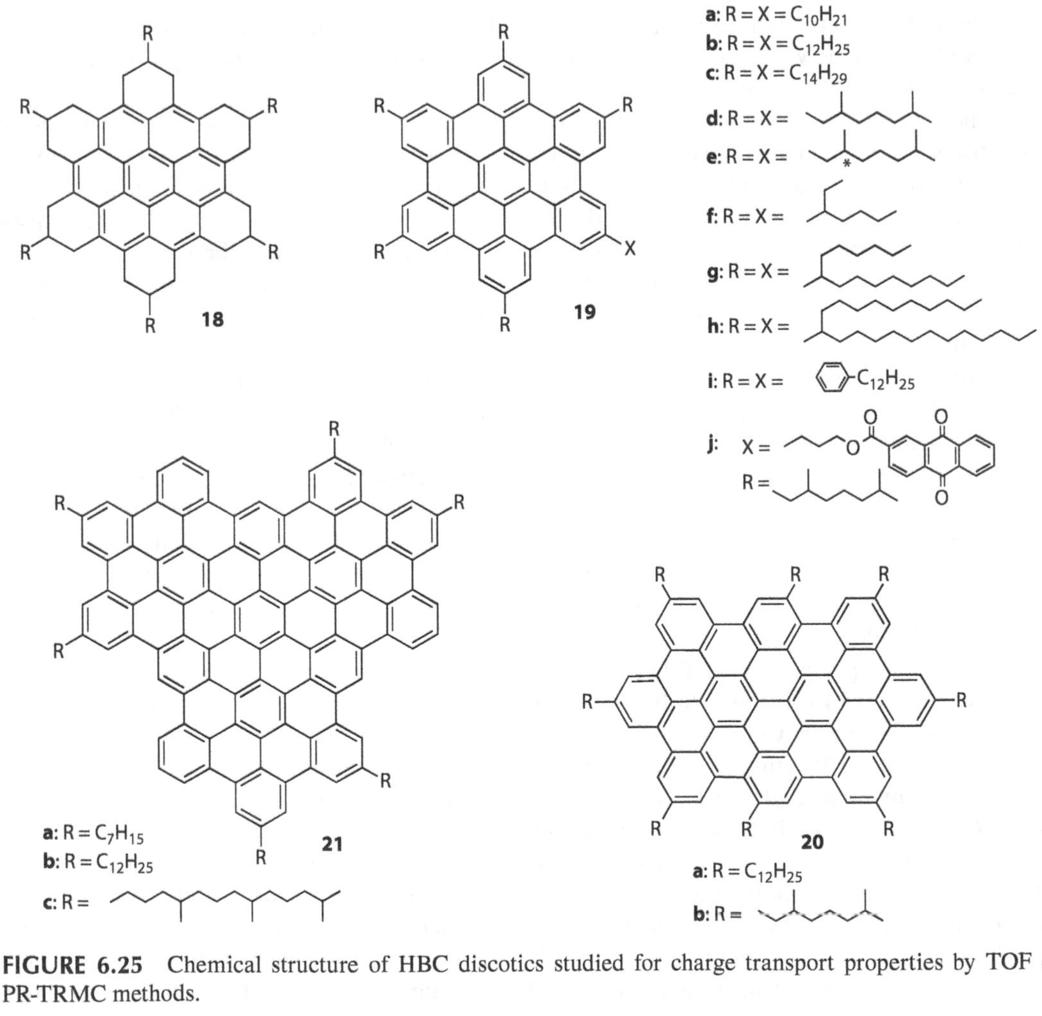

FIGURE 6.25 Chemical structure of HBC discotics studied for charge transport properties by TOF or PR-TRMC methods.

TABLE 6.2
Charge-Carrier Mobility Data of HBC and Other
Large-Core DLCs (Figure 6.25)

Compound	Mobility ($cm^2 V^{-1} s^{-1}$)		Method	Ref.
	Cr Phase	Mesophase		
18	0.72	0.1–0.2	PR-TRMC	[216]
19a	0.40	0.26	PR-TRMC	[165,216]
19b	0.70	0.38	PR-TRMC	[165,216]
19c	1.0	0.31	PR-TRMC	[165,216]
19d	0.43	0.30	PR-TRMC	[216]
19e	0.46	0.26	PR-TRMC	[216]
19f	0.26	—	PR-TRMC	[229]
19g	0.72	—	PR-TRMC	[229]
	0.0014	—	TOF	[240]
19h	0.48	—	PR-TRMC	[229]
19i	—	0.17–0.46	PR-TRMC	[165,216]
19J	0.15	0.09	PR-TRMC	[213]
20a	0.52	0.26	PR-TRMC	[216]
20b	0.45	0.29	PR-TRMC	[216]
21a	0.20	0.23	PR-TRMC	[216]
21b	0.16	0.20	PR-TRMC	[216]
21c	0.06	0.08	PR-TRMC	[216]

mobility follows the same trend with a hysteresis of about 20°C at the columnar-to-crystal phase transition. Compound **19i**, in which the alkyl chains are linked to the core via a phenyl unit, displays a columnar phase at ambient temperature and does not crystallize even at very low temperature. As there is no crystal-to-mesophase transition, this material does not show any abrupt change in the charge mobility at any temperature. Based on the charge-carrier mobility data of various discotic materials, an empirical relationship ($\Sigma\mu_{1D} = 3 \exp(-83/n)$ [$cm^2 V^{-1} s^{-1}$]) between charge mobility ($\Sigma\mu_{1D}$) and the core size was proposed [179]. However, this does not hold good for many large-core discotics. The charge-carrier mobility in very large-core discotics, such as **20** and **21**, was found to be much lower than HBC derivatives [216]. It should be noted that the charge-mobility values derived from the characteristics of FET devices or measured by the TOF technique [240] in some HBC discotics were found to be about two orders of magnitude lower than the PR-TRMC values. The anisotropy in the photoconductivity of thin film of HBC discotics was studied by the flash-photolysis time-resolved microwave conductivity technique [200,214]. The film displays a large anisotropy in photoconductivity in the direction of columnar alignment.

6.3.3.2.3 *Phthalocyanine Discotics*
The interest in phthalocyanine (Pc) derivatives for organic electronic applications originates from their broad light absorption range, interesting photophysical properties, extraordinary thermal stability, and convenient synthesis. Semiconducting properties of crystalline Pcs have been studied extensively [276–282]. The possibility of combining these properties with the self-organizing and processable properties of DLCs makes Pc-based DLCs very attractive as 1D organic semiconductors. Accordingly, a number of Pc discotics have been studied for their 1D charge migration properties. Pc-based DLCs studied by PR-TRMC and TOF techniques are shown in Figure 6.26 and the mobility data are collected in Table 6.3.

A majority of Pc discotics have been studied by PR-TRMC techniques probably because the perfect alignment of Pc molecules due to their large size was difficult. However, recently electron

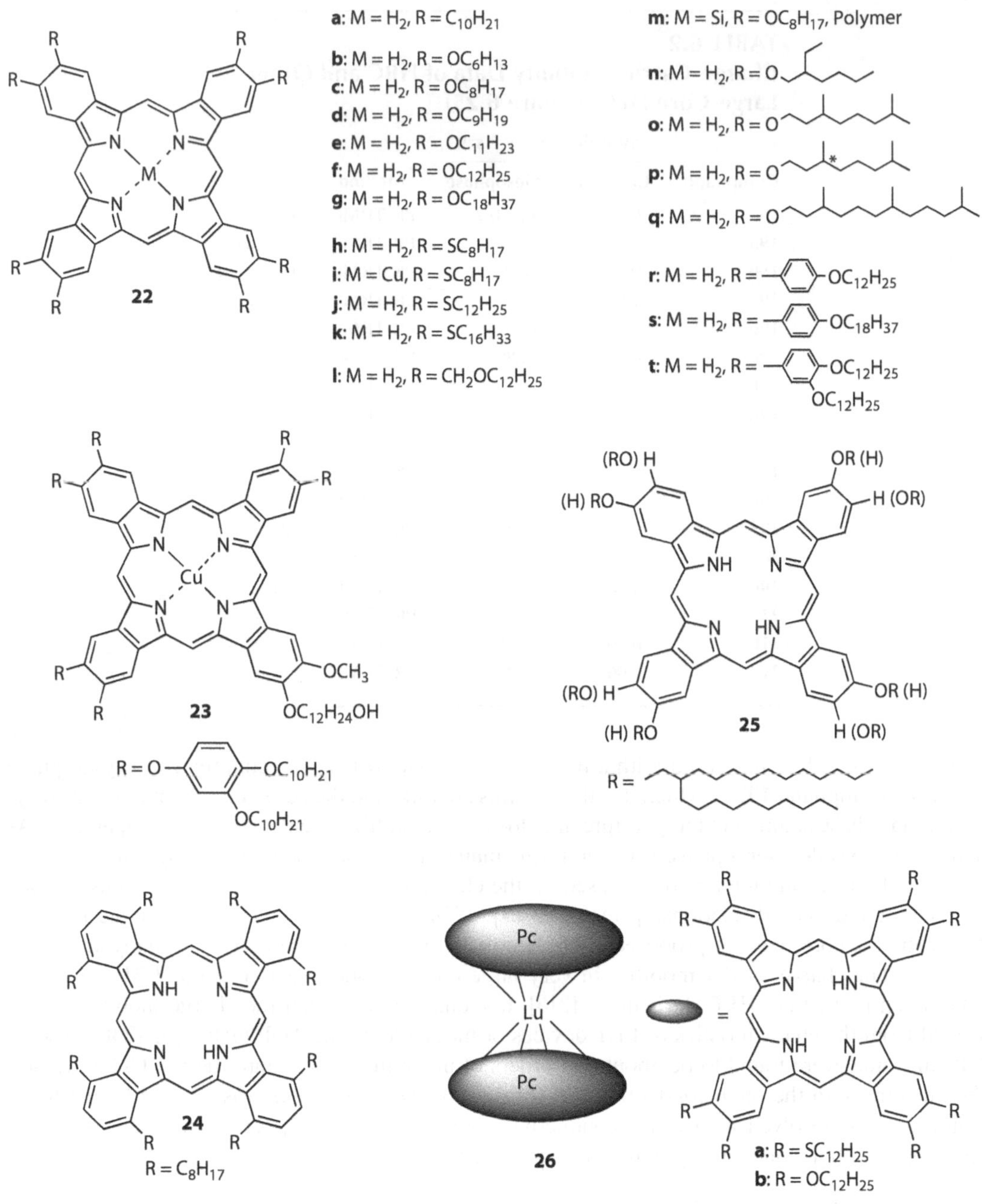
22
a: M = H2, R = C10H21
b: M = H2, R = OC6H13
c: M = H2, R = OC8H17
d: M = H2, R = OC9H19
e: M = H2, R = OC11H23
f: M = H2, R = OC12H25
g: M = H2, R = OC18H37
h: M = H2, R = SC8H17
i: M = Cu, R = SC8H17
j: M = H2, R = SC12H25
k: M = H2, R = SC16H33
l: M = H2, R = CH2OC12H25
m: M = Si, R = OC8H17, Polymer
n: M = H2, R = O
o: M = H2, R = O
p: M = H2, R = O
q: M = H2, R = O
r: M = H2, R = OC12H25
s: M = H2, R = OC18H37
t: M = H2, R = OC12H25
OC12H25
23
OCH3
OC12H24OH
R = O OC10H21
OC10H21
25
R =
24
R = C8H17
Pc
Lu
Pc
26
a: R = SC12H25
b: R = OC12H25

FIGURE 6.26 Chemical structure of phthalocyanine-based DLCs studied for charge-carrier mobility by TOF or PR-TRMC methods.

and hole mobilities in a few Pc discotics have been measured using the TOF technique. Peripherally octa-substituted Pcs **22a–j** are the simplest to prepare. The octaalkoxy derivatives **22b–g**, with varying alkyl chain length from 6 to 18 carbon atoms display similar mobility (Table 6.3). The average mobility in the columnar phase of octa-*n*-alkoxy-Pc was found to be about $0.06\,cm^2\,V^{-1}\,s^{-1}$. The mobility increases to an average value of about $0.28\,cm^2\,V^{-1}\,s^{-1}$ in the crystalline phase. Peripheral branched alkyl chain substituted Pcs **22p–q** form stable columnar phase at ambient temperature retaining the high mobility. The alkylthio-substituted Pcs **22h–k** exhibit much higher mobility than

TABLE 6.3
Charge-Carrier Mobility Data of Various Phthalocyanine-Based DLCs (Figure 6.26)

Compound	Mobility ($cm^2 V^{-1} s^{-1}$)		Method	Ref.
	Cr Phase	Col Phase		
22a	0.13	0.15	PR-TRMC	[97]
22b	0.20	0.06	PR-TRMC	[97]
22c	0.33	0.08	PR-TRMC	[97]
22d	0.30	0.08	PR-TRMC	[97]
22e	0.27	0.07	PR-TRMC	[97]
22f	0.27	0.05	PR-TRMC	[97]
22g	0.20	0.07	PR-TRMC	[97]
22h	0.55	0.16	PR-TRMC	[97]
22i	0.54	0.27	PR-TRMC	[97]
22j	0.26	0.22	PR-TRMC	[97]
22k	0.14	0.12	PR-TRMC	[97]
22l	0.14	0.04	PR-TRMC	[97]
22m	—	0.06	PR-TRMC	[97]
22n	0.17	0.19	PR-TRMC	[97]
22o	0.37	0.07	PR-TRMC	[97]
22p	—	0.03	PR-TRMC	[97]
22q	—	0.04	PR-TRMC	[97]
22r	0.33	0.18	PR-TRMC	[97]
22s	0.67	0.24	PR-TRMC	[97]
22t	—	0.21	PR-TRMC	[97]
23	0.003 (hole)	0.0022 (hole)	TOF	[212]
	0.0026 (electron)	0.0024 (electron)		
24	—	0.2 (hole)	TOF	[227]
	—	0.3 (electron)		
25	—	0.0027 (hole)	TOF	[238]
26a		0.71	PR-TRMC	[97,178]
26b		0.17	PR-TRMC	[97,178]

the corresponding alkoxy derivatives in mesophases as well as in the crystalline phase. Similarly, alkoxyphenyl-substituted Pcs **22r–t** also display higher charge mobility in the columnar phase than alkoxy derivatives. The coaxially polymerized siloxane derivative of octaalkoxy-Pc **22m** exhibits columnar phase at room temperature and slightly higher charge mobility than the corresponding octaalkoxy-Pc.

Ohta and coworkers prepared a number of octaphenoxy-substituted Pcs **23**, which spontaneously align homeotropically on a substrate [92] and, therefore, TOF measurements can be done easily. The Cu-Pc, 2-(12-hydroxydodecyloxy)-3-methoxy-9,10,16,17,23,24-hexakis(3,4-didecyloxyphenoxy) phthalocyaninato copper(II), **23**, exhibited the carrier mobility for electrons as high as for holes [212]. In the crystalline phase, the electron mobility reached to 2.6×10^{-3} $cm^2 V^{-1} s^{-1}$ and hole mobility to 2.99×10^{-3} $cm^2 V^{-1} s^{-1}$. The charge mobility was slightly lower (electron mobility 2.4×10^{-3} cm^2 $V^{-1} s^{-1}$ and hole mobility to 2.17×10^{-3} $cm^2 V^{-1} s^{-1}$) in the columnar phase. Similarly, the metal-free Pc 1,4,8,11,15,22,25-octaoctylphthalocyanine **24** also aligns homeotropically easily. The compound exhibits two columnar phases. For the Col_r phase, the hole and electron mobilities were found to be 2×10^{-1} and 3×10^{-1} $cm^2 V^{-1} s^{-1}$, respectively [227,228]. The mobilities in the Col_h phase were slightly lower (1×10^{-1} $cm^2 V^{-1} s^{-1}$) than in the Col_r phase. This could be due to the more ordered nature of the Col$_r$ phase. Interestingly even in the isotropic phase, the mobilities for positive and negative carriers were quite high (3×10^{-3} and 1×10^{-3} cm^2 V^{-1} s^{-1}, respectively). Further, the mobilities in all phases were found to be independent or very weakly dependent on temperature and field. The persistence of ambipolar charge-carrier transport at ambient conditions makes this material very interesting for device applications.

The tetraalkoxy-substituted Pc **25** (exists as a mixture of four isomers) exhibits the charge-carrier mobility up to 2.8×10^{-3} cm^2 V^{-1} s^{-1}. De Jong et al. studied the electronic coupling strength within columns of DLCs using core-level resonant photoemission spectroscopy [238,256]. The average charge-transfer times on the order of a few femtoseconds were observed. This indicates a strong electronic coupling between the Pc molecules. Therefore, the charge-carrier mobility in these molecules is limited by long-range disorder and not by nearest-neighbor coupling.

The charge-carrier mobility champion Pc discotic is the bis[octakis(alkylthio)phthalocyaninato] lutetium(III), **26** with a mobility of 0.71 cm^2 V^{-1} s^{-1} [92]. A lutetium complex of Pc (Pc$_2$Lu) is composed of a Lu^{3+} cation and two PC^{2-} anions and, therefore, it is not charge-balanced; it is a neutral radical. Unlike other discotics, including metal-free and metalled-Pcs, the free radical Pc complexes like lithium complex of Pc and Pc$_2$Lu display relatively large dark conductivities due to lower band gaps [266,268] and, therefore, have been described as "intrinsic organic semiconductors" [267]. A series of discotic liquid crystalline Pc$_2$Lu have been prepared (see Chapter 3) and some of these were studied for charge transport properties. They are derived from octaalkoxy-Pc and octaalkylthio-Pc. Surprisingly, the two series of Pc$_2$Lu discotics show marked differences in their charge transport behavior. Contrary to most discotics, the bis[octakis(dodecyloxy)phthalocyaninato]lutetium(III), **26b**, exhibits an abrupt increase in the charge mobility at the crystalline phase-to-mesophase transition. The mobility increases further at the columnar phase to higher temperature columnar phase. On the other hand, the alkylthio-substituted derivative **26a** exhibited a normal trend. In these Lu complexes, the charge-carrier mobility decreases at the transition from the crystalline state to mesophase. The crystalline phase of thioalkyl derivative **26a** display about two orders of magnitude higher mobility than the crystalline phase of alkoxy derivative **26b**. Therefore, the increase of mobility at the crystal-to-mesophase transition in the case of alkoxy derivative was attributed to a relatively low mobility in the crystalline phase rather than a jump in the mesophase. The charge-carrier mobility in the Col$_h$ phase of **26b** was 0.17 cm^2 V^{-1} s^{-1}, while it was reported to be 0.36 cm^2 V^{-1} s^{-1} in the Col$_h$ phase of **26a**. The Col$_r$ phase of **26a** displayed the highest mobility of 0.7 cm^2 V^{-1} s^{-1}.

6.3.3.2.4 Porphyrin Discotics

Though the importance of porphyrins in photosynthesis is well known, it is surprising that discotic porphyrin derivatives have not been much studied for their charge transport properties. The structures of discotic liquid crystalline porphyrins studied by PR-TRMC or by TOF methods are given in Figure 6.27 and their charge-carrier mobility data are presented in Table 6.4. While both

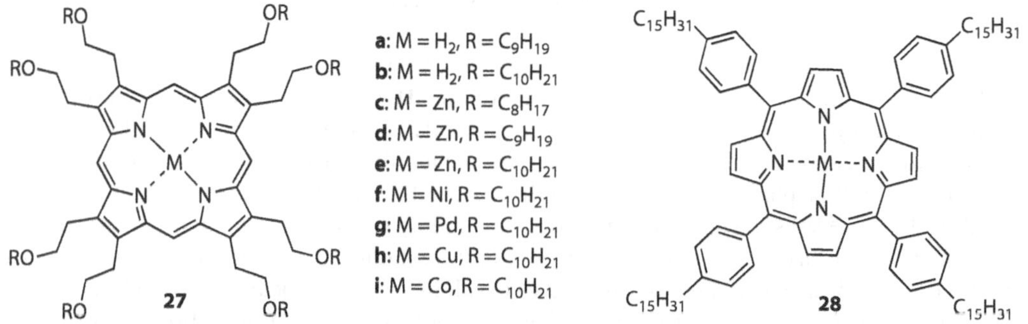

a: M = H$_2$, R = C$_9$H$_{19}$
b: M = H$_2$, R = C$_{10}$H$_{21}$
c: M = Zn, R = C$_8$H$_{17}$
d: M = Zn, R = C$_9$H$_{19}$
e: M = Zn, R = C$_{10}$H$_{21}$
f: M = Ni, R = C$_{10}$H$_{21}$
g: M = Pd, R = C$_{10}$H$_{21}$
h: M = Cu, R = C$_{10}$H$_{21}$
i: M = Co, R = C$_{10}$H$_{21}$

FIGURE 6.27 Chemical structure of porphyrin-based DLCs studied for charge-carrier mobility by TOF or PR-TRMC methods.

TABLE 6.4
Charge-Carrier Mobility Data of Various Porphyrin-Based DLCs (Figure 6.27)

| | Mobility ($cm^2 V^{-1} s^{-1}$) | | | |
Compound	Cr Phase	Mesophase	Method	Ref.
27a	0.20	—	PR-TRMC	[97]
27b	0.34	—	PR-TRMC	[97]
27c	0.01 (hole)	—	TOF	[171]
	0.008 (electron)	—	TOF	[171]
27d	0.26	0.06	PR-TRMC	[97]
27e	0.30	0.06	PR-TRMC	[97]
27f	0.35	—	PR-TRMC	[97]
27g	0.39	0.056	PR-TRMC	[97]
27h	0.36	0.065	PR-TRMC	[97]
27i	0.27	0.53	PR-TRMC	[97]
28	0.0001	0.0001	TOF	[172]

metal-free and nickel derivatives do not exhibit any mesophase, all other derivatives display a columnar mesophase.

The charge-carrier mobility measured by the PR-TRMC technique in the columnar phase of metal-free as well as metallated octakis(β-alkoxyethyl)porphyrins **27** was about 0.06 $cm^2 V^{-1} s^{-1}$ and it was about 0.3 $cm^2 V^{-1} s^{-1}$ in the crystalline phase of these derivatives. These values are comparable to the charge-carrier mobility values of phthalocyanine discotics. The nature of central metal atom does not impart much change in mobility value either in crystalline phase or in columnar phase. On the other hand, the hole and electron mobilities measured by the TOF method in the crystalline phase of zinc octakis((β-octyloxyethyl)porphyrin were reported to be about 0.01 and 0.008 $cm^2 V^{-1} s^{-1}$, respectively [171]. Monobe et al. measured the carrier mobility of discotic lamellar mesophase formed by 5,10,15,20-tetrakis(4-*n*-pentadecylphenyl)porphyrin **28** by the TOF technique. The observed values of carrier mobility were on the order of 10^{-4} $cm^2 V^{-1} s^{-1}$ [172]. The low mobility could be due to the imperfect alignment of the mesophase due to high viscosity.

6.3.3.2.5 Electron-Deficient Discotic Liquid Crystals

For many device applications, both *p*-type and *n*-type organic semiconductors are required. In *p*-type semiconductors, the major charge carriers are holes, whereas in *n*-type semiconductors, they are electrons. While *p*-type discotics are abundant and well studied for charge transport properties, *n*-type discotics are rare, and there are only a few reports on their charge transport behavior.

Perhaps the best *n*-type organic semiconducting materials are perylene derivatives. Hence, perylene derivatives are widely used as active layers in various devices such as solar cells, light emitting diodes, FETs, etc. High field-effect mobilities have been realized in *n*-type TFTs using vapor-deposited *N,N'*-dialkyl-perylene diimide, including a saturation value of 1.7 $cm^2 V^{-1} s^{-1}$ for *N,N'*-dioctadecyl-3,4,9,10-perylene diimide [283]. An efficient solar cell using a discotic liquid crystalline hole transporting HBC derivative and a nonliquid crystalline perylene derivative as electron transporting material has already been realized [284]. To gauge the potential of these materials as processable self-organizing supramolecular materials, a number of liquid crystalline perylene derivatives have been synthesized (Chapter 2) and some of these were studied for their charge transport properties (Figure 6.28). The charge-carrier mobility in the octadecylperylene **29a** was determined by the PR-TRMC method [170]. The mobility in excess of 0.1 $cm^2 V^{-1} s^{-1}$ in the liquid crystalline phase and in excess of 0.2 $cm^2 V^{-1} s^{-1}$ in the crystalline phase was observed

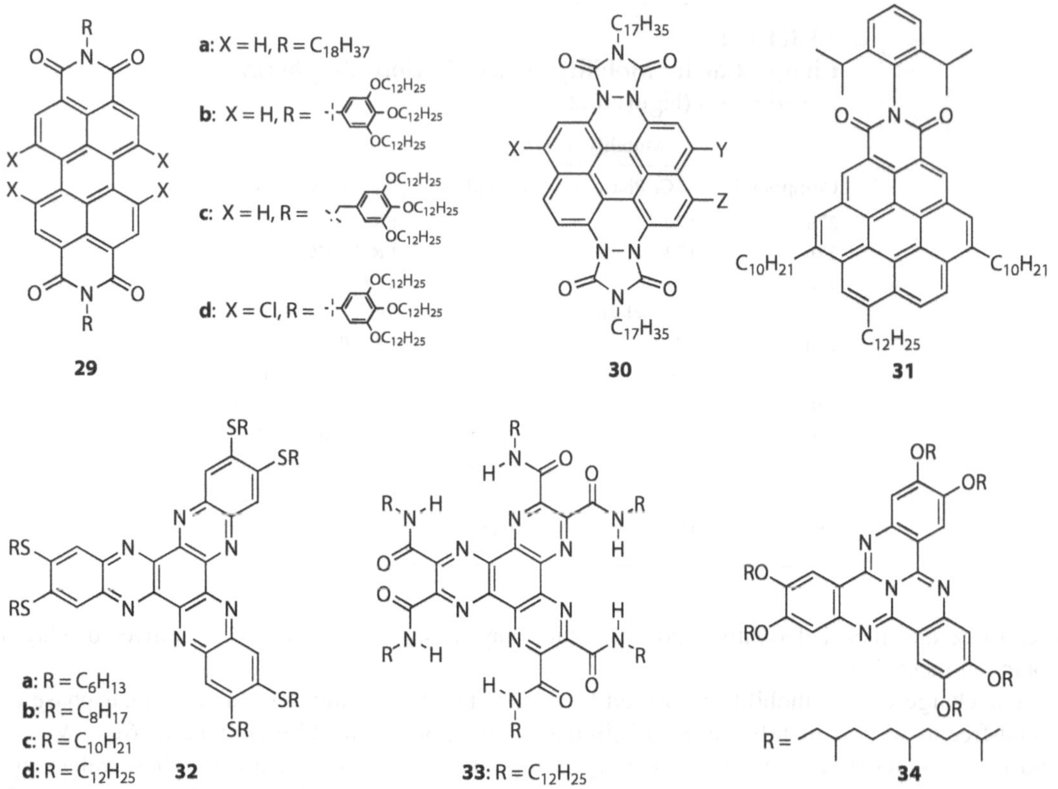

FIGURE 6.28 Chemical structure of electron-deficient DLCs studied for charge-carrier mobility by TOF or PR-TRMC methods.

in this material. The charge mobility in the perylene derivatives **30** was found to lie within the range of 0.01–0.1 cm^2 V^{-1} s^{-1} [164]. The PR-TRMC method was also used to measure the charge-carrier mobility of perylene derivatives **29b** with trialkoxyphenyl substitution and **29d** bearing four additional chlorine atoms in the bay region [222]. The charge mobility in the chlorinated derivative **29d** was found to be significantly higher than that found in the parent molecule **29b**. More importantly, a dramatic increase in charge-carrier lifetime (over 100 times) was observed upon bay substitution with chlorine atoms [222]. This enhanced lifetime persists when the material was blended with *p*-type discotic HBC derivative. As mentioned earlier, the combination of perylene and HBC derivatives has already been used to construct an efficient solar cell and, therefore, this finding may be useful in improving the device performance. The charge-carrier mobility in a room temperature discotic liquid crystalline perylene derivative **29c** was reported to be 0.011 cm^2 V^{-1} s^{-1} [223]. However, the electron mobility in the same compound measured by the steady-state SCLC technique was reported to be 1.3 cm^2 V^{-1} s^{-1} [223]. The significant difference observed in the two techniques was attributed to the different sample processing. A related coronenemono-imide **31** also exhibits a large intracolumnar charge-carrier mobility of about 0.2 cm^2 V^{-1} s^{-1} [181]. Surprisingly, the direct TOF method, which can measure both electron and hole mobilities separately, has not been explored in perylene discotics probably because of difficulties in obtaining well-aligned samples.

The hexaazatrinaphthylene (HATNA) derivatives are other well-known electron-transporting materials. Lehmann et al. prepared four liquid crystalline derivatives of HATNA **32** (Figure 6.28) and studied their charge-carrier mobility behavior by the PR-TRMC technique [224]. The mobility data are presented in Table 6.5. Unlike many other discotics where the charge-carrier mobility

TABLE 6.5
Charge-Carrier Mobility Data of Various
Electron-Deficient DLCs (Figure 6.28)

Compound	Mobility (cm² V⁻¹ s⁻¹)		Method	Ref.
	Cr Phase	Mesophase		
29a	0.2	0.1	PR-TRMC	[170]
29b	—	0.0078	PR-TRMC	[223]
	—	0.011	PR-TRMC	[222]
29c	—	0.011	PR-TRMC	[222]
30	0.01–0.1	0.01–0.1	PR-TRMC	[164]
31	—	0.2–0.3	PR-TRMC	[181]
32a	<0.01	—	PR-TRMC	[224]
32b	0.07	0.05	PR-TRMC	[224]
32c	0.59	0.26	PR-TRMC	[224]
32d	0.27	0.02	PR-TRMC	[224]
33	—	0.04–0.08	PR-TRMC	[198]
34	—	0.02 (hole)	TOF	[285]
		0.04 (electron)	TOF	[285]

in the crystalline phase is hardly dependent on peripheral chains, the HATNA homologs display dramatic variation in the mobility values. The first member of the series **32a** exhibits poor charge-carrier mobility, while the third member, **23c**, with decyl chains, displays excellent mobility value. Compounds **32b** and **32d** show intermediate behavior. In all derivatives, the mobility increases gradually with increasing temperature up to the phase transition. At the crystal-to-mesophase transition, an abrupt decrease in mobility occurs and that remains stable throughout the mesophase. Further, during the cooling cycle, the mobility values were much lower than in the heating cycle. Except in the case of **32b**, the crystalline phase that appeared from the mesophase does not show the mobility similar to the virgin material. This can only be rationalized by the different morphologies of the solvent-grown and melt-grown crystals.

The hexamides of hexaazatriphenylene **33** are very interesting electron-deficient discotics as the molecules pack very densely in the columnar phase due to intramolecular hydrogen bonding. The inter-core distance for the compound **33** was found to be 0.318 nm, the smallest inter-disk distance known in DLCs [198]. From the end-of-pulse conductivity (PR-TRMC), the sum of charge-carrier mobility for negative and positive carriers was determined to be 0.08 cm² V⁻¹ s⁻¹ at 200°C and 0.04 cm² V⁻¹ s⁻¹ at −80°C. Despite the very small core–core separation, the mobility in this compound was much lower than the mobility observed in lutetium phthalocyanine dimers having a core–core separation of 0.326 nm. This could be because compound **33** does not have a sharp isotropic phase transition. It decomposes at about 250°C and, therefore, a well-aligned columnar phase cannot be obtained. Nevertheless, the charge-carrier mobility was much higher than that observed for similar core size triphenylene derivatives.

The tricycloquinazoline (TCQ) **34** is another electron-deficient discotic core having very small core–core separation (0.329 nm) in the columnar mesophase [275]. The material has recently been studied for the charge transport properties by the TOF technique. The mobilities for the positive and negative carriers in room temperature discotic TCQ derivative **34** were estimated to be 2×10^{-2} and 4×10^{-2} cm² V⁻¹ s⁻¹, respectively, at room temperature (20°C) [285].

In another interesting approach, Percec and coworkers recently reported the charge-carrier mobility in the columnar structure of dendrimers [286]. In these molecules, the donor/acceptor groups fill the central space to form columnar phases, which can be easily aligned. TOF measurements indicate the electron and hole mobilities ranging from 1×10^{-4} to 7×10^{-3} cm^2 V^{-1} s^{-1} in the self-assembled dendrons. Recently, such supramolecular functional columnar materials derived from nonconventional discotic cores have received a great deal of attention [286–293]. However, as mentioned earlier, such non-discotic cores forming self-assembling columnar structures have not been covered here.

6.4   DISCOTIC SOLAR CELLS

Mother Nature generates chemical energy from solar energy via photosynthesis, and that is the basis of life on earth. The light-induced electron-hole generation, separation, and migration is an important process for the conversion of light to electric or chemical energy. Most of the world's current energy supplies are based on fossil energy sources, and only a little is provided by hydropower, biomass, and nuclear energy. The fossil energy sources, will in the long-term become only limitedly available; moreover, the exponential growth of CO_2 level in the atmosphere due to the burning of fossil fuels is leading to the threat of a worldwide climate change. Exploitation of nuclear energy does not appear to be socially acceptable universally due to the hazards associated with it. The problem of energy provision around the world remains unsolved. The alternative energy sources are the renewable energy sources, and solar light is the most abundant source of regenerative energy. The device that is used to convert solar energy into electric energy is called a PV solar cell. A solar cell is basically a semiconductor diode. Traditionally, inorganic semiconductors, such as Si, GaAs, CdS, etc., have been used due to their high charge-carrier mobility and stability. However, recently solar cells based on organic semiconductors have attracted much attention due to their low cost, easy processability on hard as well as on flexible substrates, and large-area application [294–304]. The power conversion efficiencies of organic solar cells have now reached to about 5.5%. These power conversion efficiencies are much lower compared to inorganic solar cells (>40% in inorganic tandem cells) or organic–inorganic hybrid dye-sensitized solar cells (~11%). From the efficiency point of view (if cost is not in consideration, e.g., in space applications), with the present state of development, organic solar cells will never be able to compete with inorganic solar cells. However, for routine applications where watt per dollar is important, organic solar cells may compete with inorganic solar cells. However, the efficiency of organic solar cells has to be improved a little more (by a factor of 2 or so) to compete with inorganic solar cells.

In the beginning, small molecules, such as anthracene, porphyrins, phthalocyanines, perylenes, etc., were mainly deposited by vacuum evaporation techniques since they possess poor solubility in organic solvents. These single layer cells generally give very low power efficiency (<0.1%). This is primarily because of the fact that the photoexcitation of organic materials produces excitons (tightly bound electron–hole pair). The potential difference of the unsymmetrical work functions of the electrodes is not sufficient to break up these excitons. The excitons diffuse within the organic layer to reach the electrode, where they may dissociate into free charges or recombine. Since the device thickness is generally much higher than the exciton lengths, most of the excitons recombine and only a few free charges are available to generate photocurrent. Much better efficiencies can be achieved by using donor–acceptor heterojunctions [305]. The bilayer cell is prepared by depositing an electron-rich and an electron-deficient organic semiconductor material. At the donor–acceptor interface, excitons separate much more efficiently and move in respective layers. Recently, attention was paid on polymeric materials, as they are commonly soluble in organic solvents. Although the vapor deposition of organic materials, generally gives higher mobility, the real technology that would impart a high impact on the manufacturing cost is the use of soluble organic semiconductors. Spin coating, spray coating, solution casting, and inkjet printing are the most commercially feasible processing techniques. In the solution-processed materials, the mobility depends strongly on film morphology, which can be controlled by the processing conditions. A combination of soluble organic polymers as hole-transporting material and C60 derivative as electron-transporting material has yielded power conversion efficiencies in excess of 5%.

Though tremendous progress has been witnessed in the field of organic electronics over the last two decades, the real commercial organic electronic devices are still not common. The reason lies in the fact that most of the organic materials lack the combination of processability and high mobility. Highly ordered organic structures like polycyclic hydrocarbons having high mobility are generally not processable. On the contrary, processable organic materials like oligomers and polymers do not display high mobility because of the low degree of order. Therefore, the use of LCs having both order and mobility is a viable solution. DLCs are highly processable self-organizing supramolecular materials with high charge-carrier mobility and, therefore, are good candidates to replace organic single crystals, polymers, and inorganic semiconducting materials. Despite these attractive features, DLCs have not been much explored for PV studies. Compared to other organic solar cells, studies on discotic solar cells are rare. Moreover, in most cases standard conditions required for PV solar cell characterization are not followed and, therefore, most of these discotic PV devices are poorly characterized. Recently, some efforts have been made to fabricate proper discotic solar cells.

In a seminal report [284], Mullen and coworkers constructed a *p/n*-type PV solar cell using discotic liquid crystalline HBC as the hole-transporting layer and a perylene dye as the electron-transporting layer. A chloroform solution of crystalline dye perylene *N,N'*-bis(1-ethylpropyl)-3,4,9,10-perylenebis(dicarboximide) and room temperature DLC hexadodecylphenyl-HBC were spin-coated onto indium tin oxide. Photodiodes were prepared by the evaporation of aluminum onto organic semiconducting blend films (Figure 6.29). The device exhibits external quantum efficiencies (EQEs) up to 34% and power efficiencies of up to ~2%. Efficient photoinduced charge transfer between the HBC and perylene and facile charge transport through vertically segregated perylene and HBC *p*-systems are primarily responsible for these efficiencies. Subsequently, a number of HBC and related large-core discotics and liquid crystalline as well as nonliquid crystalline perylene derivatives were used to prepare organic solar cells [306–312]; however, none could cross the above-mentioned power efficiency limit.

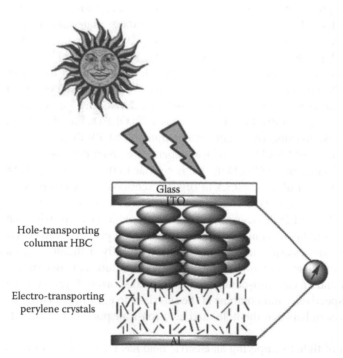

Hole-transporting columnar HBC

Electro-transporting perylene crystals

FIGURE 6.29 Schematic diagram of a bilayer PV solar cell prepared using a discotic liquid crystalline hole-transporting HBC derivative and a crystalline perylene derivative as electron-transporting materials.

Perylene derivatives were used as *n*-type semiconductor in almost all discotic devices [284,306–326]. Other *p*-type DLCs explored to construct solar cells are porphyrin [313–315], phthalocyanine [316–318], and triphenylene [319–322] derivatives. Despite very good absorption of sunlight and high charge-carrier mobilities, none of the devices show very high power-conversion efficiency. Efforts have also been made to couple donor molecules with central perylene core [323,324], but attempts to prepare PV solar cells from these liquid crystalline donor–acceptor–donor systems failed; most devices exhibited poor diode behavior. An unusual PV cell filled with a discotic liquid crystalline porphyrin in between two symmetrical ITO electrodes was reported by Gregg et al. [313]. The PV effect was observed because the concentration of excitons is much higher at the illuminated electrode interface than at the dark electrode due to strong absorption of incident light by the porphyrin and exciton dissociation at the porphyrin/ITO interface is inherently asymmetric. This leads to a strong and persistent PV effect.

The morphology of active layer plays a crucial role in charge and exciton transport. The film morphology depends on several factors such as solvent, solubility of the donor and acceptor blend, spin-coating parameters (speed, temperature), annealing, etc. Destruel et al. studied the effect of UV–ozone and argon plasma treatment of ITO on the PV parameters of devices based on triphenylene/perylene discotics [321]. The work function of ITO increases due to UV–ozone treatment, while the treatment with argon plasma decreases the ITO work function. It was observed that the open circuit voltage (V_{OC}) depends not only on the organic–organic interface but also on the electrode–organic interface. V_{OC} decreases on UV–ozone treatment of ITO, while the argon plasma treatment increases the V_{OC}.

6.5 DISCOTIC LIQUID CRYSTALS AS ORGANIC LIGHT-EMITTING DIODE MATERIALS

The phenomenon of electroluminescence is the reverse of the PV process. The generation of light using organic materials is fascinating due to their vast variety, relative ease of processing, and the ability to alter their properties by chemical means. Although organic light-emitting diodes (OLEDs) have been known for a long time, the real interest in these materials started after the patented discovery of Kodak scientists in the late 1980s. In the last 25 years, particular attention has been devoted to the development and realization of OLEDs for commercial large-area, thin, lightweight, flexible, high-density information color display, backlights, and even to general lighting applications. More than 5000 patents related to the invention of devices based on OLEDs can be found in the literature. Several companies all over the world are developing applications for the technology focusing on small, low-power displays. Displays based on OLEDs found in digital cameras, cellular phones, car audio components, and small- and medium-sized TV displays have already been commercialized and very large OLED-based "wallpapers" are under development.

The basic architecture of an OLED device is similar to that of a PV cell. In an OLED device, an organic semiconductor film or a stack of organic semiconductor layers is sandwiched between a transparent anode with high work function (usually ITO) and a low work function cathode (e.g., Al or Mg). Generally, OLED devices are single-layer, double-layer, or triple-layer structures. In a single-layer device, the fluorescent organic material should possess both *p*-type and *n*-type characteristics for hole and electron migration. This is generally difficult and, therefore, multilayer devices are more efficient. The double-layer OLED structure consists of an electron-rich hole transporting layer and an electron-deficient electron-transporting layer. Holes and electrons move in these layers, respectively, and combine at the interface to emit light. Three-layer devices consist of an emission layer in between the hole- and electron-transporting layers and are more efficient, albeit expensive.

The generation of light by applying an electric field has been reported to be highly efficient for the conjugated organic polymers such as polyphenylenevinylene having charge-carrier mobilities 10^{-6} to 10^{-8} cm^2 V^{-1} s^{-1}. High mobilities are an advantage for application in LEDs in order to obtain high current densities and hence high light intensities. As described earlier, the charge-carrier mobilities in columnar phases of DLCs are high (up to 10^{-1} cm^2 V^{-1} s^{-1}) and, therefore, they can be used in electroluminescence devices. DLCs have recently been used both as hole-transporting and electron-transporting materials to construct OLEDs [327–340]. A typical bilayer discotic OLED cell structure is shown in Figure 6.30.

Mainly triphenylene discotics have been exploited to construct OLED devices. This could be due to the fact that the work function of several triphenylene derivatives is close to the work function of an ITO electrode and they possess high charge-carrier mobility. Perylene- and pyrene-based DLCs have been used as an electron-transporting layer due to their high charge-carrier mobility in conjunction with excellent luminescence properties. Wendroff and coworkers prepared the first discotic LEDs using triphenylene monoesters **12a** and **12b** (Figure 6.23) as well as a mixture of **12b** and a tristilbene amine as the active materials [330]. The discotic material was spin coated on the ITO-covered glass surface followed by aluminum cathode deposition by the vacuum evaporation technique. The device exhibits clear diode characteristics. Annealing the film (heating the film to the isotropic state followed by slow cooling into the columnar phase) improves device characteristics. The onset voltage is reduced from 14 V to about 6 V due to the orientation of the columnar axis along the substrate normal. This corresponds to a decrease in the threshold field from 1.4×10^6 V cm^{-1} to about 6×10^5 V cm^{-1}. Double-layer and triple-layer devices using triphenylene discotic as a hole-transporting layer with tristilbene chromophore, eicosylamine as stabilizer, and a mixture of fluorinated quarterphenyl copolymer and arachidic acid methyl ester as hole-blocking materials have also been developed [329]. The diode characteristics for single-, two-, and three-layer devices are shown in Figure 6.31. The diode characteristics of these devices differ significantly. A strong reduction of the onset fields with an increasing number of layers can be seen.

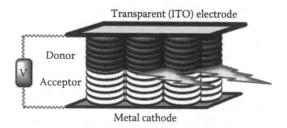

FIGURE 6.30 Schematic representation of a discotic LED device.

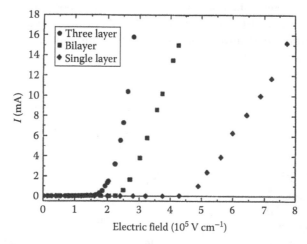

FIGURE 6.31 Diode characteristics for single-, two-, and three-layer devices with spin-coated discotic layers. (Reproduced from Stapff, I.H. et al., *Liq. Cryst.*, 23, 613, 1997. With permission.)

The threshold field decreases from 5×10^5 V^{-1} cm^{-1} for a single-layer device to 2.5×10^5 V cm^{-1} for the bilayer and finally to 2×10^5 V^{-1} cm^{-1} for the triple-layer device. Annealing the device further decreases the onset voltage to 1.5×10^5 V cm^{-1}.

Bacher et al. used various triphenylene-based monomeric, dimeric, and polymeric DLCs and well-known tris(8-quinolinolato)aluminum(III) complex (Alq$_3$) as electron transport and emitting material to construct bilayer LED devices [328]. These devices exhibit a green emission of Alq$_3$. The LED prepared with monomeric triphenylene discotic **11b** (Figure 6.23) starts to exhibit electroluminescence at a field of 5.7×10^5 V cm^{-1}. A maximum luminance of 960 cd m^{-2} at a current density of 150 mA cm^{-2} was reached at a field of 2.1×10^5 V cm^{-1} (15 V driving voltage). The LED device prepared using the discotic dimer **15a** (Figure 6.25) displayed a maximum luminance of 380 cd m^{-2} at a field of 2.4×10^6 V cm^{-1} (17 V driving voltage) at a current density of 22 mA cm^{-2}. The device starts to emit green light at 4 V and a field of 5.7×10^5 V cm^{-1}. The LED device prepared using the polymeric triphenylene discotic starts emitting green light at a voltage of 4.5 V and a field of 6.4×10^5 V cm^{-1}. A maximum luminance of 1390 cd m^{-2} at a current density of 166 mA cm^{-1} was reached at an electric field of 2.3×10^6 V cm^{-1} (16 V driving voltage). Thus, it was concluded that the polymeric discotic material is better than other discotic monomers and dimers. The LED device prepared by the photopolymerization of polymerizable triphenylene monomer on the ITO surface exhibits about one order of magnitude lower brightness compared to the device prepared using a prepolymerized polymer solution [333]. LED devices made from triphenylene diacrylate and triacrylate (which forms insoluble cross-linked networks) show reasonable performance under the applied conditions [333].

Seguy et al. prepared an OLED with both an electron-rich triphenylene-based DLC **11** (Figure 6.23) as hole-transporting layer and an electron-deficient fluorescent perylene-based DLC **35** (Figure 6.32) as an electron-transporting material [334,337]. The LED starts emitting red fluorescence above 10 V and is roughly proportional to the current density with 0.05 cd A^{-1}, corresponding to 15 cd m^{-2} at 30 mA cm^{-2} and 30 V. Perylene-based discotics are particularly very promising for LED applications as they are quite stable under ambient conditions, and even a single-layer LED emits bright orange-red luminescence above 12 V [335,336]. The device was stable for about 20 h even when operated under environmental moisture and oxygen. However, it degrades rapidly above 25 V. Pyrene and triphenylene derivatives as electron-transporting materials in conjunction with a hole-transporting material like N,N'-diphenyl-N,N'-bis(3-methylpheny)-1,1'-bipheny-4,4'diamine or poly-3,4-ethylene dioxythiophene have also been used to prepare LED devices with good characteristics [335,336]. Pure triphenylene **36**, pyrene **37**, and perylene **35** compounds display violet-blue, yellow-green, and orange-red luminescence, respectively (Figure 6.33). It has also been reported that polarized electroluminescence can be obtained from aligned perylene DLCs [341].

With the idea that a combination of hole- and electron-transporting properties within a single molecule may display better LED properties, Hanack and coworkers prepared a number of conjugated-bridged triphenylene derivatives [338]. They show bright orange to red photoluminescence, though the efficiency was not very good.

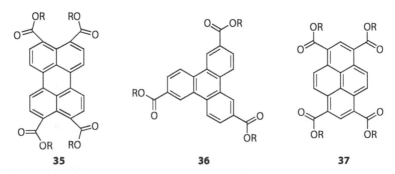

35   **36**   **37**

FIGURE 6.32   Chemical structure of some electron-deficient discotics used for OLED devices.

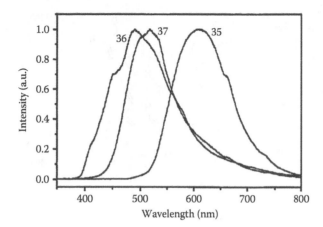

FIGURE 6.33 Electroluminescence spectra of compounds **35**, **36**, and **37**. (Reproduced from Benning, S.A. et al., *Liq. Cryst.*, 28, 1105, 2001. With permission.)

6.6 DISCOTIC FIELD EFFECT TRANSISTORS

The FET is a type of transistor commonly used for amplifying or switching weak power signals. FETs are an integral part of computer chips. Traditionally, inorganic semiconductors such as Si, GaAs, etc., have been used in TFT devices; however, recently, several efforts have been made to generate organic FETs (OFETs) due to their commercial potential in various low-cost electronic devices. Conventional transistors have three terminals: the source, the drain, and the gate electrodes. The gate controls the density of the charge carriers to migrate through the central region of a transistor, which is usually made of a semiconducting material. If the charge density is high, that is, the flow of carriers is unrestricted, current flows from the source to the drain. However, current does not flow if the charge density is low or is restricted to flow through this central region. This property allows the transistor to operate as a switch.

FETs are primarily two types: top-contact geometry and bottom-contact geometry. In a top-contact geometry, the two drain and source electrodes (usually gold deposited by vacuum evaporation) are placed on the active semiconductor layer. In a bottom-contact geometry, these electrodes are under the active semiconductor layer. The processing of a top-contact structure is much simpler and, therefore, this architecture is most commonly used.

The performance of OFETs depends largely on the active semiconducting material employed in such devices. The high anisotropic charge-carrier mobility of DLCs in conjunction with their self-assembling properties makes them an attractive candidate for OFETs. A typical top-contact structure of a discotic OFET is shown in Figure 6.34. Only a few discotic liquid crystalline semiconductors have so far been shown to have the potential for transistor devices [342–350].

Discotic HBC derivatives are attractive materials for OFETs as they possess high charge-carrier mobility and can be easily aligned parallel to the surface (planar alignment) via zone casting and other techniques. The TFT devices prepared using HBC discotic **19d** (Figure 6.25) exhibit field-effect charge-carrier mobilities of $0.5–1.0 \times 10^{-3}$ cm^2 V^{-1} s^{-1} with ON/OFF ratios of more than 10^4 and a turn-on voltage of ca. -5 to -10 V [342]. A poly(tetrafluoroethane) (PTFE) layer was used to align the discotic sample. Highly ordered thin films of HBC discotic **19i** can be obtained by applying high magnetic fields [345]. TFT devices prepared in this way also exhibit similar field-effect mobilities (10^{-3} cm^2 V^{-1} s^{-1}). Charge-carrier mobility (μ_{FE}) of up to 1×10^{-2} cm^2 V^{-1} s^{-1} with an ON/OFF ratio of an 10^4 and a turn-on voltage of ca. -15 V was observed for

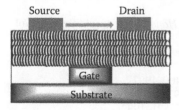

FIGURE 6.34 Schematic representation of a discotic field-effect transistor.

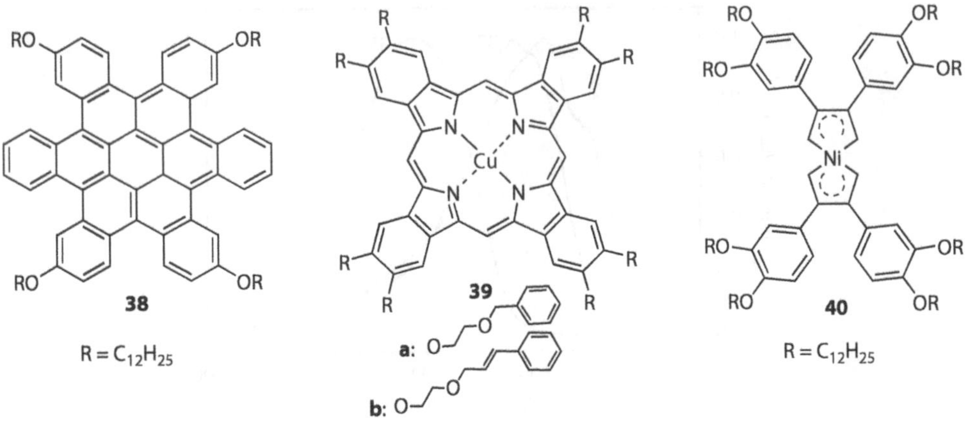

FIGURE 6.35   Chemical structure of some discotics used for OTFT devices. These structures are in addition to some other discotics presented in Figure 6.25.

a zone-cast HBC derivative [346]. Xiao et al. fabricated TFTs using a new type of HBC discotic **38** (Figure 6.35) [347]. This distorted nonplanar HBC derivative can be viewed as a hybrid of three pentacene units. It may be emphasized that pentacene is one of the most efficient organic semiconductors [351,352]. The charge-carrier mobility of $0.02\,cm^2\,V^{-1}\,s^{-1}$ with a very high ON/OFF ratio ($10^6{:}1$) was realized in this HBC device.

Donley et al. prepared bottom-contact OFETs using metal phthalocyanine discotics **39** (Figure 6.35) [347,348]. Field-effect mobilities in the range of $1–5\times10^{-6}\,cm^2\,V^{-1}\,s^{-1}$ were reported for phthalocyanine **39a**. For Pc **39b**, the mobility was reported to be $0.018\,cm^2\,V^{-1}\,s^{-1}$ at room temperature showing a field dependence interpreted with the Frenkel–Poole mobility model [347]. Another discotic metallomesogen used to fabricate OTFT is nickel bis(dithiolene)complex **40** [352]. The TFT device was prepared with an Ag source and drain electrodes. An effective mobility value of $1.3\times10^{-3}\,cm^2\,V^{-1}\,s^{-1}$ was reported in this device.

Perylene discotics are well-known n-type semiconductors and exhibit very high electron mobility. However, their charge-carrier mobilities have been studied only by the PR-TRMC and TOF methods. Efforts have not been made to study their field-effect mobility probably because of a homogenous alignment problem. Depending upon peripheral substitution, perylene derivatives can exhibit columnar or smectic phases. Electron mobilities in the smectic mesophase was reported to be about $0.1\,cm^2\,V^{-1}\,s^{-1}$ [170]. In fact, a smectic phase formed by discotic molecules (biaxial smectic) is better from an FET point of view. It is much easier to align an Sm phase homeotropically than aligning a columnar phase homogeneously. The discotic nature of central π-electronic structure may have a better overlapping of electronic orbitals and is thus expected to give better charge mobility in the discotic Sm phase.

REFERENCES

1. Bahadur, B. (ed.). *Liquid Crystals: Applications and Uses*, Vols. 1–3, World Scientific, Singapore, 1990.
2. Schadt, M. and Helfrich, W. *Appl. Phys. Lett.* 18, 127–128, 1971.
3. Demus, D., Goodby, J., Gray, G. W., Spiess, H. W., and Vill, V. (eds.). *Handbook of Liquid Crystals*, Vol. 2a, Wiley-VCH, Weinheim, Germany, 1998.
4. Dunmur, D. A., Fukuda, A., and Luckhurst, G. R. (eds.). *Physical Properties of Liquid Crystals: Nematics*, INSPEC, London, U.K., 2001.
5. Oh-E, M., Yoneya, M., Ohta, M., and Kondo, K. *Liq. Cryst.* 22, 391–400, 1997.
6. Oh-E, M., Yoneya, M., and Kondo, K. *J. Appl. Phys.* 82, 528–535, 1997.
7. Takatori, K., Sumiyoshi, K., Hirai, Y., and Kaneko, S. *Jpn. Display* 92, 521, 1992.
8. Koma, N., Yaba, Y., and Matsuoka, K. *SID Int. Symp. Digest Tech.* 869, 1995.
9. Takeda, A., Kataoka, S., Sasaki, T., Chida, H., Tsuda, H., Ohmuro, K., Sasabayashi, T., Koike, Y., and Okamoto, K. *SID Int. Symp. Digest Tech.* 29, 1077–1080, 1998.

10. Lee, S. H., Lee, S. L., and Kim, H. Y. *Appl. Phys. Lett.* 73, 2881–2883, 1998.
11. Hong, S. H., Park, I. C., Kim, H. Y., and Lee, S. H. *Jpn. J. Appl. Phys.* 39, L527–L530, 2000.
12. Lee, S. H., Lee, S. M., Kim, H. Y., Kim, J. M., Hong, S. H., Jeong, Y. H., Park, C. H., Choi, Y. J., Lee, J. Y., Koh, J. W., and Park, H. S. *SID Int. Symp. Digest Tech.* 32, 484–487, 2001.
13. Mori, H., Itoh, Y., Nishiura, Y., Nakamura, T., and Shinagawa, Y. *Jpn. J. Appl. Phys.* 36, 143–147, 1997.
14. Mori, H. *Jpn. J. Appl. Phys.* 36, 1068–1072, 1997.
15. Hoke, C. D., Mori, H., and Bos, P. J. *Jpn. J. Appl. Phys.* 38, L642–L645, 1999.
16. Mori, H. and Bos, P. J. *Jpn. J. Appl. Phys.* 38, 2837–2844, 1999.
17. Sergan, T. A., Jamal, S. H., and Kelly, J. R. *Displays* 20, 259–267, 1999.
18. Wu, L. H., Luo, S. J., Hsu, C. S., and Wu, S. T. *Jpn. J. Appl. Phys.* 39, L869–L871, 2000.
19. Okazaki, M., Kawata, K., Nishikawa, H., and Negoro, M. *Polym. Adv. Tech.* 11, 398–403, 2000.
20. Lu, M. and Yang, K. H. *Jpn. J. Appl. Phys.* 39, L412–L415, 2000.
21. Leenhouts, F. *Jpn. J. Appl. Phys.* 39, L741–L743, 2000.
22. Yamahara, M., Inoue, I., Nakai, T., Yamada, Y., and Ishii, Y. *Jpn. J. Appl. Phys.* 41, 6072–6079, 2002.
23. Nishikawa, H., Negoro, M., Kawata, K., and Okazaki, M. *J. Syn. Org. Chem. Jpn.* 60, 1190–1200, 2002.
24. Kawata, K. *Chem. Rec.* 2, 59–80, 2002.
25. Yamahara, M., Inoue, I., Sakai, A., Yamada, Y., Mizushima, S., and Ishii, Y. *Jpn. J. Appl. Phys.* 42, 4416–4420, 2003.
26. Sergan, T., Kelly, J., Yaroshchuk, O., and Chien, L. C. *Mol. Cryst. Liq. Cryst.* 409, 153–162, 2004.
27. Oka, S., Kobayashi, K., Iwamoto, Y., Toko, Y., Kimura, M., and Akahane, T. *Jpn. J. Appl. Phys.* 43, 3443–3447, 2004.
28. Mori, H. *J. Display Tech.* 1, 179–186, 2005.
29. Oikawa, T., Yasuda, S., Takeuchi, K., Sakai, E., and Mori, H. *J. Soc. Inf. Display* 15, 133–137, 2007.
30. Hwang, S. H., Lim, Y. J., Lee, M. H., Lee, S. H., Lee, G. D., Kang, H., Kim, K. J., and Choi, H. C. *Curr. Appl. Phys.* 7, 690–696, 2007.
31. Chandrasekhar, S., Prasad, S. K., Nair, G. G., Rao, D. S. S., Kumar, S., and Manickam, M. *EuroDisplay'99, The 19th International Display Research Conference Late-News Papers*, Berlin, Germany, 1999, pp. 9–11.
32. Nair, G. G., Rao, D. S. S., Prasad, S. K., Chandrasekhar, S., and Kumar, S. *Mol. Cryst. Liq. Cryst.* 397, 245–252, 2003.
33. Rahman, M. and Lee, W. *J. Phys. D: Appl. Phys.* 42, 063001, 2009.
34. Ivashchenko, A. V. *Dichroic Dyes for Liquid Crystal Displays*, CRC Press, Boca Raton, FL, 1994.
35. Yeh, P. and Paukshto. M. *Mol. Mater.* 14, 1–19, 2001.
36. Bisoyi, H. K. and Kumar, S. *Chem. Soc. Rev.* 39, 264–285, 2010.
37. Sonpatki, M. and Chien, L. C. *Mol. Cryst. Liq. Cryst.* 367, 545–553, 2001.
38. Sawade, H., Olenik, I. D., Kruerke, D., and Heppke, G. *Mol. Cryst. Liq. Cryst.* 367, 529–536, 2001.
39. Eichhorn, H., Wohrle, D., and Pressner, D. *Liq. Cryst.* 22, 643–653, 1997.
40. Sakajiri, K., Sugisaki, T., and Moriya, K. *Chem. Commun.* 3447–3449, 2008.
41. Praefcke, K., Singer, D., and Gundogan, B. *Mol. Cryst. Liq. Cryst.* 223, 181–195, 1992.
42. Boden, N., Bushby, R. J., Ferris, L., Hardy, C., and Sixl, F. *Liq. Cryst.* 1, 109–125, 1986.
43. Tam-Chang, S. W. and Huang, L. *Chem. Commun.* 1957–1967, 2008.
44. Brooks, J. D. and Taylor, G. H. *Nature* 206, 697–699, 1965.
45. Brooks, J. D. and Taylor, G. H. *Carbon* 3, 185–186, 1965.
46. Brooks, J. D. and Taylor, G. H. *Chemistry and Physics of Carbon*, Vol. 4, Marcel Dekker, New York, 1968, pp. 243–286.
47. Zimmer, J. E. and White, J. L. *Adv. Liq. Cryst.* 5, 157–213, 1982.
48. Gasparoux, H. *Mol. Cryst. Liq. Cryst.* 63, 231–248, 1981.
49. Otani, S. *Mol. Cryst. Liq. Cryst.* 63, 249–263, 1981.
50. Keinan, E., Kumar, S., Moshenberg, R., Ghirlando, R., and Wachtel, E. J. *Adv. Mater.* 3, 251–254, 1991.
51. Ho, A. S. K. and Rey, A. D. *Rheol. Acta* 30, 77–88, 1991.
52. Farhoudi, Y. and Rey, A. D. *Rheol. Acta* 32, 207–217, 1993.
53. Singh, A. P. and Rey, A. D. *J. Phys. II* 4, 645–665, 1994.
54. Hurt, R. H. and Hu, Y. *Carbon* 37, 281–292, 1999.
55. Singh, A. P. and Rey, A. D. *J. Non-Newtonian Fluid Mech.* 94, 87–111, 2000.
56. Rey, A. D. *Modell. Simul. Mater. Sci. Eng.* 8, 803–813, 2000.
57. Hu, Y. and Hurt, R. H. *Carbon* 39, 887–896, 2001.
58. Yan, J. and Rey, A. D. *Phys. Rev. E* 65, 031713, 2002.

59. Yan, J. and Rey, A. D. *Carbon* 40, 2647–2660, 2002.
60. Golmohammadi, M. and Rey, A. D. *Liq. Cryst.* 36, 75–92, 2009.
61. Yan, J. and Rey, A. D. *Mater. Res. Soc. Symp. Proc.* 702, U5.6, 2002.
62. Rey, A. D. *Mater. Res. Soc. Symp. Proc.* 709, CC8.5, 2002.
63. Yan, J. and Rey, A. D. *Carbon* 41, 105–121, 2003.
64. Grecov, D. and Rey, A. D. *Rheol. Acta* 42, 590–604, 2003.
65. Grecov, D. and Rey, A. D. *Carbon* 42, 1257–1261, 2004.
66. de Andrade Lima, L. R. P. and Rey, A. D. *J. Braz. Chem. Soc.* 17, 1109–1116, 2006.
67. Hong, S. J. and Chan, P. K. *Comput. Mater. Sci.* 36, 310–318, 2006.
68. de Andrade Lima, L. R. P. and Rey, A. D. *Chem. Eng. Commun.* 193, 1090–1109, 2006.
69. Burgess, W. A. and Thies, M. C. *Fluid Phase Equilib.* 261, 320–326, 2007.
70. Burgess, W. A., Zhuang, M. S., Hu, Y., Hurt, R. H., and Thies, M. C. *Ind. Eng. Chem. Res.* 46, 7018–7026, 2007.
71. Golmohammadi, M. and Rey, A. D. *Entropy* 10, 183–199, 2008.
72. Edie, D. D., Robinson, K. E., Fleurot, O., Jones, S. P., and Fain, C. C. *Carbon* 32, 1045–1054, 1994.
73. Jian, K. Q., Shim, H. S., Schwartzman, A., Crawford, G. P., and Hurt, R. H. *Adv. Mater.* 15, 164–167, 2003.
74. Zhi, L., Wu, J., Li, J., Kolb, U., and Mullen, K. *Angew. Chem. Int. Ed.* 44, 2120–2123, 2005.
75. Wu, J., El Hamaoui, B., Li, J., Zhi, L., Kolb, U., and Mullen, K. *Small* 1, 210–212, 2005.
76. Gherghel, L., Kubel, C., Lieser, G., Rader, H. J., and Mullen, K. *J. Am. Chem. Soc.* 124, 13130–13138, 2002.
77. Hill, J. P., Jin, W., Kosaka, A., Fukushima, T., Ichihara, H., Shimomura, T., Ito, K., Hashizume, T., Ishii, N., and Aida, T. *Science* 304, 1481–1483, 2004.
78. Jin, W., Yamamoto, Y., Fukushima, T., Ishii, N., Kim, J., Kato, K., Takata, M., and Aida, T. *J. Am. Chem. Soc.* 130, 9434–9440, 2008.
79. Chan, C., Crawford, G., Gao, Y., Hurt, R., Jian, K., Li, H., Sheldon, B., Sousa, M., and Yang, N. *Carbon* 43, 2431–2440, 2005.
80. Wang, X., Zhi, L., Tsao, N., Tomovic, Z., Li, J., and Mullen, K. *Angew. Chem. Int. Ed.* 47, 2990–2992, 2008.
81. Hanack, M. and Lang, M. *Adv. Mater.* 6, 819–833, 1994.
82. Becher, J. and Schaumburg, K. (eds.). *Molecular Engineering for Advanced Materials*, NATO Advanced Science Institutes Series, 456, 147–158, 1995.
83. Boden, N., Bissell, R., Clements, J., and Movaghar, B. *Curr. Sci.* 71, 599–601, 1996.
84. Chandrasekhar, S. In *Handbook of Liquid Crystals*, Vol. 2B, Wiley-VCH, Weinheim, Germany, 1998, Chap. VIII.
85. Boden, N. and Movaghar, B. In *Handbook of Liquid Crystals*, Vol. 2B, Wiley-VCH, Weinheim, Germany, 1998, Chap. IX.
86. Chandrasekhar, S. and Prasad, S. K. *Contemp. Phys.* 40, 237–245, 1999.
87. Hertel, D., Ochse, A., Arkhipov, V. I., and Bassler, H. *J. Imaging Sci. Technol.* 43, 220–227, 1999.
88. Boden, N., Bushby, R. J., Clements, J., and Movaghar, B. *J. Mater. Chem.* 9, 2081–2086, 1999.
89. Eichhorn, H. *J. Pophyr. Phthalocyanines* 4, 88–102, 2000.
90. Bushby, R. J. and Lozman, O. R. *Curr. Opin. Solid State Mater. Sci.* 6, 569–578, 2002.
91. Bushby, R. J. and Lozman, O. R. *Curr. Opin. Colloid Interface Sci.* 7, 343–354, 2002.
92. Ohta, K., Hatsusaka, K., Sugibayashi, M., Ariyoshi, M., Ban, K., Maeda, F., Naito, R., Nishizawa, K., van de Craats, A. M., and Warman, J. M. *Mol. Cryst. Liq. Cryst.* 397, 25–45, 2003.
93. Scott, K., Donovan, K. J., Kreouzis, T., Bunning, J. C., Bushby, R. J., Boden, N., Lozman, O. R. *Mol. Cryst. Liq. Cryst.* 2003, 397, 253–261.
94. Bunning, J. C., Donovan, K. J., Kreouzis, T., Scott, K., Bushby, R. J., Boden, N., and Lozman, O. R. *Mol. Cryst. Liq. Cryst.* 397, 263–271, 2003.
95. Pecchia, A., Siebbeles, L., and Movaghar, B. *Proc. SPIE* 4991, 253, 2003.
96. Hanna, J. I. *Proc. SPIE* 4991, 12, 2003.
97. Warman, J. M. and van de Craats, A. M. *Mol. Cryst. Liq. Cryst.* 396, 41–72, 2003.
98. Simpson, C. D., Wu, J., Watson, M. D., and Müllen, K. *J. Mater. Chem.* 14, 494–504, 2004.
99. Kumar, S. *Syn. React. Inorg. Metal-org. Nano-metal Chem.* 37, 327–331, 2007.
100. Scott, J. C. *Nature* 371, 102–103, 1994.
101. Hoeben, F. J. M., Jonkheijm, P., Meijer, E. W., and Schenning, A. P. H. J. *Chem. Rev.* 105, 1491–1546, 2005.
102. Iino, H. and Hanna, J. *Opto-Electron. Rev.* 13, 295–302, 2005.

103. Hanna, J. I. *Proc. SPIE* 5947, 594703, 2005.
104. Kumar, S. *Chem. Soc. Rev.* 35, 83–109, 2006.
105. Roncali, J., Leriche, P., and Cravino, A. *Adv. Mater.* 19, 2045–2060, 2007.
106. Sergeyev, S., Pisula, W., and Geerts, Y. H. *Chem. Soc. Rev.* 36, 1902–1929, 2007.
107. Laschat, S., Baro, A., Steinke, N., Giesselmann, F., Hagele, C., Scalia, G., Judele, R., Kapatsina, E., Sauer, S., Schreivogel, A., and Tosoni, M. *Angew. Chem. Int. Ed.* 46, 4832–4887, 2007.
108. Nguyen, T. Q., Martel, R., Bushey, M., Avouris, P., Carlsen, A., Nuckolls, C., and Brus, L. *Phys. Chem. Chem. Phys.* 9, 1515–1532, 2007.
109. Cheung, D. L. and Troisi, A. *Phys. Chem. Chem. Phys.* 10, 5941–5952, 2008.
110. Grozema, F. C. and Siebbeles, L. D. A. *Int. Rev. Phys. Chem.* 27, 87–138, 2008.
111. Piechocki, C., Simon, J., Skoulios, A., Guillon, D., and Weber, P. *J. Am. Chem. Soc.* 104, 5245–5247, 1982.
112. Chiang, L. Y., Stokes, J. P., Safinya, C. R., and Bloch, A. N. *Mol. Cryst. Liq.Cryst.* 125, 279–288, 1985.
113. Shimizu, Y., Ishikawa, A., and Kusabayashi, S. *Chem. Lett.* 1041–1044, 1986.
114. van Keulen, J., Warmerdam, T. W., Nolte, R. J. M., and Drenth, W. *Recl. Trav. Chim. Pays-Bas.* 106, 534–536, 1987.
115. Belarbi, Z., Maitrot, M., Ohta, K., Simon, J., Andre, J. J., and Petit, P. *Chem. Phys. Lett.* 143, 400–403, 1988.
116. Boden, N., Bushby, R. J., Clements, J., Jesudason, M. V., Knowles, P. F., and Williams, G. *Chem. Phys. Lett.* 152, 94–99, 1988.
117. Belarbi, Z., Sirlin, C., Simon, J., and Andre, J. J. *J. Phys. Chem.* 93, 8105–8110, 1989.
118. Simon, J. and Sirlin, C. *Pure Appl. Chem.* 61, 1625–1629, 1989.
119. van der Pol, J. F., Neeleman, E., Zwikker, J. W., Nolte, R. J. M., Drenth, W., Aerts, J., Visser, R., and Picken, S. J. *Liq. Cryst.* 6, 577–592, 1989.
120. Schouten, P. G., Warman, J. M., de Haas, M. P., Fox, M. A., and Pan, H. L. *Nature* 353, 736–737, 1991.
121. Schouten, P. G., Wenming, C., Warman, J. M., de Haas, M. P., van der Pol, J. F., and Zwikker, J. W. *Synth. Met.* 41–43, 2665–2668, 1991.
122. Schouten, P. G., Warman, J. M., de Haas, M. P., van der Pol, J. F., and Zwikker, J. W. *J. Am. Chem. Soc.* 114, 9028–9034, 1992.
123. Vaughan, G. B. M., Heiney, P. A., McCauley Jr. J. P., and Smith III, A. B. *Phys. Rev. B* 46, 2787–2791, 1992.
124. Boden, N., Borner, R., Brown, D. R., Bushby, R. J., and Clements, J. *Liq. Cryst.* 11, 325–334, 1992.
125. Schouten, P. G., Warman, J. M., and de Haas, M. P. *J. Phys. Chem.* 97, 9863–9870, 1993.
126. Fox, M. A., Bard, A. J., Pan, H. L., and Liu, C. Y. *J. Chin. Chem. Soc.* 40, 321–327, 1993.
127. Shimizu, Y., Ishikawa, A., Kusabayashi, S., Miya, M., and Nagata, A. *J. Chem. Soc., Chem. Commun.* 656–658, 1993.
128. Bengs, H., Closs, F., Frey, T., Funhoff, D., Ringsdorf, H., and Siemensmeyer, K. *Liq. Cryst.* 15, 565–574, 1993.
129. Closs, F., Siemensmeyer, K., Frey, T., and Funhoff, D. *Liq. Cryst.* 14, 629–634, 1993.
130. Catry, C., Van der Auweraer, M., De Schryver, F. C., Bengs, H., Haussling, L., Karthaus, O., and Ringsdorf, H. *Macromol. Chem.* 194, 2985–2999, 1993.
131. Liu, C. Y., Pan, H. L., Fox, M. A., and Bard, A. J. *Science* 261, 897–899, 1993.
132. Adam, D., Haarer, D., Closs, F., Frey, T., Funhoff, D., Siemensmeyer, K., Schuhmacher, P., and Ringsdorf, H. *Ber. Bunsenges. Phys. Chem.* 97, 1366–1370, 1993.
133. Adam, D., Closs, F., Frey, T., Funhoff, D., Haarer, D., Ringsdorf, H., Schuhmacher, P., and Siemensmeyer, K. *Phys. Rev. Lett.* 70, 457–460, 1993.
134. Boden, N., Bushby, R. J., and Clements, J. *J. Chem. Phys.* 98, 5920–5931, 1993.
135. Schouten, P. G., Warman, J. M., de Haas, M. P., van Nostrum, C. F., Gelinck, G. H., Nolte, R. J. M., Copyn, M. J., Zwikker, J. W., Engel, M. K., Hanack, M., Chang, Y. H., and Ford, W. T. *J. Am. Chem. Soc.* 116, 6880–6894, 1994.
136. Adam, D., Schuhmacher, P., Simmerer, J., Haussling, L., Siemensmeyer, K., Etzbach, K. H., Ringsdorf, H., and Haarer, D. *Nature* 371, 141–143, 1994.
137. Haarer, D., Adam, D., Simmerer, J., Closs, F., Funhoff, D., Haussling, L., Siemensmeyer, K., Ringsdorf, H., and Schuhmacher, P. *Mol. Cryst. Liq. Cryst.* 252, 155–164, 1994.
138. Boden, N., Bushby, R. J., and Clements, J. *J. Mater. Sci. Mater. Electron.* 5, 83–88, 1994.
139. Boden, N., Borner, R. C., Bushby, R. J., and Clements, J. *J. Am. Chem. Soc.* 116, 10807–10808, 1994.
140. Groothues, H., Kremer, F., Schouten, P. G., and Warman, J. M. *Adv. Mater.* 7, 283–286, 1995.
141. Liu, C. Y., Pan, H. L., Tang, H., Fox, M. A., and Bard, A. J. *J. Phys. Chem.* 99, 7632–7636, 1995.

142. Hiura, H., Ebbesen, T. W., and Tanigaki, K. *Adv. Mater.* 7, 275–280, 1995.
143. Warman, J. M. and Schouten, P. G. *J. Phys. Chem.* 99, 17181–17185, 1995.
144. Boden, N., Bushby, R. J., Cammidge, A. N., Clements, J., Luo, R., and Donovan, K. J. *Mol. Cryst. Liq. Cryst.* 261, 251–257, 1995.
145. Boden, N., Bushby, R. J., Clements, J., Movaghar, B., Donovan, K. J., and Kreouzis, T. *Phys. Rev. B* 52, 13274–13280, 1995.
146. Arikainen, E. O., Boden, N., Bushby, R. J., Clements, J., Movaghar, B., and Wood, A. *J. Mater. Chem.* 5, 2161–2165, 1995.
147. Boden, N., Bushby, R. J., Clements, J., and Luo, R. *J. Mater. Chem.* 5, 1741–1748, 1995.
148. Warman, J. M. and Schouten, P. G. *Appl. Organomet. Chem.* 10, 637–647, 1996.
149. van de Craats, A. M., Warman, J. M., de Haas, M. P., Adam, D., Simmerer, J., Haarer, D., and schuhmacher, P. *Adv. Mater.* 8, 823–826, 1996.
150. Haarer, D., Simmerer, J., Adam, D., Schuhmacher, P., Paulus, W., Etzbach, K. H., Siemensmeyer, K., and Ringsdorf, H. *Mol. Cryst. Liq. Cryst.* 283, 63–68, 1996.
151. Simmerer, J., Glusen, B., Paulus, W., Kettner, A., Schuhmacher, P., Adam, D., Etzbach, K. H., Siemensmeyer, K., Wendorff, J. H., Ringsdorf, H., and Haarer, D. *Adv. Mater.* 8, 815–819, 1996.
152. Adam, D., Romhildt, W., and Haarer, D. *Jpn. J. Appl. Phys.* 35, 1826–1831, 1996.
153. van de Craats, A. M., Warman, J. M., Hasebe, H., Naito, R., and Ohta, K. *J. Phys. Chem. B* 101, 9224–9232, 1997.
154. van de Craats, A. M., de Haas, M. P., and Warman, J. M. *Synth. Met.* 86, 2125–2126, 1997.
155. Yoshino, K., Nakayama, H., Ozaki, M., Onoda, M., and Hamaguchi, M. *Jpn. J. Appl. Phys.* 36, 5183–5186, 1997.
156. Etchegoin, P. *Phys. Rev. E* 56, 538–548, 1997.
157. van de Craats, A. M., Warman, J. M., Mullen, K., Geerts, Y., and Brand, J. D. *Adv. Mater.* 10, 36–38, 1998.
158. Shimizu, Y., Higashiyama, T., and Fuchita,T. *Thin Solid Films* 331, 279–284, 1998.
159. van de Craats, A. M., Siebbeles, L. D. A., Bleyl, I., Haarer, D., Berlin, Y. A., Zharikov, A. A., and Warman, J. M. *J. Phys. Chem. B* 102, 9625–9634, 1998.
160. Donovan, K. J., Kreouzis, T., Boden, N., and Clements, J. *J. Chem. Phys.* 109, 10400–10408, 1998.
161. Vaes, A., Catry, C., Van der Auweraer, M., De Schryver, F. C., Sudiwala, R. V., Wilson, E. G., Karthaus, O., and Ringsdorf, H. *J. Appl. Phys.* 84, 339–349, 1998.
162. Boden, N., Bushby, R. J., Clements, J., Donovan, K., Movaghar, B., and Kreouzis, T. *Phys. Rev. B* 58, 3063–3074, 1998.
163. Boden, N., Bushby, R. J., Clements, J., and Movaghar, B. *J. Appl. Phys.* 83, 3207–3216, 1998.
164. van de Craats, A. M., Warman, J. M., Schlichting, P., Rohr, U., Geerts, Y., and Mullen, K. *Synth. Met.* 102, 1550–1551, 1999.
165. van de Craats, A. M., Warman, J. M., Fechtenkotter, A., Brand, J. D., Harbison, M. A., and Mullen, K. *Adv. Mater.* 11, 1469–1471, 1999.
166. Shimizu, Y., Monobe, H., Mima, S., Higashiyama, T., Fuchita,T., and Sugino, T. *Mater. Res. Soc. Symp. Proc.* 559, 211–222, 1999.
167. Balagurusamy, V. S. K., Prasad, S. K., Chandrasekhar, S., Kumar, S., Manickam, M., and Yelamaggad, C. V. *Pramana* 53, 3–11, 1999.
168. Nakayama, H., Ozaki, M., Schmidt, W. F., and Yoshino, K. *Jpn. J. Appl. Phys.* 38, L1038–L1041, 1999.
169. Ochse, A., Kettner, A., Kopitzke, J., Wendorff, J. H., and Bassler, H. *Phys. Chem. Chem. Phys.* 1, 1757–1760, 1999.
170. Struijk, C. W., Sieval, A. B., Dakhorst, J. E. J., van Dijk, M., Kimkes, P., Koehorst, R. B. M., Donker, H., Schaafsma, T. J., Picken, S. J., van de Craats, A. M., Warman, J. M., Zuilhof, H., and Sudholter, E. J. R. *J. Am. Chem. Soc.* 122, 11057–11066, 2000.
171. Yuan, Y., Gregg, B. A., and Lawrence, M. F. *J. Mater. Res.* 15, 2494–2498, 2000.
172. Monobe, H., Mima, S., and Shimizu, Y. *Chem. Lett.* 1004–1005, 2000.
173. Bilke, R., Schreiber, A., Bleyl, I., Haarer, D., and Adam, D. *J. Appl. Phys.* 87, 3872–3877, 2000.
174. Kreouzis, T., Scott, K., Donovan, K. J., Boden, N., Bushby, R. J., Lozman, O. R., and Liu, Q. *Chem. Phys.* 262, 489–497, 2000.
175. Palenberg, M. A., Silbey, R. J., Malagoli, M., and Bredas, J. L. *J. Chem. Phys.* 112, 1541–1546, 2000.
176. Pecchia, A., Movaghar, B., Kelsall, R. W., Bourlange, A., Evans, S., Howson, M., Shen, T., and Boden, N. *Microelectron. Eng.* 51–52, 633–644, 2000.
177. Siebbeles, L. D. A. and Movaghar, B. *J. Chem. Phys.* 113, 1609–1617, 2000.
178. Ban, K., Nishizawa, K., Ohta, K., van de Craats, A. M., Warman, J. M., Yamamoto, I., and Shirai, H. *J. Mater. Chem.* 11, 321–331, 2001.

179. van de Craats, A. M. and Warman, J. M. *Adv. Mater.* 13, 130–133, 2001.
180. van de Craats, A. M. and Warman, J. M. *Synth. Met.* 121, 1287–1288, 2001.
181. Rohr, U., Kohl, C., Mullen, K., van de Craats, A. M., and Warman, J. *J. Mater. Chem.* 11, 1789–1799, 2001.
182. Seguy, I., Jolinat, P., Destruel, P., Mamy, R., Allouchi, H., Courseille, C., Cotrait, M., and Bock, H. *ChemPhysChem* 7, 448–452, 2001.
183. Monobe, H., Miyagawa, Y., Mima, S., Sugino, T., Uchida, K., and Shimizu, Y. *Thin Solid Films* 393, 217–224, 2001.
184. Inoue, M., Ukon, M., Monobe, H., Sugino, T., and Shimizu, Y. *Mol. Cryst. Liq. Cryst.* 365, 439–446, 2001.
185. Shimizu, Y. *Mol. Cryst. Liq. Cryst.* 370, 83–91, 2001.
186. Monobe, H., Mima, S., Miyagawa, Y., Sugino, T., Uchida, K., and Shimizu, Y. *Mol. Cryst. Liq. Cryst.* 368, 311–318, 2001.
187. Kreouzis, T., Donovan, K. J., Boden, N., Bushby, R. J., Lozman, O. R., and Liu, Q. *J. Chem. Phys.* 114, 1797–1802, 2001.
188. Bushby, R. J., Evans, S. D., Lozman, O. R., McNeill, A., and Movaghar, B. *J. Mater. Chem.* 11, 1982–1984, 2001.
189. Boden, N., Bushby, R. J., Donovan, K., Liu, Q., Lu, Z., Kreouzis, T., and Wood, A. *Liq. Cryst.* 28, 1739–1748, 2001.
190. Pecchia, A., Movaghar, B., Kelsall, R. W., Bourlange, A., Evans, S. D., Hickey, B. J., and Boden, N. *VLSI Des.* 13, 305–309, 2001.
191. Azumai, R., Ozaki, M., Nakayama, H., Fujisawa, T., Schmidt, W. F., and Yoshino, K. *Mol. Cryst. Liq. Cryst.* 366, 359–367, 2001.
192. Mizoshita, N., Monobe, H., Inoue, M., Ukon, M., Watanabe, T., Shimizu, Y., Hanabusa, K., and Kato, T. *Chem. Commun.* 428–429, 2002.
193. Wegewijs, B. R., Siebbeles, L. D. A., Boden, N., Bushby, R. J., Movaghar, B., Lozman, O. R., Liu, Q., Pecchia, A., and Mason, L. A. *Phys. Rev. B* 65, 245112, 2002.
194. Pecchia, A., Lozman, O. R., Movaghar, B., Boden, N., Bushby, R. J., Donovan, K. J., and Kreouzis, T. *Phys. Rev. B* 65, 104204, 2002.
195. Chandrasekhar, S. and Balagurusamy, V. S. K. *Proc. R. Soc. Lond. A* 458, 1783–1794, 2002.
196. Cornil, J., Lemaur, V., Calbert, J. P., and Bredas, J. L. *Adv. Mater.* 14, 726–729, 2002.
197. Pecchia, A., Kelsall, R. W., Movaghar, B., Bourlange, A., Evans, S. D., Hickey, B. J., and Boden, N. *Mater. Sci. Tech.* 18, 729–732, 2002.
198. Gearba, R. I., Lehmann, M., Levin, J., Ivanov, D. A., Koch, M. H. J., Barbera, J., Debije, M. G., Piris, J., and Geerts, Y. H. *Adv. Mater.* 15, 1614–1618, 2003.
199. Warman, J. M., Kroeze, J. E., Schouten, P. G., and van de Craats, A. M. *J. Porphyr. Phthalocyanines* 7, 342–350, 2003.
200. Piris, J., Debije, M. G., Stutzmann, N., van de Craats, A. M., Watson, M. D., Mullen, K., and Warman, J. M. *Adv. Mater.* 15, 1736–1740, 2003.
201. Haarer, D., Bilke, R., Thelakkat, M., and Jaeger, C. *Proc. SPIE* 4991, 234, 2003.
202. Donovan, K. J., Kreouzis, T., Scott, K., Bunning, J. C., Bushby, R. J., Boden, N., Lozman, O. R., and Movaghar, B. *Mol. Cryst. Liq. Cryst.* 396, 91–112, 2003.
203. Bushby, R. J., Lozman, O. R., Bunning, J. C., Donovan, K. J., Kreouzis, T., and Scott, K. *Proc. SPIE* 4991, 222, 2003.
204. Reis, F. T., Mencaraglia, D., Saad, S. O., Seguy, I., Oukachmih, M., Jolinat, P., and Destruel, P. *Synth. Met.* 138, 33–37, 2003.
205. Kats, E. I. *Mol. Cryst. Liq. Cryst.* 396, 23–34, 2003.
206. Kelsall, R. W., Pecchia, A., Bourlange, A., Movaghar, B., Evans, S. D., Hickey, B. J., and Boden, N. *Physica E* 17, 654–658, 2003.
207. Gallos, L. K., Movaghar, B., and Siebbeles, L. D. A. *Phys. Rev. B* 67, 165417, 2003.
208. Senthilkumar, K., Grozema, F. C., Bickelhaupt, F. M., and Siebbeles, L. D. A. *J. Chem. Phys.* 119, 9809–9817, 2003.
209. Boden, N., Bushby, R. J., and Lozman, O. R. *Mol. Cryst. Liq. Cryst.* 400, 105–113, 2003.
210. Bayer, A., Zimmermann, S., and Wendorff, J. H. *Mol. Cryst. Liq. Cryst.* 396, 1–22, 2003.
211. Sandhya, K. L., Prasad, S. K., Nair, G. G., and Prasad, V. *Mol. Cryst. Liq. Cryst.* 396, 113–119, 2003.
212. Fujikake, H., Murashige, T., Sugibayashi, M., and Ohta, K. *Appl. Phys. Lett.* 85, 3474–3476, 2004.
213. Piris, J., Debije, M. G., Stutzmann, N., Laursen, B. W., Pisula, W., Watson, M. D., Bjornholm, T., Mullen, K., and Warman, J. M. *Adv. Funct. Mater.* 14, 1053–1061, 2004.

214. Piris, J., Pisula, W., and Warman, J. M. *Synth. Met.* 147, 85–89, 2004.
215. Piris, J., Pisula, W., Tracz, A., Pakula, T., Mullen, K., and Warman, J. M. *Liq. Cryst.* 31, 993–996, 2004.
216. Debije, M. G., Piris, J., de Haas, M. P., Warman, J. M., Tomovi, E., Simpson, C. D., Watson, M. D., and Mullen, K. *J. Am. Chem. Soc.* 126, 4641–4645, 2004.
217. Warman, J. M., de Haas, M. P., Dicker, G., Grozema, F. C., Piris, J., and Debije, M. G., *Chem. Mater.* 16, 4600–4609, 2004.
218. Piris, J., Debije, M. G., Watson, M. D., Mullen, K., and Warman, J. M. *Adv. Funct. Mater.* 14, 1047–1052, 2004.
219. Boden, N., Bushby, R. J., Lozman, O. R., Lu, Z., McNeill, A., Movaghar, B., Donovan, K., and Kreouzis, T. *Mol. Cryst. Liq. Cryst.* 410, 13–21, 2004.
220. Reis, F. T., Mencaraglia, D., Saad, S. O., Seguy, I., Oukachmih, M., Jolinat, P., and Destruel, P. *J. Non-Cryst. Solids* 338–340, 599–602, 2004.
221. Kumar, S. and Lakshminarayanan, V. *Chem. Commun.* 1600–1601, 2004.
222. Debije, M. G., Chen, Z., Piris, J., Neder, R. B., Watson, M. M., Müllen, K., and Würthner, F. *J. Mater. Chem.* 15, 1270–1276, 2005.
223. An, Z., Yu, J., Jones, S. C., Barlow, S., Yoo, S., Domercq, B., Prins, P., Siebbeles, L. D. A., Kippelen, B., and Marder, S. R. *Adv. Mater.* 17, 2580–2583, 2005.
224. Lehmann, M., Kestemont, G., Aspe, R. G., Herman, C. B., Koch, M. H. J., Debije, M. G., Piris, J., de Haas, M. P., Warman, J. M., Watson, M. D., Lemaur, V., Cornil, J., Geerts, Y. H., Gearba, R., and Ivanov, D. A. *Chem. Eur. J.* 11, 3349–3362.
225. Jung, J. *Appl. Phys. Lett.* 87, 156101, 2005.
226. Fujikake, H., Murashige, T., Sugibayashi, M., and Ohta, K. *Appl. Phys. Lett.* 87, 156102, 2005.
227. Iino, H., Hanna, J. I., Bushby, R. J., Movaghar, B., Whitaker, B. J., and Cook, M. J. *Appl. Phys. Lett.* 87, 132102, 2005.
228. Iino, H., Takayashiki, Y., Hanna, J. I., and Bushby, R. J. *Jpn. J. Appl. Phys.* 44, L1310–L1312, 2005.
229. Warman, J. M., Piris, J., Pisula, W., Kastler, M., Wasserfallen, D., and Müllen, K. *J. Am. Chem. Soc.* 127, 14257–14262, 2005.
230. Iino, H., Takayashiki, Y., Hanna, J. I., Bushby, R. J., and Haarer, D. *Appl. Phys. Lett.* 87, 192105, 2005.
231. Inoue, M., Monobe, H., Ukon, M., Petrov, V. F., Watanabe, T., Kumano, A., and Shimizu, Y. *Opto-Electron. Rev.* 13, 303–308, 2005.
232. Iino, H., Hanna, J. I., Jager, C., and Haarer, D. *Mol. Cryst. Liq. Cryst.* 436, 217–224, 2005.
233. Iino, H., Hanna, J. I., and Haarer, D. *Phys. Rev. B* 72, 193203, 2005.
234. Bushby, R. J., Donovan, K. J., Kreouzis, T., and Lozman, O. R. *Opto-Electron. Rev.* 13, 269–279, 2005.
235. Kumar, S., Pal, S. K., and Lakshminarayanan, V. *Mol. Cryst. Liq. Cryst.* 434, 251–258, 2005.
236. Plyukhin, A. V. *Europhys. Lett.* 71, 716–722, 2005.
237. Lever, L., Kelsall, R., and Bushby, R. *J. Comput. Electron.* 4, 101–104, 2005.
238. Deibel, C., Janssen, D., Heremans, P., de Cupere, V., Geerts, Y., Benkhedir, M. L., and Adriaenssens, G. J. *Org. Electron.* 7, 495–499, 2006.
239. Rybak, A., Pfleger, J., Jung, J., Pavlik, M., Glowacki, I., Ulanski, J., Tomovic, Z., Mullen, K., and Geerts, Y. *Synth. Met.* 156, 302–309, 2006.
240. Kastler, M., Pisula, W., Laquai, F., Kumar, A., Davies, R. J., Baluschev, S., Gutierrez, M. C. G., Wasserfallen, D., Butt, H. J., Riekel, C., Wegner, G., and Mullen, K. *Adv. Mater.* 18, 2255–2259, 2006.
241. Iino, H. and Hanna, J. I. *Jpn. J. Appl. Phys.* 45, L867–L870, 2006.
242. Iino, H., Hanna, J. I., Haarer, D., and Bushby, R. J. *Jpn. J. Appl. Phys.* 45, 430–433, 2006.
243. Paraschiv, I., Giesbers, M., van Lagen, B., Grozema, F. C., Abellon, R. D., Siebbeles, L. D. A., Marcelis, A. T. M., Zuilhof, H., and Sudholter, E. J. R. *Chem. Mater.* 18, 968–974, 2006.
244. Iino, H., Hanna, J. I., Bushby, R. J., Movaghar, B., and Whitaker, B. J. *J. Appl. Phys.* 100, 043716, 2006.
245. Donovan, K. J., Scott, K., Somerton, M., Preece, J., and Manickam, M. *Chem. Phys.* 322, 471–476, 2006.
246. Duzhko, V., Semyonov, A., Twieg, R. J., and Singer, K. D. *Phys. Rev. B* 73, 064201, 2006.
247. Lever, L., Bushby, R. J., and Kelsall, R. W. *Physica E* 32, 596–599, 2006.
248. Kruglova, O., Mulder, F. M., Siebbeles, L. D. A., and Kearley, G. J. *Chem. Phys.* 330, 333–337, 2006.
249. Garcia, J. C. S. *Chem. Phys.* 331, 321–331, 2007.
250. Basova, T., Gurek, A. G., Ahsen, V., and Ray, A. K. *Org. Electron.* 8, 784–790, 2007.
251. Kumar, S., Pal, S. K., Kumar, P. S., and Lakshminarayanan, V. *Soft. Matter* 3, 896–900, 2007.
252. Pokhrel, C., Shakya, N., Purtee, S., Ellman, B., Semyonov, A. N., and Twieg, R. J. *J. Appl. Phys.* 101, 103706, 2007.
253. Monobe, H., Shimizu, Y., Okamoto, S., and Enomoto, H. *Mol. Cryst. Liq. Cryst.* 476, 31–41, 2007.
254. Mao, H., He, Z., Wang, J., Zhang, C., Xie, P., and Zhang, R. *J. Lumin.* 122–123, 942–945, 2007.

255. Marcon, V., Kirkpatrick, J., Pisula, W., and Andrienko, D. *Phys. Stat. Sol. B* 245, 820–824, 2008.

256. de Jong, M. P., Osikowicz, W., Sorensen, S. L., Sergeyev, S., Geerts, Y. H., and Salaneck, W. R. *J. Phys. Chem. C* 112, 15784–15790, 2008.

257. Andrienko, D., Kirkpatrick, J., Marcon, V., Nelson, J., and Kremer, K. *Phys. Stat. Sol. B* 245, 830–834, 2008.

258. Hirai, Y., Monobe, H., Mizoshita, N., Moriyama, M., Hanabusa, K., Shimizu, Y., and Kato, T. *Adv. Funct. Mater.* 18, 1668–1675, 2008.

259. Paraschiv, I., de Lange, K., Giesbers, M., van Lagen, B., Grozema, F. C., Abellon, R. D., Siebbeles, L. D. A., Sudholter, E. J. R., Zuilhof, H., and Marcelis, A. T. M. *J. Mater. Chem.* 18, 5475–5481, 2008.

260. Talarico, M., Termine, R., Frutos, E. M. G., Omenat, A., Serrano, J. L., Lor, B. G., and Golemme, A. *Chem. Mater.* 20, 6589–6591, 2008.

261. Kumar, P. S., Kumar, S., and Lakshminarayanan, V. *J. Phys. Chem. B* 112, 4865–4869, 2008.

262. Holt, L. A., Bushby, R. J., Evans, S. D., Burgess, A., and Seeley, G. *J. Appl. Phys.* 103, 063712, 2008.

263. Kirkpatrick, J., Marcon, V., Kremer, K., Nelson, J., and Andrienko, D. *J. Chem. Phys.* 129, 094506, 2008.

264. Cinacchi, G., Colle, R., Parruccini, P., and Tani, A. *J. Chem. Phys.* 129, 174708, 2008.

265. Kumar, P. S., Kumar, S., and Lakshminarayanan, V. *J. Appl. Phys.* 106, 093701, 2009.

266. Andre, J. J., Holczer, K., Petit, P., Riou, M. T., Clarisse, C., Even, R., Fourmigue, M., and Simon, J. *Chem. Phys. Lett.* 115, 463–466, 1985.

267. Maitrot, M., Guillaud, G., Boudjema, B., André, J. J., Strzelecka, H., Simon, J., and Even, R. *Chem. Phys. Lett.* 133, 59–62, 1987.

268. Turek, P., Petit, P., Andre, J. J., Simon, J., Even, R., Boudjema, B., Guillaud, G., and Maitrot, M. *J. Am. Chem. Soc.* 109, 5119–5122, 1987.

269. Safinya, C. R., Liang, K. S., Varady, W. A., Clark, N. A., and Andersson, G. *Phys. Rev. Lett.* 53, 1172–1175, 1984.

270. van Winkle, D. H. and Clark, N. A. *Phys. Rev. Lett.* 48, 1407–1410, 1982.

271. Teitelbaum, R. C., Ruby, S. L., and Marks, T. J. *J. Am. Chem. Soc.* 101, 7568–7573, 1979.

272. Petersen, J. L., Schramm, C. S., Stojakovic, D. R., Hoffman, B. M., and Marks, T. J. *J. Am. Chem. Soc.* 99, 286–288, 1977.

273. Law, K. Y. *Chem. Rev.* 93, 449–486, 1993.

274. Kumar, S., Naidu, J. J., and Rao, D. S. S. *J. Mater. Chem.* 12, 1335–1341, 2002.

275. Kumar, S., Rao, D. S. S., and Prasad, S. K. *J. Mater. Chem.* 9, 2751–2754, 1999.

276. Claessens, C. G., Hahn, U., and Torres, T. *Chem. Rec.* 8, 75–97, 2008.

277. Cho, C. P. and Perng, T. P. *J. Nanosci. Nanotech.* 8, 69–87, 2008.

278. Kamloth, K. P. *Chem. Rev.* 108, 367–399, 2008.

279. Di, C. A., Yu, G., Liu, Y., and Zhu, D. *J. Phys. Chem. B* 111, 14083–14096, 2007.

280. Singh, T. B. and Sariciftci, N. S. *Annu. Rev. Mater. Res.* 36, 199–230, 2006.

281. Guillaud, G., Simon, J., and Germain, J. P. *Coord. Chem. Rev.* 178–180, 1433–1484, 1998.

282. Gould, R. D. *Coord. Chem. Rev.* 156, 237–274, 1996.

283. Chesterfield, R. J., McKeen, J. C., Newman, C. R., Ewbank, P. C., da Silva Filho, D. A., Bredas, J. L., Miller, L. L., Mann, K. R., and Frisbie, C. D. *J. Phys. Chem. B* 108, 19281–19292, 2004.

284. Mende, L. S., Fechtenkotter, A., Mullen, K., Moons, E., Friend, R. H., and MacKenzie, J. D. *Science* 293, 1119–1122, 2001.

285. Shimizu, Y., Miyake, Y., and Kumar, S. *Poster* 3-*MA*53, *ILCC*. 2008.

286. Percec, V., Glodde, M., Bera, T. K., Miura, Y., Shiyanovskaya, I., Singer, K. D., Balagurusamy, V. S. K., Heiney, P. A., Schnell, I., Rapp, A., Spiess, H. W., Hudson, S, D., and Duan, H. *Nature* 419, 384–387, 2002.

287. Percec, V., Glodde, M., Peterca, M., Rapp, A., Schnell, I., Spiess, H. W., Bera, T. K., Miura, Y., Balagurusamy, V. S. K., Aqad, E., and Heiney, P. A. *Chem. Eur. J.* 12, 6298–6314, 2006.

288. Percec, V., Peterca, M., Sienkowska, M. J., Llies, M. A., Aqad, E., Smidrkal, J., and Heiney, P. A. *J. Am. Chem. Soc.* 128, 3324–3334, 2006.

289. Percec, V., Holerca, M. N., Nummelin, S., Morrison, J. J., Glodde, M., Smidrkal, J., Peterca, M., Rosen, B. M., Uchida, S., Balagurusamy, V. S. K., Sienkowska, M. J., and Heiney, P. A. *Chem. Euro. J.* 12, 6216–6241, 2006.

290. Percec, V., Won, B. C., Peterca, M., and Heiney, P. A. *J. Am. Chem. Soc.* 129, 11265–11278, 2007.

291. Percec, V., Imam, M. R., Peterca, M., Wilson, D. A., and Heiney, P. A. *J. Am. Chem. Soc.* 131, 1294–1304, 2009.

292. Kato, T., Yasuda, T., Kamikawa, Y., and Yoshio, M. *Chem. Commun.* 729–739, 2009.

293. Yasuda, T., Ooi, H., Morita, J., Akama, Y., Minoura, K., Funahashi, M., Shimomura, T., and Kato, T. *Adv. Funct. Mater.* 19, 411–419, 2009.

294. Brebee, C. J., Dgakonov, V., Parisis, J., and Sariciftei, N. S. (eds.). *Organic Photovoltoics*, Springer-Verlag, Berlin, Germany, 2003.
295. Zakeeruddin, S. M. and Gratzel, M. *Adv. Funct. Mater.* 19, 2187–2202, 2009.
296. Mishra, A., Fischer, M. K. R., and Bauerle, P. *Angew. Chem. Int. Ed.* 48, 2474–2499, 2009.
297. Krebs, F. C. *Sol. Energy Mater. Sol. Cells* 93, 394–412, 2009.
298. Armstrong, N. R., Wang, W. N., Alloway, D. M., Placencia, D., Ratcliff, E., and Brumbach, M. *Macromol. Rapid. Commun.* 30, 717–731, 2009.
299. Thompson, B. C. and Frechet, J. M. J. *Angew. Chem. Int. Ed.* 47, 58–77, 2008.
300. Bundgaard, E. and Krebs, F. C. *Sol. Energy Mater. Sol. Cells* 91, 954–985, 2007.
301. Roncali, J. *Chem. Soc. Rev.* 34, 483–495, 2005.
302. Zhu, H., Wei, J., Wang, K., and Wu, D. *Sol. Energy Mater. Sol. Cells* 93, 1461–1470, 2009.
303. Rio, Y., Vazquez, P., and Palomares, E. *J. Porphyr. Phthalocyanines* 13, 645–651, 2009.
304. de Freitas, J. N., Nogueira, A. F., and de Paoli, M. A. *J. Mater. Chem.* 19, 5279–5294, 2009.
305. Tang, C. W. *Appl. Phys. Lett.* 48, 183–185, 1986.
306. Mende, L. S., Fechtenkotter, A., Mullen, K., Friend, R. H., and MacKenzie, J. D. *Physica E* 14, 263–267, 2002.
307. Mende, L. S., Watson, M., Mullen, K., and Friend, R. H. *Mol. Cryst. Liq. Cryst.* 396, 73–90, 2003.
308. Hassheider, T., Benning, S. A., Lauhof, M. W., Kitzerow, H. S., Bock, H., Watson, M. D., and Müllen, K. *Mol. Cryst. Liq. Cryst.* 413, 461–472, 2004.
309. Jung, J., Rybak, A., Slazak, A., Bialecki, S., Miskiewicz, P., Glowacki, I., Ulanski, J., Rosselli, S., Yasuda, A., Nelles, G., Tomovic, Z., Watson, M. D., and Mullen, K. *Synth. Met.* 155, 150–156, 2005.
310. Schmidtke, J. P., Friend, R. H., Kastler, M., and Mullen, K. *J. Chem. Phys.* 124, 174704, 2006.
311. Li, J., Kastler, M., Pisula, W., Robertson, J. W. F., Wasserfallen, D., Grimsdale, A. C., Wu, J., and Mullen, K. *Adv. Funct. Mater.* 17, 2528–2533, 2007.
312. Feng, X., Liu, M., Pisula, W., Takase, M., Li, J., and Mullen, K. *Adv. Mater.* 20, 2684–2689, 2008.
313. Gregg, B. A., Fox, M. A., and Bard, A. J. *J. Phys. Chem.* 94, 1586–1598, 1990.
314. Gregg, B. A. *Mol. Cryst. Liq. Cryst.* 257, 219–227, 1994.
315. Li, L., Kang, S. W., Harden, J., Sun, Q., Zhou, X., Dai, L., Jakli, A., Kumar, S., and Li, Q. *Liq. Cryst.* 35, 233–239, 2008.
316. Petritsch, K., Friend, R. H., Lux, A., Rozenberg, G., Moratti, S. C., and Holmes, A. B. *Synth. Met.* 102, 1776–1777, 1999.
317. Levitsky, I. A., Euler, W. B., Tokranova, N., Xu, B., and Castracane, J. *Appl. Phys. Lett.* 85, 6245–6247, 2004.
318. Kim, J. Y. and Bard, A. J. *Chem. Phys. Lett.* 383, 11–15, 2004.
319. Seguy, I., Mamy, R., Destruel, P., Jolinat, P., and Bock, H. *Appl. Surf. Sci.* 174, 310–315, 2001.
320. Oukachmih, M., Destruel, P., Seguy, I., Ablart, G., Jolinat, P., Archambeau, S., Mabiala, M., Fouet, S., Bock, H. *Sol. Energy Mater. Sol. Cells* 85, 535–543, 2005.
321. Destruel, P., Bock, H., Seguy, I., Jolinat, P., Oukachmih, M., and Pereira, E. B. *Polym. Int.* 55, 601–607, 2006.
322. Wang, J., He, Z., Zhang, C., Sun, J., and Hui, G. *Proc. SPIE* 6841, 68411, 2007.
323. Peeters, E., van Hal, P. A., Meskers, S. C. J., Janssen, R. A. J., and Meijer, E. W. *Chem. Euro. J.* 8, 4470–4474, 2002.
324. van Herrikhuyzen, J., Syamakumari, A., Schenning, A. P. H. J., and Meijer, E. W. *J. Am. Chem. Soc.* 126, 10021–10027, 2004.
325. Archambeau, S., Bock, H., Seguy, I., Jolinat, P., and Destruel, P. *J. Mater. Sci. Mater. Electron.* 18, 919–923, 2007.
326. Podhajecka, K., Matejicek, P., Vohlidal, J., Masuda, T., and Pfleger, J. *Synth. Met.* 158, 775–781, 2008.
327. Christ, T., Stumpflen, V., and Wendorff, J. H. *Macromol. Rapid. Commun.* 18, 93–98, 1997.
328. Bacher, A., Bleyl, I., Erdelen, C. H., Haarer, D., Paulus, W., and Schmidt, H. W. *Adv. Mater.* 9, 1031–1035, 1997.
329. Stapff, I. H., Stumpflen, V., Wendorff, J. H., Spohn, D. B., and Mobius, D. *Liq. Cryst.* 23, 613–617, 1997.
330. Christ, T., Glusen, B., Greiner, A., Kettner, A., Sander, R., Stumpflen, V., Tsukruk, V., and Wendorff, J. H. *Adv. Mater.* 9, 48–52, 1997.
331. Christ, T., Geffart, F., Glusen, B., Kettner, A., Lussem, G., Schafer, O., Stumpflen, V., Wendorff, J. H., and Tsukruk, V. V. *Thin Solid Films* 302, 214–222, 1997.
332. Stuempflen, V., Stapff, L., and Wendorff, J. H. *Proc. SPIE* 3281, 164, 1998.
333. Bacher, A., Erdelen, C. H., Paulus, W., Ringsdorf, H., Schmidt, H. W., and Schuhmacher, P. *Macromolecules* 32, 4551–4557, 1999.

334. Seguy, I., Destruel, P., and Bock, H. *Synth. Met.* 111–112, 15–18, 2000.

335. Hassheider, T., Benning, S. A., Kitzerow, H. S., Achard, M. F., and Bock, H. *Angew. Chem. Int. Ed.* 40, 2060–2063, 2001.

336. Benning, S. A., Hassheider, T., Baumann, S. K., Bock, H., Sala, F. D., Frauenheim, T., and Kitzerow, H. S. *Liq. Cryst.* 28, 1105–1113, 2001.

337. Seguy, I., Jolinat, P., Destruel, P., Farenc, J., Mamy, R., Bock, H., Ip, J., and Nguyen, T. P. *J. Appl. Phys.* 89, 5442–5448, 2001.

338. Freudenmann, R., Behnisch, B., and Hanack, M. *J. Mater. Chem.* 11, 1618–1624, 2001.

339. Nguyen, T. P., Ip, J., Jolinat, P., and Destruel, P. *Appl. Surf. Sci.* 172, 75–83, 2001.

340. Fouet, S. A., Dardel, S., Bock, H., Oukachmih, M., Archambeau, S., Seguy, I., Jolinat, P., and Destruel, P. *ChemPhysChem* 4, 983–985, 2003.

341. Benning, S. A., Oesterhaus, R., and Kitzerow, H. S. *Liq. Cryst.* 31, 201–205, 2004.

342. van de Craats, A. M., Stutzmann, N., Bunk, O., Nielsen, M. M., Watson, M., Mullen, K., Chanzy, H. D., Sirringhaus, H., and Friend, R. H. *Adv. Mater.* 15, 495–499, 2003.

343. Cherian, S., Donley, C., Mathine, D., LaRussa, L., Xia, W., and Armstrong, N. *J. Appl. Phys.* 96, 5638–5643, 2004.

344. Donley, C. L., Zangmeister, R. A. P., Xia, W., Minch, B., Drager, A., Cherian, S. K., LaRussa, L., Kippelen, B., Domercq, B., Mathine, D. L., O'Brien, D. F., and Armstrong, N. R. *J. Mater. Res.* 19, 2087–2099, 2004.

345. Shklyarevskiy, I. O., Jonkheijm, P., Stutzmann, N., Wasserberg, D., Wondergem, H. J., Christianen, P. C. M., Schenning, A. P. H. J., de Leeuw, D. M., Tomovic, Z., Wu, J., Mullen, K., and Maan, J. C. *J. Am. Chem. Soc.* 127, 16233–16237, 2005.

346. Pisula, W., Menon, A., Stepputat, M., Lieberwirth, I., Kolb, U., Tracz, A., Sirringhaus, H., Pakula, T., and Mullen, K. *Adv. Mater.* 17, 684–689, 2005.

347. Xiao, S., Myers, M., Miao, Q., Sanaur, S., Pang, K., Steigerwald, M. L., and Nuckolls, C. *Angew. Chem. Int. Ed.* 44, 7390–7394, 2005.

348. Cho, J. Y., Domercq, B., Jones, S. C., Yu, J., Zhang, X., An, Z., Bishop, M., Barlow, S., Marder, S. R., and Kippelen, B. *J. Mater. Chem.* 17, 2642–2647, 2007.

349. Shimizu, Y., Oikawa, K., Nakayama, K. I., and Guillon, D. *J. Mater. Chem.* 17, 4223–4229, 2007.

350. Tsao, H. N., Rader, H. J., Pisula, W., Rouhanipour, A., and Mullen, K. *Phys. Stat. Sol. A* 205, 421–429, 2008.

351. Schön, J. H., Kloc, C., and Batlogg, B. *Science* 288, 2338–2340, 2000.

352. Schön, J. H., Berg, S., Kloc, C., and Batlogg, B. *Science* 287, 1022–1023 2000.

Index